ECOLOGICAL PHYTOGEOGRAPHY IN THE NINETEENTH CENTURY

This is a volume in the Arno Press collection

HISTORY OF ECOLOGY

See last pages of this volume for a complete list of titles.

ECOLOGICAL PHYTOGEOGRAPHY IN THE NINETEENTH CENTURY

ARNO PRESS

A New York Times Company

New York / 1977

Editorial Supervision: LUCILLE MAIORCA

Reprint Edition 1977 by Arno Press Inc.

HISTORY OF ECOLOGY
ISBN for complete set: 0-405-10369-7
See last pages of this volume for titles.

Manufactured in the United States of America

Library of Congress Cataloging in Publication Data
Main entry under title:

Ecological phytogeography in the nineteenth century.

(History of ecology)
CONTENTS: Humboldt, A. von. On isothermal lines and the distribution of heat over the globe.—De Candolle, A. Géographie botanique.—Hooker, W. J. Geography considered in relation to the distribution of plants. England, botany.—[etc.]
1. Phytogeography—History—Sources. 2. Botany—Ecology—History—Sources. I. Arno Press. II. Series.
QK101.E26 581.9 77-74218
ISBN 0-405-10388-3

CONTENTS

Humboldt, Alexander von
ON ISOTHERMAL LINES, AND THE DISTRIBUTION OF HEAT OVER THE GLOBE (Reprinted from *Edinburgh Philosophical Journal,* Volume III-V), Edinburgh, 1820-1821

De Candolle, Auguste
GÉOGRAPHIE BOTANIQUE (Reprinted form *Dictionnaire des Sciences Naturelles,* Volume XVIII), Strasbourg, 1820

[Hooker, William Jackson]
GEOGRAPHY CONSIDERED IN RELATION TO THE DISTRIBUTION OF PLANTS *and* ENGLAND, BOTANY (Reprinted from *An Encyclopaedia of Geography),* London, 1834

Grisebach, Heinrich Rudolf August
UEBER DEN EINFLUSS DES CLIMAS AUF DIE BEGRENZUNG DER NATÜRLICHEN FLOREN (ON THE INFLUENCE OF CLIMATE ON THE RANGE OF THE NATURAL FLORA) (Reprinted from *Linnaea,* Volume XII), 1838

Grisebach, Heinrich Rudolf August
REPORT ON THE CONTRIBUTIONS TO BOTANICAL GEOGRAPHY, DURING THE YEAR 1842 *and* REPORT ON THE CONTRIBUTIONS TO BOTANICAL GEOGRAPHY DURING THE YEAR 1843 (Reprinted from *Ray Society, Reports and Papers on Botany,* I), London, 1846

Grisebach, Heinrich Rudolf August
REPORT ON THE PROGRESS OF GEOGRAPHICAL BOTANY, DURING THE YEAR 1844 *and* REPORT ON THE PROGRESS OF GEOGRAPHICAL AND SYSTEMATIC BOTANY, DURING THE YEAR 1845 (Reprinted from *Ray Society, Reports and Papers on Botany,* II), London, 1849

Lippincott, James S[tarr]
GEOGRAPHY OF PLANTS (Reprinted from *Report of the Commissioner of Agriculture for the Year 1863),* Washington, D.C., 1863

ON ISOTHERMAL LINES AND THE DISTRIBUTION OF HEAT OVER THE GLOBE

Alexander von Humbolt

THE

EDINBURGH

PHILOSOPHICAL JOURNAL.

ART. I.—*On Isothermal Lines, and the Distribution of Heat over the Globe.* By Baron ALEXANDER DE HUMBOLDT *.

THE distribution of heat over the Globe belongs to that kind of phenomena, of which the general circumstances have been long known, but which were incapable of being rigorously determined or submitted to exact calculation, till experience and observation furnished data from which the theory might obtain the corrections and the different elements which it requires. The object of this memoir is to facilitate the collection of these data, to present results drawn from a great number of unpublished observations, and to group them according to a method which has not yet been tried, though its utility has been recognised for more than a century in the exposition of the phenomena of the variation and dip of the magnetic needle. As the discussion of individual observations will be published in a separate work, I shall at present limit myself to a simple sketch of the distribution of heat over the globe, according to the most recent and accurate data. Although we may not be able to refer the complex phenomena to a general theory, it

---

* As this interesting and valuable Memoir, the original of which was published in the *Memoires D'Arcueil*, tom. iii. p. 462, has never appeared in our language, and as it must be constantly referred to in all subsequent speculations on Meteorology, and should be familiar to every person who pursues this important study, we have resolved to present a translation of it to our readers. A small part of the memoir appeared in an English journal; but almost all the reductions from the Centigrade to Fahrenheit's scale were so erroneous, that the numbers cannot be trusted. We have added various notes, which will be distinguished from those of the Author by affixing ED. to the former, and H to the latter.—ED.

will be of considerable importance to fix the numerical relation by which a great number of scattered observations are connected, and to reduce to empirical laws the effects of local and disturbing causes. The study of these laws will point out to travellers the problems to which they should direct their principal attention, and we may entertain the hope, that the theory of the distribution of heat will gain in extent and precision, in proportion as observations shall be more multiplied, and directed to those points which it is of most importance to illustrate.

As the phenomena of geography and of vegetables, and in general the distribution of organised beings, depend on the knowledge of the three co-ordinates of Latitude, Longitude, and Altitude, I have been occupied for many years in the exact valuation of atmospherical temperatures; but I could not reduce my own observations without a constant reference to the works of Cotte and Kirwan, the only ones which contain a great mass of meteorological observations obtained by instruments and methods of very unequal precision. Having inhabited for a long time the most elevated plains of the New Continent, I availed myself of the advantages which they present for examining the temperature of the superincumbent strata of air, not from insulated data, the results of a few excursions to the crater of a volcano, but from the collection of a great number of observations made day after day and month after month in inhabited districts. In Europe, and in all the Old World, the highest points of which the mean temperatures have been determined, are the Convent of Peissenberg in Bavaria, and the Hospice of St Gothard *. The first of these is placed at 3264, and the second at 6808 feet above the level of the sea. In America a great number of good observations have been made at Santa Fe de Begota and at Quito, at altitudes of 8,727 and 9,544 feet. The town of Huancavelica, containing 10,000 inhabitants, and possessing all the resources of modern civilisation, is situated in the Cordilleras of the southern hemi-

* The mean temperature of the air at the Convent of the Great St Bernard, the height of which is 7,960 feet, is not determined. There are several villages in Europe placed at more than 5000 feet of altitude; for example, St Jacques de Ayas at 5,479, and Trinita Nuova, near Grasfoney, at 5,315 feet.—H.

sphere at 12,310 feet of absolute elevation; and the mine of Santa Barbara, encircled with fine edifices, and placed a league to the south of Huancavelica, is a place fit for making regular observations, at the height of 14,509 feet, which is double that of the Hospice of St Gothard.

These examples are sufficient to prove how much our knowledge of the higher regions of the atmosphere, and of the physical condition of the world in general, will increase, when the cultivation of the sciences, so long confined to the temperate zone, shall extend beyond the tropics into those vast regions, where the Spanish Americans have already devoted themselves with such zeal to the study of physics and astronomy. In order to compare with the mean heat of temperate climates, the results which M. Bonpland and I obtained in the equinoctial regions from the plains to the height of 19,292 feet, it was necessary to collect a great number of good observations made beyond the parallels of 30° and 35°. I soon perceived how vague such a comparison was, if I selected places under the meridian of the Cordilleras, or with a more eastern longitude, and I therefore undertook to examine the results contained in the most recent works. I endeavoured to find, at every 10° of latitude, but under different meridians, a small number of places whose mean temperature had been precisely ascertained, and through these, as so many fixed points, passed my *isothermal lines* or *lines of equal heat*. I had recourse, in so far as the materials have been made public, to those observations the results of which have been published, and I found, in the course of this easy, but long and monotonous labour, that there are many mean temperatures pointed out in meteorological tables, which, like astronomical positions, have been adopted without examination. Sometimes the results were in direct contradiction to the most recent observations, and sometimes it was impossible to discover from whence they were taken.

Many good observations were rejected, solely because the absolute height of the place where they were made was unknown. This is the case with Asia Minor, Armenia and Persia, and of almost all Asia; and while the equinoctial part alone of the New World presents already more than 500 points, the greater number of which are simple villages and hamlets, determined by barometrical levelling, we are still ignorant of the height of Erzeroum, Bagdad, Aleppo, Teheran, Ispahan, Delhi and Lassa, above the level of the neighbouring seas. Notwithstanding the intimate relation in which we have lately stood with Persia and Candahar, this branch of knowledge has not made any progress in the last fifty years.

We are not authorised, however, on account of the decrease of temperature in the upper regions of the atmosphere, to confound the mean temperatures of places which are not placed on the same level. In the Old World, good observations, which can alone be used for establishing empirical laws, are confined to an extent between the parallels of 30 and 70 degrees of latitude, and the meridians of 30° east longitude, and 20° of west longitude. The extreme points of this region are the island of Madeira, Cairo, and the North Cape. It is a zone which is only a thousand nautical leagues, (1 7th of the circumference of the globe,) from east to west, and which, containing the Basin of the Mediterranean, is the centre of the primitive civilisation of Europe. The extraordinary shape of this part of the world, the interior seas and other circumstances, so necessary for developing the germ of cultivation among nations, have given to Europe a particular climate, very different from that of other regions placed under the same latitude. But as the physical sciences almost always bear the impress of the places where they began to be cultivated, we are accustomed to consider the distribution of heat observed in such a region, as the type of the laws which govern the whole globe. It is thus that, in geology, we have for a long time attempted to refer all volcanic phenomena to those of the volcanoes in Italy. In place of estimating methodically the distribution of heat, such as it exists on the surface of continents and seas, it has been usual to consider as real exceptions every thing which differs from the adopted type, or, by pursuing a method still more dangerous in investigating the laws of nature, to take the mean temperatures for every five degrees of latitude, confounding together places under different meridians. As this last method appears to exclude the influence of extraneous causes, I shall first discuss it briefly before I proceed to point out the method, essentially different, which I have followed in my researches.

The temperature of the atmosphere, and the magnetism of the globe, cannot, like those phenomena which depend on one cause, or on a single centre of action, be disengaged from the influence of disturbing circumstances, by taking the averages of many observations in which these extraneous effects are mutually destroyed. The distribution of heat, as well as the dip and variation of the needle, and the intensity of the terrestrial magnetism, depend, by their nature, on local causes, on the constitution of the soil, and on the particular disposition of the radiating surface of the globe. We must, however, guard against confounding under the name of extraneous and disturbing causes, those on which the most important phenomena, such as the distribution and the more or less rapid developement of organic life, essentially depend. Of what use would it be to have a table of magnetic dips, which, in place of being measured in parallels to the magnetic equator, should be the mean of observations made on the same degrees of terrestrial latitude, but under different meridians? Our object is to ascertain the quantity of heat which every point of the globe annually receives, and, what is of most importance to agriculture, and the good of its inhabitants, the distribution of this quantity of heat over the different parts of the year, and not that which is due to the solar action alone, to its altitude above the horizon, or to the duration of its influence, as measured by the semidiurnal arcs.

Moreover, we shall prove, that the method of means is unfit for ascertaining what belongs exclusively to the sun, (inasmuch as its rays illuminate only one point of the globe,) and what is due both to the sun and to the influence of foreign causes. Among these causes may be enumerated the mixture of the temperatures of different latitudes produced by winds;—the vicinity of seas, which are immense reservoirs of an almost invariable temperature;—the shape, the chemical nature, the colour, the radiating power and evaporation of the soil;—the direction of the chains of mountains, which act either in favouring the play of descending currents, or in affording shelter against particular winds;—the form of lands, their mass and their prolongation towards the poles;—the quantity of snow which covers them in winter, their temperature and their reflection in summer;—and, finally, the fields of ice, which form, as it were, circumpolar continents, variable in their extent, and whose detached parts dragged away by currents modify in a sensible manner the climate of the temperate zone.

In distinguishing, as has long been done, between the *solar* and the *real climate*, we must not forget, that the local and multiplied causes which modify the action of the sun upon a single point of the globe, are themselves but secondary causes, the effects of the motion which the sun produces in the atmosphere, and which are propagated to great distances. If we consider separately (and it will be useful to do this in a discussion purely theoretical) the heat produced by the sun, the earth being supposed at rest and without an atmosphere, and the heat due to other causes regarded as disturbing ones, we shall find that this latter part of the total effect is not entirely foreign to the sun. The influence of small causes will scarcely disappear by taking the mean result of a great number of observations; for this influence is not limited to a single region. By the mobility of the aerial ocean, it is propagated from one continent to another. Every where in the regions near the polar circles, the rigours of the winters are diminished by the admixture of the columns of warm air, which, rising above the torrid zone, are carried towards the poles: Every where in the temperate zone, the frequent west winds modify the climate, by transporting the temperature of one latitude to another *. When we reflect, besides, on the extent of seas, on the form and prolongation of continents, either in the two hemispheres, or to the east and west of the meridians of Canton and of California, we shall perceive, that even if the number of observations on the mean temperature were infinite, the compensation would not take place.

It is, then, from the theory alone that we must expect to determine the distribution of heat over the globe, in so far as it depends on the immediate and instantaneous action of the sun. It does not indicate the degrees of temperature expressed by the dilatation of the mercury in a thermometer, but the ratios between the mean annual heat at the equator, at the parallel of 45°, and under the polar circle; and it determines the ratios between the solstitial and equinoctial heats in different zones. By comparing

* Raymond, *Memoire sur la Formule Baromet.* p. 108 and 113.

the results of calculation, not with the mean temperature drawn from observations made under different longitudes, but with that of a single point of the earth's surface, we shall set out with that which is due to the immediate action of the sun, and to the whole of the other influences, whether they are solar or local, or propagated to great distances. This comparison of theory with experience will present a great number of interesting relations.

In the year 1693, previous to the use of comparable thermometers, and to precise ideas of the mean temperature of a place, Halley laid the first foundations of a theory of the heating action of the sun under different latitudes . He proved that these actions might compensate for the effect of the obliquity of the rays. The ratios which he points out, do not express the mean heat of the seasons, but the heat of a summer day at the equator and under the polar circle, which he finds to be as 1.834 to 2.310 †. According to Geminus ‡, Polybius among the Greeks had perceived the cause why there should be less heat at the equator than under the tropic. The idea also of a temperate zone, habitable and highly elevated in the midst of the torrid zone, was admitted by Esatosthenes, Polybius, and Strabo.

In two memoirs §, published at long intervals in 1719 and 1765, Mairan attempted to solve the problems of the solar action, by treating them in a much more extended and general manner. He compared, for the first time, the results of theory with those of observation; and as he found the difference between the heat of summer and winter much less than it ought to be by calculation, he recognised the permanent heat of the globe and the effects of radiation.

Without mistrusting the observations he employed, he conceived the strange theory of central emanations which increase the heat of the atmosphere from the equator to the pole. He supposes that these emanations decrease to the parallel of 74°, where the solar summers attain their maximum, and that they then increase from 74° to the pole. Lambert ¶, with that

---

* *Phil. Trans.* 1693, p. 878. † This should be 2.339.—Ed.

‡ *Isag. in Aratum*, cap. 13.; Strabo, *Geogr.* lib. ii. p. 97.

§ *Mem. de l'Acad.* 1719, p. 133; and 1765, p. 145. and 210.

¶ *Pyrometrie oder Vom Maasse des Feuers*, 1779. p. 342.

sagacity which distinguishes all his mathematical researches, has pointed out in his *Pyrometrie* the error of Mairan's theory. He might have added, that this geometer confounds a quantity of heat which a point of the globe receives under the latitude of 60° during the three months of summer, with the maximum to which the inhabitants of these northern regions see their thermometers rising in a clear day. The mean temperatures of the summers, far from decreasing from the pole to the tropics, are under the equator, under the parallel of 45°, and under that of Stockholm, Upsal, or St Petersburg, in the ratio of 81°.86; 69°.8; 61°.16 of Fahrenheit's scale. Reaumur had sent his new thermometers to the torrid zone, to Syria, and to the north. As it was then reckoned sufficient to mark the warmest days, an idea was formed of an *universal summer*, which is the same in all parts of the globe. It had been remarked, and with reason, that the extreme heats are more frequent, and even more powerful, in the temperate zone in high latitudes, than under the torrid zone. Without attending to the mean temperature of months, it was vaguely supposed, that in these northern regions the summers followed the ratio of the thermometrical extremes. This prejudice is still propagated in our own day, though it is well established, that in spite of the length of the days in the north, the mean temperatures of the warmest months at Petersburg, Paris, and the Equator, are 65°.66; 69°.44, and 82°.4. At Cairo, according to the observations of Nouet, the three months of summer are 84°.74, and consequently 19° warmer than at St Petersburg, and 15° warmer than at Paris. The summer heats of Cairo, are almost equal to those I have experienced at Cumana and La Guayra between the tropics.

With regard to the *central emanation* of the system of Mairan, or to the quantity of heat which the earth gives to the ambient air, it is easy to conceive that it cannot act in all seasons. The temperature of the globe at the depths to which we can reach, in general differs little from the mean annual temperature of the atmosphere. Its action is of great importance for the preservation of vegetables; but it does not become sensible in the air, unless where the surface of the globe is not entirely covered with snow, and during those months, whose mean temperature is below that of the whole year. In the south of France, for example, the radiation of the earth may act upon the atmosphere in the five months which precede the month of April. We speak here of the proper heat of the globe, of that which is invariable at great depths, and not of the radiation of the surface of the globe, which takes place even at the summer solstice, and the nocturnal effects of which have furnished M. Prevost with an approximate measure of the direct action of the sun

Mairan had found, that in the temperate zone the heat of the solar summer is to that of the solar winter as 16 to 1. M. Prevost admits for Geneva 7 to 1. Good observations have given me for the mean temperature of the summers and the winters at Geneva 34°.7; 64°.94; and at St Petersburg 46°.94 and 62°.06. These numbers neither express ratios nor absolute quantities, but thermometrical differences considered as the total effect of the calorific influences; the ratios furnished by theory separate the solar heat from every other indirect effect. Euler was not more successful than Mairan in his theoretical essays on the solar heat. He supposes that the negative sines of the sun's altitude during the night give the measure of the nocturnal cooling, and he obtains the extraordinary result †, that under the equator the cold at midnight ought to be more rigorous than during winter, under the poles. Fortunately, this great geometer attached but little importance to this result, and to the theory from which it is deduced. The second memoir of Mairan, without adding to the problems which had been attempted since the time of Halley, has at least the advantage of containing some general views on the real distribution of heat in different continents. It is true, that the extreme temperatures are there constantly confounded with the mean temperatures; but previous to the works of Cotte and Kirwan, it was the first attempt to group the facts, and to compare the most distant climates.

Dissatisfied with the route followed by his predecessors, Lambert, in his Treatise on Pyrometry, directed his attention to two very different objects. He investigated analytical expressions for the curves, which express the variation of temperature in a

* *Du Calorique rayonnant*, p. 271. 277. 292. † *Comment. Petrop.* tom. ii. p. 98.

place where it had been observed, and he resumed in its greatest generality the theorem of solar action. He gives formulæ, from which we may find the heat of any day at all latitudes; but being perplexed with the determination of the nocturnal dispersion of the acquired heat, or the subtangents of the nocturnal cooling *, he gives tables of the distribution of heat under different parallels, and in different seasons †, which deviate so much from observation, that it would be very difficult to ascribe these deviations to the heat radiating from the globe, and to disturbing causes. We are struck with the slight difference which the theory indicates between the mean annual temperatures of places situated under the equator and the polar circle, and between the summers of the torrid zone and those of the temperate zone. It cannot be expected, indeed, that analysis is capable of determining the distribution of heat such as it exists on the surface of the globe. Without employing empirical laws, and deducing the data from actual observation, the theory can subject to calculation only a part of the total effect, or that which belongs to the immediate action of the solar rays; but after the recent successful applications of analysis to the phenomena of the radiation of surfaces, the transmission of heat through solid bodies, and the cooling of these bodies in media of variable density, we may still expect to be able to perfect the theory of solar action, and to compute the distribution of the heat received into the exterior crust of our planet.

In discussing what may be expected from the purely theoretical labours of Geometers, I have not spoken of a celebrated, but very concise Memoir of Mayer, the reformer of the Lunar Tables. This work, written in 1755, was published twenty years afterwards, in his *Opera Inedita* ‡. It is a method, and

* *Pyrometrie*, p. 141. 179. † *Id.* 318. 339.

‡ *De Variationibus Thermometri accuratius definiendis*, (*Opera inedita*, vol. i. p. 3—10.) M. Daubuisson, in a note inserted in the *Journal de Physique*, tom. lxii. p. 449. has given a formula which accords better with observation than that of Mayer. He admits that the temperature increases from the pole to the equator, as the cosine of the latitude raised to the power of $2\frac{1}{2}$°; but he judiciously adds, that this formula is applicable only to a zone of the Old World, near the Northern Atlantic Ocean.—H.

not a theory : It is an essay essentially different from those we have quoted, and, as its learned author calls it, a determination of the mean heat found empirically by the application of co-efficients furnished by observation. The method of Mayer is analogous to that which Astronomers pursue with so much success, when they correct by small steps the mean place of a planet, by means of the inequalities of its motion: It does not present the result of the solar action disengaged from the influence of foreign circumstances; but, on the contrary, it estimates the temperatures such as they are distributed over the globe, whatever be the cause of that distribution. The mean heat of two places situated under different latitudes being given, we find by a simple equation the temperature of every other parallel *. The calculations of Mayer, according to which the temperatures decrease from the equator to the poles, as the squares of the sines of the latitudes, give results sufficiently precise, when the place does not differ much in longitude from that of the regions where the empirical co-efficients have been obtained. But, even in the northern hemisphere, when we apply the formula to places situated 70° or 80° to the east or west of the meridian of Paris, the calculated results no longer agree with observation. The curve which passes through those points whose temperature is 32°, does not coincide with any terrestrial parallel. If, in the Scandinavian Peninsula, we meet with this curve under the 65th or 68th de-

* The formula given by Mayer was $T = 24 \cos^2$ Lat.; or $T = 12 + 12 \cos 2$ Lat. for Reaumur's scale; and $T = 84 - 52 \sin^2$ Lat., or $T = 58 + 26 \cos 2$ Lat. for Fahrenheit's scale. Since the publication of Humboldt's memoir, M. Daubuisson has resumed the subject of the earth's temperature in his *Traité de Geognosie*, tom. i. p. 424. Paris, 1819. He gives the following formula, which is almost the same as that of Mayer, for finding the mean temperature, according to the Centigrade scale, viz. $T = 27° \cos^2$ Latitude. This formula, which is superior in accuracy to Mayer's, gives all the temperatures in *defect* for latitudes below 42°, and in *excess* for all the higher latitudes, as appears from Daubuisson's table. It is therefore obviously defective. M. Daubuisson, however, considers it as applicable principally between the parallels of 30° and 60° of N. lat. It ought to be remarked, that in the above formula, 27° has been *assumed as the mean temperature* of the equator, in order to make the results agree with observations made in the temperate regions, whereas the mean temperature of the equator, as ascertained by Humboldt, is 27°.5; and if this were used in Daubuisson's formula, it would make the differences still more in excess.—Ed.

gree of latitude, it descends, on the contrary, in North America, and Eastern Asia, to the parallel of from 53° to 58°. But the direction and the inflexions of this curve of 32° of temperature influences the neighbouring isothermal lines in the same manner as the inflexions of the magnetic equator modify the lines of inclination. To demand what is the mean temperature, or what is the magnetic inclination under a particular degree of latitude, is to propose problems equally indeterminate. Though, even in high latitudes the magnetic and the isothermal lines are not rigorously parallel to the magnetic equator, and to the curve of 32° of temperature; yet it is the distance of any place from this curve which determines the mean temperature, as the inclination of the needle depends on the magnetic latitude.

These considerations are sufficient to prove, that the empirical formulæ of Mayer require the introduction of a co-efficient, which depends upon the longitude, and consequently on the direction of the isothermal lines and their nodes with the terrestrial parallels. Mayer had no intention of disengaging the results which he obtained from the influence of all disturbing causes: He limited himself to the determination of the effects of altitude above the level of the sea, and those of the seasons, and the length of the day. He wished to point out the way which philosophers ought to pursue in imitating the method of astronomers. His Memoir was written at a time when we did not know the mean temperature of three points on the globe; and the corrections which I propose after tracing the isothermal lines, so far from being incompatible with the method of Mayer, are, on the contrary, among the number of those which this geometer seems to have indistinctly foreseen.

Kirwan, in his work on Climates, and in a learned Meteorological Memoir, inserted in the eighth volume of the Memoirs of the Irish Academy, attempted at first to pursue the method proposed by Mayer, but, richer in observations than his predecessors, he soon perceived, that after long calculations, the results agreed ill with observation *. In order to try a new method, he selected, in the vast extent of sea, those places

* Kirwan's *Estimate of the Temperature of the Globe*, chap. iii.

whose temperature suffered no change but from permanent causes. These were in the part of the great ocean commonly called the Pacific Ocean, from 40° of South to 45° of North latitude, and in the part of the Atlantic Ocean, between the parallels of 45° and 80°, from the coasts of England to the Gulf Stream, the high temperature of which was first determined by Sir Charles Blagden. Kirwan tried to determine for every month the mean temperature of these seas at different degrees of latitude; and these results afforded him terms of comparison with the mean temperatures observed on the solid part of the terrestrial globe. It is easy to conceive, that this method has no other object, but to distinguish in climates that is in the total effect of *calorific influences*, that which is due to the immediate action of the sun on a single point of the globe. Kirwan first considers the earth as uniformly covered with a thick stratum of water, and he then compares the temperatures of this water at different latitudes, with observations, at the surface of continents indented with mountains, and unequally prolonged towards the poles.

This interesting investigation may enable us to appreciate the influence of local causes, and the effect which arises from the position of seas, on account of the unequal capacity of water and earth for absorbing heat. It is even better fitted for this object than the *Method of Means* deduced from a great number of observations made under different meridians; but in the actual state of our physical knowledge, the method proposed by Kirwan cannot be followed. A small number of observations made far from the coasts, in the course of a month, fixes, without doubt, the mean annual temperature of the sea at its surface, and, on account of the slowness with which a great mass of water follows the changes of the temperature of the surrounding air, the extent of variations in the course of a month is smaller in the ocean than in the atmosphere: But it is still greatly to be desired *, that we should be able to indicate by direct experience, for every parallel, and for every month, the mean temperature of the ocean under the temperate zone. The scheme which Kirwan has formed for the extent of the seas, that ought to

* See my *Relation Historique*, tom. i.

form the term of comparison, is founded only in a small degree upon the observations of navigators, and to a great degree on the theory of Mayer. He has also confounded experiments made on the superficial temperature of the ocean with the results of meteorological journals, or the indications of the temperature of the air which rests upon the sea: He has obviously reasoned in a circle, when he modified, either by theoretical suppositions or by observations made on the air upon the coasts of continents, the table of the temperature of the ocean, in order to compare afterwards with these same results, partly hypothetical, those which observation alone furnished in the interior of the earth. After the works of Kirwan, we must notice those of Cotte, which are merely laborious, though useful, compilations, which, however, ought not to be used without much circumspection. A critical spirit has rarely presided over the reduction of the observations, and they are not arranged so as to lead to general results.

In detailing the actual state of our knowledge on the distribution of heat, I have shewn how dangerous it is to confound the results of observation with theoretical deductions. The heat of any point of the globe depends on the obliquity of the sun's rays, and the continuance of their action, on the height of the place, on the internal heat and radiation of the earth in the middle of a medium of variable temperature; and, in short, upon all those causes which are themselves the effects of the rotation of the earth, and the inequal arrangement of continents and seas. Before laying the foundation of a system, we must group the facts, fix the numerical ratios, and, as I have already pointed out, submit the phenomena of heat, as Halley did those of terrestrial magnetism, to empirical laws. In following this method, I have first considered whether the method employed by meteorologists for deducing the mean temperature of the year, the month and the day, is subject to sensible errors. Assured of the accuracy of the numerical averages, I have traced upon a map the isothermal lines, analogous to the magnetic lines of dip and variation. I have considered them at the surface of the earth in a horizontal plane, and on the declivity of mountains in a vertical plane. I have examined

2

the increase of temperature from the pole to the equator, which is inequal under different meridians; the distribution of the same quantity of heat over the different seasons, in the same isothermal parallel, and under different latitudes; the curve of perpetual snows, which is not a line of equal heat; the temperature of the interior of the earth, which is a little greater towards the north, and in high mountains, than the mean temperature of the atmosphere under the same parallel; and, lastly, the distribution of heat in the ocean, and the position of those bands, which may be called Bands of the warmest waters. As the limits of this extract will not permit me to enter, in a detailed manner, upon these different discussions, I shall confine myself solely to the principal results.

It was formerly the custom to take the *maximum* and *minimum* of temperature observed in the course of a year, and to consider half the sum as the mean temperature of the whole year. This was done by Maraldi, De la Hire, Muschenbröek, Celsius, and even Mairan, when they wished to compare the very warm year of 1718 with the excessively cold years of 1709 and 1740. De la Hire was struck with the identity between the uniform temperature of the caves of the Observatory of Paris, and the mean of the observed annual extremes. He appears to have been the first who had an idea of the mean quantity of heat which a point of the globe receives; and he adds, "We may regard the air of the caves as the mean state of the climate *." Reaumur followed also the method of a *maximum*, though he confessed that it was incorrect †. He noticed the hours at which it was necessary to make observations; and after 1735, he published in the Memoirs of the Academy the extremes of daily temperature: he even compared the produce of two harvests with the sum of the degrees of heat to which during two consecutive years the crops had been exposed. When he treated, however, of the mean temperature of the month, he contented himself, as Duhamel did thirty years afterwards, with recording three or four thermometrical extremes. In order to have some idea of the errors of this imperfect method, I may

* *Mem. de l'Acad.* 1719, p. 4. † *Id.* 1735, p. 559.

state, that even in 1777, the mean temperature of Toulon was estimated by Cotte * at 78°.08, though he afterwards found, by employing the whole mass of observations, that it was not more than 60°.26.

In order to diminish the errors of the method of annual extremes, it was perceived, though very late, that it was necessary to subdivide the curve which expresses the variation of temperature. Twenty-four extremes divided among twelve months of the year, give an annual mean more exact than the two extremes of all the observations. The ordinates do not increase uniformly and uninterruptedly up to the maximum of the year, and there are partial inflexions sufficiently regular. The more we subdivide, and the more we know the terms in the series, the more will these terms approximate, and the less error will there be in the supposition of an arithmetical progression, and in that of the equidistance of the different *maxima* and *minima* of temperature. These considerations enable us to appreciate the three methods according to which observations are at present made. 1. Observations are made *three times* a-day, at sunrise and sunset, and at two o'clock in the afternoon. This was done at Geneva during the three years 1796, 1797 and 1798. In the observations, the hour of mid-day was preferred to that of sunset. 2. Observations are made *twice* every day, at the two periods which are supposed to give the maximum and the minimum, namely, at sunrise, and at two o'clock in the afternoon. 3. Observations are made *once* a-day, at an hour which, in different seasons, has been found to represent the mean temperature of the day. It is thus that M. Raymond, by a judicious induction, has proved, that the height of the barometer, at mid-day, gives, in our climates, the mean atmospherical pressure, corrected for the diurnal variation.

In calculating † a great number of observations made between the parallels of 46° and 48°, I have found, that a single observation at sunset, gives a mean temperature which differs only some tenths of a degree from that which is deduced from observations made at sunrise and at two o'clock. The devia-

* *Mem. de la Soc. Royale de Med.* 1777, p. 101. † *De la Formule Barom.* p. 213.

tions of different months, do not exceed 1°.8, and they are very regularly positive or negative, according to the order of the seasons. M. Arago * has examined for seven years the observations of noon. They give for Paris 5°.4 more than the mean temperature of the whole year. Upon high mountains in the temperate zone, the difference is scarcely 1°8 †. By the application of coefficients, variable according to the latitude and the elevation, we may deduce the true mean temperatures from observations made at any particular period of the day, nearly in the same manner as we can ascertain the latitude of a place from altitudes of the sun, taken out of the meridian.

If we do not stop at two observations of the maximum and minimum, but add a third observation, we commit an error more or less serious, if we divide simply by three the sum of the observations, without attending to the duration of the partial temperatures, and to the place which the third observation occupies between the last terms of the series ‡. Experience proves, that the mean temperatures of the year, obtained by two or three observations, do not differ sensibly, if the intermediate observation is sufficiently distant (*four* or *five* hours) from the observation of the *maximum* and *minimum*. Whenever, therefore, we do not take into account the duration of the intermediate temperatures, we should prefer the two observations of the extreme temperature, which is the method most generally adopted. We shall content ourselves with pointing out the errors to which

* The mean of the observations at noon at Paris was 56°.84; at Clermont in Auvergne (elevation 1348 feet), 56°.30; at Strasburg (elevation 453 feet), 55°.22. *Bulletin de la Soc. Philom.* 1814, Oct. p. 95.—H.

† At the Hospice of St Gothard. *Ephem. Soc. Pal.* 1785, p. 47.

‡ *Example.*—On the 13th June, at 4^h in the morning, 46°.4; at 2^h in the afternoon, 55°.4; and at 11^h in the evening, 50°, (erroneously 46°.4, or 8° Centig. in the original). In calculating by the duration, we have

50°.9 the mean for	10^h of interval,	= 509°.0
52.7	9	= 474.3
48.2	5	= 241.0

The true mean of which is 51°.0. The mean of the three observations, as commonly taken, is 50°.6. If we stop at the two extreme temperatures, we shall have for their half sum 50°.45.—H.

it is liable. In our climates, the two extreme terms do not divide the series of twenty-fours into two equal parts. The maximum is an epoch nearly fixed: the rising of the sun retards or hastens it three hours. As we ought to take into account the duration of the partial temperature, in order to find the quantity of heat divided between the night and the day, we must couple the *maximum* of one day with the *minimum* of the day following, and not be content with taking half the sum of all the *maxima* and *minima* of a month. In the ordinary method, we determine only the mean temperature of the part of the day comprehended between the rising of the sun and two o'clock in the afternoon; and we take it for granted that the mean temperature is the same * from two o'clock to sunrise next day. This double error, of want of equi-distance and of the coupling of observations, does not in general produce errors of more than some tenths of degrees, sometimes in excess, and sometimes in defect, since the warm and cold days are mixed †.

All the calculated results will err in defect, if the 365 ordinates, through which the curve of the year passes, do not express an arithmetical progression, and if the partial irregularities do not sensibly compensate one another. It is only on this supposition that we can judge by the extreme terms of the series, of the sum of the terms, that is, of the partial temperatures. It is very obvious, that near the maximum, the increase ought to be more slow than in other points of the curve, and that this increase in the temperature of the air ought to depend on the sine of the sun's altitude, and on the emission of the radiant heat of the globe.

---

* *Example.*—At sunrise at 6^h, 50°; at 2 o'clock in the afternoon, 62°.6. At sunset, 51°.8; at 2^h, 66°.2; at sunrise, 50°. The true means will be for the first 24 hours 56°.9, and for the second 59°.0, for we shall have

For 8^h, $\frac{1}{2}(50°.0 + 66°.2) \times 8 = 450°.4$ for $8^h = 472°.0$

16 $\frac{1}{2}(51°.8 + 62°.6) \times 16 = 915°.2$ for $16^h = 929°.6$

The method commonly employed gives $\frac{1}{2}(50° + 62°.6) = 56°.3$, and $\frac{1}{2}(66°.2 + 51°8) \doteq 58°.1$. The errors being — 0°.6 and + 0.9, sometimes positive and sometimes negative.—H.

† The error disappears when days of equal temperature succeed one another. It amounts to 1°.8, if the mean temperatures of two successive days differ from 7° to 9°, which however very rarely happens.—H.

It appeared to me very important to establish, by observations made at every hour, at different periods of the year, and under different latitudes, the degree of confidence that can be placed in those results which are called *Mean Temperatures.* I have selected from the registers of the Royal Observatory at Paris clear and calm days, which offered at least ten or twelve observations. Under the equator, I have spent whole days in determining the horary increments and decrements of temperature, in marking the thermometer both in the shade and in the sun, and also the progress of evaporation and humidity; and in order to avoid calculation, I measured with a quadrant the altitude of the sun at each observation. I chose days and nights completely calm, and when the heavens were entirely free from clouds, because the mass of vesicular vapours interrupts the radiation from the earth. The result of these experiments has been very satisfactory, and proves, what had already been deduced from the coincidence between the temperature of the earth and the mean of daily observations, and from the regular progress of the mean temperatures of months in different years, that the effects of small disturbing causes may be compensated by a great number of observations *. I have obtained analogous results by taking, for several months, the mean of 9 o'clock in the morning, of sunrise and midnight. I have computed the temperatures by the distance of the *maximum* expressed in time, and on the supposition of an arithmetical progression. I have found that, under the Torrid Zone, the morning curve from sunrise to the *maximum*, differs very regularly from the even-

* On the 3d and 4th September 1811, lat. 48°.50′.

Sum of the temperatures during 24 hours.	True mean temperature, taking into account the duration.	Half sum of the two extreme temperatures.
625°.71 Fahr.	57°.92 Fahr.	58°.28 Fahr.
672.49	59.90	61.88
834.67	66.74	65.12
834.67	66.74	68.00
835.37	66.74	63.50
	63.61 Mean.	63.35 Mean.

The three last days shew an equality of temperature, which is very surprising, and which does not appear but in the true means.—H.

ing curve. In the morning the true mean temperature, such as we find by taking the duration into account, is a little greater than half the sum of the extremes *. In the evening the error is in a contrary direction, and the series of temperatures approaches more to a progression by quotients. The differences do not in general exceed half a degree, and calculation proves that the compensation is regular. It would be curious to examine the effect which the radiation of the earth has on these horary effects, as the changes of temperature at the surface do not follow the geometrical progression, in so far as they take place in a medium of uniform temperature.

In order to avoid the use of an arbitrary measure, astronomers express the diameters of the planets by taking that of the earth for unity. In like manner, I express the mean temperatures not in parts of the equatorial heat, but by the arithmetical ratios which subsist between this heat and that of the other parallels. This method frees us from the want of uniformity, which arises from the use of different thermometers. Instead of saying that in Europe, under the parallel of 45°, the mean temperature is 13°.4 Centigrade, or 56°.12 of Fahrenheit, we say that it is = 1.0°,487, and in lat. 55° = 1.0°,29. These arithmetical ratios inform us of what is most interesting in the theory of the distribution of heat, that in thermometers whose *zero* is the point of melting ice, the mean temperatures under the latitude of 45° and 55° are, in our regions, the *half* and the *third* nearly of the equatorial temperature, which I suppose to be 81°.5.

(To be continued.)

---

* *Example.*—Latitude 10°.25′.

		Calculation of a true mean by the duration.	Supposition of an arithmetical progression.
Before the *maximum*,	11th September 1799,	70°.52 Fahr.	69°.44 Fahr.
	14th	69.26	68.00
	18th	71.24	70.34
After the *maximum*,	18th August,	68.72	69.80
	20th	70.16	71.24
	27th	68.72	69.26
Before the *maximum*,	17th August,	69.26	68.00
After the *maximum*,	17th August	65.48	66.02
Total effect,	17th August.	67.37	67.01

H.

Art. VIII.—*On Isothermal Lines, and the Distribution of Heat over the Globe.* By Baron Alexander de Humboldt. (Continued from Vol. III. p. 20.)

Having discussed the method of taking averages, and of reducing temperatures to general expressions, we shall now proceed to trace the course of the *Isothermal Lines* on the surface of the Globe, and at the level of the sea. From a slight attention to the difference of climates, it has been remarked, more than a century ago, that the temperatures are not the same under the same parallels; and that in advancing 70° to the east or the west, the heat of the atmosphere suffers a sensible diminution. In pursuance of our method, we shall reduce these phenomena to numerical results, and shew that places situated under the same latitudes do not differ, in America and Europe, by the same number of degrees of temperature, as has been vaguely stated. This assertion would make us suppose that the isothermal lines are parallel in the temperate zone.

		Lat.	Mean Temp.
I. Parallels of Georgia, of the State of Mississippi, of Lower Egypt, and Madeira.	*Natches*, -	31° 28′	64° 8′
	Funchal, -	32 37	68 7
	Orotava, -	28 25	69 8
	Rome, -	41 53	60 4
	Algiers, -	36 48	70 0
	Difference,	7 0	4 1

		Lat.	Mean Temp.
II. Parallels of Virginia, Kentucky, Spain, and the South of Greece.	*Williamsburg,*	38° 8′	58° 0′
	Bourdeaux,	44 50	56 5
	Montpellier,	43 36	59 4
	Rome, -	41 53	60 4
	Algiers, -	36 48	70 0
	Difference,	7 0	7 7
III. Parallels of Pennsylvania, Jersey, Connecticut, Latium, and Romelia.	*Philadelphia,*	39 56	54 9
	New-York,	40 40	53 8
	St Malo, -	48 39	54 5
	Nantes, -	47 13	54 7
	Naples, -	40 50	63 3
	Difference,	7 0	9 5
	Ipswich, -	42 38	50 0
	Cambridge (Amer.)	42 25	50 4
	Vienna, -	48 13	50 5
	Manheim, -	49 29	51 3
	Toulon, -	43 7	62 1
	Rome, -	41 53	60 4
	Difference,	6 30	11 0
IV. Parallels of Canada, Nova Scotia, France, and the South of Germany.	*Quebec,* -	46 47	41 9
	Upsal, -	59 51	41 9
	Padua, -	45 24	57 7
	Paris, -	48 50	51 4
	Difference,	13 0	12 6
V. Parallels of Labrador, the South of Sweden, and Courland.	*Nains,* -	57 0	26 4
	Okak, -	57 20	29 8
	Umea, -	63 50	33 3
	Enontekies,	68 30	27 0
	Edinburgh,	55 58	47 8
	Stockholm,	59 20	42 3
	* Difference,	11 0	17 1

This table † indicates the difference of climates, expressed by that of the mean temperature, and by the number of degrees in latitude which it is necessary to go northward in Europe, in

The differences under the column of latitudes, is the difference of the latitude of a place in Europe and a place in America, which have the same mean temperature; and the differences under the column of mean temperatures, is the difference of the mean temperatures of a place in Europe and a place in America, which have the same latitude.—Ed.

† See my *Prolegomena de distributione geographica plantarum, secundum cœli temperiem et altitudinem montium.* p. 68.—H.

order to find the same quantity of annual heat as in America As a place could not be found in the Old World, whose mean temperature was 48° the same as that of Williamsburg, I have supplied it with an interpolation between the latitudes of two points whose mean temperatures are 56°.5 and 59°.4. By an analogous method, and by employing only good observations, I have found that

1. The *isothermal line* of 32° (0° centig.) passes between Ulco and Enontekies in Lapland (lat. 66° to 68°; East long. from London 19° to 22°), and Table Bay in Labrador (lat. 54° 0', west long. 58°.)

2. The *isothermal line* of 41° (5° centig.) passes by near Stockholm (lat. 60° east long. 18°) and the Bay of St George in Newfoundland (lat. 48°, and long. 59°.)

3. The *isothermal line* of 50° (10° centig.) passes by Belgium (lat. 51°, east long. 2°) and near Boston (lat. 42° 30', west long. 70° 59.)

4. The *isothermal line* of 59° (15° centig.) passes between Rome and Florence (lat. 43° 0', east long. 11° 40') and near Raleigh in North Carolina (lat. 36° 0', and west long. 76° 30'.)

The direction of these lines of equal heat, gives for the two systems of temperature, which we know by precise observations, viz. part of the middle and west of Europe, and that of the coast of America, the following differences:

Latitude.	Mean Temp. of the west of the Old World.	Mean Temp. of the east of the New World.	Difference.
30	70°.52	66°.92	3°.60
40	63.14	54.50	8.64
50	50.90	37.94	12.96
60	40.64	23.72	16.92

If we call the mean equatorial temperature 1, we shall have the half of this temperature in the Old World at 45°, and in the east of the New World, at 39° of lat *.

The mean temperatures decrease

Latitude.		Temp.		Temp.
From 0°—20°	In the Old World,	3°.6	In the New World,	3°.6
20 —30		7.2		10.8
30 —40		7.2		12.6
40 —50		12.6		16.2
50 —60		9.9		13 3
0 —60		40.5		56.5

In both continents, the zone in which the mean temperature decreases most rapidly is comprehended between the parallels of

* This observation relates to the Centigrade scale. If we count the temperatures from 32°, it applies also to Fahrenheit's scale.—Ed.

40° and 45°. Observation here presents a result entirely conformable to theory, for the variation of the square of the cosine, which expresses the law of the temperature, is a *maximum* towards 45° of latitude. This circumstance ought to have a favourable influence on the civilization and industry of the people who inhabit the regions under this mean parallel. It is the point where the regions of the vines touch those of the olives and the citrons. On no other part of the globe, in advancing from north to south, do we observe the temperatures increase more sensibly, and no where else do vegetable productions, and the various objects of agriculture, succeed one another with more rapidity. But a great difference in the productions of contiguous countries, gives activity to commerce, and augments the industry of the cultivators of the soil.

We have traced the direction of the isothermal lines from Europe to the Atlantic Provinces of the New World. We have seen them approach one another from parallelism towards the south, and converge towards the north, particularly between the thermometric curves of 41° and 50°: We shall now endeavour to pursue them to the west. North America presents two chains of mountains, extending from N. E. to S. W., and from N. W. to S. E. forming almost equal angles with the meridian, and nearly parallel to the coasts which are opposite to Europe and Asia, viz. the chain of the *Alleghanys* and the *Rocky Mountains*, which divide the waters of the Missouri and the Columbia. Between these chains stretch the vast basin of the Mississippi, the plains of Lousiana, and of the Tenessee, and the states of Ohio, the centre of a new civilization. It is generally believed in America that the climate is more mild to the west of the Alleghany Mountains, than under the same parallels in the Atlantic States *. Mr Jefferson, has estimated the difference at 3° of latitude; and the *Gleditsia monosperma*, the *Catalpa*, and the *Aristolochia Sypho*, and other vegetable productions, are found so many degrees farther to the north, in the basin of the Ohio, than on the coast of the Atlantic †.

* This is true also of the Columbian Valley. See Warden's *Account of the United States*, vol. iii. p. 169.—Ed.

† See my *Essai sur la Geographie des Plantes*, p. 154.

M. Volney has endeavoured to explain these phenomena by the frequency of the south-west winds, which drive back the warm air of the Gulf of Mexico towards these regions. A series of good observations, made for *seven* years by Colonel Mansfield at Cincinnati, on the banks of the Ohio, and recently published by Mr Drake, in an excellent treatise on American meteorology*, has removed the doubts which obscured this point. The thermometrical *means* prove that the isothermal lines do not rise again in the regions of the west. The quantity of heat which each point of the globe receives under the same parallels, is nearly equal on the east and the west of the Alleghany range, the winters being only a little milder to the west, and the summers a little warmer †. The migrations of vegetables towards the north are favoured in the basin of the Mississippi, by the form and the direction of the valley which opens from the north to the south. In the Atlantic Provinces, on the contrary, the valleys are transverse, and oppose great obstacles to the passage of plants from one valley to another.

If the isothermal lines remain parallel, or nearly so, to the equator, from the Atlantic shores of the New World to the east of the Mississippi and the Missouri, it cannot be doubted that they rise again beyond the *Rocky Mountains*, on the opposite coast of Asia, between 35° and 55° of latitude. To the consi-

* *Natural and Statistical View or Picture of Cincinnati and the Miami Country*, 1 vol. 8vo. Cincinnati.—H. See Warden's *Account of the United States*, vol. ii. 236 for an abstract of Mr Drake's results.—Ed.

† The following comparison of the mean temperatures has been deduced with great care.

Cincinnati.		Philadelphia.	
Lat. 36° 6′, west long. 84° 24′.		Lat. 39° 56′, west long. 75° 16′.	
Winter,	32°.9 Fahr.	Winter,	32°.2 Fahr.
Spring,	54.1.	Spring,	51.4
Summer,	72.9	Summer,	73.9
Autumn,	54.9	Autumn,	56.5
Mean,	53.7	Mean,	53.5

I have taken for Philadelphia the means between the observations of Coxe and Rush. I have also referred for correction to the observations made by M. Legaux at Spring-Mill upon the Schuylkill, to the north of Philadelphia. As Cincinnati is 512 feet above the level of the sea, it's mean temperature is 1°.4 too low.—H.

derations which I pointed out in my work on Mexico *, are to be added the observations of Captain Lewis, and some other Anglo-American travellers, who have passed the winter on the banks of the Columbia. In New California, they cultivate with success the olive, along the canal of Santa Barbara, and the vine from Monterey to the north of the parallel of 37°, which is that of Chesapeake Bay. At Nootka, in the Island of Quadra and Vancouver, and almost in the latitude of Labrador, the smallest rivers do not freeze before the month of January. Captain Lewis saw the first frosts near the embouchure of the Colombia, only on the 7th of January, and the rest of the winter was rainy. Through 122° 40′ of west long. the isothermal line of 50° Fahr. appears to pass almost as in the Atlantic part of the Old World, at 50° of lat. The western coasts of the two worlds resemble one another to a certain point †. But these returns of the isothermal lines do not extend beyond 60°. The curve of 32° Fahr. is already found to the south of the Slave Lake, and it comes still farther south in approaching Lakes Superior and Ontario.

In advancing from Europe towards the east, the isothermal lines again descend ‡, the number of fixed points being few. We can only employ those which are made in places whose known elevation allows us to reduce the mean temperatures to the level of the sea. The few good materials which we possess, have enabled us to trace the curves of 32° and 55°.4. We know even the nodes of the latter curve round the whole globe. It

* *Essai Politique sur la Nouvelle Espagne*, tom. ii. p. 440, 478, 509.

† On account of the influence of west and south-west winds. See Dalton's *Meteor. Observ.* p. 125.

‡ In comparing places from the west to the east, and nearly under the same parallel, we find,

West.	Lat.	Mean Temp.	East.	Lat.	Mean Temp.
St Malo,	48° 39′	54°.5	Vienna,	48° 13′	50°.5
Amsterdam,	52 21	53 .4	Warsaw,	52 14	48 .6
Naples,	40 50	63 .3	Pekin,	39 54	54 .9
Copenhagen,	55 41	45 .7	Moscow,	55 46	40 .1
Upsal,	59 52	41 .9	Petersburgh,	59 56	38 .8

The elevation of Pekin is inconsiderable. That of Moscow is 984 feet. The absolute temperature of Madrid, to the west of Naples, is 59°; but the city is elevated 1978 feet above the level of the sea.—H.

passes to the N. of Bourdeaux, (lat. 45° 46′, W. long. 0° 37′,) near Pekin, (lat. 39° 54′, E. long. 116° 27′.,) and Cape Foulweather to the S. of the embouchure of the Colombia, (lat. 44° 40′, W. long. 104°.) Its nodes are distant at least 162° of longitude. We have here pointed out only the empirical laws. under which are ranged the general phenomena, and the variations of the temperature which embrace at once a vast extent of the globe. There are *partial inflexions* of the isothermal lines, which form, so to speak, particular systems modified by small local causes; such as the strange inflexion of the thermometric curves on the shores of the Mediterranean, between Marseilles, Genoa, Lucca and Rome *, and those which determine the difference between the climate of the western coast and the interior of France. These last depend much less on the quantity of heat received by a part of the globe during the whole year, than upon the unequal distribution of heat between winter and summer. It will one day be useful to have upon particular charts the partial inflections of the isothermal lines, which are analogous to the lines of soundings or of equal declivity. The employment of graphical representations will throw much light upon phenomena, which are deeply interesting to agriculturists. If, in place of geographical charts, we possessed only tables containing the co-ordinates of latitude, longitude, and altitude, a great number of curious facts relative to the configuration and the superficial inequalities of continents would have remained forever unknown.

We have already found, that towards the north, the isothermal lines are neither parallel to the equator nor to one another; and it is on account of the want of parallelism, that we have, in order to simplify such complicated phenomena, traced round the whole globe the curves of equal heat. The position of the line of 32° acts like the magnetic equator, whose inflexions in the South Sea modify the inclinations at great distances. We may even believe that, in the distribution of climates, the line of 32° determines the position of the *curve of greatest heat*,

	Lat.	Mean Temp.		Lat.	Mean Temp.
Bologna,	44° 29′	56°.3	Marseilles,	43° 17′	58°.8
Genoa,	44 25	60.6	Rome,	41 53	60.4

H.

which is as it were the isothermal equator, and that in America and Asia through 78° of west, and 102° of east longitude, the torrid zone commences more to the south of the tropic of Cancer, or that it there presents temperatures of less intensity. An attentive examination of the phenomena proves that this is not the case. Whenever we approach the torrid zone below the parallel of 30°, the isothermal lines become more and more parallel to one another, and to the earth's equator. The great colds of Canada and Siberia do not extend their action to the equatorial plains. If we have long regarded the Old World as warmer between the tropics than the new world, it is, *first*, Because till 1760, travellers used thermometers of spirit of wine, coloured, and affected by light; *2d*, Because they observed it either under the reflection of a wall, or too near the ground, and when the atmosphere was filled with sand; and, *3d*, Because in place of calculating the true *mean*, they used only the thermometric *maximum* and *minimum*. Good observations give,

Old World.	Lat.	Mean Temp.	New World.	Lat.	Mean Temp.
Senegambia,	15° 0	79°.07	Cumana,	10° 27′	81°.86
Madras,	13 5	80 .42	Antilles,	17 0	81.05
Batavia,	6 12	80 .42	Vera Cruz,	19 11	78.08
Manilla,	14 36	78 .08	Havannah,	23 10	78.08

The mean temperature of the equator cannot be fixed beyond 81½°. Kirwan values it at 84°, but only two places of the earth were known, viz. Chandernagor and Pondicherry, to which old travellers attributed annual temperatures above 81°½. At Chandernagor, in latitude 21°.6, the mean temperature, according to Cotte, is 91°.9, but the Jesuite Boudier marked only the days when the thermometer was above 98°.6, and below 57°.2. And at Pondicherry, in latitude 11° 55′, the mean temperature, according to Cotte, is 85°.3, and according to Kirwan, 88°; but M. de Cossigny observed with a spirit-of-wine thermometer.

The distribution of heat over different parts of the year differs, not only according to the decrease of the mean annual temperatures, but also in the same isothermal line. It is this unequal division of the heat which characterises the two systems of climate of Europe and Atlantic America. Under the torrid zone, a small number of months are warmer in the Old World than in the New. At Madras, for example, according to Dr

Roxburgh, the mean temperature of June is 89°.4; at Abusheer 93°.2, but at Cumana I have found it only 84°.6.

With respect to the temperate zone, it has been long known, that from the parallel of the Canary Isles to the Polar Circle, the severity of the winter augments in a progression much more rapid than the summers diminish in heat. It is also known, that the climate of the islands and the coasts differs from that of the interior of continents, the former being characterised by mild winters and less temperate summers. But it is the heat of summer particularly which affects the formation of the amylaceous and saccharine matter in fruits, and the choice of the plants that ought to be cultivated. As the principal object of this memoir is to fix, after good observations, the numerical relations between the unequal quantities of heat distributed over the globe, we shall now compare the mean temperatures of three months of winter and summer under different latitudes, and shew how the inflections of the isothermal lines modify these relations. In following the curves of equal heat from west to east, from the Basin of the Mississippi to the eastern coasts of Asia, through an extent of 4000 leagues, we are struck with the great regularity which appears in the variations of the winter temperature.

I. *Differences of the Seasons from the Equator to the Polar Circle.*

Isothermal Lines of	Cisatlantic Region. Long. 1° W. and 17° E.			Transatlantic Region. Long. 58°— 72° W.		
	Mean Temperature.			Mean Temperature.		
	Winter.	Summer.	Diff.	Winter.	Summer.	Diff.
68°	59°.0	80°.6	21°.6	53°.6	80°.6	27°.0
59	44.6	73.4	28.8	39.2	78.8	39.6
50	35.6	68.0	32.4	30.2	71.6	41.4
41	24.8	60.8	36.0	14.0	66.2	52.2
32	14.0	53.6	39.6	1.4	55.4	54.0

This table shews the increase of the difference between the winters and summers from 28° and 30° to the parallels of 55° and 65°. The increase is more rapid in the Transatlantic Zone, where the isothermal lines of 32° and 50° approach one another very much; but it is remarkable, that in the two zones which form the two systems of different climates, the division of the annual temperature between winter and summer is made in such a manner, that, upon the isothermal line of 32°, the difference of the two seasons is almost double of that which is observed on the isothermal line of 68°.

Cisatlantic Region. Long. 31° E. and 22° W.				
Places.	Latitude.	Mean Temperature.		
		Whole Year.	Winter.	Summer.
(Pondicherry)	11°.55	85°.3	77°.0	90°.5
Cairo,	30 02	72.7	57.7	84.7
Funchal,	32 37	68.7	63.9	72.5
Rome.	41 53	60.4	45.9	55.2
Bourdeaux,	44 50	56.5	42.1	70.7
Paris,	48 50	51.4	38.3	66.2
Copenhagen,	55 41	45.7	30.7	64.6
Stockholm,	59 20	42.3	25.5	61.9
Drontheim,	63 24	39.9	24.7	61.3
Umeo,	63 50	33.3	12.9	54.9

Transatlantic Region. Long. 69° E. and 99° W				
Places.	Latitude.	Mean Temperature.		
		Whole Year.	Winter.	Summer.
Cumana,	10°.27	81°.9	81°.7	83°.7
Havannah,	23.10	78.1	71.2	83.3
Natchez,	31.28	64.8	48.6	79.2
Cincinnati,	39.06	53.6	32.9	72.9
Philadelphia,	39 56	54.9	32.2	73.9
New York,	40.40	53.8	29.8	79.2
Cambridge,	42.25	50.4	34.0	70.5
Quebec,	46.47	41.9	14.2	68.0
Nain,	57.10	26.4	0.6	48.4
Fort Churchhill,	59.02	25.3	6.8	52.2

If, instead of the mean temperatures of the seasons, we consider, I do not say the days of the *maxima* and *minima* of the year, which are the ordinates of the concave and convex summits of the entire curve, but the mean temperatures of the warmest and the coldest month, the increase of the differences becomes still more perceptible. We request the reader to compare in the following Table only the places which belong to regions bounded by the same meridians, and consequently to the same system of climate; as for example, to the region of Eastern America to that of Western Europe, and that of Eastern Asia. We must also attend to the changes of temperature produced by the monsoons in a part of the equinoctial regions, and distinguish under the temperate zone between the climate of the interior, or the continental climate, and that of islands and coasts.

Places.	Lat.	Mean Temperature. Coldest Month.	Mean Temperature. Warmest Month.	Difference.	Observations.
Cumana,	10° 27	80°.1	84°.4	4°.3	Uninterrupted trade winds.
Pondicherry,	11 55	76.1	91.4	15.3	Monsoons. Radiation of the sands.
Manilla,	14 36	68.0	86.9	18.9	Monsoons.
Vera Cruz,	19 11	70.0	81.7	11.7	North winds in winter.
Cape Français,	19 46	77.0	86.0	9.0	Uninterrupted trade winds.
Havannah,	23 10	70.0	83.8	13.8	North winds in winter.
Funchal,	32 37	64.0	75.6	11.6	Insular climate.
Natchez,	31 28	46.9	78.8	31.9	Transatlantic region. Interior.
Cincinnati,	39 6	29.6	74.4	44.8	Same system of climate.
Pekin,	39 54	24.8	84.2	59.4	Region of eastern Asia.
Philadelphia,	39 56	29.8	77.0	47.2	Transatlant. region. Eastern coasts
New York,	40 40	25.3	80.8	55.5	*Idem.*
Rome,	41 53	42.1	77.0	34.9	Cisatlantic region.
Milan,	45 28	33.8	55.2	21.4	Interior land.
Buda,	47 29	27.7	71.6	43.9	*Idem.*
Paris,	48 50	35.1	69.8	34.7	Nearer the western coast.
Quebec,	46 47	14.0	73.4	59.4	Transatlant. region. Eastern coasts
Dublin,	53 21	37.6	60,3	22.7	Region of the west of Europe. Insular climate.
Edinburgh,	55 58	38.3	59.4	21.1	*Idem.*
Warsaw,	52 14	27.1	70.3	43.2	Interior land.
Petersburg,	59 56	8.6	65.7	57.1	East of Europe.
North Cape,	71 0	22.1	46.6	24.5	Climate of coasts and islands.

We may conclude in general, that for any given place in the curves which express the annual temperatures, the ordinates of the concave and convex summits differ the more from one another, as the temperatures diminish. In the New World, under 40° of latitude, we find a greater difference between the warmest and coldest months of the year than in the Old World, at Copenhagen and Stockholm under 56°—59° of latitude. At Philadelphia the thermometer descends to 50° or 59° below the freezing point, while under the same parallel in Europe it descends scarcely 30°.6 below it.

I have endeavoured to shew, in another work, how this circumstance which characterises the regions which Buffon indicates by the name of *Excessive Climates*, influences the physical constitution of the inhabitants. In the United States, the Europeans, and indeed all the natives, are, with great difficulty, inured to the climate. After winters that have been very rigorous, not from the general temperature, but from an extreme depression of the thermometer, the irritability of the nervous system is prodigiously increased by the excessive heat of summer; and it is undoubtedly to this cause that we must, in a great measure, ascribe the difference in the propagation of the yellow fever, and the different forms of the marsh fever, under the equator, and in the temperate zone of the New World *. On high mountains in islands of little extent, and along the shores, the lines of annual temperature take nearly the same form as in warm climates, having only a less degree of curvature. The difference between the seasons, too, becomes smaller. At the North Cape, in 71° of latitude, and in the isothermal line of 32°, it is almost 11° greater than at Paris, in 49° of latitude, and in the isothermal line of 50°. The sea-breezes and the fogs which render the winters so temperate, diminish at the same time the heats of summer †. The characteristic of any climate is not the difference between the winters, expressed in degrees of the thermometer;—it is this difference, compared with the absolute quantities indicated by the mean temperature of the seasons.

II. *Difference between the Winters and Summers, in following the same Isothermal Line from West to East.*

The differences between the seasons of the year are less great near the convex summits of the isothermal curves, where these curves rise again towards the North Pole, than near the concave summits. The same causes, which affect the inflexion or the greatest curvature of the isothermal lines, tend also to equalise the temperatures of the seasons.

The whole of Europe, compared with the eastern parts of America and Asia, has an insular climate, and, upon the same isothermal line, the summers become warmer, and the winters colder, in proportion as we advance from the meridian of Mont Blanc towards the east or the west. Europe may be considered as the western prolongation of the old continent; and the western parts of all continents are not only warmer at equal latitudes than the eastern parts, but even in the zones of equal annual temperature, the winters are more rigorous, and the summers hotter on the eastern coasts than upon the western coasts of the two continents. The northern part of China, like

---

* *Political Essay on the Kingdom of New Spain*, tom. iv. p. 528.

† Leopold von Buch's *Travels in Lapland*, tom. ii.

the Atlantic region of the United States, exhibits *excessive climates*, and seasons strongly contrasted, while the coasts of New California, and the embouchure of the Colombia, have winters and summers almost equally temperate. The meteorological constitution of these countries in the N. W. resembles that of Europe as far as 50° or 52 of latitude; and without wishing to ascribe the great revolutions of our species solely to the influence of climate, we may affirm that the difference between the eastern and western shores of continents, has favoured the ancient civilisation of the Americans of the west,—facilitated their migrations towards the south, and multiplied those relations with eastern Asia, which appear in their monuments, their religion, traditions, and the division of the year. In comparing the two systems of climates, the concave and convex summits of the same isothermal lines, we find at *New York* the summer of *Rome* and the winter of *Copenhagen*;—at *Quebec* the summer of *Paris* and the winter of *Petersburg*. In China, at *Pekin* for example, where the mean temperature of the year is that of the coasts of Britanny, the scorching heats of summer are greater than at Cairo, and the winters as rigorous as at Upsal.

The mean temperature of the year being equal to the fourth part of the winter, spring, summer and autumnal temperatures, we shall have upon the same isothermal line of 53° 6′ (12° cent.)

At the *concave* summit in *America*, 74° 40′ west long. $\Big\}\ 53°.6 = \dfrac{32° + 52°.3 + 75°.6 + 54°.5}{4}$

At the *convex* summit in *Europe*, 2° 20′ west long. $\Big\}\ 53°.6 = \dfrac{40°.1 + 51°8 + 68°.4 + 54°.1}{4}$

At the concave summit in Asia, 116° 20′ east long. $\Big\}\ 53°.6 = \dfrac{-24°.8 + 54°.7 + 80°.6 + 54.3}{4}$

This analogy between the eastern coasts of Asia and America, sufficiently proves that the inequalities of the seasons, of which we have endeavoured to fix the numerical relations, depend on the prolongation and enlargement of continents towards the pole; of the size of seas in relation to their coasts, and on the frequency of the N. W. winds, which are the *Vents de Remous* of the temperate zone, and not on the proximity of some plateau or elevation of the adjacent lands. The great plateaus of Asia do not stretch beyond 52° of latitude; and in the interior of the New Continent, all the immense basin bounded by the Allegany Range, and the rocky mountains, and covered with secondary formations, is not more than from 656 to 920 feet above the level of the ocean, according to the levels taken in Kentucky, on the banks of the Monongahela, at Lake Erie *.

The following table indicates for all the habitable temperate zone the division of the same quantity of annual heat between the two seasons of winter and summer. The numbers which it contains, are either the result of direct observations, or of interpolations between a great number of observations made in neighbouring places, and situated under the same meridian. We have followed each isothermal curve from west to east, giving the preference to places situated near the summits of the curve, as presenting at the same time the greatest differences in the distribution of the annual heat. The longitudes are reckoned from the Observatory of Greenwich.

Isothermal Lines from 32° to 68°.

	Long.	Lat.		Mean Temperature. Winter.	Mean Temperature. Summer.
Isoth. Line of 68°.	82° 10′ W.	29° 30	*Florida,*	53°.6	80°.6
	16 56 W.	32 37	*Madeira,*	63.5	72.0
	3 0 E.	36 48	*North Africa,*	59.0	80.6
Isoth. Line of 63°.5	89 40 W.	32 30	*Mississippi,*	46.4	77.0
	14 11 E.	40 50	*Italy,*	50.0	77.0
Isoth. Line of 59°.	84 10 W.	35 30	*Basin of the Ohio,*	39.2	77.9
	3°—4° E.	43 30	*Middle of France,*	44.6	75.2
Isoth. Line of 54°.5	84 40 W.	38 30	*America, W. of Allegh.*	34.7	75.2
	74 10 W.	40 0	*America, E. of ditto.*	32.5	77.0
	1 32 W.	47 10	*West of France,*	39.0	68.0
	9 20 E.	45 30	*Lombardy,*	34.7	73.4
	116 20 E.	40 0	*East of Asia,*	26.6	82.4
Isoth. Line of 50°.	84 20 W.	41 20	*America W. of Allegh.*	31.1	71.6
	71 10 W.	42 30	*America E. of ditto.*	30.2	73.4
	6 40 W.	52 30	*Ireland,*	39.2	59.5
	0 40 W.	53 30	*England,*	37.4	62.6
	2 20 E.	51 0	*Belgium,*	36.5	63.5
	19 0 E.	47 30	*Hungary,*	31.1	69.8
	116 20 E.	40 0	*Eastern Asia,*	23.0	78.8
Isoth. Line of 45°.5.	71 0 W.	44 42	*America E. of Allegh.*	23.9	71.6
	2 10 W.	57 0	*Scotland,*	36.1	56.5
	12 35 E.	55 40	*Denmark,*	30.3	62.6
	21 20 E.	53 5	*Poland,*	28.0	66.2
Isoth. Line 41°.	71 10 W.	47 0	*Canada,*	14.0	68.0
	9 20 E.	62 45	*West of Norway,*	24.8	62.6
	17 20 E.	60 30	*Sweden,*	24.8	60.8
	24 20 E.	60 0	*Finland,*	23.0	63 5
	36 20 E.	58 30	*Central Russia,*	22.1	68.0
Isoth. Line of 36°.5.	71 40 W.	50 0	*Canada,*	6.8	60.8
	18 5 E.	62 30	*West coast of Gulf of Bothnia,*	17.6	57.2
	22 20 E.	62 50	*East coast of ditto.*	16.5	59.0
Isoth. Line of 32°.	57 40 E.	53 0	*Labrador,*	3.2	51.8
	19 50 E.	65 0	*Sweden,*	11.3	53.6
	25 20 E.	71 0	*North Extremity of Norway,*	23.9	43.7

* Drake's *Nat. and Statist. View of Cincinnati*, p. 63.

When we consider that the annual temperature of a place is nothing more than the numerical expression of the mean of the ordinates, we may imagine an infinity of entirely dissimilar curves, in which the twelve ordinates of the months have exactly the same mean. This consideration should not lead us to believe, that a place which has the winter of the south of France, that is, where the mean temperature of winter is 44°.6, may, by the compensation of a summer and an autumn, much less warm, have the mean temperature of Paris. It is true, that the constant ratio which is observed in the same parallel, between the solstitial heights of the sun and the semidiurnal arcs, is differently modified by the position of a place in the centre of a continent or upon the coast, by the frequency of certain winds, and by the constitution of an atmosphere more or less favourable to the transmission of light, and of the radiating caloric of the earth. But these variations, which travellers have often exaggerated, have a maximum which nature never oversteps. It is impossible to examine the preceding table without observing, that the division of the annual heat between summer and winter, follows on each isothermal line a determinate type; that the deviations of that type are contained between certain limits, and that they obey the same law in the zones which pass by the concave or convex summits of the isothermal lines, for example, by 58°—68° of West Long., and by 5°—7°, and 116° of East Longitude.

The following table shews the oscillations, or the *maxima* and *minima*, observed in the division of the heat between the seasons. I have added the means of the winters and summers found at different degrees of longitude, and under the same isothermal line.

Isoth. Lines.	Degrees of Long. examined.	Oscillations observed in the Means.				Means calculated.	
		Winters.		Summers.		Winters.	Summers.
32°	83	3°.2 to	24°.8	51°.8 to	53°.6	14°.0	52°.7
41	107	14.0	24.8	62.6	68.0	19.4	65.3
50	200	23.0	37.4	62.6	78.8	30.2	70.7
59	87	39.0	44.6	75.2	77.0	41.9	75.2
68	84	53.6	59.0	71.6	80.6	56.3	77.9

The deviations round the mean, that is, the inequality of the winters on the same isothermal line, increase in proportion as the annual heat diminishes, from Algiers to Holland, and from Florida to Pennsylvania. The winters of the curve of 68° are not found upon that of 51°, and the winters of 51° are not met with on the curve of 42°. In considering separately what may be called the same *system of climate*, for example, the European Region, the Transatlantic Region, or that of Eastern Asia, the limits of the variations become still more narrow. Wherever in Europe, in 40° of longitude the mean temperature rises

Mean temperature	The winters are from		and the summers from	
To 59°.0	44°.6 to	46°.4	73°.0 to	75°.2
54.5	36.5	41.0	68.0	73.0
50.0	31.1	37.4	62.6	69.8
45.5	28.4	36.1	57.2	68.0
41.0	20.3	26.8	55 4	66.2

In tracing *five* isothermal lines between the parallels of Rome and Petersburg, the coldest winter presented by one of these lines is not found again on the preceding line. In this part of the globe, those places whose annual temperature is 54° 5, have not a winter below 32°, which is already felt upon the isothermal line of 50. If, in place of stopping at the most rigorous winter which each curve presents, we trace the *lines of equal winter temperature*, (or the *Isocheimal lines*,) these lines, instead of coinciding with the lines of equal annual heat, oscillate round them. As the *Isocheimal lines* unite points placed on different isothermal lines, we may examine to what distance their summits extend. In considering always the same system of climates, for example, the European region, we shall find that the lines of equal winter cut isothermal lines, which are 9° distant. In Belgium * (in latitude 52°, and in isothermal latitude 51° 8,) and even in Scotland, (in latitude 57°, and isothermal latitude 45° 5,) the winters are more mild than at Milan, (in latitude

* Throughout all Holland, 90 days of winter have a mean temperature of from 36°.7 to 38°.7. At Milan, at Padua, and at Verona, the same season is only from 34°·7 to 36°.7. The observations made in Belgium and Holland, offer also a very remarkable example of an equal quantity of heat distributed in the space of a year over a vast extent of territory. The mean temperatures scarcely vary from Paris to Franecker, over 3½ degrees of latitude, which, in the interior of a continent, should produce a difference of 3¼ degrees of annual temperature. The canal of the Channel opens towards the north. The west winds blew therefore, over a great

45° 28′, and isothermal latitude 55° 8′,) and in a great part of Lombardy. Farther to the north, in the Scandinavian Peninsula, we meet with three very different systems of climate, viz. 1. The region of the west coasts of Norway to the west of the mountains. 2. The region of the eastern coasts of Sweden, to the east of the mountains. And, 3. The region of the west coasts of Finland, along the Gulf of Bothnia. Baron Von Buch has made us acquainted with the atmospherical constitution of these three different regions, in which the slowest increase of the winter cold is felt from Drontheim to the North Cape, on the west and north-west coasts. At the Isle of Mageroe, (in north lat. 32°,) at the northern extrémity of Europe, under the parallel of 71°, the winters are still 7°.2 milder than at St Petersburg, (in north lat. 38°.8,) but the mean heat of the summers never reaches that of the winters of Montpellier, (in north lat. 59°.4). At the Faroe Isles, under 62° of north lat. the lakes are very seldom covered with ice, and to so temperate a winter succeeds a summer, during which snow often falls upon the plains. Nowhere without the tropics is the division of the annual heat among the seasons more equal. In the temperate zone, under parallels nearer to our own, Ireland presents an ex-

---

part of the ocean, and during a long rainy winter, with the sky almost always clouded, the surface of the earth is less cooled by radiation than farther to the east, in the interior of the country, where the atmosphere is pure and dry.

	Lat.	Mean Temperature.		
		Year.	Winter.	Summer
Franecker,	52° 36′	51°.8	36°.7	67°.3
Amsterdam,	52 22	53.4	36.9	65.8
Hague,	52 3	51.8	38.3	65.5
Rotterdam,	51 54	51.1	36.9	64.9
Middelburg,	51 30	49.6	36.1	64.0
Dunkirk,	51 2	50.5	38.5	64.0
Brussels,	50 50	52.0	36.7	66.2
Arras,	50 17	49.4	35.8	63.3
Cambray,	50 10	52.0	39.0	66.6

The mean duration of the observations at each place is from eight to nine years, and 52,000 partial observations have been employed to obtain nine mean temperatures. A similar harmony in the results is also found in Lombardy.

	Mean Temp.		Mean Temp.
Milan,	55°.8	Bologna,	56°.3
Padua,	56.3	Venice,	56.5
Verona,	55.9		

H.

ample still more striking of the union of very mild winters, with cold and moist summers. Notwithstanding a difference of 4° of latitude, the winters there are as mild as in Britain, while the mean temperature of the summers is three degrees less. This is the true maritime climate. The month of August, which on the same isothermal line, in the east of Europe *, (in (Hungary) has the temperature of 71°.6, reaches only 60°.8 at Dublin. The month of January, whose mean temperature at Milan, and in a great part of Lombardy, is only 35°.6, rises in Ireland to 5°.4, and 7°.2. On the coasts of Glenarm, also (in north lat. 54° 56′,) under the parallel of Konigsberg, the myrtle vegetates with the same strength as in Portugal †. It scarcely freezes there in winter, but the heat of summer is not capable of ripening the vine.

These examples are sufficient to prove, that the isocheimal lines deviate much more than the isothermal lines from the terrestrial parallels. In the system of European climates, the latitudes of two places that have the same annual temperature cannot differ more than from 4° to 5°, while two places whose mean winter temperature is the same, may differ more than 9° or 10 in latitude. The farther we advance to the east, the more rapidly do these differences increase.

The lines of equal summer, or *isotheral curves*, follow a direction exactly contrary to the *isocheimal lines*. We find the same summer temperature at Moscow, in the centre of Russia, and towards the mouth of the Loire, notwithstanding a difference of 11° of latitude. Such is the effect of the radiation of the earth on a vast continent deprived of mountains. It is sufficiently remarkable, that the inflexions of the isothermal lines, and the division of lands and seas are such upon the globe, that every where in North America, in Europe, and in Eastern Asia, the mean temperature of the summers does not denote more than 36° in the parallels of from 45° to 47°. The same causes which in Canada, and in the north of China, sink the curves of equal annual heat, where the isothermal lines (those of

* Wahlenberg *Flora Carpath.* p. 90.

† *Irish Transactions*, tom. viii. p. 116, 203, 269.

51°.8, and 53°.6,) corresponding to the parallels of 45° and 47°, tend to raise the lines of equal summer or the *isotheral curves.*

However great be the influence of the unequal division of the heat between the seasons, on the physical condition of nations, on the developement of agricultural industry, and the selection of plants for culture, I would not recommend the tracing upon the same chart the isothermal lines, and the winter and summer curves. This combination would not be more fortunate than the lines of declination, inclination, and equal intensity of the magnetic forces, which, however, all depend upon one another. Instead of multiplying the intersection of the curves, it will be sufficient to add to the isothermal lines, near their summits, the indication of the mean temperatures of summer and winter. In this way, by following the line of 50°, we shall find marked in America, to the west of Boston, $\left(\frac{30°.2}{73°.4}\right)$, in England $\left(\frac{37°.4}{62°.6}\right)$, in Hungary $\left(\frac{31°.1}{69°.8}\right)$, and in China $\left(\frac{23°.0}{78°.8}\right)$

(To be continued.)

Art. III.—*On Isothermal Lines, and the Distribution of Heat Over the Globe.* By Baron Alexander de Humboldt. (Continued from Vol. III. p. 274.)

After what has already been stated respecting the limits between which the annual heat divides itself on the same isothermal curve, it will be seen how far we are authorised to say, that the *Coffee-tree*, the *Olive*, and the *Vine*, in order to be productive, require mean temperatures of 64°.4; 60°.8, and 53°.6 Fahr. These expressions are true only of the same system of climate, for example, of the part of the Old World which stretches to the west of the meridian of Mont Blanc; because in a zone of small extent in longitude, while we fix the annual temperatures, we determine also the nature of the summers and the winters. It is known likewise, that the olive, the vine, the varieties of grain, and the fruit-trees, require entirely different constitutions of the atmosphere. Among our cultivated plants, some, slightly sensible of the rigours of winter, require very warm but not long summers; others require summers rather long than warm; while others, again, indifferent to the temperature of summer, cannot resist the great colds of winter. Hence, it follows, that, in reference to the culture of useful vegetables, we must discuss three things for each climate,—the mean temperature of the entire summer,—that of the warmest month,—and that of the coldest month. I have published the numerical results of this discussion in my *Prolegomena de Distributione Geographica Plantarum, secundum Cœli Temperiem*; and I shall confine myself at present to the limits of culture of the olive and the vine. The olive is cultivated in our continent between the parallels of 36° and 44°, wherever the annual temperature is from 62°.6 to 58°.1, where the mean temperature of the coldest month is not below from 41°.0 to 42°.8, and that of the whole summer from 71°.6 to 73°.4 *. In the New World, the division of heat between the seasons is such, that on the isothermal line of 58°.1, the coldest month is

* In cases like the present, we have not used the round numbers of Fahrenheit, as is done in the original with the Centigrade scale, but have given the real value of the degrees used by the author, that his exact numbers may always be ascertained.—Ed.

35°.6, and that the thermometer sometimes sinks there even during several days from 14° to 10°.4. The region of potable wines extends in Europe between the isothermal lines of 62°.6 and 50°, which correspond to the latitudes of 36° and 48°. The cultivation of the vine extends, though with less advantage, even to countries whose annual temperature descends to 48°.2 and to 47°.48; that of winter to 33°.8, and that of summer to 66°.2 and 68°. These meteorological conditions are fulfilled in Europe as far as the parallel of 50°, and a little beyond it. In America, they do not exist farther north than 40°. They have begun, indeed, some years ago to make a very good red wine to the west of Washington, beyond the first chain of mountains, in the valleys which do not extend beyond 38° 54′ of Lat. On the Continent of Western Europe, the winters, whose mean temperature is 32°, do not commence till on the isothermal lines of 48°.2 and 50°, in from 51° to 52° of latitude; while in America, we find them already on the isothermal lines of from 51°.8 to 53°.6, under from 40° to 41° of latitude.

If, instead of considering the natural inflexions of the isothermal lines, that is to say, those that propagate themselves progressively at great intervals of longitude, we direct our attention to their partial inflexions, or to *particular systems of climates* occupying a small extent of country, we shall still find the same variations in the division of the annual heat between the different seasons. These partial inflexions are most remarkable,

1*st*, In the Crimea, where the climate of Odessa is contrasted with that of the S.W. shores of the Chersonesus, sheltered by mountains, and fit for the cultivation of the olive and the orange tree.

2*dly*, Along the Gulf of Genoa, from Toulon and the Hieres Isles to Nice and Bordighera, (*Annales du Museum*, tom. xi. p. 219.), where the small maritime palm-tree, *Chamærops*, grows wild, and where the date-tree is cultivated on a large scale, not to obtain its fruit, but the palms or etiolated leaves.

3*dly*, In England, on the coast of Devonshire, where the port of Salcombe has, on account of its temperate climate, been called the Montpellier of the North, and where (in South Hams) the Myrtle, the *Camellia Japonica*, the *Fuchsia coccinea*, and the

*Buddleia globosa**, pass the winter in the open ground, and without shelter.

4*thly*, In France, on the western coasts of Normandy and Brittany. In the Department of Finisterre, the arbutus, the pomegranate-tree, the *Yucca gloriosa* and *aloifolia*, the *Erica Mediterranea*, the *Hortensia*, the *Fuchsia*, the *Dahlia*, resist in open ground the inclemency of a winter which lasts scarcely fifteen or twenty days, and which succeeds to a summer by no means warm. During this short winter, the thermometer sometimes falls to 17°.6. The sap ascends in the trees from the month of February; but it often freezes even in the middle of May. The *Lavatera arborea* is found wild in the isle of Glenans, and opposite to this island, on the continent, the *Astragalus Bajonensis*, and the *Laurus nobilis* †.

From observations made in Brittany for twelve years, at St Malo, at Nantes, and at Brest, the mean temperature of the peninsula appears to be above 56°.3. In the interior of France, where the land is not much elevated above the sea, we must descend 3° of latitude in order to find an annual temperature like this.

It is known from the researches of Arthur Young ‡, that in spite of the great rise of the two isothermal lines of 53°.6 and 55°.4 on the western coast of France, the lines of culture (those of the olive, and of the maize and vine,) have a direction || quite opposite, from S.W. to N.E. This phenomenon has been ascribed §, with reason, to the low temperature of the summers

* Knight, *Trans. Hort. Soc.* vol. i. p. 32. In 1774, an Agave flowered at Salcombe, after having lived twenty-eight years without being covered in winter. On the coast of England, the winters are so mild, that orange trees are seen on espaliers, which are sheltered, as at Rome, only by means of a matting.—H.

† Bonnemaison, *Geogr. Botan. du Depart. du Finisterre*, (*Journal de Botan.* tom. iii. p. 118.)

‡ *Travels in France*, vol. ii. p. 91.

|| The line which limits the cultivation of the vine, extends from the embouchure of the Loire and of the Vilaine, by Pontoise, to the confluence of the Rhine and the Moselle. The line of the olive trees commences to the west of Narbonne, passes between Orange and Montelimart, and carries itself to the N.E. in the direction of the Great St Bernard.—H.

§ Decandolle, *Flor. Franç.* 2d edit. tom. ii. pl. viii. xi. Lequinio, *Voy. dans le Jura*, tom. ii. p. 81.—91.

along the coast; but no attempt has been made to reduce to numerical expressions the ratios between the seasons in the interior and on the coast. In order to do this, I have chosen *eight* places, some of which lie under the same geographic parallels, and others in the prolongation of the same isothermal line. I have compared the temperatures of winter, of summer, and of the warmest months; for a summer of uniform heat excites less the force of vegetation, than a great heat, preceded by a cold season. The terms of comparison have been along the Atlantic; the coasts of Brittany, (from St Malo and St Brieux to Vannes and Nantes); the sands of Olonne; the Isle of Oleron; the embouchure of the Garonne and Dax, in the department of the Landes; and in the interior, corresponding to the same parallel, Chalons sur Marne, Paris, Chartres, Troyes, Poitiers, and Montauban. Farther south, from 44½° of Lat. the comparisons become incorrect, because France, locked between the Ocean and the Mediterranean, presents, along this last basin, in the fine region of the olives, a system of climate of a particular kind, and very different from that of the western coast.

Places in the Interior.	Latitude.	Mean Temperature			
		Of the Year.	Of Winter.	Of Summer.	Of the Warmest Month.
		Fahr.	Fahr.	Fahr.	Fahr.
Chalons sur Marne, -	48°.57	50°.5	36°.1	66°.6	67°.5
Paris, - - -	48.50	51.1	38.7	65.3	67.5
Chartres, - -	48.26	50.7	37.0	64.6	65.7
Troyes, - -	48.18	52.2	38.3	67.3	68.4
Chinon, - -	47.26	53.4	38.7	69.1	70.2
Poitiers, - -	46.39	54.3	39.7	67.1	69.3
Vienne, - -	45.31	55.0	38.7	71.6	73.4
Montauban, -	44.01	55.6	42.6	69.3	71.4
Places on the Coast.					
St Malo, - -	48°.39′	55°.5	42°.4	66°.9	67°.5
St Brieux, - -	48.31	52.3	41.7	64.4	67.1
Vannes, - -	47.39	51.8	39.7	64.4	65.8
Nantes, - -	47.13	54.7	40.5	68.5	70.5
La Rochelle, - -	46.14	53.1	40.3	66.6	67.1
Oleron, - -	45.56	58.1	44.6	68.5	72.1
Bourdeaux, - -	44.50	56.5	42.1	70.9	71.4
Dax, - -	43.52	54.1	44.4	67.3	68.9

These results are deduced from 127,000 observations, made with sixteen thermometers, of, no doubt, unequal accuracy. In supposing, on the theory of probabilities, that in such a number of observations, the errors, in the construction and exposure of the instruments, and in the hours of observation, will, in a great measure, destroy one another, we may determine, by interpolation, either under the same parallel, or upon the same isothermal line, the *mean winters* and *summers* of the interior and of the coast of France. This comparison gives,—

			Mean Winter.	Mean Summer.	Annual Temp.
I. Isothermal Lines of	52°.7	Coast,	40°.6	65°.1	
		Interior,	38.5	68.0	
	51°.7	Coast,	41.4	67.3	
		Interior,	39.2	68.4	
I. Parallels of	47° to 49°	Coast,	41°.0	66°.7	53°.0
		Interior,	37.8	66.6	51.6
	45° to 46°	Coast,	42.3	67.8	55.8
		Interior,	39.2	69.3	54.7

As the isothermal lines rise again towards the western coasts of France; that is to say, as the mean temperature of the year becomes there greater than under the same latitude in the interior of the country, we ought to expect, that in advancing from east to west under the same parallel, the heat of the summers would not diminish. But the rising, again, of the isothermal lines, and the proximity of the sea, tend equally to increase the mildness of the winters; and each of these two causes acts in an opposite manner upon the summers. If the division of the heat between these seasons was equal in Brittany and in Orleannois, in the *climate of the coast*, and the *continental climates*, we ought to find the winters and summers warmer in the same latitude along the coast. In following the same isothermal lines, we readily observe, in the preceding table, that the winters are colder in the interior of the country, and the summers more temperate upon the coasts. These observations confirm in general the popular opinion respecting the climate of coasts; but in recollecting the cultivation and the developement of vegetation on the coasts and in the interior of France, we should expect differences of temperature still more considerable. It is surprising that these differences between the winter and the summer should not exceed 1°.8, or nearly a quarter of the difference between the mean temperature of the winters or the summers of Montpellier and Paris. In speaking of the limits of the cultivation of plants upon mountains, I shall explain the true cause of this apparent contradiction. In the mean time, it may be sufficient to remark, that our meteorological instruments do not indicate the quantity of heat, which, in a clear and dry state of the air, the direct light produces in the more or less coloured parenchyma of the leaves and fruits. In the same mean temperature of the atmosphere, the developement of vegetation is retarded or accelerated, according as the sky is foggy or serene, and according as the surface of the earth receives only a diffuse light, during entire weeks, or is struck by the direct rays of the sun. On the state of the atmosphere, and the degree of the extinction of light, depend, in a great measure, those phenomena of vegetable life, the contrasts of which surprise us in islands, in the interior of continents, in plains, and on the summit of mountains. If we neglect these photometrical considerations, and do not appreciate the production of heat in the interior of bodies, and the effect of nocturnal radiation in a clear or a cloudy sky, we shall have some difficulty in discovering, from the numerical ratios of the observed summer and winter temperatures of Paris and London, the causes of the striking difference which appears in France and England in the culture of the vine, the peach, and other fruit-trees *.

When we study the organic life of plants and animals, we must examine all the stimuli or external agents which modify their vital actions. The ratios of the mean temperatures of the months are not sufficient to characterise the climate. Its influence combines the simultaneous action of all physical causes; and it depends on heat, humidity, light, the electrical tension of vapours, and the variable pressure of the atmosphere. It is the last cause which, on the tops of mountains, modifies the perspiration of plants, and even increases the exhaling organs. In making known the empirical laws of the distribution of heat

* Young's *Travels in France*, vol. ii. p. 195.

over the globe, as deducible from the thermometrical variations of the air, we are far from considering these laws as the only ones necessary to resolve all the problems of climate. Most of the phenomena of nature present two distinct parts, one which may be subjected to exact calculation, and another which cannot be reached but through the medium of induction and analogy.

Having considered the division of heat between winter and summer on the same isothermal line, we shall now point out the numerical ratios between the mean temperature of spring and winter, and between that of the whole year and the warmest month. From the parallel of Rome to that of Stockholm, and consequently between the isothermal lines of 60°.8 and 41°, the difference of the months of April and May is everywhere 10°.8 or 12°.6, and all the successive months are those which present the most rapid increase of temperature. But, as in northern countries, in Sweden, for example, the month of April is only 37°.4, the 10°.8 or 12°.6 which the month of May adds *, necessarily produces there a much greater effect on the developement of vegetation than in the south of Europe, where the mean temperature of April is from 53°.6 to 55°.4. It is from an analogous cause, that in passing from the shade to the sun, either in our climates in winter, or between the tropics on the back of the Cordilleras, we are more affected by the difference of temperature than in summer and in the plains, though in both cases the thermometrical difference is the same, for example from 5°.4 to 7°.2. Near the polar circle, the increase of the vernal heat is not only more sensible, but it extends equally to the month of June. At Drontheim, the temperatures of April and May, like those of May and June, differ not 10°.8 or 12°.6, but 14°.4 or 16°.2.

In distinguishing upon the same isothermal line the places which approach its concave or convex summits, in the same sys-

* In calculating for Europe, from 46° to 48° of Lat. for ten years the mean temperatures of every ten days, we find, that the decades which succeed one another, differ near the summits of the annual curve only 1°.44, while the differences rise in autumn from 3°.6 to 5°.4, and in spring from 5°.4 to 7°.2.—H.

tem of climates in the northern and southern regions, we shall find,

1*st*, That the increase of the vernal temperature is great, (from 14°.4 or 16°.2, in the space of a month), and equally prolonged, wherever the division of the annual heat between the seasons is very unequal, as in the north of Europe, and in the temperate part of the United States.

2*dly*, That the vernal increase is great, (at least above 9° or 10°.8), but little prolonged, in the temperate part of Europe.

3*dly*, That the increase of the vernal temperature is small, (scarcely 7°.2), and equally prolonged, wherever there is an insular climate.

4*thly*, That in every system of climates, in the zones contained between the same meridians, the vernal increase is smaller, and less equally prolonged, in low than in high latitudes.

The isothermal zone from 53°.6 to 55°.4, may serve as an example for confirming these different modifications of spring. In Eastern Asia, near the concave summit, the differences of temperature between the four months of March, April, May and June, are very great, and very equal, (15°.7, 13°.3, and 13°.9). In advancing westward towards Europe, the isothermal line rises again, and in the interior of the country, near the convex summit, the increase is still greater, but little prolonged; that is to say, that of the four months which succeed one another, there are only two whose difference rises to 13°: they are 9°.4; 13°.3; 4°.1. Farther west, on the coasts, the differences become small and equal, viz. 3°.6; 6°.5; 5°.6. In crossing the Atlantic, we approach the western concave summit of the isothermal line of 53°.6. The increase of vernal temperature shews itself anew, and almost as great, and as much prolonged, as near the Arctic concave summit. The differences of the four months are 10°.4; 13°.9; and 10°.8. In the curve of annual temperature, the spring and autumn mark the transitions from the *minimum* and the *maximum*. The increments are naturally slower near the summits than in the intermediate part of the curve. Here they are greater, and of longer continuance, in proportion to the difference of the extreme ordinates. The autumnal decrease of temperature is less rapid than the vernal increase, because the surface of the earth acquires the maximum of heat slower than the atmosphere, and because, in spite of the serenity of the air which prevails in autumn, the earth loses slowly, by radiation, the heat which it has acquired. The following Table will shew how uniform the laws are which I have just established.

Names of Places.	Latitude.	March.	April.	May.	June.	Differences of Temperature of the Four Months.			Mean Temp. of the Year.
I. Group,—*Concave Summits in America.*									
Natchez, - - -	31° 28′	57°.9	66°.2	72°.7	79°.5	8°.3	6°.1	7°.2	64°.8
Williamsburg, - -	37 18	46.4	61.2	66.6	77.7	14.8	5.4	11.2	58.1
Cincinnati, - -	39 0	43.7	57.4	61.2	70.9	13.7	3.6	9.7	53.8
Philadelphia, - -	39 56	44.1	53.6	62.1	72.3	9.5	8.5	10.3	53.6
New York, - -	40 40	38.7	49.1	65.8	80.2	10.4	16.7	14.4	53.8
Cambridge, - -	42 25	34.5	45.5	56.8	70.2	11.0	11.3	13.3	50.4
Quebec, - - -	46 47	23.0	39.6	54.7	63.9	16.6	15.1	41.2	41.7
Nain, - - -	57 0	6.8	27.5	37.0	43.3	20.7	9.5	8.1	26.4
II. Group,—*Convex Summits in Europe.*									
1. *Continental Climate :*									
Rome, - - -	41 53	50.4	55.4	66.9	72.3	5.0	11.5	5.4	60.4
Milan, - - -	45 28	47.8	51.1	65.1	70.5	7.7	9.5	5.4	55.8
Geneva, - - -	46 12	39.6	45.5	58.1	62.2	6.1	12.4	4.1	49.3
Buda, - - -	47 29	38.3	49.1	64.8	68.4	10.8	15.7	3.6	51.1
Paris, - - -	48 50	42.3	48.2	60.1	64.4	8.5	11.9	4.3	51.1
Gottingen, - -	51 32	34.2	44.2	57.7	62.2	10.1	13.5	4.5	46.9
Upsal, - - -	59 51	29.5	39.7	48.7	57.9	10.3	9.0	9.2	41.9
Petersburg, - -	59 56	27.5	37.0	50.2	59.4	9.5	13.1	9.2	38.8
Umeo, - - -	63 50	23.0	34.2	43.7	55.0	11.2	9.5	11.3	33.3
Uleo, - - -	65 0	14.0	26.2	41.0	55.0	12.2	14.8	14.0	33.1
Enontekies, - -	68 30	11.5	26.6	36.5	49.5	15.1	9.9	13.0	27.0
2. *Climate of the Coast :*									
Nantes, - - -	47 13	50.0	53.6	60.1	65.7	3.6	6.5	5.6	54.7
London, - - -	51 30	44.2	49.8	56.5	63.1	5.6	6.7	6.7	51.6
Dublin, - - -	53 21	41.9	45.3	51.8	55.6	3.4	6.5	4.0	48.4
Edinburgh, - -	55 57	41.4	47.3	50.5	57.2	5.8	3.2	6.7	47.8
North Cape, - -	71 0	25.0	30.0	31.0	40.1	5.2	4.0	6.1	32.0
III. Group,—*Concave Summit of Asia.*									
Pekin, - - -	39 54	41.4	57.0	70.3	84.2	15.7	13.3	13.9	54.9

In all places whose mean temperature is below 62°.6, the revival of nature takes place in spring, in that month whose mean temperature reaches 42°.8 or 46°.4. When a month rises to,

41°.9, the Peach-tree *(Amygdalus Persica)* flowers.
46°.8, the Plum-tree *(Prunus domestica)* flowers.
51°.8, the Birch-tree * *(Betula alba)* pushes out its leaves.

At Rome, it is the month of March, at Paris the beginning of May, and at Upsal the beginning of June, that reaches the mean temperature of 51°.8. Near the Hospice of St Gothard, the birch cannot vegetate, as the warmest month of the year there scarcely reaches 46°.5. Barley, in order to be cultivated advantageously, requires †, during ninety days, a mean temperature of from 47°.3 to 48°.2. By adding the mean temperatures of the months above 51°.8, that is, the temperatures of those in which trees vegetate that lose their foliage, we shall have a sufficiently exact mean of the strength and continuance of vegetation. As we advance towards the north, vegetable life is confined to a shorter interval. In the south of France, there are 270 days of the year in which the mean temperature exceeds 51°.8; that is to say, the temperature which the birch requires to put forth its first leaves. At St Petersburgh, the number of these days is only 120. These two *cycles of vegetation*, so unequal, have a mean temperature which does not differ more than 5°.4; and even this want of heat is compensated by the effects of the direct light, which acts on the parenchyma of plants in proportion to the length of the days. If we compare, in the following *Table*, Eastern Asia, Europe, and America, we shall discover, by the increase of heat during the cycle of vegetation, the points where the isothermal lines have their concave summits. The exact knowledge of these cycles, will throw more light on the problem of Agricultural Geography, than the examination of the single temperatures of summer.

Cotte, *Meteorologie*, p. 418.—Wahlenberg, *Flor. Lap.* Pl. 51.

† Playfair, *Edin. Trans*, vol. v. p. 202.—Wahlenberg in Gilbert's *Annalen*, tom. xli. p. 282.

Lines of Equal Heat.	Names of Places.	Latitude.	Mean Temp. of the Year.	Sum of the Mean Temp. of the Months that reach 51°.8.	Number of those Months.	Mean Temperat. of the days which reach 51°.8.	Mean Temperature of the warmest Months.	Observations.
Isothermal Line of 59°.0,	Rome,	41° 53′	60°.4	585°	9	64°.8	77°.0	Basin of the Mediterranean.
	Nismes,	43 50	60.3	593	9	65.8	78.3	*Idem.*
Isothermal Line of 53°.6,	Pekin,	39 54	54.9	499	7	71.2	84.2	Eastern concave summit.
	Poitiers,	46 34	54.3	426	7	60.8	69.3	Convex summit.
	Nantes,	47 13	54.7	438	7	62.6	69.8	*Idem*, coasts.
	St Malo,	48 39	53.8	431	7	61.5	68.4	*Idem.*
	Philadelphia,	39 56	53.4	463	7	66.2	77.0	Western concave summit.
	Cincinnati,	39 6	53.8	458	7	65.5	74.3	*Idem.*
Isothermal Line of 50°.0,	London,	51 30	51.8	364	6	60.6	66.6	Insular climate.
	Paris,	48 50	51.1	381	6	63.5	69.8	Near the coasts.
	Buda,	47 29	51.1	323	5	64.6	72.0	Interior.
Isothermal Line of 48°.2,	Geneva,	46 12	49.3	311	5	62.2	66.6	Interior.
	Dublin,	53 21	48.7	282	5	56.5	60.8	Climate of the coasts.
	Edinburgh,	55 57	47.8	279	5	55.8	59.4	*Idem,*
Isothermal Line of 41°.0,	Upsal,	59 51	41.9	229	4	57.2	61.9	Convex summit.
	Quebec,	46 47	41.7	318	5	63.7	73.4	Western concave summit.
Isothermal Line of 32°.0,	Petersburgh,							
	Umeo,	59 56	38.8	236	4	59.0	65.7	East of Europe.
		53 50	33.3	118	2	59.0	62.6	E. Coast of Gulf of Bothnia.
	North Cape,	71 0	32.0	0	0	0	46.6	Interior climate.
	Enontekies,	68 30	27.0	116	2	58.1	59.5	Continental climate.

In the system of European climates, from Rome to Upsal, between the isothermal lines of 59° and 41°, the warmest month adds from 16°.2 to 18° to the mean temperature of the year. Farther north, and also in eastern Asia, and in America, where the isothermal lines bend towards the equator, the increments are still more considerable.

As two hours of the day indicate the temperature of the whole day, there must also be two days of the year, or two decades, whose mean temperature is equal to that of the whole year. From the mean of ten observations, this temperature of the year is found at Buda in Hungary from the 15th to the 20th of April, and from the 18th to the 23d of October. The ordinates of the other decades may be regarded as functions of the mean ordinates. In considering the temperatures of entire months, we find, that to the isothermal line of 35°.6, the temperature of the month of October coincides (generally within a degree) with that of the year. The following Table proves that it is not the month of April, as Kirwan affirms, *(Estimate*, &c. p. 166.), that approaches nearest to the annual temperature.

Names of Places.	Mean Temperature			Names of Places.	Mean Temperature		
	Of the Year.	Of October.	Of April.		Of the Year.	Of October.	Of April.
Cairo, -	72°.3	72°.3	77°.9	Gottingen,	46°.9	47°.1	44°.4
Algiers, -	69.8	72.1	62.6	Franeker,	52.3	54.9	50.0
Natchez,	65.0	68.4	66.4	Copenhagen,	45.7	48.7	41.0
Rome, -	60.4	62.1	55.4	Stockholm,	42.3	42.4	38.5
Milan,	55.8	58.1	55.6	Christiania,	42.6	39.2	42.6
Cincinnati,	53.6	54.9	56.8	Upsal, -	41.7	43.3	39.7
Philadelphia,	53.4	54.0	53.6	Quebec,	41.9	42 8	39.6
New York,	53.8	54.5	49.1	Petersburgh,	38.8	39.0	37.0
Pekin, -	54.7	55.4	57.0	Abo, -	41.4	4 .0	40.8
Buda, -	51.1	52.3	49.1	Drontheim,	39.9	39·2	34.3
London, -	51.8	52.3	49.8	Uleo, -	33.1	37.9	34.2
Paris, -	51.1	51.3	48.2	Umeo, -	23.3	37.8	34.0
Geneva, -	49.3	49.3	45.7	North Cape,	32.0	32.0	30.2
Dublin, -	48.6	48.7	45 3	Enontekies,	27.0	27.5	26.6
Edinburgh,	47.8	48.2	46.9	Nain, -	26.4	33.1	27.5

As travellers are seldom able to make observations for giving immediately the temperature of the whole year, it is useful to know the constant ratios which exist in each system of climates, between the vernal and autumnal temperatures, and the annual temperature.

The quantity of heat which any point of the globe receives, is much more equal during a long series of years than we would be led to believe from the testimony of our sensations, and the variable product of our harvests. In a given place, the number of days during which the N.E. or S.W. winds blow, preserve a very constant ratio, because the direction and the force of these winds, which bring warmer or colder air, depend upon general causes,—on the declination of the sun,—on the configuration of the coast,—and on the lie of the neighbouring continent. It is less frequently a diminution in the mean temperature, than an extraordinary change in the division of the heat between the different months, which occasions bad harvests. By examining between the parallels of 47° and 49° a series of good meteorological observations, made during ten or twelve years, it appears, that the annual temperatures vary only from 1°.8 to 2°.7; those of winter from 3°.6 to 5°.4; those of the months of winter from 9° to 10°.8. At Geneva, the mean temperatures of twenty years were as follows:

Years.	Mean Temp.	Years.	Mean Temp.
1796,	49°.3	1806,	51°.4
1797,	50.5	1807,	49.3
1798,	50.0	1808,	46.9
1799,	48.7	1809,	48.9
1800,	50.5	1810,	51.1
1801,	51.1	1811,	51.6
1802,	50.9	1812,	47.8
1803,	50.4	1813,	48.6
1804,	51.1	1814,	48.2
1805,	47.8	1815,	50.0
		Mean of 20 Years,	49°.67

If, in our climates, the thermometrical oscillations are a sixth part of the annual temperature, they do not amount to one twenty-fifth part under the tropics. I have computed the thermometrical variations, during eleven years, at Paris, for the whole year, the winter, the summer, the coldest month, the warmest month, and the month which represents most accurately the annual mean temperature; and the following are the results which I obtained:

Observations of M. Bouvard.	Mean Temperature					
	Of the Year.	Of Winter.	Of Summer.	Of January.	Of August.	Of October.
Paris, 1803, -	51°.1	36°.7	67°.6	34°.3	67°.6	50°.5
1804, -	52.0	41.0	65.5	43.9	64.6	52.7
1805, -	49.5	36.0	63.1	34.9	64.8	49.3
1806, -	53.4	40.6	65.3	43.0	64.6	51.8
1807, -	51.4	42.3	67.8	36.1	70.5	54.3
1808, -	50.5	36.7	66.2	36.3	66.6	48.2
1809, -	50.9	40.5	62.4	40.8	64.2	49.6
1810, -	50.9	36.5	63.3	30.6	63.7	52.9
1811, -	52.7	39.2	65.1	26.6	63.7	57.6
1812, -	49.8	39.6	63.1	34.7	64.2	51.1
1813, -	49.8	36.1	61.7	32.5	62.6	53.1
Mean of these 11 years,	51°.1	38°.7	64°.0	36°.6	65°.1	51°.9

At Geneva, the mean temperatures of the summers were, from 1803 to 1809,—

Years.		Mean Temp of Summers.
1803,	-	67°.3
1804,	-	65.0
1805,	-	62.2
1806,		65.7
1807,	-	68.2
1808,	-	62.9
1809,	-	63.0
Mean of seven years,		64°.9

M. Arago has found, that in the two years 1815 and 1816, the last of which was so destructive to the crops in a great part of France, the difference of the mean annual temperature was only 2°, and that of the summer 3°.2. The summer of 1816 at Paris was 59°.9, 4°.7 below the mean of the former. From 1803 to 1813, the oscillations round the mean did not go beyond —2°.9, and +3°4.

In comparing places which belong to the same system of climates, though more than eighty leagues distant, the variations seem to be very uniform, both in the annual temperature and that of the seasons, although the thermometrical quantities are not the same.

Years.	Paris.		Geneva.		Paris.		Geneva.		Paris.		Geneva.	
	Mean Annual Temperature.	Difference between Mean Ann. Temp. and that for 12 years, 51°.1	Mean Annual Temperature.	Difference between Mean An. Temp. and that of 12 years, 49°.6.	Mean Temperature of Winter.	Difference with the mean Winter Temperature of 12 years, 38°.7.	Mean Temperature of Winter.	Difference with the Mean Winter Temperature of 12 years, 34°.9.	Mean Temperature of Summer.	Difference with the Mean Temperature of Summer for 12 years, 64°.6.	Mean Temperature of Summer.	Difference with the Mean Temperature of Summer for 12 years, 64°.9.
1803,	51°.1	0	50°.4	+ 0.8	36°.7	— 2.0	32°.2	— 2°.7	67°.6	+ 3°.0	67°.6	+ 2.7
1804,	52.0	+ 0.9	51.1	+ 1.5	41.0	+ 2.3	38.3	+ 3.4	65.5	+ 0.9	66.2	+ 1.3
1805,	49.5	— 1.6	47.8	— 1.8	36.0	— 2.7	33.8	— 1.1	63.1	— 1.5	63.0	— 1.9
1806,	53.4	+ 2.3	51.4	+ 1.8	40.6	+ 1.9	38.5	+ 3.6	65.3	+ 0.7	64.6	— 0.3
1807,	51.4	+ 0.3	49.3	— 0.3	42.3	+ 3.6	35.8	+ 0.9	67.8	+ 3.2	68.2	+ 3.3
1808,	50.5	— 0.6	46.8	— 2.8	36.7	— 2.0	33.8	— 1.1	66.2	+ 1.6	63.5	— 1.4
1809,	50.9	— 0.2	48.7	— 0.9	40.5	+ 1.8	35.1	+ 0.2	62.4	— 2.2	63.1	— 1.8
1810,	50.9	— 0.2	51.1	+ 1.5	36.5	— 2.2			63.3	— 1.3		
1811,	52.7	+ 1.6	51.8	+ 2.2	39.2	+ 0.5			65.1	+ 0.5		
1812,	49.8	— 1.3	47.8	— 1.8	39.6	+ 0.9			63.1	— 1.5		
1813,	49.8	— 1.3	48.6	+ 1.0	36.1	— 2.6			61.7	— 2.9		

(To be continued in next Number.)

ART. VI.—*On Isothermal Lines, and the Distribution of Heat over the Globe.* By Baron ALEXANDER DE HUMBOLDT. (Continued from Vol. IV. p. 37.)

ALL the ratios of temperature which we have hitherto fixed, belong to that part of the lower strata of the atmosphere which rests on the solid surface of the globe in the northern hemisphere. It now remains for us to discuss the *temperature of the southern hemisphere.* In few parts of natural philosophy, have naturalists differed so widely in opinion. From the beginning of the 16th century, and the first navigations round Cape Horn, an idea prevailed in Europe, that the southern was considerably colder than the northern hemisphere. Mairan and Buffon * combated this opinion by inaccurate reasonings of a theoretical nature. Æpinus † established it anew. The discoveries of Cook made known the vast extent of ice round the South Pole; but the inequality in the temperature of the two hemispheres was then exaggerated. Le Gentil, and particularly Kirwan ‡, had the merit of having first demonstrated, that the influence of the circumpolar ice extended much less into the temperate zone than was generally admitted. The less distance of the sun from the winter solstice, and his long continuance in the northern signs, act in an opposite manner || on the heat in the two hemispheres; and as (after the theorem of Lambert) the quantity of light which a planet receives from the sun, increases in proportion to the true anomaly, the inequality in the temperature of the two hemispheres is not the effect of unequal radiation. The southern hemisphere receives the same quantity of light; but the accumulation of heat in it is less §, on account of the emission of the radiant heat which takes place during a long winter. This hemisphere being also in a great measure covered with

* *Theorie de la Terre*, tom. i. p. 312.—*Mémoires de l'Acad.* 1765, p. 174.

† *De Distributione Caloris*, 1761.

‡ *Estimate*, &c. p. 60.—*Irish Trans.* vol. viii. p. 423.—Le Gentil, *Voyage dans l'Inde*, vol. i. p. 73.

|| Mairan, *Mem. Acad.* 1765, p. 166.—Lambert, *Pyrometrie*, p. 310.

Prevost, *De la Chaleur Rayonnante*, 1809, p. 329. & 367. § 280,—306.

water, the pyramidal extremities of the continents have there an irregular climate. Summers of a very low temperature are succeeded, as far as 50° of south latitude, by winters far from rigorous. The vegetable forms also of the torrid zone, the arborescent ferns, and the orchideous parasites, advance towards 38° and 42° of S. Latitude. The small quantity of land * in the southern hemispheres, contributes not only to equalise the seasons, but also to diminish absolutely the annual temperature of that part of the globe. This cause is, I think, much more active than that of the small eccentricity of the earth's orbit. The continents during summer radiate more heat than the seas, and the ascending current which carries the air of the equinoctial and temperate zones towards the circumpolar regions, acts less in the southern than in the northern hemisphere. That cap of ice which surrounds the pole to the 71st and 68th degree of south latitude, advances more towards the equator, whenever it meets a free sea; that is, wherever the pyramidal extremities of the great continents are not opposite to it. There is reason to believe, that this want of dry land would produce an effect still more sensible, if the division of the continents was as unequal in the equinoctial as in the temperate zones †.

Theory and experience prove, that the difference of temperature between the two hemispheres, cannot be great near the limit which separates them ‡. Le Gentil had already observed, that the climate of Pondicherry is not warmer than that of Madagascar at the Bay of Antongel in 12° of S. Lat. Under the parallels of 20 the Isle of France has the same annual temperature, viz. 80°.1, as Jamaica and St Domingo. The Indian Sea between the east coasts of Africa, the Isles of Sonde and New Holland, form a kind of gulf which is shut up to the north by Arabia and Hindostan. The isothermal lines there appear to go back to the South Pole; for farther to the west in the open sea between Africa and the New World, the cold of the southern hemisphere already causes itself to be felt from the 22d degree, on account of its insulated mountains and particular loca-

* The dry lands in the two hemispheres are in the ratio of 3 to 1.

† The dry lands between the tropics, are in the two hemispheres as 5 to 4. and without the tropics as 13 to 1.

‡ Prevost, p. 343.

lities. I shall not mention the island of St Helena, Lat. 15° 55′ whose mean temperature, according to the observations of M. Beatson, at the sea side, does not exceed 71°6 or 73°4. It is the eastern coast of America, which, in the observations of a Portuguese astronomer, M. Benito Sanchez Dorta *, present us with the S. Lat. of 22° 54′, almost at the limit of the equinoctial region with a plan, of which we know the climate by more than 3500 thermometrical and barometrical observations made every year, to ascertain the horary variations in the heat and pressure of the air. The mean temperature of Rio Janeiro is only 74°.3, whilst, notwithstanding the north winds which bring the cold air of Canada during winter into the Gulf of Mexico, the mean temperatures of Vera Cruz, (Lat. 19° 11′,) and of the Havannah, (Lat. 23° 10′,) are 77°.9. The differences of the two hemispheres become more sensible in the warmest months.

Rio Janeiro.		Havannah.	
	Mean Temp.		Mean Temp.
June,	68°.0	December,	71°.8
July,	70 2	January,	70 2
January,	79 2	July,	83 3
February,	80 6	August,	83 8

The great equality in the division of the annual heat in 34° of N. and S. Lat. is very surprising. If we attend to the three continents of New Holland, Africa and America, we shall find, that the mean temperature of Port Jackson, (Lat. 33° 51′,) is, after the observations of MM. Hunter, Peron, and Freycinet, - - - 66°.7
That of the Cape of Good Hope, (Lat. 33° 53′,) 66 9
That of the city of Buenos Ayres, (Lat. 34° 36′,) 67 5

In the northern hemisphere 60°.8 or 69°.8 of annual temperature corresponds to the same latitude in the northern hemisphere, according as we compare the American system of climates † or the Mediterranean one;—the concave or the convex

* *Mem. de l'Acad. de Lisbonne*, tome ii. p. 348. 369.

	Latitude.	Mean Temp.
† Natchez,	31° 28′,	64°.8
Cincinnati,	39 06,	53.8.

parts of the isothermal lines. At Port Jackson, where the thermometer descends sometimes below the freezing point, the warmest month is 77°.4, and the coldest 56°.8. We find here the summer of Marseilles and the winter of Cairo *. In Louisiana 2½ of Lat. nearer the Equator, the warmest month is 79°.7, and the coldest 46°.9. In Van Diemen's Land, corresponding nearly in latitude to Rome, the winters are more mild than at Naples; but the coldness of the summers † is such, that the mean temperature of the month of February appears to be scarcely 64°.4, or 66°.2, whilst at Paris, under a latitude more distant from the Equator by 7°, the mean temperature of the month of August is also from 64°.4 to 66°.2', and at Rome above 77°. Under the parallel of 51° 25' south, the mean temperature of the Malouine Isles is well ascertained to be 47°.3. At the same Lat. N. we find the mean temperature in Europe from 50° to 51°.8, and in America scarcely from 35°.6 to 37°.4. The warmest and the coldest months are at London 66°.2 and 35°.6; at the Malouine Isles 55°.8 and 37°.4. At Quebec, the mean temperature of the water is 14°; at the Malouine Isles 39°.6, though those isles are 4° of Lat. farther from the Equator than Quebec. These numerical ratios prove, that, to the parallels of 40° and 50°, the corresponding isothermal lines are almost equally distant from the Pole in the two hemispheres; and that, in considering only the system of transatlantic climates between 70° and 80° of W. Long., the mean temperatures of the year, under the corresponding geographical parallels, are even greater in the southern than in the northern hemisphere.

The division of the heat between the different parts of the year, gives a particular character to southern climates. In the

	Latitude.	Mean Temp.
* Cairo,	30° 2'	72°.3
Funchal,	—— 32 37	68 5
Algiers,	—— 36 48	70 0

† In Van Diemen's Land the thermometer descends in February, in the morning, to 45°.5. The mean of mid-day is 60°.8. At Paris it is in August 73°.4. In Van Diemen's Land, in February the mean of the maxima is 78°8.; of the minima 54°.5. At Rome these means are 86° and 64°.5.—D'Entrecasteaux, *Voyage*, tom. i. p. 205. and 542.

southern hemisphere on the isothermal lines of 46°.4 and 50°.0, we find summers which in our hemisphere belong only to the isothermal lines of 35°.6 and 40°. The mean temperature is not precisely known beyond 51° of S. Lat. Navigators do not frequent those regions when the sun is in the northern signs, and it would be wrong to judge of the rigour of winter, from the low temperature of the summer. The eternal snows which in 71° of N. Lat. support themselves at the height of 2296 feet above the sea, descend even into the plains, both in South Georgia * and in Sandwich Land in 54° and 58° of S. Lat. But these phenomena, however striking they may appear, do not by any means prove that the isothermal line of 32° is 5° nearer the South Pole than the North Pole. In the system of transatlantic climates, the limit of eternal snow is not at the same altitude as in Europe; and in order to compare the two hemispheres, we must take into account the difference of longitude. Besides, an equal altitude of the snows, does not by any means indicate an equal mean temperature of the year. This limit depends parcularly † on the coldness of summer, and this again on the quick condensations of the vapour caused by the passage of the floating ice. Near the poles the foggy state of the air diminishes in summer the effect of the solar irradiation, and in winter that of the radiation of the globe. At the Straits of Magellan, MM. Churruca and Galeano have seen snow fall in 53° and 54° of S. Lat. in the middle of summer; and though the day was 18 hours long, the thermometer seldom rose above 42°.8 or 44°.6 and never above 51°.8.

The inequal temperature of the two hemispheres, which, as we have now proved, is less the effect of the eccentricity of the

* It is the more surprising to find in the Island of Georgia snow on the banks of the ocean, because 2° 39′ nearer the Equator at the Malouine Isles, the mean temperature of the summers is 53°.1, or 9° greater than at the point in our hemisphere in 71° of Lat. where the limit of perpetual snow exists at 2296 feet of absolute elevation. But we must recollect, 1st, That the Malouine Isles are near a continent which is heated in summer; 2d, That Georgia is covered with mountains, and is placed not only in a sea open to the north, but under the influence of the perennial ices of Sandwich Land; and, 3dly, That in Lapland, 20° of Lat. produce in certain local circumstances 10°.8 of difference in the temperatures of the summers.

† Baron Von-Buch's *Travels in Lapland*, vol. ii. p. 393,—420.

earth's orbit, than of the unequal division of the continents, determines * the limit between the N. E. and S. E. Trade Winds. But as this limit is much more to the north of the Equator in the Atlantic Ocean, than in the South Sea, we may conclude that, in a region between 130° and 150° of W. Long. the difference of temperature between the two hemispheres, is less great than farther to the east in 20° or 50° of longitude. It is indeed under this region in the Great Ocean, that, as far as the parallel of 60°, the two hemispheres are equally covered with water, and equally destitute of dry land, which, radiating the heat during summer, sends the warm air towards the poles. The line which limits the N. E. and S. E. Trade Winds, approaches the Equator, whereas the temperature of the hemispheres is different; and if, without diminishing the cold of the southern atmosphere, we could increase the inflexion of the isothermal lines in the system of transatlantic climates, we should meet the S. E. winds in 20° and 50° of W. Long. to the north, and in 130° and 150° of W. Long. to the south of the Equator †.

The low strata of the atmosphere which rest upon the aqueous surface of the globe, receive the influence of the temperature of the waters. The sea radiates less absolute heat than continents; it cools the air upon the sea, by the effect of evaporation; it sends the particles of water cooled and heavier towards the bottom; and it is heated again, or cooled, by the currents directed from the Equator to the Poles, or by the mixture of the superior and inferior strata on the sides of banks.

It is from these causes combined, that, between the tropics, and perhaps as far as 30° of Lat., the mean temperatures of the air next the sea, are 3°.6 or 5°.4 lower than that of the continental air. Under high latitudes, and in climates where the atmosphere is coolest in winter, much below the freezing point, the isothermal lines rise again towards the Poles, or become convex when the continents pass below the seas ‡.

With respect to the temperature of the ocean, we must distinguish between four very different phenomena. 1*st*, The tem-

---

* Prevost, *Journ. de Phys.* tom. xxxviii. p. 369.—Irish *Trans.* vol. viii. p. 374.

† Humboldt's *Relat. Histor.* tom. i. p. 225, 237. ‡ *Id.* p. 67, 230. 242.

perature of the water at the surface corresponding to different latitudes, the ocean being considered at rest, and destitute of shallows and currents. *2d*, The decrease of heat in the superimposed strata of water. *3d*, The effect of billows on the temperature of the surface water. *4th*, The temperature of currents, which impell with an acquired velocity, the waters of our zone across the immoveable waters of another zone. The region of warmest waters no more coincides with the Equator, than the region in which the waters reach their maximum of saltness. In passing from one hemisphere to another, we find the warmest waters between 5° 45′ of N. Lat., and 6° 15′ of S. Lat. Perrins found their temperature to be 82°.3; Quevedo 83°.5; Churruca 83°.7, and Rodman 83°.8. I have found them in the South Sea to the east of the Galapagos Isles 84°.7. The variations and the mean result do not extend beyond 1°.3. It is very remarkable that in the parallel of warmest waters, the temperature of the surface of the sea is from 3°.6 to 5°.4 higher than that of the superincumbent air. Does this difference arise from the motion of the cooled particles towards the bottom, or the absorption of light, which is not sufficiently compensated by the free emission of the radiant coloric. As we advance from the Equator to the Torrid Zone, the influence of the seasons on the temperature of the surface of the sea becomes very sensible; but as a great mass of water follows very slowly the changes in the temperature of the air, the means of the months do not correspond at the same epochs in the ocean and in the air. Besides, the extent of the variations is less in the water than in the atmosphere, because the increase or decrease in the heat of the sea takes place in a medium of variable temperature, so that the *minimum* and the *maximum* of the heat which the water reaches, are modified by the atmospherical temperature of the months which follow the coldest of the warmest months of the year. It is from an analogous cause, that in springs which have a variable temperature, for example, near Upsal *, the extent of the variations of temperature is only 19°.8, while the same extent in the air from the month of January to August, is 39°.6. In the parallel of the Canary Islands,

* Gilbert's *Annalen*, 1812, p. 129.

Baron Von Buch found the minimum of the temperature of the water to be 68°, and the maximum 74°.8. The temperature of the air in the warmest of the coldest months, is, in that quarter, from 64°.4 to 75°.2. In advancing towards the north, we find still greater differences of winter temperature between the surface of the sea and the superincumbent air. The cooled particles of water descend till their temperature reaches 39°.2; and hence in 46° and 50° of Lat. in the part of the Atlantic which is near Europe, the *maximum* and *minimum of heat* are

In the water at its surface,	68°.0 and 41°.9
In the air from the mean of warmest and coldest months,	66.2 and 35.6

The excess in the mean temperature of the water over that of the air, attains its maximum beyond the polar circle, where the sea does not wholly freeze. The atmosphere is cooled to such a degree in these seas, (from 63° to 70° of Lat., and 0° of Long.) that the mean temperature of several months of winter descend on the continents to 14° and 10°.4, and on the coasts to 23° and 21°.2, while the temperature of the surface of the sea is not below 32° or 30°.2. If it is true, that even in those high latitudes the bottom of the sea contains strata of water which, at the maximum of their specific gravity, have 39°.2 or 41° of heat, we may suppose that the water at the bottom contributes to diminish the cooling at the surface. These circumstances have a great influence on the mildness of countries in continents separated from the Pole by an extensive sea.

Hitherto we have attended to the distribution of heat on the surface of the globe at the level of the sea. It only remains for us to consider the variations of temperature in the higher regions of the atmosphere, and in the interior of the earth.

The decrease of heat in the atmosphere, depends on several causes, the principle of which, according, to Laplace and Leslie *, is the property of the air to increase its capacity for heat by its rarefaction. If the globe was not surrounded by a mixture of elastic and aëriform fluids, it would not be sensibly colder at the height of 8747 yards than at the level of the sea. As each

* *Essay on Heat and Moisture*, p. 11.; and *Geometry*, p. 495.

part of the globe radiates in every direction, the interior of a spherical envelope which would rest on the top of the highest mountains, would receive the same quantity of radiant heat as the lower strata of the atmosphere. The heat, it is true, will be spread over a surface a little greater; but the difference of temperature will be insensible, since the radius of the spherical envelope will be to that of the earth as 1.001 to 1.

Considering the earth as surrounded with an atmospherical fluid, it is obvious, that the air heated at its surface will ascend, dilate itself, and be cooled, either by dilatation, or, by a more free radiation across the other strata that are equally rarified. These are the ascending and descending currents, which keep up the decreasing temperature of the atmosphere *.

The cold of mountains is the simultaneous effect, 1*st*, Of the greater or less vertical distance of the strata of air at the surface of the plains and of the ocean. 2*d*, Of the extinction of light, which diminishes with the density of the superincumbent strata of air †; and, 3*d*, Of the emission of radiant heat, which is favoured by air very dry ‡, very cold, and very clear. The mean temperature of our present plains would be lowered, if the seas should experience a considerable diminution. The plains of continents would then become *plateaux*, and the air which rested on them would be cooled by the circumjacent strata of air, which, at the same level, would receive but a small portion of the heat emitted from the dry bottom of the seas.

The following Table contains the results of observations which I have made near the Equator, on the Andes of Quito, and towards the northern extremity of the torrid zone, in the Cordilleras of Mexico. These results are true means, given either by stationary observations made during several years, or by insulated observations. In these last, we have taken into account the hour of the day,—the distance of the solstices,—the direction of the wind,—and the reflection from the plains.

* *Essay on Heat and Moisture*, p. 11.; and *Geometry*, p. 495.

† Humboldt on Refraction below 10°, *Observ. Astron.* tom. i. p. 126.

‡ Wells on *Dew*, p. 50.

HEIGHT above the LEVEL of the SEA.	CORDILLERAS OF THE ANDES. From 10° of North Lat. to 10° of South Lat.		MOUNTAINS OF MEXICO. From 17° of North Lat. to 21° of North Lat.	
	Mean Temp. of the Year.	EXAMPLES, which may serve as a Type.	Mean Temp. of the Year.	EXAMPLES, which may serve as a Type.
0 Toises. 0 Feet. (Comparative heights in Europe have been added for every 1000 metres.)	Fahr. 81°.5	*Cumana*, 33 *feet*. Temp. of day, 78°.8—86° ——— night, 71.6—74.3 *Maximum*, 90.9 *Minimum*, 70.2 Mean, - 81.9	78°.80	*Vera Cruz*, 0 *feet*. Temp. of day, 80°.6—86° —— night, in summer, 78.26—82.4 —— night, in winter, 66.2—75.2 / 64.4—71.6 Mean Temp. 77.72
500 Toises. 3197 Feet. *Vesuvius* 3870 feet.	71°.24	*Caraccas*, 2906 *feet*. Temp. of day, 64°.4—73°4 ——— night, 60.8—62.6 *Maximum*, 78.3 *Minimum*, 54.5 Mean, - 69.4 *Guaduas*, 3769 feet. Temp. Mean, 67°.5	67°.64	*Xalapa*, 4330 *feet*. Temp. Mean in winter, 64°.76 —— of day, 57°.2—59 *Chilpantzingo*, 4523 *feet*, on a plateau which radiates. Mean Temp. 69°.08
1000 Toises. 6394 Feet. *Hospice of St Gothard*, 6806 feet.	64°.4	*Popayan*, 5815 *feet*. Temp. of day, 66°.2—75°.2 —— night, 62.6—64.4 Mean, - 65.66 *Santa Fé de Bogota*, 8721 *feet*. Temp. of day, 59°—64°.4 ——— night, 50—53.6 *Minimum*, 36.5 Mean, - 57.74	64°.4	*Valladolid de Mechoachan*, 6396 *feet*. Mean Temp. 66°.2—68°. *Mexico*, 7468 *feet*. Temp. of day, 60°.8—69°.8 —— night, 55.4—59 Warmest months, 52.7—59 Coldest months, 32—44.6 Mean, 62°.6
1500 Toises. 9591 Feet. *Canigou*, 9118 feet	57°.74	*Quito*, 9538 *feet*. Temp. of day, 60°.08-66°.74 —— night, 48.2--51.8 *Maximum*, 71.6 *Minimum*, 42.8 Mean, - 57.92	57°.2	*Toluca*, 8823 *feet*. Temp. Mean, - 59° *At the Nevado de Toluca*, 11,178 *feet*. Temp. of spring. 48°.2
2000 Toises. 12,789 Feet. *Peak of Teneriffe*, 12,169 feet.	44°.6	*Micuipampa*, 11,867 *feet*. Temp. of day, 41°—48°.2 ——— night, 35.6—31.28 *Les Paramos*, 11,480 feet. Mean Temp. in gen. 47°.12	45°.5	*At the Nivado de Toluca*, 12,178 *feet*. Temp. in Sept. at noon, 52°.7 *At Coffre de Perote*, 12,136 *f*. In February, at 9h, 50°.36
2500 Toises. 15,985 Feet. *Mont Blanc*, 15,662 feet.	37°.7	*At the Inferior Limit of Perpetual Snows*, 15,774 *feet*. Temp. of day, 39°.2—46°.4 ——— night, 28.4—21.2 *Chimboraso*, 19,286 feet. In June, at 1 o'clock, I have seen the therm. at 29°.12.	33°.8	*At the Pic del Fraille*, 15,157 *feet*. I have seen the thermometer in September at 39°.74.

The means given by the Mexican observations are a little different from those given by the observations on the Cordilleras. When the differences and the coincidences amount to about a degree of Fahrenheit. they may be regarded as purely accidental. The length of the day is more unequal in the 20th degree of latitude, but the perpetual snows do not descend 656 feet lower than under the Equator. As the Cordilleras of New Granada, Quito, and Peru, present a great number of points where stationary observations have been made, I shall collect here the mean temperatures which M. Caldas * and I have determined with some certainty, and which all belong to a zone bounded by the parallels of 10° N. and 10° S. Lat.

	Alt. in Feet.	Mean Temp.		Alt. in Feet.	Mean Temp.
Coasts of Cumana,	0	{ 80°.6 82.4	Alausi, - -	7970	59°.00
			Pasto, - -	8308	58.28
Tomependa, Amazons R.	1279	78.44	Santa Rosa, -	8459	57.74
Tocayma, - -	1581	81.5	Cuenca, - -	8633	60.08
Antioquia, - -	1666	77.00	Santa Fé de Bogota,	8721	57.74
Neiva, - -	1702	77.00	Hambato, - -	8849	60.44
Caraccas, - -	2906	69.44	Caxamarca, -	9381	60.80
Caripe, - -	2959	65.3	Llactacunga, -	9473	59.00
Carthago, - -	3149	74.84	Riobamba Nuevo,	9482	61.16
La Plata, - -	3437	74.66	Tunja, - -	9522	56.66
Guaduas, - -	3772	67.46	Quito, - -	9538	57.92
La Meya, - -	4225	72.50	Malbasa, - -	9971	54.50
Medellin, - -	4858	68.9	Plateau de los Pastos,	10099	54.50
Estrella, - -	5645	65.84	Les Paramos, -	11480	47.30
Popayan, - -	5815	65.66	At the Inferior Limit of Perpetual tual Snow,	15744	34.88
Loxa, - -	6855	64.4			
Almaguer, - -	7413	62.6			
Pamplona, -	8016	61.16			

These thirty-two points are not insulated points, as balloons would be if they were fixed in the atmosphere at a perpendicular height of 16,400 feet. They are stations taken on the declivity of mountains, upon that part of the solid mass of the globe which, in the form of a wall, rises into the higher regions of the atmosphere. These mountains, too, have at each height parti-

* I have used the mean temperature and barometrical measurements published at Santa Fé de Bogota by MM. Caldas and Restrepo in the *Semanario del N. R. de Granada*, tom. i. p. 273.; tom. i. p. 93.—341.

cular climates, modified by the radiation of the plateaus on which they stand,—upon the slope of the ground,—the nakedness of the soil,—the humidity of the forests,—and the currents which descend from the neighbouring summits.

Without knowing the localities themselves, the effect of disturbing causes will be readily seen, by comparing in the preceding Table the mean temperatures which correspond to the same elevations; and the discussion of these observations would prove, also, that the extent of the variations is much less than is generally believed. If we examine thirty-two temperatures, upon the hypothesis that a degree of cooling corresponds to an altitude of 200 metres (656 feet), we shall deduce the temperature of the plains (from 30°.6 to 82°.4) twenty-six times from that of elevated places. For the other six deductions, the temperatures differ only about 3°.6; and the errors of observation are here combined with the effects of localities. The air which rests on the plains of the Andes mixes itself with the great mass of the free atmosphere, in which there prevails under the torrid zone a surprising stability of temperature. However enormous be the mass of the Cordilleras, it acts but feebly on the strata of air which are unceasingly renewed. On the other hand, if the plains are heated during the day, they radiate as much during the night; for it is principally on the plains elevated 8856 feet above the sea, that the sky is most clear and uniformly serene. At Peru, for example, the magnificent plateau of Caxamarca, in which the wheat yields the eighteenth, and barley the sixtieth grain, has an extent of more than twelve square leagues: it is smooth like the bottom of a lake, and sheltered by a circular wall of mountains free from snow. Its mean temperature is 60°.8, yet the wheat is often frozen during the night; and in a season where the thermometer fell before sunrise to 46°.4, I have seen it rise in the day to 77° in the shade. In the vast plains of Bogota, which are 656 feet less elevated than that of Caxamarca, the mean temperature, as established by the fine observations of Mutis, is scarcely 57°.74.

In comparing towns situated on elevated plains with those which are placed on the declivity of mountains, I have found for the first an augmentation of temperature, which, on account of the nocturnal radiation, does not exceed from 2°.7 to 4°.14. This augmentation is a little greater in the lower regions of the Andes, in those large valleys whose smooth bottoms reach the height of from 1312 to 1640 feet, principally in the valley of La Madalcine, between Neiva and Houda. It is singular to find in the middle of mountains heats which equal those of the plains, and which are more insupportable, as the air of the valleys is almost never agitated by the winds. If we compare, however, the mean temperatures of these same places with those of the strata of the true atmosphere, or on the declivity of mountains, we shall find them only from 3°.6 to 5°.4.

On these grounds, we may place some confidence on the four results which we have deduced from such a great number of observations, for the perpendicular heights of 1000, 2000, 3000, and 4000 metres. I have confined myself to a simple arithmetical mean, and to the fortuitous compensation of irregularities; for I could not have avoided employing an hypothesis on the decrease of heat, if I had wished to reduce to a standard height those heights which approach it the most. I have added the observations with which an intimate knowledge of localities has furnished me.

1. *For* 1000 *Metres* (3280 *feet.*) *of Elevation.*	Alt. in Feet.	Temp. Fahr.
Convent of Caripe, (thick and damp forests),	2959	65°.3
Caraccas, (a foggy sky, valley of small extent,)	2906	69.44
La Plata, (very warm, valley communicating with that of L'Alto Magdalena,)	3437	74.66
Carthago, (very warm valley of Cauca),	3149	74.84
2. *For* 2000 *Metres* (6560 *feet*) *of Elevation.*		
Loxa, (a plateau of small extent),	6855	64.4
Almaguer, declivity covered with very thick vegetation),	7413	62.6
Popayan, (small plateau, a little elevated above the valley of Cauca,)	5815	65.66
3. *For* 3000 *Metres* (9840 *feet*) *of Elevation.*		
Caxamarca, (very extensive plateau, sky serene,)	9381	60.80
Quito, (at the foot of Pinchincha, a narrow valley,)	9538	57.92
Tunja, (mountains of New Grenada),	9522	56.66
Malvasa, (elevated plains, cooled by the snows of the volcano of Puracé,)	9971	54.50
Los Pastos, (very cold plateau, from which rise snow covered summits,)	10099	54.50
Llactacunga, (temperate valley),	9473	59.0
Riobamba Nuevo, (arid plains of Tupia, covered with pumice-stone,)	9482	61.16

Between the tropics, the Cordilleras form the centre of the civilization and industry of Spanish America. They are inhabited to the height of 4000 metres, (13,120 feet); and a small number of observations made on the back of the Andes, gives a sufficiently accurate idea of the mean temperature of the year. In Europe, on the contrary, in the temperate zone, the high mountains are in general little inhabited. The descent of the isothermal line of 32°, causes to cease the cultivation of crops of grain, at the point where they begin in the Cordilleras. Stationary habitations are rare above 2000 metres (6560 feet) of elevation; and in order to judge with any precision of the mean temperature of the superincumbent beds of air, we must unite at least 730 thermometrical observations made in the course of a year

---

* Elevations of 400 metres (1312 feet) appear to have a very sensible influence on the mean temperature, even when great portions of countries rise progressively. In order to establish this point, I have examined the temperatures of places situated almost on the level of the sea, and under the same parallels.

	Lat.	Elevation in Feet.	Mean Temp.
Buda, - - -	47°.29	512	51°.08
Paris, - - -	48.50	116	51.08
Vienna, - - -	48.12	551	50.54
Manheim, - -	49.20	384	50.18

Whence, in the longitudes of Paris and Buda, and between the latitudes of 47° and 48°, and almost at the level of the sea, the mean temperature is from 50°.9 to 51°.44.

Under the same longitudes, we have,—

	Elevation in Feet.	Mean Temp.
Geneva, - - -	1177	49°.28
Zurich, - - -	1437	47.84
Munich, - - -	1711	50.74
Berne, - - -	1755	49.28
Marschling, - - -	1834	51.98 *
Coire, - - -	1991	48.92 †

By taking the means of these results, we cannot mistake the influence of *small elevations*, or of very *extensive plateaus*, on the decrease of the mean temperature.

* Heated by the winds of Italy.

† In spite of the winds of Italy.

Places situated between 46°—47° of North Lat.	Elevations.		Mean Temperatures.		
	Metres.	Feet.	Of the Year.	Of the Coldest Months.	Of the Warmest Months.
Level of the sea,	0		53°.60	36°.32	69°.80
Geneva,	359	1177	49.64	34.16	66.56
Tegernsée,	744	2440	42.44	22.10	59.36
Peissenberg,	995	3264	41.00	20.84	57.02
Chamouni,	1028	3372	39.20		55.40
Hospice de St Gothard,	2076	6809	30.38	15.08	46.22
Col de Géant,	3436	11270	21.20		36.50

In comparing the mean temperature of superincumbent beds of air, I find that the isothermal line of 41°, which, in the parallel of 45°, is found at the height of 1000 metres, (3280 feet), makes the equatorial mountains of an absolute elevation of 4250 metres, (13,940 feet). It had, however, been long believed, after Bouguer, that the inferior limits of perpetual snows characterised every where a bed of air, whose mean temperature was 32°; but I have shewn in a Memoir read to the Institute in 1808 *, that this supposition is contrary to experience. By uniting good observations, I have found, that at the limit of perpetual snows, the mean temperature of the air is,—

	Metres.	Feet.	Mean Temp. of Limit of Perpetual Snows.
At the Equator,	4800	15,744	34°.70
In Temperate Zone,	2700	8,856	25.34
In Frigid Zone, in Lat. 68°—69°,	1050	3,444	21.20

As the heat of the higher regions of the atmosphere depends on the radiation of the plains, we may conceive, that, under the same geographical parallels, we cannot find, in the transatlantic climates, (on the declivities of rocky mountains), the isothermal lines at the same height above the level of the sea as in European climates. The inflexions which these lines experience, when traced on the surface of the globe, necessarily influence their position in a vertical plane, whether we unite in the atmosphere points placed under the same meridians, or consider only those that have the same latitude.

Hitherto we have attempted to determine the mean temperatures which correspond under the Equator and in Lat. 45° and

* *Observations Astronomiques*, tom. i. p. 136.

47° to beds of the atmosphere equally elevated. This determination is founded on stationary observations, and indicates the mean state of the atmosphere. General physics has its numerical elements, as well as the system of the world; and these elements, so important in the theory of barometrical measurements and in that of refractions, will be perfected in proportion as natural philosophers shall direct their attention to the study of general laws.

Height, in		Equatorial Zone, from 0°——10°.		Temperate Zone, from 45°——47°.	
Metres.	Feet.	Mean Temp.	Diff.	Mean Temp.	Diff.
0	0	81°.50	10°.26	53°.60	12°.60
974	3195	71.24	6.12	41.00	9.36
1949	6393	65.12	7.38	31.64	8.28
2923	9587	57.74	13.14	23.36	
3900	12792	44.60	9.90		
4872	15965	34.70			

This Table proves, in conformity with the deductions of theory, that in the mean state of the atmosphere, the heat does not decrease uniformly in an arithmetical progression. In the Cordilleras, (and the fact is extremely curious), we observe the decrease getting less and less between 1000 and 3000 metres, particularly between 1000 and 2500 metres of elevation, and then increasing anew from 3000 to 4000 metres. The strata, where the decrease attains its *maximum* and its *minimum*, are in the ratio of 1 to 2. From the height of the Caraccas to that of Popayan and Loxa, 1000 metres produce a difference of 6°.3. From Quito to the height of Paramos, the same 1000 metres change the mean temperature more than 12°.6. Do these phenomena depend only on the configuration of the Andes, or are they the effect of the accumulation of clouds in the aërial ocean? In considering that the Andes form an enormous mass, 3600 metres (11,808 feet) high, from which rise peaks or domes insulated and covered with snow, we may conceive how, from the point where the mass of the chain diminishes so rapidly, the heat decreases also with rapidity. It is not easy, however, to explain, by an analogous cause, why the progressive cooling diminishes between 1000 and 2000 metres. The great plateaus of the Cordilleras commence only at the height of 2600 or 2900 metres, (8528 or 9512 feet); and I am of opinion, that the slowness with which the heat decreases in the stratum of air between 1000 and 2000, is the triple effect of the extinction of light, or the absorption of the rays in the clouds,—of the formation of rain,—and the obstacle which the clouds oppose to the free passage of radiant heat. The bed of air of which we speak, is the region in which are suspended the large clouds which the inhabitants of the plains see above their heads. The decrease of temperature, which is very rapid from the plains to the region of clouds, becomes less rapid in that region; and if this change is less sensible in the temperate zone, it is no doubt because at the same height, the effect of radiation there is less sensible than above the burning plains of the equinoctial zone. In these zones, too, the cooling appears to follow the same law in the beds of air of equal temperature; but the force of radiation varies with the temperature of the radiating beds.

The results which we have now discussed, deserve the preference over those which are deduced from observations made during excursions to the tops of some lofty mountains. The first give for the

	Metres.	Cent.		Fahr.		Metres.
Equinoctial Zone,	0—4900	1°	or	1°.8	for	187 *
Temperate Zone,	0—2900	1°	or	1°.8	for	174

* This is the mean result or the measure of the distribution of heat in the whole column of air. The partial results are from the back of the Andes.

Heights in Metres.	Cent.		Fahr.		Metres.
0—1000	1°	or	1°.8	for	170
1000—2000	1	or	1.8	for	294
2000—3000	1	or	1.8	for	232
3000—4000	1	or	1.8	for	131
4000—5000	1	or	1.8	for	180

In these numbers, we recognise, as in the above Table, the influence of the region of clouds upon the decrease of heat. In order to shew the utility of these numerical ratios, I shall give here the approximate calculation of the height of the plain of Thibet, deduced from the mean temperature of the month of October, which, according to the former, is 42°.26. As the latitude of Tissoolumbo 29°, gives 69°.8 for the mean temperature of the plain; and as at Mount St Gothard, the mean

The last give for the

		Cent.		Fahr.	Metres.
Equinoctial zone,	-	1°	or	1°.8	190
Parallels of 45°—47°,	-	1°	or	1°.8	160—172 *

This agreement is no doubt very remarkable, and the more so, as, in comparing stationary with insulated observations, we confound the mean state of the atmosphere in the course of a whole year with the decrease which corresponds to a particular season, or a particular hour of the day. M. Gay-Lussac found, in his celebrated aëronautical voyage from 0 to 7000 metres, (0 to 22,960 feet), a centigrade degree for 187 metres, near Paris, at a period when the heat of the plains was nearly equal to that of the equinoctial region. It is on account of this observed equality in the decrease of heat, in reckoning from the standard temperature of the plains, that the astronomical refractions corresponding to angles below 10°, have been found the same under the equator and in temperate climates. This result, contrary to the theory of Bouguer, is confirmed by observations which I have made in South America, and by those of Maskelyne at Barbadoes, calculated by M. Oltmanns.

We have seen, that between the tropics, on the back of the Cordilleras, we find, at 2000 metres of elevation, I will not say the climate, but the mean temperature of Calabria and of Sicily. In our temperate zone, in 46° of Lat. we meet at the same elevation with the mean temperature of Lapland †. This comparison

---

temperature of October is even a little below that of the whole year, it is probable that the height of the plain of Great Thibet exceeds from 2900 to 3000 metres.—See my Memoir on the Mountains of India in the *Ann. de Chim. et de Phys.* 1817.

Note by the Editor.

As the cold meridian of the globe passes through the plains of Great Thibet, we conceive that the mean temperature of Lat. 29° in that plain, when reduced to the level of the sea, will be about 65°, and therefore that the height of the plain of Great Thibet will not exceed 2800 metres or 9184 feet.—D. B.

* Saussure gives for the summer 160 metres, (525 feet); for winter 230, (754 feet); and for the whole year 195, (640 feet). M. Ramond gives 165, (538 feet). M. D'Aubuisson 173 metres, (567 feet).—See *Journ. de Phys.* tom. lxxi. p. 37.; *De la Formul. Barometr.* p. 189.; and my *Recueil d'Obs. Astron.* tom. i. p. 129.

† As the temperature varies very little in the course of a whole year in the equinoctial zone, we may form a pretty correct idea of the climate of the Cordil-

leads us to an exact knowledge of the numerical ratios between the elevations and the latitudes, ratios which we find indicated with little precision in works on physical geography.

The following are the results which I have obtained from exact data in the temperate zone, from the plains to 1000 metres of elevation. Every hundred metres of perpendicular height, diminishes the mean temperature of the year, by the same quantity that a change of 1° of latitude does in advancing towards the Pole. If we compare only the mean temperature of summer, the first 1000 metres are equivalent to 0°.81 Fahr. From 40° to 50° of latitude, the mean heat of the plains in Europe decreases in Europe 12°.6 of Fahr.; and this same decrease of temperature takes place on the declivity of the Swiss Alps from 0 to 1000 metres of elevation.

Differences of Latitude, Compared with Differences of Elevation.	Mean Heat of the Year.	Mean Heat of Summer.	Mean Heat of Autumn.
I. *At the Level of the Sea.*			
a. Latitude, 40°, -	63°.14	77°.00	62°.60
b. Latitude, 50°, -	50.54	64.40	
II. *On the Declivity of Mountains.*			
a. At the foot in 46° of Lat.	53.60	68.00	51.80
b. At an elevation of 1000 metres,	41.00	58.46	42.80

These numerical ratios are deduced from observations made on the temperature of the air. We cannot measure the quantity of heat produced by the solar rays on the parenchyma of plants, or in the interior of fruits which receive their colour in ripening. The fine experiment of MM. Gay-Lussac and Thenard, the combustion of chlorine and hydrogen, proves what a powerful action direct light exercises on the molecules of bodies. But as the extinction of light is less upon the mountains in dry and rarified air, maize, fruit-trees, and the vine, still flourish at heights which, according to our thermometrical observations made in the air, and far from the ground, we ought to suppose

leras, by comparing them to the temperature of certain months in France or in Italy. We find in the plains of Orinoco the month of August of Rome; at Popayan, (2988 feet), the month of August of Paris; at Quito (4894 feet), the month of May; in the Paramos, (5904 feet), the month of March at Paris.

too cold for the cultivation of plants useful to man. M. De Candolle, indeed, to whom the geography of vegetables owes so many valuable observations, has seen the vine cultivated in the south of France at 800 metres (2624 feet) of absolute height, when, under the same meridian, this same cultivation went on with difficulty at 4° of latitude farther north; so that if we consider only the ratios in France, an elevation of 100 metres, (328 feet), appears to correspond, not to 1°, but to half a degree of latitude *.

(To be concluded in next Number.)

ART. V.—*On Isothermal Lines, and the Distribution of Heat over the Globe.* By Baron ALEXANDER DE HUMBOLDT. (Concluded from Vol. IV. p. 281.)

I SHALL now conclude this Memoir by the enumeration of the most important results which have been obtained by Baron Von Buch, M. Wahlenberg, and myself, on the distribution of heat in the interior of the earth, from the Equator to 70° of N. Lat., and from the plains to 3600 metres (11,808 feet) of elevation. I shall limit myself to an enumeration of the facts. The theory by which these facts are connected, will be found in the fine analytical work with which M. Fourier will soon enrich natural philosophy.

The interior temperature of the earth is measured either by the temperature of subterraneous excavations, or by that of springs. This kind of observation is very liable to error, if the traveller does not pay the most minute attention to local circumstances, which are capable of altering the results *. The air, when cooled, accumulates in caverns, which communicate with the atmosphere by perpendicular openings. The humidity of rocks depresses the temperature by the effect of evaporation. Caverns that have little depth are more or less warmed, according to the colour, the density, and the moisture, of the strata of stone in which nature has hollowed them. Springs indicate too low a temperature, if they descend rapidly from a considerable height upon inclined strata. There are some under the torrid zone and in our climate, which do not vary in their temperature throughout the whole year more than half a degree; and there are others which shew the mean temperature of the earth only by observing them every month, and taking the mean of all the observations. From the Polar circle to the Equator, and from the tops of mountains towards the plains, the progressive increase of the temperature of springs diminishes with the

* Baron von Buch, in the *Bibl. Brit.* tom. xix. p. 263.; Saussure, *Voyages*, sect. 1418.; Wahlenberg, *De Veget. Helvet.* Pl. 77.—84.; Gilbert, *Annalen*, 1812, p. 150. 160. 277.; Lambert, *Pyrometrie*, p. 296. Dr Roebuck appears to have been the first who entertained exact notions on the temperature of springs, and upon their relation to the mean temperature of the air; Phil. Trans. 1775, vol. lxv, p. 461.

mean temperature of the ambient air. The temperature of the interior of the earth is, at

	Lat.	Temp. Fahr.		Lat.	Temp. Fahr.
Vadso,	70°.0	35°.96	Paris,	48° 50′	53.°6
Berlin,	52.31	49.28	Cairo,	30 2	72.5

In equinoctial America, I have found it in the plains from 77° to 78°.8.

The following are examples of the decrease of temperature from the plains to the tops of mountains.

	Alt. in Feet.	Temp. Fahr.
Spring of Utliberg, near Zurich, -	1532	48°.92
Ditto of Rossbaden at St Gothard, -	7016	38.30

Between the Tropics I have found,

	Alt. in Feet.	Temp. Fahr.
Springs of Cumanacoa, - -	1,148	72°.5
Ditto Montserrate, above Santa Fé de Bogota,	10,680	59.9
Ditto in the Mine of Hualgayoc in Peru,	11,759	53.24

In the plains, and to the height of 3280 feet, between the parallels of from 40° to 45°, the mean temperature of the earth is nearly equal to that of the ambient air; but very accurate observations by Baron Von Buch and Wahlenberg tend to prove, that in high latitudes, towards the top of the Swiss Alps, for example, beyond the height of 1400 or 1500 metres (4592 or 4920 feet), the springs and the earth are 5°.4 warmer than the air.

Zone of 30°—55°.	Lat.	Mean Temp. of Air, Fahr.	Temp. of the Interior of the Earth.
Cairo, - -	30° 2′	72°.68	72°.50
Natchez, - -	31 28	64.76	64.94
Charlestown, - -	33 0	63.14	63.50
Philadelphia, - -	39 56	53.42	52.16
Geneva, - - -	46 12	49.28	50.74
Dublin, - - -	53 21	49.10	49.28
Berlin, - - -	52 31	47.30	49.28
Kendal, - -	54 17	46.22	47.84
Keswick, - -	54 33	48.02	48.56
Zone of 55°—70°.			
Carlscrona, - -	56 6	46.04	47.30
Upsal, - -	59 51	41.90	43.70
Umeo, - -	63 50	33.26	37.22
Vadso, - -	70 0	29.66	35.96

At Enontekies, in 68½° of Lat. the difference between the mean temperatures of the earth and the air, is so great as 7°.74. Analogous differences are observed on the back of the Alps, at the altitude of 1400 metres (4592 feet).

In the following small table, I have added the mean temperature of the atmosphere, by supposing, with M. Ramond, that there is a decrease of 1° centigrade for 164 metres (1° Fahr. for 300 feet nearly), and by placing the temperature of 32° (according to observations made at the Hospice of St Gothard), at 1950 metres (6396 feet) of elevation.

	Alt. in Feet.	Temperature. Springs.	Air.
Rigi Kaltebad, -	4717	43°.7	38°.12
Pilate, - -	4858	41.0	37.40
Blancke Alp, -	5786	37.4	35.78
Rossbaden, - -	7016	38.3	31.38

It may be objected, that in the Alps of Switzerland, the temperature of springs has only been observed from the beginning of June to the end of September, and that the differences between the air and the interior of the earth would almost entirely disappear, if we knew the temperature of the springs during the whole year. It must not be forgotten, however, that the springs of the Alps did not vary in the space of *four* months at the time of the observations of M. Wahlenberg;—that among the small number of scanty springs which indicate changes of temperature in different seasons, these variations amount from June to September to 11° or 15°;—and that several springs, particularly those which are very copious, do not vary during a whole year more than half a degree of Fahrenheit.

It appears to me, therefore, sufficiently certain, that where the earth is covered with a thick stratum of snow, while the temperature of the air descends to 15° or — 4° of Fahrenheit, the temperature of the earth is above the mean temperature of the air.

When we consider what a large portion of the globe is covered with the sea, and examine the temperature of the deepest waters *, we are constrained to admit, that in islands, along

* At Funchal in Madeira, the temperature of caverns appears to be 61°.16, and consequently 7°.2 *below* that of the air.—*Phil. Trans.* 1778, p. 372.

coasts, and perhaps even in continents of small extent, the interior heat of the earth is modified by the proximity of the strata of rocks on which the waters of the ocean rest.

I have considered successively in this memoir, the distribution of heat,

1. At the surface of the globe.
2. On the declivity of mountains.
3. In the ocean.
4. In the interior of the earth.

In explaining the theory of isothermal lines and their inflexions, which determine the different systems of climates, I have endeavoured to reduce the phenomena of temperature to empirical laws. These laws will appear much more simple, when we shall have multiplied and rectified by degrees the numerical elements which are the results of observation.

In the following general Table of the distribution of heat, the temperatures are expressed in degrees of Fahrenheit; the longitudes are reckoned from east to west of the meridian of the observatory of Greenwich. The mean temperatures of the seasons have been calculated, so that those of the months of *December*, *January*, and *February*, form the mean temperature of *Winter*. An asterisk (*) is prefixed to those places whose mean temperatures have been most accurately determined, and in general by means of 8000 observations. The isothermal lines have a convex summit in Europe, and two concave summits in Asia and Eastern America

Isothermal Bands.	Names of Places.	Position. Lat.	Position. Long.	Position. Height in Feet.	Mean Temp. of the Year.	Distribution of Heat in the different Seasons. Mean Temp. of Winter.	Mean Temp. of Spring.	Mean Temp. of Summer.	Mean Temp. of Autumn.	*Maximum* and *Minimum*. Mean Temp. of Warmest Month.	Mean Temp. of Coldest Month.	
Isothermal Bands from 32° to 41°.	Nain, -	57° 8′	61°20′ W	0	26°.42	— 0°.60	23°.90	48°.38	33°.44	51°.80	— 11°.20	1
	* Enontekies,	68 30	20 47 E	1356	26.96	+ 0.68	24.98	54.86	27.32	59.54	— 0.58	2
	Hospice de St Gothard,	46 30	8 23 E	6390	30.38	18.32	26.42	44.96	31.82	46.22	15.08	3
	North Cape,	71 0	25 50 E	0	32.0	23.72	29.66	43.31	32.08	46.58	22.10	4
	* Uleo, -	65 3	25 26 E	0	35.08	11.84	27.14	57.74	35.96	61.52	7.70	5
	* Umeo, -	63 50	20 16 E	0	33.26	12.92	33.80	54.86	33.44	62.60	11.48	6
	* St Petersburg,	59 56	30 19 E	0	38.84	17.06	38.12	62.06	38.66	65.66	8.60	7
	Drontheim,	63 24	10 22 E	0	39.92	23.72	35.24	61.21	40.10	64.94	19.58	8
	Moscow, -	55 45	37 32 E	970	40.10	10.78	44.06	67.10	38.30	70.52	6.08	9
	Abo, -	60 27	22 18 E	0	40.28	20.84	38.30	61.88	40.64	—	—	10
Isothermal Bands from 41° to 50°.	* Upsal, -	59 51	17 38 E	0	42.08	24.98	39.38	60.26	42.80	62.42	22.46	11
	* Stockholm, -	59 20	18 3 E	0	42.26	25.52	38.30	61.88	43.16	64.04	22.82	12
	Quebec, -	46 47	71 10 W	0	41.74	14.18	38.84	68.00	46.04	73.40	13.81	13
	Christiania, -	59 55	10 48 E	0	42.8	28.78	39.02	62.60	41.18	66.74	28.41	14
	* Convent of Peyssenburg,	47 47	10 34 E	3066	42.98	28.58	42.08	58.46	42.98	59.36	30.20	15
	* Copenhagen,	55 41	12 35 E	0	45.68	30.74	41.18	62.60	48.38	65.66	27.14	16
	* Kendal, -	54 17	2 46 W	0	46.22	30.86	45.14	56.84	46.22	58.10	34.88	17
	Malouin Islands,	51 25	59 59 W	0	46.94	39.56	46.58	53.06	48.46	55.76	37.40	18
	* Prague, -	50 5	14 24 E	0	49.46	31.46	47.66	68.90	50.18	—	—	19
	Gottingen, -	51 32	9 53 E	456	46.94	30.38	44.24	64.76	48.74	66.38	29.66	20
	* Zurich, -	47 22	8 32 E	1350	47.84	29.66	48.20	64.04	48.92	65.66	26.78	21
	* Edinburgh, -	55 57	3 10 W	0	47.84	38.66	46.40	58.28	48.56	59.36	38.30	22
	Warsaw, -	52 14	21 2 E	0	48.56	28.76	47.48	69.08	49.46	70.34	27.14	23
	* Coire, -	46 50	9 30 E	1876	48.92	32.36	50.00	63.32	50.36	64.58	29.48	24
	Dublin, -	53 21	6 19 W	0	49.10	39.20	47.30	59.54	50.00	61.16	35.42	25
	Berne, -	46 5	7 26 E	1650	49.28	32.00	48.92	66.56	49.82	67.28	30.56	26
	* Geneva, -	46 12	6 8 E	1080	49.28	34.70	47.66	64.94	50.00	66.56	34.16	27
	* Manheim, -	49 29	8 28 E	432	50.18	38.80	49.64	67.10	49.82	68.72	33.44	28
	Vienna, -	48 12	16 22 E	420	50.54	32.72	51.26	69.26	50 54	70.52	26.60	29

Band	Place	Lat.	Long.	Height								No.
Isothermal Bands from 50° to 59°.	* Clermont, -	45 46	3 5 E	1260	50.00	34.52	50.54	64.40	51.26	66.20	28.04	30
	* Buda, -	47 29	19 1 E	494	51.08	33.98	51.08	70.52	52.34	71.60	27.78	31
	Cambridge, (U.S.)	42 25	71 3 W	0	50.36	33.98	47.66	70.70	49.82	72.86	29.84	32
	* Paris, -	48 50	2 20 E	222	51.08	38.66	49.28	64.58	51.44	65.30	36.14	33
	* London, -	51 30	0 5 W	0	50.36	39.56	48.56	63.14	50.18	64.40	37.76	34
	Dunkirk, -	51 2	2 22 E	0	50.54	38.48	48.56	64.04	50.90	64.76	37.75	35
	Amsterdam,	52 22	4 50 E	0	51.62	36.86	51.62	65.84	51.62	66.92	35.42	36
	Brussels, -	50 50	4 22 E	0	51.80	36.68	53.24	66.20	51.08	67.28	35.60	37
	* Franeker, -	52 36	6 22 E	0	51.80	36.68	51.08	67.28	54.32	69.08	32.90	38
	Philadelphia,	39 56	75 16 W	0	53.42	32.18	51.44	73.94	56.48	77.00	32.72	39
	New York,	40 40	73 58 W	0	53.78	29.84	51.26	79.16	54.50	80.78	25.34	40
	* Cincinnati, -	39 6	82 40 W	510	53.78	32.90	54.14	72.86	54.86	74.30	30.20	41
	St Malo, -	48 39	2 1 W	0	54.14	42.26	52.16	66.02	55.76	66.92	41.74	42
	Nantes, -	47 13	1 32 W	0	54.68	40.46	54.50	68.54	55.58	70.52	39.02	43
	Pekin, -	39 54	116 27 E	0	54.86	26.42	56.30	82.58	54.32	84.38	24.62	44
	* Milan, -	45 28	9 11 E	390	55.76	36.32	56.12	73.04	56.84	74.66	36.14	45
	Bourdeaux, -	44 50	0 34 W	0	56.48	42.08	56.48	70.88	56.30	73.04	41.00	46
Isothermal Band from 59° to 63°.	Marseilles, -	43 17	5 22 E	0	59.00	45.50	57.56	72.50	60.08	74.66	44.42	47
	Montpellier,	43 36	3 52 E	0	59.36	44.06	56.66	75.74	60.98	78.08	42.08	48
	* Rome, -	41 53	12 27 E	0	60.44	45.86	57.74	75.20	62.78	77.00	42.26	49
	Toulon, -	43 7	5 50 E	0	62.06	48.38	60.80	75.02	64.40	77.00	46.40	50
	Nangasacki,	32 45	129 55 E	0	60.80	39.38	57.56	82.94	64.22	86.90	37.40	51
	* Natchez, -	31 28	90 30 W	180	64.76	48.56	65.48	79.16	66.02	79,70	46.94	52
Isothermal Band from 68° to 77°.	* Funchal, -	32 37	16 56 W	0	68.54	64.40	65.84	72.50	72.32	75.56	64.04	53
	Algiers, -	36 48	3 1 E	0	69.98	61.52	65.66	80.24	72.50	82.76	60.08	54
Isothermal Bands above 77°.	* Cairo, -	30 2	31 18 E	0	72.32	58.46	73.58	85.10	71.42	85.82	56.12	55
	* Veracruz, -	19 11	96 1 W	0	77.72	71.96	77.90	81.50	78.62	81.86	71.06	56
	* Havannah, -	23 10	82 13 W	0	78.08	71.24	78.98	83.30	78.98	83.84	69.98	57
	* Cumana, -	10 27	65 15 W	0	81.86	80.24	83.66	82.04	80.24	84.38	79.16	58

[1] Coast of Labrador. Two years of observations. Floating ice towards the east. A transatlantic climate. Mean temp. of Oct. about 32°.72; Nov. 26°.68.

[2] Centre of Lapland. A European climate. Fine vegetation. June, 49°.46; July, 59°.54; Aug. 55°.94; Sept. 41°.74; Oct. 27°5; Nov. 12°.38. Inland situation. Specimen of a continental climate.

[3] Eleven years of observations, calculated afresh in decads by Wahlenberg. Thermometer verified by Saussure. Mean temp. of seven months of the year below 32°. Winds blow from Italy in the winter. *Minimum* observed in the winter —0°.4; in Aug. at noon, in the shade, *maximum* 54°.5; the nights in Aug. frequently from 33°.8 to 29°.8; the mean temp. of Oct. 31°.1 represents that of the whole year; at the Col de Géant, 10,598 feet high, the mean temp. of July is 36°.5. We find 32° to be the mean temp. in Europe in 45° of latitude, at 5,400 feet high; at the parallel of the Canaries, at 12,300 feet; in the Andes, under the Equator, at 16,500 feet.

[4] Buch, *Voy. en Norw.* ii. 416. Specimen of the climate of the islands and coasts in the north of Europe. April, 30.°02; May, 33°.98; Oct. 32°; Nov. 25°.88. At Alten, Lat. 70°, mean temp. of July, 63°.5; a continental climate.

[5] Finland, eastern coast. May, 40°.82; June, 55°.04; July, 61°.52; Aug. 56°.66; Sept. 46°.58; Oct. 38°.66; Nov. 24°.62. Julin and Buch.

[6] Eastern coast of Western Bothnia. Dr Nœzen. March, 23°.18; April, 33°.98; Oct. 38°.12; Nov. 24°.62.

[7] Euler. Mean temp. of the year, 37°.94. Inochodzow. *Acta. Petr.* xii. 519,—533.

[8] Two years. Berlin, in the *Mem. de l'Acad. de Drontheim*, iv. 216. April, 34°.34; May, 50°.74; Oct. 39°.2; Nov. 27°.68. Climate of the west coast of Europe.

[9] Four years. *Journal de Phys.* xxxix. 40. A continental climate. Winter colder, and summer warmer than at Petersburg. Eastern part of Europe; height as taken from Stritter. (Chamouni, Lat. 46° 1′; Long. 6° 18′ E.; height, 3,168 feet; mean temp. 39°.2.)

[10] Twelve years. Kirwan. Cotte, mean of the year, 41°.18; of the summer, 67°.46; too high. West coast of Finland.

11 Observations from 1774 to 1804, made by Mallet, Prosperin, Holmquist, and Schleling, calculated by M. De Buch, *Voy. de Norw.* ii. 309. It is, perhaps, the place the mean temp of which is the best determined. Winters more serene than at Stockholm; colder on account of the radiation of the ground and the air.

12 Thirty-nine years of observations, 15 of which are very good. Wargentin. Cotte, mean temp. of the year, 44°.24. Five months below 32° as at Petersburg.

13 Four years. A transatlantic climate.

14 Buch, two years. Mean temp. of the winter often scarcely 31°.1. West coast.

15 Alps of Bavaria. Six years' observations, calculated by M. Wahlenberg. Many fruit trees. Convent of Tegernsée, in Bavaria, height of, 2,292 feet; mean temp. of 1785, 42°.44; Peyssenberg, 41°.

16 Bugge. Three months below 32°. Under the Equator, mean temp. of 44°.6, at an elevation of 18,000 feet.

17 Dalton. West of England. Climate of islands; springs 47°.84. Keswick, Lat. 54° 33′, Long. 3° 3′ W.; mean temp 48°.02; springs, 48°.56.

18 Kirwan. Scarcely two years' observations. Southern latitude.

19 Strnadt. Fifteen years. Climate of the continent of Europe.

20 Maier.

21 Six years' observations of M. Escher, calculated by Wahlenberg. The town is situated in a hollow, to which those warm winds cannot penetrate, that render the winters more temperate in the other parts of Switzerland.

22 The calculation has been made from six years of excellent observations, by Professor Playfair; during this time the thermometer was never seen above 75°.74 *. Vegetation continues from March 20. to Oct. 20.; mean temp. of these seven months is from 55°.76 to 50°.90, according as the years are more or less fruitful; wheat does not ripen if the mean temp. descends to 47°.66.

* See *Edinburgh Transactions*, vol. ix. p. 209.—Ed.

[23] Guittard. Only three years. Mean temp. a little too high. Eastern part of Europe. A continental climate.

[24] Four years of observations, by M. de Salis Sewis, calculated by M. Wahlenberg. Mountains of the Grisons.

[25] Kirwan. *Irish Trans.* viii. 203. and 269. Specimen of the climate of the islands. Coldest days, 23°; interior of the ground, 49°.28. Hamilton.

[26] The climate of Berne is a continental climate, in comparison with that of Geneva; there is no lake near it.

[27] Seven years of observations. Saussure. Mean temp. 50°74. *Voy.* § 1418. I find the mean temp. from 1796—1815, 49°.7. Interior of the earth, 51°.98. Pictet, *Bibliotheque Brit.* 1817, iv. 109.

[28] Six years.

[29] Austria. Berlin, Lat. 52° 31′; mean temp. probably 46°.4 to 47°.3; according to Beguelin, 48°.74; springs, 49°.28. Ratisbon, Lat. 49°; height, 1,104 feet; mean temp. 47°.66. Munich, Lat. 48° 8′; height, 1,608 feet; mean temp. 50°.74.

[30] Ramond. Seven years of excellent observations. The mean of the months, at noon, well ascertained; winter, 39°.92; spring, 57°.02; summer, 70°.88; autumn, 57°.92. *Mem. Inst.* 1812, p. 49. Cotte, mean temp. 51°.26.

[31] Wahlenberg, *Flor. Carp.* p. 90. Continental climate. Height of the observatory, 474 feet.

[32] Two years, near Boston, in New England. Transatlantic climate. The thermometer sometimes descends to 0°.

[33] Eleven years (1803—1813) of observations made at the observatory. A greater number of years will, perhaps, give the mean temp. a little higher. Vaults, 53°.06. Kirwan finds for Paris, from seven years of observations of unequal value, 51°.62; he fixes upon 52°.7. Cotte, from 29 years of observations, (*Journ. de Phys.* 1782, July), 53°.24. Cotte, for 33 years, (1763—1781, *Mem. Instit.* iv. 266.), 52°.34. The extraordinary year of 1816 offers the mean temp. of 48°.74; winter, 37°.04; spring, 48°.92; summer, 59°.54; autumn, 50°: the preceding year, 1815, offers a mean temp. of 50°.74; winter, 37°.04; spring, 52°.7; summer, 62°.78; autumn, 50°.74. Arago. Mean temp. of Montmorency, for 33 years, 50°.74; height, 498 feet. Cotte,

Strasburg, Lat. 48° 34′; height, 480 feet; mean temp. 49°.28. Herrenschneider.

[34] Dr Young. Mean temp. varies from 47°.84 to 51°.62, (*Lectures*, ii. 453.) Cavendish, (*Trans*. 1788, p. 61.), 48°.74, Roebuck, Hunter, and Kirwan, 51°.62. Horsley, 51°.26 According to Kirwan, the four seasons in London are, 39°.56, 50°.9, 64°.76, 51°.98; at Paris, 36°.68, 51°.08, 65°.84, 52°.52; from which results, London, 51°.62; Paris, 51°.44. Cotte (*Journ. de Phys*. xxxix. 36.) thinks London is 51°.26, and Paris, 52°.34. The difference which we observe in cultivated plants depends less upon mean temp. than upon direct light, and the serenity of the atmosphere.

[35] Seven years. Cotte. Lisle, 48°.38; Rouen, 51°.44; Cambray, 51°.98; Soissons, 53°.42; Rethel, 53°.24; Metz, 52°.88; Nancy, 51°.98; Etampes, 51°.08; L'Aigle, 50°.9; Brest, 54°.14; Mayenne, 51°.98.

[36] Mohr, and Van Swinden. Five years.

[37] Thirteen years. Temperature rather too high?

[38] Eleven years. Van Swinden. From 1771—1783. Mean temp. 51°.26.

[39] Concave transatlantic summit. Seven years of observations give 54°.86; for the four seasons, 33°.98, 53°.06, 75°.2, 56°.12. Rush, 52°.52, (Drake's *View of Cincin*. p. 116.) Coxe, 54°.14. M. Legaux finds for 17 years, for Springmill on the Schuylkill, Lat. 40° 50′; mean temp. 53°.42. Springs, near Philadelphia, 54°.86, Warden.

[40] Two years only. Retif de la Serve. The thermometer sometimes descends to — 4° in the parallel of Naples! Springs, 54°.86. Ipswich, Lat. 42° 38′; mean temp. 50°. Williamsburg, in Virginia, 58°.1. Cotte and Kirwan. Transatlantic climates.

[41] Transatlantic climates west of the Alleghanys. Good observations, from 1806—1813. Col. Mansfield, (Drake, p. 93.) *Minimum* of the winter, from 5° to — 9°.4; Jan. 1797, as low as — 16°.6, for 39° latitude. *Maximum* 89°.6 to 107°.6 in the shade, without reflection; $\frac{1}{3}$ of all the winds SW.; springs near Cincinnati, 54°.32. Little snow falls; but it is abundant between Lat. 40° and 42°.

[42] Three years only. Bougourd. Dijon, height, 810 feet;

Lat. 47° 19′; mean temp. 50°.9. Besançon, height, 804 feet; Lat. 47° 14′; mean temp. 51°.26.

[43] Six years. Duplessis, and Boudan. Temperature of the summer too high? Rochelle, 53°.06; Poitiers, 52°.7.

[44] Amyot. Six years. Concave. Asiatic summit. Three months below 32°, as at Copenhagen; the summer like that at Naples.

[45] One of the best determined points. The years 1789—1812 are calculated in decads of days. Observations of the Astronomer Reggio, April, 55°.76; Oct. 58°.1. The two decads which approach the nearest to the mean temp. of the year, are, the first of April, 53°.24; and the last of Oct. 54°.68. The mean temps. for January have varied in 10 years from 24°.98 to 38°.48; those of July, from 71°.42 to 78°.44; the mean of the years, from 54°.5 to 57°.2. (Reggio, taking only 24 *maxima* and *minima* in a year for 1763—1798; mean temp. 55.4, *Ephem. Mil.* 1779, p. 82.)

[46] Ten years. Guyot. Lyons, 528 feet, 55°.76. Mafra, near Lisbon, Lat. 38° 52′; height, 600 feet; mean temp. 56°.3, too small. *Mem. de Lisbonne,* ii. 105—158.

[47] Seven years, (1777—1782). St Jacques de Sylvabelle. The thermometer sometimes descends to 23°. Cotte, *Traité de Met.* ii. 420.) 34 years (Raymond, in *Mem. de la Soc. de Med.* 1777, p. 86.) give 62°.06. Cotte (*Journ. de Phys.* xxxix. 21.) fixes it at 58°.64. Kirwan, at 61°.24. The observations made at the Royal Observatory of Marseilles can alone decide.

[48] Ten years. Nismes, 60°.26; Perpignan, 59°.54; Tarascon, 59°.9; Arles, 59°; Rieux, 57°.2: Montauban, 55°.58; Tonains, 54°.86; Dax, 54°.14; Rodez, 57°.02; Aix, 56°.66. Under the equator, 57°.74, at 9,000 feet of elevation.

[49] William Humboldt. Calandrelli, 60°.08. The thermometer sometimes descends to 24°.5, and rises to 99°.5. Naples, 67°.1; Toaldo, probably 63°.5; Florence, 61°.52; Tartini, too high; Lucca, 60°.44; Genoa, 60°.26; Bologna, 56°.3; Verona, 55°.76; Venice, 56°.48; Padua, 56°.3. Kirwan regards it as an established fact, that in Europe the mean temp. in Lat. 40°, is 61°.88; in Lat. 50°, 52°.52.

[50] Only two years. Barberet, and d'Angos. Sheltered by mountains. Estimate a little too high.

[51] Japan. A single year. *Voy. de Thunberg*, p. 121. Climate of islands. Under the Equator, 64°.4, at a height of 6000 feet.

[52] West of the Alleghanys, in Louisiana. Four years. Dunbar. Transatlantic climate.

[53] Madeira. Heberden. Climate of islands. St Croix, of Teneriffe, 71°.42. The remainder of the island of Teneriffe, in the plains, 69°.26. Buch.

[54] Old observations of Tartebout. They appear good. Bagdad, Lat. 33° 19′; according to Beauchamps, 73°.76. The four seasons, 50°.74; 74°.64; 92°.66; 77°; but there was reflection from a house. The thermometer falls to 29°.44. Under the Equator, at 3,000 feet high; mean temp. 71°.24.

[55] The calculations are made from the observations of Nouet, (*Decade*, ii. 213.) The following are the mean temps. of the 12 months: 58°.1; 56°.12; 64°.58; 77°.9; 78°.26; 83°.66; 85°.02; 85°.82; 79°.16; 72°.32; 62°.96; 61°.24. (Niebuhr, 72°.2.) Temp. of Joseph's Well, 72°.5. Catacombs of Thebes, 81°.5. Well of the great pyramid, surrounded by sand, 88°.16. Jomard. Bassora, on the Persian Gulf; mean temp. 77°.9; winter, 64°.04; summer, 90°.86; July, 93°.2.

[56] Orta. Humboldt, *Nouv. Esp.* iv. 516. Jamaica, coast, 80°.6. Blagden.

[57] Ferrer, 1810—1812. *Con. des Tems*, 1817, p. 338. Wells of 10 feet deep; air, 75°.92; water, 74°.48; in 1812, *maximum*, Aug. 14. 86°; *minimum*, Feb. 20. 61°.52. Grottos, 81°.5. Humboldt, *Observ. Astron.* i. 134.

[58] Humboldt. Pondicherry, 85°.1; Madras, 80°.42; Manilla, 78°.08; Isle de France, coast, 80°.42.

GÉOGRAPHIE BOTANIQUE

Auguste De Candolle

GÉOGRAPHIE BOTANIQUE. (*Bot.*) On désigne sous le nom de *géographie botanique* l'étude méthodique des faits relatifs à la distribution des végétaux sur le globe, et des lois plus ou moins générales qu'on en peut déduire. Cette branche des connoissances humaines n'a pu exciter l'attention des observateurs que depuis que la géographie et la botanique, enrichies par un grand nombre de faits, ont su s'élever à des idées générales. Les anciens naturalistes avoient fort négligé l'étude et même l'indication des patries des plantes. Linnæus est le premier qui ait pensé à les indiquer dans les ouvrages généraux ; il est le premier qui ait donné et le précepte et le modèle de la manière de rédiger les Flores ; il est le premier surtout qui ait distingué avec soin les *habitations*, c'est-à-dire les pays dans lesquels les plantes croissent, et les *stations*, c'est-à-dire la nature particulière des localités dans lesquelles elles ont coutume de se développer. C'est donc de Linnæus que sont réellement sorties les premières idées de géographie botanique.

Depuis cette époque, tous les botanistes ont indiqué avec plus de précision la patrie des plantes, et quelques-uns même ont fait de cette étude l'objet de leurs recherches spéciales. Ainsi Giraud-Soulavie, dans son Histoire naturelle de la France méridionale, publiée en 1783, et Bernardin de Saint-Pierre, dans ses élégantes Études de la nature, ont présenté à cet égard quelques considérations intéressantes, mais dépourvues de cette exactitude qui fixe l'attention des savans et qui seule constate la vérité. M. Link[1], en 1789, a fait connoître les plantes qui lui paroissent propres aux terrains calcaires. M. Stromeyer[2], en 1800, a présenté sur la géographie botanique le plan d'un travail qui fait connoître toute l'étendue de la science, et qui fait regretter qu'elle n'ait pas été étudiée plus tôt. M. Lavy[3], en 1801, a classé les plantes du Piémont relativement à leur ordre géographique.

---

1 Link, *Floræ Gœttingensis specimen;* in-8.° *Gœttingæ*, 1789.

2 Stromeyer, *Commentatio sistens historiæ vegetabilium geographicæ specimen;* in-8.° *Gœttingæ*, 1800.

3 Lavy, *Stationes plantarum Pedemontio indigenarum;* in-8.° *Taurini*, 1801.

M. Kielman[1], en 1804, a publié quelques observations intéressantes sur la végétation des Alpes. J'ai moi-même, puisque l'ordre chronologique me force à me citer ici, exposé d'une manière abrégée, dans la Flore françoise[2], quelques observations générales déduites de l'étude des plantes de France, et j'ai, depuis, ajouté à cette base quelques détails ultérieurs, soit dans les rapports de mes voyages[3], soit dans l'article *Géographie botanique et agricole* du Dictionnaire d'agriculture[4], soit enfin dans le 3.ᵉ volume des Mémoires de la société d'Arcueil, publié en 1817. M. Bossi a fait à la Lombardie l'application de la méthode que j'avois proposée pour la France[5]. Mais l'ouvrage le plus précieux que nous possédions sur la géographie des plantes, le seul, peut-être, qui l'ait fait entrevoir dans toute son étendue, est la Géographie des plantes que M. de Humboldt a publiée dans son Tableau physique des régions équatoriales[6], auquel on doit joindre quelques développemens insérés dans ses élégans Tableaux de la nature[7]; ouvrages remarquables par le grand nombre de faits qu'ils font connoître, et par leur heureuse liaison avec les lois les plus importantes des sciences physiques. Dès-lors la géographie botanique prit une marche plus assurée. M. Wahlenberg, dans sa Flore de Laponie[8], et ensuite dans ses Essais sur la végétation de la Suisse[9] et des monts Carpathes[10], a développé l'histoire générale des végétaux de ces trois pays avec une sagacité remarquable.

---

1 Kielman, *Dissertatio de vegetatione in regionibus Alpinis*; in-8.° *Tubingæ*, 1804.

2 Flore françoise, 3.ᵉ édition, 1805, vol. 2, p. 1, avec une carte géographique.

3 Rapports des voyages botaniques et agronomiques dans les départemens de la France, imprimés parmi ceux de la Société d'agriculture de Paris; 1808 — 1814.

4 Dictionnaire d'agriculture, chez Déterville, à Paris, 6 vol., 1809.

5 *Giornale della società d'incoragemento del regno d'Italia*, *n.°* 7.

6 Essai sur la géographie des plantes; 1 vol. in-4.°, Paris, 1807.

7 Tableaux de la nature, traduits par Eyries; 2 vol. in-12. Paris, 1808.

8 *Flora Laponica*, 2 vol. in-12. *Berolini*, 1812.

9 *De vegetatione et climate Helvetiæ tentamen*; in-8.° *Tiguri*, 1813.

10 *Flora Carpathorum principalium*; in-8.° *Gœttingæ*, 1814.

M. Robert Brown a fait connoître plusieurs généralités piquantes sur la géographie botanique de la Nouvelle-Hollande [1] et de la partie d'Afrique voisine du Congo [2], et a, dans ses divers Mémoires, comme c'est le propre de son talent, ouvert aux botanistes une nouvelle route. M. Schouw [3] a cherché à démêler, au milieu des faits nombreux et divers qui semblent se contredire, si l'on pouvoit admettre que chaque espèce de plante eût pris naissance dans un seul lieu; il prépare, sur la géographie des plantes de l'Italie, un travail que les botanistes attendent avec impatience. M. Boué [4] a publié quelques considérations utiles sur la manière d'étudier la Flore d'un pays donné, et les a appuyées sur l'exemple de l'Écosse. M. Winch [5] a fait un travail presque analogue sur quelques parties de l'Angleterre. M. Léopold de Buch, après avoir indiqué, dans son Voyage en Norwége, plusieurs faits curieux de géographie botanique, a publié un travail très-intéressant sur la distribution des plantes dans les îles Canaries [6], résultat de ses propres recherches et de celles de son ami Chr. Smith, dont la botanique a pleuré depuis la mort misérable. Enfin, M. de Humboldt a recueilli, avec son talent ordinaire, tout ce que l'on connoît sur les bases de la géographie des plantes, et, en le combinant avec ses propres recherches, en a tracé, dans les Prolégomènes de la Flore d'Amérique [7], le tableau le plus fidèle et le plus brillant.

A ces divers ouvrages il faut, pour avoir une idée complète de l'état actuel de nos connoissances, joindre cette multitude immense de notes relatives à la patrie des plantes

---

1 *General geographical remarks on the botany of Terra australis;* in-4.° *London*, 1814.

2 *Observations on the herbarium collected by prof. Chr. Smith, in the vicinity of Congo;* in-4.° *London*, 1818.

3 *De sedibus plantarum originariis sectio prima. Havniæ*, 1816, in-8.°

4 *De methodo Floram cujusdam regionis conducendi;* in-8.° *Edinburgi*, 1817.

5 *Essai on the geographical distribution of plants through the counties of Northumberland, etc.;* in-8.° *New-Castle*, 1819.

6 *Allgemeine Uebersicht der Flora auf den Canarischen Inseln. Berlin*, 1819; in-4.°

7 Humboldt, Bonpland et Kunth, *Nova plantarum genera et species Americæ, etc.;* in-4.° Paris, 1015 et suiv.

qu'on trouve éparses dans les écrits des voyageurs, dans les collections des naturalistes, dans les Flores et les ouvrages généraux de botanique; j'oserai peut-être encore ajouter ici, que, par la manière dont j'ai récapitulé ces notes dans le Système universel du règne végétal, elles deviendront plus utiles dans l'avenir à l'étude de la distribution des plantes sur le globe.

Tels sont les ouvrages qui constituent la bibliothèque de la géographie botanique, et dont cet article doit être le résumé : j'y joindrai les considérations qui m'ont été fournies par l'examen attentif que j'ai fait, pendant sept années de voyages en France, de la distribution des plantes sur le sol qui nous entoure.

Je me propose de publier sous peu la statistique végétale de la France, qui contiendra, entre autres résultats de mes voyages, l'ensemble des faits observés sur la distribution des plantes sauvages et cultivées sur la surface de la France. L'article actuel peut être considéré comme l'introduction de cet ouvrage.

Toute la science me paroît se classer sous trois chefs généraux :

1.° L'influence que les élémens extérieurs exercent sur les végétaux, et les modifications qui résultent, pour chaque espèce, du besoin qu'elle a de chaque substance, ou des moyens par lesquels elle peut échapper à son action ;

2.° Les conséquences qui résultent de ces données générales pour l'étude des stations;

3.° L'examen des habitations des plantes, et les conséquences qui en résultent relativement à l'ensemble de la science.

1.re Partie. *Influence des élémens ou agens extérieurs sur les végétaux.*

Nous devons examiner ici l'influence de la température, de la lumière, de l'eau, du sol et de l'atmosphère, et ne pas perdre de vue que, quoique pour la clarté de l'exposition nous devions les séparer, elles agissent cependant presque toutes à la fois.

A. *Influence de la température.*

De toutes ces influences la plus prononcée est la température. Cette action est tellement claire qu'elle est connue de tout le monde, et qu'en l'analysant je ne puis que classer des faits la plupart triviaux.

La température influe sur les végétaux, ou par une action purement physique sur leurs liquides et leurs solides, ou par une action physiologique sur leur force vitale.

Considérée dans son action purement physique, la température dilate ou condense les parties des plantes, comme celles de tous les corps. L'influence sur les solides est peu manifeste; celle sur les liquides est tellement évidente, qu'on peut établir en principe que l'action physique de la température sur les végétaux ou les parties de végétaux est sensiblement proportionnée à la quantité de liquides aqueux qu'ils renferment. Ainsi, les organes qui ne renferment point de liquides, sont comme insensibles aux extrêmes du froid et du chaud : tels sont les bois à leur état parfait, et les graines complétement mûres. De là vient que les graines peuvent être transportées par des causes occasionelles dans des climats entièrement différens des leurs, et y conservent leur vie là où les plantes elles-mêmes périroient.

Mais, pour analyser les effets de la température sur les liquides des végétaux, il faut distinguer ceux qui sont hors du végétal et destinés à y pénétrer, et ceux qui sont déjà introduits dans son tissu.

Toutes les matières dont les végétaux se nourrissent, sont ou de l'eau, ou des substances dissoutes ou suspendues dans l'eau. Si la température est au-dessous de la congélation, l'eau, devenue solide, ne peut pénétrer dans le tissu, et la végétation est suspendue : si la température est trop élevée, le terrain se dessèche et ne fournit plus d'alimens. La première cause de stérilité s'observe au pôle et dans les hautes montagnes; la seconde, dans les lieux très-chauds. Mais l'action de la température est très-sensible à la surface du sol, et l'est moins à une certaine profondeur : d'où il résulte, 1.° que, dans un terrain donné, les plantes à racines profondes résistent mieux aux extrêmes de la température que celles à racines superficielles ; 2.° qu'une plante donnée résiste mieux aux extrêmes de la température dans un terrain plus compacte, ou moins bon conducteur du calorique, ou moins doué de la faculté rayonnante, que dans un sol ou trop léger ou bon conducteur, ou rayonnant fortement le calorique. 3.° La nature des plantes et celle du sol étant données, les plantes résistent mieux au froid dans une atmosphère sèche, et à la chaleur dans une atmosphère humide.

Quant aux liquides renfermés dans le tissu même du végétal, ils sont soumis aux lois générales de la physique. Le froid peut les atteindre au point de les congeler; et comme cette congélation est toujours accompagnée de dilatation, celle-ci, lorsqu'elle est brusque, rompt les parois des cellules ou des vaisseaux, et détermine ainsi la mort partielle des plantes. Si, au contraire, la chaleur est extrême, elle détermine une trop forte évaporation, d'où suit la flétrissure et le desséchement. Voyons par quels mécanismes les plantes peuvent plus ou moins résister à ces effets.

Leur résistance contre la congélation se fonde sur la marche de leur nutrition. Leurs racines sont plongées dans un sol dont la température est en hiver plus chaude que celle de l'air : elles absorbent donc, quoiqu'en petite quantité, un liquide qui, en s'introduisant dans leur tissu, tend à le réchauffer au point que l'intérieur des gros arbres est en général au même degré de température que celle indiquée par un thermomètre placé à la profondeur moyenne de leurs racines. Cette action s'étend jusqu'aux sommités, parce que les liquides ne se communiquent pas leur chaleur de molécule à molécule, et qu'ils ne peuvent la transmettre qu'avec lenteur aux substances ligneuses et mauvaises conductrices qui les entourent. Il s'établit ainsi une lutte entre le froid extérieur de l'atmosphère et la chaleur interne de la séve. Les différences d'un arbre à l'autre tiennent essentiellement à la facilité plus ou moins grande avec laquelle la chaleur de celle-ci peut se dispenser. Ainsi, 1.°, plus le nombre des couches interposées et distinctes par des zones d'air captif sera grand entre l'aubier (qui, renfermant plus d'humidité, est plus susceptible de gel) et l'extérieur, plus les arbres pourront résister au froid : c'est ainsi que les vieux arbres résistent mieux au froid que les jeunes[1]; c'est ainsi que les bouleaux, dont l'écorce présente un grand nombre d'épidermes superposés, résistent à des froids étonnans; c'est ainsi que la plupart des arbres monocotylédones, étant privés d'écorce, vivent moins bien dans les climats froids que les dicotylédones; c'est ainsi que les jeunes pousses résistent bien mieux au froid lorsque, dans leur premier développement, elles sont abritées par des bourgeons écailleux que lorsqu'elles sont à nu, etc.

2.° Plus les couches extérieures sont dépourvues d'eau et abondamment munies de matières charbonneuses ou résineuses, plus aussi les végétaux résistent au froid : ainsi les plantes grasses gèlent assez facilement; ainsi les conifères résistent à des froids très-vifs, tandis que les arbres verts non résineux gèlent à des degrés de froid peu intenses; ainsi les jeunes pousses, imbibées d'eau au printemps, gèlent à des degrés de froid qu'elles supportent en automne, lorsqu'elles sont moins aqueuses; ainsi les arbres gèlent moins facilement après un été bien chaud, qui a, comme disent les jardiniers, parfaitement *aoûté* leurs pousses, qu'après un été froid et pluvieux, où les pousses n'ont pas acquis toute leur dureté.

Toutes ces causes combinées, soit entre elles, soit avec l'état particulier de chaque organe, soit avec la nature du tissu intime de chaque végétal, expliquent assez bien la diversité d'action d'un même degré de froid sur des végétaux divers. Si nous examinons de la même manière l'action d'une température trop élevée, nous verrons que certains végétaux, tels que les bois très-durs, y résistent, parce que, renfermant peu de sucs aqueux, ils offrent peu de matière à évaporer; d'autres, comme les plantes grasses, parce qu'elles sont douées d'un très-petit nombre d'organes évaporatoires; d'autres, comme les herbes des lieux humides, parce qu'elles pompent promptement une quantité d'eau suffisante pour suppléer aux effets de l'évaporation.

---

1 L'azédarach, jeune, gèle souvent, à Montpellier, à 3 ou 4 degrés, et je l'ai vu, plus âgé, supporter sans périr un froid de 15° (therm. centigr.) dans le jardin botanique de Genève.

Quoique ce soit par des causes très-complexes que les végétaux résistent aux actions extrêmes du froid et du chaud, et que par leur réunion on pût peut-être expliquer complétement pourquoi telle plante gèle là où une autre très-semblable ne gèle pas; il seroit, je pense, impossible d'expliquer, par ces simples considérations de physique, pourquoi, entre les limites mêmes où la végétation est possible, des plantes différentes requièrent des degrés de chaleur différens, en sorte que telle graine germe à 5 ou 6°, et que telle autre en exigera 20 ou 30° pour se développer. Cette diversité, qu'on retrouve dans les animaux, doit, très-probablement, dans les deux règnes organiques, être rapportée à l'intensité de l'excitabilité de la fibre ou du tissu de chaque espèce. Le problème se complique donc de causes physiques appréciables et de causes physiologiques que nous sommes obligés d'admettre, quoique nous ne puissions en rendre compte avec la même précision.

L'influence de la température sur la géographie des plantes doit être étudiée sous trois points de vue : 1.° la température moyenne de l'année ; 2.° les extrêmes de la température, soit en froid, soit en chaud ; 3.° la distribution de la température dans les différens mois de l'année.

La température moyenne, qui pendant long-temps a été l'objet presque unique des physiciens, est en réalité la donnée la moins importante pour la géographie des plantes : à ne la considérer que comme une indication vague, elle est d'un emploi assez commode; mais la même température moyenne peut être déterminée par des circonstances tellement différentes, que les conséquences et les analogies qu'on en voudroit déduire sur la végétation, seroient très-erronées.

On tire des résultats plus bornés, mais plus exacts, de l'étude des points extrêmes de la température : ainsi toute localité qui, ne fût-ce que de loin en loin, présente ou un froid ou une chaleur d'une certaine intensité, ne peut présenter à l'état sauvage les végétaux incapables de supporter ce degré extrême. Lorsque ces températures exagérées ne reviennent qu'à de longs intervalles, l'homme peut maintenir dans le pays la culture d'un végétal qui ne sauroit s'y maintenir sauvage, soit parce que, à chaque fois qu'il est détruit par la rigueur exagérée de la saison, il le rétablit par des graines ou des plantes tirées de pays plus tempérés; soit parce que, dans ces momens critiques, il l'abrite contre l'intempérie de l'air; soit, enfin, parce que l'agriculteur ne demande pas toujours aux plantes qu'il cultive de porter des graines fertiles. C'est ainsi que la vigne, l'olivier et la plupart de nos plantes cultivées végètent très-bien pour notre usage dans des climats dont il seroit impossible qu'elles supportassent les hivers, si elles étoient livrées à elles-mêmes: c'est une des causes qui établit une différence absolue entre la géographie agricole et la géographie botanique.

Dans cette dernière, qui nous occupe essentiellement ici, les plantes ne peuvent s'établir à demeure dans un pays que lorsque ce pays ne présente pas, même de loin en loin, des causes de destruction complète. Ainsi, quelle que soit la température moyenne, une plante ne peut vivre sauvage dans un climat où, ne fût-ce que tous les vingt ans, elle viendroit à geler; ou, si quelques graines y sont portées par des causes accidentelles, elles n'ont jamais le temps de s'y établir d'une manière fixe. Les plantes annuelles, qui n'ont d'autre moyen de réproduction que leurs graines, sont complétement exclues de toute localité où une intempérie quelconque peut, ou les tuer, ou seulement empêcher la production de leurs graines: aussi sont-elles exclusivement bornées aux régions tempérées. Les végétaux vivaces peuvent encore vivre sauvages dans des climats qui ne leur permettent pas toujours de produire des graines; celles qui sont douées de moyens particuliers de réproduction par les racines, peuvent vivre même dans des climats où elles ne sauroient presque jamais donner des graines fertiles.

Sous ces divers rapports, et sous plusieurs autres, la distribution de la température dans les mois de l'année est la partie la plus importante de cette étude.

Il est des climats éminemment uniformes, dans lesquels une certaine température moyenne est produite par un hiver doux et un été frais: tels sont en général tous les pays maritimes; ce qui tient à ce que leur température est continuellement ramenée près de la moyenne par la mer, ce vaste réservoir de température constante, qui les rafraîchit l'été et les réchauffe l'hiver : telles sont encore, sans qu'on en connoisse bien les raisons, les parties occidentales des deux continens de l'hémisphère boréal, et, jusques à un certain point, la presque-totalité de l'hémisphère austral. Au contraire, une même température moyenne peut être produite par la combinaison d'hivers très-froids avec des étés très-chauds : c'est ce qu'on observe dans les pays continentaux comparés aux pays maritimes, dans les parties orientales des continens comparées aux occidentales, dans l'hémisphère boréal comparé à l'hémisphère austral.

Les plantes annuelles, qui ont absolument besoin de chaleur pendant l'été pour mûrir leurs graines, et qui peuvent passer l'hiver endormies, pour ainsi dire, à l'état de graines et indifférentes au froid de l'hiver, préfèrent les climats de la seconde série; les plantes vivaces, qui peuvent mieux se passer de mûrir leurs graines, et qui redoutent les grands froids de l'hiver, préfèrent ceux de la première. Parmi celles-ci, les plantes qui perdent leurs feuilles s'accommodent mieux des climats inégaux, et les plantes toujours vertes préfèrent les climats égaux.

Si de ces données générales on descend dans les détails, on concevra facilement comment la température de chaque saison en particulier, comment la durée de la chaleur dans certaines époques de l'année ou de la journée (durée que nos tableaux météorologiques ne représentent que d'une manière imparfaite), peuvent exclure tel ou tel végétal de chaque localité. Je n'ai pu, pressé par l'espace, indiquer ici que les principes et la marche du raisonnement : ceux qui voudront étudier ce sujet curieux d'une manière approfondie, doivent lire et méditer le beau travail de M. de Humboldt sur les lignes isothermes, inséré dans le 3.e volume des Mémoires de la société d'Arcueil.

B. *Influence de la lumière.*

L'influence de la lumière solaire sur la végétation est presque aussi importante que celle de la température, et, quoiqu'elle influe un peu moins que la précédente sur la distribution géographique des végétanx, elle mérite cependant une mention très-particulière.

La lumière est l'agent qui opère le plus grand nombre des phénomènes de la vie végétale. 1.° Elle détermine une grande partie de l'absorption de la séve; les plantes pompent peu d'humidité pendant la nuit et à l'obscurité. 2.° Elle détermine complétement l'émanation aqueuse des parties vertes des plantes; celles-ci n'exhalent point ou presque point d'eau pendant la nuit ou à l'obscurité, tandis que cette exhalaison est très-considérable de jour et surtout aux rayons directs du soleil. 3.° La lumière détermine, sinon absolument dans tous les cas, au moins dans presque tous ceux qu'on connoît bien et qui nous intéressent le plus; la lumière détermine, dis-je, dans le parenchyme des parties vertes, la décomposition de l'acide carbonique, et conséquemment la fixation du carbone dans les végétaux, la coloration des parties vertes, le degré de leur consistance et de leur alongement, l'intensité des propriétés sensibles, et, enfin, la direction de plusieurs organes. 4.° Elle est une des causes principales, et peut-être l'unique, des mouvemens singuliers connus sous le nom de sommeil des feuilles et des fleurs. 5.° Pendant l'absence de la lumière les parties vertes absorbent une certaine quantité de gaz oxigène, déterminée pour chacune d'elles dans un temps donné.

Quoique ces diverses influences s'exercent sur presque tous les végétaux, elles ne s'exercent pas sur toutes les espèces au même degré, et c'est de cette diversité même que naît le besoin qu'a chaque végétal d'une dose particulière de lumière.

A considérer le globe dans sa totalité, la lumière est en moyenne plus également répartie que la chaleur; mais elle offre des disparates importantes dans son mode de répartition. Dans les pays situés près de l'équateur, une lumière intense, parce qu'elle agit plus perpendiculairement, éclaire les végétaux à peu près également toute l'année pendant douze heures chaque jour. A mesure qu'on s'éloigne de l'équateur et qu'on s'approche du pôle, l'intensité des rayons devenus plus obliques va en diminuant; mais, par la distribution de ces rayons, la lumière manque presque complétement pendant l'hiver, où l'absence de végétation la rendroit presque inutile aux plantes, et est presque continue pendant la durée de la végétation, de sorte que sa continuité compense en tout ou en partie son intensité. Quoique les conséquences de la continuité de la lumière n'aient pas encore été suffisamment étudiées, on voit déjà, d'après cette donnée générale, qu'indépendamment de ce qui tient à la température, les plantes qui perdent leurs feuilles peuvent mieux supporter les pays septentrionaux, et que celles à végétation continue doivent avoir un plus grand besoin des régions méridionales. Les plantes dont les feuilles et les fleurs conservent habituellement la même position, peuvent vivre dans les climats du nord, où la lumière est presque continue en été; tandis que c'est dans les climats méridionaux qu'on trouve et qu'on doit trouver les espèces qui sont remarquables par le sommeil et le réveil alternatif de leurs feuilles ou de leurs fleurs, mouvement qui est en rapport avec l'alternative des jours et des nuits.

Dans les pays situés au niveau de la mer, les rayons solaires ne parviennent aux végétaux qu'au travers d'une épaisse atmosphère, qui éteint, pour ainsi dire, une partie de leur éclat; à mesure que l'on s'élève sur les sommités des montagnes, l'action de ces rayons est plus intense, parce que l'atmosphère est moins épaisse : d'où il résulte que, sous chaque latitude donnée, les espèces qui ont besoin en proportion de plus de lumière que de chaleur, doivent occuper le sommet des montagnes, et celles qui veulent plus de chaleur que de lumière doivent demeurer dans les plaines. Tous ceux qui ont tenté de cultiver les plantes des Alpes dans les plaines, savent combien il est difficile d'imiter cette station et de leur donner de la clarté sans trop de chaleur.

Enfin, dans chaque pays déterminé, les plantes se distribuent entre les diverses localités, d'après le besoin qu'elles ont d'une certaine quantité de lumière, et le point auquel chacune d'elles peut, sans trop souffrir, supporter un certain degré d'obscurité. Ainsi, toutes les plantes à feuilles très-aqueuses, qui ont besoin de beaucoup d'évaporation; toutes les plantes grasses qui, ayant très-peu d'organes évaporatoires, ont besoin d'un stimulant pour déterminer sûrement leur action; toutes celles qui sont d'un tissu très-abondant en carbone, ou qui ont des sucs très-résineux ou huileux, ou qui offrent une grande étendue de surfaces vertes, etc., ont besoin de beaucoup de lumière et se trouvent dans les lieux découverts : les autres, selon qu'elles s'écartent davantage de ces conditions, vivent ou à l'ombre légère des buissons, ou à celle plus forte des haies ou des murs, ou à celle des forêts (qui varient entre elles selon la nature des arbres), ou, comme le font certains champignons, dans les cavernes et à l'obscurité totale. On a encore peu étudié les végétaux relativement à la dose de lumière dont ils ont besoin ; mais je ne doute pas qu'il n'y ait, à cet égard, de grandes diversités, et qu'elles ne puissent expliquer celles des stations : ainsi j'ai vu des fougères rester vertes dans des caves où les autres plantes étoient toutes étiolées ; ainsi j'ai vu la lumière artificielle des lampes produire des effets très-divers sur différens végétaux exposés à son action. Ce sujet seroit digne des recherches de quelques observateurs exacts. Les époques même où une certaine dose de lumière parvient aux végétaux, quoique moins variables que ce qui tient à la température, présentent encore quelque intérêt. Ainsi, par exemple, les mousses et les arbustes toujours verts, comme le houx, qui végètent principalement en hiver, vivent très-bien dans les forêts d'arbres qui perdent leurs feuilles, où ne pourroient vivre des plantes qui végètent surtout pendant l'été.

C. *Influence de l'eau.*

Tout le monde connoît l'absolue nécessité de l'eau pour la végétation, et les physiologistes ne se sont, à cet égard, distingués du vulgaire, que parce que quelques-uns, tels que Van-Helmont, ont eu l'art d'exagérer encore un effet si puissant. Si nous nous bornons d'abord à l'examen de l'eau en tant que faisant partie du sol lui-même, nous savons qu'elle est le véhicule universel qui apporte aux végétaux tous leurs alimens, et qu'elle-même fait partie de la nourriture qui se fixe dans les plantes et accroît leurs parties solides. Sous ce double rapport les végétaux peuvent différer, et quant à la quantité absolue d'eau qu'ils acquièrent, et quant au mode de son absorption, et quant au besoin qu'a chaque espèce de trouver certaines matières dissoutes dans l'eau qu'elle absorbe. Montrons, en peu de mots, l'influence de ces différences sur la géographie botanique.

La quantité diverse de l'eau absorbée par chaque espèce offre les disparates les plus prononcées, et chacun sait que c'est une des causes qui influent le plus puissamment sur la distribution topographique des végétaux.

Ceux qui ont besoin d'absorber une grande quantité d'eau, savoir, ceux à tissu lâche et spongieux; ceux qui ont des feuilles larges, molles et surtout munies d'un grand nombre de pores corticaux; ceux qui ne portent que peu ou point de poils à leur superficie; ceux dont la végétation est rapide; ceux qui forment peu de matériaux huileux ou résineux; ceux dont les parties ne sont pas susceptibles d'être altérées ou corrompues par l'humidité; ceux, enfin, dont les racines sont très-nombreuses, ont en général besoin d'absorber beaucoup d'eau et ne peuvent vivre que dans des lieux où ils en trouvent naturellement une grande proportion.

Ceux, au contraire, qui ont le tissu serré et compacte, qui ont les feuilles petites, dures ou munies d'un petit nombre de pores; ceux qui ont beaucoup de poils; ceux dont la végétation est lente; ceux qui forment dans le cours de leur végétation beaucoup de matériaux huileux ou résineux; ceux dont le tissu est susceptible d'être altéré ou corrompu par trop d'humidité; ceux, enfin, dont les racines sont peu nombreuses, ont besoin d'une petite quantité d'eau et choisissent de préférence pour leur station naturelle les lieux les plus secs.

Le degré d'action de chacune des causes que je viens d'énumérer, et leur combinaison mutuelle, déterminent pour chaque espèce le besoin d'une quantité d'eau à peu près déterminée. Mais, quelque compliquées que soient ces causes, il faut encore les combiner avec d'autres : ainsi, plus la température est élevée, et plus la lumière est intense dans une époque et un lieu donnés; plus aussi, toutes choses étant d'ailleurs égales, les plantes ont besoin d'absorber une plus grande quantité d'eau, parce qu'elles en combinent et en éliminent davantage. De là vient le besoin qu'ont certaines plantes de trouver plus ou moins d'eau à certaines époques de leur vie, ou dans certaines localités, ou dans certains modes de culture.

Si je suivois dans les détails cette marche de raisonnement, je pourrois montrer assez clairement comment les végétaux, par des causes diverses, ont besoin d'une quantité d'eau déterminée, et, par conséquent, doivent prospérer chacun dans la localité qui répond à ses besoins. Mais les exemples sont trop faciles à trouver pour qu'il vaille la peine de les présenter à l'attention du lecteur. Les conséquences mêmes des lois générales que je viens d'indiquer, sont généralement connues : ainsi on sait que les plantes à racines profondes prospèrent mieux dans les pays sujets à de longues sécheresses, parce que le fond de la terre végétale présente toujours un peu d'humidité; celles à racines très-superficielles ne peuvent vivre que dans des climats où l'humidité est plus continue, etc.

Mais la nature de l'eau absorbée par les plantes présente encore de grandes diversités : moins l'eau est chargée de principes nutritifs, plus il est nécessaire que les végétaux en absorbent dans un temps donné pour suffire à leur nourriture; plus, au contraire, l'eau est chargée de principes qui altèrent sa fluidité ou sa transparence, et qui, en tant que molécules solides, tendent à obstruer l'orifice des pores ou à gêner l'absorption par leur viscosité, moins aussi les végétaux en absorbent dans un temps donné.

La nature même des molécules dissoutes ou suspendues dans l'eau influe beaucoup sur la distribution topographique des plantes. Ces matières dissoutes sont : 1.° de l'acide carbonique; 2.° de l'air atmosphérique; 3.° des matières solubles, végétales ou animales; 4.° des principes alcalins ou terreux. On conçoit facilement que, quoique les besoins spéciaux des plantes soient beaucoup moins différens que ceux des animaux comparés entre eux, il doit y avoir à cet égard des diversités remarquables. Quoique cet objet ait été moins étudié que les autres parties de la physiologie végétale, on peut déjà entrevoir bien des faits qui s'y rapportent : ainsi, les végétaux dont le tissu doit contenir beaucoup de carbone, tels que les arbres à bois dur, redoutent plus que d'autres les eaux extrêmement pures et qui renferment peu de gaz acide carbonique.

Les plantes qui présentent beaucoup de matières azotées dans leur composition chimique, telles que les crucifères et les champignons, recherchent de préférence les terrains qui renferment beaucoup de matières animales en solution; les plantes qui présentent à leur analyse chimique une quantité notable de certaines substances terreuses, telles que la silice dans les monocotylédones, le gyps dans les légumineuses, etc., ont besoin d'en trouver dans le sol où elles croissent, et, s'il en manque, l'agriculteur a soin d'en ajouter artificiellement. Les espèces qui offrent, lorsqu'on les brûle, une quantité de substances alcalines plus considérable qu'à l'ordinaire, ne peuvent vivre que là où ces matières sont accumulées : ainsi, toutes celles qui ont un besoin absolu de carbonate de soude, ne peuvent prospérer que près de la mer ou des sources salées; quelques-unes peuvent suppléer à ce besoin de leur nature par l'absorption du carbonate de potasse, et alors elles peuvent vivre indifféremment près et loin de la mer. Ainsi la nature diverse des matières dissoutes dans les eaux est évidemment une des nombreuses causes qui déterminent les stations des espèces végétales.

Jusqu'ici je n'ai examiné l'eau qu'en tant qu'elle est destinée à être absorbée par les plantes; mais l'eau agit encore sous un autre rapport : lorsqu'elle est amassée en quantité plus considérable que la plante ne peut en absorber, elle réagit sur son tissu, et tend à le décomposer, à le dissoudre ou à le corrompre. Parmi les plantes qui ont besoin d'absorber une grande quantité d'eau, il en est qui ne peuvent pas résister long-temps à cette action, pour ainsi dire, extérieure de l'eau accumulée : ainsi, les plantes à racines très-charnues, comme les bulbes succulentes ou les racines bulbeuses du *protea argentea*, ou les tubercules charnus des cyclamens, sont assez facilement altérés par l'humidité, et ces plantes ne peuvent vivre par conséquent dans des lieux aquatiques ou marécageux. Au contraire, les tiges et les feuilles de certaines plantes sont naturellement douées de moyens par lesquels elles peuvent résister à l'action de l'eau extérieure. Ainsi, les unes ont la faculté de sécréter une matière visqueuse qui les enveloppe et les protège contre l'eau; c'est ce qu'on voit très-bien dans les *batrachospermum*, par exemple; d'autres, telles que plusieurs potamogétons, suintent à leur superficie une espèce de vernis qui empêche l'eau de les toucher, et qui agit pour les en défendre, précisément comme l'huile dont sont enduites les plumes des oiseaux aquatiques. Enfin, les plantes monocotylédones, dont la surface présente un tissu remarquablement siliceux et par conséquent très-peu altérable par l'humidité, résistent mieux que les dicotylédones à l'action de l'eau extérieure. Aussi voyons-nous un plus grand nombre de plantes aquatiques parmi les monocotylédones que parmi les dicotylédones; aussi certaines plantes, même charnues, telles que les aloès, peuvent vivre plusieurs mois sous l'eau sans en être sensiblement altérées.

Me seroit-il permis de faire remarquer ici, en passant, que c'est à cause de cette quantité de silice et de cette inaltérabilité qui en est la suite, que la plupart des peuples du monde ont choisi des monocotylédones pour couvrir leurs maisons? Les septentrionaux ont employé le chaume d'après le même principe par lequel les peuples des tropiques emploient les feuilles des palmiers.

Ce que je viens de dire de l'eau accumulée à l'état de liquide autour des racines ou des feuilles des plantes, seroit applicable, avec de légères modifications, à l'eau dissoute ou suspendue dans l'air : c'est ce que nous verrons tout à l'heure, en parlant de l'influence de l'atmosphère; mais je dois auparavant dire quelques mots de l'influence du sol.

D. *Influence du sol.*

Cette influence est peut-être plus compliquée encore que toutes les précédentes; on peut cependant la réduire à trois considérations principales.

1.° Le sol sert de point d'appui aux végétaux, et par conséquent sa consistance doit lui donner, sous ce rapport, une aptitude particulière pour soutenir, plus ou moins bien, des plantes douées de formes diverses. Ainsi, les terrains de sable très-mobile ne peuvent servir de point d'appui qu'aux végétaux ou assez bas et couchés pour que le vent ne les renverse pas, ou aux arbres munis de racines assez profondes et assez ramifiées pour les fixer dans cette matrice mobile; encore ces deux effets seront-ils modifiés dans leurs résultats selon qu'il s'agira de pays plus ou moins sujets à l'action impétueuse des vents, selon qu'il s'agira d'arbres qui vivent naturellement isolés, ou de ceux qui, croissant en sociétés nombreuses, se protègent réciproquement.

Les règles inverses se trouvent vraies pour les terrains compactes : les plantes à petites racines peuvent y être suffisamment fixées, et celles-là seules peuvent y vivre ; car les racines très-grandes ne sauroient pénétrer dans des terrains trop tenaces.

Enfin, les deux termes extrêmes de cette série présentent également des terrains stériles : les sables trop mobiles, ou les eaux trop courantes ; les argiles trop compactes, ou les rochers trop durs, sont, par des causes inverses, presque entièrement dépourvus de végétation.

2.° La nature chimique des terres ou des pierres qui composent le terrain, influe aussi sur le choix des végétaux qui peuvent le peupler ou y prospérer ; mais c'est ici encore un effet qui, quoiqu'en apparence simple, est en réalité très-complexe.

Les différentes terres agissent sur la végétation par des circonstances physiques, ainsi, par exemple, selon qu'elles sont plus ou moins douées de la force hygroscopique, ou, en d'autres termes, selon qu'elles absorbent l'eau ambiante plus ou moins facilement, qu'elles la retiennent avec plus ou moins de force, ou l'abandonnent plus ou moins facilement. Les plantes qui exigent plus ou moins d'humidité, peuvent prospérer dans tel ou tel terrain ; mais cet effet, évident en lui-même, se complique avec d'autres circonstances : ainsi, Kirwan a montré par l'analyse comparée des terres réputées bonnes pour le froment dans divers pays, qu'elles contiennent d'autant plus de silice que le climat est plus sujet à la pluie, d'autant plus d'alumine que le climat est moins pluvieux ; ou, en d'autres termes, que le terrain, pour être bon pour un végétal donné, doit être plus hygroscopique dans un climat sec, moins hygroscopique dans un climat humide : d'où résulte évidemment que, dans des localités différentes, on peut trouver les mêmes espèces de végétaux dans des terrains différens.

3.° Chaque nature de roche a un certain degré de ténacité et une certaine disposition à se déliter ou à se pulvériser : de là résulte la facilité plus ou moins grande de certains terrains à être formés ou de sable ou de gravier, et à être composés de fragmens de forme ou de grandeur à peu près déterminée. Certains végétaux, par les causes ci-dessus indiquées, pourront préférer tel ou tel de ces sables ou de ces graviers; mais la nature propre de la roche n'agit ici que médiatement : ainsi, lorsqu'on rencontre des roches calcaires qui se délitent comme les schistes argileux, on y trouve les mêmes espèces de végétaux. Les deux considérations que je viens d'indiquer sont très-particulièrement applicables aux lichens des rochers.

4.° Les roches, selon leur couleur ou leur nature, sont plus susceptibles d'être réchauffées par les rayons directs du soleil, et, par conséquent, elles peuvent un peu modifier la température d'un lieu donné; par conséquent aussi influer, quoique légèrement, sur le choix des plantes susceptibles d'y prospérer.

Mais, indépendamment de toutes ces causes physiques, la nature chimique des roches a-t-elle une influence sur les végétaux? On ne peut, sans doute, le nier absolument; mais on doit convenir que cette action a été en général fort exagérée. Il faut remarquer, en effet, que les plantes ne vivent pas en général sur le roc pur, mais dans un détritus de ces mêmes roches; que les roches d'un pays même assez borné présentent souvent des natures très-diverses; que la terre végétale n'est pas seulement formée par les roches qui l'entourent immédiatement, mais encore par le mélange des molécules terreuses chariées par les eaux, transportées par les vents et déposées dans un lieu donné par les débris des animaux ou des végétaux qui y ont vécu précédemment. Il résulte de toutes ces causes que les terres végétales diffèrent beaucoup moins entre elles que les roches qui leur servent de support, et que la plupart des plantes trouvent dans la plupart des terrains les alimens terreux qui leur sont nécessaires; aussi, après sept années de voyages en France, j'ai fini par trouver presque toutes les plantes naissant spontanément dans presque tous les terrains minéralogiques. Lorsqu'il s'agit d'une localité peu étendue et par conséquent d'un même climat, on trouve bien quelquefois certaines plantes qui s'arrêtent à la limite d'un terrain; mais, lorsqu'on étend ses recherches sur un espace plus étendu, on voit souvent cette même plante vivre, sous un climat différent, dans ce terrain qu'elle dédaignoit ailleurs. Je pourrois citer une foule d'exemples à l'appui de ces diverses assertions : ainsi on dit que le buis ne croît que dans les terrains calcaires, et il est vrai qu'il paroît les préférer; mais je l'ai trouvé en abondance dans les schistes argilo-calcaires des Pyrénées, et il n'est complétement exclu ni des granites de la Bretagne, ni des terrains volcaniques de l'Auvergne. On dit que le châtaignier ne croît point dans les pays calcaires, et il y est en effet plus rare qu'ailleurs; cependant on trouve de beaux châtaigniers des deux côtés du lac de Genève, au pied des montagnes calcaires du Jura et du Chablais. M. Carradori a trouvé, par des expériences de laboratoire, que la magnésie pure est un poison pour la plupart des plantes; et M. Dunal, ayant été, à ma demande, visiter un point des environs de Lunel où le sol présente une grande quantité de magnésie presque pure, y a trouvé les mêmes plantes que dans le calcaire environnant, et leurs racines prospéroient dans les fentes de cette roche magnésienne. Sans nier donc entièrement l'influence de la nature chimique des terres (et j'ai, plus haut, en parlant des matières dissoutes dans l'eau, cité quelques exemples qui la prouvent), je pense qu'elle ne doit jamais être séparée des influences purement physiques, et qu'on lui a en général attribué une importance exagérée.

E. *Influence de l'atmosphère.*

Plus nous avançons dans la carrière que nous nous sommes proposée, plus nous trouvons que tout est compliqué, qu'aucun effet ne peut être produit par une cause unique, qu'aucun agent n'opère d'une manière simple. Ainsi l'atmosphère peut agir ou simultanément ou séparément par sa composition accidentelle, c'est-à-dire par l'eau et les autres matières qu'elle renferme, suspendues ou dissoutes; par son mouvement, par sa transparence et par sa densité. Je ne parle pas de sa composition primitive, car les expériences les plus exactes ont prouvé que les proportions d'azote et d'oxigène sont constamment les mêmes dans l'atmosphère; mais des matières qui n'en font pas partie intégrante et nécessaire, s'y mêlent dans certains lieux, et la rendent plus ou moins propre à certaines espèces de végétaux. Ainsi, comme cela a lieu dans certaines grottes ou certaines mines, les quantités de gaz acide carbonique ou d'hydrogène peuvent être assez considérables pour empêcher la végétation de toutes les plantes, ou pour ne permettre que celle de quelques-unes, ou plus robustes, ou plus avides de ces substances. Ainsi l'air chargé des émanations salines de la mer nuit à certains végétaux, et favorise au contraire le développement de ceux qui ont besoin de carbonate de soude, comme on le voit dans les vallées du midi de l'Europe, où l'on trouve des plantes maritimes, et où l'on peut cultiver de la soude à une grande distance de la mer, pourvu qu'elles soient ouvertes de son côté et exposées au vent marin.

Mais ces effets divers sont bornés à des localités peu étendues; l'influence la plus générale que l'atmosphère exerce sous le rapport des substances qu'elle renferme, est son influence hygroscopique. Elle est habituellement chargée d'eau, ou invisible et simplement appréciable par l'hygromètre, ou visible et à l'état de vapeur. On n'a encore qu'un petit nombre d'observations ou d'expériences exactes pour connoître, 1.° si ces deux états de l'eau atmosphérique agissent d'une manière bien différente sur les végétaux; 2.° pour déterminer l'influence sur les plantes d'une certaine quantité habituelle ou momentanée, continue ou variable, d'humidité atmosphérique. Les expériences, un peu vagues il est vrai, des cultivateurs, et les observations déduites de la distribution des plantes sur le globe, tendent à prouver que cette influence est assez importante : tel végétal prospère mieux, à égal degré de température, dans un air modérément humide; tel autre dans un air très-humide ou très-sec. C'est une des circonstances que la culture en plein air ne peut point imiter, que la culture des serres n'imite que d'une manière imparfaite, et qui influe, par conséquent, sur les difficultés que nous éprouvons à transporter les végétaux d'un pays dans l'autre. Par conséquent, elle doit agir aussi sur la géographie des plantes, et mérite plus d'attention que les voyageurs ne lui en ont accordé jusqu'ici. C'est en partie à cette cause que tient la différence de la végétation des pays maritimes et des pays continentaux, des montagnes et des plaines, etc. Les brouillards empêchent la fécondation des fleurs, et, par conséquent, telle plante ne pourroit prospérer habituellement dans un climat qui seroit trop souvent nébuleux à l'époque de sa floraison.

L'influence de l'agitation de l'air est bien connue dans les cas extrêmes, mais n'a pas encore été appréciée dans les détails. Tout le monde sait que les vents trop impétueux brisent ou déracinent les arbres, et leur effet est grave dans les pays où ces accidens sont intenses ou fréquens; il l'est d'autant plus que la nature du sol est plus sablonneuse, et qu'il s'agit d'arbres à tiges plus élevées, à branches plus ramifiées, à bois plus fragile, à feuilles plus larges, à fruits plus gros. Mais la stagnation absolue de l'air paroît aussi nuisible à la végétation : déjà plusieurs jardiniers avoient observé qu'on se trouve bien d'établir un peu de mouvement dans l'air des serres; et récemment M. Knight a prouvé que des arbres retenus immobiles croissent moins dans un temps donné que ceux qui sont soumis à l'action du vent. Quoiqu'on n'ait point encore assez apprécié cet effet pour savoir s'il agit sur la distribution des végétaux, je ne crois pas devoir le passer entièrement sous silence.

Mais, de toutes les influences de l'atmosphère, la plus difficile peut-être à réduire à sa véritable valeur est l'action de sa densité, ou, ce qui est la même chose, l'influence de la hauteur absolue sur la végétation. J'ai déjà cherché à analyser cette influence de la hauteur dans un Mémoire qui fait partie du troisième volume de ceux de la société d'Arcueil, et je me bornerai à indiquer les bases générales du phénomène.

La hauteur peut agir sur les végétaux, parce qu'elle a une action très-prononcée, et sur la température, et sur l'intensité de la lumière solaire, et sur l'humidité ambiante, et sur la rareté de l'air atmosphérique.

A mesure qu'on s'élève dans l'atmosphère, la température va en diminuant, d'après des lois aujourd'hui assez bien connues des physiciens, et qui paroissent dépendre de ce que l'air rare a plus de capacité pour la chaleur que l'air dense. Les faits qui prouvent que l'abaissement de la température dans les hautes montagnes est une des causes qui influent le plus sur la distribution des végétaux, sont les suivans.

1.° La fixité de la croissance naturelle de chaque plante à une élévation déterminée au-dessus du niveau de la mer, est d'autant plus grande qu'il s'agit de pays plus voisins de l'équateur, d'autant moindre qu'il s'agit de pays plus tempérés : ce qui tient à ce que, plus on s'éloigne de l'équateur, plus l'exposition d'un lieu donné a d'influence sur sa température.

2.° Dans les pays tempérés, comme la France, par exemple, les plantes qui sont peu affectées par la température et qui croissent à toutes les latitudes, croissent aussi à toutes les hauteurs où le terrain n'est pas couvert de neiges éternelles, depuis le niveau de la mer jusqu'au sommet des montagnes. J'ai recueilli environ sept cents exemples de cette loi : ainsi la bruyère commune, le genévrier, le bouleau, etc., croissent indifféremment au niveau de la mer et à 3000 mètres de hauteur.

3.° Si des plantes qui, selon leur constitution, redoutent une température trop chaude ou trop froide, croissent à des latitudes diverses, on observe que c'est à des hauteurs telles que l'effet de l'élévation puisse compenser celui de la latitude : ainsi, les plantes des plaines du Nord croissent dans le Midi sur les montagnes.

4.° Les plantes cultivées en grand suivent des lois tout-à-fait correspondantes aux précédentes : celles qu'on cultive à toutes latitudes, végètent aussi à toutes hauteurs ; celles qu'on ne trouve qu'à des latitudes déterminées, s'arrêtent aussi à des hauteurs proportionnelles : la pomme de terre, qui vient si bien dans nos plaines, se cultive, au Chili, jusqu'à 3600 mètres d'élévation ; l'olivier, qui n'atteint nulle part 44° de latitude, ne s'élève pas au-dessus de 400 mètres de hauteur.

5.° L'élévation au-dessus du niveau de la mer établit, dans la comparaison de la température des saisons, des effets assez analogues à ceux qui résultent de la distance de l'équateur, de sorte que les effets sur la végétation en sont d'autant plus analogues dans les deux cas.

A mesure qu'on s'élève dans une ligne verticale, il résulte de la diminution de la densité de l'air, que l'intensité de la lumière solaire va en augmentant cet effet est représenté dans la ligne des distances à l'équateur, parce que la continuité de la lumière pendant la durée de la végétation est d'autant plus grande qu'il s'agit d'une latitude plus élevée.

A mesure qu'on s'élève dans les montagnes, on voit l'hygromètre, par sa marche descendante, annoncer que l'humidité de l'air va en diminuant : le même effet général a lieu à mesure qu'on va de l'équateur au pôle.

Dans les montagnes couvertes de neiges éternelles et où les plantes sont arrosées habituellement avec de l'eau glacée, celles qui craignent les températures trop chaudes peuvent vivre à des hauteurs inférieures à celles que, sous la même latitude, elles supportent lorsqu'elles ne sont pas arrosées par de l'eau de neige.

Il semble donc que, sous tous ces rapports, l'espèce de fixité des plantes à de certaines hauteurs tient éminemment à l'abaissement de la température d'après l'élévation. Le seul point de vue, purement théorique, d'après lequel on pourroit croire que la rareté de l'air a par elle-même une action directe sur la végétation, c'est le besoin qu'ont les végétaux d'absorber une quantité plus ou moins grande de gaz oxigène pendant la nuit par leurs parties vertes, et jour et nuit par leurs parties colorées. Il n'est pas douteux qu'il y auroit un terme d'élévation où l'atmosphère, devenue trop rare, ne présenteroit pas assez d'air pour satisfaire à ce besoin des plantes; mais partout les montagnes se trouvent couvertes de neige avant que cet effet devienne sensible. Aussi voyons-nous les plantes qui ont besoin de la plus grande dose d'oxigène, tout comme celles qui ont besoin de la moindre, croître indifféremment dans les plaines et dans les montagnes. Si cette influence entre donc pour quelque chose dans la station des plantes à certaines hauteurs, elle ne me paroît pas appréciable au milieu de l'influence prédominante de la température de la lumière et de l'humidité.

La diminution de la pression de l'air peut encore, selon M. de Humboldt, agir en favorisant et en augmentant l'évaporation. Cet effet est certain en théorie; mais je ne connois pas de moyens, dans les connoissances actuelles, pour en apprécier l'influence réelle.

Pour prouver combien, dans les climats tempérés, l'influence de la hauteur est moindre qu'on ne pourroit le croire, j'ai coté, dans une suite de tableaux qui font partie du Mémoire cité plus haut, les *maxima* et *minima* des hauteurs où j'ai trouvé une même espèce de plantes. Ces tableaux, où j'ai presque toujours négligé à dessein les exemples où la différence ne va pas à mille mètres, prouvent que l'influence des hauteurs est beaucoup moins grande qu'on ne l'avoit cru.

2.e PARTIE. *Des stations.*

Nous venons d'analyser l'influence générale des agens extérieurs sur les végétaux, et d'entrevoir comment la structure propre à chaque plante, combinée avec cette influence générale, détermine pour chaque espèce, ou la possibilité de vivre dans un lieu déterminé, ou sa plus grande prospérité dans une certaine localité. Nous devons maintenant appliquer ces données générales aux stations et aux habitations des plantes. C'est sur cette distinction fondamentale que me semblent reposer tous les moyens de mettre quelque exactitude dans la généralisation des faits connus.

On exprime par le terme de *station*, la nature spéciale de la localité dans laquelle chaque espèce a coutume de croître, et par celui d'*habitation*, l'indication générale du pays où elle croît naturellement. Le terme de station est essentiellement relatif au climat, au terrain d'un lieu donné; celui d'habitation est plus relatif aux circonstances géographiques et même géologiques. La station de la salicorne est dans les marais salés, celle de la renoncule aquatique est dans les eaux douces et stagnantes; l'habitation de ces deux plantes est en Europe, celle du tulipier dans l'Amérique septentrionale. L'étude des stations est, pour ainsi dire, la topographie, et celle des habitations la géographie botanique.

La confusion de ces deux classes d'idées est une des causes qui ont le plus retardé la science, et qui l'ont empêchée d'acquérir quelque exactitude. Nous voyons très-évidemment que dans une région bornée les plantes se distribuent uniquement par le besoin que chacune d'elles a, d'après sa structure, de certaines combinaisons des milieux où elle doit vivre.

La même cause détermine-t-elle les habitations ? C'est une des questions fondamentales de la science, et même pour la discussion des faits il importe de ne pas confondre ceux qui sont relatifs à ces deux classes d'idées. Nous nous bornerons d'abord à l'examen des stations des plantes d'une même région. Les lois relatives aux stations paroissent applicables à toutes les régions; mais on ne doit comparer que les exemples réellement comparables, c'est-à-dire, déduits d'une même région.

Toutes les plantes d'un pays, toutes celles d'un lieu donné, sont dans un état de guerre les unes relativement aux autres. Toutes sont douées de moyens de réproduction et de nutrition plus ou moins efficaces. Les premières qui s'établissent par hasard dans une localité donnée, tendent, par cela même qu'elles occupent l'espace, à en exclure les autres espèces: les plus grandes étouffent les plus petites; les plus vivaces remplacent celles dont la durée est plus courte; les plus fécondes s'emparent graduellement de l'espace que pourroient occuper celles qui se multiplient plus difficilement.

Dans cette lutte perpétuelle il se passe deux phénomènes principaux. 1.° Certaines plantes, d'après leur organisation, ont besoin de certaines conditions d'existence: l'une ne peut pas vivre là où elle ne trouve pas une certaine quantité d'eau salée; l'autre, là où elle n'a pas, à telle époque de l'année, telle quantité d'eau ou telle intensité de lumière solaire, etc. Il résulte de ce besoin de certaines circonstances, que certaines plantes ne peuvent pas se développer dans certaines localités: première cause de la distribution locale des végétaux. 2.° Les conditions d'existence de chaque espèce ne sont pas rigoureusement fixes, mais admettent une certaine latitude entre des limites. On pourroit, pour chaque espèce, déterminer le point qui convient le mieux à sa nature, relativement à la dose de chaleur, de lumière, d'humidité, etc., qu'elle doit recevoir pour être dans le plus grand degré de prospérité possible: ce point une fois déterminé, on ne tarde pas à reconnoître que chaque espèce peut s'en écarter en plus ou en moins dans des limites quelconques. Lorsque ces limites sont très-rapprochées, la plante est plus délicate; elle ne peut vivre que dans un petit nombre de localités, et ne peut, par le même motif, ni se naturaliser au loin ni se cultiver facilement : telles sont, par exemple, les bruyères, les *pinguicula*, les *brunia*, etc. Lorsque ces limites sont larges et plus elles sont larges, plus aussi la plante est robuste ; plus elle peut vivre dans des localités diverses ; plus aussi elle est facile à cultiver et à naturaliser au loin : telles sont la plupart des graminées, les plantains, les centaurées, etc. On trouve tous les degrés de délicatesse ou de force entre ces deux extrêmes.

Mais, à mesure que la localité dans laquelle une plante se développe est plus contraire à sa nature, à mesure aussi elle y croît plus foible; de sorte que telle espèce, le *carex arenaria*, je suppose, qui, dans un terrain sablonneux acquiert tout son développement et étouffe toutes ses voisines, pourra bien dans un terrain compacte être à son tour étouffée par ces mêmes espèces qu'elle auroit domptées dans son sol de prédilection. Ce que le terrain produit dans l'exemple que je viens de citer, pourroit être, dans d'autres cas faciles à remarquer, produit par la température, la lumière, l'eau ou l'atmosphère; bien plus, les mêmes plantes, dans les mêmes localités, luttent les unes avec les autres, et avec des succès différens selon leur âge. Ainsi, dans la culture des dunes des landes, on sème pêle-mêle du genêt et du pin : le genêt, qui pousse très-rapidement, domine et protège les jeunes pins, et quand il se trouve trop serré, il les étouffe quelquefois ; le pin, lorsqu'il échappe à ce danger, grandit plus que les genêts, il les dépasse et finit par les étouffer à son tour. Le même effet peut être produit par des maladies ou des accidens, par la nature diverse des couches de terre à différentes profondeurs, par les intempéries plus dangereuses pour une espèce que pour l'autre, et enfin par l'action de l'homme.

On peut conclure de ces faits, que je me contente d'indiquer, vu que la plupart sont très-bien connus; on peut, dis-je, conclure que dans chaque localité, parmi les plantes qui y sont semées naturellement et qui peuvent réellement y vivre, celles qui y prospèrent davantage tendent à s'emparer de l'espace et à en exclure celles qui y sont plus languissantes : seconde cause de la distribution locale des végétaux, et de la tendance naturelle de chacun d'eux à vivre dans le terrain qui lui convient le mieux.

On peut facilement, de ces considérations générales, déduire l'explication d'un fait observé dès long-temps, mais plus méthodiquement par M. de Humboldt, savoir, qu'il est des espèces dont on trouve le plus souvent les individus épars et égrenés, et d'autres, qu'on a nommées plantes *sociales*, dont les individus naissent rapprochés et comme en sociétés nombreuses. Ainsi, pour citer des extrêmes de ces deux manières de vivre, le *cypripedium calceolus* ou l'*orchis hircina* vit presque toujours isolé, tandis que les bruyères de l'ouest, les rhododendrons des Alpes, les potamogétons, etc., vivent le plus souvent en sociétés nombreuses. Cet effet est dû à des causes diverses. Ainsi, lorsqu'un terrain donné est d'une nature tellement particulière qu'il convient très-bien à certaines espèces et mal à la plupart des autres, celles qui y prospèrent finissent par s'en emparer entièrement. C'est ainsi qu'on trouve des plantes sociales dans tous les terrains spéciaux: telles sont l'*elimus arenarius* dans les sables, les *sphagnum* dans les lieux tourbeux, les rhododendrons sur les pentes élevées des Alpes, les bruyères dans les landes, etc. Toutes ces plantes sont sociales, parce qu'elles ne vivent que dans des localités déterminées.

Au contraire, lorsqu'un terrain convient, au même degré, à un grand nombre de végétaux différens, ceux-ci luttent ensemble, à forces égales, pour s'y établir, et y vivent alors mélangées. C'est ainsi que dans nos terrains cultivés toutes les mauvaises herbes prospèrent pêle-mêle lorsqu'on leur en laisse la liberté; c'est ainsi que les forêts des régions fertiles des tropiques présentent un mélange de plusieurs arbres, tandis que celles des pays tempérés, moins favorisées du climat, présentent d'ordinaire une essence dominante.

Enfin, les espèces éminemment robustes, qui par cela même sont le plus souvent dispersées, deviennent quelquefois sociales: c'est ce qui a lieu, par exemple, dans les très-mauvais terrains, où ces plantes robustes peuvent vivre, tandis que toutes les autres périssent; c'est ainsi que les individus de l'*eryngium campestre* sont égrenés dans certains pays, et vivent souvent en sociétés dans les sables à demi fixés du bord des mers.

A ces causes générales, déduites du mode de nutrition, il faut joindre les causes qui dépendent de la réproduction des plantes : celles qui se propagent par des racines, des tiges ou des jets rampans, comme la piloselle ; celles qui produisent un grand nombre de graines, et dont les graines ne peuvent pas être facilement emportées au loin par les vents, vivent plus rapprochées entre elles que celles d'organisation analogue d'ailleurs, mais à graines peu nombreuses ou très-volatiles.

La disposition ou le rapprochement des individus d'une même espèce est donc une conséquence immédiate de la théorie générale des stations, telle que nous l'avons développée ci-dessus.

La classification des stations des plantes, qui, à la manière dont elle est exposée dans la plupart des livres, semble fort simple, est en réalité fort compliquée et peu susceptible d'une exactitude rigoureuse. Nous avons vu, dans la première partie de cet article, combien une seule des circonstances qui influent sur la végétation présente de modifications, la plupart simultanées : or, une station est une espèce de résultat moyen produit par la combinaison variée et inégale de toutes ces circonstances : ainsi, un marais est différent de lui-même, selon qu'il est alimenté d'eau douce ou d'eau salée ; qu'il est sur un sol d'argile ou sur du sable, dans la plaine ou sur une montagne, dans un climat chaud ou froid, etc. Quoique cette difficulté soit évidente, il existe cependant des données générales dans les stations, de sorte qu'il est utile de les distinguer, lors même qu'on ne peut le faire avec rigueur.

Voici les classes qui paroissent les moins incertaines, savoir :

1.° Les plantes *maritimes* ou *salines*, c'est-à-dire celles qui, sans croître plongées dans l'eau salée et sans flotter à sa surface, ont cependant besoin de vivre près des eaux salées pour en absorber une portion nécessaire à leur nourriture. Il faut distinguer ici celles qui, comme la salicorne, vivent dans les marais salés, et qui paroissent absorber des matières salines par leurs racines et leurs feuilles ; celles qui, semblables au *roccella fuciformis*, vivent sur les rocs exposés à l'air marin, et ne semblent absorber que par leurs feuilles ;

et, enfin, les plantes, telles que l'*eryngium campestre*, qui n'ont pas besoin d'eau salée, mais qui vivent sur les bords de la mer comme ailleurs, parce qu'elles sont assez robustes pour ne pas trop redouter l'action du sel.

2.° Les plantes *marines*, appelées récemment *thalassiophytes* par M. Lamouroux, qui croissent, ou plongées dans l'eau salée, ou flottantes à sa surface. Ces plantes se distribuent dans le fond de la mer ou des eaux salées, d'après le degré de salure de l'eau; d'après le degré habituel de son agitation, la continuité ou l'intermittence de leur immersion, le degré de tenacité du sol, et peut-être l'intensité de la lumière.

3.° Les plantes *aquatiques*, qui vivent plongées dans les eaux douces, soit entièrement immergées, comme les conferves; soit flottantes à la surface, comme les stratiotes; soit fixées dans le sol par leurs racines, avec le feuillage dans l'eau, comme plusieurs potamogétons; soit enracinées dans le sol, et venant ou flotter à la surface, comme les *nymphæa*, ou s'élever au-dessus de la surface, comme l'*alisma plantago*. Cette dernière sous-division se rapproche beaucoup de la classe suivante.

4.° Les plantes des *marais* d'eau douce et des lieux très-humides, parmi lesquelles on doit distinguer principalement celles des terrains tourbeux, des prairies marécageuses, du bord des eaux courantes; et, enfin, celles des terrains inondés pendant l'hiver et plus ou moins desséchés pendant l'été.

5.° Les plantes des *prairies* et des pâturages, dans l'étude desquelles il faut distinguer celles qui, par leur réunion sociale, soit naturelle, soit artificielle, forment le fond de la prairie, et celles qui croissent entre elles avec plus ou moins de fréquence et de facilité. Ces plantes des prairies ne diffèrent que par le degré d'humidité de celles des prairies marécageuses.

6.° Les plantes des *terrains cultivés*. Cette classe est tout-à-fait due à l'action de l'homme : les plantes qui croissent dans nos terres cultivées sont celles qui, dans l'état sauvage, se plaisent dans les terrains légers et substantiels; plusieurs d'entre elles ont été transportées d'un pays à l'autre avec les graines mêmes des plantes cultivées. Celles qu'on trouve dans les champs, les vignes et les jardins, quoique souvent les mêmes, présentent souvent aussi un choix particulier déterminé par le mode de culture.

7.° Les plantes des *rochers*, desquelles on passe, par des nuances insensibles, à celles des murailles, des lieux rocailleux et pierreux, et jusques à celles des graviers, qui, à mesure que la masse des fragmens va en diminuant, nous conduisent, par de nombreuses nuances, jusqu'à la classe suivante. L'étude des plantes des rochers présente des diversités remarquables, d'après la nature propre de chaque roche.

8.° Les plantes des *sables* ou des terrains très-meubles, pour la classification desquelles on éprouve quelque difficulté : car celles des sables maritimes se confondent avec les plantes salines; celles des terrains meubles avec les espèces des terrains cultivés; et celles des sables grossiers ne diffèrent pas de celles des graviers.

9.° Les plantes des *lieux stériles*, à raison de ce qu'ils sont trop compactes, comme le sont les terrains argileux, ou ceux dont la superficie se durcit par la sécheresse ou la chaleur, ou ceux qui sont fortement tassés par l'homme ou les animaux. Cette classe hétérogène renferme des végétaux peu tranchés.

10.° Les plantes des *décombres*, ou qui naissent voisines des habitations humaines : ces espèces, en petit nombre, semblent déterminées dans le choix de leur station, les unes par le besoin qu'elles ont des sels nitreux, d'autres peut-être par le besoin de matières azotées.

11.° Les plantes des *forêts*, parmi lesquelles il faut distinguer les arbres qui, par leur réunion, composent la forêt, et les végétaux qui peuvent avec plus ou moins de facilité croître sous leur abri. Parmi les végétaux habitans des bois, leur distribution dans des forêts de diverses essences se détermine d'après le degré d'obscurité plus ou moins grand que chaque espèce peut supporter, soit toute l'année, comme dans les forêts d'arbres verts; soit pendant tout l'été, dans les forêts d'arbres qui perdent leurs feuilles.

12.° Les plantes des *buissons* et des *haies*. Les arbustes qui composent cette station, diffèrent des végétaux des forêts par leurs moindres dimensions et par la légèreté de leur ombrage : les espèces qui croissent entre eux sont plus particulièrement les herbes grimpantes.

13.° Les plantes *souterraines*, qui vivent, soit dans les cavernes plus ou moins obscures, comme les byssus; soit dans le sein même de la terre, comme les truffes. Ces plantes peuvent se passer de l'action de la lumière, et plusieurs d'entre elles ne peuvent même la supporter. Les espèces qui naissent dans les cavités des vieux troncs, ont de grands rapports avec celles des cavernes.

14.° Les plantes des *montagnes*, parmi lesquelles on pourroit admettre comme sous-divisions toutes les autres stations. On a coutume de classer comme plantes montagnardes celles qui, dans nos climats, ne se trouvent qu'à une hauteur absolue de plus de 500 mètres; mais cette limite est tout-à-fait arbitraire. La division la plus importante à établir parmi les plantes montagnardes, est celle des espèces qui croissent dans les montagnes alpines où la neige persiste pendant tout l'été, et où l'arrosement est non-seulement continu, mais d'autant plus abondant et plus froid qu'il fait plus chaud; et des espèces qui croissent dans les montagnes dépouillées de neige pendant l'été, et où, par conséquent, l'arrosement cesse au moment où il seroit le plus nécessaire. Ces dernières sont évidemment plus robustes que les premières, et sont beaucoup plus faciles à soumettre à la culture.

15.° Les plantes *parasites*, c'est-à-dire, qui sont dépourvues de la faculté, ou de pomper leur nourriture du sol, ou de l'élaborer complétement, et qui ne peuvent vivre qu'en absorbant la séve d'un autre végétal : on en trouve dans toutes les stations précédentes. On doit distinguer parmi les plantes parasites : 1.° celles qui naissent à la surface des végétaux, et s'y implantent pour vivre à leurs dépens, telles que le gui et la cuscute; et 2.° les parasites intestines, qui se développent dans l'intérieur même des plantes vivantes, et percent le plus souvent l'épiderme pour paroître au dehors, telles que les urédos et les æcidium.

16.° Les plantes *fausses-parasites*, c'est-à-dire, qui vivent ou sur des végétaux morts ou sur des végétaux vivans, mais sans en pomper la séve. Cette classe, qui a souvent été confondue avec la précédente, présente trois sous-divisions assez distinctes. La première, qui se rapproche des vraies parasites, comprend des plantes cryptogames, dont les germes, apportés probablement pendant l'acte de la végétation, se développent à l'époque où soit la plante, soit l'organe qui la recèle, commence à dépérir, et qui vivent de sa substance pendant son agonie ou après sa mort; telles sont les némaspores et plusieurs sphéries : ce sont de fausses parasites *intestines*. La seconde comprend des végétaux, soit cryptogames, comme les lichens et les mousses, soit phanérogames, comme les épidendrums, qui vivent sur les arbres vivans sans pomper leur séve, et en se nourrissant ou de l'humidité superficielle de l'écorce, ou de celle de l'air : ce sont de fausses-parasites *superficielles;* plusieurs peuvent vivre sur les rochers, les arbres morts ou le sol. La troisième comprend les fausses-parasites *accidentelles*, comme le sont les herbes qu'on voit naître çà et là dans les cavités des troncs.

Ces seize classes admettent assez tolérablement la totalité des végétaux connus; mais, comme j'en ai prévenu, elles ne doivent point être considérées d'une manière rigoureuse. Les unes se rapportent à l'influence du sol, d'autres à celle de l'eau, d'autres à celle de l'air ou de la lumière; et dans chacune d'elles on a pris un élément prédominant pour base de la division, et on a négligé momentanément tous les autres. Cette méthode est peu logique; mais on est forcé de s'en contenter là où des causes très-nombreuses se compliquent ensemble.

L'influence de la température, quoique très-puissante sur les végétaux, a été négligée dans la classification des stations; nous la verrons, au contraire, tenir le premier rang dans le peu qui est appréciable pour nous dans la théorie des habitations, dont nous allons maintenant nous occuper.

3.e Partie. *Des habitations.*

Si l'étude des stations nous a déjà présenté bien des parties vagues et peu susceptibles d'appréciations rigoureuses, celle des habitations nous offre cette incertitude à un degré plus éminent encore. Une partie du phénomène de la distribution des végétaux dans les pays divers, paroît bien tenir à l'influence appréciable de la température ; mais il est encore une partie de faits qui échappe à toutes les théories actuelles, parce qu'elle se lie à l'origine même des êtres organisés, c'est-à-dire au sujet le plus obscur de la philosophie naturelle.

Tous ou presque tous les végétaux, livrés à eux-mêmes, tendent à occuper sur le globe un espace déterminé ; c'est la détermination des lois d'après lesquelles se fait cette circonscription végétale, qui constitue l'étude des habitations. Si l'on se contente de connoissances relatives aux espèces, on peut assez bien déterminer, pour chacune d'elles, les limites en latitude, en longitude et en hauteur, qu'elle n'a pas coutume de franchir. La collection de ces faits de détail est la base de la science. Lorsqu'on les aura tous réunis avec exactitude, peut-être en pourra-t-on déduire des lois générales et rigoureuses; mais nous ne connoissons probablement pas la moitié des espèces du globe, et parmi celles que nous connoissons il en est à peine la moitié dont l'habitation soit déterminée avec précision. Les généralités que nous tentons d'établir en ce moment, sont donc évidemment provisoires; mais elles tendent, tout imparfaites qu'elles sont, à faire connoître l'ensemble de la végétation, et à diriger les voyageurs dans le choix de leurs observations ultérieures : c'est sous ce double rapport qu'elles ont déjà un intérêt réel.

L'influence de la température est manifeste lorsque l'on compare la nature, le nombre et le choix des végétaux qui croissent dans les pays divers, à différentes latitudes et à différentes hauteurs. Cette influence paroît plus grande encore lorsqu'on réfléchit que ces élémens se compensent de manière à procurer aux individus d'une même espèce une température à peu près semblable dans les localités diverses où elle se trouve. Il se passe ici le même phénomène que pour les stations; savoir, que les espèces délicates, qui ont besoin d'une température bien déterminée (soit quant à l'intensité, soit quant à l'époque), n'habitent que dans un seul pays, tandis que les espèces plus robustes, qui s'accommodent de divers degrés de froid et de chaud, peuvent se rencontrer à des distances très-considérables. La température des eaux présentant de moindres diversités que celles de l'air, il est probable que les plantes aquatiques doivent être, moins que toutes les autres, bornées à un climat déterminé: c'est aussi ce que les botanistes croient avoir observé; mais je ne suis pas bien certain que ce résultat probable soit fondé sur des comparaisons assez nombreuses et assez exactes.

Le nombre des espèces diverses d'un espace donné va en augmentant à mesure qu'on avance vers les pays chauds, et en diminuant vers les pays froids. Cette loi est évidente dans les montagnes, qui ont bien moins de plantes à leur sommet qu'à leur base; mais plusieurs autres causes concourent avec la température pour produire ce résultat, qui est plus clair en comparant les pays soumis à des latitudes diverses. Ainsi M. de Humboldt compte 4000 espèces seulement dans l'Amérique tempérée et 13000 dans l'Amérique équinoxiale entre les tropiques, 1500 dans l'Asie tempérée et 4500 dans l'Asie équinoxiale. Ces nombres ne peuvent être que très-approximatifs, vu que les différens pays sont très-inégalement connus.

On peut atteindre à une précision un peu plus grande, en comparant, sous d'autres rapports, le choix des végétaux du Nord et du Midi. En général, si l'on part des régions tempérées, on voit évidemment,

1.° Que le nombre proportionnel des plantes dicotylédones va en augmentant à mesure que l'on approche de l'équateur, et en diminuant vers le pôle;

2.° Que le nombre des acotylédones ou cellulaires suit une règle inverse, c'est-à-dire qu'il va en augmentant vers le pôle, et en diminuant vers l'équateur;

3.° Que celui des monocotylédones, parmi lesquelles je comprends les fougères, souffre peu de variations comparativement aux deux classes précédentes, et forme environ un sixième de la Flore totale de chaque pays, comme du monde entier.

Ces trois propositions peuvent se déduire du tableau suivant.

1.er *Tableau, indiquant le nombre proportionnel des trois grandes classes de végétaux dans divers pays.*

Laponie. Latit. bor. 66 — 69°.

		Soit à la totalité comme
D'après M. Wahlenberg. Nombre total des plantes,	1087.	
Dicotylédones	340	1 : 3
Monocotylédones	186	1 : 6
Acotylédones	557	1 : 2

Islande. Lat. bor. 63 — 67°.

D'après M. Hooker : nombre total	642.	
Dicotylédones	239	1 : 3
Monocotylédones	135	1 : 5
Acotylédones	268	1 : 2 ½

Allemagne. Latit. bor. 45 — 54°.

Principalement d'après M. Hoffmann : nombre total,	3650.	
Dicotylédones	1466	1 : 2 ½
Monocotylédones *	483	1 : 7 ½
Acotylédones **	1700	1 : 2 ⅙

France. Latit. bor. 42 — 51°.

D'après la Flore françoise et le Supplément :

Nombre total	5966.	
Dicotylédones	2997	1 : 2
Monocotylédones	798	1 : 7 ½
Acotylédones	2171	1 : 2 ½

Barbarie. Lat. bor. 34 — 37°.

D'après M. Desfontaines : nombre total	1577.	
Dicotylédones	1200	1 . 1 ¼
Monocotylédones	316	1 : 5
Acotylédones ***	61	1 : 26

Égypte. Latit. bor. 24 — 32°.

D'après M. Delille : nombre total	1030.	
Dicotylédones	776	1 : 1 ⅓
Monocotylédones	192	1 : 6 ⅓
Acotylédones	62	1 : 16

* Ce nombre est plus fort que celui de Hoffmann, parce que j'ai supputé les graminées d'après la partie publiée de la Flore de M. Schrader.

** Ce nombre paroît au-dessous de la vérité. Je n'ai pu noter les algues et les champignons que par approximation.

*** Ce nombre est au-dessous de la vérité. L'auteur s'est moins occupé de cryptogames que du reste du règne végétal.

Jamaïque. Latit. bor. 18°.

D'après M. Lunan : nombre total	1335.	Soit à la tot. comme
Dicotylédones	801	1 : 1 5/8
Monocotylédones	412	1 : 5 1/3
Acotylédones	122	1 : 11

Guiane françoise. Latit. bor. 1 — 4°.

D'après Aublet : nombre total	1209.	
Dicotylédones	960	1 : 1 1/4
Monocotylédones	226	1 : 6
Acotylédones***	23	1 : 57

Amérique équinoxiale entre les tropiques.

D'après M. de Humboldt : total des espèces observées,	4160.	
Dicotylédones	3226	1 : 1 1/4
Monocotylédones	654	1 : 6 1/3
Acotylédones	280	1 : 15

Nouvelle-Hollande. Latit. austr. 10 — 43°.

D'après M. Rob. Brown : total des espèces connues,	4160.	
Dicotylédones	2900	1 : 1 1/2
Monocotylédones	860	1 : 4 7/8
Acotylédones	400	1 : 10

Tristan da Cunha. Latit. austr. 37°.

D'après MM. du Petit-Thouars et Dugald-Charmichael.

Total des espèces	113.	
Dicotylédones	18	1 : 6
Monocotylédones	37	1 : 3
Acotylédones	58	1 : 2

Globe, dans sa totalité.

D'après M. Persoon, en 1805 et 1806.

Total des espèces	27000.	
Dicotylédones	17670	1 : 1 1/2
Monocotylédones	4560	1 : 5 9/10
Acotylédones, environ	4770	1 : 5 6/10

Ou les Dicotylédones font du nombre total environ	4/6
Monocotylédones	1/6
Acotylédones	1/6

Ce genre de calculs ne peut pas être fort exact, 1.° parce qu'on y compare des Flores faites d'après des principes divers et avec un soin inégal; 2.° parce que les acotylédones sont beaucoup moins bien connues que les deux autres classes, et manquent même complétement dans plusieurs Flores.

Sous ce dernier rapport on atteint à une précision plus

grande en comparant seulement les rapports numériques des dicotylédones et des monocotylédones. C'est dans ce but que sont rédigés les deux tableaux suivans.

2.e *Tableau, indiquant le nombre des dicotylédones et des monocotylédones dans diverses Flores non consignées dans le premier tableau.*

États-unis de l'Amérique septentrionale.

D'après M. Pursh :	Vasculaires	2891
	Dicotylédones	2253
	Monocotylédones	638

Isles Britanniques.

D'après M. Smith :	Vasculaires	1485
	Dicotylédones	1078
	Monocotylédones	407

Suisse.

D'après Haller	Vasculaires	1712
	Dicotylédones	1315
	Monocotylédones	397

Venise.

D'après M. Moricand :	Vasculaires	757
	Dicotylédones	568
	Monocotylédones	189

Crimée et Caucase.

D'après M. Marschall de Bieberstein :

Vasculaires	2413
Dicotylédones	2000
Monocotylédones *	413

Royaume de Naples.

D'après M. Tenore :	Vasculaires	2537
	Dicotylédones	2001
	Monocotylédones	536

Isles Canaries.

D'après l'ouvrage et les notes manuscrites de M. de Buch.

Vasculaires	371, ou, en comptant les plantes acclimatées,	533
Dicotylédones	308	419
Monocotylédones	63	114

Sainte-Hélène (île de).

D'après M. Roxburgh :	Vasculaires	61
	Dicotylédones	31
	Monocotylédones	30

* Les fougères sont comptées d'après une note fournie par M. Steven.

3.e *Tableau. Nombres proportionnels des monocotylédones et des dicotylédones, tels qu'ils résultent des deux tableaux précédens.*

1.re CLASSE. Continens ou îles voisines des continens.

			Monocot. sont aux dicotyl. comme
Laponie.............	lat. 66 — 69°	lat. moy. 67°,30′	100 : 183
Islande.............	63 — 67°	— 65°	100 : 170
Isles Britanniques....	50 — 59°	— 54°,30′	100 : 265
Allemagne..........	45 — 54°	— 49°,30′	100 : 304
Suisse..............	46 — 48°	— 47°	100 : 331
France.............	42 — 51°	— 46°,30′	100 : 375
Venise-.............	45 — 46°	— 45°,27′	100 : 300
Royaume de Naples..	38 — 42°	— 40°	100 : 392
États-unis d'Amérique.	31 — 47°	— 39°	100 : 353
Barbarie............	34 — 37°	— 35°,30′	100 : 379
Nouvelle-Hollande...	10 — 43°	— 34° *	100 : 337
Isles Canaries.......	28 — 30°	— 29°	100 : 490
Égypte..............	24 — 32°	— 28°	100 : 404
Guiane françoise....	1 — 4°	— 2°,30	100 : 424
Amérique équinoxiale entre les tropiques....		0′	100 : 493

2.e CLASSE. Isles éloignées des continens.

Jamaïque.................	lat. bor. 18°	100 : 194
Sainte-Hélène.............	lat. austr. 15°55′	100 : 103
Tristan da Cunha..........	lat. austr. 37°	100 : 49

Il résulte des tableaux précédens, que,

1.° Si l'on se borne aux continens ou aux grandes îles très-voisines des continens, le nombre des monocotylédones, comparé aux dicotylédones, va en augmentant vers le pôle et en diminuant vers l'équateur, avec assez de régularité.

2.° Dans les îles éloignées des continens le nombre proportionnel des dicotylédones est plus petit que leur latitude ne paroît le comporter. Ainsi, dans la Jamaïque, où selon l'analogie la proportion devroit être =1 : 4, elle se trouve être =1 : 1,94; à Sainte-Hélène, où la proportion devroit être aussi à peu près =1 : 4, elle se trouve =1 : 1,03; à Tristan da Cunha, où la proportion devroit être =1 : 3,6, elle se trouve =1 : 0,49.

Ce double résultat, et surtout le dernier, pourroit tenir en partie à ce que les monocotylédones ont généralement

* Moyenne des lieux suffisamment explorés.

besoin de plus d'humidité que les dicotylédones : aussi voyons-nous les régions très-sèches, comme les Canaries, la Crimée, le royaume de Naples, présenter moins de monocotylédones que l'analogie de leur latitude ne l'indique, tandis que la Guiane, les environs de Venise, qui sont fort humides, en ont un peu plus que la moyenne des pays situés aux mêmes latitudes.

Des calculs analogues, qu'il seroit trop long de rapporter en détail, montrent que le nombre des arbres, qui, proportionnellement aux herbes, est très-petit près du pôle, va sans cesse en augmentant à mesure qu'on approche de l'équateur, et comme le plus grand nombre des arbres appartient à la classe des dicotylédones, ce résultat est tout-à-fait conforme aux précédens. Pour donner une idée de cette disproportion, je dirai qu'on compte en Laponie 11 arbres et 24 arbustes qui s'élèvent au-dessus de deux pieds : on trouve en France 74 espèces d'arbres sauvages et 195 arbustes s'élevant au-dessus de deux pieds. La Flore de la Guiane, pays mal connu, mais situé sous les tropiques, offre 225 arbres et un nombre très-grand d'arbrisseaux, c'est-à-dire que la proportion des arbres à la totalité de la végétation est en Laponie......... $1/100$,
en France.......... $1/80$,
à la Guiane....... $1/5$.

Ce plus grand nombre de végétaux ligneux qu'on observe dans les pays chauds, se retrouve même en comparant la distribution sur le globe des espèces de chaque famille. Ainsi les fougères en arbre ne vivent que sous les tropiques : les palmiers, qu'on peut regarder comme des liliacées en arbre, ne sortent guère de cette zone : les malvacées fournissent, sous les tropiques, les plus grands arbres du monde, et ne présentent que des herbes dans les pays les plus septentrionaux où elles parviennent ; on en peut dire autant des rubiacées, des composées, etc.

Jusqu'ici nous voyons la végétation de la zone tempérée tenir le milieu entre celle de la zone glaciale et de la zone torride ; mais il est un point de vue sous lequel elle présente un caractère qui lui est propre, c'est qu'elle est la patrie de prédilection des herbes annuelles et bisannuelles. Ainsi, en négligeant les acotylédones, la Laponie ne présente que

36 especes d'herbes, qui ne fructifient qu'une seule fois; on n'en connoît à la Guiane que 73, et la France en compte 1073: de sorte qu'en comparant ces nombres absolus avec la totalité des végétaux de chaque pays, on trouve que le nombre proportionnel des plantes annuelles est en Laponie $\frac{1}{30}$, à la Guiane $\frac{1}{17}$, en France au-delà de $\frac{1}{6}$. Les extrêmes de la température produisent ici des effets analogues: les herbes délicates ne peuvent réussir que dans ces heureuses zones tempérées où l'homme, qui à bien des égards est l'un des êtres les plus délicats de la nature, a lui-même éminemment prospéré; ce n'est que dans ces fortunés climats que l'œil est récréé chaque printemps par cette verdure nouvelle dont la fraîcheur est inconnue et aux habitans de la zone polaire, et à ceux qui vivent sous le soleil brûlant de l'équateur.

Ce que nous venons d'esquisser pour les classes, on devra le faire un jour pour toutes les familles; mais la plupart des Flores étrangères sont encore trop incomplètes pour qu'on puisse donner une grande importance aux résultats qu'on obtiendroit aujourd'hui de recherches longues et minutieuses à faire sur des documens imparfaits. M. de Humboldt a tenté ce beau travail pour quelques grandes familles, et a lui-même consigné les résultats curieux auxquels il est parvenu, dans un article qu'il a bien voulu me communiquer et qui se trouvera à la suite de celui-ci : ceux qui désireront poursuivre ce genre de recherches autant que le comporte l'état actuel de la science, devront aussi étudier avec soin et les Prolégomènes du grand ouvrage botanique de M. de Humboldt, et les notes de géographie botanique qu'il a placées à la fin des principales familles des plantes, et les Mémoires de M. Brown sur la Nouvelle-Hollande et le Congo, que j'ai déjà cités plus haut. L'espace me manque pour donner ici tous les faits de détail; je m'attache surtout à faire connoître la marche du raisonnement qui me paroît propre à la science que quelques botanistes philosophes travaillent à créer.

Toutes les lois que, selon la précision des documens, nous venons d'établir avec plus ou moins de probabilité sur la distribution des plantes, relativement aux degrés de latitude, on devroit les chercher relativement aux hauteurs absolues au-dessus de la mer; mais le nombre des plantes dont l'habitation a été constatée sous ce rapport, est trop borné pour oser l'entreprendre : on peut déjà cependant entrevoir que les mêmes lois s'y représentent avec assez de précision. Les classes, les familles ou les genres qui s'approchent le plus du pôle, tendent à s'élever plus haut sur les montagnes, tandis que celles qui restent dans les zones voisines de l'équateur sont aussi celles qui dans les pays tempérés restent dans les plaines. A mesure qu'on avance vers l'équateur, on retrouve sur les montagnes un choix de végétaux analogues, quant aux genres et aux familles, à ceux des plantes des pays tempérés; et comme les montagnes des pays équinoxiaux sont plus hautes que les nôtres, on y retrouve même des plantes de genres et de familles analogues à nos plantes montagnardes.

Mais, quoique la latitude et la hauteur soient les causes dominantes de la température moyenne d'un lieu, il est encore d'autres causes que j'ai indiquées plus haut, et qui influent principalement sur la distribution de la chaleur dans les diverses époques de l'année : tels sont le voisinage ou la distance de la mer, la forme générale des continens, la direction des vents, etc. Ces causes modifient continuellement les résultats précédens, et établissent de certains rapports de végétation entre des localités éloignées.

Pour achever ce qui est relatif à cette espèce d'arithmétique botanique, comme l'appelle M. de Humboldt, et pour montrer jusqu'à quel point elle peut peindre l'aspect général de la végétation des pays divers, je dirai encore qu'on a tiré quelque parti de la comparaison du nombre proportionnel des espèces et des genres d'un pays. Plus le nombre moyen des espèces de chaque genre ou de chaque famille est borné, plus l'aspect de la végétation présente de variété; plus, au contraire, ce nombre est grand, plus le coup d'œil du pays présente de monotonie dans les formes. Le tableau suivant fait connoître ces résultats pour quelques pays; mais il est nécessaire de faire observer ici combien peu ces résultats offrent de certitude réelle. Ils sont, en effet, modifiés par la tendance plus ou moins grande des auteurs à diviser davantage les genres, ou à distinguer plus d'espèces; ils le sont encore par cette autre circonstance, que, dans les pays souvent étudiés, les espèces ont été toutes distinguées, tandis qu'on confond plus souvent les unes avec les autres lorsqu'il est question de plantes étrangères. Au milieu des incertitudes de ce genre de calcul, il est difficile de ne pas remarquer que c'est dans les îles isolées que le nombre des espèces de chaque genre est proportionnellement le plus petit : fait que je me borne à consigner ici, en attendant des résultats plus exacts.

4.e *Tableau. Nombre proportionnel des genres et des espèces de divers pays.*

	Espèces.	Genres.	Moyenne des espèces par genre.
France	5966	830	7 1/5
Allemagne	4100	608	6 2/3
Cap (10e classe du Prodr. de Thunb.)	1300	265	5
États-unis	2891 *	739	4
Laponie	1087	320	3 1/2
Isles Britanniques	1485 *	458	2 1/3
Barbarie	1577	504	3 1/7
Islande	642	211	3
Jamaïque	1335	504	2 3/5
Égypte	1030	426	2 1/2
Guiane	1209	566	2 1/7
Tristan da Cunha	113	55	2
Sainte-Hélène	61 *	35	1 2/3
Canaries	371 *	212	1 1/2

J'ai cherché à prouver jusqu'ici que les habitations considérées dans leur ensemble paroissent déterminées par la température. Sans doute, il faut combiner avec elle les considérations déduites des stations; car il est clair que, plus un pays sera sablonneux, plus on y trouvera de plantes des sables, etc. Mais, lors même que l'on donne à ces causes toute la latitude qu'on peut leur attribuer, peut-on parvenir à rendre complétement raison des faits les mieux connus? C'est ce dont je doute, et ce qui exige une nouvelle discussion.

---

* Les nombres marqués d'un astérisque se rapportent aux plantes vasculaires seulement.

Il ne seroit peut-être pas difficile de trouver deux points dans les États-Unis et l'Europe, ou dans l'Amérique et l'Afrique équinoxiale, qui présentent toutes les mêmes circonstances, savoir, une même température, une même hauteur, un même sol, une dose égale d'humidité; cependant, presque tous, peut-être tous les végétaux seroient différens dans ces deux localités semblables : on pourroit bien trouver une certaine analogie d'aspect et même de structure entre les plantes de ces deux localités supposées; mais ce seroient en général des espèces différentes. Il semble donc que d'autres circonstances que celles qui déterminent aujourd'hui les stations, ont influé sur les habitations. Avant de discuter cette question, établissons d'abord les faits indépendamment de toute théorie.

Lorsque l'on compare entre elles les diverses parties du monde séparées par de vastes mers, on trouve de grandes différences dans le choix des végétaux; mais il y en a aussi quelques-uns de communs. S'il s'agit de l'hémisphère boréal, on trouve de ces espèces communes à plusieurs régions, principalement vers le pôle, où tous ces pays se réunissent ou se rapprochent beaucoup. On en retrouve encore çà et là dans le reste des deux continens; mais, si l'on fait abstraction des espèces qui paroissent avoir été transportées par l'homme, leur nombre va toujours en diminuant à mesure qu'on approche des régions australes, où la distance des continens devient plus grande : ainsi, sur 2891 espèces phanérogames décrites par Pursh dans les États-Unis, on en trouve 385 qui se retrouvent dans l'Europe boréale ou tempérée, et sur ce nombre, comme l'observe M. de Humboldt, il en est plusieurs qu'il est difficile de croire transportées par l'homme; telles sont le *satyrium viride*, le *betula nana*, etc. Au contraire, MM. de Humboldt et Bonpland n'ont trouvé, dans tous leurs voyages dans l'Amérique équinoxiale, qu'environ vingt-quatre espèces (toutes cyporacées ou graminées) qui fussent communes à l'Amérique et à quelque partie de l'ancien monde. Le nombre des acotylédones commun aux deux continens est plus considérable (autant du moins que la difficulté de distinguer les espèces dans cette classe permet de l'affirmer). Mais les proportions paroissent les mêmes, c'est-à-dire qu'il y a plus d'espèces communes aux deux continens vers le nord que vers le sud.

Si l'on compare la Nouvelle-Hollande avec l'Europe, on trouve, d'après M. Brown, que sur 4100 espèces connues dans cette terre australe il y en a 166 qui lui sont communes avec l'Europe. Sur ce nombre, 15 sont dicotylédones, 32 monocotylédones, et 119 acotylédones. Parmi les deux premières classes, il en est plusieurs qu'on peut soupçonner avoir été transportées par l'homme; mais il en est quelques-unes, telles que les potamogétons, sur lesquelles ce soupçon paroîtroit peu fondé.

Le nombre des espèces communes aux parties de l'ancien continent fort éloignées les unes des autres est peut-être un peu plus considérable que dans les deux exemples que je viens de citer; mais il est encore très-borné : il faut en effet se défier beaucoup, dans les recherches de ce genre, des Flores un peu anciennes; ce n'est que depuis quelques années que les botanistes ont senti toute l'importance de cette question, et ont apporté à l'examen de ces plantes dites communes à divers pays une suffisante attention. Les premiers voyageurs croyoient toujours retrouver dans les pays lointains les plantes de leur patrie, et se plaisoient à leur en donner les noms. Dès qu'ils en ont rapporté des échantillons en Europe, l'illusion s'est dissipée pour le plus grand nombre : lorsque la vue des échantillons secs a laissé encore des doutes, la culture dans les jardins a contribué à les lever, et il reste aujourd'hui (sauf les plantes transportées par l'influence de l'homme) un bien petit nombre d'espèces phanérogames communes à des continens divers. Ainsi, la Nouvelle-Hollande a $1/80$, l'Amérique équinoxiale $1/134$ de ses espèces communes avec l'Europe, et moins encore avec le reste du monde.

Avant d'attacher quelque degré d'importance à ce petit nombre d'espèces communes à des régions fort éloignées, il convient d'examiner quels sont les divers moyens par lesquels les graines peuvent se transporter d'un pays dans un autre.

S'il s'agit d'un transport de proche en proche, il suffit que les circonstances nécessaires à la vie de l'espèce ne soient pas interrompues, ou, en d'autres termes, qu'il ne se rencontre pas sur la route des espaces dans lesquels la végétation de telle ou telle espèce devient impossible. Ces barrières naturelles au transport des plantes sont de divers genres.

1.° Les mers sont des obstacles à la propagation des plantes d'autant plus puissans qu'elles sont plus étendues. Ainsi les plantes des îles participent à la végétation des continens dont elles sont voisines, à peu près en proportion inverse de leur distance : par exemple, en faisant exception des végétaux évidemment naturalisés, on trouve que, sur 1485 végétaux vasculaires qui croissent dans les îles britanniques, il n'y en a que 43 ou $\frac{1}{34}$ qui n'aient pas encore été retrouvées en France ; sur 533 espèces, les îles Canaries en offrent 310, soit environ $\frac{28}{34}$, qui n'ont pas été retrouvées sur le continent d'Afrique, et la Flore de Sainte-Hélène présente à peine deux ou trois espèces qui aient été retrouvées dans l'un des deux continens voisins. Les mers arrêtent le transport des plantes par leur étendue et par l'influence délétère de l'eau salée sur les graines soumises à son action. Ainsi les graines du *lodoicea* des îles Sechelles, transportées par les courans aux Maldives, comme l'a vu M. Labillardière, ou celles du *mimosa scandens* et du *dolichos urens*, transportées des Antilles aux Hébrides, comme je l'ai appris de M. Louis Necker, arrivent dans ces pays lointains privées de la faculté de germer. Mais, quand nous avons des exemples prouvés de graines transportées régulièrement à de telles distances, quand nous avons de fortes probabilités pour croire que l'action délétère de l'eau salée n'agit pas au même degré sur toutes les graines, quand nous voyons les îles offrir si souvent des végétaux semblables à ceux des côtes voisines, pouvons-nous douter qu'un certain nombre d'espèces ne puissent avoir été et être ainsi transportées par la mer d'une région à l'autre, et prospérer, lorsque les plantes y rencontrent un climat conforme à leurs besoins ? Ce transport, qui est très-difficile quand les mers sont très-vastes, devient plus facile lorsqu'il se trouve entre deux continens quelques séries d'îles qui servent aux graines comme de points d'étapes : c'est ainsi que les îles Aleutiennes établissent une communication entre le nord de l'Asie et de l'Amérique ; aussi presque toutes les plantes recueillies jusques à présent dans ces îles sont du nombre des espèces communes à l'ancien et au nouveau continent.

Il est des mers qui semblent avoir moins que les autres arrêté le passage des végétaux; telle est, par exemple, la mer Méditerranée, qui présente sur ses deux bords une végétation presque semblable : sur 1577 espèces observées par M. Desfontaines en Barbarie, il y en a seulement 300 environ, soit à peine $\frac{1}{5}$, qui n'aient pas été retrouvées en Europe. Ce phénomène peut tenir ou à la multitude des îles qui sont dispersées dans cette mer, ou à ce qu'elle est depuis plus long-temps que toute autre parcourue par les navigateurs, ou peut-être à ce qu'elle a dû son origine à quelque irruption de l'océan postérieure à l'origine de la végétation.

2.° La seconde sorte de limites naturelles pour le transport des végétaux est déterminée par les déserts assez vastes et assez continus pour que les graines ne puissent être qu'avec peine transportées d'un côté à l'autre : c'est ainsi que les sables arides et brûlans du Sahara offrent une barrière presque impossible à franchir, et établissent une grande différence entre les végétaux des deux parties de l'Afrique séparées par le désert. Hors les plantes transportées évidemment par l'homme, on peut à peine trouver dans la Flore atlantique quelques espèces qui aient été observées au Sénégal. Les steppes salés de l'Asie occidentale produisent un effet analogue, mais d'une manière moins prononcée, parce qu'ils sont plus interrompus, et moins générale, parce qu'il est un certain nombre d'espèces végétales qui peuvent encore vivre dans cette eau saumâtre.

3.° Une troisième sorte de limites est déterminée par les grandes chaînes de montagnes : celles-ci peuvent influer, ou parce qu'étant couvertes de neiges éternelles elles offrent un obstacle à la propagation des graines, ou parce que la différence brusque de température déterminée par leur élévation empêche certaines espèces de se propager d'un côté à l'autre. Mais il faut remarquer que ce genre de limites est très-imparfait, comparé aux deux précédens. Les chaînes de montagnes sont toujours coupées par des fissures plus ou moins profondes, qui permettent aux plantes de s'étendre d'un côté à l'autre : ainsi on remarque très-bien en France que quelques plantes du Midi s'échappent au travers des gorges des Alpes ou des Cevennes, et se trouvent sur le revers septentrional de ces deux chaînes, principalement dans les lieux où elles sont plus basses ou plus interrompues.

Enfin, tout obstacle continu à la végétation d'une espèce quelconque l'empêche de s'étendre dans une certaine direction : un grand marais est une limite pour les plantes qui craignent l'eau ; une grande forêt, pour celles qui craignent l'ombre ; un changement de latitude ou d'élévation, pour celles qui craignent le froid.

Les plantes sont, à des degrés inégaux, douées de la faculté de franchir ces limites, et il importe beaucoup, pour la question qui nous occupe, de prendre une idée générale de ces moyens de transport, soit naturels, soit factices.

1.° Les mouvemens des eaux transportent fréquemment les graines des plantes riveraines ; j'en ai déjà dit quelques mots en parlant de celles que les courans de la mer charient avec eux : mais les rivières produisent cet effet d'une manière plus sûre, parce que l'eau douce nuit moins que l'eau salée à la faculté germinative ; ainsi on voit souvent des plantes alpines se développer le long du cours des rivières qui descendent des Alpes.

Mais, en donnant à ce transport des graines par les eaux toute l'importance possible, on ne peut guères expliquer comment les graines des plantes aquatiques peuvent s'être transportées d'un bassin dans un autre. Comment, par exemple, l'aldrovanda peut-il se trouver dans le bassin du Pô et dans celui du Rhône ? Si ces faits étoient rares, on pourroit admettre quelques causes accidentelles ; mais les plantes aquatiques, qui, moins que toutes les autres, peuvent être transportées par le vent, l'homme ou les animaux, sont la plupart dispersées dans diverses régions. Ce fait ne seroit-il point une conséquence et une preuve nouvelle des inondations ou déluges qui, en recouvrant d'eau une partie quelconque des terres, ont pu jadis transporter et déposer çà et là les graines des plantes aquatiques ? Il est difficile de comprendre autrement l'existence des poissons et autres animaux d'eau douce dans des lacs privés de toute communication entre eux ; et la même explication, en s'appliquant aux deux règnes organisés, devient plus probable pour l'un et pour l'autre, et moins gigantesque relativement au fait spécial auquel je l'avois d'abord appliquée.

Ainsi les eaux, soit dans leur état actuel, soit dans des états anciens dont d'autres phénomènes attestent la réalité, contribuent à expliquer la dispersion de certaines espèces de plantes.

2.° L'atmosphère peut aussi contribuer au même phénomène : nous en avons la preuve directe dans certaines trombes, qui transportent quelquefois à de grandes distances des graines de végétaux divers ; nous voyons tous les jours les vents transporter çà et là les graines qui, par leur petitesse, ou par les ailes et les aigrettes dont elles sont munies, se prêtent facilement à leur action. Mais, outre les faits de ce genre, tellement triviaux que personne ne songe à les contester, il en est d'autres qui doivent peut-être se rapporter à la même cause. Les graines ou germes des cryptogames sont d'une dimension si petite et d'un poids si léger, que nous les voyons emportés dans l'air, comme ces molécules de poussière impalpable qui flottent sans cesse dans l'atmosphère. On peut concevoir que ces graines sont ainsi transportées à d'immenses distances, sans que cette hypothèse contrarie les lois de la physique ni même celle des simples probabilités. Ainsi les vents qui soufflent long-temps dans de certaines directions, devront transporter avec eux certaines espèces de cryptogames ; j'oserois presque en citer un exemple : la côte de Bretagne est habituellement battue par les vents de sud-ouest, et j'ai trouvé sur les arbres de la promenade de Quimper-Corentin deux lichens (le *sticta crocata* et le *physcia flavicans*) qui n'avoient encore été trouvés qu'à la Jamaïque et qu'on ne retrouve point dans le reste de la France.

3.° Les animaux concourent encore au transport des graines d'une région dans l'autre. Les semences qui, comme le *xanthium spinosum* ou le *galium aparine*, sont munies de crochets ou de piquans, s'attachent aux poils des animaux, et sont ainsi chariées hors de leur terre natale ; celles qui se trouvent entourées par des péricarpes charnus, dont certains oiseaux font leur nourriture, résistent souvent à l'effet de la digestion, et sont semées çà et là avec les excrémens de ces oiseaux : la manière dont les grives sèment le gui, peut donner un exemple de ce fait. Les migrations des oiseaux. à des distances considérables, et même au travers des mers, peuvent, dans quelques cas, transporter des graines au loin.

4.° Enfin, l'homme joue un rôle si important et si actif sur le globe, qu'il en modifie continuellement la surface, et que son action, soit volontaire, soit involontaire, se fait sentir sur la plupart des corps de la nature. Il s'est répandu dans le monde entier, et a transporté partout avec lui les végétaux qu'il cultive pour ses besoins. Lorsque l'introduction de ces cultures est récente, on n'a point de doute sur leur origine; mais, lorsqu'elle est ancienne, on ignore la vraie patrie de ces plantes nourricières. Ainsi personne ne conteste l'origine américaine du maïs ou de la pomme de terre, non plus que l'origine dans l'ancien monde du café ou du froment. Mais il est certains objets cultivés de très-ancienne date entre les tropiques, tels, par exemple, que le bananier, dont l'origine n'est pas avérée : tantôt l'un des continens l'a fourni à l'autre ; tantôt tous les deux possédoient des espèces analogues, qui se confondent aujourd'hui sous le nom de variétés. On peut voir, dans le beau Mémoire de M. Brown sur les plantes du Congo, par quel genre de raisonnemens et d'analogies on peut démêler la vérité relativement à ces anciennes naturalisations.

Parmi celles qui sont plus récentes, il en est encore de difficiles à constater : c'est ainsi que les Nègres arrachés de l'Afrique par l'avide activité des Européens et transportés dans les colonies américaines, y ont porté avec eux quelques-uns des arbres fruitiers et des végétaux utiles de leur patrie ; c'est ainsi que nous avons vu de nos jours des armées porter çà et là des graines et des procédés de culture d'une extrémité de l'Europe à l'autre, et nous montrer ainsi comment dans des temps plus anciens les conquêtes d'Alexandre, les expéditions lointaines des Romains et ensuite les croisades ont pu transporter plusieurs végétaux d'une partie du monde à l'autre.

Mais, outre les plantes qu'il cultive, l'homme en charie sans cesse avec lui, qu'il répand sans s'en douter et quelquefois contre son gré dans le monde entier : ainsi toutes les mauvaises herbes qui croissent au milieu de nos céréales et que peut-être nous avons reçues d'Asie avec elles, nous les avons nous-mêmes introduites dans toutes les parties du globe ; ainsi, avec les blés de Barbarie, les habitans du midi de l'Europe sèment depuis plusieurs siècles les plantes d'Alger et de Tunis; ainsi, avec les laines et les cotons de l'Orient ou de la Barbarie, on apporte fréquemment en France des graines de plantes exotiques, dont quelques-unes se naturalisent. J'en citerai un exemple frappant. Il est à la porte de Montpellier une prairie consacrée à faire sécher les laines étrangères après qu'elles ont été lavées: il ne se passe presque point d'année qu'on ne trouve dans ce pré aux laines des plantes étrangères naturalisées; j'y ai cueilli la *psoralea palæstina*, l'*hypericum crispum*, le *centaurea parviflora*, etc. On voit de même, dans quelques villes maritimes, les plantes étrangères naturalisées par les lests des batimens : Bonamy en cite plusieurs semées de cette manière dans les environs de Nantes; le *datura stramonium*, le *senebiera pinnatifida*, etc., pourroient bien avoir été introduits en Europe de cette manière. Enfin, les jardins de botanique, où l'on réunit tant de végétaux divers, deviennent autant de centres de naturalisation: ainsi l'*erigeron canadense*, le *phytolacca decandra*, etc., qui paroissent en être sortis, sont aujourd'hui plus communs en Europe que bien des plantes indigènes; ainsi nous avons vu dernièrement, aux portes de Genève, le *veronica filiformis* se naturaliser autour d'un jardin particulier de botanique.

Dans nos pays anciennement civilisés, médiocrement favorables à la végétation et sans cesse débarrassés des plantes inutiles par l'agriculture, ces sortes de naturalisation de hasard ne se font qu'avec lenteur, et un grand nombre de plantes ainsi propagées périssent sans postérité; mais dans les pays chauds et mal cultivés ces naturalisations deviennent très-faciles. Ainsi M. Burchell a vu le *chenopodium ambrosioides*, qu'il avoit lui-même semé dans un point de l'île Sainte-Hélène, se multiplier en quatre ans au point d'y être une des mauvaises herbes les plus communes. On trouve une preuve expérimentale de ces naturalisations que l'homme fait à son insçu, dans la comparaison même des plantes qui se retrouvent à de grandes distances : ainsi, dans la Nouvelle-Hollande, dans l'Amérique, au cap de Bonne-Espérance, on trouve plus d'espèces originaires d'Europe que d'aucune autre partie du monde ; d'où l'on voit que l'influence de l'homme l'emporte dans ce cas sur celle des causes purement physiques. Les pays dans lesquels on aborde pour la première fois, ne présentent en général que les espèces véritablement indigènes, et, à mesure que les relations de commerce se multiplient, on voit s'accroître le nombre des plantes européennes ou communes à divers continens. Hâtons-nous donc, pendant qu'il en est temps encore, de faire les Flores exactes des pays lointains ; recommandons surtout aux voyageurs celles des îles peu fréquentées par les Européens : c'est dans leur étude que doit se trouver la solution d'une foule de questions de géographie végétale.

Si l'on réfléchit maintenant à l'action perpétuelle des quatre causes de transport de graines que je viens d'indiquer, les eaux, les vents, les animaux et l'homme, on trouvera, je pense, qu'elles sont bien suffisantes pour expliquer ce petit nombre de végétaux qu'on retrouve semblables dans des continens divers. La première s'applique particulièrement aux plantes aquatiques, la seconde aux cryptogames, les deux dernières aux phanérogames ordinaires. Leur action, lente, simultanée, continue et inaperçue, tend sans cesse à transporter les plantes en tous sens, et celles-ci se naturalisent là où elles rencontrent des circonstances favorables à leur existence.

De l'ensemble de ces faits on peut donc déduire qu'il existe des *régions botaniques :* je désigne sous ce nom des espaces quelconques qui, si l'on fait exception des espèces introduites, offrent un certain nombre de plantes qui leur sont particulières et qu'on pourroit nommer véritablement *aborigènes.* Les plantes d'une région s'y distribuent, d'après leur nature, dans les localités qui leur conviennent, et elles tendent avec plus ou moins d'énergie à dépasser leurs limites et à se répandre dans le monde entier ; mais elles sont la plupart arrêtées, ou par des mers, ou par des déserts, ou par des changemens de température, ou seulement parce qu'elles viennent à rencontrer des espaces déjà occupés par les plantes d'une autre région. Il y a donc des régions parfaitement circonscrites et déterminées ; il en est d'autres qu'on ne peut apprécier que par un certain ensemble ou une certaine masse de végétaux communs.

Nous sommes encore loin de pouvoir appliquer ces principes avec quelque exactitude ; mais on peut cependant déjà entrevoir quelques-unes de ces régions de manière à éveiller sur ces recherches l'attention des voyageurs. Voici à peu près celles qui se présentent à moi dans l'état actuel de nos connoissances.

1.° La région *hyperboréenne*, qui comprend les extrémités boréales de l'Asie, de l'Europe et de l'Amérique, et qui se confond trop avec la suivante.

2.° La région *européenne*, qui comprend toute l'Europe moyenne, sauf les parties voisines du pôle, et celles qui entourent la Méditerranée : elle s'étend à l'est jusqu'à peu près aux monts Altaï.

3.° La région *sibérienne*, où je comprends les grands plateaux de la Sibérie et de la Tartarie.

4.° La région *méditerranéenne*, qui comprend tout le bassin géographique de la Méditerranée ; savoir : la partie d'Afrique en-deçà du Sahara, et la partie d'Europe qui est abritée du nord par une chaîne plus ou moins continue de montagnes.

5.° La région *orientale*, ainsi désignée relativement à l'Europe australe, et qui comprend les pays voisins de la mer Noire et de la mer Caspienne.

6.° L'Inde avec son archipel.

7.° La Chine, la Cochinchine et le Japon.

8.° La Nouvelle-Hollande.

9.° Le cap de Bonne-Espérance, ou l'extrémité australe de l'Afrique, hors des tropiques.

10.° L'Abyssinie, la Nubie et les côtes du Mosambique, sur lesquelles on manque de documens suffisans.

11.° Les environs du Congo, du Sénégal et du Niger, ou l'Afrique équinoxiale et occidentale.

12.° Les îles Canaries.

13.° Les États-Unis de l'Amérique septentrionale.

14.° La côte ouest de l'Amérique boréale tempérée.

15.° Les Antilles.

16.° Le Mexique.

17.° La partie de l'Amérique méridionale située entre les tropiques.

18.° Le Chili.

19.° Le Brésil austral et Buénos-Ayrès.

20.° Les terres Magellaniques.

Enfin, il faudroit joindre à cette indication générale chacune des îles qui est assez écartée de tout autre continent pour présenter un choix de végétaux qui lui est propre.

Les botanistes savent qu'en général les plantes de ces vingt régions sont différentes les unes des autres, de sorte que, lorsqu'on trouve dans les écrits des voyageurs des plantes de l'une de ces régions qu'on dit avoir été retrouvées dans une autre, on doit, avant d'admettre cette proposition, étudier les échantillons venus des deux pays avec un soin tout particulier. A ne considérer cette division du globe que comme une précaution dans la synonymie et la détermination des espèces, elle auroit déjà quelque utilité; mais elle sert surtout à pouvoir exprimer sous une forme un peu plus générale la multitude immense des faits relatifs aux patries des plantes.

Parmi les phénomènes généraux que présente l'habitation des plantes, il en est un qui me paroît plus inexplicable encore que tous les autres : c'est qu'il est certains genres, certaines familles, dont toutes les espèces croissent dans un seul pays (je les appellerai, par analogie avec le langage médical, *genres endémiques*), et d'autres dont les espèces sont réparties sur le monde entier (je les appellerai, par un motif analogue, *genres sporadiques*). Ainsi, quoique très-nombreuses, toutes les espèces des genres *Hermannia, Manulea*, *Borbonia*, *Cluytia*, *Antholiza*, *Gorteria*, etc., sont originaires du cap de Bonne-Espérance; celles de *Banksia*, de *Styphelia*, de *Goodenia*, etc., de la Nouvelle-Hollande; celles de *Mutisia*, de *Cinchona*, de *Fuchsia*, de *Cactus*, de *Tillandsia*, etc., de l'Amérique équatoriale : tandis qu'au contraire la plupart des genres ont des espèces qui croissent spontanément dans des pays très-divers. Quelques familles mêmes semblent affecter certaines régions : ainsi les hespéridées sont toutes de l'Inde ou de la Chine; les labiatiflores, de l'Amérique méridionale; les épacridées, de l'Australasie. Mais rien ne paroît cependant bien régulier dans cette disposition des espèces sur le globe. Ainsi, par exemple, nous possédons en Europe certaines espèces de genres très-nombreux, et dont toutes les autres espèces sont originaires de quelque autre région. Toutes les *passiflora* habitent l'Amérique, sauf une, découverte il y a peu de temps dans l'extrémité australe de l'Afrique par M. Burchell. Tous les *mesembryanthemum* habitent le cap de Bonne-Espérance, excepté les *M. nodiflorum* et *copticum*, qu'on trouve en Corse et en Barbarie; tous les *ixia*, excepté l'*ixia bulbocodium*, commun sur nos côtes méridionales; tous les *gladiolus*, excepté le *gladiolus communis*, si commun dans nos moissons; toutes les bruyères, au nombre de deux ou trois cents espèces, excepté cinq à six qu'on trouve en Europe; presque toutes les oxalis, excepté trois espèces sauvages en France et quelques-unes en Amérique. Ces espèces égrenées, qu'on compareroit volontiers à des soldats séparés de leurs régimens, ont été les causes pour lesquelles les botanistes ont pendant si long-temps négligé l'étude des ordres naturels: il falloit que la botanique exotique fût très-avancée pour qu'on pût reconnoître leurs affinités; car elles sembloient échapper à toutes les règles, lorsque ces règles n'étoient établies que sur les familles européennes. Au reste, cette disposition plus ou moins régulière des espèces et des familles sur le globe est un fait avéré, mais qu'il est aujourd'hui tout-à-fait impossible de réduire à quelque théorie. Un autre fait assez remarquable qui se présente dans la comparaison des régions, c'est que certains pays qui n'offrent point ou presque point d'espèces semblables, donnent naissance à des espèces analogues, c'est-à-dire appartenant aux mêmes genres. Ainsi, par exemple, les États-Unis d'Amérique présentent un grand nombre de genres semblables à ceux de l'ancien continent: tantôt les espèces sont partagées entre les États-Unis et l'Europe, comme, par exemple, dans les genres *Fraxinus*, *Populus*, *Pinus*, *Tilia*; tantôt entre les États-Unis et l'Asie, comme dans les genres *Juglans*, *Magnolia*, *Vitis*; quelquefois entre les trois régions, comme pour les genres *Acer*, *Salix*, *Delphinium*, etc. Ce phénomène se présente d'une manière plus piquante lorsqu'il s'agit de genres très-peu nombreux en espèces: ainsi, par exemple, nous ne connoissons, dans le monde entier, que deux liquidambars, deux panax, deux platanes, deux *stillingia*, deux *planera;* l'une des espèces de chaque genre habite l'Asie orientale, l'autre l'Amérique septentrionale : nous ne connoissons que deux *majanthemum*, deux *vallisneria*, deux *ostrya*, deux châtaigniers, deux *hipophae*, l'une des espèces en Europe, l'autre aux États-Unis : nous ne connoissons que trois espèces de *larix*, de *carpinus*, de *trollius*, l'une en Europe, la seconde en Sibérie, la troisième aux États-Unis. Ce que je viens de dire des trois régions principales de la partie tempérée de l'hémisphère boréal, est également vrai des trois régions équatoriales; ainsi on trouve entre les tropiques, en Asie, en Afrique et en Amérique, des espèces analogues, mais jamais semblables entre elles : par exemple, les espèces des genres *Cratæva*, *Bertiera*, *Elæis*, etc., sont partagées entre l'Amérique et l'Afrique équatoriales; celles des genres *Sagus*, *Strophranthus*, etc., entre l'Asie et l'Afrique équatoriales; celles des genres *Psychotria*, *Begonia*, etc., entre l'Amérique et l'Asie équatoriale; celles des genres *Melastoma*, *Stercutia*, *Jussieua*, entre les trois régions équatoriales. Nous ne connoissons dansle monde entier que deux *cytinus*, l'un dans la région méditerranéenne, l'autre au Mexique; deux *sphenoclea*, l'un au Malabar, l'autre au Mexique; deux *melothria*, l'un en Guinée, l'autre aux Antilles; deux *gyrocarpus*, l'un dans l'Inde, l'autre aux Antilles; deux *sauvagesia*, l'un à Cayenne, l'autre à Madagascar, etc. La même analogie s'aperçoit aussi entre les régions de l'hémisphère austral, mais d'une manière moins marquée, soit parce que les mers en occupent une partie proportionnément plus grande, soit surtout parce que nous connoissons moins les détails de leur botanique locale.

Si nous comparons les régions analogues des deux hémisphères, nous y trouverons de même quelques rapports assez remarquables : ainsi, les espèces des genres *Caltha*, *Empetrum*, etc., se trouvent dans les parties les plus froides des deux hémisphères, et manquent dans tout l'espace intermédiaire; les espèces des genres *Oxalis*, *Passerina*, etc., se trouvent dans les régions tempérées des deux hémisphères, et manquent dans les espaces intermédiaires; les *hypoxis* offrent même ceci de singulier, qu'une partie des espèces croît dans la région tempérée australe de l'ancien monde, et l'autre seulement dans la région tempérée boréale du nouveau.

Enfin, certaines régions présentent des analogies plus particulières encore, et que je dirois volontiers plus mystérieuses. Par exemple, certains genres assez nombreux en espèces sont partagés entre le cap de Bonne-Espérance et le cap de Van-Diémen ; tels sont les *pelargonium*, les *protea*, etc. La région des Canaries et celle de l'Europe offrent un grand nombre de genres semblables, mais qui ont cette particularité que les espèces herbacées sont en Europe, et les espèces ligneuses aux Canaries : ainsi, on trouve dans cette région des *sonchus*, des *prenanthes*, des *convolvulus*, des *echium*, qui sont des arbrisseaux et presque des arbres ; l'île de Sainte-Hélène, dont les forêts sont des espèces de *solidago*, est, sous ce rapport, analogue aux Canaries.

Il semble au premier coup d'œil, et cette idée est si séduisante qu'elle est presque populaire, que ces espèces sont les mêmes que les nôtres, devenues ligneuses par leur séjour dans un climat chaud ; mais il n'en est rien : les espèces ligneuses des Canaries restent ligneuses dans nos climats plus froids ; nos espèces herbacées ne deviennent point ligneuses dans les pays chauds, ou du moins celles qui en sont légèrement susceptibles ne le deviennent pas plus aux Canaries qu'ailleurs. Observons, en effet, pour faire mieux sentir ce caractère particulier de la végétation des Canaries, que d'autres régions également chaudes ont de même des espèces communes avec l'Europe, mais qui y sont herbacées comme chez nous : ainsi, les *sonchus* et les *echium* d'Égypte, les *convolvulus* d'Égypte et de l'Inde, sont herbacés et non ligneux comme aux Canaries. Ces rapports de certains pays les uns avec les autres tiennent sans doute à des ressemblances de localités quelquefois appréciables, quelquefois inconnues ; mais, même dans ce dernier cas, elles peuvent servir de guides dans les naturalisations. Au reste, tout ce que nous venons de dire des régions ne doit s'entendre que des plantes sauvages ; car, dès que les graines d'une espèce trouvent, où que ce soit, un climat et un terrain convenables, elles peuvent s'y développer comme dans leur sol natal. Ce fait nous amène à l'idée déjà indiquée plus haut, savoir, que les stations tiennent uniquement à des causes physiques agissant actuellement, et que les habitations pourroient bien avoir été en partie déterminées par des causes géologiques qui n'existent plus aujourd'hui. Dans cette hypothèse on concevroit facilement pourquoi certaines plantes ne se trouvent jamais sauvages dans des lieux où elles viennent parfaitement dès qu'on les y apporte. Mais cette théorie participe, il faut l'avouer, à l'incertitude de toutes les idées relatives à l'état ancien de notre globe et à l'origine primitive des êtres organisés.

Sous le premier rapport, on pourroit se demander, avec quelques physiciens, si les parties les plus élevées du globe, ayant été les premières découvertes par les eaux, n'ont pas dû être les premières peuplées de végétaux, et servir comme de centres d'où les plantes se seroient dispersées de tous côtés. Cette hypothèse seroit assez d'accord avec l'idée des régions; mais la différence de température des plaines et des montagnes, aussi bien que la circonstance, observée plus haut, que certaines chaînes de montagnes semblent plutôt servir de limites que de centres de végétation, empêche de pouvoir donner trop d'importance à cette idée, que le célèbre Willdenow paroissoit avoir admise.

Dira-t-on, avec quelques autres naturalistes, que les terrains primitifs ont dû les premiers se couvrir de végétaux, ceux-ci ayant dû précéder le développement des animaux, et par conséquent la formation des terrains secondaires? Dans cette idée, les parties primitives du globe devroient être les centres des régions; mais, outre qu'il est difficile de reconnoître des traces de cette dispersion, il est très-douteux que les espèces de plantes qui végètent aujourd'hui soient les mêmes que celles qui ont dû exister avant les terrains secondaires, et dont nous trouvons des empreintes ou des débris dans ces terrains. Cette étude curieuse, commencée il y a peu de temps, au moins avec quelque exactitude, par M. de Sternberg, et que M. Adolphe Brongniart, tout jeune qu'il est, paroît déjà destiné à perfectionner; cette étude, dis-je, semble indiquer que nos espèces végétales sont différentes des espèces antédiluviennes, et que par conséquent il y a eu développement d'une nouvelle végétation depuis la formation des terrains secondaires.

Que seroit-ce, si de ces considérations purement géologiques nous passions à celles qui tiennent aux bases, et je dirois volontiers à la métaphysique de l'histoire naturelle? Toute la théorie de la géographie botanique repose sur l'idée que l'on se fait de l'origine des êtres organisés et de la permanence des espèces. Je n'entreprendrai point de discuter ici ces deux questions fondamentales et peut-être insolubles; mais je ne puis me dispenser de faire remarquer leurs rapports avec l'étude de la distribution des végétaux.

Tout l'article qu'on vient de lire est rédigé en suivant l'opinion que les espèces des êtres organisés sont permanentes, et que tout individu vivant provient d'un autre être semblable à lui : j'ai cherché à montrer qu'en suivant cette opinion, à laquelle tous les faits certains nous conduisent, et qu'on n'attaque qu'en combinant les conséquences de faits douteux ou ambigus, on pouvoit se rendre raison de la plus grande partie de la géographie des plantes. Que si l'on vient à dire que la permanence des espèces n'est pas prouvée, je répondrai qu'elle l'est au moins dans certaines limites : si l'on vient à trouver que deux ou trois plantes voisines, prises pour des espèces, sont des variétés, nous étendrons seulement les bornes qui circonscrivent telle ou telle espèce; mais l'idée même d'espèce n'en sera pas altérée. De ce que les botanistes ont quelquefois admis trop d'espèces, parce qu'ils ont mis trop d'importance à des caractères déduits des parties les plus visibles, mais les moins essentielles, des organes de la végétation, peut-on raisonnablement conclure que les organes de la fructification participent à la même incertitude, et qu'il n'existe pas d'espèces fixes? Je ne le pense pas, et je ne vois pas que ceux-mêmes qui soutiennent ces idées, se conduisent d'après elles. La plupart sont obligés de convenir qu'au moins dans les êtres d'organisation compliquée, lorsqu'une fois les types des espèces sont fixés, ils sont constans dans des limites données : c'est ce qu'on observe dans tous les êtres des deux règnes organisés dont l'anatomie est bien connue. Mais quelle preuve a-t-on qu'il en soit autrement dans les êtres à organes moins distincts et moins bien connus? On auroit facilement soutenu, avant Hedwig, qu'il n'existoit point d'espèces constantes dans les mousses : aujourd'hui on est obligé de se rejeter dans les champignons, dans les algues, pour citer des exemples qu'on ne puisse pas arguer d'erreur dès le premier examen. Singulière logique que celle où l'on néglige à dessein les conséquences de tous les faits bien connus, pour établir les théories générales sur des faits mal connus et bornés à un petit nombre d'êtres!

L'identité plus fréquente des cryptogames, dans divers pays éloignés, a paru un argument en faveur de leur production par les élémens extérieurs; mais nous avons vu qu'on peut l'expliquer par l'agitation permanente de l'atmosphère; et les partisans des formations spontanées me sembleroient, au contraire, dans l'impossibilité d'expliquer le fait général et incontestable, qu'un grand nombre d'espèces bien déterminées ne trouvent que dans une région, et ne se rencontrent pas sauvages dans des pays où toutes les circonstances leur sont favorables et où elles vivent très-bien lorsqu'on les y sème.

Jusqu'à présent les variétés des végétaux paroissent se ranger sous deux chefs généraux : celles qui sont produites par les élémens extérieurs actuels et qui sont modifiables par des circonstances contraires, et celles qui sont formées par l'hybridité et que les circonstances extérieures ne paroissent pas altérer. Les différences constantes des végétaux nés dans diverses régions ne semblent se rapporter ni à l'une ni à l'autre de ces classes: on ne peut les attribuer aux circonstances externes, car d'autres circonstances ne les détruisent pas; on ne peut les attribuer à l'hybridité, car l'hybridité ou le croisement des races suppose nécessairement le rapprochement des êtres analogues. Je comprends très-bien, quoique je ne partage pas complétement cette opinion, je comprends et j'admets, dans quelques cas, que, dans un pays où se trouvent rapprochées plusieurs espèces des mêmes genres, il peut se former des espèces hybrides, et je sens qu'on peut expliquer par là le grand nombre d'espèces de certains genres qu'on trouve dans certaines régions; mais je ne conçois pas comment on pourroit soutenir la même explication pour des espèces qui vivent naturellement à de grandes distances. Si les trois mélèzes connus dans le monde vivoient dans les mêmes lieux, je pourrois croire que l'un d'eux est le produit du croisement des deux autres : mais je ne saurois admettre que, par exemple, l'espèce de Sibérie ait été produite par le croisement de celles d'Europe et d'Amérique. Je vois donc qu'il existe, dans les êtres organisés, des différences permanentes qui ne peuvent être rapportées à aucune des causes actuelles de variations; ce sont ces différences qui constituent les espèces: ces espèces sont distribuées sur le globe en partie d'après des lois qu'on peut immédiatement déduire de la combinaison des lois connues de la physiologie et de la physique, en partie d'après les lois qui paroissent tenir à l'origine des choses et qui nous sont inconnues.

Tel est, en résumé, le point où la géographie botanique est obligée de s'arrêter. Ne perdons pas de vue que cette science n'a pu commencer que lorsque l'étude des espèces a été assez avancée pour lui fournir des faits nombreux et constatés, et que, d'un autre côté, il importe de l'étudier beaucoup, afin d'en fixer les bases avant que les rapports de commerce, les naturalisations, les voyages, les cultures dans les jardins, aient achevé de confondre toutes les régions les unes avec les autres, et quelquefois même aient lié les espèces entre elles par des productions intermédiaires.

Pour donner une idée, et du degré réel de confiance qu'on peut accorder aux résultats des connoissances acquises aujourd'hui, et du nombre des espèces qui restent à découvrir pour pouvoir établir la géographie des plantes sur la connoissance réelle des espèces, je terminerai cet article en rappelant un calcul approximatif, que j'ai mentionné ailleurs[1], sur le nombre proportionnel des espèces connues et de celles qui restent à découvrir sur le globe.

Le catalogue le plus complet du règne végétal que nous possédions aujourd'hui, l'*Enchiridium* de M. Persoon, contient 21,000 espèces, sans compter les cryptogames, qu'on peut estimer à 6000. Depuis lors les grands ouvrages de MM. Brown, de Humboldt, Pursh, etc., en ont fait connoître plusieurs milliers, et il existe, dans les collections des naturalistes, un nombre très-considérable de plantes qui, quoique non décrites, ne peuvent pas être considérées comme inconnues. Pour avoir une idée approximative du nombre

1 Biblioth. univ. des sciences, vol. 6, p. 119.

total des espèces, soit décrites, soit réunies dans les collections, j'ai comparé le nombre des espèces des familles dont j'ai été en dernier lieu appelé à faire des monographies, avec le nombre que les mêmes genres présentent dans Persoon; voici le résultat de cette comparaison.

	Dans Persoon.	Dans le Syst. univ.
Renonculacées	268	509
Dilléniacées	21	90
Magnoliacées	21	37
Anonacées	44	105
Menispermées	37	80
Berbéridées	23	50
Podophyllées	4	6
Nymphæacées	13	30
Papavéracées	27	53
Fumariacées	32	49
Crucifères	504	970
Capparidées	70	215
	1064	2194

Si divers botanistes faisoient donc simultanément le même travail sur toutes les familles du règne végétal, les 27,000 espèces indiquées dans l'ouvrage de Persoon se trouveroient portées à 56,000. Il n'est, en effet, nullement probable qu'il y ait eu dans les livres et dans les collections modernes plus d'augmentations dans ces douze familles que dans toutes les autres; la plus grande portion de ce calcul repose même sur deux familles européennes et qu'on croyoit des mieux connues. En me bornant à dire que le nombre des espèces décrites ou observées dans les collections est de 56,000, je suis probablement au-dessous et non au-dessus de la vérité.

Mais quelle proportion du nombre réel des végétaux du globe représentent ces cinquante-six mille espèces déjà acquises pour la science? Si l'on calcule que c'est depuis trente ans que le plus grand nombre a été recueilli; si l'on compare le nombre proportionnel des espèces européennes et étrangères; si, enfin, l'on cherche à se faire une idée de l'étendue des pays peu ou point parcourus par les botanistes et du nombre des végétaux qu'ils doivent renfermer, on arrive par ces voies diverses à ce même résultat, qu'il est probable que nous n'avons encore recueilli que la moitié des végétaux du globe, et que par conséquent le nombre total des espèces peut être évalué entre 110,000 et 120,000 : nombre immense, qui tend à prouver l'admirable fécondité de la nature; qui démontre la nécessité de perfectionner, autant que possible, les méthodes de classification naturelle; qui doit, enfin, montrer aux voyageurs et aux botanistes qu'il reste beaucoup à recueillir et à observer dans tous les pays du monde.

On voit par ce qui précède, que les lois de la géographie botanique ne sont guères établies que sur la connoissance souvent incomplète d'un quart des végétaux du globe. Ce nombre, tout borné qu'il est, peut suffire pour donner une idée de la théorie des stations, parce que l'étude d'une seule région suffit pour expliquer une foule de faits communs à toutes; mais, quant à la théorie des habitations, nous avons besoin de recherches nombreuses et exactes. Les travaux qui, pour l'avancement de cette partie de la science, me paroissent les plus dignes d'être recommandés aux observateurs, sont les suivans.

Il importe d'abord de multiplier les Flores locales dans différens points du globe, en ayant soin de mettre plus de précision, qu'on ne l'a fait généralement, aux limites géographiques de l'espace dont on décrit la végétation, aux élévations absolues auxquelles les plantes vivent dans diverses localités, et à l'état habituel des milieux ou élémens qui peuvent influer sur la végétation.

Les Flores des îles offrent en particulier un intérêt réel, soit par les bizarreries qu'elles présentent, soit parce que le travail, étant circonscrit, peut être fait avec exactitude.

Il importe que les voyageurs ne se contentent pas seulement de noter qu'ils ont trouvé telle espèce connue dans tels lieux, mais qu'ils rapportent des échantillons qui puissent en constater l'identité. Il est encore à désirer qu'ils notent avec soin les circonstances locales qui peuvent faire présumer si l'espèce est réellement indigène, ou si elle a été naturalisée; si elle vit en société ou éparse, si elle est abondante ou rare dans le pays : en un mot, des détails précis et variés sur les stations et les habitations des plantes sont absolument nécessaires pour donner une marche plus certaine à la géographie botanique. J'ose recommander ces recherches aux voyageurs : il est, je le répète, instant de les faire avant que la civilisation ait trop changé la surface du globe.

Quant aux botanistes sédentaires, leur rôle pour l'avancement de la géographie botanique est de comparer tous les résultats obtenus par les voyageurs, pour en déduire les généralités. Il seroit fort précieux, pour faciliter ce travail, que quelque savant exact et laborieux voulût bien compulser toutes les Flores déjà publiées, et les ranger dans l'ordre des familles naturelles, afin de pouvoir profiter, sans trop de perte de temps, des documens déjà acquis par la laborieuse activité des naturalistes. Je ne doute point qu'un pareil travail ne fasse naître dans l'esprit de celui qui l'entreprendroit une foule d'idées nouvelles et de rapprochemens ingénieux.

Il seroit encore singulièrement utile et à ce genre de recherches, et à plusieurs autres branches des sciences, qu'il se publiât enfin un résumé exact et complet des connoissances acquises sur l'état actuel de la géographie physique et de cette partie de la physique générale qui fait réellement partie de la géographie. Assez long-temps, dans les livres élémentaires consacrés à cette étude, nous n'avons vu que les divisions politiques et les travaux des hommes; il est temps que nous possédions quelque recueil, soit méthodique, soit même alphabétique, sur la nature même des pays divers. Si, en formant ces vœux, je pouvois déterminer quelque savant à exécuter ces travaux, j'aurois sans doute plus contribué à l'avancement de la géographie botanique que par l'esquisse bien imparfaite que je viens d'en présenter. (De Cand.)

Sur les lois que l'on observe dans la distribution des formes végétales; par Alexandre de Humboldt.[1]

Les rapports numériques des formes végétales peuvent être considérés de deux manières très-distinctes. Si l'on étudie les plantes, groupées par familles naturelles, sans avoir égard à leur distribution géographique, on demande quels sont les types d'organisation d'après lesquels le plus grand nombre d'espèces sont formées? Y a-t-il plus de Glu-

1 Cet article est tiré de la seconde édition, *inédite*, de la Géographie des plantes de M. de Humboldt.

macées que de Composées sur le globe? ces deux tribus de végétaux font-elles ensemble le quart des Phanérogames? quel est le rapport des Monocotylédonées aux Dicotylédonées? Ce sont là des questions de phytologie générale, de la science qui examine l'organisation des végétaux et leur enchaînement mutuel. Si l'on envisage les espèces qu'on a réunies d'après l'analogie de leur forme, non d'une manière abstraite, mais selon leurs rapports climatériques ou leur distribution sur la surface du globe, les questions que l'on se propose offrent un intérêt beaucoup plus varié. Quelles sont les familles de plantes qui dominent sur les autres Phanérogames plus dans la zone torride que sous le cercle polaire? les Composées sont-elles plus nombreuses, soit à la même latitude géographique, soit sur une même bande isotherme, dans le nouveau continent que dans l'ancien? Les types qui dominent moins en avançant de l'équateur au pôle, suivent-ils la même loi de décroissement à mesure qu'on s'élève vers le sommet des montagnes équatoriales? Les rapports des familles entre elles ne varient-ils pas sur des lignes isothermes de même dénomination, dans les zones tempérées au nord et au sud de l'équateur? Ces questions appartiennent à la géographie des végétaux proprement dite; elles se lient aux problèmes les plus importans qu'offrent la météorologie et la physique du globe en général. De la prépondérance de certaines familles de plantes dépend aussi le caractère du paysage, l'aspect d'une nature riante ou majestueuse. L'abondance des Graminées qui forment de vastes savanes, celle des Palmiers ou des Conifères, ont influé puissamment sur l'état social des peuples, sur leurs mœurs, et le développement plus ou moins lent des arts industriels.

En étudiant la distribution géographique des formes, on peut s'arrêter aux espèces, aux genres et aux familles naturelles (Humboldt, *Prolog. in Nov. Gen.*, tom. 1, p. XIII, LI et 33). Souvent une seule espèce de plantes, surtout parmi celles que j'ai appelées *sociales*, couvre une vaste étendue de pays. Telles sont, dans le nord, les bruyères et les forêts de pins; dans l'Amérique équinoxiale, les réunions de Cactus, de Croton, de Bambusa et de Brathys de la même espèce. Il est curieux d'examiner ces rapports de multiplication et de développement organique on peut demander quelle espèce, sous une zone donnée, produit le plus d'*individus;* on peut indiquer les familles auxquelles, sous différens climats, appartiennent les espèces qui dominent sur les autres. Notre imagination est singulièrement frappée de la prépondérance de certaines plantes que l'on considère à cause de leur facile reproduction, et du grand nombre d'*individus* qui offrent les mêmes caractères spécifiques, comme les plantes les plus vulgaires de telle ou telle zone. Dans une région boréale où les Composées et les Fougères sont aux Phanérogames dans les rapports de 1 : 13 et de 1 : 25 (c'est-à-dire, où l'on trouve ces rapports en divisant le nombre total des Phanérogames par le nombre des espèces de Composées et de Fougères), une seule espèce de fougères peut occuper dix fois autant de terrain que toutes les espèces de Composées ensemble. Dans ce cas, les Fougères dominent sur les Composées par la *masse*, par le nombre des *individus* appartenant aux mêmes espèces de Pteris ou de Polypodium; mais elles ne dominent pas, si l'on compare à la somme totale des espèces de Phanérogames les formes différentes qu'offrent les deux groupes de Fougères et de Composées. Comme la multiplication de toutes les espèces ne suit pas les mêmes lois, comme toutes ne produisent pas le même nombre d'*individus*, les quotiens obtenus en divisant le nombre total des Phanérogames par le nombre des espèces des différentes familles ne décident pas seuls de l'aspect, je dirois presque du genre de monotonie de la nature dans les différentes régions du globe. Si le voyageur est frappé de la répétition fréquente des mêmes espèces, de la vue de celles qui dominent par leur masse, il ne l'est pas moins de la rareté des individus de quelques autres espèces utiles à la société humaine. Dans les régions où les Rubiacées, les Légumineuses ou les Térébinthacées composent des forêts, on est surpris de voir combien sont rares les troncs de certaines espèces de Cinchona, d'Hæmatoxylum et de Baumiers.

En s'arrêtant aux espèces, on peut aussi, sans avoir égard à leur multiplication et au nombre plus ou moins grand des individus, comparer sous chaque zone, *d'une manière absolue*, les espèces qui appartiennent à différentes familles.

Cette comparaison intéressante a été faite dans le grand ouvrage de M. De Candolle (*Regni vegetabilis Systema Naturæ*, t. 1, p. 128, 396, 439, 464, 510). M. Kunth l'a tentée sur plus de 3300 Composées déjà connues jusqu'à ce jour (*Nov. gen.*, t. 4, p. 238). Elle n'indique pas quelle famille domine au même degré sur les autres Phanérogames indigènes, soit par la masse des individus, soit par le nombre des espèces; mais elle offre les rapports numériques entre les espèces d'une même famille appartenant à différens pays. Les résultats de cette méthode sont généralement plus précis, parce qu'on les obtient sans évaluer la masse totale des Phanérogames, après s'être livré avec soin à l'étude de quelques familles isolées. Les formes les plus variées, des Fougères, par exemple, se trouvent sous les tropiques; c'est dans les régions montueuses, tempérées, humides et ombragées de la région équatoriale, que la famille des Fougères renferme le plus d'espèces. Dans la zone tempérée, il y en a moins que sous les tropiques; leur nombre absolu diminue encore en avançant vers le pôle : mais comme la région froide, par exemple, la Laponie, nourrit des espèces de Fougères qui résistent plus au froid que la grande masse des Phanérogames, les Fougères, par le nombre des espèces, dominent plus sur les autres plantes en Laponie qu'en France et en Allemagne. Les *rapports numériques* qu'offre le tableau que j'ai publié dans mes *Prolegomena de distributione geographica plantarum*, et qui reparoît ici perfectionné par les grands travaux de M. Robert Brown, diffèrent entièrement des rapports que donne la *comparaison absolue* des espèces qui végètent sous les zones diverses. La variation qu'on observe en se portant de l'équateur aux pôles, n'est par conséquent pas la même dans les résultats des deux méthodes. Dans celle des fractions que nous suivons, M. Brown et moi, il y a deux variables, puisqu'en changeant de latitude, ou plutôt de zone isotherme, on ne voit pas varier le nombre total des Phanérogames dans le même rapport que le nombre des espèces qui constituent une même famille.

Lorsque des *espèces* ou des *individus* de même forme qui se reproduisent d'après des lois constantes, on passe aux divisions de la *méthode naturelle* qui sont des *abstractions diver-*

sement graduées, on peut s'arrêter aux genres, aux familles, ou à des sections plus générales encore. Il y a quelques genres et quelques familles qui appartiennent exclusivement à de certaines zones, à une réunion particulière de conditions climatériques; mais il y a un plus grand nombre de genres et de familles qui ont des représentans sous toutes les zones et à toutes les hauteurs. Les premières recherches qui ont été tentées sur la distribution géographique des formes, celles de M. Treviranus, publiées dans son ingénieux ouvrage de *Biologie* (tom. 2, pag. 47, 63, 83, 129), ont eu pour objet la répartition des genres sur le globe. Cette méthode est moins propre à présenter des résultats généraux, que celle qui compare le nombre des espèces de chaque famille ou des grands groupes d'une même famille à la masse totale des Phanérogames. Dans la zone glaciale, la variété des formes génériques ne diminue pas au même degré que la variété des espèces : on y trouve plus de genres dans un moindre nombre d'espèces (De Candolle, *Théorie élément.*, p. 190; Humboldt, *Nova gen.*, tom. 1, pag. XVII et L). Il en est presque de même sur le sommet des hautes montagnes, qui reçoivent des colons d'un grand nombre de genres que nous croyons appartenir exclusivement à la végétation des plaines.

J'ai cru devoir indiquer les points de vue différens sous lesquels on peut envisager les lois de la distribution des végétaux. C'est en les confondant que l'on croit trouver des contradictions qui ne sont qu'apparentes, et que l'on attribue à tort à l'incertitude des observations (*Berliner Jahrbücher der Gewächskunde*, *Bd.* 1, p. 18, 21, 30). Lorsqu'on se sert des expressions suivantes : « cette forme ou « cette famille se perd vers la zone glaciale; elle a sa vé« ritable patrie sous tel ou tel parallèle; c'est une forme « australe; elle abonde dans la zone tempérée; » il faut énoncer expressément si l'on considère le nombre absolu des espèces, leur fréquence absolue croissante ou décroissante avec les latitudes, ou si l'on parle des familles qui dominent, au même degré, sur le reste des plantes phanérogames. Ces expressions sont justes; elles offrent un sens précis, si l'on distingue les différentes méthodes d'après lesquelles on peut étudier la variété des formes. L'île de

Cuba (pour citer un exemple analogue et tiré de l'économie politique) renferme beaucoup plus d'individus de race africaine que la Martinique; et cependant la masse de ces individus domine bien plus sur le nombre des blancs dans cette dernière île que dans celle de Cuba.

Les progrès rapides qu'a faits la géographie des plantes depuis douze ans, par les travaux réunis de MM. Brown, Wahlenberg, De Candolle, Léopold de Buch, Parrot, Ramond, Schouw et Hornemann, sont dus, en grande partie, aux avantages de la méthode naturelle de M. de Jussieu. En suivant, je ne dirai pas les classifications artificielles du système sexuel, mais les familles établies d'après des principes vagues et erronés (*Dumosæ*, *Corydales*, *Oleraceæ*), on ne reconnoît plus les grandes lois physiques dans la distribution des végétaux sur le globe. C'est M. Robert Brown qui, dans un mémoire célèbre sur la végétation de la Nouvelle-Hollande, a fait connoître le premier les véritables rapports entre les grandes divisions du règne végétal, les Acotylédonées, les Monocotylédonées et les Dicotylédonées (Brown, dans *Flinder's voyage to Terra australis*, tom. 2, p. 538; et *Observ. syst. et geographical on the herbar. of the Congo*, p. 3). J'ai essayé, en 1815, de suivre ce genre de recherches, en l'étendant aux différens ordres ou familles naturelles. La physique du globe a ses *élémens numériques*, comme le système du monde, et l'on ne parviendra que par les travaux réunis des botanistes voyageurs à reconnoître les véritables lois de la distribution des végétaux. Il ne s'agit pas seulement de grouper des faits; il faut, pour obtenir des approximations plus précises (et nous ne prétendons donner que des approximations), discuter les circonstances diverses sous lesquelles les observations ont été faites. Je pense, comme M. Brown, qu'on doit préférer, en général, aux calculs faits sur les inventaires incomplets de toutes les plantes publiées, les exemples tirés de pays considérablement étendus, et dont la Flore est bien connue, tels que la France, l'Angleterre, l'Allemagne et la Laponie. Il seroit à désirer qu'on eût déjà une Flore complète de deux terrains de 20,000 lieues carrées, dépourvus de hautes montagnes et de plateaux, et situés entre les tropiques dans l'ancien et le nouveau monde. Jusqu'à ce que ce vœu soit accompli, il faut se contenter des grands herbiers formés par des voyageurs qui ont séjourné dans les deux hémisphères. Les habitations des plantes sont si vaguement et si incorrectement indiquées dans les vastes compilations connues sous les noms de *Systema vegetabilium* et de *Species plantarum*, qu'il seroit très-dangereux de s'en servir d'une manière exclusive. Je n'ai employé ces inventaires que subsidiairement, pour contrôler et modifier un peu les résultats obtenus par les Flores et les herbiers partiels. Le nombre des plantes équinoxiales que nous avons rapportées en Europe, M. Bonpland et moi, et dont notre savant collaborateur, M. Kunth, aura bientôt terminé la publication, est peut-être numériquement plus grand qu'aucun des herbiers formés entre les tropiques : mais il se compose de végétaux des plaines et des plateaux élevés des Andes. Les végétaux alpins y sont même beaucoup plus considérables que dans les Flores de la France, de l'Angleterre et des Indes, qui réunissent aussi les productions de différens climats appartenant à une même latitude. En France, le nombre des espèces qui végètent exclusivement au-dessus de 500 toises de hauteur, ne paroît être que $\frac{1}{9}$ de la masse entière des Phanérogames (De Cand., dans les *Mém. d'Arcueil*, t. 3, p. 295).

Il sera utile de considérer un jour la végétation des tropiques et celle de la région tempérée, entre les parallèles de 40° et de 50°, d'après deux méthodes différentes, soit en cherchant les rapports numériques dans l'ensemble des plaines et des montagnes qu'offre la nature sur une grande étendue de pays, soit en déterminant ces rapports dans les plaines seules de la zone tempérée et de la zone torride. Comme nos herbiers sont les seuls qui font connoître, d'après un nivellement barométrique, pour plus de 4000 plantes de la région équinoxiale, la hauteur de chaque station au-dessus du niveau de la mer, on pourra, lorsque notre ouvrage des *Nova genera* sera terminé, rectifier les rapports numériques du tableau que je publie aujourd'hui, en défalquant des 4000 Phanérogames que M. Kunth a employés à ce travail (*Prolegom.*, pag. XVI) les plantes qui croissent au-dessus de mille toises, et en divisant le nombre total des plantes non alpines de chaque famille par celui des végétaux qui viennent dans les régions froides et tempérées de l'Amérique équinoxiale. Cette manière d'opérer doit affecter le plus, comme nous le verrons tantôt, les familles qui ont des espèces alpines très-nombreuses, par exemple, les Graminées et les Composées. A 1000 toises d'élévation, la température moyenne de l'air est encore, sur le dos des Andes équatoriales, de 17° cent., égale à celle du mois de Juillet à Paris. Quoique sur le plateau des Cordillères on trouve la même température annuelle que dans les hautes latitudes (parce que la *ligne isotherme* de 8°, par exemple, est la trace marquée dans les plaines par l'intersection de la *surface isotherme* de 8° avec la surface du sphéroïde terrestre), il ne faut pas trop généraliser ces analogies des climats tempérés des montagnes équatoriales avec les basses régions de la zone circompolaire. Ces analogies sont moins grandes qu'on ne le pense; elles sont modifiées par l'influence de la distribution partielle de la chaleur dans les différentes parties de l'année (*Proleg.*, p. LIV, et mon *Mémoire sur les lignes isothermes;* p. 137). Les *quotiens* ne changent pas toujours en montant de la plaine vers les montagnes, de la même manière qu'ils changent en approchant du pôle : c'est le cas des Monocotylédonées considérées en général; c'est le cas des Fougères et des Composées. (*Proleg.*, pag. LI et LII; Brown, *on Congo*, pag. 5.)

On peut d'ailleurs remarquer que le développement des végétaux de différentes familles et la distribution des formes ne dépendent ni des latitudes géographiques seules, ni même des latitudes isothermes; mais que les quotiens ne sont pas toujours semblables sur une même ligne isotherme de la zone tempérée, dans les plaines de l'Amérique et de l'ancien continent. Il existe sous les tropiques une différence très-remarquable entre l'Amérique, l'Inde et les côtes occidentales de l'Afrique. La distribution des êtres organisés sur le globe dépend non-seulement de circonstances climatériques très-compliquées; mais aussi de causes géologiques qui nous sont entièrement inconnues, parce qu'elles ont rapport au premier état de notre planète. Les grands Pachydermes manquent aujourd'hui dans le nouveau monde, quand nous les trouvons encore abondamment, sous des climats analogues, en Afrique et en Asie. Dans la zone équinoxiale de l'Afrique la famille des Palmiers est bien peu nombreuse, comparée au grand nombre d'espèces de l'Amérique méridionale. Ces différences, loin de nous détourner de la recherche des lois de la nature, doivent nous exciter à étudier ces lois dans toutes leurs complications. Les lignes d'égale chaleur ne suivent pas les parallèles à l'équateur; elles ont, comme j'ai tâché de le prouver ailleurs, des *sommets convexes* et des *sommets concaves*, qui sont distribués très-régulièrement sur le globe, et forment différens systèmes le long des côtes orientales et occidentales des deux mondes, au centre des continens et dans la proximité des grands bassins des mers. Il est probable que, lorsque des physiciens-botanistes auront parcouru une plus vaste étendue du globe, on trouvera que souvent les lignes des *maxima d'agroupement* (les lignes tirées par les points où les fractions sont réduites au dénominateur le plus petit) dévient des lignes isothermes. En divisant le globe par bandes longitudinales comprises entre deux méridiens, et en en comparant les rapports numériques sous les mêmes latitudes isothermes, on reconnoîtra l'existence de différens *systèmes d'agroupement*. Déjà, dans l'état actuel de nos connoissances, nous pouvons distinguer quatre systèmes de végétation, ceux du nouveau continent, de l'Afrique occidentale, de l'Inde et de la Nouvelle-Hollande. De même que, malgré l'accroissement régulier de la chaleur moyenne du pôle à l'équateur, le *maximum* de chaleur n'est pas identique dans les différentes régions par différens degrés de longitude, il existe aussi des lieux où certaines familles atteignent un développement plus grand que partout ailleurs : c'est le cas de la famille des Composées dans la région tempérée de l'Amérique du nord, et surtout à l'extrémité australe de l'Afrique. Ces accumulations partielles déterminent la physionomie de la végétation, et sont ce que l'on appelle vaguement les traits caractéristiques du paysage.

Dans toute la zone tempérée les Glumacées et les Composées font ensemble plus d'un quart des Phanérogames. Il résulte de ces mêmes recherches, que les formes des êtres organisés se trouvent dans une dépendance mutuelle. L'unité de la nature est telle, que les formes se sont limitées les unes les autres d'après des lois constantes et immuables. Lorsqu'on connoît sur un point quelconque du globe le nombre d'espèces qu'offre une grande famille (p. ex., celle des Glumacées, des Composées ou des Légumineuses), on peut évaluer avec beaucoup de probabilité, et le nombre total des plantes phanérogames, et le nombre des espèces qui composent les autres familles végétales. C'est ainsi qu'en connoissant, sous la zone tempérée, le nombre des Cypéracées ou des Composées, on peut deviner celui des Graminées ou des Légumineuses. Ces évaluations nous font voir dans quelles tribus de végétaux les Flores d'un pays sont encore incomplètes : elles sont d'autant moins incertaines que l'on évite de confondre les *quotiens* qui appartiennent à différens *systèmes de végétation*. Le travail que j'ai tenté sur les plantes, sera sans doute appliqué un jour avec succès aux différentes classes des animaux vertébrés. Dans les zones tempérées il y a près de cinq fois autant d'oiseaux que de mammifères, et ceux-ci augmentent beaucoup moins vers l'équateur que les oiseaux et les reptiles.

La géographie des plantes peut être considérée comme une partie de la *physique du globe*. Si les lois qu'a suivies la nature dans la distribution des formes végétales étoient beaucoup plus compliquées encore qu'elles ne le paroissent au premier abord, il ne faudroit pas moins les soumettre à des recherches exactes. On n'a pas abandonné le tracé des cartes lorsqu'on s'est aperçu des sinuosités des fleuves et de la forme irrégulière des côtes. Les lois du magnétisme se sont manifestées à l'homme dès que l'on a commencé à tracer les lignes d'égale déclinaison et d'égale inclinaison, et que l'on a comparé un grand nombre d'observations qui paroissoient d'abord contradictoires. Ce seroit oublier la marche par laquelle les sciences physiques se sont élevées progressivement à des résultats certains, que de croire qu'il n'est pas encore temps de chercher les *élémens numériques* de la géographie des plantes. Dans l'étude d'un phénomène compliqué, on commence par un aperçu général des conditions qui déterminent ou modifient le phénomène ; mais, après avoir découvert de certains rapports, on trouve que les premiers résultats auxquels on s'est arrêté, ne sont pas assez dégagés des influences locales : c'est alors qu'on modifie et corrige les *élémens numériques*, qu'on reconnoît de la régularité dans les effets mêmes des perturbations partielles. La critique s'exerce sur tout ce qui a été annoncé prématurément comme un résultat général, et cet esprit de critique, une fois excité, favorise la recherche de la vérité et accélère le progrès des connoissances humaines.

Acotylédonées. Plantes cryptogames (Champignons, Lichens, Mousses et Fougères); Agames celluleuses et vasculaires de M. De Candolle. En réunissant les plantes des plaines et celles des montagnes, nous en avons trouvé sous les tropiques $\frac{1}{9}$; mais leur nombre doit être beaucoup plus grand. M. Brown a rendu très-probable que dans la zone torride le rapport[1] est pour les plaines $\frac{1}{15}$, pour les montagnes $\frac{1}{5}$ (*Congo*, p. 5). Sous la zone tempérée, les Agames sont généralement aux Phanérogames comme 1 : 2; dans la zone glaciale, elles atteignent le même nombre, et le surpassent souvent de beaucoup.

En séparant les Agames en trois groupes, on observe que les Fougères sont plus fréquentes (le dénominateur de la fraction étant plus petit) dans la zone glaciale que dans la zone tempérée (*Berl. Jahrb.*, *B.* 1, p. 32). De même les Lichens et les Mousses augmentent vers la zone glaciale. La distribution géographique des Fougères dépend de la réunion de circonstances locales d'ombre, d'humidité et de chaleur tempérée. Leur *maximum* (c'est-à-dire le lieu où le dénominateur de la fraction normale du groupe devient le plus petit possible) se trouve dans les parties montagneuses des tropiques, surtout dans des îles de peu d'étendue, où le rapport s'élève à $\frac{1}{3}$ et au-delà. En ne séparant pas les plaines et les montagnes, M. Brown trouve pour les Fougères de la zone torride $\frac{1}{20}$. En Arabie, dans l'Inde, dans la Nouvelle-Hollande et dans l'Afrique occidentale (entre les tropiques), il y a $\frac{1}{26}$: nos herbiers d'Amérique ne donnent que $\frac{1}{38}$; mais les Fougères sont rares dans les vallées très-larges et les plateaux arides des Andes, où nous avons été forcés de séjourner longtemps (*Congo*, pag. 43, et *Nov. gen.*, tom. 1, pag. 33). Dans la zone tempérée, les Fougères sont $\frac{1}{70}$; en France $\frac{1}{73}$; en

---

1 Dans cet article, les fractions $\frac{1}{9}$, $\frac{1}{15}$, $\frac{1}{5}$, indiquent le rapport entre les espèces d'une famille et la somme des Phanérogames qui végètent dans le même pays. Les abréviations *Trop.*, *Temp.*, *Glac.*, désignent les trois zones, torride, tempérée et glaciale.

Allemagne, d'après des recherches récentes, $\frac{1}{71}$ (*Berl. Jahrb.*, *B.* 1, pag. 26). Le groupe des Fougères est extrêmement rare dans l'Atlas, et manque presque entièrement en Égypte. Sous la zone glaciale, les fougères paroissent s'élever à $\frac{1}{25}$.

Monocotylédonées. Le dénominateur devient progressivement plus petit en allant de l'équateur vers le 62.e de latitude nord ; il augmente de nouveau dans des régions plus boréales encore, sur la côte du Groenland, où les Graminées sont très-rares (*Congo*, pag. 10). Le rapport varie de $\frac{1}{5}$ à $\frac{1}{6}$ dans les différentes parties des tropiques. Sur 3880 Phanérogames de l'Amérique équinoxiale que nous avons trouvées, M. Bonpland et moi, en fleur et en fruit, il y a 654 Monocotylédonées et 3226 Dicotylédonées : donc la grande division des Monocotylédonées seroit $\frac{1}{6}$ des Phanérogames. D'après M. Brown, ce rapport est dans l'ancien continent (dans l'Inde, dans l'Afrique équinoxiale et dans la Nouvelle-Hollande), $\frac{1}{5}$. Sous la zone tempérée on trouve $\frac{1}{4}$ (France 1 : $4\frac{2}{5}$; Allemagne, 1 : $4\frac{1}{2}$; Amérique boréale, d'après Pursh, 1 : $4\frac{1}{2}$) ; Royaume de Naples, 1 : $4\frac{1}{5}$; Suisse, 1 : $4\frac{1}{4}$; Isles britanniques, 1 : $3\frac{3}{4}$). Sous la zone glaciale, $\frac{1}{3}$.

Glumacées (les trois familles des Joncacées, des Cypéracées et des Graminées, réunies).= *Trop.*, $\frac{1}{11}$.—*Temp.*, $\frac{1}{8}$.—*Glac.*, $\frac{1}{4}$.

L'augmentation vers le nord est due aux Joncacées et aux Cypéracées, qui sont beaucoup plus rares, relativement aux autres Phanérogames, sous les zones tempérées et sous la zone torride. En comparant entre elles les espèces appartenant aux trois familles, on trouve que les Graminées, les Cypéracées et les Joncacées sont sous les tropiques comme 25, 7, 1 ; dans la région tempérée de l'ancien continent, comme 7, 5, 1 ; sous le cercle polaire, comme $2\frac{2}{5}$, $2\frac{3}{5}$, 1. Il y a en Laponie autant de Graminées que de Cypéracées : de là vers l'équateur les Cypéracées et les Joncacées diminuent beaucoup plus que les Graminées ; la forme des Joncacées se perd presque entièrement sous les tropiques (*Nov. gen.*, T. Ier, p. 240).

Joncacées seules. = *Trop.*, $\frac{1}{400}$. — *Temp.*, $\frac{1}{90}$. — *Glac.*, $\frac{1}{25}$ (Allemagne, $\frac{1}{94}$; France, $\frac{1}{86}$).

Cypéracées seules. = *Trop.* Amérique, à peine $\frac{1}{57}$; Afrique occidentale, $\frac{1}{18}$; Inde, $\frac{1}{25}$; Nouvelle-Hollande, $\frac{1}{14}$ (*Congo*, p. 9). — *Temp.*, peut-être $\frac{1}{20}$ (Allemagne, $\frac{1}{19}$; France, toujours d'après les travaux de M. De Candolle, $\frac{1}{27}$; Danemarck, $\frac{1}{16}$). — *Glac.*, $\frac{1}{9}$. C'est le rapport trouvé en Laponie et au Kamtschatka.

Graminées seules. = *Trop.* J'ai admis jusqu'ici $\frac{1}{15}$. M. Brown trouve pour l'Afrique occidentale $\frac{1}{12}$, pour l'Inde $\frac{1}{12}$ (*Congo*, p. 41). M. Hornemann s'arrête pour cette même partie de l'Afrique à $\frac{1}{10}$ (*De indole plant. Guineensium*, 1819, p. 10). — *Temp.* Allemagne, $\frac{1}{13}$; France, $\frac{1}{13}$. — *Glac.*, $\frac{1}{10}$.

Composées. En confondant les plantes des plaines avec celles des montagnes, nous avons trouvé dans l'Amérique équinoxiale $\frac{1}{6}$ et $\frac{1}{7}$; mais, sur 534 Composées de nos herbiers, il n'y en a que 94 qui végètent depuis les plaines jusqu'à 500 toises (hauteur à laquelle la température moyenne est encore de 21°, 8; égale celle du Caire, d'Alger et de l'île de Madère). Depuis les plaines équatoriales jusqu'à 1000 toises de hauteur (où règne encore la température moyenne de Naples), nous avons recueilli 265 Composées. Ce dernier résultat donne le rapport des Composées, dans les régions de l'Amérique équinoxiale au-dessous de 1000 toises, de $\frac{1}{9}$ à $\frac{1}{10}$. Ce résultat est très-remarquable, puisqu'il prouve qu'entre les tropiques, dans la région très-basse et très-chaude du nouveau continent il y a moins de Composées, dans les régions subalpines et tempérées plus de Composées, que sous les mêmes conditions dans l'ancien monde. M. Brown trouve pour le Rio-Congo et Sierra-Léone, $\frac{1}{23}$; pour l'Inde et la Nouvelle-Hollande, $\frac{1}{16}$ (*Congo*, p. 26; *Nov. gen.*, t. IV, p. 239). Quant à la zone tempérée, les Composées font en Amérique $\frac{1}{6}$ (c'est peut-être aussi dans l'Amérique équinoxiale le rapport des Composées des très-hautes montagnes à toute la masse des Phanérogames alpins); au cap de Bonne-Espérance, $\frac{1}{5}$; en France, $\frac{1}{7}$ (proprement $\frac{2}{15}$); en Allemagne, $\frac{1}{8}$. Sous la zône glaciale les Composées sont, en Laponie, $\frac{1}{13}$; au Kamtschatka, $\frac{1}{13}$. (Hornemann, p. 18; *Berl. Jahrb.*, *B. I*, p. 29.)

Légumineuses. = *Trop.* Amérique, $\frac{1}{12}$; Inde, $\frac{1}{9}$; Nouvelle-Hollande, $\frac{1}{9}$; Afrique occidentale, $\frac{1}{8}$ (*Congo*, p. 10). — *Temp.* France, $\frac{1}{16}$; Allemagne, $\frac{1}{20}$; Amérique boréale, $\frac{1}{19}$; Sibérie, $\frac{1}{14}$ (*Berl. Jahrb.*, *B. I*, p. 22). — *Glac.*, $\frac{1}{35}$.

Labiées. = *Trop.*, $\frac{1}{40}$. — *Temp.* Amérique boréale, $\frac{1}{40}$; Allemagne, $1/26$; France, $1/24$. — *Glac.*, $1/70$. La rareté des Labiées et des Crucifères dans la zone tempérée du nouveau continent est un phénomène très-remarquable.

Malvacées. = *Trop.* Amérique, $1/47$; Inde et Afrique occidentale, $1/34$ (*Congo*, p. 9); dans la seule côte de Guinée, $1/20$ (Hornemann, p. 20). — *Temp.*, $1/200$. — *Glac.*, o.

Crucifères. = Presque point sous les tropiques, en faisant abstraction des montagnes au-dessus de 1200 à 1700 toises (*Nov. gen.*, p. 16). France, $1/19$; Allemagne, $1/18$; Amérique boréale, $1/62$.

Rubiacées. Sans diviser la famille en plusieurs sections, on trouve pour les tropiques, en Amérique $1/29$, dans l'Afrique occidentale $1/14$; pour la zone tempérée, en Allemagne $1/70$, en France $1/73$; pour la zone glaciale, en Laponie $1/80$. M. Brown sépare la grande famille des Rubiacées en deux groupes qui offrent des rapports climatériques très-distincts. Le groupe des *Stellatæ* sans stipules interposées appartient principalement à la zone tempérée : il manque presque entre les tropiques, excepté sur le sommet des montagnes. Le groupe des Rubiacées à feuilles opposées et à stipules appartient très-particulièrement à la région équinoxiale. M. Kunth a divisé la grande famille des Rubiacées en huit groupes, dont un seul, celui des Cofféacées, renferme dans nos herbiers un tiers de toutes les Rubiacées de l'Amérique équinoxiale. (*Nov. gen.*, t. III, p. 341.)

Euphorbiacées. = *Trop.* Amérique, $1/35$; Inde et Nouvelle-Hollande, $1/30$; Afrique occidentale, $1/28$ (*Congo*, p. 25). — *Temp.* France, $1/70$; Allemagne, $1/100$. — *Glac.*, Laponie $1/500$.

Éricinées et Rosages. = *Trop.* Amérique, $1/130$. — *Temp.* France, $1/125$; Allemagne, $1/90$; Amérique boréale, $1/36$. — *Glac.* Laponie, $1/25$.

Amentacées. = *Trop.* Amérique, $1/800$. — *Temp.* France, $1/50$; Allemagne, $1/40$; Amérique boréale, $1/25$. — *Glac.* Laponie, $1/20$.

Ombellifères. = Presque point sous les tropiques au-dessous de 1200 toises; mais, en comptant dans l'Amérique équinoxiale les plaines et les hautes montagnes, $1/100$: sous la zone tempérée beaucoup plus dans l'ancien que dans le nouveau continent. France, $1/34$; Amérique boréale, $1/57$; Laponie, $1/60$.

En comparant les deux mondes, on trouve en général

dans le nouveau, sous la zone équatoriale, moins de Cypéracées et de Rubiacées, et plus de Composées; sous la zone tempérée, moins de Labiées et de Crucifères, et plus de Composées, d'Éricinées et d'Amentacées, que dans les zones correspondantes de l'ancien monde. Les familles qui augmentent de l'équateur vers le pôle (selon la *méthode des fractions*), sont les Glumacées, les Ericinées et les Amentacées; les familles qui diminuent du pôle vers l'équateur, sont les Légumineuses, les Rubiacées, les Euphorbiacées et les Malvacées; les familles qui semblent atteindre le *maximum* sous la zone tempérée, sont les Composées, les Labiées, les Ombellifères et les Crucifères.

J'ai réuni les résultats principaux de ce travail dans un seul tableau; mais j'engage les physiciens à recourir aux éclaircissemens sur les diverses familles, chaque fois que les nombres partiels leur paroissent douteux. Les *quotiens* des tropiques sont modifiés de telle manière qu'ils ont rapport aux régions dont la température moyenne est de 28° à 20° (de 0 à 750 toises de hauteur). Les quotiens de la zone tempérée sont adaptés à la partie centrale de cette zone, entre 13° et 10° de température moyenne. Dans la zone glaciale la température moyenne est de 0° à 1°. A ce tableau des quotiens ou des fractions, qui indique les rapports de chaque famille à la masse totale des phanérogames, on pourroit ajouter un tableau dans lequel seroient comparés entre eux les nombres absolus des espèces. Nous en donnerons ici un fragment qui n'embrasse que les zones tempérées et glaciales.

	France.	Amérique boréale.	Laponie.
Glumacées	460	365	124
Composées	490	454	38
Légumineuses	230	148	14
Crucifères	190	46	22
Ombellifères	170	50	9
Caryophyllées	165	40	29
Labiées	149	78	7
Rhinanthées	147	79	17
Amentacées	69	113	23

Ces nombres absolus sont tirés des ouvrages de MM. De Candolle, Pursh et Wahlenberg. La masse des plantes décrites en France est à celle de l'Amérique boréale dans le rapport de $1\frac{1}{3}$: 1; à celle de Laponie, dans le rapport de 7 : 1.

GROUPES FONDÉS SUR L'ANALOGIE DES FORMES.	RAPPORTS A TOUTE LA MASSE DES PHANÉROGAMES. Zone équatoriale; lat. 0° — 10°	Zone tempérée; lat. 45° — 52°	Zone glaciale; lat. 67° — 70°	SIGNES indiquant la direction de l'accroissement.
Agames (Fougères, Lichens, Mousses, Champignons.)	Plaines .... 1/15 Montagnes 1/5	1/2	1/1	↗
Fougères seules	Pays peu montueux .... 1/20 Pays très-montueux .. 1/3 à 1/8	1/70	1/25	← →
Monocotylédonées	Ancien continent .... 1/5 Nouveau continent 1/6	1/4	1/3	↗
Glumacées (Joncacées, Cypéracées, Graminées).	1/11	1/8	1/4	↗
Joncacées seules	1/400	1/90	1/25	↗
Cypéracées seules	Ancien continent .... 1/22 Nouveau continent 1/50	1/20	1/9	↗
Graminées seules	1/14	1/12	1/10	↗
Composées	Ancien continent .... 1/18 Nouveau continent 1/12	Ancien continent .... 1/8 Nouveau continent 1/6	1/13	→ ←
Légumineuses	1/10	1/18	1/35	↙
Rubiacées	Ancien continent .... 1/14 Nouveau continent 1/25	1/60	1/80	↙
Euphorbiacées	1/32	1/80	1/500	↙
Labiées	1/40	Amérique .... 1/40 Europe 1/25	1/70	→ ←
Malvacées	1/35	1/200	0	↙
Éricinées et Rosages	1/130	Europe .... 1/100 Amérique 1/36	1/25	↗
Amentacées	1/800	Europe .... 1/45 Amérique 1/25	1/20	↗
Ombellifères	1/500	1/40	1/60	→ ←
Crucifères	1/800	Europe .... 1/18 Amérique 1/60	1/24	→ ←

Explication des signes: ↗ le dénominateur de la fraction diminue de l'équateur vers le pôle nord; ↙ le dénominateur diminue vers l'équateur; → ← le dénominateur diminue du pôle nord et de l'équateur vers la zone tempérée; ← → le dénominateur diminue vers l'équateur et vers le pôle nord.

GEOGRAPHY CONSIDERED IN RELATION TO THE DISTRIBUTION OF PLANTS

and

ENGLAND, BOTANY

[William Jackson Hooker]

BOOK III.

GENERAL PRINCIPLES OF GEOGRAPHY UNDER ITS RELATION TO ORGANISED AND LIVING BEINGS.

1031. *In considering the extensive range of subjects* which this book embraces, we have arranged them as they successively rise above the scale of inanimate nature.

1032. The *first* chapter treats of geography, in its relation to botany, or to the distribution of plants over the surface of the globe.

1033. The *second* chapter considers it in its relation to zoology, or the distribution of animals, including man viewed simply as to his physical condition.

1034. The *third* chapter views geography in reference to human society, to man in his political, moral, and social condition.

Chap. I.

GEOGRAPHY CONSIDERED IN RELATION TO THE DISTRIBUTION OF PLANTS.

1035. *In proportion as our knowledge increases relative to any of the sciences,* we find a more intimate relation and connection between them. Formerly geography was only studied as it regarded the surface of the earth itself, its figure, the constitution of the several regions and countries, their boundaries, &c.; and botany has had too many votaries who devoted their attention almost exclusively to determining the generic and specific names of plants, neglecting the more beautiful and philosophical parts of the science. Of late years, indeed, our systems of geography have, in some instances, contained a meagre catalogue of the vegetable productions of the different regions, but nothing that could give the least information with respect to the laws of their general distribution: and now that some of the most able naturalists and philosophers of our day have, by their labours, thrown new light upon this interesting subject, we should feel that our work would ill merit the character which we hope it may obtain with the public, were we to omit a notice of it. At the same time, the limits of our publication will permit us to give only a sketch of what indeed must be considered as still in its infancy; and those who have most devoted their attention to botanical geography will most readily join with Mirbel in declaring that "we are even yet far from having arrived at that period when it will be possible to write a good history of this subject. What we do know of climates and of vegetation, is little, in comparison with what we have yet to learn; and hence it would be rash in us to form an estimate of what we do not know by what we are already acquainted with. The surest way is to confine ourselves to collecting and arranging facts, leaving, to those who may follow us, the charge of discovering and developing the theory."

1036. *To exhibit the present state of botanical science, we shall endeavour to put together the more interesting facts,* collected principally from the writings of our most authentic travellers and naturalists; and, devoting this memoir to vegetable geography in its more enlarged and general sense, shall afterwards, in the different countries, under the head of botany, point out some of the most striking and important productions of their respective regions. As the nature of the present work does not permit us to enter minutely into the subject in all its bearings, we shall give a popular view of it, as little encumbered as possible with technical terms.

1037. *That certain vegetables are confined to certain districts or limits,* depending in a great measure, but by no means altogether, upon soil and climate, must be familiar to the most careless enquirer into the works of nature. In regard to climate, the two extremes are represented by the country within the tropics, and that which approaches the poles. In the one, nature exhibits herself in her most lovely and her most magnificent and exuberant form, and the earth is covered with vegetables which indicate a never-ending summer; whilst in the others a brief summer, a few days of freedom from frost and snow, call into existence a thinly scattered vegetation of small and stunted flowering plants, which scarcely rise above the mosses and lichens that surround them; and the intermediate zones will be found to be occupied by other races, gradually, however, increasing in difference as they approach to one or other of these extremities. The same gradation exists, we know, upon a lofty mountain situated within the tropics. At its base may be seen those plants which are peculiar to the tropics; and the beauty, the grandeur and perpetual verdure will gradually diminish in the ascent, until a soil and climate be found on the higher summits similar in respect to climate and productions to those in the vicinity of the poles.

1038. *In regard to climate and vegetable productions,* our globe has been aptly compared, in its two hemispheres, to two immense mountains, placed base to base, the circumference of which at the foot is constituted by the equator, and the two poles represent the summits, crowned with perpetual glaciers.

1039. *That almost every country possesses a vegetation peculiar to itself,* is also well known; and this is particularly the case with countries whose natural boundaries are formed by mountains, seas, or deserts, even in the same or different degrees of latitude. Europe exhibits a widely different class of plants from that part of North America which lies immediately opposite to it. The botany of Southern Africa has little or no resemblance to that of the same parallels in South America, or to that of New Holland. Nay, in our own country of Britain, we have some plants that are confined to the eastern and some to the western side of the kingdom. In Scotland, the *Tutsane* and the *Isle of Man Cabbage* are never found but on the western side of the country, and the same is the case with the *pale Butterwort* (Pinguicula Lusitanica), both in England and Scotland. Nature has constituted the barrier, for by art they may be cultivated as well on one as the other side of the island.

1040. *Botanical geography is constituted* by considering plants in relation to their *habitation,* region, or the country in which they grow, and in regard to their locality or particular *station,* and forming a collection of facts, deduced from these circumstances, from which general laws may be derived: nor is this a science destitute of advantages; such, we mean, as are immediately manifest; for there are few, in the present age, who will be disposed to deny that the study of the works of nature, like every thing that can exalt and refine the mind, is highly deserving of our attention. Vegetable geography is intimately connected with horticulture. Our gardens will be better stocked with vegetables and fruits, our forests with trees, our fields with corn, and our pastures with grasses, in proportion to our knowledge of the relation of plants with the exterior elements. Nay, Schouw has justly observed, that a good chart of the distribution of the vegetable forms over any given country will afford a far more correct idea of the productive strength of that country than many statistical tables. The systematic botanist may thence derive benefit; for by it he will be better able to determine whether certain kinds of plants are species or varieties; he will consider that a different local situation produces different effects upon them; that those growing in wet places are less hairy or downy than those growing in dry; that at great elevations plants are more dwarf in their stature, with fewer leaves, but with larger and more brilliant flowers than those found at lesser heights. The station, too, of certain plants, or groupes of plants, frequently leads to a discovery of characters diverse from other individuals of other countries with which they had been associated. Thus the *Canadian Strawberry* and the *Canadian chickweed Wintergreen* (Trientalis), though long confounded with our *Strawberry* and with our *Trientalis,* are found to be quite distinct. The regions, too, and the limits of those regions, of very important medicinal drugs, are determined by vegetable geography.

Sect. I. *Progress of Botanical Geography.*

1041. *This branch of science had been, however, for a long time, wholly neglected.* Linnæus, indeed, with whom originated so many improvements in botany, besides what related to systematic arrangement, was the first writer who gave stations for plants, as he called them, or rather habitations, or frequently both combined, and this plan has been followed by every succeeding systematic botanist. Yet although these stations or habitations are frequently consulted in the geographical arrangement of plants, they are too vague and uncertain to be generally depended upon; and they must be employed with caution. We are not, however, to forget, with regard to Linnæus, that he has, in some of his writings, entered far more fully into the subject than any of his predecessors. In his beautiful memoir entitled *Stationes Plantarum*, he divides all, primarily, into *Aquatic, Alpine*, those growing in *shady places* (*Umbrosæ*), those growing in *open ground* (*Campestres*), *Mountain-plants*, and *Parasitical;* and these again are subdivided into *fresh-water, maritime, marsh-plants, heath-plants*, &c.; and in his oration entitled "*De telluris habitabilis incremento*," he has published that ingenious if not correct theory of the origin and distribution of all plants as well as animals; namely, that, at the beginning, the whole earth was covered with sea, except one island, large enough to contain all the animals and vegetables; that this island was situated within the tropics, and was crowned with a very high mountain, which, in consequence of its varied soil and climate, might contain one or more of every vegetable and animal. He then proceeds to show how, by means of the amazing fertility of plants from seed, they might, at an after period, be scattered all over the globe in the soil and situation best adapted to them; and, lastly, in his *Flora Lapponica*, he pays particular attention to the differences of plants as influenced by the variety of stations, and especially by their altitude above the level of the sea. Tournefort, indeed, is said previously to have found upon Mount Ararat, at its base, the plants of Armenia; a little way up, those of Italy; higher again, those which grow about Paris; afterwards, the Swedish plants; and lastly, on the top, the Lapland alpine ones. De Saussure, who so assiduously studied vegetable physiology, was particularly attentive, on that account, to the elevation at which plants grow above the level of the sea; and appears to have been the first to ascertain *that* elevation barometrically. St. Pierre did not leave this subject unnoticed in his *Studies of Nature;* but he was so led away by his genius, that he believed there was not a square league upon the surface of the globe which did not include within its area some one vegetable, at least, which does not exist any where else. Mr. Young, the celebrated agriculturist, in his *Travels upon the Continent*, determined with considerable accuracy the northern boundaries of several of the most important cultivated plants, the *Olive*, the *Vine*, and the *Maize;* whilst Girand Soulavie, in the south of France, has characterised the limits of them, and of the *Orange* and *Chestnut;* and these two men have, no doubt, excited a spirit of enquiry into the geographical distribution of other vegetable productions. These, and several other authors of less note, prepared the way, during the last century, for the more important labours of the present, when the study has begun to rank as a science. Stromeyer, in 1800, described, to a certain extent, the boundaries of the vegetable kingdom, in a work entitled "*A Specimen of the History of Vegetable Geography*," which appeared at Göttingen. The work of Kielmann, edited at Tubingen, 1804, entitled "*A Dissertation concerning Vegetation in the Alpine Regions*," was followed by that of Treviranus, named "*Biologie*," which seems to be the first wherein attention was paid to the distribution of plants according to their natural families; the latter author dividing the globe into regions or distinct Floras; and De Candolle, about the same time, partitioned France into regions in the same way, and wrote on the influence of height upon vegetation. To the celebrated M. de Humboldt, however, we are indebted for the most valuable writings on vegetable geography, and those which have first given it the true character of a science. His "*Essai sur la Géographie des Plantes*, in 1807, and his beautiful "*Tableau de la Nature*," contained his first ideas on the subject; while his celebrated "*Prolegomena de distributione geographica Plantarum secundum cœli temperiem et altitudinem montium*," forming the introductory chapter to the botanical part of his travels; his invaluable memoir on isothermal lines and the distribution of heat over the globe, published in the Mémoires d'Arcueil, and translated into Brewster's and Jameson's Philosophical Journal, vol. iii. p. 1. &c.; together with his latest work on the subject, "*New Enquiries into the Laws which are observed in the Distribution of Vegetable Forms*," likewise inserted in the Edin. Phil. Journal, vol. vi. p. 273: these may justly be considered as the most important dissertations on a comprehensive scale that have yet appeared. In the mean time, other eminent naturalists, by their well-directed labours, contributed materially to extend the science: Wahlenberg, for example, in his admirable *Flora Lapponica*, and in that of a portion of Switzerland, and of the Carpathian Alps; whilst Von Buch, in his travels in Norway, detailed many curious facts respecting the distribution of vegetables in that climate, and also in his interesting *Voyage to the Canaries*, made in company with the lamented Professor Christian Smith. Mr. R. Brown, who treats with the most masterly hand whatever is connected with the study of botany, has much increased his already eminent fame by those memoirs in which this subject is entered upon; and his profound knowledge of the natural orders, his possession of the richest herbarium and the finest library in the world, both of which he has turned to the best possible account, together with the advantages arising from his extensive travels; these have enabled him to publish memoirs which will be studied with infinite advantage, and which will rank among the most valuable that have ever appeared. We particularly allude to his "*Remarks, Geographical and Systematic, on the Botany of Terra Australis*, 1814," and "*Observations on the Herbarium collected by Professor Christ. Smith, in the Vicinity of the Congo*, 1818." Mr. Boué, a foreigner, but for a long time a resident in Scotland, published a pamphlet at Edinburgh, entitled "*De Methodo Florum cujusdam regionis conducendi,*" and Mr. Winch of Newcastle, an "*Essay on the Geographical Distribution of Plants through the County of Northumberland.*" Taking advantage of what others had done, Dr. Schouw compiled, in 1824, an admirable history of the science, of which some portions have been translated into Brewster's and Jameson's Journals; and we regret that, to our knowledge at least, this valuable work exists, in an entire state, in no other language but the Danish and the German. It has the merit of being accompanied by an Atlas of several maps of the world; each exhibiting the geographical extent of certain tribes or families of vegetables, indicated by different colours. This plan of the maps has been successfully followed by Dr. Von Martius, in his History of the natural order *Amaranthaceæ,* in the thirteenth volume of the Nova Acta Acad. Cæsar.-Leopold.; so that we see, at one view, upon a plan of the world, the countries in which these plants are found, their boundaries, and the comparative abundance, indicated by the greater or less depth of colour employed. De Candolle, again, in the "*Nouveau Dictionnaire des Sciences Naturelles,*" has given an admirable *résumé* of all these writers, and has added much important original information. A somewhat similar plan is adopted by M. Brongniart in the "*Dictionnaire Classique d'Histoire Naturelle,*" vol. vii. p. 275. Mr. Allan Cunningham, both in Mr. Barron Field's "*Memoirs of New South Wales,*" and in the second volume of "*Captain King's Survey of the Intertropical Coasts of Australia,*" has furnished some further and very excellent remarks upon the distribution of vegetables, especially of the less frequented parts of New Holland. The "*Mémoires du Muséum d'Histoire Naturelle*" contain some important papers on this subject, particularly that of Mirbel, "*Sur la Géographie des Conifères,*" a tribe of plants valuable for its economical uses; and his "*Recherches sur la Distribution Géographique des Végétaux phanérogames dans l'ancien Monde, depuis l'Equateur jusqu'au Pôle Arctique:* and, lastly, we shall only name a useful little manual, entitled a "*Lecture on the Geography of Plants,*" by Mr. J. Barton.

1042. We might yet have swelled this list to a considerable extent; but enough has been said to show that the subject has merited and received the attention of some of the ablest naturalists; and that it is a science which gives an increased degree of interest both to geography and botany.

Sect. II. *On the Influence of the Elements on Plants.*

1043. In regarding the limits to which certain plants are circumscribed upon the surface of the globe, we shall see that it is with them as with the mighty ocean; they are equally subject to that fiat of the Almighty, "Thus far shalt thou go, and no farther." The Palms, the Tree-Ferns, the parasitical Orchideæ, are ever confined to the tropics; the Cruciferous and Umbelliferous plants almost exclusively to the temperate regions; while the Coniferous plants, and many of the Amentaceous tribes, flourish in those of the north: and since these are all affected by physical agents, we must consider, before proceeding any farther, the influences which the *elements* or *exterior agents* exercise upon plants. These M. de Candolle considers to be *Heat, Light, Moisture, Soil, Atmosphere.*

Subsect. 1. *On the Influence of Heat.*

1044. *Heat* is the most obvious and powerful agent in affecting the existence and growth of plants: and of this we have continual experience before our eyes. In winter all vegetation is at a stand, and we can only cultivate those plants which are in a continued state of vegetation, by artificial heat. Plants are nourished either by water alone, or by substances dissolved or suspended in the water. Hence vegetation is arrested when the temperature is below the freezing point; for the water, becoming solid, cannot enter the vegetable tissue. Again, as in the great deserts of many countries, the heat may be so great that the earth is dried up, and cannot part with its nutritive properties. These effects, however, it is but reasonable to suppose, are more remarkable upon the surface of the earth than at a considerable depth: hence it happens that trees which have long tap-roots resist both the extremes of temperature better than those whose roots are nearer to the surface; their fibres penetrate into a soil whose temperature is greater in winter than that of the outer air, so that the fluids imbibe and keep the interior of large trees, as has been ascertained by experiment, at a degree of heat pretty nearly the same as that indicated by a thermometer placed at the roots of such trees. Hence, the greater the thickness of the stem or branch, and the greater the number of layers interposed between the pith (the softest part being the moistest and the most susceptible of cold) and the exterior air, the better are they able to resist the severity of the cold. It is a well known fact that a shrub or tree as it grows older becomes more hardened against frost. De Candolle relates, that at Montpellier the *Pride of India* (Melia Azedarach) when young is destroyed by a moderate degree of cold; but that when it attains a more advanced age it will endure, in the garden at Geneva, an intensity of atmosphere four times as severe as that which killed the young plant in the south of France.

1045. Again, *in proportion as the exterior layers are deprived of sap or watery fluid, and for-*

tified by a deposit of carbon and resinous matter, the more powerfully they withstand the cold. Every gardener and cultivator is acquainted with the fact, that in cold and wet summers, when the sun and heat have been insufficient to produce good bark upon the new shoots of the fruit-trees, they are liable to be affected by a very moderate frost in the ensuing winter. *Succulent plants* and *Monocotyledonous plants*, in general, which have no distinct bark, are highly susceptible of cold; whilst the *Birch*, which is fenced around with numerous layers of old and dry bark, and the *Fir*, whose bark abounds with resin, endure an intense degree of it without injury. At Fort Enterprise, in North America, lat. 64° 30″, Dr. Richardson has ascertained that the *Banksian Pine* (Pinus Banksiana), the *white*, the *red*, and *black Spruce*, the *small-fruited Larch*, and other Amentaceous trees, bear a degree of cold equal to 44° below zero of Fahrenheit; and in Siberia, lat. 65° 28″, the *common Larch*, the *Siberian Stone Pine*, the *Alder*, *Birch*, and *Juniper*, &c. attain their greatest size, and are not affected by the extremest cold of that severe climate.

1046. *Powerful summer heats* are capable of causing trees and shrubs to endure the most trying effects of cold in the ensuing winter, as we find in innumerable instances; and *vice versâ*. Hence, in Britain, so many vegetables, fruit-trees in particular, for want of a sufficiently powerful sun in summer, are affected by our comparatively moderate frosts in winter; whilst upon continents in the same degrees of latitude the same trees arrive at the highest degree of perfection. Even in the climate of Paris the *Pistacia tree* and the *Oleander* will not bear the winter. Yet the winters there are mild in comparison with those which prevail in the environs of Pekin, where the Oleander was found by Lord Macartney to remain abroad the whole year; and at Casbin in Persia, where Chardin assures us that the Pistacia Nuts, produced in the open air, are larger than those of Syria. On the other hand, the heat of these two countries in summer is infinitely greater than that at Paris; the summer temperature of Pekin especially nearly equals that of Cairo, and surpasses that of Algiers. For the same reason, too, the *Weeping Willow* becomes a large tree in England; while in Scotland, where the winters are at least as mild, but where the summer affords much less warmth, this beautiful tree can only be cultivated in highly favoured situations, and even there its vegetation is exceedingly languid: its young shoots, not ripened by the summer sun, are destroyed even by a slight frost.

1047. Hence *the influence of temperature* upon the geography of plants is ably pointed out by M. de Candolle under *three* points of view: — 1. The mean temperature of the year. 2. The extreme of temperature, whether in regard to cold or heat. 3. The distribution of temperature in the different months of the year.

1048. *The mean temperature*, that point which it has for a long time been the great object of the physician to ascertain, is in reality what is of the least importance in regard to the geography of plants. In a general view, it may be useful to take it into consideration; but the mean temperature is often determined by circumstances so widely different, that the consequences and the analogies to be deduced from it relative to vegetables would be very erroneous.

1049. *By attending to the extreme points of temperature*, results more limited, but far more exact, are to be obtained. Thus, every locality which, though at only short intervals, affords a degree of cold or heat of certain intensity, cannot but produce plants which are capable of supporting those extreme degrees. When, however, these widely different temperatures recur at very long intervals, man may cultivate in such a country a vegetable which cannot exist in a wild state; either because, when destroyed by the rigour of the season, he restores it by seeds or by plants derived from a more temperate country; or because he shelters it from the inclemency of the air; or, as is too often the case in our climate, because he is satisfied with the product of the plant, although it should not bring its seeds to perfection. And thus it is that, in the south of Europe, the *Vine*, and *Olive*, and *Orange* trees often vegetate exceedingly well for all the purposes for which they are required, though, if left to themselves, they could not propagate themselves, nor sustain the winter. Thus we see a wide difference in the geography of plants; between those in a state of nature, and those individuals whose growth is artificially encouraged by man.

1050. This, indeed, is a subject closely connected with the *acclimatation of plants*, or the power which man is supposed to exert over them in inuring them by degrees to a climate not originally natural to them. This power is, however, denied by very able vegetable physiologists. Mirbel, in particular, declares that he has known many species indeed whose wants have been, to a certain degree, artificially supplied; but not one whose constitution has been changed. "If," he says, "from time to time, exotics mingle themselves with our indigenous tribes, propagate as they do, and even dispute the very possession of the soil with the native inhabitants; this, assuredly, is not the work of man, but it is the *climate* which dispenses this faculty of naturalisation." Cultivators, however, maintain that seedlings from *Myrtles*, which had ripened their fruit in Devonshire in the open air, are better able to endure the cold of our climate than those seeds perfected by artificial heat, or that have come from the warmer parts of Europe. It is true, the power of so acclimating itself already exists in the vegetable; but it is man that calls it into action, for naturally the myrtle would never extend itself to our latitudes. Nay, something of the same kind M. Mirbel himself allows, where he says, "When we consider that the *Vine* is cultivated in the plains of Hindostan and Arabia, between the 13th and 15th parallels; that it is cultivated on the banks of the Rhine and Maine, in lat. 51°; in Thibet, at an elevation above the level of the sea of from 9,000 to nearly 11,000 feet, under the 32d degree of latitude; what astonishes and interests us the most is, not that the vine inhabits countries so remote from one another, or that it grows at so great an elevation above the sea, but that it possesses in so eminent a degree the *property of accommodating itself to different climates;* a property, indeed, much more restricted in a great number of vegetables, which extend from the equator to the tropics on both sides, without ever crossing them; for notwithstanding the greater distance between the 23d degree of southern parallel and the 23d degree of northern parallel, the climatic differences are much less from one tropic to the other than from the plains of Hindostan to the banks of the Maine."

1051. The *distribution of heat at different months of the year* is what we shall find to be of the most importance in regard to vegetable geography. Some climates are eminently uniform; a certain mean temperature is produced by a mild winter and a moderate degree of warmth in summer. This is frequently the case on the sea-coasts, because the extremes of heat are continually modified by the sea; that vast reservoir of nearly equal temperature, which therefore imparts heat in winter and cold in summer, and enables even tropical plants to subsist in some situations of the temperate zone. Such are the western shores of Europe and America, and a great portion of the southern hemisphere. A similar mean temperature may indeed be produced by a combination of very severe winters and very hot summers, as in the great continents compared with islands; or the shores of those continents, or the eastern side of continents, as compared with the western; or the northern with the southern hemisphere; but these two climates, as may be expected, will produce a very different vegetation.

1052. *Annual plants,* which require heat during the summer to ripen their seeds, and which pass the winter, so to say, in torpidity, in the state of grain, indifferent to the intensity of cold*, abound most in those regions where the extremes are the greatest; whilst the *perennial plants,* which can better dispense with the maturing of their seeds, and which are injured by the severities of winter, affect the temperate climates. Of these, again, those kinds which have deciduous leaves accommodate themselves best to unequal temperatures; whilst the individuals on which the foliage remains, or *evergreens,* give the preference to districts where the temperature is more constantly equal.

1053. *Mirbel reckons* that there are about 150 or 160 natural groups or families of plants in the Old World, types of all which exist in the tropical parts of it. Beyond these limits, a great number become gradually extinct. In the 48th degree of latitude, scarcely one half of that number appear; in the 65th, not 40; and but 17 in the vicinity of the polar regions. He further estimates, that within the tropics the proportion of woody species, trees and shrubs, equals, if it does not exceed, that of herbaceous, annual, biennial, and perennial plants. The relative number of the woody species with the herbaceous, annual, biennial, and perennial, decreases from the equator to the poles; but, as an equivalent, the proportion of perennial to annual or biennial plants goes on increasing. Near the extreme limits of vegetation these are, at least, as twenty to one.

1054. *We must, however, by no means conclude* that the same elevation in corresponding degrees of latitude is necessarily suited to the vegetation of the same plants. A number of circumstances may exist to modify the degree of heat at the same elevation. In Switzerland, for example, an investigation of the temperature that prevails in districts situated at only a short distance from each other among the mountains will produce very dissimilar results; and on this subject we find some interesting remarks, extracted from Kasthofer, "*Voyage dans les petits Cantons et dans les Alpes Rhétiennes,*" in the first number of the *Foreign Review and Continental Miscellany.* The elevation of the valley of Untersee, we are there told, is the same as that of Gestein; yet the thermometer, in 1822–3, fell only to 8° below zero in the former spot; whereas at Gestein it fell to $10\frac{1}{2}$°, and at Berne to 16°. The depth of the valleys influences vegetation; the deeper they are, the more intense is the cold on the summits of the surrounding mountains. Thus, the pine does not thrive on the Bragel, at a height of 5100 feet; whereas it succeeds perfectly, at the same elevation, on the Rhetian Alps, the valleys of the Linth, the Muotta, and Kloen being deeper than those of the latter districts. In like manner, in the valley of the Davos, agricultural produce is certain in places much more elevated than the Bernese valleys, because the latter are deeper. The warm winds from Italy have a perceptible power over the vegetation of the contiguous parts of Switzerland; but the degrees of that influence depend upon circumstances. In the valley of the Inn, barley and flax are cultivated with success at an elevation of more than 5400 feet; whereas at Laret, in the valley of Davos, though the height is only 4900 feet,

* Seeds being, in general, furnished with few organs which abound in moisture, are in a degree insensible to the extremes of heat and cold; whence it arises that, in conveying them from one country to another, they pass through a variety of climate uninjured.

no grain will thrive. Yet these valleys are alike in most respects, and are surrounded by mountains of similar altitudes; they are both sheltered from the north-east wind; their soil is of the same nature; but in the valley of the Inn, the warm winds from Italy are intercepted only by a single chain of mountains, whereas two chains lie between Italy and the valley of Davos: and, besides, the latter being of smaller extent than the former, it admits of the reception of less solar heat. In the Oberland of Berne, an increase in height of 2000 feet diminishes the crop one third.*

Subsect. 2. *On the Influence of Light.*

1055. *The influence of the solar light* upon vegetation De Candolle considers to be as important as that of temperature; and although it acts less powerfully upon the geographical distribution of plants, it nevertheless merits a particular notice.

1056. *Light is that agent* which operates in producing the greatest number of phenomena in vegetable life. It determines, in a great measure, the absorption; for plants imbibe less humidity during the night and in darkness. It completely influences the watery exhalations of the green parts of plants; for these parts do not exhale during the night or in obscurity, whilst these exhalations are very considerable during the day, and especially under the direct influence of the rays of the sun. The light affects, in most cases, the decomposition of the carbonic acid; and consequently the deposition of carbon in vegetables, their substance and their growth, the intensity of their sensible properties, and the direction of many organs. It is the principal, and perhaps the only, cause of those singular movements known by the name of the *sleep of plants;* and, lastly, during the absence of light the green parts absorb a certain quantity of oxygen gas. Although these different causes affect all vegetables, yet they are not affected in the same degree.

1057. *Light is more equally distributed than heat* upon the surface of the globe; but its mode of diffusion induces some very important consequences. In the countries situated under the equator, an intense light, since it acts more perpendicularly, influences vegetables nearly equally, during twelve hours each day, throughout the whole year. In proportion as we recede from the equator and approach the poles, the intensity of the more oblique rays gradually diminishes; but in regard to the distribution of these rays, the light is completely wanting during the winter, when the absence of vegetation indeed renders it nearly useless to plants; and it is continued during almost the whole period of vegetation, in such a manner that its lengthened influence compensates wholly or in part for its want of intensity. Thus we see that, independently of what concerns the temperature, plants which lose their leaves can better exist in northern countries, and that those whose vegetation is continued have need of the southern regions. And another beautiful and just remark is made by De Candolle, in reference to the distribution of light; namely, that those plants whose foliage and flowers maintain habitually and constantly the same position, can live in northern climates, where the light is almost continued in summer; whilst it is in the regions of the south that we find, as might naturally be expected, those species which are remarkable for the alternate closing and expanding, or sleeping and waking, of their flowers, a motion which has an intimate connexion with the alternation of days and nights. Thus we see why it is found so difficult in our country to cultivate many of the tropical vegetables, or, at any rate, to bring them to perfection. M. de Humboldt has proved that it is less owing to the absence of heat than to the want of sufficient solar light that the *Vine* does not ripen its fruit beneath the foggy skies of Normandy; and M. Mirbel has satisfied himself that the uninterrupted action of the sun's rays, during a great number of days, is the cause of the astonishingly rapid developement of alpine plants in high northern regions.† Dr. Richardson, too, states that the sugar-boilers in the Canadian forests observe that the flow of sap in the *Sugar Maple* (Negundo fraxinifolium) is not so immediately influenced by a high mean temperature as by the power of the direct rays of the sun. The greatest quantity of sap is collected when a smart frost during night is succeeded by a warm sunshiny day.

* The same author, too, mentions a curious fact of vegetation resting upon a basis of ice. The glacier of Roccosecco, which forms one of the branches of the Berneria, has on its summit a nearly horizontal valley filled with ice; and on this the avalanches have brought down masses of earth. This earth produces a number of alpine plants, that afford abundant and nourishing food to the flocks of the inhabitants of Samaden. There are documents which prove that this singular pasture has been used ever since the year 1556.

† "Vegetables," says M. Mirbel, in his *Elémens de Physiologie Végétale*, "when secluded from the light, send out long, thin, and whitish shoots; their substance becomes lax, and without firmness; in fact, they are bleached. The operation of the luminous beams on these organised bodies consists chiefly in separating the constituent parts of water and carbonic acid which they contain, and in disengaging the oxygen of the latter. The carbonic acid, with the hydrogen and oxygen of the water, produce those gums, resins, and oils, which flow in the vessels and which fill the cells. These juices nourish the membranes, and bring them into the ligneous state; a result which becomes more marked as the light is strongest and its action most protracted. Darkness and light produce, therefore, diametrically opposite effects on vegetation. Darkness, by keeping up the softness of the vegetable parts, favours their increase in length; light, by ministering to their nourishment, consolidates them, and arrests their growth. Hence it follows that a fine state of vegetation, such as unites in just proportions size and strength, must depend, in a measure, on the nicely balanced alternation of day and night. Now, the hyperborean plants spring up at a period when the sun is constantly above the horizon, and the light which incessantly acts upon them confirms and perfects them before they have time to attain a considerable degree of length. Their vegetation is active, but soon over; they are robust, but small."

Again, Humboldt assures us, that in all places where the mean temperature is below 62° 6′, the revival of nature takes place in spring in that month whose mean temperature reaches 42° 8′, or 46° 4′. At Cumberland House, Dr. Richardson found vernation to begin in May, when the mean temperature was only 49°, nearly 3° below that which Baron Humboldt considered necessary for the evolution of deciduous leaves; but he adds, "*the influence of the direct rays of the sun was at this time very great*, and the high temperature of the last decade of the month compensated for the first." We can imitate the native climes of many of the delicate exotics, as far as regards temperature; and in summer, when our days are long, we see them flourish almost as if they were in their natural situations; but in winter they languish, and often die, especially the more tender species, such as the *Hedysarum gyrans*, and the *humble plant* (Mimosa pudica). It is evident that they want that distribution of light which is most congenial to them.

1058. Plants, then, are arranged in their different localities, according to the certain quantity of light which they may require. All those with very watery leaves, which evaporate much, which are of a succulent nature, which, having few pores or organs of evaporation, need a stimulus to determine their action, all which have a tissue abounding in carbon, or which contain very resinous or oily juices, or which offer a great extent of green surface, require much light, and are generally found in exposed places; the rest, according as they are more or less distinguished by these properties, exist either under the slight shadow of bushes, or beneath the more powerful shelter of hedges and walls, or of forests; or, as is the case with many *Fungi*, in caves and darkness. These last are, indeed, destitute of any green colour; but *Mosses, Ferns*, and even some evergreens, such as the *Ivy*, flourish best beneath the shade of dense forests, if the trees of those forests have deciduous leaves; and in situations where plants that vegetate only during the summer could scarcely live.

1059. The subject, however, of the *action of light upon vegetation, has not yet received the attention* which it deserves. Many more observations and experiments are required before we can employ it with certainty in connection with botanical geography.

Subsect. 3. *On the Influence of Moisture.*

1060. *Water being the vehicle by means of which nourishment is conveyed into the plant*, and, indeed, itself yielding a large proportion or even the whole of the nutriment of many vegetables, it follows that this element is not only of the highest importance in vegetable economy, but one of the causes which affects most powerfully the geographical distribution of plants upon the surface of the globe.

1061. *Those vegetables, in particular, necessarily absorb a great quantity of water*, which have a large and spongy cellular tissue; those which possess broadly expanded soft leaves, furnished with a great number of cortical pores; those having few or no hairs on their surface; those whose growth is very rapid, which deposit but little oily or resinous matter; those of which the texture is not subject to be changed or corrupted by humidity; those, in fine, whose roots are very numerous, generally need to absorb much moisture, and cannot live but in places where they find naturally a large proportion of it. On the other hand, those plants which are of a firm and compact cellular tissue, which have small or rigid leaves, furnished with very few pores, which are abundantly clothed with hairs, of which the growth is slow, and which deposit, during the progress of their vegetation, much oily or resinous matter; those whose cellular tissue is liable to be changed and decayed by too much moisture, and of which the roots are not numerous, require little water, and prefer, for their natural situation, dry places. Great differences, however, are produced, according to the nature of the water that is absorbed; the less it is charged with the nutritive principle, the more necessary is it that the vegetable shall absorb, in a given time, enough to suffice for its support. Again, the more the water abounds with substances which alter its fluidity or transparency, and which, inasmuch as they are solid particles, tend to obstruct the orifices of the pores, or to impede absorption by their viscosity, the less do such vegetables imbibe in a given time.

1062. *The very nature even of those substances dissolved or suspended in the water* has a great influence upon the topographical distribution or the locality of plants. The matters so dissolved are, 1. Carbonic acid. 2. Atmospheric air. 3. Animal and vegetable substances. 4. Alkaline principles or earths. Those plants whose cellular tissue is found to contain much carbon, such as trees producing hard wood, avoid, more than others, the vicinity of waters which are extremely pure, and which contain but little carbonic acid gas. Plants which exhibit much azote in their chemical composition, such as the *Cruciferous Plants* and the *Fungi*, seek those spots where there is much animal matter in solution. Those, again, which present, when chemically analysed, a considerable quantity of certain earthy substances, such as silica* in the *Monocotyledonous Plants*, gypsum in the *Leguminosæ*, &c., will require it in a

* This silica, we know, abounds in the grasses, as well as in other monocotyledonous plants; and M. de Candolle observes, that it is in consequence of its existence in the grasses, &c., and of the comparative indissolubility which is the result, that it is preferred by almost all nations of the world for a covering to their houses. The people of the North thus employ straw for that purpose on the same principle that those of the Tropics use the leaves of the palms.

greater or less proportion in the soil where they grow; and if it does not exist there naturally, the agriculturist must supply it artificially; and those species which yield, when burned, a more abundant portion of alkaline substances than usual, can only flourish or even live where these matters abound. The species which have need of carbonate of soda will only grow successfully near the sea or saline lakes or springs. Thus the different property of the substances dissolved in the water is evidently one of the many causes which determine the stations of the vegetable species.

Subsect. 4. *On the Influence of the Soil.*

1063. The influence of soil M. de Candolle considers as perhaps more complicated than that of the preceding agents. He reduces it to three principal heads: —

1064. (1.) *The soil serves as a means of support to vegetables,* and consequently its *consistence* or tenacity ought to possess, in this point of view, a peculiar fitness for sustaining, in a greater or less degree, plants exhibiting very various forms. Thus, soils composed of blowing sand can only serve as a support to vegetables which are of very humble stature and prostrate growth, so that the winds may not overturn them; or to trees, furnished with very deep and branching roots, which may attach them into this moveable matrix. The contrary holds good in regard to very compact soils. Small-rooted plants may thus be firmly enough fixed, and they may subsist; but the very large roots are incapable of penetrating into soils that are very tenacious. The two extremes of these soils present an equally sterile vegetation. Sands which are not sufficiently stationary (as those very remarkable ones on the northern shores of the Moray Firth), water which is subject to very rapid currents, clay of an extremely compact nature, or rocks of great hardness, are equally unfriendly to the growth of plants.

1065. (2.) *The chemical nature of the earths or stones* of which the soil is composed, affects the choice of vegetables, as regards their flourishing in such situations. But this subject, simple as it appears at first sight, is in reality very complicated. For the different earths act upon vegetation by physical circumstances; as, for example, according as they absorb the surrounding water with more or less facility, retain it with more or less force, or part with it more or less easily. Now, the celebrated Kirwan ascertained by a comparative analysis of earths which were reckoned excellent for the growth of wheat in various countries, that they contain more silica if the climate is more subject to rain, more alumine if the contrary be the case; in short, that the soil, to be good for any given vegetable, ought to have the power of absorbing more moisture in a dry climate, less in an humid atmosphere: whence it is plain that in different localities the same species of vegetable may be found in different soils.

1066. (3.) *Every kind of rock has a certain degree of tenacity,* and a certain disposition to decompose or become pulverised: whence results the greater or less facility of particular soils to be formed either of sand or gravel, and to be composed of fragments of a nearly determined form and size. Certain vegetables, from causes which we shall presently indicate, will prefer such or such of this sand or gravel; but the peculiar nature of the soil does not act here immediately; thus, when we find calcareous rocks which decompose like argillaceous schist, the same species of vegetation is observed. These two considerations are particularly applicable to lichens.

1067. (4.) *Rocks, according to their colour or their nature,* are more susceptible of being heated by the direct rays of the sun; and consequently they may, in some degree, modify the temperature of a given place; and influence also, though slightly, the choice of plants capable of succeeding upon them.

1068. But, independently of all these physical causes, it may be asked, whether the *chemical nature of rocks* has any effect upon vegetables? It is generally considered to be so; but it must be allowed that this action has been frequently very much exaggerated. Bory de St. Vincent, indeed, has assured us that calamine, or native carbonate of zinc, in the vicinity of Aix-la-Chapelle, is always indicated, to a certainty, by particular plants; and the fact is confirmed by a little work, since published, called A Flora of the Environs of Spa. The *yellow heartsease,* a small variety of the *common Eyebright* (Euphrasia officinalis), the *white Campion* (Silene inflata), a *Sandwort* (Arenaria), a shrubby *Lichen,* a species of *Bromus* (Brome-grass), constitute this poor but constant vegetation. These, however, no doubt, grow in greater abundance and perfection in other soils: the wonder is that they do not altogether perish here; for even the gallinaceous birds, which eat gravel to triturate their food, die from swallowing fragments of calamine. It must be remarked, in reality, that plants do not often live upon pure rock, but among the decomposed matter of that rock; that the rocks, even though very circumscribed, often present very different natures: that vegetable mould is not only formed by the rocks which immediately surround it, but also by the admixture of earthy substances carried by the waters, transported by the winds, or by the remains of animals and vegetables which have before existed there. Hence it will be understood how the vegetable earths differ much less in themselves, than the rocks which produce them or serve to support them; and that the greater number of plants yield, in most situations, the alimentary earths which are necessary for them. Indeed, after various botanical journeys made through France, M. de Candolle has found nearly the same plants vegetating spontaneously in almost all the different rocky substances. It has been said that the *Box* (Buxus sempervirens) grows only in calcareous soils, and it certainly prefers them; but it is found abundantly in the argillaceous calcareous schistose rocks of the Pyrenees; and it is even seen among the granite of Brittany and upon the volcanic parts of Auvergne. The *Chestnut* has been said to avoid a calcareous country; but there are beautiful chestnuts on both sides of the Lake of Geneva, at the foot of the calcareous mountains of Jura and Chablais.

1069. *Pure magnesia*, M. Carradori has found, by chemical experiment, acts as a poison on most plants; yet M. Dunal, in visiting a portion of the environs of Lunel, where the soil presents a great quantity of almost pure magnesia, found there the same plants as in the surrounding calcareous soil, and the roots flourishing in the clefts of this magnesian rock. Thus we must be careful not to attach too much importance to the nature of the earth, which is frequently acted upon by causes purely physical.

Subsect. 5. *Atmospheric Influence.*

1070. *The atmosphere*, taken in its pure state, we know to be composed, at all times, of the same proportions of *azote* and *oxygen;* and in such cases we may suppose its action to be similar upon all vegetables. But the atmosphere also is of different degrees of transparency or density; it holds in solution other matters or substances, which mix with it in certain places, and render it more or less suitable to certain species of plants. In mines, for instance, the quantity of carbonic acid gas, or of hydrogen, may be so great as to preclude vegetation altogether; or to allow only of the growth of such individuals as are very strong and vigorous, or particularly absorbent of these substances. Then, too, the air charged with saline emanations from the sea injures some plants, and on the other hand encourages the developement of such as require carbonate of soda; as may be seen in the valleys of the south of Europe, where maritime plants affording soda may be cultivated at a considerable distance from the ocean, provided that they lie open towards the sea, and are exposed to the winds that blow from it.

1071. *We cultivate in our inland gardens*, languidly and but for a year or two, many of the *maritime plants*, such as the *Lithospermum maritimum*. The *Nitraria Schoberi* is improved by employing salt where it is grown. Many of the *Statices* may be, however, easily cultivated, and one of them, the *common Thrift* (S. Armeria) even succeeds in crowded towns, whence its English name; yet its native country is either on the shores of the sea or in salt marshes, or upon the summits of the highest mountains.

1072. *The most general influence*, however, exercised by the atmosphere, is its power of containing and parting with moisture, or its hygroscopic action. The atmosphere is habitually charged with moisture; sometimes in such a manner as to be invisible, and then only ascertainable by the hygrometer; at other times visible in a state of vapour or dew; and we find that vegetables in general succeed better in a climate where, at a given degree of temperature, the air is moderately moist, than in another where it is either too much saturated with moisture or too dry. This is a circumstance which cannot well be imitated in the cultivation of plants in the open air: but in our stoves, and especially by the aid of steam, the various degrees of humidity necessary to a vigorous vegetation may be produced to the greatest nicety. Thus, in the magnificent stoves of the Messrs. Loddiges, at Hackney, near London, by an ingenious but simple contrivance, a gently falling dew is imitated; and with what success will best be seen by an inspection of the healthy and beautiful state of those plants which delight in a warm and moist atmosphere.

1073. *The agitation or movement* of the air by winds and other causes exercises some power over vegetation; but we are too little acquainted with this subject to be able to deduce any particular theory from it.

1074. *Of all the atmospheric influences*, the most difficult to reduce to its proper value is that of *density;* or, what is the same thing, the influence of height or elevation above the level of the sea. This M. de Candolle has made the subject of a memoir in the volume of the Society of Arcueil, and we shall here give his general ideas upon it.

1075. *In proportion as we are elevated in the air*, the temperature as well as the moisture continue to diminish; a circumstance which appears to depend upon this, that the rare air has more capacity for heat than dense air. The facts that go to prove that the diminution of the temperature upon high mountains is one of the causes which most affect the distribution of vegetables, are the following:—

(1.) The natural situation of each plant at a determined elevation above the level of the sea is so much the greater in proportion as the country is nearer the equator, and less in more temperate regions; that is to say, the farther we recede from the equator, the greater influence has the exposure upon the temperature.

(2.) In temperate climates, as France, for instance, those plants which are but little affected by temperature, and which grow in all its latitudes, are found also at all those elevations where the earth is not covered by eternal snows; from the level of the sea to the summits of the mountains. M. de Candolle has detected about 700 examples of this law; the *common Heath*, the *Juniper*, the *Birch*, &c. grow indifferently at the level of the sea, and at a height of 10,000 feet.

(3.) If plants which, according to their nature, avoid either too high or too low a degree of temperature, yet grow at different latitudes, we may observe that it is at heights where the effect of elevation may compensate that of latitude: thus the native plants of the northern plains will be seen to grow upon the mountains of the south.

(4.) Plants which are cultivated upon a large scale are guided by laws which entirely correspond with the preceding: those which are cultivated in various latitudes will grow indifferently at various heights; those which are only found at certain latitudes will extend no farther than to proportional elevations. The *potatoe*, which succeeds so well in our plains, is cultivated in Peru at an elevation of 10,000 feet above the level of the sea: the *olive*, which nowhere passes 44° north latitude, will not grow at a height exceeding 1250 feet.

(5.) The elevation above the level of the sea, when we compare the temperature of the seasons, establishes effects very analogous to those which result from the distance from the equator; so that there is the more analogy between the results on vegetation in the two cases. *In proportion as we rise in a direct line*, it follows, from the lessened density of the air, that the intenseness of the solar light continues to increase; this effect is represented in the line of distances from the equator, because the perpetuity of light during the continuance of vegetation is so much the greater in proportion as the latitude is more elevated.

(6.) *In proportion to the greater height upon the mountains*, so will the hygrometer be seen to indicate a less degree of humidity; the same general effect takes place as we recede from the equator towards the poles.

1076. *On mountains*, covered with perpetual snow, where the plants are constantly moistened with water in a freezing state, those species, to which a warm temperature is unfriendly, will live at inferior heights to those which they brave in the same latitude, when they are not watered from those cold sources.

1077. *It would appear*, therefore, from all these considerations, that the situation or fixed locality of plants at certain heights depends mainly on the fall of the temperature attributable to that elevation. Now, the only purely theoretical point of view, says M. de Candolle, according to which we can comprehend how the rarefaction of the air bears in itself a direct influence upon vegetation, is this; that plants require to absorb a greater or less degree of oxygen gas in their green or their coloured parts. It cannot be doubted that there is a certain point of elevation where the atmosphere becomes too much rarefied to supply the wants of plants; but where this is the case the mountains are always clothed with snow. M. de Humboldt, too, inclines to think that the pressure of the air may act in encouraging and increasing the quantity of evaporation. But we must say that direct experiment is still wanting to confirm these opinions (and this is perhaps unattainable in the present state of science), in order that we may form a conclusive judgment on their value.

Sect. III. *Station and Habitation of Plants.*

1078. *The station and habitation of plants* must next engage a portion of our attention. They are both important: the former implies their situation as regarding local circumstances, and the action of physical causes upon vegetables; the latter implies the geographical position. When we say, that such a plant is found in marshes, on the sea-shore, in woods, or upon mountains, in England, in France, in North America; by the marshes, shore, woods, or mountains, we mean what we here term the *station;* and by England, France, or North America, the *habitation:* such is the sense, at least, in which we shall here use the terms; for in systematic botanical writings the meaning is by no means always thus restricted.

1079. *The seeds of plants*, by varied and beautiful means, *are widely dispersed* by the liberal hand of nature: whilst some, however, fall upon barren ground, or a soil unfit for the nature of that particular vegetable, others take root in situations, both with regard to the earth and surrounding medium, which are in harmony with their growth, and produce, "some thirty, some sixty, and some an hundred-fold." There are, again, tribes which, under these circumstances, increase so prodigiously that they destroy vegetables of a less vigorous growth, and, to the exclusion of others, appropriate to themselves a great extent of the surface of the earth. Such are termed by M. de Humboldt "plantes sociales," plants living in society. In this way, and notwithstanding the extreme poverty of the soil, the *Seaside Sedge* (Carex arenaria), the *upright Sea Lymegrass* (Elymus arenarius), and the *Sea-reed* or *Marram** (Arundo arenaria), occupy a prodigious surface of our sandy shores, almost to

* The Celtic name of this plant is *Maraim*. A village upon the sea-coast of Norfolk is named Marham, from the circumstance of the great abundance in which the *Arundo arenaria* grows in its vicinity.

the exclusion of other vegetation; their long, creeping, and entangled roots serving to bind the sands together, and thus forming a barrier to the encroachments of the sea. Thus it is with the heaths in our country, where our sterile moors are purple with the blossoms of the heath.

1080. The flowers of the *Gentians* cover, as with a carpet of the most brilliant ultramarine blue, the sides of the alpine hills in Switzerland and the south of Europe. Our fields are often, too often, red with *Poppies*, and the marshes are whitened with the "snowy beard" of the *Cottongrass*, and our pastures with the blossoms of the *Cardamine pratensis*, so that they appear at a distance as if covered with linen laid out for bleaching, whence arises the vulgar English name* of the latter plant. Some of these plants thus living in society are continually striving with their neighbours, till the strongest obtain the victory. Many low perennial and herbaceous vegetables are overpowered by a colony of taller shrubs; such as the *Whin* or *Furze* and the *Broom*: and these in their turns must occasionally give place to trees and shrubs of a larger and stronger growth. Mr. Brown has, however, noticed a curious fact in regard to the *Field Eryngo* (Eryngium campestre), and the *Star-thistle* (Centaurea Calcitrapa), which cover much cultivated ground upon the continent, that these two engrossers are never mixed together indiscriminately, but that each forms groups of partial masses, placed at certain distances from their rivals.

1081. *On the other hand*, there are plants, which, from the circumstance of their not increasing much by root, or bearing few seeds, or such seeds as from their light and volatile nature are much dispersed, and which are not particular in their choice of soil, do not form groupes, but lie scattered (*Plantes éparses, egrenées*, or *rares*, of the French).

1082. *The former kind*, or "*social plants*," are those which it will be most important for us to consider in relation to Botanical Geography.

1083. *The stations of plants* being thus, as we have already mentioned, liable to the influence of physical agents, it becomes necessary to define them by terms which are calculated at once to point out the places and the circumstances in which they grow. This, however, is a task of no small difficulty; for, without swelling the list to an immeasurable length, it will be impossible to define the various local situations of plants. There are many situations which produce only one or two kinds: for example, the snow, in the highest arctic regions to which our enterprising travellers have attained, has been found to nourish and to bring to the greatest perfection that highly curious vegetable, the *Red Snow* (Protococcus nivalis).† The *truffle* (Tuber cibarium) is found entirely hid beneath the surface of the earth. Some *fungi* are detected upon the dead horns and hoofs of animals (no plant exists upon living bodies ‡), and upon dead chrysalides; and both *fungi* and *mosses* grow on the dung of animals. Paper nourishes the minute *Conferva dendroidea*: the glass of windows, and the glass table of the microscope, if laid by in a moist state for a certain length of time, produce the *Conferva fenestralis*. Wine casks in damp cellars give birth to the *Racodium cellare*: and Dutrochet has detected living vegetables in Madeira wine and in Goulard water, (a solution of Saturn). These, however, and many others that might be noticed, may be numbered among the extraordinary stations, and they principally affect cryptogamic vegetables. In a popular view of the subject, though we cannot altogether omit the notice of such minute yet curious vegetable productions, we shall mainly direct our attention to the more conspicuous plants; and they may be thus divided. 1. *Maritime or saline plants.* These are terrestrial, but growing upon the borders of the ocean or near salt lakes; as the *Saltworts* (Salsolæ) and *Glassworts* (Salicorniæ), &c. Hence these plants abound in the interior of Africa and the Russian dominions, where there are saltpans, as well as on the shores. 2. *Marine plants.* This tribe is indeed mostly cryptogamic, and comprises the *Algæ, Fuci, Ulvæ*, &c. The phænogamous, or perfect marine plants, are the *Sea-wracks* (Ruppia and Zostera), and a few others allied to them. 3. *Aquatic plants.* Growing in fresh water. Both stagnant pools and running streams in various situations, abound in plants. Some are entirely submerged, but in this case, with the rare exception of the little *Awlwort* (Subularia aquatica), the flowers rise to the surface of the water for the purpose of fructification.§ 4. *Marsh or swamp plants.* 5. *Meadow and pasture plants.* 6. *Field plants.*

* *Lady's Smock.* Not Ladies' Smock. Such plants were in olden time dedicated to Our Lady the Virgin Mary.

† More will be said of this plant when we come to treat of the plants of the high northern regions.

‡ Schouw, indeed, has a tribe of plants which he calls "*Plantæ Epizoæ*," *attached to living animals.* Thus, he says, *Fuci* and other *Algæ* are attached to whales, mussels, and barnacles. But in this case the plants manifestly adhere to a dead portion of the animal; like those vegetables which exist upon the outer and dead part of the bark of trees. I have seen a bee which was brought from Sicily. It had been captured alive, with a substance attached to its head which had all the appearance of a fungus, and from its shape, to that particular genus of Fungi called *Clavaria.* Many persons were deceived into a belief that this vegetable substance was growing on the living insect; but an eminent naturalist, well acquainted with the structure of the parts of fructification in the Orchideous plants, immediately detected that the supposed fungus was nothing more than the pollen-mass, which had attached itself to the nose of the bee by means of its glutinous gland, while the busy insect was thrusting its head into the flower to gather its honeyed sweets.

§ Raymond certainly observed, in the Pyrenees, a species of *Crowfoot*, the *Water Crowfoot* (Ranunculus aquatilis), producing its flower and fruit wholly under water; but, upon a closer investigation of the phenomenon, he found that in these cases the calyx enclosed a globule of air, with which this important function of fertilisation was performed. The curious aquatic, *Vallisneria spiralis*, has a still more wonderful contrivance

This tribe often includes such as, introduced with the grain sown in those districts, are equally placed there by the hand of man. 7. *Rock plants*, which may include the natives of very stony spots, and such as grow upon walls. Walls, although artificial structures, are known to produce many plants in greater perfection than natural rock; yet we must not suppose that any vegetable is exclusively confined to this habitat. The *Holosteum umbellatum* and *Draba muralis* may be cited as examples of this tribe in England; and amongst mosses, the *Grimmia pulvinata, Tortula muralis*, &c. 8. *Sand plants*. 9. *Plants of dry moors*, where our *heaths* (Ericæ) abound. These seem to be included among the *plantes des lieux stériles* of De Candolle; a very heterogeneous groupe it must be confessed, and by no means easy to characterise. 10. *Plants which attach themselves to the vicinity of places inhabited by man*. Such are the *Dock, Nettle*, &c.; these species follow every where the human footsteps, even to the huts and cabins of the highest mountains; encouraged, perhaps, by the presence of animal substances, and the azote which in such substances is known to abound. 11. *Forest plants*, consisting of such trees as live in society. 12. *Plants of the hedges*, as are many of our climbing plants, the *Honeysuckle*, the *Traveller's joy*, the *Bryony*, &c. 13. *Subterranean plants*. Those that live in mines and caves, and which, though tolerably numerous and important, are yet mostly cryptogamous. One species, a fungus, yields a pale phosphoric light of considerable intensity. 14. *Alpine or mountain plants*, for it is very difficult to draw the limit, and indeed they will depend much upon latitude. A plant which grows upon a hill of inconsiderable elevation in Norway, Lapland, and Iceland, will of course inhabit the loftiest Alps of the south of Europe. Again, upon mountains that have no perpetual snow lying on them, alpine plants will be found much higher than on such as have continued streams of cold snow-water descending, which affects the state of the atmosphere at much lower regions. 15. *Parasitic plants*, such as the *Missletoe*, the various species of *Loranthus*, &c., and that most wonderful of all vegetable productions, the *Rafflesia Arnoldii*: these, as their name implies, derive nourishment from a living portion of the vegetable to which they attach themselves. This is the case, too, with many Fungi which subsist upon the living foliage of plants; some exclusively on the upper, others as invariably on the lower side of these leaves; and, lastly, the name of 16. *Pseudo-parasites* has been given to a very extensive tribe, which subsist upon the decayed portions of the trunk or branches of the trees to which they are attached, as many of the *Lichens, Mosses*, &c.; or which are simply attached by the surface of their roots to tropical trees, obtaining no nourishment from them, but from the surrounding element. Among this number may be reckoned that numerous and singular family of the *Orchideæ*, called, from their nature and property, "*air plants*." Greatly as this list might be swelled, we shall find that even here there is a gradation and an approximation of one tribe to another; but these are amply sufficient for our purpose.

1084. *We have been able to account in some measure for the stations of plants*, affected as these are by local circumstances; but the study of the succeeding part, which refers to their *habitations*, considered in their most extensive scale, for instance, as belonging to certain regions or countries, we shall find to be much more difficult; and we must frequently be content to study and to admire the amazing variety of vegetable forms which the beneficent hand of nature has scattered over the different parts of our world, without being able to account for these important phenomena. In New Holland we find, almost exclusively, all the species of *Banksia, Goodenia*, and *Epacris*, and the curious *Acaciæ* without leaves, but with petioles so much enlarged as to amusse the shape and perform the functions of leaves. At the Cape of Good Hope, the *Fig Marigolds* (Mesembryanthema), the *Stapeliæ*, the numerous kinds of *Ixia, Gladiolus, Pelargonium*, and *Protea* abound. The *Aurantiaceæ*, the family of plants to which the *Orange* and *Lemon* belong, are of Asiatic origin; as the *Camellia* and *Thea* are of Chinese. Those curious plants, the *Mutisiæ*, the various species of *Fuchsia*, the *Cinchonæ* or *medicinal barks*, the *Cacti*, are all peculiar to South America. If a few of them are found in other countries, such circumstances are of very rare occurrence, and do not overturn the general laws for the exclusive existence of many plants in certain countries. We have in the temperate parts of Europe one species of *Ixia*, one of *Gladiolus*, and in the north of Africa and south of Europe a few kinds of *Fig Marigold*. Within the tropics the genera of plants, throughout Asia, Africa, and America, are similar, but rarely are the species the same. This rule nearly holds good on the opposite continents in temperate climates. We have the *Oriental Plane* (Platanus orientalis) in the old world, and the Americans have the *Occidental Plane* (P. occidentalis). Even in the two hemispheres, in similar parallels of latitude, the genera of plants have a great affinity: the southern extremity of the great continent of

---

for bringing the male and female flowers in contact. The plant is diœcious. The female flower is attached to the parent plant by means of a very long stalk, spirally twisted like a corkscrew, so that when it is in perfection, it rises to the surface by the untwisting of the stalk. The male flowers, upon a separate plant, are almost sessile, borne on a very short straight stem, which never could reach the surface without detaching themselves from the plant. This they do at the proper season; they float upon the top of the water along with the female flowers, scatter their pollen, and die. The female blossoms, on the contrary, by the spiral twisting of their stalks, retire, and ripen their seeds under water.

America has many in common with the north of Europe; and the plants of our regions, transported thither, succeed extremely well.

1085. *To what extent plants migrate,* unaided by man, it is not easy to say; but that such migration is going on, by various means and causes, cannot be questioned. Islands which lie near to continents, and which evidently appear at one period to have been joined with them, as England for example, although they may contain a vegetation similar to that of the neighbouring continental shores, have always a smaller number of species; and this can only be accounted for by the interruption which straits or seas occasion to the progress of the seeds.

1086. *The Field Eryngo* (Eryngium campestre), to which we have already alluded, the *Venus's looking-glass* (Campanula Speculum), and many other plants of France and Germany, seem to stop at the line formed by the sea in approaching our shores; yet these, and many other vegetables of France, reach a limit upon the same continent more northern than any part of England.

1087. *The transmigration of plants* may be reckoned to be facilitated by the following causes. 1. *The sea and its currents,* but to a very limited extent; for if the seed be of such a nature that the water penetrates its integuments and reaches the embryo, life is destroyed. Yet to such a distance are they carried by this medium, that upon the coasts of Britain, of Iceland, and Norway, the seeds of the West Indies are frequently cast, and it is said sometimes even in a fit state for vegetation. 2. *Rivers,* by the continual movement of their waters, convey many plants to a considerable distance from their original place of growth; and the banks of streams are generally adorned with a vegetation of a more varied kind than the districts remote from them. Thus, too, the different species of *Saxifrage* and other alpine plants are, in mountainous regions, brought down from the higher situations, and flourish in the valleys. 3. *Winds,* which waft the light, winged, and pappose seeds to immense distances, and by means of which they are widely dispersed. 4. *Animals,* which, in wandering from place to place, often carry on their coats those seeds which have hooked bristles, &c. 5. *Birds,* which, swallowing berries and other fruits, pass the seeds in a perfect state, and, it is even said, sometimes better fitted for germination than before. In this manner the seeds are often deposited in the places necessary for their growth, and to which they could not otherwise have reached; of which a familiar instance is found in the *Misseltoe.*

1088. *Man is however the most active agent in the dispersion of plants,* and we must not overlook the important consequences. Sometimes, indeed, the causes are accidental, but more frequently intentional. The shipwreck of a vessel on the island of Guernsey, having some bulbs on board from the Cape of Good Hope, caused a plant to propagate in the sands upon the shores of that mild climate, to which has been since given the name of *Amaryllis Sarniensis* or *Guernsey Lily,* and a branch of trade of some importance is carried on in the sale of this very root. At Buenos Ayres, a species of *Artichoke* (Cynara Carduncus) has increased so much by seeds imported from Europe, that Mr. Head, in his amusing "Sketches of a Journey across the Pampas," &c. tells us that "there are three regions of vegetation between Buenos Ayres and the base of the Cordilleras; a space of 900 miles: the first of which is covered, for 180 miles, with clover and *thistles.* This region," the author continues, "varies with the seasons of the year in a most extraordinary manner. In winter, the leaves of the thistles * are large and luxuriant, and the whole surface of the country has the rough appearance of a turnep field. The clover in this season is extremely rich and strong; and the sight of the wild cattle grazing in full liberty on such pasture is very beautiful. In spring the clover has vanished, the leaves of the thistles have extended along the ground, and the country still looks like a rough crop of turneps. In less than a month the change is most extraordinary; the whole region becomes a luxuriant wood of enormous thistles, which have suddenly shot up to a height of ten or eleven feet, and are all in full bloom. The road or path is hemmed in on both sides; the view is completely obstructed; not an animal is to be seen; and the stems of the thistles are so close to each other, and so strong, that, independent of the prickles with which they are armed, they form an impenetrable barrier. The sudden growth of these plants is quite astonishing; and though it would be an unusual misfortune in military history, yet it is really possible that an invading army, unacquainted with this country, might be imprisoned by these thistles before it had time to escape from them. The summer is not over before the scene undergoes another rapid change: the thistles suddenly lose their sap and verdure, their heads droop, the leaves shrink and fade, the stems become black and dead; and they remain, rattling with the breeze, one against another, until the violence of the pampero or hurricane levels them with the ground, when they rapidly decompose and disappear, the clover rushes up, and the scene is again verdant."

1089. *The strong-scented Everlasting* (Elichrysum fœtidum, vid. Bot. Mag. t. 1987), a

* From specimens in our Herbarium, we have ascertained that this *thistle* is the *Cardoon* (Cynara Carduncus), introduced no doubt from Europe as an article of food, but now growing wild, useless, and pernicious.

native of the Cape of Good Hope, has found a soil and climate equally suited to its growth on the shores of Brest, where it covers a great portion of the sands, to the exclusion of the aboriginal natives of the soil. *Wheat* is supposed to be indigenous to Barbary. The *potatoe*, first found in South America, is now cultivated all over the world. *Rice*, from Asia, is grown to an immense extent in America, &c.; these, and many other plants similarly circumstanced, which we could mention, together with those that adorn our gardens, often owe their wide diffusion to having escaped into uncultivated places, and become to a certain degree naturalised there.

1090. But *there are limits to migration*, for some of which we can account, and for others we cannot. Even many of our garden plants, which, escaping by accident, or designedly placed in uncultivated spots so as to appear wild, have only for a time maintained a languid existence, and then have disappeared altogether. Thus we know that the beautiful *Gentianella* (Gentiana acaulis) cannot have a title to a place in the British Flora, nor can some others, which are mere outcasts from gardens. Some plants are wholly confined to particular spots, and can be found nowhere else. The *Tree-Pink* (Dianthus arboreus) grows still on the single rock in the island of Crete, where Prosper Alpinus first detected it; and the *Double Cocoa-nut* of the isle Praslin, one of the little groupe of islands called the Seychelles, notwithstanding the annual migration of its nuts for many thousands of miles, has never established itself in any other place. Nature has planted the *common Thrift* (Statice Armeria), the *Scurvy Grasses* (Cochlearia anglica and danica), and the *Rose-root* (Rhodiola rosea), in rocky and stony places, upon our shores and on the tops of the highest mountains; yet these plants are never found in any intermediate places.

1091. *The visible obstacles to the migration of plants* are —

(1.) *The sea*, which, though we have introduced as a means of extending the habitations of plants, is yet a far greater impediment, by the injury it does to the seeds, and the difficulty of their being conveyed to distant countries in a sufficiently short time to prevent the natural death of the seed. It must be observed, too, that the greater number of seeds have a specific gravity heavier than that of water when in a living state. The *Double Cocoa-nut*, when found floating, has always lost its vegetative property. The living nut is immensely heavy, and would inevitably sink.

(2.) *Dry and burning deserts*. These, in spite of their oases, which have been happily assimilated to the isles of the ocean, prove a powerful obstacle to the transport of seeds. Thus, those districts of Africa which are separated from one another by the scorching sands of Sahara exhibit a great dissimilarity in their vegetation. The plants of Morocco and the northern parts of Africa have little resemblance to the indigenous growth of Senegal; whilst the affinity of the vegetables brought by Caillaud from Upper Egypt to those collected by Palisot de Beauvois in Oware and Benin would in themselves lead to the conclusion that no very great and continued deserts intervene between these far distant countries.

1092. *Mountain ranges*. The barriers which these present would almost be insurmountable, were it not for the defiles which here and there occur, forming passages for men and animals, as well as for plants. Thus, the plants on the Italian side of the Alps are quite different from those on the Switzerland side; those of the Spanish Pyrenees from those of the French Pyrenees; and it was a subject of peculiar regret to the enterprising Drummond, when he reached the summits of the Rocky Mountains in North America, that his commission did not allow him to penetrate farther into the western side of that great continent, where he found, every step he took, a vegetation very different from what had been presented to him by the eastern side.

1093. *A knowledge of the Natural Orders of plants* is in no department of botany so important as in treating of their geographical distribution. The system of Linnæus, or the Artificial Arrangement, does not, as we know, regard the habits and affinities of vegetables, but simply and beautifully points out to us, by certain characters, the means of arriving at the knowledge of any given species. The natural method, which owes so much to the labours of Jussieu, Decandolle and Brown, has a higher object in view, that of grouping plants together according to their natural affinities; and by such an arrangement we are often led to other and very important results. The primary divisions of the Natural Method are, first, ACOTYLEDONES, or plants which have no cotyledons to the seed: these are synonymous to the Cryptogamia, and include the *Mosses, Lichens, Sea-weeds, Fungi, Ferns*, &c.; secondly, MONOCOTYLEDONES; those whose seeds have one cotyledon, such as the *Grasses, Liliaceous Plants*, the *Rushes, Sedges*, the *Palms*, &c.; and, thirdly, DICOTYLEDONES, or the plants which have two or rarely more cotyledons to the seed, such as our *Shrubs* and *Trees*, and very many *Herbaceous Plants*. Each of these possesses external characters which, though not very easily defined in words, yet cannot fail to strike the observer who devotes his attention, even for a little while, to the subject; and we find that, in a great proportion of instances, they have not only a peculiar station, but that their geographical distribution is different.

1094. *The* ACOTYLEDONOUS *plants increase in number in proportion to the other great classes*, as we recede from the equator to the poles; with the exception, however, of the *Ferns*. The latter abound more within the tropics than any where else: not, however, so much in open plains as in the sheltered, moist, and hilly countries; so that their maximum is in the mountainous part of the tropics. The island of Martinique afforded to the Abbé Plumier a rich and abundant harvest of ferns; and some isles of small extent are said to have one third of their vegetation composed of this kind of plants.

1095. *Among the* MONOCOTYLEDONOUS *Plants,* the *Palms* are exclusively confined to the tropics: the *Liliaceous plants* abound there and in the warm zones; the three families of *Grasses, Sedges* (Cyperaceæ), and *Rushes* (Junci), present some important differences in regard to a comparison with the phænogamous or flowering plants. The disparity between these latter and the *grasses* is not great in each of the zones; whilst the two other families, the *Cyperaceæ* and *Junci,* diminish near the equator and increase towards the north. Nevertheless, there are exceptions to this rule; for the *grasses* are very rare upon the coasts of Greenland. In what we have now said, we allude to the grasses, &c. in a wild state; having no reference to those regions where so many of the grass tribe, as the *Wheat, Barley, Oat, Maize, Rye, Rice,* &c., are found simply in a state of cultivation.

1096. *The* DICOTYLEDONOUS *plants are the most extensively distributed,* and we must offer some further remarks upon them. The *Compound* or *Syngenesious plants* (Compositæ), as every one knows, form a very extensive natural family. They are diffused throughout the whole earth, but they are most abundant in the temperate and tropical climates. Fewer, however, of them are found in the warm regions of equinoctial America than in the subalpine and temperate districts of the same country. At the Congo and Sierra Leone in Africa, in the East Indies and New Holland, they exist in comparatively smaller numbers than in other regions situated in similar parallels, but which afford situations more congenial to their growth. Again, in the frozen zone, in Kamtschatka and Lapland, the relative proportion of plants of this family is one half less than in the temperate climates.

1097. *The Leguminous plants* (*to which the Pea, the Bean, &c. belong,* and such as bear papilionaceous flowers,) abound most in the equinoctial regions: they diminish gradually in each hemisphere in diverging from the equator, except indeed in certain countries where particular genera, by the multiplicity of their species, give a peculiar feature to the vegetation, as in Siberia and the vast provinces of Russia, where so many *Astragali* or Bitter-vetches are found.

1098. Mr. Brown has judiciously separated the natural order of *Rubiaceæ* into two groups: those with verticillate leaves and no stipules (the *Stellatæ* of Linnæus), to which belong our *Goosegrass* (Galium), *Madder* (Rubia), &c., and which are almost peculiar to the temperate zones; and the true *Rubiaceæ,* with opposite pairs of leaves, and two opposite stipules (which are in fact abortive leaves, and thus show their affinity with the *Stellatæ*), to which belong the real *medicinal barks* (or Cinchonæ), and some other nearly related plants possessing similar virtues: these latter are almost wholly confined to the equinoctial regions.

1099. The two well known and extensive natural families, the *Umbelliferous* and *Cruciferous plants,* are very rare in the tropics, if we except the mountains. They abound in the south of Europe, and especially about the valley or basin of the Mediterranean.

Sect. IV. *View of Botanical Regions.*

1100. *To divide the globe into botanical regions or districts* will not be difficult, seeing that certain countries possess a peculiar vegetation, and that numerous impediments prevent emigration; seeing, too, that certain forms or tribes are incompatible with certain climates. M. De Candolle has constituted twenty of those regions; but although each is, to a certain degree, peculiar in its vegetable productions, it would require more space than we can devote to such a subject to characterise them. We must, therefore, content ourselves with giving a bare list. 1. *Hyperborean region.* This district includes the northern extremity of Asia, Europe, and America; and gradually merges into the following. 2. *European region;* comprising all Europe, except the part bordering upon the pole, and the southern districts approaching the Mediterranean. To the east it extends to the Altaic mountains. 3. *Siberian region,* comprehending the great plains of Siberia and Tartary. 4. *Mediterranean region;* comprising all the basin of this great inland sea; that is, Africa on this side the Sahara, and that part of Europe which is sheltered from the north by a more or less continued range of mountains. 5. *Oriental region;* thus called relatively to southern Europe, and containing the countries bordering upon the Black and Caspian Seas. 6. *India,* with its archipelago. 7. *China,* Cochinchina, and Japan. 8. *New Holland.* 9. *The Cape of Good Hope,* or southern extremity of Africa, beyond the tropics. 10. *Abyssinia,* Nubia, and the Mozambique Coast (imperfectly known). 11. *Equinoctial Africa;* viz. the neighbourhood of the Congo, the Senegal, and Niger. 12. *The Canary Isles.* 13. *The United States of North America.* 14. *The Western and Temperate Coasts of North America.* 15. *The West Indian Isles.* 16. *Mexico.* 17. *Tropical South America.* 18. *Chili.* 19. *Southern Brazil and Buenos Ayres.* 20. *The Straits of Magellan.*

1101. *Many of the productions of these regions will be considered somewhat at large in other parts of this work;* and we shall conclude our introductory sketch of Botanical Geography by a notice of Professor Schouw's *Phyto-Geographic* or General Botanical Division of the

Globe. This is illustrated by a map, which accompanies this memoir. Unlike M. De Candolle, Professor Schouw characterises the regions by the most remarkable feature of their vegetation, adopting commonly used geographical terms only where he conceives that a certain division of the earth ought to constitute a distinct region, but is not sufficiently acquainted with its productions to determine and define their forms. He makes the characteristic feature of his regions to depend on these facts: first, that at least one half of the species should be peculiar to that region; secondly, that at least a quarter of the genera should belong exclusively to it, or at least have there a decided maximum, so that their species in other districts might merely be considered as their representatives; and, thirdly, that individual families of plants be either peculiar to the region, or else have their *maxima* there; nevertheless, when this last characteristic is wanting, while the difference in genera and species is very considerable, it may yet be admitted as a region.

Professor Schouw in this manner reckons twenty-two regions: —

1102. (1.) *Region of Saxifrages and Mosses*, or the *Alpine Arctic Flora.* — This corresponds with De Candolle's first region, and comprehends all the countries within the polar circle; namely, Lapland, the north of Russia and Siberia, Kamtschatka, Canada, Labrador, Greenland, and Iceland; but Professor Schouw adds to it, with much propriety, part of the Scottish and Scandinavian mountains, as far as they fall within the alpine region, as also the mountains in the southern and central parts of Europe, inasmuch as they are related to the alpine regions. It is characterised by the abundance of *mosses* and *lichens*, the presence of the *Saxifrages, Gentians, Chickweed tribe* (Alsineæ), *Sedges*, and *Willows*; an entire absence of tropical families; a considerable decrease of the peculiar forms of the temperate zone; by the forests of *beech* or *fir*, or else the total want of trees; the scarcity of animals, and the prevalence of cæspitose plants, whose blossoms are large in proportion, and generally of a pale colour.

1103. (2.) *Region of the Umbelliferous and Cruciferous plants.* — This tribe takes in the whole of Europe, except what belongs to the preceding division, from the Pyrenees, the mountains of the south of France, of Switzerland, and the north of Greece, to the greater part of Siberia, and the country about Mount Caucasus. Schouw has characterised it by the *cruciferous* and *umbelliferous* plants, because they form a larger portion of the total number than any other kinds, and because it may thus be best separated from the vegetation of North America in the same parallel. It is not easily distinguished from the next region; but it may be said of it, that *Fungi* abound more, that the *Rosaceous* family and the *Crowfoots* (Ranunculaceæ), the *Amentaceous* and *Coniferous* tribes (Pines), form rather a large proportion; that it bears a resemblance to many of the polar forms, especially in the abundance of its *Sedges* (Cyperaceæ); that its meadows are most flourishing, and that almost all the trees are deciduous in winter. In the northern part of this region, the *Cichoraceæ* (a tribe of the Compositæ or syngenesious plants, including the *Endive, Lettuce, Dandelion*, &c.) much prevail; while in its southern division, or in northern Asia, the *Cynarocephalæ* (Artichoke and Thistle tribes), together with the *Butter-vetches* (Astragali), and *Saline plants* (Sea-worts and Glass-worts), seem to have their maximum.

1104. (3.) *Region of the Labiate flowers* and *Caryophylleæ* (to which the *Pink*, the *Catch-fly*, the *Sandworts*, &c. belong); or the *Mediterranean Flora.*—This is bounded on the north by the Pyrenees, the Alps of Switzerland and of the south of France, and the north of Greece, and thus includes the three peninsulas of southern Europe, namely, Spain, Italy, and Greece; on the west by Asia Minor and its islands; on the south it takes in Egypt and all the north of Africa as far as the deserts; and, lastly, it includes the Canary Islands, Madeira, and the Azores. It is marked especially by the two families above mentioned, which are much rarer both to the north and south of the countries just enumerated, and in the corresponding parallels in North America. The *Compositæ*, the *Stellatæ* (*Goosegrass, Madder*, &c.), and the rough-leaved plants (*Asperifoliæ*), are here in considerable numbers, as well as in the similar latitudes. A few tropical plants, or individuals allied to them, now appear; one or two *Palms*, the *Laurels*, the *Arum* tribe, the *Terebinthaceæ* (Pistacia, &c.), some tropical grasses and true *Cyperaceæ*. *Nightshades* (Solaneæ), *Leguminous* plants, the *Mallow* and *Nettle* tribes, and the *Spurges* (Euphorbiaceæ), increase; *Evergreens* are numerous; vegetation never entirely ceases, but verdant meadows are more rare. This region may be subdivided into provinces: of the *Cisti*, Spain and Portugal; of the *Sage* and *Scabious*, the south of France, Italy, and Sicily; of the shrubby *Labiatæ*, the Levant, Greece, Asia Minor, and the southern part of the Caucasian country; and of *Houseleeks* (Semperviva), the Canary Isles, probably also the Azores, Madeira, and the north-west coast of Africa. Many *Sempervivæ*, some succulent plants, *Spurges*, and *Cacaliæ*, characterise especially this province.

1105. (4.) *The Japanese region.*—The eastern temperate part of the old continent, namely, Japan, the north of China, and Chinese Tartary, probably forms a peculiar region; but we are too little acquainted with the botany of these countries to admit it with certainty, and still less are we able to define correctly the characteristics of its Flora. Of the 358 genera found in Japan, 270 occur in Europe and the north of Africa, and about the same number in North America; so that its Flora seems to occupy a middle place between those of the old and new worlds. Its vegetation, indeed, approaches more to the tropical than to the European; for we meet with the *Cycas* family, the *Scitamineæ* (to which belong the *Ginger*, *Cardamom*, &c.), the *Bananas*, the *Palms*, the *Anonæ* or *Custard-apples*, and the *Sapindaceæ*; so that there is a considerable affinity, as might be expected from its situation, to the flora of India. The families of the *Buckthorns* (Rhamni) and *Honeysuckles* are found in a relatively considerable number, and they exhibit some peculiar genera; thus, perhaps, this region might be correctly termed that of the *Rhamni* and the *Caprifoliaceæ*.

1106. (5.) *Region of the Asters and Solidagos* (Michaelmas-daisies and golden-rods).—The eastern part of North America, with the exception of such as belongs to the first or arctic district, comprehends without doubt two regions; for amongst 417 genera in Walter's Flora of Carolina, 117 are wanting in Barton's Flora of Philadelphia. The northern divisions of the United States have, indeed, but few genera which do not occur also in the southern; but this only shows that a similar relation exists here to what takes place between the north and south of Europe. The southern region will include Florida, New Orleans, Georgia, and Carolina; the northern contains the other states of North America. What characterises this region is (besides the number of species of the genera *Aster* and *Solidago*), the great variety of *Oaks* and *Firs*; the very few *Cruciferæ* and *Umbelliferæ*, *Cichoraceæ* and *Cynarocephalæ*; the total absence of the genus *Erica*, and the presence of more numerous species of the allied family of *Vaccinium* (Whortleberries) than are to be met with in Europe.

1107. (6.) *Region of Magnoliæ.*—This, which comprises the most southern parts of North America, is separated from the preceding region by the number of tropical forms which here appear, and which show themselves more frequently than on the similar parallel of the old continent (such, for instance, as the *Scitamineæ*, *Cycadeæ*, *Anonaceæ*, *Sapindaceæ*, *Melastomeæ*, *Cacti*, &c.). From the old world, too, in corresponding latitudes, it is still further distinguished by a smaller proportion of *Labiatæ* and *Caryophylleæ*; and by having more trees of broad shining foliage and splendid blossoms (the *Magnolias*, the *Tulip-tree*, the *Horse-chestnut*, &c.) and with pinnated leaves (the *Gleditschiæ*, *Robiniæ*, *Acaciæ*, &c.).

1108. (7.) *Region of the Cacti, Peppers, and Melastomas*: a very extensive region, including the lower districts of Mexico, the West Indies, New Granada, Guiana, and Peru, perhaps also a part of Brazil; in short, all intertropical America. The three families here mentioned appear peculiarly to characterise these countries; for the first belongs exclusively to America, and of the other two there exist comparatively few species out of these districts. *Palms*, the *Rubiaceæ*, the *Solaneæ* (in which are classed the *Nightshades* and *Potatoe*), the *rough-leaved plants* (Boragineæ), the *Passion-flowers* and *Compositæ*, are here very common. It may admit of several provinces, as that of the *Ferns* and *Orchideæ* (in the West India islands); of the *Palms* (the continent of South America). Brazil ought certainly to constitute a peculiar province, if indeed it be not a distinct region; and the works of Spix and Martius, St. Hilaire, the Prince de Neuwied, &c., will soon enable us to characterise its vegetable forms. The *Melastomæ* and *Palms* appear to belong to the more numerous inmates of this region.

1109. (8.) *Region of the Cinchonæ* (or Medicinal Barks).—It appears from Humboldt's works that the middle districts (such at least in respect to their altitude) of South America should form a distinct region from that last mentioned, as they differ considerably from the low lands; and the name now proposed seems to be characteristic of their vegetation, at least of Peru and New Granada, though certainly not of Mexico, where the species of *Cinchona* are wanting.

1110. (9.) *Region of Escalloniæ*, *Vaccinia* (Whortleberries), and *Winteræ* (Winter's Barks). —These, according to Humboldt, occupy the highest parts of South America. Besides the plants mentioned, there belong to this region many species of *Lobelia*, *Gentian*, *Slipperwort* (*Calceolaria*), *Sage*, several European genera of *Grasses*, *Brome*, *Festuca* and *Poa*, the *Cichoraceæ*, as *Hypochæris* and *Apargia*; as well as the more strictly speaking alpine plants (*Saxifrages*, *Whitlow-grasses*, *Sandworts*, and *Sedges*). Perhaps also those parts of the high lands where the species of *Oak* and *Fir* flourish belong to the same region, though in all probability they constitute a peculiar province.

1111. (10.) *Chilian region.*—It appears that Chili should form a distinct region; for amongst the genera which appear there, not one half are found in the low districts of South America. Its character, perhaps, most resembles that of the mountainous country, in its *Slipperworts*, *Escalloniæ*, *Weinmanniæ*, *Bæa*, *Bellflowers*, and *Buddleæ*; but yet the difference is scarcely sufficient to constitute it a province. The Flora of this country appears to be essentially distinct from that of New Holland, the Cape, and New Zealand; though an approach to them is observable in *Goodenia*, *Araucaria* (Chilian pine), the *Protea* family, *Gunnera*, and *Ancistrum*.

1112. (11.) *Region of arborescent Compositæ* (syngenesious plants with tree-like stems).—This takes in Buenos Ayres, and in general the eastern side of the temperate part of South America. It has been already remarked, that the Flora of this district of the world agrees to a considerable degree with that of Europe: amongst 109 genera, 70 are likewise European, and 85 in the north temperate zone. On the other hand, it differs considerably from the Floras of the Cape and of New Holland, for the *Proteas*, the *Myrtle tribe*, and the *Mimosas* are either wholly wanting, or are seen but sparingly; and there are no *Epacrideæ*, *Heaths*, *Irideæ*, *Mesembryanthema*, or *Geraniums*. Nor can it be compared with the Flora of the north-west coast of America; for amongst 189 genera mentioned, only 35 are found in Chili. The characteristics of this region seem to lie in the great number of *Arborescent Syngenesiæ* (particularly of the sub-family *Boopideæ*), which, however, do not exclusively appertain to it, but are also seen at the Cape.

1113. (12.) *Antarctic region.* — This includes the countries near the Straits of Magellan. There is a considerable affinity between the vegetation here and what is seen in the north temperate zone; for, amongst 82 known genera from thence, there are 59 of them which have species in the northern hemisphere. The arctic polar forms also appear, such as *Sedges* (Carices), *Saxifrages*, *Gentians*, *Arbutus*, and *Primroses*. Some resemblance to the highlands of South America and to Chili is also shown in the *Slipperworts*, *Ourisia*, *Bæa*, *Bolax*, *Wintera*, *Escallonia*; to the Cape, in the genera *Gladiolus*, *Witsenia*, *Gunnera*, *Ancistrum*, *Oxalis*; and to New Holland, in *Proteaceæ* and *Mniarum*.

1114. (13.) *Region of New Zealand.*—This well deserves to be characterised as a separate region, although its vegetation be a mixture of what prevails on the nearest continents, as South America, Southern Africa, and New Holland. It has, in common with South America, *Ancistrum*, *Weinmannia*, *Wintera*; with Southern Africa, the *Fig Marigolds*, *Gnaphalium*, *Xeranthema* (Everlastings), *Tetragonia* (the famous New Zealand Spinach), *Woodsorrel*, and *Passerina*; and with New Holland, the *Epacris*, *Melaleuca*, *Myoporum*; with both the latter, the families of *Proteaceæ* and *Restiaceæ*: some species also are common both to New Holland and Van Diemen's Land, for instance *Mniarum biflorum*, *Samolus littoralis*, *Gentiana montana*; the first also a native of the Straits of Magellan.

1115. (14.) *Region of Epacrides and Eucalypti*: comprehending the temperate parts of New Holland, together with Van Diemen's Land.—This region is very marked. The families of *Stackhouseæ* and *Tremandreæ* are quite peculiar to New Holland, the *Epacrideæ* nearly so. *Proteaceæ*, *Acaciæ*, *Aphyllæ*, and the greater number of the *Myrtle* family (especially of the genera *Eucalyptus*, *Leptospermum*, *Melaleuca*); the *Stylideæ*, *Restiaciæ*, *Casuarineæ*, *Diosmeæ*, separate it from other regions. The tropical part of New Holland, according to Brown, can hardly be united to this, but must be either a particular region, whose Flora resembles that of India, or else a province of this latter region.

1116. (15.) *Region of Fig-Marigolds* (Mesembryanthema) and *Stapelias*.—This comprehends the southern extremity of Africa, the Flora of which is distinguished by a high degree of peculiarity. By the families *Proteaceæ*, *Restiaceæ*, *Polygalæ* (Milkworts), *Diosmeæ*, it may be recognised from most others, except New Holland, and from this it is distinguished by the two numerous genera *Mesembryanthemum* and *Stapelia*, and by the family *Ericeæ*, which is here more abundant than any where else. Further characteristics of this region may be found in the many *Irideæ*, *Geraniæ*, *Oxalideæ*, and the extremely large proportion of *Compositæ*. On the other hand, there exist in this district, as in New Holland, but very sparingly, those peculiar forms of the northern temperate zones, the *Cruciferæ*, *Ranunculaceæ*, *Rosaceæ*, *Umbelliferæ*, *Caryophylleæ*.

1117. (16.) *Region of Western Africa.* —We are only acquainted with Guinea and Congo, the vegetation of which, as we have already remarked, possesses but few peculiarities, and is a mixture of the Floras of Asia and America, though most resembling the former. The American tropical families of *Cacti*, *Peppers*, *Palms*, *Passion-flowers*, are either absent entirely, or they occur in small numbers. *Leguminosæ* are more numerous than in America. Above two thirds of the genera and some of the species of Guinea are found also in the East Indies. On the other hand, this region approximates to America, in possessing many *Rubiaceæ*, as also in the genera *Schwenkia*, *Elais* (a palm), *Paullinia*, *Malpighia*, and several more which are wanting in Asia, and in several species which it has in common with America. A considerable proportion of *Grasses* and *Sedges* (Cyperaceæ), with the peculiar genus *Adansonia* (the Baobab, which is the largest known tree in the world), belong to the characteristics of this country. The interior of Africa is unknown to us.

1118. (17.) *Region of Eastern Africa.* — Of the coast of this side of Africa and the adjacent islands our knowledge is imperfect. We are tolerably acquainted with the islands of Bourbon and France; of Madagascar we know but little; and of the east coast itself scarcely any thing. The Flora of the two first-named islands has a considerable resemblance to that of India. Amongst 290 known genera, 196 of them (equal to two thirds) are found also in India; and of the species, not a few are likewise Indian; many of these, however, may have been introduced by the constant intercourse that takes place between these two parts of the globe. The genera *Eugenia*, *Ficus* (fig), *Urtica* (nettle), *Euphorbia* (spurge), *Hedysarum*, *Panicum*, *Andropogon*, *Sida*, *Pandanus* (screw-pine), *Dracæna* (dragon-wood) *Conyza*, are very numerous in species, as are the same genera in India. In *ferns*, these islands are peculiarly rich. Again, their flora differs considerably from the South African; an analogy existing, however, in their possessing single representatives of the Cape genera *Erica*, *Ixia*, *Gladiolus*, *Bleria*, *Mesembryanthemum*, *Seriphium*, and several arborescent *Synge-*

nesiæ. Still less is the affinity to the extra-tropical parts of New Holland. The similarity is stronger to the tropical portion of that country, of which the flora also approaches that of India. Single genera are all that it seems to possess in common with America; for instance, *Melicocca, Ruizia, Dodonæa, Dichondra*. The following are, perhaps, peculiar to this region, *Latania, Hubertia, Poupartia, Tristemma, Fissilia, Cordylina, Assonia, Fernalia, Lubinia*, and others. The flora of Madagascar seems very peculiar. It agrees with the islands last mentioned; and several genera are seen no where else than in them and Madagascar; for example, *Danais, Ambora, Dombeya, Dufourea, Didymomeles, Senacea;* several species also are common to both; as *Didymomeles Madagascariensis, Danais fragrans, Cinchona Afro-inda*. Still, among the 161 known genera from Madagascar, 54 only are found in the Isles of France and Bourbon; so that there might be good grounds for forming a separate region of the first; unless, perhaps, the east coast of Africa should come under the same. With New Holland and the Cape, Madagascar has probably still less in common than the two other islands.

1119. (18.) *Scitaminean region* (of the *Turmeric, Zedoary, Cardamom, Indian-shot*, &c), or the Indian Flora. — To this appertain India, east and west of the Ganges, together with the islands between India and New Holland; perhaps, also, that division of New Holland which falls within the tropics. The *Scitamineæ* are here in far greater numbers than in America; also, though to a less degree, the *Leguminosæ, Cucurbitaceæ, Tiliaceæ*. The previously mentioned South American forms are rare, or else wanting. This region should be separated into several provinces; but as yet we know too little to undertake such a division with any degree of certainty.

1120. (19.) *The Indian highlands* ought to form one or perhaps two regions, their vegetation being very dissimilar to that of the lowlands: in the middle region, *Melastomæ, Orchideæ*, and *Filices*, appear to prevail; in the higher, the vegetation is more like the European and North Asiatic, and probably the Japanese: these districts perhaps constitute one region with the whole of Central Asia; but of all these countries we shall know much more when the Flora of India by Roxburgh and Wallich is completed.

1121. (20.) *The Flora of the South of China and of Cochinchina* partly resembles that of India, especially in regard to families; but still Loureiro's Flora contains a great many peculiar genera. It is true that perhaps the number of these genera might be reduced; but even then, the vegetation of this tract will probably prove sufficiently peculiar to constitute a distinct region.

1122. (21.) *The region of the Cassiæ and Mimosæ*, which prevail particularly in Arabia and Persia, seems likewise to have a good right to be separated from India, as it is already sufficiently distinct from the Mediterranean region (No. 3.); for, of 281 genera mentioned by Forskäl, 109 only are found in the south of Europe. It is more probable that the Flora of Nubia and part of Central Africa appertains to this region. Abyssinia perhaps forms a distinct region, its elevated parts possessing such a different climate.

1123. (22.) *The islands in the South Sea* which lie within the tropics form perhaps a separate region; though with but a slender degree of peculiarity. Among 214 genera, 173 are found in India; most of the remainder are in common with America; for instance, *Chiococca, Weinmannia, Guajacum*. Of the species which exist equally in them and Asia, are *Zapania nodiflora, Kyllingia monocephala, Fimbristylis dichotoma, Tournefortia argentea, Plumbago zeylanica, Morinda umbellata, Sophora tomentosa*. In common with America, *Dodonæa viscosa, Sapindus saponaria* (soap-berry): with both, *Rhizophora Mangle* (mangrove tree): it has also some in common with New Holland, as *Daphne indica* (a species of *Spurge Laurel*). Peculiar families, or such as have there a decided maximum, can scarcely be cited; though, on the other hand, most of the species are peculiar. The *Bread-fruit* is among the characteristics of these islands; though this tree is not confined to the South Seas.

1124. *The limit of the present essay* does not allow of the intended introduction of the geographical situation of many of our more useful and important plants, which Professor Schouw has so ably delineated; such as that of the *Beech*, the *Vine*, the *Fir tribes*, the *Heaths, Corn*, and such fruits or vegetables as are employed as bread: the *Palms*, the *Proteaceæ*, which form so remarkably striking a feature in the Cape of Good Hope and in New Holland; the *Compositæ*, which are perhaps more universally diffused than any other kind of plant; the *Cruciferæ*, to which the *Cabbage, Turnep, Mustard, Scurvy-grass*, &c. appertain; and the *leguminous tribes*, whose seeds (as the *Pea* and *Bean*) are so valuable for man, and whose foliage, as the *Lupine* and *Trefoil*, &c. affords most of the nourishment to cattle. We must endeavour to incorporate these with the vegetation of the various regions where they are found in the greatest abundance.

Subsect. 2. *Botany.*

1429. *The botany* of the different parts of the British empire is so similar, that it is proposed to treat under one head that of England, Scotland, Ireland, and their adjacent islands.

1430. *The British Islands*, as regards their vegetable productions, are possessed of considerable interest, partly on account of their situation, their variety of soil and exposure, their great extent of coast, the height of many of the mountains; which, though of small elevation, as compared with those of the continent of Europe, yet, from their northern latitude, almost attain the elevation at which perpetual snow is found; and, above all, on account of the careful manner in which their botany has been explored. Were certain tribes of the Cryptogamia, especially the Fungi and the Algæ, more satisfactorily studied and described, we should possess, in the works of Sir James Smith, the most correct flora, of any country in the world. Yet, with all these advantages, few countries have been less known in respect to the distribution or peculiar localities of these plants. Turner and Dillwyn, in their work called the *Botanist's Guide*, have, indeed, done much towards making known the particular stations, especially of the rare plants of England and Wales. M. Boué, in his inaugural thesis, printed at Edinburgh in 1817, on the botany of Scotland, has been the first to offer any remarks on a geographical arrangement, however partial, of the vegetables of that portion of the kingdom; and this was followed by the very useful Essay of Mr. Winch, on the geographical distribution of plants throughout the counties of Northumberland, Cumberland, and Durham; and by Mr. Atkinson on that of Yorkshire, published in the *Transactions of the Wernerian Society*. Aided by these, and some valuable communications that have been made to us by our friends, and by our personal observations, made during various excursions in different parts of the British dominions, we shall endeavour to throw together such remarks as may, at some future time, lay the foundation for a geographical distribution of our vegetables, upon a more extended scale.

1431. *The principal of these islands* in question, Great Britain, although its breadth from E. to W. scarcely averages 8°, yet extends from lat. 50° in the S. to 58° 40′, or, if the extreme northern point, where it is broken into a number of smaller islands, be taken into consideration, to N. lat. 61°: it thus, necessarily, even in its plains, includes a considerable variety of climate, but every where, more or less, tempered by the surrounding ocean; so as, in no part of the island, except on the mountains, or high table-lands, can the temperature be compared, in point of climate, to similar latitudes upon the European, much less upon the American continents. Yet, from the proximity of the whole group to the former, the vegetation is, with a very few exceptions, hereafter to be mentioned, altogether similar to that of the adjacent districts of Europe. In one point of view, indeed, our insular situation affects vegetation in a way which could hardly be expected in so high a latitude; and if, on the one hand, in consequence of our unfavourable summers, the frequent obscurity of the sun, our damp and foggy atmosphere, we are not permitted, without artificial heat and protection, to bring many of the fruits of more favoured climates to perfection; yet the mildness of our winter enables us to introduce and to naturalise plants of much more southern latitudes: so that our gardens, our parks, our shrubberies, and even our forests, are adorned with a vegetation the most varied that can be imagined, or producing the most beautiful flowers, or the most valued of the timbers.

1432. *If there be a spot in our island* which may be designated "a land of the cypress and myrtle, the cedar and vine," it is on the *extreme southern coast of England and Ireland*, yet not assuredly "where the trees ever blossom, the beams ever shine;" but in a spot, certainly, where the native vegetables of the warmer temperate zone are successfully grown in the open air, and where they come to considerable perfection. We have ourselves seen, in the south of Devonshire, the orange and lemon trees loaded with fruit of the finest kind; trained, indeed, to a wall, but without protection, and only provided with it during a very short portion of the winter months; the Lemon-scented Vervain (*Lippia citriodora*, formerly called *Verbena triphylla*) becomes quite a tree, without any artificial protection; the American Agave, the Creeping Cereus, the Prickly Pear, the former having twice flowered in the grounds of the same gentleman, Mr. Yates of Woodville, near Salcombe. The account of the latter of these is recorded in the *Transactions of the Horticultural Society of London*, for 1822; and the plant itself is of that extraordinary character and the circumstance of its blossoming at such an early age is so characteristic of the climate, that we shall here record some of the most important particulars respecting it. It was planted in the open ground, in 1804, when only three years old, and about 6 inches high, within a few yards of the sea-shore, at an elevation of about 40 or 50 feet above the level of the sea; and it was never, in any way, artificially sheltered, nor aided by manure or any cultivation. In eight years from that period, it had attained the height of 5 feet, and during the summer of 1812, it grew nearly $\frac{1}{8}$ of an inch daily. In 1820, it was 11 feet high, and covered a space, the diameter of which was 16 feet; the leaves, close to the stem, being nearly 9 inches thick. During that year it threw up its flowering stem, at first resembling a great head of asparagus, which, for six weeks, rose at the rate of 3 inches a day. By the end of September the flower-stalk attained the height of 27 feet, sending out side branches, and blossoming throughout the whole of the months of September and October. The number of flowering branches was 40, the lowest projecting 2 feet in horizontal length from the stem, gradually diminishing upwards to about a foot or 9 inches at the top; each of them had between 300 and 400 flowers; and in all, there were above 16,000 blossoms. The base of the flowering stalk was 20 inches in circumference.*

1433. *Myrtles*, from the south of Europe, the Tea, Camellias and other Chinese and Japanese plants, thrive well in the open air, as well as the Magnolias, and many other trees, from the southern states of North America, whose native latitudes lie many degrees nearer to the tropics than the highly favoured line of coast in question.

1434. *A beautiful picture of the exotic vegetation* of the island of Guernsey is given by Dr. Macculloch, in the *Memoirs of the Caledonian Horticultural Society*. He justly states, that the Guernsey Lily (*Amaryllis sarniensis*) is not a native of that island, but, as it is supposed, was introduced by accident or design from Japan, and cultivated there to a vast extent to supply the London market. The climate seems to be peculiarly favourable to its growth and increase, as it is for those of other species of the same genus; the Amaryllis vittata, Belladonna undulata and formosissima natives of the Cape or South America. The Magnolia grandiflora is as certain to flower yearly, as it is remarkable for the luxuriance of its growth. The Garden Hydrangea (*H. hortensis*), the Scarlet Fuchsia (*F. coccinea*), the Horseshoe Geranium (*G. zonale, G. inquinans, G. radula,* and *G. glutinosum,*) emulate in summer the luxuriance which they display in their native climates. Every rustic cottage is covered with Geraniums, and adorned with numerous kinds of Pinks, such as are rarely seen in the rest of Britain, except among careful florists. The Lemon-scented Vervain here, too, flourishes perfectly exposed, and grows to 20 feet high and upwards, spreading in a circle of which the diameter equals its height, with its long branches reaching down to the ground on all sides. Its growth, indeed, is so luxuriant, that it is necessary to keep it from becoming troublesome, by perpetually cutting it about the root, from which fresh shoots, 14 feet long, resembling those of the Osier Willow, are annually produced. The Celtis micrantha, a stove plant, flourishes, with little care, out of doors; the double and single Camellias, some species of Olive, as well as many Proteas, several kinds of Cistus, particulary C. crispifolius and formosus, Yuccas, the Azorian Jessamine (*Jasminum azoricum*), the Oleander (*Nerium Oleander*), the Daphne odora, Clethra arborea, Corræa alba, Melaleuca hypericifolia, Galardia rigens, with numerous Ixias and Heaths, all equally requiring the protection of the green-house during the severer winters of our own island, and many of them liable to perish, in spite of our greatest care, — all these, as well as the Myrtle, brave the utmost rigour of a Guernsey climate, and flourish in the greatest luxuriance.

1435. *The Indian Shot* (*Canna indica*) has become thoroughly habituated to the climate of Guernsey, and scatters its seeds, every year, so as to prove a weed in the gardens where it has been introduced. The effect of this delightful climate is visible, too, on fruits of various kinds. Both the Seville and the Sweet Oranges, standing under the shelter of a wall, and only protected by a mat in winter, produce fruit in abundance every year.

1436. *What we have been stating* of the climate of the extreme south of Britain is, however, a remarkable exception to that of the rest of this kingdom; and even in that part such productions are only witnessed upon the very verge, as it were, of the ocean. A few miles inland, and the climate, rather than the natural vegetation, undergoes a considerable change. The lines which constitute the northern boundaries in Europe of the Maize and the Vine (the former especially), are considerably to the south of England; nor is there a chance of these two valuable products ever being cultivated with success in such high latitudes as the British isles; and especially when the fickle character of the summer is taken into account. It is only by art that the fruits of southern climes can, generally speaking, be brought to perfection with us; and such are the skill and assiduity of our gardeners, and such the encouragement given to them by public institutions, and by the nobility and men of large fortune, that in no part of the world is better fruit to be met with, either with regard to size or to flavour.

* In Spain, especially near the shores of the Mediterranean, the flowering stems of the Agave americana grow to 30 feet in height, in a period of ten days, as we are assured by a friend who has recently returned from that country; and they have all the appearance of a forest of ships' masts. In South America, they are applied to many valuable purposes; cut into slices, they are employed in various parts of the country as razor strops, and are sold at Rio Janeiro for that purpose. Like many other monocotyledonous plants, these stems probably abound in silex, which acts upon the blade of the knife or razor. The young stalks are so hard that they cannot be cut with a large and strong knife; and the fibres of the leaves are manufactured, at Perpignan, into ropes and cordage. In Catalonia, the Agave americana flowers in its ninth or tenth year; in the south of France so much more rarely, that it is there called "the plant that flowers as seldom as an Englishman smiles." In our country it is outvied, we hope, by the smiles of the natives; for, as we have before stated, it blossomed, in Devonshire, in its twenty-first year; whereas, in the rest of England, in the green-house, or even stove, it flowers so unfrequently, that is called "the plant that flowers once in a hundred years."

1437. *Before, however, we speak further* of the exotic productions, which are affected by the different latitudes or by the different elevations in Great Britain, we must be permitted to offer a few remarks on its native vegetation; which (exclusive of cryptogamia) has been so admirably described by Sir James E. Smith, in his *English Flora*, the last, and, we may add, too, the best, of the many excellent works of that learned botanist.

1438. *The general nature of the vegetation of the British isles* will be best seen by a table of the different natural families, accompanied by columns, referring to the number of species in each of these groups; and the proportion which each of these latter bears to the whole.

1439. *So widely different as is our flora*, and, indeed, that of the greater part of Europe, from that of America, in the same latitude, the comparison of the vegetable productions of the two countries would be attended with no interesting results; nor do we think that, to institute a comparison between that of Britain and of any country on the continent of Europe, either to the north or south of our own islands, would repay the labour of the investigation; but, did our space allow us, we should gladly have scrutinised the relation between the vegetation of our own islands, and that of an equal space of territory, as far as it would be possible, in the same degrees of latitude; in Germany and Denmark, and a part of Sweden, for example. But this would, doubtless, be attended with great labour. Many floras would have to be consulted, and only such plants selected as belong to the given space of country. Nor would the comparison be exactly fair, unless the two districts contained, and in nearly the same parallels of latitude, similar mountains and similar plains. A great obstacle, too, to the obtaining of correct results would arise from the very different opinions which different botanists entertain as to what are species, and what varieties; and what are aboriginal natives, and what imported from other countries; so that no two naturalists, even publishing at the same time the flora of the same region, would at all approximate in what concerns the number of native species in that region, more especially if each uses his own eyes, and is unbiassed by the opinion of his fellow-labourer in the same studies. Nees von Esenbeck and Hornschuch's *Bryologia Germanica*, contains, in the first volume, but few genera, and comprises 68 species: the same group of genera in Mohr's *Flora Germanica* has only half that number. In our own country, Smith reckons 17 saxifrages of Scotland, 59 willows, and 14 roses; Hooker, in his *Flora Scotica*, enumerates but 13 saxifrages, 10 roses, and 43 willows; and the same holds good with many other genera in different writers. At present, we can only guess, from a tolerable acquaintance with the vegetable productions, at the relative proportions; and whatever may be our partiality to our native land, we must, nevertheless, express our opinion that the result of such an investigation would be greatly in favour of the Continent. The soil, we apprehend, is more varied, the quantity of uncultivated ground is greater, and the very circumstance of its continuity of land, extending north and south and eastward, would naturally aid the migration of plants from great distances, which must materially increase the number of species. Thus it is, that, on landing on the shores of France or Holland, we find many plants which have reached their limits in those directions, simply in consequence of the barrier afforded by the sea; whereas the same individual species passes on to Germany, Holland, and even to Denmark and Sweden, to enrich those more northern floras.

1440. *The only two floras of Great Britain*, which are so complete as to demand our particular attention, are Smith's work, above alluded to the (*English Flora*), and Gray's *Arrangement of British Plants;* the former classed according to the Linnæan system, extending, however, only to the end of the class Polygamia, and the first order of the class Cryptogamia Filices. Gray's *Flora* includes the whole of the British vegetables, arranged according to the natural method, and is the only one that approaches, however deficient it may still be, to any thing like a catalogue of our present state of knowledge of the Cryptogamia. Among the Phænogamous plants, however, Mr. Gray has included a great number that are only known in a state of cultivation, as has been done by De Candolle, in his *Flore Française*, and many other continental botanists. We have, therefore, deemed it convenient thus to give a list of the plants, according to each of these authors; and the increased number in the columns of species, according to Mr. Gray, will be thus easily accounted for.

COMPARATIVE TABLE.

A List of the Number of Species of British Plants, arranged according to the Classes and principal Families to which they belong; exhibiting the relative proportion which these latter bear to the whole of the respective Classes.*

Names of the Natural Families.	Species in Smith's *English Flora.*			Proportion of Natural Orders to Phænogamous, per Smith.	Species in Gray's *Arrangement of British Plants.*			Proportion of Natural Orders to Phænogamous, per Gray.	Average Proportion of Natural Orders to Phænogamous Plants.
Fungi	- -	- -	- -	- -	- -	800	- -	1 to 2	1 to 2
Algæ	- -	- -	- -	- -	- -	400	- -	4	4
Lichenes	- -	- -	- -	- -	- -	400	- -	4	4
Hepaticæ, by Hooker	- -	90	- -	$16\frac{1}{10}$	- -	97	- -	$16\frac{8}{9}$	$16\frac{7}{9}$
Musci, by Hooker	- -	290	- -	$5\frac{1}{5}$	- -	290	- -	$5\frac{6}{10}$	$5\frac{4}{10}$
Filices	- -	58	- -	26	- -	58	- -	$28\frac{1}{5}$	$27\frac{1}{10}$
ACOTYLEDONES	- -	- -	- -	- -	- -	- -	2045	$\frac{4}{5}$	$\frac{4}{5}$
Gramineæ	121	- -	- -	$12\frac{1}{2}$	170	- -	- -	$9\frac{3}{5}$	11
Cyperaceæ	92	- -	- -	17	91	- -	- -	18	$17\frac{1}{2}$
Junceæ and Restiaceæ	32	- -	- -	47	33	- -	- -	$49\frac{1}{2}$	$48\frac{1}{4}$
Glumaceæ	- -	245	- -	$6\frac{1}{8}$	- -	294	- -	$5\frac{1}{2}$	$5\frac{3}{4}$
Orchideæ	- -	37	- -	$40\frac{2}{3}$	- -	33	- -	$49\frac{1}{2}$	45
Monocotyledones cæteræ	- -	73	- -	- -	- -	89	- -	- -	
MONOCOTYLEDONES	- -	- -	355	$4\frac{1}{4}$	- -	- -	416	$3\frac{9}{10}$	4
Coniferæ	- -	4	- -	376	- -	7	- -	234	305
Amentaceæ	- -	78	- -	$19\frac{1}{4}$	- -	72	- -	$22\frac{3}{4}$	21
Euphorbiaceæ	- -	16	- -	94	- -	16	- -	$102\frac{1}{4}$	98
Scrophul. and Orobancheæ	- -	52	- -	$28\frac{1}{2}$	- -	55	- -	$29\frac{4}{5}$	$29\frac{1}{4}$
Labiatæ and Verbenæ	- -	56	- -	$26\frac{4}{5}$	- -	69	- -	$23\frac{3}{4}$	$25\frac{1}{4}$
Boragineæ	- -	23	- -	1 to $65\frac{1}{3}$	- -	23	- -	1 to $71\frac{1}{8}$	1 to $68\frac{1}{5}$
Ericineæ and Pyroleæ	- -	22	- -	67	- -	22	- -	$74\frac{1}{3}$	$70\frac{2}{3}$
Campanulaceæ	- -	14	- -	$107\frac{1}{3}$	- -	15	- -	109	$108\frac{1}{6}$
Compositæ	- -	137	- -	11	- -	144	- -	$11\frac{1}{3}$	$11\frac{1}{6}$
Rubiaceæ	- -	21	- -	$71\frac{1}{2}$	- -	19	- -	86	$78\frac{3}{4}$
Umbelliferæ	- -	64	- -	$23\frac{1}{2}$	- -	69	- -	$23\frac{3}{4}$	$23\frac{5}{8}$
Rosaceæ	- -	81	- -	$18\frac{1}{2}$	- -	81	- -	$20\frac{1}{5}$	$19\frac{1}{3}$
Leguminosæ	- -	66	- -	$22\frac{4}{5}$	- -	69	- -	$23\frac{3}{4}$	$23\frac{1}{4}$
Malvaceæ	- -	6	- -	$250\frac{1}{2}$	- -	6	- -	$272\frac{2}{3}$	$261\frac{1}{2}$
Caryophylleæ	- -	59	- -	$25\frac{1}{2}$	- -	60	- -	$27\frac{3}{10}$	$26\frac{4}{10}$
Cruciferæ	- -	71	- -	$21\frac{1}{6}$	- -	73	- -	$22\frac{5}{7}$	$21\frac{7}{10}$
Ranunculaceæ	- -	36	- -	$41\frac{3}{4}$	- -	42	- -	39	$40\frac{3}{8}$
Dicotyledones cæteræ	- -	342	- -	- -	- -	378	- -	- -	
DICOTYLEDONES	- -	- -	1148	$1\frac{1}{3}$	- -	- -	1220	$1\frac{1}{3}$	$1\frac{1}{3}$

1441. *Scotland*, comprising the northern extremity of the island, commencing in latitude 55°, and its flora having been tolerably satisfactorily investigated, we here subjoin a similar list, drawn up by Mr. Arnott. The Phænogamous plants are taken from the *Flora Scotica* of Dr. Hooker; the Cryptogamia from the *Flora Scotica*, the *Flora Edinensis*, and *Cryptogamic Flora* of Dr. Greville, and from the late Captain Carmichael's manuscripts.† It will be observed that the Musci, Filices, Lichenes, and Hepaticæ have of late years received comparatively few additions; while the Algæ and Fungi have increased so enormously, that more are now reckoned for each order than Gray has stated as natives of the whole island.

---

* Kindly drawn up for this work by G. A. W. Arnott, Esq. of Edinburgh.

† The proportions in the Cryptogamia will be found probably much more correct for Scotland than those given in the British table are for the whole of Britain; and this owing to the very accurate researches which have been made in that tribe by Dr. Greville, and by the late Captain Dugald Carmichael; particularly by the latter in the Fungi and Algæ; so that the discoveries of that gentleman alone in those two groups, in one small district (Appin) in the west highlands of Scotland, amount to more species than were previously described as inhabiting the whole of the British dominions.

Names of the Natural Families.	Species.			Proportion of Natural Orders to Phænogamous Plants.	Names of the Natural Families.	Species.			Proportion of Natural Orders to Phænogamous Plants.
Fungi - -	-	974	-	1 to $1\frac{1}{9}$	Scrophularinæ -	-	37	-	1 to $29\frac{1}{4}$
Algæ - -	-	465	-	$2\frac{1}{3}$	Labiatæ - -	-	39	-	$28\frac{1}{5}$
Lichenes -	-	260	-	$4\frac{1}{6}$	Boragineæ -	-	18		
Hepaticæ -	-	73	-	$14\frac{6}{7}$	Ericineæ - -	-	18		
Musci - -	-	264	-	$4\frac{1}{7}$	Campanulaceæ -	-	9		
Filices - -	-	48	-	$22\frac{1}{2}$	Compositæ -	-	105	-	$10\frac{1}{3}$
Acotyledones	-	-	2084	$\frac{2}{3}$	Rubiaceæ - -	-	16		
Gramineæ -	94				Umbelliferæ -	-	44	-	$24\frac{3}{5}$
Cyperaceæ -	66				Rosaceæ - -	-	52	-	$20\frac{4}{5}$
Junceæ - -	28				Leguminosæ -	-	43	-	25
Glumaceæ -	-	188	-	$5\frac{7}{9}$	Malvaceæ -	-	5		
Orchideæ -	-	19			Caryophylleæ -	-	45	-	24
Monocotyl. cæteræ	-	53			Cruciferæ - -	-	56	-	$19\frac{1}{3}$
Monocotyled.	-	-	260	$4\frac{1}{6}$	Ranunculaceæ -	-	25		
Coniferæ - -	-	8			Dicotyled. cæter.	-	245		
Amentaceæ -	-	56			Dicotyledones	-	-	823	$1\frac{3}{10}$
Euphorbiaceæ -	-	7							

1442. *It must be remarked*, that in *Cyperaceæ*, *Junceæ*, *Salix*, *Saxifraga*, *Rosa*, *Rubus*, and some others, the *species* are not formed on the same rules as in Smith's *English Flora*; and therefore, before drawing a parallel between these orders in Scotland, and in the whole of Britain, a considerable number of species ought to be added. To make this comparison, then, about twenty species may be added to the *Monocotyledones*, and about fifty (say forty-seven), to the *Dicotyledonous* plants, making these two, 280 and 870; whence the Monocotyledones of Scotland are to the whole of those in the British dominions as one to one and a quarter, or as four to five; and the Dicotyledones as eight to eleven.

1443. *Ireland* possesses a flora which partakes of the nature of those of England and Scotland. A list of the phænogamous plants has been recently published by Mr. J. T. Mackay, of the Dublin College Botanic Garden. It exhibits a much poorer vegetation than its sister island, including only 934 species; of which there are,

41 Filices. 211 Monocotyledones, and 682 Dicotyledones.

So that the proportion of Filices to Phænogamous plants is as - 1 : $21\frac{3}{4}$
Monocotyledones to Phænogamous plants 1 : $4\frac{1}{4}$
Dicotyledones to Phænogamous plants - 1 : $1\frac{1}{3}$

1444. *The proportion* of Irish Monocotyledones to British Monocotyledones (according to the species of Smith) is as - - 1 to $1\frac{2}{3}$, or as 3 to 5.
Of Irish Dicotyledones - 1 to $1\frac{2}{3}$, or as 3 to 5.

1445. *If we take into consideration* the plants of the northern part of England, according to the species in Mr. Winch's list of the plants of Northumberland, Cumberland, and Durham, we shall find that there are 47 *Filices*; other *Cryptogamia*, 1206; *Monocotyledones*, 249; and *Dicotyledones*, 788; so that the total Cryptogamous to Phænogamous plants are as - - 1 to $\frac{4}{5}$
Filices to ditto - - - - 1 to 22
Monocotyledones to ditto - - - 1 to $4\frac{1}{6}$
Dicotyledones to ditto - - - 1 to $1\frac{1}{3}$
Or, if compared with the whole of the British empire: —
Monocotyledones to British Monocotyledones (by Smith's *Flora*) as 1 to $1\frac{2}{5}$, or as 5 to 7.
Dicotyledones to British Dicotyledones (by Smith's *Flora*) as 1 to $1\frac{1}{2}$, or as 2 to 3 nearly.

1446. *Proportion* of the Classes of Plants to the Phænogamous: —

Names of the Natural Families.	Great Britain and Ireland. (Smith.)	Scotland. (Hooker, &c.)	Ireland. (Mackay.)	N. of England. (Winch.)
Dicotyledones - -	1 to $1\frac{1}{3}$	1 to $1\frac{3}{10}$; or nearly 1 to $1\frac{1}{3}$	1 to $1\frac{1}{3}$	1 to $1\frac{1}{3}$
Monocotyledones -	1 to $4\frac{1}{4}$	1 to $4\frac{1}{6}$	1 to $4\frac{1}{4}$	1 to $4\frac{1}{6}$
Acotyledones (including Filices) -	1 to $\frac{4}{5}$	1 to $\frac{1}{2}$	unknown.	1 to $\frac{4}{5}$
Filices - -	1 to 26	1 to $22\frac{1}{2}$	1 to $21\frac{3}{4}$	1 to 22

1447. *Relative proportion* of species in —

Names of the Natural Families.	Great Britain and Ireland. (Smith.)	Scotland. (Hooker.)	Ireland. (Mackay.)	N. of England. (Winch.)
Dicotyledones - -	28	20	17	19
Monocotyledones -	36	26	21	25

1448. *Few, indeed, of the* 3300 (taking them in round numbers) species of plants now enumerated as natives of England, Scotland, and Ireland, and the adjacent islets, can be considered as exclusively belonging to these countries. For though there are many which are not referred to as species in the works of other authors, yet they are, for the most part, among such families as are not well understood, and about which there will always exist a difference of opinion; as among the Grasses, Willows, Brambles, &c. The Scottish and Cornish Lovages (*Ligusticum scoticum* and *cornubiense* Sm.) were long considered to be peculiar to the British dominions; the one in the extreme south, the other in the north; but the former is now well known to be an inhabitant of maritime places in all the arctic and subarctic regions, so that it has only in Scotland and the north of England found its southern limits; whilst the latter, which we have received from Italy, and which grows also in Greece and Portugal, has, in Cornwall, reached its extreme northern boundary. The place of the Bird's-eye Primrose (*Primula farinosa*), so abundant in pastures in the north and west of England, is supplied by the Primula scotica in the northern districts of Scotland, and in the Orkney Islands: it is unquestionably a distinct species; but we are confident, from the excellent representation of it in the *Svensk Botanik*, that it is also a native of Sweden, though there confounded with P. farinosa. These instances will suffice to show with what great caution we ought to look upon any plant which may have been considered to belong exclusively to the British dominions.

1449. *Many plants* have indeed reached their northern limits in the south of England and Ireland. We must particularly mention the Strawberry Tree (*Arbutus Unedo*, *fig.* 107.), which forms so charming a feature in that most beautiful of all scenery, the Lake of Killarney. Some have, indeed, supposed that it was introduced into Ireland by the monks of Mucruss Abbey, at some very remote period. Its appearance is, however, altogether that of an aboriginal native, coming to a great size*, perfecting its bright scarlet berries, which are disseminated over the rocks and islands in every direction. The *Erica vagans*, or Cornish Heath (*fig.* 108. *a*), is found no where in Britain except Cornwall; and the same may be said of the newly-discovered *E. ciliaris* (*b*), and the following, of great beauty or rarity: *Lobelia Dortmanna*, *Phyteuma orbicularis* and *P. spicata*, *Sibthorpia europæa* and *Isnardia palustris*, are quite southern plants in the British dominions.

107

STRAWBERRY TREE.

1450. *The Water-Soldier* (*Stratiotes aloides*); the Water Violet (*Hottonia palustris*); the small Maidenhair Grass (*Briza minor*); the Sweet Violet (*Viola odorata*); several Mulleins, the Primrose-peerless (*Narcissus poeticus* and *biflorus*); the common Snake's Head (*Fritillaria meleagris*); the Agrostis setacea, the Star of Bethlehem (*Ornithogalum pyrenaicum*); the two species of Squill (*Scilla autumnalis* and *bifolia*); the Mountain Spiderwort (*Anthericum serotinum*); the Solomon's Seal (*Convallaria polygonatum*); and Sweet Sedge (*Acorus Calamus*); the Yellow-wort (*Chlora perfoliata*); the Mezereum (*Daphne Mezereum*); the Flowering Rush (*Butomus umbellatus*); the Yellow Marsh Saxifrage (*Saxifraga Hirculus*); though on the Continent a very arctic plant, the Clove Pink (*Dianthus caryophyllus*); and D. prolifer, several Catchflys (*Silene*); Euphorbias, Cistuses, Anemones, the Traveller's Joy (*Clematis Vitalba*); the Ground Pine (*Ajuga Chamæpitys*); the Wood-Sage (*Teucrium Scorodonia*); the crested and field Cow-wheat (*Melampyrum cristatum* and *arvense*); some Orobanches, the Vella annua, Draba aizoides, and Iberis amara, some Fumitories (*Fumaria solida, lutea,* and *parviflora*); the yellow and crimson Vetchlings (*Lathyrus Aphaca* and *Nissolia*); the Vicia hybrida, lævigata, and bithynica, Hippocrepis comosa; Orchis Morio†, pyramidalis, ustulata, fusca, militaris, tephrosanthos, hircina; Aceras anthropophora, Herminium monorchis; all the species of Ophrys, Epipactis rubra, Malaxis Loeselii; the beautiful and rare Lady's Slipper (*Cypripedium Calceolus*); the Birthwort (*Aristolochia Clematitis*); the Roman Nettle (*Urtica pilulifera*); the Xanthium strumarium and Amaranthus Blitum; the Spanish Chestnut Tree (*Fagus castanea*); and Misseltoe (*Viscum album*); the Sea Buckthorn (*Hippophae rhamnoides*); and White Poplar (*Populus canescens*): these are some among the most striking of the British plants, which do not reach the middle of the kingdom, and fail below the south of Scotland.

108

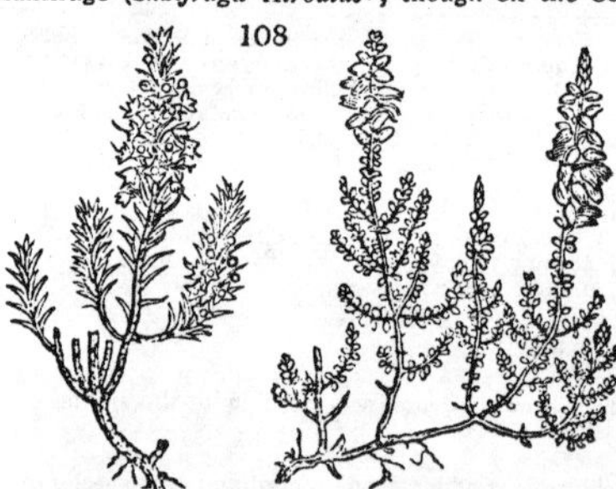

CORNISH HEATH. CILIATED HEATH.

1451. *The most interesting of the Scottish plants* are, principally, such whose types are found on the continent of Europe, in high northern latitudes, or in the extreme arctic regions of both Asia and America; such as Veronica fruticulosa, saxatilis, and alpina, several alpine grasses, and other glumaceous plants; such as Phleum alpinum and Alopecurus alpinus, Eriophorum alpinum; Juncus castaneus, arcticus, and biglumis; and Luzula arctica, Primula scotica (*fig.* 109. *a*), the Myosotis alpestris (*d*), Azalea procumbens, Gentiana nivalis (*c*), Sibbaldia procumbens, Convallaria verticillata, Epilobium alpinum, Arbutus alpina, Pyrola uniflora (*b*), Saxifraga nivalis and rivularis, Stellaria scapigera (the latter is exclusively British), Arenaria rubella and fastigiata, the Cherleria sedoides, Lychnis Viscaria and alpina, Spergula saginoides, Potentilla opaca, Nuphar Kalmiana, Ranunculus alpestris, Ajuga pyramidalis, Cardamine bellidiflora, Orobus niger, Astragalus tralensis and campestris, Erigeron alpinum, Corallorhiza innata, Achillæa tomentosa, Goodyera repens; the most alpine Carices and Salices, and the Dwarf Birch (*Betula nana*).

109

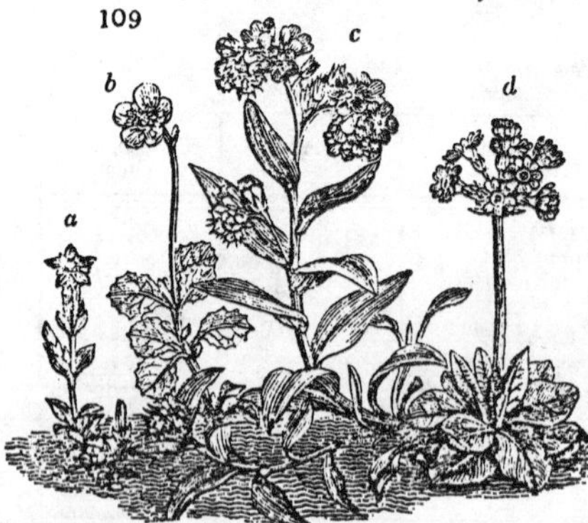

a, SCOTTISH PRIMROSE. b, ROCK SCORPION GRASS.
c, SINGLE-FLOWERED WATER-GREEN. d, SMALL ALPINE GENTIAN.

1452. *There are two plants* which deserve particular notice, as natives of Great Britain, and found nowhere else in Europe; but these are again met with in North America; the one is Potentilla tridentata (*fig.* 110. *a*) abundant in arctic America and upon the Rocky and White

* Mr. Mackay measured a trunk of this fine evergreen tree on Rough Island, nearly opposite O'Sullivan's Cascade, which, in 1805, was 9¼ feet in girth, at a foot from the ground.

† On the authority of Lightfoot, indeed, this plant, so abundantly found in England, is given as a native of Scotland; but no living botanist, that I am aware of, has ever seen it there.

Mountains, the other the Eriocaulon septangulare (*fig.* 110. *b*). This latter genus is mostly tropical, or a native of the warm temperate zones in America, the East Indies, and Australia. The only exceptions to this rule are the Eriocaulon pellucidum of Michaux, and the plant in question; the former being found in North America as high as Canada; and, upon examination, the two species prove identical. In these instances, the Eriocaulon and the Potentilla seem to have overcome many obstacles in their migration, and to have reached their eastern boundary. The Eriocaulon is confined to a few lakes in the Hebrides, where we have been surprised in the month of September at the high temperature of the water, which probably never freezes; and in some spots in the south and west of Ireland: the Potentilla is only found on one hill in Angusshire.

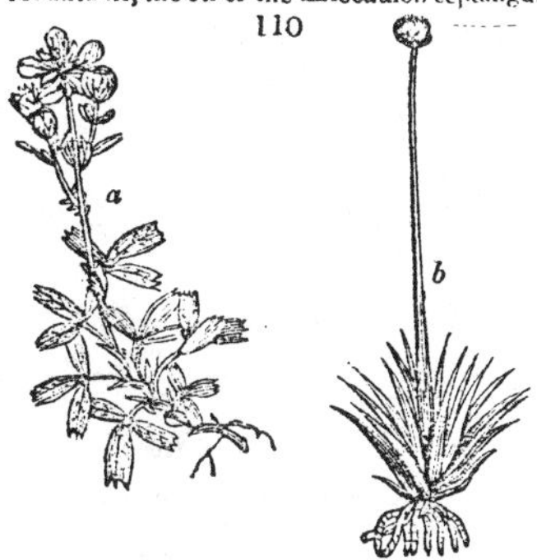

a, TRIFID-LEAVED CINQUEFOIL.
b, JOINTED PIPEWORT.

1453. *It is worthy of remark*, that the genus Pedicularis, which is so numerous in species, in the eastern and southern parts of Europe, almost wholly disappears in Britain; for, notwithstanding the vast numbers of it which are found in Siberia, the south of Russia, Switzerland, extending even to the Pyrenees, and Germany, our country possesses but two, which are equally abundant upon the Continent; and although almost wholly an alpine genus, yet our mountains possess not one really alpine species. It would appear that the climate is peculiarly unsuited to their nurture; for in North America, in the same and especially in still higher northern latitudes, they again become abundant.

1454. *Ireland*, again, exhibits a few striking peculiarities in some of its vegetable productions. Besides the Strawberry tree (*Arbutus unedo*) already mentioned, it can boast of Pinguicula grandiflora (*fig.* 111. *a*), a beautiful flower, native of France and the Pyrenees; Menziesia polifolia (*b*), a species belonging to the latter country and to Spain, and found in a wild state in no other parts of the world; it is, too, a most lovely one: also St. Patrick's Cabbage (*Saxifraga umbrosa*) and the London Pride (*S. Geum*, *c*) and their varieties, which are scarcely known to exist but in Switzerland and the Pyrenees; Arenaria ciliata, a native of the mountains on the continent of Europe; and to these rarities have lately been added, by Professor Giesecké, the Yellow Poppy (*Papaver nudicaule*, *d*), and the Ledum palustre

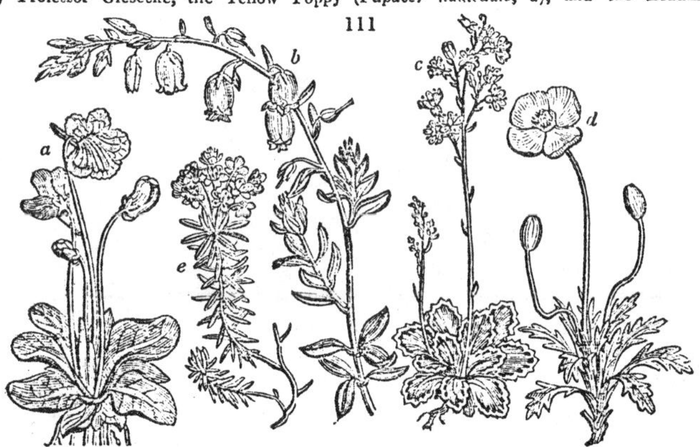

a, LARGE-FLOWERED BUTTERWORT. *b*, IRISH MENZIESIA. *c*, KIDNEY-LEAVED SAXIFRAGE.
d, NAKED-STALKED YELLOW POPPY. *e*, MARSH LEDUM.

(*e*,) both of them peculiarly arctic productions, and plentiful on the northern extremity of America and Greenland; and with these we must be permitted to number, though Cryptogamic plants, the Trichomanes brevisetum (*fig.* 112. *b*), which scarcely grows any where else in the world but in Madeira and in Yorkshire (if it be not now extinct in the latter habitat), the Adiantum Capillus Veneris (*a*), whose only locality in the British dominions is the west of Ireland, and one spot in Wales, but which is frequent in the south of Europe, and even in the tropical parts of America; and two mosses, Hookeria latevirens, and Daltonia splachnoides, entirely peculiar to Ireland.

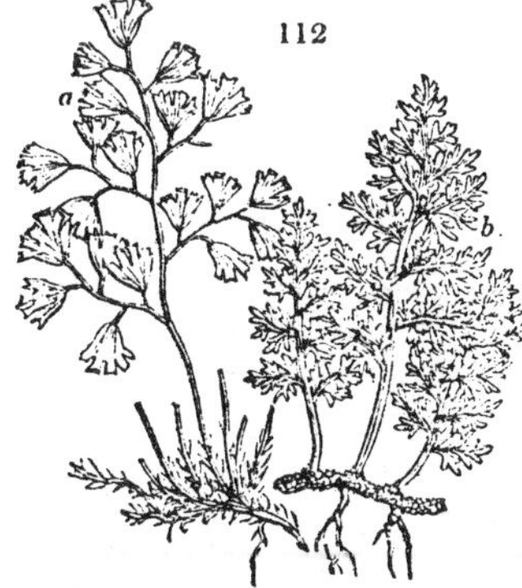

a, TRUE MAIDEN-HAIR.
b, SHORT-STYLED BRISTLE-FERN.

1455. *Hibernia*, again, possesses three remarkable vegetable productions, which are now, we believe, pretty generally distributed in our gardens and shrubberies throughout the kingdom, and are universally known by the names of the Irish Broom, Irish Furze, and Irish Yew. The former, we believe to be the Sparticum patens of Linnæus, a Spanish species, with very hairy pods; and it is, probably, not wild in that country. The Irish Furze has an appearance very different from that of the European or Dwarf Furzes (*Ulex europæus* and *nanus*), having very erect short branches, and closely placed spines; so that the whole plant has a remarkably dense and compact habit, appearing almost as it it were kept close clipped with shears. It blossoms rarely, but we have seen both flowers and seed-vessels, which do not differ in any material point from those of Ulex nanus. In some gardens it is called U. europæus var. strictus; but Mr. Mackay, perhaps with justice, considers it to be quite a distinct species, and he has called it, in his "Catalogue of the Indigenous Plants of Ireland," Ulex strictus. Still, the only stations for this plant are in the Marquess of Londonderry's park and shrubberies, at Mount Stewart, county of Down, where there are some very large bushes; but whence it came, no one can tell. This would, however, be a very valuable plant to the agriculturist; for, according to Mr. Stewart Murray, it has been planted (for it increases readily by cuttings) in dry hilly pastures in the north of Scotland, and in the early spring throws up an abundant crop of succulent shoots, which are greedily eaten by the sheep, when there is little or no grass to support them.

113

IRISH YEW.

1456. *The third Irish botanical curiosity* is the Irish Yew (*fig.* 113), or Florence-Court Yew, as it is called in that country, from its being first discovered at Florence Court, the seat of Lord Enniskillen. Mr. Mackay does not consider it to be wild; but Mr. Hervey, in the *Agricultural Magazine* for October, 1828, says, that it is an undoubted native, and plentiful in the neighbourhood of Antrim, where there are trees at least a century old. It is distinguished by its upright branches, which give the whole plant somewhat the habit of a Cypress; by the leaves growing, not in a distichous manner, but from all sides of the stem: the drupe or berry, too, Mr. Mackay says, is of a different form from that of the common Yew.

1457. *If we had room here for its insertion*, we would gladly have drawn up a full list of those plants which are exclusively peculiar to the east or to the west sides of this kingdom, or which have, at least, their maximum in either one, and are very rare or altogether wanting on the opposite side. We are sure it would be productive of highly interesting results. The hasty catalogue which follows, though made without the accurate care which the subject would require, if it received the scrupulous attention which it merits, will, nevertheless, suffice to show that the plants of the east coast, generally, have the character of those which are natives of dryer and warmer climates; whereas those of the west are, in most instances, types of the subalpine plants of more southern latitudes, or exhibit a vegetation analogous to that of the more southern coasts of France and the Mediterranean. It must be observed, however, that it is extremely difficult to define the limits of the verge of the plants on the east or west coast, for want of more full stations than we are yet possessed of, and more ample local floras; and it must be clearly understood that the annexed list must be considered only as an approximation to the truth, even upon its own limited scale.

1458. *Plants* which are either wholly confined to the *east* side of England, or have their maximum there, rarely extending to the centre of the kingdom: —

- Veronica triphylla
- ——— verna
- Dactylis stricta
- Polypogon monspeliensis
- ——— littoralis
- Alopecurus bulbosus
- Panicum sanguinale
- Agrostis Spica venti
- Dipsacus pilosus
- Holosteum umbellatum
- Gentiana verna
- Tillæa muscosa
- Bupleurum tenuissimum
- ——— rotundifolium
- Menyanthes nymphæoides
- Pimpinella magna
- Myosurus minimus
- Verbasca
- Campanula patula
- ——— Rapunculus
- Thesium linophyllum
- Salsola fruticosa
- Statice reticulata (differs from the plant so called of Scotland and Wales)
- Tofieldia palustris
- Acorus Calamus
- Luzula Forsteri
- Fritillaria meleagris
- Tulipa sylvestris
- Ornithogalum nutans
- ——— umbellatum
- ——— luteum
- Trientalis europæa
- Dianthus caryophyllus
- Sedum album
- Silene quinquevulnera
- —— conica
- —— anglica
- —— noctiflora
- Lythrum hyssopifolium
- Adonis autumnalis
- Anemone Pulsatilla
- Glaucium violaceum
- Stratiotes aloides
- Ajuga alpina?
- —— Chamæpitys
- Melampyrum cristatum
- ——— arvense
- ——— sylvaticum
- Orobanche minor
- ——— ramosa
- Turritis glabra?
- Orchis fusca
- —— militaris
- —— taphrosanthos
- —— ustulata
- —— hircina
- Ophrys muscifera
- —— apifera
- —— arachnites
- —— aranifera
- Aristolochia clematitis
- Hippophae rhamnoides
- Urtica pilulifera
- Atriplex pedunculata.

1459. *Plants* which are either wholly confined to the *western* side of England, or to Wales, or which have their maximum there, rarely extending beyond the centre of the kingdom: —

- Pinguicula lusitanica
- Polycarpon tetraphyllum
- Knappia agrostidea
- Scirpus Holoschænus
- Rubia peregrina
- Lysimachia thyrsiflora (Yorkshire and Scotland)
- Campanula hederacea
- Lobelia Dortmanna
- ——— urens (but not so in Scotland)
- Pulmonaria maritima, (found as far south as Devonshire*; not confined to either coast of Scotland)
- Primula farinosa (most abundant in Cumberland; in Scotland confined to the east, and rare)
- Impatiens noli tangere
- Gentiana campestris
- Corrigiola littoralis
- Exacum filiforme
- Eryngium campestre
- Sium verticillatum
- Sibthorpia europæa
- Bupleurum Odontites
- Illecebrum verticillatum
- Anthericum serotinum
- Scilla verna
- —— autumnalis
- Rumex digynus
- Juncus filiformis
- Saxifraga nivalis (Wales and Yorkshire)
- Cotyledon Umbilicus
- Oxalis corniculata
- Andromeda polifolia
- Arenaria tenuifolia
- Cotoneaster vulgaris
- Papaver cambricum
- Cistus polifolius
- —— guttatus
- —— marifolius
- Actæa spicata
- Melittis melissophyllum
- ——— grandiflorum
- Bartsia viscosa
- Brassica oleracea
- ——— monensis
- Draba aizoides
- Cheiranthus sinuatus (Anglesea)
- Arabis hispida
- Erodium moschatum
- Vicia bithynica
- Lotus diffusus
- Chrysocoma aure
- Euphorbia Peplis
- Asplenium septentrionale (Denbighshire; in Scotland, east?)
- Adiantum Capillus Veneris.

1460. *In general, with regard to such of the plants* above mentioned as are found in Scotland, the same rule, as to locality, holds good in that country also. There are, however, some differences, and a few unexpected peculiarities, which we shall shortly notice; and Scotland being of so much smaller extent, and perhaps more thoroughly investigated, than England, we can speak with more certainty as to the range of these vegetables.

1461. *Plants* found on the *west* side of Scotland almost exclusively: —

- Pinguicula lusitanica
- Lysimachia Nummularia
- Convolvulus Sepium
- Campanula hederacea
- Jasione montana
- Selinum palustre
- Sison verticillatum
- Œnanthe pimpinelloides
- Statice Limonium
- Drosera longifolia
- Epilobium tetragonum
- Dianthus deltoideus
- Cotyledon Umbilicus
- Lythrum Salicaria
- Glaucium luteum
- Thalictrum flavum
- Bartsia viscosa
- Brassica monensis
- Ervum tetraspermum
- Hypericum Androsæmum
- ——— elodes
- Picris hieracioid
- Inula crithmoides
- Epipactis latifolia
- Gymnadenia conopsea (very rare on the east coast).

* According to Turner and Dillwyn's *Botanist's Guide*.

1462. *Plants* found on the *east* side of Scotland almost exclusively: —

Phleum arenarium
Festuca loliacea
Echium vulgare (doubtfully wild on the west)
Primula veris
——— elatior
Convolvulus arvensis
Gentiana amarella
Saxifraga tridactylites
Silene noctiflora
Euphorbia Esula
——— exigua
Arenaria verna
Potentilla verna
——— argentea
Spiræa Filipendula
Marrubium vulgare
Lavatera arborea
Astragalus glycyphyllus
——— hypoglottis
Trifolium fragiferum
——— ornithopodioides
——— scabrum
Vicia lutea
Cnicus Eriophorus
Onopordon Acanthium
Artemisia maritima
Epipactis palustris
Asplenium septentrionale.

1463. *The trees that are aboriginal natives of Britain* are the Oak (two species); the Elm (five species); the Beech, the Ash, the Maple, Sycamore, Hornbeam, Lime (three species, according to Smith); the Spanish Chestnut (?); the Alder, Birch, Poplar, (four species); and the Scotch Fir: to which may be added the Mountain Ash, which in some parts of Scotland attains to a great size. Of these, then, the Oak, the Beech, Birch, and Scotch Fir live in society, —

> " In grouping beauty rise the loftiest trees,
> Secure in friendship from the blighting breeze;" —

forming vast forests, almost to the exclusion of other trees. The finest forests of Oak and Beech are unquestionably to be seen in the southern parts of England; the latter flourishing, to an extraordinary degree, in the chalk and deep clay soils of Sussex and some of the neighbouring counties. In Scotland, the oak, though there may be some exceptions, generally forms copse woods, and is mostly confined to the valleys. Its northern limit is scarcely within the British dominions. It extends to lat. 60° on the continent in Russia, and 64° in Norway; and if in Scotland oaks are not found in the extreme north, it is rather owing to want of shelter and of suitable soil, than to any other circumstance.

1464. *The Pine* (*Pinus sylvestris*, *fig.* 114.) constitutes noble forests among the mountainous districts of North Britain, filling the valleys, and ascending, probably, to the height of 2500 feet upon the hills, among the northern Grampians, especially in the districts of Braemar, Glenmore, and Rothiemurcus; and exhibiting individual specimens of such size and beauty as no one can form an idea of, who has not seen this tree in its native situations. About 20 years ago, the greater part of the noble forest of Glenmore, belonging to his Grace the Duke of Gordon, was cut down, and floated to the sea at Speymouth; where several ships were built of the excellent timber which it afforded. One individual was then felled, which afforded a plank, which we have seen at Fochabers: it is 5½ feet in diameter, and free from every kind of imperfection.

114

SCOTCH FIR.

1465. *Of the fruit trees* which are successfully cultivated in the open air, the number is limited. In the south, exclusively, or, perhaps, as far as the centre of the kingdom, under favourable circumstances, the Vine, the Fig, the Quince, the Mulberry, Chestnut, Walnut, and Medlar may be advantageously planted. The Apple, Pear, the Plum of various kinds, the Peach, Nectarine, and Apricot; all, according to soil, exposure, and other local circumstances, ripen their fruit in the open air, if afforded the protection of a wall, as high north as Inverness, and some of the most hardy ones much higher: but the want of sun must ever be a hinderance to the thorough perfecting of good fruit in the north of Scotland.

1466. *Of the various kinds of Corn*, which are used as food for man or cattle, Wheat, Barley, Bere, Bigg, Oats, and Rye are the universal crops; and these all succeed in situations not too much elevated above the level of the sea, as far to the northward as Inverness, beyond which the wheat becomes a very uncertain crop; and even considerably south of Inverness, to the north of the Forth and Clyde, in lat. 56°, the cultivation of wheat is almost wholly confined to the eastern side of the country, the west being the district for pasture.

1467. *In regard to the height* at which certain plants will grow above the level of the sea, the southern and midland parts of Great Britain do not contain mountains upon a sufficiently lofty scale to render their investigation particularly interesting. The northern parts of England posssess mountains of considerable elevation, upwards of 3000 feet; and it has fortunately happened that Mr. Winch's " Essay on the Geographical Distribution of Plants throughout the Counties of Northumberland, Cumberland, and Durham," of which the lat. 55° may be considered the medium, embraces a very great proportion of this very country, which, from its situation, may, in point of climate, be considered as intermediate between the more northern and southern floras of Great Britain; and though we cannot devote the room that would be necessary to introducing the whole of his remarks upon the humble and less known vegetables, yet we gladly select from the writings of so accurate an observer what concerns the more valuable and more striking vegetable productions.

1468. *The Oak*, in lat. 55°, attains a large size in the valleys; it ascends the hills, but gradually becomes of stunted growth in Weardale and Teesdale, to the elevation of 1600 and 1700 feet.

1469. *The Common Elm* (*Ulmus campestris*) is not indigenous north of the Tees;

its place being taken by the Wych Elm (*U. montana*), which skirts the mountains at a height of 2000 feet.

1470. *The Beech and Aspen* are truly natives, flourishing beautifully in the low sheltered spots, but not climbing the hills to equal heights with the oak. The White and Black Poplars (*Populus alba* and *nigra*) are doubtful natives of the north of England, as of Scotland; though the White Poplar is remarkable for withstanding the north-easterly winds, which are so destructive to vegetation in the counties of Northumberland and Durham. The Lime (*Tilia europæa*), the Chestnut (*Castanea vesca*), and the Hornbeam (*Carpinus Betulus*), stand in the same predicament.

1471. *Holly trees* are among the chief ornaments of the woods in Durham, Northumberland, and Cumberland, as is the Yew (*Taxus baccata*). The Birch (*Betula alba*) equals in size the birches of Norway and Sweden; but is not found on the mountains at a greater elevation than the Sycamore (*Acer Pseudo-platanus*), which in the subalpine regions seems to be as vigorous, and to attain as great a size as it does near the sea-coast. The Mountain Ash (*Pyrus aucuparia*) is found on the hills; the White Beam (*Pyrus Aria*) may be traced from the High-Force of the river Tees to the coast; the Alder (*Alnus glutinosa*) and the Guelder Rose (*Viburnum Opulus*) accompany our streams; and the Hazel, Black Cherry (*Prunus Cerasus*), Bird Cherry (*Prunus Padus*), the Spindle-tree (*Euonymus europæus*), the Raspberry (*Rubus idæus*), and the common Elder (*Sambucus nigra*), are found in all the woods from the sea-shore to those situated on an elevation of 1600 feet: but the common Maple (*Acer campestris*) occurs only in the hedges, in some parts of the flat country.

1472. *The Ash tree* (*Fraxinus excelsior*), the White Thorn (*Mespilus Oxyacantha*), the Crab tree, or Wild Apple (*Pyrus Malus*), and Black Thorn (*Prunus spinosa*), abound throughout the district in question. The Bullare (*Prunus insititia*) is rare; and the Plum-tree (*Prunus domestica*), Pear (*Pyrus communis*), Red Currant (*Ribes rubrum*), the Berberry (*Berberis vulgaris*), and Gooseberry (*Ribes Grossularia*), though of frequent occurrence, appear not to be original natives of the soil. But the Rock Currant (*Ribes petræum*), the Acid Mountain Currant (*Ribes spicatum*), Alpine Currant (*Ribes alpinum*), Black Currant (*Ribes nigrum*), and Privet (*Ligustrum vulgare*), are indigenous, and not unfrequent.

1473. *On the elevated moors*, and even on the mountain of Crossfell, at an elevation of nearly 3000 feet, the roots and trunks of very large Pines (*Pinus sylvestris*) are seen protruding from the black peat moss: but the tree is no longer indigenous to the country; nor does the same tree now attain the size of the ancient Pine, though planted in similar moorland situations, though the young trees be protected, and even the plantations situated at a lower level.

1474. *The Furze* (*Ulex europæus*) attains to an elevation of 2000 feet in sequestered spots, accompanied by the Bramble. Juniper may be traced from the coast to the height of 1500 feet. The Cloudberry (*Rubus Chamæmorus*), the Bearberry (*Arbutus Uva ursi*), and Sand Willow (*Salix arenaria*), attain the same elevation; while the Dwarf Willow (*Salix herbacea*), but without its usual attendant the Reticulated Willow (*S. reticulata*), reaches to the tops of the loftiest mountains, upwards of 3000 feet above the level of the sea.

1475. *Coarse Grasses, Sedges, and Rushes* too often cover the wet moors with a scanty and almost useless vegetation. To the agriculturist the different Heaths are scarcely more acceptable; but they are unquestionably among the most beautiful of our native plants, and their abundance and the vast extent of ground which they clothe, give a peculiar character to very many parts of Great Britain, especially in the North. In the districts in question, the common Heather (*Calluna vulgaris*), the Fir-leaved Heath (*Erica cinerea*), and the Cross-leaved Heath (*Erica Tetralix*), the latter, however, less fragrant, and preferring moist situations; these flourish in various situations, from 1000 to 3000 feet above the level of the sea, but never in calcareous soil; which circumstance, Mr. Winch observes, occasions the striking difference between the heaths of Durham and Northumberland, and the Yorkshire Wolds as they are called, where the substratum is chalk.

1476. *The most considerable elevation* to which the cultivation of wheat extends in the north of England does not exceed 1000 feet above the level of the sea. Oats grow at nearly double that height; but in unfavourable years the sheaves may often be seen standing among the snow, which not uncommonly covers the tops of the mountains in October, and is never later in falling than the middle of November. The limits of Barley and Rye are between those of wheat and oats; but Bigg, a more hardy kind of grain than either of the former, is no longer cultivated. Turnips, though of small size, and Potatoes, grow at the same height as Oats. On the soil of the moors being ploughed for the first time, and lime applied, White Clover (*Trifolium repens*) comes up in abundance; and the same circumstance is noticed by Mr. Pursh to occur in North America when the woods are cleared away, and the ground first broken up.

1477. *Observations, similar to those we have now exhibited*, in a suitable part of Scotland were still a desideratum; and we have been fortunate enough, through the medium of Dr. Alexander Murray of Aberdeen, to obtain them from the Rev. James Farquharson, minister of Alford, Aberdeenshire; a gentleman, as his remarks will show, of much scientific knowledge, and great and patient research. It is to be hoped he may be induced to continue his investigations on these interesting subjects: and this may encourage others possessed of similar acquirements, and equal facilities, in different parts of Scotland, to collect a series of observations, which would be of the greatest service in forming a complete system of the *Vegetable Geography* of these islands. With this communication of Mr. Farquharson, which we shall give in his own words, we shall close our remarks, already perhaps too much extended for the nature of this work, on *British Botany*.

1478. "*The district*," Mr. Farquharson writes, "to which the following observations chiefly refer, is near the middle of Aberdeenshire, about lat. 57° 15′; and may be termed a table land, elevated 400 to 600 feet above the sea, studded with many irregular ridges and groups of mountains, of different elevations, up to 1800 feet from the level of the sea. The substratum is every where what is named primitive rock, granite, gneiss, and micaceous slate; the soil generally a friable, dark, vegetable mould, passing in some places into a lighter-coloured and more tenacious clay, and all containing many fragments of the primitive rock. A debris of clay and broken stones sometimes intervenes between the rock and the soil; but the many operations of road-making, draining, enclosing, and quarrying, which have of late years been performed within the district, show that this debris is not very extensive nor very deep, excepting at the edges of some of the more precipitous elevations. A valley in this district, about eight miles long and seven broad, nearly rectangular, and surrounded by ridges of the mountains, has furnished most of the phenomena here stated. The mean temperature of the air has not been determined; but two fine perennial springs, at an elevation of 500 feet above the level of the sea, on the 20th of November, 1828, gave 47° of Fahrenheit.

1479. *List of plants*, stating in feet the elevation above the sea which they reach, in the district of Alford, Aberdeenshire; and beyond which they do not grow, or cannot be cultivated successfully. Latitude 57° 15′; 25 miles inland: —

"1. 400 feet. Wheat. This grain has been repeatedly cultivated in the valley, from this height up to 600 or 650 feet, within the last thirty years, by persons from the south country, well acquainted with its mode of culture, and on a large scale; but the result, owing to its frequently failing to ripen well, has not justified the extension, or even the continuance, of its growth here; and it is now rarely attempted. When it did ripen, in favourable seasons, the crop was abundant. It succeeds tolerably well nearer to the coast, about 150 feet lower.

"2. 400 feet. Barley (*Hordeum distichum*). This has often been attempted, in place of the Bigg (*Hordeum hexastichum*).* It proved late, and frequently failed.

"An early variety, named Corsican Barley, succeeded admirably in situations as elevated as the Bigg, over which it had the great advantage of not being liable to lodge; but its culture has been discontinued, owing to the impolitic regulations of the Excise, which prohibit the malting of Bigg and Barley on the same premises; and as the maltsters generally use Bigg, our common grain, they could not, owing to this regulation, purchase Barley.

"3. 950 feet. Bigg succeeds quite well up to this height, and often ripens in the more inland districts about 100 feet higher, where oats frequently fail.

"4. 950 feet. Oats fail so frequently to perfect their grain higher than this, as to render their cultivation inexpedient at a greater elevation; and as they form the most important crop of our known rotations of cropping, no other kind is attempted at more elevated districts.

115

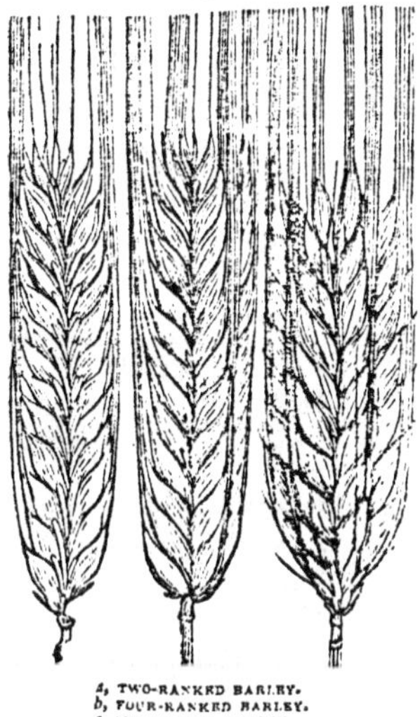

a, Two-ranked barley.
b, Four-ranked barley.
c, Six-ranked barley.

"5. 950 feet. Avena strigosa follows the cultivation of both Bigg and Oats to this height; and as its increase is greater in proportion than theirs, it might probably succeed a good deal higher.

"N.B. The introduction of comparatively early varieties, as the Early Angus and Hardy Red Oats, which has already greatly benefited the farmer in

* I am not even sure that our Bigg is the Hordeum hexastichum. We have another kind of barley, called Packman-rich, which is as early as the Bigg, and more evidently distinctly six-rowed in the ear; the grains being also more closely packed in the rachis. It is seldom sown, being not quite so productive as the common Bigg. I suspect it may be the true Hordeum hexastichum. The former is doubtless the Hordeum vulgare of Linnæus, or four-ranked barley (*fig.* 115. *b*), the latter the Hordeum hexastichum *Linn.*, or six-ranked barley (*c*). In both there are six rows or ranks of seeds; but in the Hordeum vulgare (*Hordeum hexastichum* of Gray's *British Flora*), two opposite pairs of rows have the seeds more erect, imbricated, and compressed, appearing to constitute but two single rows; the other two or lateral rows of seeds are much more prominent. In Hordeum hexastichum the seeds arrange themselves in six distinct and prominent rows. These are the two most hardy kinds, and such as are generally cultivated in the northern and mountainous regions. In the south of Scotland, and in England generally, the Hordeum distichum of Linnæus, common or two-ranked barley (*a*), is cultivated. In this kind the two opposite double rows are abortive, and nothing but small empty bushels appear instead of seeds. The name of Bigg, or Bere, or Bear, is usually given to the four-sided barley, and that of six-sided Bigg or Bere, and Barley Bigg, to six-sided; but in common language they have often no distinguishing appellation. — *Ed.*

this valley, will, perhaps, enable him to push his cultivation a little higher; where, at many points, there is yet good soil untouched.

" 6. 950 feet. White Norfolk Pea. I introduced these in the lower part of the valley, about seven years ago, where they have always done well; and two years since I gave some for sowing in some of the highest cultivated lands of the district, where they have been equally successful, ripening before the Bigg. Grey Peas so often fail to ripen here, that they are seldom sown.

" 7. 600 feet. English Winter Tares (*Vicia sativa*). I began to plant these eight years ago, as a spring crop, and they have answered admirably in damp land. I have always been enabled from a portion of the crop to secure seed for a succeeding one. When sown in May they are heaviest to be cut from September to November.

" 8. 950 feet. Potatoes succeed with the oats here; but often fail to come to perfection in the Highland glens, even a very little more elevated than this.

" 9. 950 feet. Turnips become a valuable crop. The Red and White varieties are only for autumn use, as they often perish in winter; but we have a hardy Yellow variety, which I have never seen materially injured by cold, except in the unusually severe frosts of 1813 and 1814. Rutabaga is not planted; and as its produce is of much less weight, and as we have this valuable yellow sort, we do not require any which is more hardy.

" 10. 950 feet. Rye Grass (*Lolium perenne*).

" 11. 950 feet. Red Clover (*Trifolium pratense*).

" 12. 950 feet. White Clover (*Trifolium repens*).

" These three plants form our usual sown-grass fields for hay and pasture, and they answer well at this height. The last is a native, even higher. Thus we see that, to the most elevated points where oats, the principal crop, ripen, the cultivator gets up along with them all the plants necessary for a judicious rotation of crops: Turnips, Bigg, Rye-grass, and the Red and White Clovers.

" 13. 450 feet. Sainfoin (*Hedysarum Onobrychis*). I made an experiment on a small scale, three years ago, to introduce this on a piece of very dry, sandy, but rich and well-limed land, at this altitude: the plants came up freely, and many of them were remaining on the second year, but in the third year they had almost entirely disappeared, none having ever flowered.

" 14. 400 feet. Field Bean (*Vicia Faba*) does not ripen: a broad Spanish bean comes to perfection in gardens often.

" 15. 450 feet. Jerusalem Artichoke (*Helianthus tuberosus*) never flowers, but is propagated by tubers.

" 16. 450 feet. The Squash, recently introduced, named Vegetable Marrow (*Cucurbita Succado*, in Loudon's *Encyclopædia of Gardening*), perfects its seeds when sown under a hand-glass, till the second leaf expands, and then it may bear transplantation into the open ground.

" 17. 670 feet. Nasturtium (*Tropæolum majus*) has sometimes perfect seeds.

" 18. 950 feet. Flax (*Linum usitatissimum*) grows as high up as the Bigg.

" 19. 650 feet. Carraway-seed (*Carum Carui*) propagates itself by seed on a piece of enclosed grass.

" N.B. Many plants, as Papaverum somniferum, Verbascum Thapsus, Dipsacus fullonum, Cheiranthus Cheiri, Dianthus barbatus, not being natives of this part of the kingdom, yet propagate themselves by seed, without any attention being paid to sowing them, on the cultivated ground of a garden at 450 feet from the level of the sea: and at New, in Strathdow, about 800 feet high, a great variety of kitchen-garden herbs come to perfection in the open air, such as Strawberries, Beet, Carrot, Onions, Chives, Leeks, Parsley, Cresses, Lettuces, Radish, &c. The Cabbages and Greens are good at all the highest points to which cultivation is extended.

Trees and Shrubs.

" 20. 700 feet. Oak (*Quercus Robur*).* Half-rotten trunks of this tree, of only moderate dimensions, occasionally dug out of the ground (of which I have seen several), prove this tree to have been, recently, a native of this district; and a tradition, which may be relied on, points out the spots which furnished the timber of the old church at Forbes, situated within the valley, the only building constructed of oak in the neighbourhood. The native Oaks have all disappeared; unless a solitary small tree, remote from any others of equal age, and growing amidst an unprotected Hazel copse, within a mile of the above-mentioned spot, and on similar ground, may be considered as such. There are a great many planted Oaks in the valley, some of them at the above elevation, but their autumnal shoots (this tree sending out two shoots in a season) are generally killed in the succeeding winter.

" 21. 750 feet. Beech (*Fagus sylvatica*) becomes a very fine tree in the lower part of the valley, but a few reaching this elevation are miserably stunted.

" 22. 500 feet. Hazel (*Corylus Avellana*), is wild here, and ripens its nuts well: this cannot be considered as the farthest limit of its growth; but the progress of cultivation has probably exterminated it in higher spots.

" 23. 950 feet. Alder (*Alnus glutinosa*), in a natural state, skirts every where the streams in the low grounds: a few very small dwarf bushes of it are found in one place at this height.

" 24. 1050 feet. Scotch Fir (*Pinus sylvestris*) ceases to be of any value at a greater height, though it struggles a little farther still, in stunted mis-shapen bushes.

" 25. 1100 feet. White Birch (*Betula alba*) advances but a little farther than the Scotch Fir, as appears at one spot, where they were planted and have grown together.

---

* Probably Quercus sessiliflora *Smith.*

"26. 1200 feet. Larch (*Pinus Larix*). This native of the Alps is now extensively planted, in all parts of the valley, with the greatest success; and at this height some extensive shrubberies, of eight or ten years' standing, mingled with Scotch Firs, appear in a prosperous state. One plantation has been carried as high as 1450 feet; but the Larches for 250 feet downwards are dwarf and sickly; and it appears likely that but few will exist long, beyond 1200 feet of altitude. The Larch was introduced here about sixty years ago; but being then deemed a delicate tree, none were tried on very high ground. The growth of the Larch is much more vigorous than that of the Scotch Fir; and being from the primitive regions of the Alps, the soil of Aberdeenshire proves very congenial to it, being chiefly primitive rock. The timber which it affords is strong and very durable. Sixteen years ago I had two gates made of Larch only forty years old; and although these have been constantly exposed ever since, the timber shows very little appearance of decay.

"27. 420 feet. Silver Fir (*Pinus Picea*). One plant of this native of the Pyrenees, the only one of any considerable age in the valley, has attained a large size, being 75 feet high, and containing 230 cubical feet of very healthy and growing timber.

"28. 800 feet. Hawthorn (*Mespilus Oxyacantha*) has, at New, at Strathdow, formed fine hedges at this elevation. In the valley, there are two fine solitary trees of it, at an altitude of 650 feet; a few good hedges are also seen in the lower grounds.

"29. 900 feet. Ash (*Fraxinus excelsior*) grows tolerably well at this height.

"30. 900 feet. Sycamore (*Acer Pseudo-platanus*).

"31. 500 feet. Elm (*Ulmus montana*) becomes here a fine tree: it has not been tried higher.

"N.B. Tilia europæa, Populus alba and tremula, Salix alba, Æsculus Hippocastanum, Sambucus nigra, Carpinus Betulus, Cytisus Laburnum, Pyrus Aucuparia, all thrive well in the lower districts of the valley, but have not here been planted nearly so high as the extreme limits of their growth. The Populus balsamifera, at 450 feet, appears sickly, and would not probably succeed any farther up.

"32. 800 feet. Cherries produced very fine and abundant crops, at New, in Strathdow.

"33. 800 feet. Apples are here a very uncertain crop; but at Castle Forbes, 500 feet high, they do not often fail. Auchintoul, the Ribston Pippin generally ripens well.

"34. 800 feet. Pears, the Jargonelle ripens here; but at 450 feet, the Auchen Pear seldoms comes to perfection.

"35. 900 feet. Gooseberries. This valley is the region of most delicious gooseberries; and when our best kinds, especially the Red Walnut, are in season, we need not envy the fruits of the most favoured regions. I have heard, from good authority, that Bishop Forbes of Aberdeen, who died in 1635, had remarked, that gooseberries became more delicious as they approached the upper limit of their cultivation, and that the Scots, in his day, possessed the finest varieties of this fruit; and that at his own castle of Carse, about 750 feet high, he had them of a higher flavour than he knew them to have any where else. Gooseberry bushes cease to become productive near the limits of our highest cultivation.

"36. 800 feet Raspberry (*Rubus Idæus*) is indigenous in the woods, at 650 feet.

"37. 1000 feet. Broom (*Spartium scoparium*) is found here in a stunted state, but bearing flowers.

"38. 850 feet. Whin or Furze (*Ulex europæus*). Several patches, at this elevation, seem to have no tendency to grow any higher up, although they spread laterally.

"39. 700 feet. Honeysuckle (*Lonicera Periclymenum*) will flower, if nailed to a wall.

"40. 600 feet. Evergreen Thorn (*Mespylus Pyracantha*), when attached to a wall, flowers in June, and ripens its dense branches of fine scarlet berries in October, which continue till it is again in blossom. It is one of our most ornamental shrubs.

"41. 600 feet. The Lilac (*Syringa vulgaris*) flowers here, but the inflorescence is always abortive.

"42. 450 feet. Evergreen oak (*Quercus Ilex*). Some years ago I obtained a few plants of this: they died down to the ground in winter; sending up new shoots from the root in spring, for two or three years, and then perished.

"43. 450 feet. Common osier (*Salix viminalis*) was given to me several years ago; but it does not thrive, dying very frequently in winter, especially the preceding year's shoots, and often indeed the whole plant.

"44. 450 feet. Berberry (*Berberis vulgaris*) ripens its fruit.

"45. 450 feet. Daphne Mezereum ripens its berries.

"46. 450 feet. The Hop (*Humulus Lupulus*) rarely produces any catkins.

"N.B. Ligustrum vulgare, Cornus sanguinea, and Lonicera Xylosteum blossom at 450 feet; but I have not seen them produce berries, except in the favourable season of 1827, succeeding the still warmer one of 1826. The Ligustrum and Cornus bore many berries on that year, and even the Lonicera had a few, all of which appeared to be fully ripe. Artemisia Abrotanum never flowers."

To some further queries which we addressed to Mr. Farquharson, respecting the influence of soil and situation, he obligingly communicated to us the following replies: —

"It is quite evident, from many indications, that the richness or poverty of the soil possesses considerable influence in extending or limiting the upper boundaries of the growth of all vegetables. In regard to the Herbaceous Plants and Gramineæ, the power of these circumstances becomes quite apparent in any field where a portion has been put into better order than the rest, by recent manuring, &c. In the manured part, the crops are much more luxuriant, and always somewhat earlier. This advantage would be again counteracted by a slightly increased elevation. Another condition of soil, its wetness or dryness, has much more effect in checking or extending the boundaries of these vegetables. It is only on dry lands that the different grains ripen at their respective highest elevations; and the retarding influence of a wet soil is so great as to make the crops in some of the lowest districts of the valley perhaps equally late with those in the highest. In regard to trees, it is apparent at many points in this country, that poverty of soil has prevented them from attaining their usual limit of altitude. The effect of a wet or dry soil varies upon trees, according to the habits of the different kinds, and cannot be defined.

"As to exposure, I have never been able to ascertain that this modifies the attainable elevation of any of the herbaceous or annual tribes.* Almost every variety of situation,

* "As the position, that exposure does not modify the attainable elevation of the herbaceous and annual tribes, may appear rather startling; it seems necessary, in support of it, to refer to several localities, where the most satisfactory proofs of it occur. And first, in respect of aspect. In the lower part of the valley, the three points where the earliest crops uniformly occur almost every season are, Tulloch, in the parish of Keig; Montgarry, in the parish of Tullynessle; and some north fields in the Gallowhill, in the parish of Alford. The first of these has a direct west aspect; the second, a direct south one; and the last has an aspect within one point of the north, on a steep descent. The earliness is owing to the dryness of the soil. Next in order to these, in point of earliness, are some fields on Upper Auchintoul, having an aspect a little to the south of east. In the higher cultivated lands, the south-lying fields of Upper Balfour, in the parish of Forbes; the east-lying fields of Tiverchindy; and the west-lying fields, on the opposite side of the dingle at the same place, in the parish of Alford; and the fields near the top of the hill at Campfield, parish of Lum-

both of aspect and shelter, occurs within this valley at all elevations; and it does not appear that in sheltered grounds the crops are earlier than on those which are open, or the contrary; or that the produce of fields having a northerly aspect is earlier or later than that of ground lying to the south. The circumstance of elevation, with respect to these plants, quite counterbalances the conditions both of aspect and shelter. It is true, indeed, that in the spring the vegetation is often much earlier in fields lying open to the south, than those which face the north; and pastures which have the former exposure are valuable on account of their early herbage. But for the final maturing of the ordinary crops, the disadvantages attendant on a north-lying field is compensated when the sun reaches the northern tropic; and these situations then enjoy all the benefit of his morning and especially his evening rays, which is generally considerable. In exact conformity with this view, it does not appear that farther inland, where the glens are narrower, and more protected by high mountains, the grain ripens at higher elevations than here. Certainly cultivated spots are found somewhat higher than any in Alford: but then the culture is not conducted solely with a view to the grain, but chiefly for the winter fodder necessary to support numerous herds of small cattle, which feed on the extensive mountain pastures in summer; and it is a common saying among persons possessed of these lofty spots, that if their grain ripens once in three years, it is all that they expect.

"On Trees and Shrubs, however, which are of loftier growth than the annual and herbaceous plants, and have to endure the severity of winter storms, the exposure has a most powerful influence. The numerous and extensive plantations in this valley furnish the most decisive evidence on this point. Many stations occur where the Scotch Fir, for instance, is dwarfish, and even killed, much below its natural limit; owing evidently to an unfavourable exposure, and not to any deficiency or unfitness of the soil; for in some of these places, corn crops of good growth are mingled with the plantations, or pass along the sides of them, and the trees have succeeded quite well in spots that are incidentally sheltered. It is, indeed, from its circumstances of shelter, and not of aspect, that exposure exercises such a powerful influence; for if the trees, planted in ground having a westerly or north-westerly aspect generally suffer most, and are most limited in their elevations, it is because our severest storms come from the west and north-west, and the trees so situated are exposed to their full violence. It may be concluded, therefore, that if we could command more shelter, trees, at least our native ones, would succeed at higher points than they here reach; and accordingly, I have been given to understand that at Braemar, in the upper part of this country, which is sheltered by the loftiest and most extensive masses of mountains in all Scotland, attaining a height of 3000 and 4000 feet and upwards, the Scotch Fir and White Birch grow at a much higher elevation than any I have named. Of this, however, I cannot give the exact particulars, but I believe the general fact.

"I can say nothing, from personal knowledge, of the different circumstances attending the extension of plants on the east and west coasts of Scotland; and I would not, therefore, have referred to the subject, except to bring forward a remark made in a work, not professedly scientific, and therefore, perhaps, liable to be overlooked: it is in Dr. Macculloch's interesting *Letters from the Highlands and Western Islands of Scotland*. He states the fact, that grain ripens better in the outer Hebrides than in the islands nearer the mainland, or on the west coast of the mainland itself; and the cause which he assigns for this striking circumstance is, that in these outer islands there is not sufficient elevation or continuity of land to lift up the atmosphere, and thus arrest and condense the vapours with which our prevailing westerly winds come loaded from the Atlantic. These islands, therefore, often, in summer, enjoy clear weather, when the inner islands and mountains are deluged with rain, or obscured by a heavy cloud."

UEBER DEN EINFLUSS DES CLIMAS AUF DIE BEGRÄNZUNG DER NATÜRLICHEN FLOREN

Heinrich Rudolf August Griesbach

Ueber

den Einfluss des Climas auf die Begränzung der natürlichen Floren.

Vom

Dr. *A. Grisebach*,

Privatdocenten in Göttingen.

Es kann als eine allgemein anerkannte Thatsache angesehen werden, dass die Erdoberfläche in eine Anzahl von natürlichen Floren zerfalle, die die Natur sowohl nach der Polhöhe, wie nach Meridianen begränzt. Dass an ihren Gränzen verhältnissmässig nur sehr schmale Mittelgebiete liegen, keineswegs aber, wie Herr Philippi neulich behauptet hat *), allmählige stetige Uebergänge, z. B. zwischen der süd- und mittel-europäischen Flora, beobachtet werden, lehrt die Erfahrung jedes Reisenden, der in den Alpen auf den plötzlichen Vegetationswechsel an der untern Rhododendron-Gränze achtet, oder der in der Gegend von Montélimart eine solche Naturgränze in der Ebene aufsucht, wo ihm nicht bloss neue Culturpflanzen begegnen, von denen allein Herr P. die Physiognomie Süd-Europa's ableiten will sondern wo neben an-

*) In Wiegmann's Archiv für Naturgeschichte. 1836.

dern fremdartigen Eindrücken auch die Wiesen aufhören und keine Wälder mehr aus einer einzigen, sonstige Vegetation ausschliessenden Art gebildet werden. Die von den Familienquotienten aber entlehnten Gründe für die entgegenstehende Ansicht verschwinden bei einer verbesserten Berechnungsmethode, und es lässt sich mit Schärfe nachweisen, dass wenigstens zwei grosse Familien im südlichen Europa ein wesentlich verschiedenes Verhältniss zur ganzen Vegetation haben: die Leguminosen und die Cyperoideen.

Zur Characteristik der natürlichen Floren kann von botanischer Seite eine zwiefache Methode angewendet werden, deren jede für die genügend bekannten Floren zu gleichen Resultaten, d. h. zu identischen Gränzbestimmungen derselben führt; es ist das wichtigste Problem der Pflanzengeographie, auch für jede natürliche Flora ausschliessliche climatologische Charactere aufzufinden.

Die erste Methode, deren Anwendung schon eine sehr oberflächliche Kenntniss einer Gegend gestattet, geht von der Physiognomie ihrer Vegetation, von der Gruppirung ihrer Individuen im Grossen aus, sei es, dass sie durch grosse Verbreitung hervortreten, oder durch ihre Gestaltung auffallen. Ich möchte eine Gruppe von Pflanzen, die einen abgeschlossenen physiognomischen Character trägt, wie eine Wiese, ein Wald u. s. w., eine *pflanzengeographische Formation* nennen. Sie wird bald durch eine einzige gesellige Art, bald durch einen Complex von vorherrschenden Arten derselben Familie characterisirt, bald zeigt sie ein Aggregat von Arten, die, mannigfaltig in ihrer Organisation, doch eine gemeinsame Eigenthümlichkeit haben, wie die Alpentriften fast nur aus perennirenden Kräutern bestehen. Bei einer übersichtlichen Darstellung der Formationen einer Flora würde es darauf ankommen, die *Characterpflanzen* derselben nachzuweisen, die Arten zu bestimmen, denen sie ihre physiognomischen Eigenthümlichkeiten verdanken, die keineswegs subjectiv sind: eine Aufgabe, die Reisenden um so mehr empfohlen werden muss, als sie leicht und gründlich auszuführen ist. Diese Formationen nun wiederholen sich überall nach localen Einflüssen, aber sie finden mit der natürlichen Flora, die sie constituiren, ihre absolute, ihre climatische Gränze. So weit Wälder von Pinus sylvestris, oder mit Calluna vulgaris bedeckte Ebenen reichen, findet man sich im Gebiete der mittel-europäischen Flora. Mag die einzelne Art aus einer Flora in die andere übergreifen, die in ihrer Gruppirung characterisirende Art kommt nicht zugleich in zwei Floren vor: eine jede Formation, deren Character und deren Glieder mit Schärfe dargestellt sind, eignet sich daher zur Gränzbestimmung ihrer natürlichen Flora. Entgegenstehende Erfahrungen sind mir noch nicht bekannt geworden: es braucht indessen kaum erinnert zu werden, dass in dieser Wissenschaft jede Thatsache nur mit grösster Vorsicht verallgemeinert werden darf, und jeder ausgesprochene Grundsatz stillschweigend berichtigende Thatsachen voraussieht. In diesem Sinne nur mag dasjenige mitgetheilt werden, worauf die vergleichende Untersuchung leitete.

Hr. Brown*) bemerkte, dass die Flora von Congo 9 Familien enthalte, die über die Hälfte der von Smith daselbst gesammelten Arten einschlössen. Dies ist eine Thatsache, die für alle Floren gilt, und es liegt in dieser Hinsicht ein bestimmter Begriff zum Grunde, wenn man von den 8 — 15 *vorherrschenden Familien* einer Flora spricht. Da man indessen hier eine willkührliche Gränze feststellen muss, so schlage ich vor, dazu die Familien zu rechnen, welche über 4 Procente der ganzen phanerogamischen Vegetation enthal-

*) In Tuckey Narrative p. 425.

ten: sie werden dann in den meisten Fällen zusammen $^2/_3$ der Gesammtvegetation bilden. Nur von diesen vorherrschenden Familien gilt nach meinen Untersuchungen das Humboldt'sche Gesetz, dass die Summe der Arten einer jeden derselben dividirt in die Summe aller Phanerogamen gleiche Quotienten an jedem Orte innerhalb derselben natürlichen Flora giebt. Man kann vielleicht mit Grund behaupten, dass, da jene vorherrschenden Familien grösstentheils die natürlichsten sind, diese Verschiedenheit zwischen kleineren und grösseren Familien auf der Ungleichförmigkeit des Begriffs beruhe, den die systematische Pflanzenkunde bei der Aufstellung jener natürlichen Gruppen befolgt hat, dass, wenn alle Pflanzen der Erde nach einem einfachen natürlichen Princip in gleichförmige Familien getheilt werden könnten, *alle* Pflanzenarten einer Flora ein bestimmtes statistisches Verhältniss zeigen würden. Da wir aber das natürliche System als eine unvollendete Bemühung ansehen müssen, die Typen der vegetabilischen Organisation aufzustellen und sie nach dem Plane, der ihnen zu Grunde liegt, zu ordnen, so wird es nicht auffallend erscheinen, dass jene Verhältnisse nur bis zu einem Grade gelten, der unserer Erkenntniss entsprechen mag. Wenn man nach dem Reichthume an Arten für jede einzelne Flora eine Reihe ihrer vorherrschenden Familien bildet, so findet man, dass zwischen je zwei natürlichen Floren sowohl die Glieder der Reihe, als die Folge derselben verschieden sind, dass aber in verschiedenen Bezirken derselben Flora sich diese Reihefolge nie ändere. Man hatte Anfangs eine solche Congruenz der Familien-Quotienten für ganze Zonen angenommen, später, als diese Annahme sich unhaltbar zeigte, dieselbe in derselben Flora für alle Familien nachweisen zu können geglaubt: die Abweichungen in den kleinern Familien, die sich aus vervielfältigten Berechnungen ergaben, veranlassten Beilschmied zu dem irrigen Schlusse, dass man nur gleich grosse Gebiete vergleichen dürfe, während es sich leicht darthun lässt, dass z. B. jene Reihefolge für ganz Deutschland diesseits der Alpen dieselbe ist, wie für eine deutsche Ortsflora, oder bei Moskau dieselbe, wie von ganz England. So gültig die Beweise waren, die Humboldt für einige der grossen Familien geltend gemacht hatte, so glaubte man doch jene Abweichungen damit nicht vereinigen zu können, und meinte nur noch von allmäliger Zunahme und Abnahme der Familien in gewissen Richtungen der Erdoberfläche sprechen zu dürfen, was jede Begränzung der natürlichen Floren von dieser Seite aufhebt. Eine Nachweisung über specielle Untersuchungen, die zu den Humboldtschen Sätzen zurückgeführt haben, ist von mir an einem andern Orte versucht worden. Herr Meyen ist der Ansicht, dass die Eintheilung der Erde nach pflanzenphysiognomischen Momenten eine ganz andere sei, als nach der Verbreitung der Pflanzenfamilien; ausserdem weist er für jetzt die Berechnungen über die letzteren zurück, weil die Erde noch nicht gleichförmig untersucht und hinreichend bekannt sei, so dass man nur zu zufälligen Resultaten gelange, die in der Natur keine Geltung haben. Es giebt indessen eine leichte und einfache Methode der Beweisführung: man braucht nur die Quotienten der vorherrschenden Familien eines Landes zu berechnen, von dem wir eine Reihe von Floren besitzen, die zu verschiedenen Zeiten erschienen sind und in der Summe der Arten je nach dem Standpuncte der Kenntniss bedeutend unter sich abweichen. Eine solche Reihefolge gewähren die Floren der vereinigten Staaten von Walter, Michaux, Pursh, Nuttall, Beck: man findet für die grossen Familien unter einigen Cautelen gleiche Resultate. Ein ähnlicher, anderswo mitgetheilter Beweis liegt in zwei der Humboldtschen Verzeichnisse der Andenflora aus ziemlich entfernt liegenden Orten, die eine analoge Familienreihe ergeben, so gross auch übrigens der Gegensatz in den vorherrschenden Familien tropischer Floren ist. Es verhält sich hier, wie mit den Berechnungen der mittlern Wärme: je näher dem Aequator, desto weniger Beobachtungen sind zur Bestimmung derselben erforderlich; je grösser die Familien einer Flora, desto früher tritt unserer Kenntniss ihre Bedeutsamkeit entgegen.

Der andere Einwurf des Herrn Meyen scheint eine gewisse Gleichförmigkeit der Pflanzen*formen* in den ganzen Zonen im Sinne zu haben, womit indessen die Physiognomie der Vegetation noch nicht erschöpft ist. Die Tropen haben überall ihre Palmen, ihre Bananen, ihre Pandanus-Arten, ihre fiederblättrigen Dicotyledonenbäume; in den gemässigten Erdstrichen der nördlichen Hemisphäre begegnen wir unter allen Meridianen Wäldern von Coniferen, Wiesen von dichtem Graswuchse. Gleichwohl finden wir in Nord-America, dessen Eichenwälder und Asterngebüsche im Süden unmittelbar von tropischen Formen, von Yuccà-Bäumen, begränzt werden, nichts, was man mit der südeuropäischen Flora vergleichen könnte. Viel entschiedener aber ist dieser Gegensatz in der südlichen Hemisphäre: die Eucalyptus- und Acacien-Wälder nur in Neuholland, die Ericoiden, die Liliaceen und Irideen dem südlichen Africa eigenthümlich, nichts von dem im extratropischen Südamerica, dessen Formen mehr an europäische erinnern. Da indessen die Gleichartigkeit der Physiognomie eines Landes keineswegs bloss auf der Gestaltung der vorherrschenden Formen, sondern auch wesentlich auf ihrer Gruppirung beruht, so ergaben sich hieraus nicht minder bedeutende Differenzen unter den Floren derselben Zone. So kommt Pinus Cembra, die ausgedehnte Waldungen am Altai bildet, auf den Carpaten und Alpen nur in einzelnen Individuen vor. Es ist endlich eine willkührliche Bestimmung, dass man in der Physiognomik der Natur nur so allgemein von Pflanzenformen reden will, dass man etwa Laubholzwälder, Nadelholz und Palmen unterscheidet; jede Art, die in grossen Massenverhältnissen auftritt, hat vielmehr ihren eigenthümlichen Character, der dem Sinne des Malers nicht entgeht und zur Characteristik der Floren benutzt werden kann; es ist schon oben bemerkt, dass die vorherrschenden Glieder der Formationen sich in zwei Floren nicht wiederholen, und dass man also von diesem Gesichtspuncte allerdings zu denselben Gränzbestimmungen derselben gelangt, wie durch statistische Rechnungen.

Ausser diesen beiden Eigenthümlichkeiten, die den Character jeder Flora bezeichnen, treten bei der Vergleichung derselben in der Natur, die hier allein zu Resultaten führen kann, noch einige Verschiedenheiten unter denselben auf, die aber nicht zu Bestimmungen ihrer Gebiete benutzt werden können. Die erste hierauf bezügliche Bemerkung betrifft die absolute Zahl der Pflanzenarten einer Flora. Vergleicht man z. B. die Anzahl der Pflanzen, die Perrottet und Leprieur auf vieljährigen, mit besonderm Sammlertalent ausgeführten Reisen in der ganzen Ausdehnung der Küste von Senegambien bis tief in das Innere des tropischen Africa's gesammelt haben, und die kürzlich in einer sorgfältig bearbeiteten Flora von Guillemin und Richard herausgegeben sind, so wundert man sich, dass sie nicht höher ist (nach einer Schätzung etwa 1200 Sp.), als die Artenanzahl eines kleinen Bezirks der deutschen Flora, während unter gleicher Breite, bei gleichen climatischen Verhältnissen, Reisende im tropischen America auch fern von den Anden über die dreifache Zahl in viel kürzerer Zeit gesammelt haben, während am Cap der guten Hoffnung in einem weit trocknern Clima, auf einem kleinern und gleichartigeren Terrain einzelne Sammler über 6000 Arten zusammenzubringen im Stande gewesen sind. Sei es, dass diese Verschiedenheiten geologische Ursachen haben, sei es, dass sie in der Natur des Zusammenlebens der Arten jeder Flora begründet sind (— beide Hypothesen sind für die Wissenschaft werthlos, da sie unbekannte Grössen berühren —): genug sie existiren zwischen je zwei Floren und sie entziehen sich climatischen Bestimmungen. Eine andere Differenz der Floren, die man gleichfalls nicht aus climatischen Ursachen ableiten kann, liegt in der Differenz der Formen selbst. Warum fehlen uns die Proteaceen, die Eucalyptus, die Restiaceen, die unter ähnlichen climatischen Verhältnissen auf der südlichen Hemisphäre wachsen? warum kommt im tropischen America ursprünglich fast keine einzige Art vor, die in den Tropen des alten Continents ihre Heimath hat? Eine weitere Ausführung würde in's Unbegränzte gehen, aber im Allgemeinen wird hier diese Erscheinung erwähnt, um anzudeuten, wie wenig naturgemäss es zu sein scheine, die Floren nur nach willkührlichen climatischen Bestimmungen zu begränzen, wie es der Erfahrung auf jedem Schritte widerspreche, dass gleiche climatische Bedingungen auch gleiche Pflanzenarten produciren.

Es soll uns hier die Frage beschäftigen, ob überhaupt climatische Bestimmungen der natürlichen Floren möglich sind, oder ob man sich mit ihrer Begränzung von botanischer Seite begnügen müsse. Wir wollen mit einigen Bemerkungen über den bisherigen Gang dieser Untersuchung beginnen. Die Abhängigkeit des Pflanzenlebens von Temperaturextremen *) musste eine der ältesten Erfahrungen des Menschen sein: südliche Culturpflanzen ertrugen ein rauheres Clima nicht, andere waren dagegen unempfindlicher; aber wenn aus dieser Beobachtung sich ergiebt, dass jede Pflanzenart ihre eigen-

*) Der Einfluss der Wärme auf die Pflanzen innerhalb dieser Extreme gehört nicht der Pflanzengeographie, sondern der Physiologie an.

thümliche Temperatursphäre habe, eine Thatsache, die für die Pflanzengeographie unfruchtbar ist, deren Object die climatische Bedingung einer ganzen Flora, also vieler Arten, die eine gemeinsame Abhängigkeit von physikalischen Einflüssen haben sollen, nachzuweisen fordert: so musste, wie dies der Grundgedanke vergleichender Wissenschaften ist, erst ein grösserer, aus allgemeinern Anschauungen geschöpfter Maassstab an diese Verhältnisse gelegt werden, unter denen die Verschiedenheit in den Wärmesphären einzelner Arten derselben Flora ein verschwindendes Moment wird. Dazu gehörte zuerst die Beobachtung der Wiederkehr einer ganzen Flora aus höhern Breiten im Gebirge. Wie nun dieser Erscheinung die Abnahme der Temperatur nach der Höhe und nach der Entfernung vom Aequator, die durch ihr arithmetisches Mittel *gemessen* wird, parallel geht, so folgerte man daraus folgenden Satz: Finde sich gleich, dass einige Pflanzenarten grosse Temperatur-Differenzen ertragen können und daher über einen grossen Theil des Erdbodens sich zu verbreiten im Stande sein, so gelte doch bei Weitem von der Mehrzahl der Pflanzen, und somit als pflanzengeographisches Gesetz, dass nur eine mittlere Jahreswärme unter engen Gränzen einer jeden Flora entspreche, und dass die Arten derselben überall da auftreten können wo diese mittlere Wärme vorkomme. Darüber ist hier vorläufig zu bemerken, dass man zwar die jährliche Temperaturcurve auf ein arithmetisches Mittel zurückführen könne, nicht aber den Process der vegetabilischen Entwickelung, in der jede Stufe ein nothwendiges Glied bildet, das seine besondere Bedingungen, seine besondere Temperatursphäre hat. Sodann entsprachen weitere Erfahrungen jener Hypothese nicht. Nirgends zeigt sich eine grössere Mannichfaltigkeit der vegetabilischen Formen, eine engere Begranzung der natürlichen Floren, als unter dem Aequator, wo dagegen die mittlere Jahreswärme nur zwischen

27° und 29° differirt; Nord-America und Mittel-Europa haben nur einen kleinen Theil von Pflanzen gemeinschaftlich, aber alle mittleren Temperaturen von Mittel-Europa kommen in den vereinigten Staaten vor; Moskau hat eine mittlere Temperatur = 3,°9, Warschau = 8,°2, und doch gehören beide Städte zu derselben Flora u. s. w. Ferner ist die Thatsache, dass auf den Gipfeln der Alpen Pflanzen wachsen, die in Lappland und dem arktischen Asien und America wiederum vorkommen, viel zu sehr verallgemeinert worden. Schon auf dem Aetna, auf dem Pic von Teneriffa hört diese Uebereinstimmung in den Arten auf; im tropischen America aber wiederholt sich auf den Anden keine der Arten *), die in Nord-America bei entsprechender mittlerer Jahreswärme wachsen, und wenn Eichen und Tannen in der Cordillere von Mexico wiederkehren, so sind es einmal von den nordischen verschiedne Arten, und „die Gruppirung derselben zu einem Ganzen nimmt dort den verschiedendsten Character an **)."

Da diese einer gangbaren Hypothese entgegenstehenden Thatsachen theilweise hie und da ausgesprochen wurden, der Zusammenhang des Climas und der pflanzengeographischen Phänomene aber im Allgemeinen nicht geläugnet werden konnte, so versuchte man, die Theorie dadurch zu verbessern, dass man aussprach: nur die mittlere Wärme des Sommers sei es, von der man bei solchen Untersuchungen ausgehen müsse. Man erkennt den Fortschritt der Erkenntniss, der die gleichgültige Temperatur während des Winterschlafs der Pflanzen von der wirksamen während ihrer Vegetation absondert, aber man sieht auch, wie willkührlich es sei, in dieser Rück-

*) Befaria paniculata Mich., die man zu diesem Zwecke anführt, ist von den Arten in den Anden specifisch verschieden.

**) A. v. Humboldt Ansichten der Natur p. 175.

sicht 6 Monate für alle Zonen der Erde festzustellen, während auf den Alpen das thätige Pflanzenleben nur halb so lange dauert, in Italien fast 9 Monate, dass man ferner diese Bestimmung auf tropische Länder gar nicht anwenden könne, wo das Pflanzenleben in feuchten Gegenden gar nicht unterbrochen ist, in trocknen dagegen nur während der Regenzeit flüchtig erscheint und bald wieder verschwindet. Es erhellt aus diesen Gesichtspuncten hinlänglich, dass überhaupt diese Erscheinungen viel complicirter sind, wie es den Anschein hatte, und dass man daher, um zu sichern Principien zu gelangen, zuerst alle climatischen Momente einzeln untersuchen, ihre Beziehung zu den Pflanzen und zu den Floren würdigen müsse, und dann erst erfahren könne, wie gross der Antheil jedes einzelnen sei, wenn man die Flora als ein Product mehrfacher Bedingungen anzusehn sich gedrungen sähe, und in welchem Gesetze die Abhängigkeit der Floren von dem Clima begründet sei, oder ob sich ein solches überall nachweisen lasse. Um diese Frage indessen bestimmter zu stellen, müssen wir zuerst die Art der Schlussfolge näher zu bezeichnen suchen, die die climatischen und pflanzengeographischen Erscheinungen in Zusammenhang zu bringen sucht: 1) Welche climatische Differenzen, fragt es sich zuerst, lassen sich innerhalb des Gebiets *einer* natürlichen Flora, so wie sie botanisch begränzt ist, nachweisen, und müssen daher von den climatischen Bedingungen *ganzer* Floren ausgeschlossen werden? sie können höchstens für die Vertheilung der Formationen in der Flora wirksam sein, ohne für sich einen Einfluss auf die Gränzbestimmung derselben äussern zu können. So findet sich, dass diejenige Differenz in der jährlichen Temperaturcurve, die das Inselclima von England dem Continentalclima von Mittel-Europa gegenüberstellt, keinen solchen pflanzengeographischen Werth habe, um eine eigenthümliche Flora zu begründen: denn sowohl die Formationen Englands sind den unsrigen gleich, als die Quotienten der vorherrschenden Familien übereinstimmen. 2) Welche climatische Momente bestimmen den Umfang einer Flora? Dies kann nur von den climatischen Grossen behauptet werden, die an den botanischen Gränzen der Flora gleichfalls eine wesentliche Modification erleiden, innerhalb derselben aber eine grossere Gleichartigkeit zeigen, als anderswo. 3) Giebt es climatische Differenzen zwischen den Hauptzonen der Erde, die es nicht gestatten, dass dieselbe Flora aus einer Zone in die andere übergreife? Da nun, wie wir sehen werden, wesentliche Differenzen dieser Art zwischen dem Clima der Tropenländer und dem der übrigen stattfinden, so ist deren pflaugengeographische Bedeutung wiederum empirisch festzustellen, ob nämlich ein solches Uebergreifen derselben Flora über die Wendekreise hinaus beobachtet werde. Gehen wir nun zu einer speciellen Betrachtung der climatischen Momente selbst über, die auf die Verbreitung der Pflanzen von Einfluss sein könnten.

Da die *mechanische Zusammensetzung der Atmosphäre* auf der ganzen Erde und auf allen der organischen Welt zugänglichen Höhen nach ihren beiden Hauptbestandtheilen dieselbe ist, so kann sie die Pflanzengeographie, die nur auf Differenzen in den Zuständen der Atmosphare ihr Augenmerk richtet, nicht interessiren: ebenso wenig der Reichthum an Kohlensäure, deren Quantitätsverhältnisse keine allgemeine Beziehungen enthalten.

Man konnte die Verminderung des *Drucks der Atmosphäre* für die climatische Ursache der Eigenthümlichkeiten alpiner Floren halten: aber die Wiederkehr von vielen dieser Pflanzen am Pol beweist das Unhaltbare einer solchen Hypothese, die noch entschiedner durch den Umstand widerlegt wird, dass man in botanischen Gärten *alle* jene Gewächse mit dem besten Erfolge cultivirt.

Man kann den Einfluss der *atmosphärischen Feuchtigkeit* auf pflanzengeographische Erscheinungen auf eine dreifache Weise betrachten, indem man 1) von der Tension des Wasserdampfs ausgeht und die mittlern Werthe desselben für verschiedene Floren vergleicht; oder indem man 2) die Geschwindigkeit voranstellt, mit der an einem Orte die Aggregatzustände des Wassers in der Atmosphäre wechseln, wozu man nur die meist unsichern Angaben über die Mengen des niedergeschlagenen und verdunsteten Wassers benutzen kann; oder indem man 3) die Vertheilung beider Werthe auf das Jahr, ihre Intensität in den verschiedenen Vegetationsperioden für die einzelnen Floren untersucht. Eine einfache Betrachtung des Pflanzenlebens lehrt indessen, dass psychrometrische Werthe gegen die atmosphärischen Niederschläge für die Vegetation verschwindende Grössen sind, und dass ferner die absolute Menge des niedergeschlagenen Wassers viel bedeutungsloser ist, als die Häufigkeit und gleichmässige Vertheilung der Niederschläge: ein Grundsatz, der bei der Wiesenbewässerung und bei künstlichen Begiessungen seine practische Anwendung findet, und der darauf beruht, dass das Wasser hier nur als Nahrungsmittel der Pflanzen zu betrachten ist, oder vielmehr dass ihr Leben von der Geschwindigkeit abhängt, mit der liquides Wasser von der Wurzel absorbirt und gasförmiges von der Epidermis ausgeschieden wird, also auch von *der Geschwindigkeit, mit der die Circulation des Wassers durch die Atmosphäre vor sich geht*, die der Wurzel das liquide Wasser zuführt. Mag auch der organische Verdunstungsprocess theilweise von der Tension des Wasserdampfs der Atmosphäre abhängen, so kann man darin doch nur ein sehr untergeordnetes Moment erkennen: andererseits aber hängen bekanntlich die Niederschläge nicht allein von der Menge des gasförmigen Wassers ab, das z. B. durch herrschende Winde in andere Länder fortgeführt werden mag und also den Organismen verloren geht, die es zum Theil producirten. Ferner lässt sich nicht nachweisen, dass die Fülle oder die Art der Vegetation in einem Verhältnisse zu der jährlichen Regenmenge stehe: vielmehr beweisen die grossen Differenzen, die hierin an nahe liegenden Orten sich gezeigt haben, z. B. in Frankreich, und die Gränzen, innerhalb deren auf der ganzen Erde die Mengen atmosphärischer Niederschläge sich bewegen, in derselben Flora fast berühren, dass wir darin keine climatischen Charactere der natürlichen Floren zu suchen haben. Einen ganz andern Gesichtskreis aber eröffnet sich die Untersuchung, sobald sie die Länder absondert, in denen die atmosphärischen Niederschläge auf bestimmte Perioden des Jahrs beschränkt sind.

Einige Tropenländer zeigen, höhern Breiten gegenüber, einen umgekehrten Gegensatz in ihren Feuchtigkeits- und Wärmeverhältnissen. Wenn in den Tropen die jährliche Temperatur-Curve sich einer geraden Linie nähert, so wird der Gegensatz zwischen Sommer und Winter desto grösser, je mehr man in das Innere gemässigter Continente eindringt; während in diesen die atmosphärischen Niederschläge sich gleichförmig über das ganze Jahr verbreiten, so erreicht der Gegensatz zwischen trocknen und feuchten Jahreszeiten in tropischen Ländern sein Maximum. Aber die Gleichförmigkeit der Temperatur im ganzen Jahre gilt für alle Tropenländer, die ungleiche Vertheilung der Niederschläge nur für einen Theil, und hierin liegt das wichtigste Moment für eine climatische Characteristik tropischer Floren.

Abgesehen von den meteorologischen Wechselwirkungen selbst (von der Tension des vorhandenen Wasserdampfs und der Wärme), hängt die Intensität der Verdunstung von der Menge ihres Materials ab: von der Grösse der Oberfläche des vorhandenen liquiden Wassers und von den Processen der Vegetation. Vorausgesetzt, dass nicht Seewinde oder Aequatorial-Strömungen *) fremde Feuchtigkeit herführen und niederschlagen, wird in Ländern, in denen die Verdunstung das Gleichgewicht der Wassercirculation stört und das vorhandene Liquidium mindert, allmälich die wesentlichste Bedingung des Pflanzenlebens, die stete Gegenwart liquiden Wassers an allen Puncten des Erdbodens, verschwinden; dasselbe Resultat werden zweitens herrschende Polarströmungen hervorbringen, die den Wasserdampf in andere Länder füh-

*) Seit den Dove'schen Untersuchungen ist man gewohnt, Aequatorial-Strömungen und Regen bringende Polar-Strömungen und trockene für identisch anzusehen, weil jene sich in ihrer Bewegung abkühlen, diese erwärmen. Man ist hier ohne Zweifel von einer horizontalen Richtung der Winde ausgegangen, für die allein jene Annahme gültig sein kann, da die Erwärmung oder Abkühlung einer nicht horizontalen Strömung wesentlich von dem Winkel abhängt, den sie mit der Erdoberfläche macht; da indessen die Dove'sche Theorie sich auf Beobachtungen stützt, so kann man daraus den Schluss ziehen, dass die meisten Winde (wenigstens an der Erdoberfläche, wo jene Beobachtungen grösstentheils gemacht wurden) eine horizontale Richtung haben. Man darf indessen den angeführten Unterschied nie aus den Augen verlieren, und würde z. B. sehr irriger Weise die Feuchtigkeit der Aequatorial-Strömungen verallgemeinern, wenn man sie auch von dem rückkehrenden Passate behaupten wollte. Einmal hat er einen grossen Theil seines Wasserdampfs durch die Niederschläge verloren, die von der Abkühlung des Courant ascendant abhängen, und zweitens weht er in einer so beträchtlichen Höhe, dass er sich in seinem Fortschreiten, bis er den Erdboden wieder in der Nähe der Wendekreise errreicht, fortwährend erwärmen muss, *obgleich* er vom Aequator kommt. Die Niederschläge jenes Berührungspunctes entstehen daher nicht von der Abkühlung eines, hier ohnehin trockenen, Aequatorialstroms, sondern von der Erkältung der hier vorhandenen feuchten Luft durch jenen, was mit Hutton's Ansicht übereinstimmt.

ren, ohne ihn niederzuschlagen. Es ist bekannt, welche climatische Verschiedenheiten in den Tropen-Ländern von diesen Verhältnissen abhängen, und welche Mittel die Natur anwendet, so entschieden dem organischen Leben feindlichen Einflüssen zu begegnen. Abgesehen von diesen Floren erschaffenden Perturbationen, wozu besonders die Solstitial-Bewegung gehört, würden die Passatwinde die Tropen-Länder in fünf scharf gesonderte Zonen theilen, von denen zwei ohne Feuchtigkeit und ohne Vegetation wären: eine Aequatorialzone mit einer Wassercirculation von grösster Geschwindigkeit, durch die ununterbrochenen Niederschläge des in der Höhe abgekühlten Courant ascendant bedingt, also von seitlichen, eben deswegen langsamer ihre Temperatur verändernden Strömungen unabhängig; zwei Passatzonen, durch ihre perennirenden Polarwinde zu ewiger Trockenheit und Sterilität bestimmt; zwei Zonen der Polargränzen der Passate, wegen steter Vermischung ungleich erwärmter und relativ ungleich gesättigter Luftschichten nie ohne Niederschläge, jedoch durch den untern Passat und durch seitliche Strömungen nach und aus den gemässigten Zonen in ihrer Wassercirculation störend afficirt. Die Einflüsse, die diese Gleichförmigkeit zu modificiren bestimmt sind, und die theils der Solstitial-Bewegung, theils der Configuration des Erdbodens und seiner einmal vorhandenen organischen Decke angehören, haben durchaus die Tendenz, einen Gegensatz zwischen trocknen Jahreszeiten und Regenzeiten hervorzurufen, in jenen ewig trocknen Zonen wenigstens eine periodische Vegetation möglich zu machen und an den Wendekreisen gleichfalls einen durch Trockenheit bedingten Winterschlaf in die Pflanzenwelt einzuführen: hier bewirkt dies die Verschiebung der Passate, die in die einzelne Flora ihre feuchte Polargränze nur während einer *Jahreszeit* setzen, dort die Wanderung des Courant ascendant in die Passatzone, der bis zum 17° der Sonne gegen die Wendekreise folgt, oder in der alten Welt die periodische Umkehrung des obern und untern Passats, die die Aequatorialzone selbst in Ostafrica zu afficiren scheint. Vergleichen wir nun die tropischen Floren mit diesen climatischen Bestimmungen, aus denen sich bei weiterer Ausführung ergiebt, dass man zwischen wahren und periodischen Wüsten und Ländern mit perennirender Wassercirculation unterscheiden müsse, so finden wir denselben Gegensatz in dem Character der Vegetation auf das Entschiedendste ausgesprochen, und sehen uns befugt, die natürlichen Floren innerhalb der Wendekreise danach zu characterisiren und einzutheilen. Wir finden zuerst einige Floren in der Nähe des Aequators, in denen die Blüthezeit der Pflanzen auf das ganze Jahr vertheilt ist, in denen ausser den Parasiten die meisten Pflanzenarten holzig werden, wo nichts Periodisches bemerkt wird, was für die ganze Flora Geltung hätte, wo sich der grösste Formenreichthum zeigt und die grösste Intensität des vegetabilischen Lebens kund giebt: diesem entsprechen grosse Reservoirs von süssem Wasser, die der Verdunstung keine Schranke setzen, tägliche Niederschläge aus dem Courant ascendant, der durch die Passate nie dauernd gestört wird, angemessene Neigung des Bodens, um das niedergeschlagene Wasser möglichst vielen Puncten zuzuführen und vor Seebildung zu sichern, hinreichend nahe Gebirge, um aus zersetzten Fossilien ununterbrochen der Pflanze ihre festen Elemente zuzuführen und die Humusdecke zu binden. Solche Floren, die auf die Nachbarschaft des Aequators beschränkt sind, treten in ihrer ganzen Fülle nur in dem Theile von Süd-America auf, der östlich von den Anden liegt und vielleicht auf diesen selbst; ferner scheint die Flora von einigen ostindischen Inseln hieher zu gehören, wenigstens die von Java, wo auch für alle Monate Blüthezeiten angegeben werden, bei mangelnden Nachrichten ein entscheidendes Criterium, sobald sich in dieser

Hinsicht eine gleichförmige Vertheilung über das Jahr herausstellt: ausser diesen Floren, also der von Nord-Brasilien von Guiana und von Java kenne ich keine auf der Erde ohne gemeinsamen Winterschlaf, womit ohne Zweifel das gleichzeitige Aufsteigen des Frühlingssaftes in den holzigen Gewächsen und die deutliche Ausbildung der Jahresringe bei den Dicotyledonen zusammenhängen. Bestimmte Beobachtungen über diese interessanten Puncte habe ich in den Schriften der Naturforscher, die tropische Länder untersucht haben, vergeblich gesucht. Aublet hat indessen in Guiana, Meyen auf den Philippinen das Bluten der Lianen in längern Perioden beobachtet.

Wo eine periodische Regenzeit in einem so scharfen Gegensatze gegen die trockene Jahreszeit steht, dass in der letztern die Wassercirculation durch die Atmosphäre aufhört, wird sich der Einfluss dieses Gegensatzes auf die Pflanzenwelt nach der verschiedenen Natur ihrer Organismen auf eine dreifache Art äussern. Man kann nämlich die Pflanzen nach ihrer Abhängigkeit von jener Circulation in drei Classen eintheilen: 1) *Pflanzen mit Knospen producirenden Stammbildungen*, Organen, die die Botaniker nach einem organographischen, nicht aber physiologischen Eintheilungsprincip bald Holzstamm, bald Rhizom, Knollen, Zwiebeln u. s. w. nennen; diese Organe, die gemeinsame Eigenthümlichkeit der perennirenden Gewächse mit Ausnahme einiger Saftpflanzen, sind unter der Vegetation ungünstigen Einflüssen, zum Winterschlaf, zu einer Unterbrechung ihrer vitalen Functionen für eine von jenen Einflüssen abhängige Zeit befähigt und theilen diese Eigenthümlichkeit mit den Saamen der Gewächse, beide zur Erhaltung der vegetabilischen Schöpfung bestimmt, beide durch Entwickelung ihrer Knospen ihr Erwachen bezeichnend und neue krautartige Individuen erzeugend. Wenn nun das Aufhören der grossen Circulation in der Atmosphäre auch das

Aufhören der kleinen Circulation des Wassers durch die Pflanze bedingt, so beginnen die Stammbildungen ihren Winterschlaf und stellen in jenen Ländern während der trockenen Jahreszeit dasselbe Bild der schlafenden Natur dar, was bei uns die gesunkene Temperatur hervorruft; die krautartigen Theile, die wahren Individuen der Pflanze, die Herr Schleiden den Stammgebilden gegenüber treffend mit den Polypen am Polypenstock vergleicht, können ohne dauernde Ernährung und Entwickelung nicht bestehen und sterben ab, wenn sie selbst oder ihre Stammbildungen nicht zu der folgenden Classe gehören. 2. *Saftpflanzen* in einem weitern Sinne nenne ich diejenigen Gewächse, die durch ein Uebergewicht der Wurzelabsorption über die organische Verdunstung, durch die Langsamkeit ihrer eigenen Wassercirculation ein Réservoir von Wasser in ihrem Parenchym bilden und sich dadurch eine kürzere oder längere Zeit von der Circulation der Atmosphäre unabhängig machen. Zu dieser Unabhängigkeit gelangen nicht bloss die eigentlichen Succulenten (diese vielmehr nur zum Theil), sondern auch die immergrünen Gewächse der Länder, die eine trockene Jahreszeit haben; sie sind während dieser Periode die einzigen Repräsentanten vegetabilischer Lebensprocesse, wichtig für die Physiognomik des Landes in dieser Zeit, der einzige Pflanzenschmuck, wenn alles Andere abgestorben scheint. Aber sie bilden stets nur einen kleinen Theil der Pflanzenarten ihrer Flora, und heben daher den Gegensatz derselben gegen die Tropenfloren mit ununterbrochener Feuchtigkeit nicht auf. 3. *Annuelle Pflanzen*, die während der trocknen Jahreszeit nur in Saamen existiren. Ihre Vegetationszeit ist oft kürzer, als die Regenzeit, und hängt von dem Zeitpuncte ihrer Saamenreife ab, aber sie kann nie einen Wassermangel überdauern. Wiewohl es allgemeiner Character aller tropischen Floren ist, arm an jährigen Pflanzen zu sein, so zeigt sich doch schon ein Gegensatz in ihrer Zahl, z. B. zwischen Senegambien und den Floren mit einer Vegetatio continua, in denen die Vegetationsfülle auch dadurch ausgesprochen erscheint, dass die meisten Gewächse sich zur Stammbildung erheben.

Die Unterscheidung dieser Grade der Abhängigkeit des Pflanzenlebens von der Vertheilung der atmosphärischen Niederschläge auf das Jahr hervorzuheben, schien deshalb nöthig, um die Verschiedenheit des botanischen Characters der feuchten Aequatorial-Floren von den Passatfloren, in denen ein Winterschlaf auftritt, schärfer zu bestimmen, die zwar nur die Gewächse mit Stammbildungen, aber damit bei Weitem die Mehrzahl aller Pflanzenarten betrifft. Da die Verschiebung der Passate bekanntlich an den Wendekreisen nur periodische Regenzeiten gestattet, so werden Floren dieser Art überall auftreten können, wo die Passatwinde herrschen, und es scheint keinen Unterschied in ihrem Character zu begründen, ob die Regenzeit in die heisseste oder in eine andere Jahreszeit fällt. In America zeigt sich der Character der Passatfloren nirgends auffallender, als an der Westküste, wo die periodischen Nebel nur während weniger Monate eine flüchtige Vegetation hervorrufen *); diesseits der Anden ist der Gegensatz, den die periodische Flora der Llanos, deren Character der Griffel des Meisters in diesen Wissenschaften gezeichnet hat, der Flora der grossen Ströme gegenüber bildet, gerade wegen ihrer Nachbarschaft und wegen des Mangels einer natürlichen Gränze entscheidend, wenn die Bewegung der Atmosphäre nicht eben diese Gränze darböte. In der alten Welt gehören zu den Passatfloren Senegambien, Abyssinien und die beiden Halbinseln von Hindostan nebst

*) Vgl. über das Vorkommen der Passatfloren Meyen's Pflanzengeographie pag. 10. 13. und dessen Mittheilungen über Canton in Nov. Act. Acad. Caesar. Vol. 17. p. 2.

dem südlichen China. Man wird ohne Zweifel in der Folge manche Eigenthümlichkeiten dieser Floren auf die Dauer ihrer Regenzeiten beziehen können, so wie es jetzt schon bemerkenswerth erscheint, dass in der alten Welt die Regenzeiten grösstentheils von Monsoons abhängen, und daher fast 6 Monate dauern, in der westlichen Hemisphäre an der Aequatorialgränze, z. B. in Peru, viel kürzere Zeit und an den Wendekreisen weniger entschieden von der trockenen Jahreszeit gesondert sind.

Endlich bleibt unter den Tropen noch die Reihe von Ländern zu betrachten übrig, in denen die Wassercirculation niemals für die Vegetation ausreicht, und die daher höchstens Saftpflanzen zu produciren im Stande sind. Herr von Humboldt hat die Ursachen im Zusammenhange entwickelt, denen die Sterilität der Sahara zuzuschreiben ist, und die wahrscheinlich auch zum Theil auf gewisse Theile von Asien angewendet werden können: Mangel an Wasser, das verdunsten könnte, ist neben herrschenden Polarströmungen als das Wichtigste jener Momente zu betrachten, so dass hier die reine Wirkung der Passatwinde in Erscheinung tritt. Das Product, worin sie sich äussert, zeigt sich schon am Saume des Landes in dem Mangel an Flüssen, die es verlassen: dieselbe Thatsache in Neuholland lässt auch hier auf eine wahre Wüste innerhalb der Passatzone schliessen, da die Steppen mit periodischer Vegetation in America, dem nordwestlichen Asien und Europa stets von grossen Flüssen durchströmt werden, und da wir für eine ringförmige Erhebung in so grossem Maassstabe, dass ein die Ströme eines Welttheils aufnehmendes Binnenmeer dadurch bedingt werden könnte, kein Analogon auf der Erde finden.

Es ist schon oben erwähnt worden, dass dieser Unterschied unter den tropischen Floren, der von ihren Feuchtigkeitsverhältnissen abhängt, in allen den Ländern verschwindet, in denen keine Winde mehrere Monate lang herrschen und damit die Regelmässigkeit in der Vertheilung der Niederschläge auf das Jahr aufhört. Die extratropischen Floren verhalten sich daher in dieser Rücksicht wie die Aequatorialfloren, nur durch die Geschwindigkeit der Circulation unterschieden, dennoch, abgesehen von der Wärme, zu einer Vegetatio continua befähigt; es ist nämlich zu bemerken, dass die beiden Regenzeiten von Süd-Europa, die Herr Dove nachgewiesen hat, zu wenig von den übrigen Jahreszeiten gesondert sind, um einen durch Trockenheit bedingten Winterschlaf der Vegetation zu bewirken. Ebenso wenig sind, um diesen wichtigen Satz zu wiederholen, andere Differenzen in den Feuchtigkeitsverhältnissen in diesen Zonen für die Begränzung der Floren von Wichtigkeit. Zwei der folgereichsten Thatsachen in der Pflanzengeographie, die stets vorangestellt zu werden verdienen, sind die Identität der alpinen Flora von Mittel- und Nord-Europa, und die von England mit dem nördlichen und mittlern Russland. Aus den Untersuchungen des Herrn Kämtz über die Aufstellung des Wasserdampfs in verschiedenen Höhen der Atmosphäre ergeben sich die complicirten Verhältnisse, nach denen die Alpenflora grösstentheils in eine Region fällt, die die geringste Tension des Wasserdampfs, aber die reichsten Niederschläge darbietet, Eigenthümlichkeiten, die der Atmosphäre nur in verticaler Richtung, nicht in der Richtung vom Aequator zum Pol, zukommen, und zu keiner climatischen Analogie zwischen der alpinen und lappländischen Flora führen. Das Gebiet der mittel-europäischen Flora bietet, wie mehrfach erwähnt wurde, einen ebenso lehrreichen Beweis von dem Nicht-Einflusse der Differenzen in seinen Feuchtigkeitsverhältnissen dar.

Da wir also in den periodischen Regenzeiten einen strengen climatischen Character des grössten Theils der tropischen

Floren im Gegensatz zu den extratropischen finden, so haben wir endlich noch zu untersuchen, in wie weit natürliche Florengränzen mit den Polargränzen der Passate zusammenfallen. Diese Betrachtung bezieht sich auch auf die von den Wärmerverhältnissen abhängige Differenz beider Zonen, die in climatischer Hinsicht zu ziemlich analogen Gränzbestimmungen führt. Im Allgemeinen fällt die Polargränze der Passate auf der nördlichen Halbinsel mit dem Wendekreise zusammen *). In America entspricht dieser Gränze der Gegensatz zwischen den Floren von Cuba und Florida, zwischen Vera Cruz und Texas (nach den Sammlungen Drummonds: vgl. Hooker's Journal of Bot. vol. I.), zwischen Acapulco (nach Humboldt's Verzeichniss) und Californien (nach den daher neuerlich in England beschriebenen, zahlreichen Arten, unter denen sich keine tropischen Formen finden). In Africa bildet die Sahara eine breite Scheidewand zwischen den Floren der Nordküste und den durch Regenzeiten periodischen Floren von Senegambien und Abyssinien. Aus Asien haben die von Decaisne beschriebenen **) und die während der Euphrat-Expedition gesammelten Pflanzen***) im Gegensatze zu Forskal's Flora einen Character, der sich nicht

*) Nach Dove (Poggend. Ann. XXV. pag. 193.) reichen die Passate im atlantischen Meere im September bis zum 24° L. B., im stillen Meere durchschnittlich für das ganze Jahr bis zum Wendekreise. Es kann uns hier nur die äussere Gränze während der nördlichen Verschiebung angehen, da davon die Absonderung einer Regenzeit abhängt.

**) In den Annales des sciences 1834 etc.

***) Diese Sammlung wurde von Herrn Lindley dem königl. Herbarium in Berlin mitgetheilt, und durch die grosse Liberalität, mit der dessen Benutzung gestattet wird, hatte ich Gelegenheit, jene Pflanzen kennen zu lernen.

wesentlich von dem der südeuropäischen Flora entfernt; wahrscheinlich verläuft von da eine Florengränze durch den südlichen Theil von Persien und an der Südgränze von Afghanistan und Lahore; weiter nach Osten werden die tropischen Floren vom Himalajah ebenso begränzt, wie die Monssons sich an diesen Höhen brechen; die Gegend von Canton gehört endlich nach Meyen's Beobachtungen zu den Floren, deren Winterschlaf von unterbrochener Feuchtigkeit abhängt. Man erkennt hieraus, dass auf der nördlichen Hemisphäre nirgends eine Flora aus den passatlosen Ländern in die tropischen übergreife, aber man muss in einer Wissenschaft, deren Wesen es ist, die Verhältnisse allgemein aufzufassen, nicht eine Genauigkeit der Angaben fordern, die weder der extensiven Grösse unserer botanischen und climatologischen Kenntnisse entsprechen, noch dem Grade einer physikalischen Forschung zukommen würde, in der zu viele und zu wenig in ihrem relativen Einflusse gekannte Factoren in Betracht gezogen werden müssen. Auf der südlichen Halbkugel würde eine Linie, die die tropischen von den extratropischen scheidet, erst gezogen werden können, wenn die Floren von Süd-America und des tropischen Australiens genauer bekannt sein werden: in den Sammlungen von Sello zeigt sich der erwähnte Gegensatz des botanischen Characters zwischen den Pflanzen aus Brasilien und aus Montevideo; Chile steht durchaus unter dem Einflusse periodischer Niederschläge, ebenso nach den Beobachtungen von Drège die Südspitze von Africa, so dass in der südlichen Hemisphäre die tropischen Floren unstreitig in weit höhere Breiten reichen, wie in der nördlichen. Die Anwesenheit tropischer Formen und solcher Familien, die entschieden extratropischen Ländern fehlen, ist indessen ein trügerisches Criterium für einen Begriff, der nur von climatischer Seite bestimmt ist: es wird noch nicht behauptet, dass alle Passat-Floren einige gemeinschaftliche botanische Merkmale haben, sondern nur, dass die äussere Polargränze der Passatwinde überall mit irgend einer Florengränze zusammenfalle. Ferner müsste sich nachweisen lassen, dass die Zone der Calmen oder des perennirenden Courant ascendant ihre eigenthümlichen Floren habe, aber die Gränzen derselben sind zu wenig bekannt.

Die *Bewegungen der Atmosphäre*, die nur einen mittelbaren Einfluss auf das Leben der Pflanze äussern können, haben in den bisherigen Erörterungen schon eine nähere Betrachtung veranlasst; der andere Theil ihrer Wirksamkeit, ihre, wenngleich früherhin überschätzte, Bedeutung für die Temperatur, kann uns gleichfalls nur in ihren Wirkungen interessiren. Wenden wir uns nun zu den imponderabeln Bestandtheilen der Atmosphäre.

Dieselbe Schlussfolge, die früher gegen den Einfluss des Luftdrucks auf die Florengränzen geltend gemacht wurde, findet auch seine Anwendung gegen die *Lichterscheinungen*, von deren Intensität man die Eigenthümlichkeiten alpiner Floren hat ableiten wollen, indem man insbesondere, nach Wahlenbergs Andeutungen, einen Unterschied zwischen den arktischen und alpinen Pflanzen in ihren habituellen Characteren darzustellen sich bemühte, wie sie mit einer eindringlichen Beobachtungsgabe von Schouw *) für die Alpenflora waren aufgefasst worden. Wie wenig diese Unterscheidung in der Natur gegründet sei, zeigt eine unbefangene Vergleichung des Einzelnen: dasselbe Vorherrschen der Rhizom-Kräuter, dieselben reinfarbigen, grossen Blumen, im Allgemeinen dieselben vorherrschenden Familien: Schouw selbst sprach sich in gleichem Sinne aus **). Man darf bei diesen Untersuchun-

*) Pflanzen-Geogr. pag. 460.

**) Ebendas. pag. 469.

gen nie die *vier verschiedenen Grade* aus den Augen verlieren, in denen die Abhängigkeit des Pflanzenlebens von physikalischen Einflüssen von Lebensreizen steht: 1) Reize, von denen das Leben des Individuums abhängt, die die Physiologen integrirende Reize zu nennen pflegen, wie die Gegenwart liquiden Wassers, einer Humusdecke, atmosphärischer Luft, eine bestimmte Temperatursphäre; diese haben nur in Rücksicht auf den Winterschlaf ein pflanzengeographisches Interesse. 2) Reize, die auf die Qualität des Individuums einwirken, alternirende Reize, wie gewisse Wärmegrade innerhalb der Temperatursphäre der Art, die Intensität des Lichts, die Menge der Feuchtigkeit: diese Reize, die nur innerhalb der Gränzen, in denen sich das Leben jeder Art bewegt, variiren, bilden für die Pflanzengeographie gar kein Object. 3) Reize, von denen das Leben der Art abhängt, deren Umfang auch die Lebenssphäre der Art ist, wie die Temperatur-Maxima und Minima, die für jede Art besonders gefunden werden müssen, deren Ueberschreiten die Individuen der Art tödten und die für die einzelnen Epochen der Vegetation derselben Art verschieden sein können. 4) Die physikalischen Bedingungen einer ganzen Flora, die gleichfalls Gränzbestimmungen ihres Lebens sein sollen, die nur für einen Theil ihrer Arten, aber für die Gruppirung *aller* gelten. Aus diesem Gesichtspuncte wird man weder von dem pflanzengeographischen Einflusse des Lichts, noch des Magnetismus, noch der Electricität reden können: liegen darin Momente, die erforscht werden können, so gehören sie wenigstens nicht dem jetzigen Standpuncte unserer Erkenntniss an, auf dem wir uns vergeblich bemühen würden, die Erdoberfläche nach Differenzen in diesen Grössen einzutheilen.

Von den verschiedenen Beziehungen, in die man die *Wärme* der Atmosphäre mit den Gränzen der natürlichen Floren gestellt hat, ist schon im Eingange Einiges erwähnt worden: diese Beziehungen sind jetzt umständlicher zu erörtern. Bevorwortet kann werden, dass man bei der Bestimmung des solaren Einflusses auf die Pflanzen nur auf die in dieser Hinsicht nicht genügenden Thermometer-Beobachtungen hingewiesen ist, bei denen die Geschwindigkeit, mit der das Thermometer steigt und fällt, und darin die für das Pflanzenleben ohne Zweifel wichtige Potenz, die von der directen Sonnenwärme ihr Maass erhält, verloren geht.

Jedes organische Wesen hat ein Maximum und ein Minimum der Temperatur, innerhalb deren Gränzen es allein fähig ist zu existiren. Vergleicht man indessen die Temperatursphäre, in der sich die Keimkraft eines Getreidekorns erhält*), mit der weit beschränktern Skale, die die vegetirende Pflanze später in Anspruch nimmt: so ergiebt sich daraus die Verschiedenheit ihrer Abhängigkeit von der Wärme in verschiedenen Lebensperioden. Wenden wir dies Gesetz von der einzelnen Pflanze auf den Lebensprocess der ganzen Flora an, so wird das Clima derselben nicht bloss durch die mittlere Temperatur, nicht bloss durch die Temperaturextreme des ganzen Jahres, sondern durch die Temperatursphäre jeder einzelnen Periode des Pflanzenlebens bestimmt werden müssen. Bei dem Mangel an Jahrescurven, bei dem fühlbaren Mangel an Beobachtungen über die Perioden des Pflanzenlebens einer ganzen Flora, wofür man noch keine Methode der Beobachtung angegeben hat, kann es nicht befremden, dass eine Untersuchung dieser Art für verschiedene Floren für jetzt nur zu wenigen und ungewissen Resultaten führen kann: indessen glaube ich das wichtige Gesetz nachweisen zu können, dass *an allen Puncten der mittel-europäischen Flora die mittlere Temperatur des Zeitraums*

*) Vergl. Annales des sc. naturelles. Nouv. Sér. pag. 257—70.

der vegetirenden krautartigen Axe (bestimmter vom Aufsteigen des Frühlingssaftes in den Bäumen bis zum Abfallen ihrer Blätter) = 13° *C.* ist, während Isothermen und Isotheren grosse Differenzen zeigen und jener Zeitraum variabel ist. Ich stelle dies Gesetz, das ich als für jede natürliche Flora auf einen constanten, climatischen Character führend halten möchte, an die Spitze dieser Bemerkungen, um ihren Gang und ihr Resultat zu bezeichnen, während ich sie mit einigen Nachweisungen über die Beobachtungen, auf die es sich stützt, beschliessen werde. Zuvörderst stehen die vorhin erwähnten Abhängigkeitsgrade von der Wärme mit folgenden pflanzengeographischen und climatischen Phänomenen in Verbindung:

1) Wir finden eine absolute Gränze des Pflanzenlebens nach der Polhöhe und Erhebung über dem Meere, also es giebt eine Temperatursphäre für alle Floren und somit für alle Pflanzen.

2) Wir finden einige Pflanzenarten, deren Temperatursphäre fast ebenso gross ist. Kann dies zwar mit Sicherheit bis jetzt für natürliche Standorte nur von einigen Cryptogamen *) behauptet werden, so gilt es desto entschiedener von einigen Culturpflanzen, z. B. der Kartoffel, dem Roggen, Medicago sativa **). Sehr viele Pflanzenarten haben eine Ausdehnung, die zwei oder mehrere natürliche Floren übergreift: so kommen die meisten mittel-europäischen Arten auch in Süd-Europa, viele in Süd-Sibirien vor. Die einzelne Art ist nicht an die Temperatursphäre ihrer Flora gebunden.

3) Enontekis ist von Wäldern umgeben; das Bernhardhospiz liegt mehr als 2000′ über der Baumgränze: dort ist

*) v. Humboldt de distributione etc. pag. 60.

**) Meyen Reise I. pag. 401.

die mittlere Temperatur = — 2°,9 C., hier = — 1°,1 C. Also wird die Temperatursphäre der Formationen so wenig, als der Arten durch die mittlere Jahrestemperatur gemessen.

4) Unter den Tropen weicht sowohl in der Ebene, wie in den durch Erhebung kalten Regionen die Wärme irgend eines Zeitraums nur wenige Grade von der mittlern Temperatur ab. Daher wird die Temperatursphäre tropischer Floren durch die Temperaturextreme und die mittlere Wärme hinreichend genau bestimmt.

5) Ausserhalb der Wendekreise zeigt sich überall ein Winterschlaf der Floren, der von einer gesunkenen Temperatur abhängt und in seiner Dauer verchieden ist. Das bekannte physiologische Experiment *), wodurch man einen Baum künstlich zum Erfrieren bringt, wenn man einen Ast desselben in ein Treibhaus leitet und dadurch zum Ausschlagen nöthigt, beweist die Verschiedenheit der Temperatursphäre der Pflanzen während ihres Winterschlafs. Da wir keine Unterschiede ähnlicher Art zwischen den übrigen Vegetationsepochen für ganze Floren nachweisen können, so sind wir zu dem Satze berechtigt, dass die Temperatursphäre einer natürlichen Flora durch die mittlere Temperatur und durch die Temperaturextreme *ausserhalb ihres Winterschlafs* gemessen werde. Curven nämlich, deren Gesetz unbekannt ist, können nur auf diese Weise verglichen und somit benutzt werden: man kann sie zwar eintheilen und die mittlern Werthe der Theile vergleichen, aber dies würde nur in dem Falle zu Resultaten für den Character der Floren führen, wenn das Eintheilungsprincip von dem Wesen des Pflanzenlebens hergenommen wäre. Da ausserdem durch die Untersuchun-

*) Mustel Traité de la végétation II. pag. 326.

gen des Herrn Kämtz *) wahrscheinlich wird, dass die Vertheilung der Wärme in der Jahrescurve und der Eintritt der Wendepuncte auf der ganzen Erde gleichförmig ist, so würde bei Vergleichungen die mittlere Temperatur des Jahres oder eines Theiles desselben das climatische Moment der Curve vollständig enthalten.

Es frägt sich nun zunächst, ob der Vegetationsprocess der krautartigen Axe einer Art an einen absoluten Zeitraum geknüpft sei, und somit die Dauer des Winterschlafs einer Flora in ihrem Gebiete constant sein müsse; ferner ist zu untersuchen, ob man Mittel habe, die nöthigen Zeitbestimmungen über diese beiden Perioden des Pflanzenlebens zu machen. Die erste Frage ist kürzlich durch Herrn Boussaingault negativ beantwortet, indem er das für die Cultur überaus wichtige Gesetz nachwies, dass die mittlere Temperatur irgend einer Vegetationsperiode multiplicirt mit der Zahl der Tage, die darüber verflossen sind, für dieselbe Art stets dasselbe Product giebt, während *beide* Grössen innerhalb gewisser Gränzen variabel sind. Dies erklärt Schübler's Beobachtungen **) über die verschiedenen Blüthezeiten derselben Pflanzenart unter verschiedenen Breiten, die er nach einer irrigen Theorie auf die mittleren Temperaturen der verglichenen Orte bezog, während die einfache Thatsache des verschiedenzeitigen Blühens einer Art an demselben Orte in verschiedenen Jahren mit Schärfe darauf hinweist, dass die mittlere Wärme dabei ein gleichgültiger Factor sei. Das richtige Princip hatte bekanntlich schon Adanson angegeben, aber auf eine unangemessene Weise ausgeführt; Herr De Candolle ***)

---

*) Meteorol. I. pag. 127.

**) Flora 1830. I. pag 353.

***) Physiologie végétale 2. pag. 476.

wies die Unzulässigkeit seines Verfahrens nach, kam aber selbst nur zu dem negativen Resultate, dass sich die climatische Ursache des verschiedenzeitigen Ausschlagens von Aesculus nicht ermitteln lasse. Freilich hängt es von der Ausbildung der Knospe im vorigen Herbste ab, aber diese hat keinen Einfluss auf die Dauer des Zeitraums vom Ausschlagen bis zum Abfallen der Blätter. Auf dem jetzigen Standpuncte der Untersuchung ergeben sich nun folgende Gesetze, die für die Temperatursphären einzelner Pflanzenarten Gültigkeit haben.

1) Im Allgemeinen kann die Vegetationszeit gewisser Pflanzen bei einer höhern Temperatur verkürzt werden, aber es findet hier eine bestimmte Gränze statt, die von der Natur der Pflanze abhängt, und somit tritt die Aequatorial-Gränze einer Pflanze mit dem Minimum der Zeit ein, in der sie sich bei einem Maximum der Temperatur entwickeln kann. Der eine Factor ihrer Temperatursphäre, die Anzahl der Tage, in denen sie sich entwickelt, kann nicht unbegränzt in demselben Sinne verringert werden, als der andere, das erhaltene Wärme-Quantum sich vermehrt. Deshalb gebraucht man unter den Tropen bei der Cultur extratropischer Pflanzen kühlende Vorrichtungen, die ihre Vegetation verlängern sollen.

2) Eine Verschiebung der Entwickelungszeiten ist die gewöhnlichere Erscheinung, in der das obige Gesetz in Wirksamkeit tritt. In der Provence beginnt die Weinerndte einen Monat früher, als am Rhein, aber das Ausschlagen des Weinstocks tritt dort gleichfalls früher ein. Dadurch wird eine mehrfache Erndte mancher Culturpflanzen unter den Tropen möglich.

3) Die Polargränze einer Pflanze tritt mit dem Maximum der Zeit ein, in der sie sich bei einem Minimum der Temperatur entwickeln kann. Sie wird auf eine doppelte Weise bestimmt werden können, indem sich entweder die Constante

(das Product aus Entwickelungszeit und mittlerer Temperatur derselben) in der ganzen Curve nicht mehr hervorbringen lässt, oder wenn die Factoren der Constante nicht mehr in die Temperatur- und Entwickelungs-Sphäre der Art fallen.

Welche Anwendung können wir aus diesem Gesetze, das eine von der Temperatur abhängige Veränderlichkeit der Vegetationszeit derselben Pflanzenart nachweist, für die climatischen Gränzbestimmungen der natürlichen Floren machen? Man erkennt leicht, dass dasselbe für diese letztere nicht gilt, und dass diese Nichtgültigkeit desselben eben die climatische Ursache der möglichen Verbreitung einzelner Arten durch verschiedene Floren ist. Der Beweis dafür liegt in dem Umstande, dass der Winterschlaf der mittel-europäischen Flora von Süden nach Norden allmälig länger wird, ohne dass die Sommerwärme in gleichem Sinne zunimmt. Dies ist näher nachzuweisen, und man muss dabei zuerst die Methode, nach der man die Dauer des Winterschlafs bestimmen kann, erörtern.

Zeitbestimmungen dieser Art sind wohl hauptsächlich deswegen nicht versucht, weil man glaubte, nur das Erwachen bestimmter Arten beobachten zu können, nicht das gleichzeitige Erwachen des grössten Theils der ganzen Flora, und weil man dazu oft Frühlingspflanzen wählte, die eben als Ausnahmen keine allgemeine Schlüsse gestatten. Wenn man in landwirthschaftlichen Journalen die Saat- und Erndte-Zeiten verschiedener Jahre vergleicht*), so findet man eine höchst auffallende Uebereinstimmung unter den dazwischen verflosse-

*) Vgl. einige Angaben hierüber in meinen Genera et species Gentianearum pag. 32., in denen auch ein Beweis für die Unabhängigkeit des Pflanzenlebens von der Temperatur während des Winterschlafs enthalten ist.

nen Zeiträumen, eine Uebereinstimmung, viel grösser, als unter irgend welchen meteorologischen Beobachtungen, aus denen man mit Erfolg arithmetische Mittel zieht. Hätte man also für die ganze Flora ein solches zeitbestimmendes Criterium, wie die Saatzeit der Kornarten ist, so würde man dadurch für jeden Ort die mittlere Dauer des Winterschlafs im Gegensatz zu der Vegetationszeit bestimmen können, alsdann aber in der mittlern Temperatur dieser letztern die Temperatursphäre der ganzen Flora erhalten. Solche Momente des Pflanzenlebens nun liegen ohne Zweifel im Aufsteigen des Frühlingssaftes durch die dicotyledonischen Stammbildungen, da das Wachsthum dieser letztern (im Gegensatze zu den Knospen, d. h. krautartigen Organen) eine abgeschlossene, von frühern Bildungen unabhängige Periodicität zeigt; zweitens in der herbstlichen Blattentfärbung, das das Aufhören des Assimilationsprocesses im Zellensafte und wahrscheinlich seiner Cyclose bezeichnet. Da mir über diese beiden Perioden des vegetabilischen Lebens, die wegen ihrer grösseren Gleichzeitigkeit für die ganze Flora einen wesentlichen Vorzug vor andern Objecten der Beobachtung behaupten, keine vergleichbare Zeitbestimmungen bekannt geworden sind, so sah ich mich genöthigt, auf eine Genauigkeit in den Resultaten zu verzichten, die die Natur in diesen Verhältnissen zu beobachten scheint. Eine schätzenswerthe Zusammenstellung von Beobachtungen über die Zeiten des Ausschlagens und Blühens verschiedener Gewächse an mehreren Orten in Europa findet sich in der Regensburger botanischen Zeitung für 1836, aber ich habe sie nur wenig benutzen können, da, wie oben erwähnt wurde, die Zeit des Ausschlagens von dem Entwickelungsgrade der Knospe mit abhängt, das denselben bedingende Clima des verflossenen Jahres aber nicht mit in Rechnung gebracht werden kann, da der Zeitpunct des ersten Entstehens der Knospe zwischen Holz- und Rinden-System unbekannt bleiben muss.

Da sich ferner verschiedene Arten in Hinsicht auf die Ausbildung ihrer Knospen im Herbste sehr verschieden verhalten, so erklären sich hieraus die bedeutenden Zeitunterschiede, die die Beobachtung der Zeit des Ausschlagens verschiedener Holzgewächse an demselben Orte liefert, z. B. in Neapel für Sambucus nigra der mittlere Werth = 14. Januar, für die Eiche = 31. März. Besser schien mir hingegen die Blüthezeit von Pflanzen benutzt werden zu können, deren Blüthen zu der Zeit sich entfalten, wenn der Frühlingssaft in den Bäumen anfängt sich zu zeigen, und zugleich die Ordinaten der Jahrescurve am stärksten wachsen. Keine Pflanze dürfte sich hiezu mehr eignen, als Primula elatior Jacq. und, so wie es sich hier nur um angenäherte Werthe handeln kann, so sind den nachfolgenden Angaben über die Anfangsperiode der Vegetationszeit Beobachtungen über die Blüthezeit jener Pflanze zu Grunde gelegt. Da für den Anfang des Winterschlafs solche Beobachtungen nicht zu benutzen waren, so habe ich statt dessen nach folgendem Raisonnement die Jahrescurven selbst benutzt: der Eintritt, wie das Aufhören des Winterschlafs hängt von der Temperatur ab; die Ordinaten der Jahrescurven nehmen zu beiden Seiten der Jahrescurve nahe gleichförmig ab; wenn also das Aufhören des Winterschlafs von der Grösse einer gewissen Ordinate abhängt, so wird der Wiederanfang desselben eintreten, wenn im Herbste dieselbe Ordinate wiederkehrt. Aus diesen Gesichtspuncten ist die folgende Tafel zur Beurtheilung der Temperatursphäre der Flora von Mitteleuropa entstanden, über deren Zusammenstellung erst einige Bemerkungen nöthig scheinen.

Die Angaben über die Blüthezeit von Primula elatior sind theils einigen unter den directen Beobachtungen, die in der Flora (1830 und 1836) mitgetheilt sind, entnommen, theils nach dem von Schübler eingeführten und für geringe Breitenunterschiede gültigem Verfahren bestimmt, wonach zwischen dem 44sten und 54sten Grade eine Verzögerung von etwa 4 Tagen einem Breitengrade entspricht, in höhern Breiten dagegen diese Verzögerung geringer wird und nur zu 2 — 3 Tagen angenommen werden kann. Es musste indessen dabei auch auf den Lauf der Isothermen, auf das Verhältniss der Isotheren und Isochimenen gegen einander, die Höhe u. s. w. Rücksicht genommen werden, so dass die Angabe jeder Blüthezeit als ein Mittel, aus zum Theil widersprechenden Betrachtungen hervorgegangen, anzusehen ist. Es würde zu weit führen, diese Methode in ihrer Anwendung auf die einzelnen Fälle nachzuweisen; wie es wesentlich darauf ankam, welche directe Beobachtung jedesmal zu Grunde zu legen war, so schien es genügend zu sein, diese einer jeden abgeleiteten Angabe in Parenthese beizufügen. So erschien es aus einleuchtenden Gründen zweckmässiger, Cuxhafen auf die Beobachtungen in England zu beziehen, als auf die deutschen. Die Temperaturangaben rühren grösstentheils aus den im zweiten Theile von Kämtz's Meteorologie zusammengestellten Beobachtungsreihen her; die von Berlin, Manchester und Upsala sind bekannten Jahrescurven entnommen. Die mittlere Temperatur der Vegetationszeit, die *Phytoisotherme*, ist durch das arithmetische Mittel aus dem Temperatur-Maximum und der Temperatur der beiden Endpuncte bestimmt, welches bekanntlich der wirklichen mittlern Temperatur hinreichend genau entspricht.

	Aufsteigen des Frühlingssaftes. (Bläthezeit von Primula elatior.)	Abfall der Blätter. (Wiederkehr der Frühlings - Ordinate.)	Temperatur dieser Ordinate.	Vegetationsdauer.	Mittl. Temp. während der Vegetationszeit. (Phytoisotherme)	Temperatur-Maximum.
1) Petersburg	15. Mai (Greifsw. 27. Apr.)	27. September	+ 7°,5 C.	4,4 Monate.	13°,2 C.	19° C.
2) Stockholm	12. Mai. (=)	1. October	+ 8° C.	4,6 =	13°,0	18°,1 C.
3) Upsala	6. Mai.*) (Beobachtet.)	1. Oct. (Mittel aus den Angaben in Wahlenberg's Flora Upsal.)	+ 8° C.	4,7 =	13°,0	18° C.
4) Königsberg	28. Apr. (Greifswalde.)	7. October	+ 7°,6 C.	5,3 =	12°,7	17°,8 C.
5) Berlin	20. Apr. (Beobachtet.)	10. October	+ 8° C.	5,6 =	13°,0	18°,1 C.
6) Manchester	13. Apr. (Norfolksh. 11. Apr.)	20. October	+ 8° C.	6,1 =	12°,5	17° C.
7) Cuxhafen	13. Apr. (=)	25. October	+ 7°,6 =	6,4 =	13°,4	19°,3 C.
8) Zwanenburg	9. Apr. (=)	1. November	+ 8° =	6,7 =	13°,8	19°,6 =
9) London	5. Apr. (=)	29. October	+ 8° =	6,7 =	13°,0	18° =
10) Frankfurt	5. Apr. (Heidelberg 1 Apr.)	1. November	+ 7°,6 =	6,8 =	13°,5	19°,4 =
11) Manheim	1. Apr. (=)	27. October	+ 7°,4 =	6,9 =	14°,1	20°,9 =
12) Carlsruhe	29. März. (=).	27. October	+ 8°, =	7 =	14°,2	20°,5 =
13) Paris	26. März. (Zürich 11. M.)	6. November	+ 7°,7 =	7,3 =	13°,5	19°,3 =
14) La Rochelle	18. März. (=)	6. November	+ 7°,5 =	7,5 =	13°,9	20°,3 =

*) Zu berichtigen ist die Angabe von Schübler (a. a. O.), nach der die Frühlingspflanzen bei Upsala erst zu Anfang Junius blühen sollen. Wahlenberg theilt Beobachtungen über die Perioden der dortigen Flora mit. (Praefat. in Flor. Ups.)

Ich mache zuerst auf die völlige Uebereinstimmung der Phytoisotherme von Stockholm, Berlin und London aufmerksam, die theils wegen des Gegensatzes in der Vegetationsdauer merkwürdig ist, theils als ein starkes Argument gegen diejenigen gebraucht werden kann, die eine Flora von Nord-Europa oder eine Flora der europäischen Seeküste unterscheiden wollen. Im Besondern ergeben sich aus dieser Uebersicht folgende Puncte:

1) Bei einer Verschiedenheit von 3 Monaten in der Dauer der Vegetationszeit differiren die mittlern Temperaturen derselben nicht um anderthalb Grade, eine Differenz, die man, da sie keine stetige Zunahme mit der Polhöhe zeigt, mit Recht auf die ungenaue Bestimmung der Endpuncte der Periode beziehen darf.

2) Das Aufhören des Winterschlafs entspricht an den verschiedensten Orten der mittel-europäischen Flora einer analogen Ordinate in der Jahrescurve, so dass man als Thatsache aussprechen darf, dass der Frühlingssaft anfängt zu steigen, wenn die Curve sich über 7,°5 C. erhebt. So sehr die Werthe dieser Ordinate sich indessen nähern, so möchte vielleicht eine weitere Untersuchung dahin führen, dass diese Temperatur gegen Süden abnimmt: wenigstens würde bei vorausgesetzter Identität der Phytoisotherme die geringe Zunahme der Temperatur-Maxima dafür sprechen.

3) Die Dauer der Vegetationszeit steht in directem Verhältnisse mit der mittlern Jahreswärme; und in indirectem Verhältnisse mit der Krümmung der Jahrescurve, d. h. ein Seeclima bedingt einen kürzern Winterschlaf. Hierdurch wird auch vom climatologischem Gesichtspuncte die im Anfange dieses Aufsatzes aus botanischen Gründen behauptete Identität der europäischen Küstenfloren mit der Vegetation des Conti

nents gerechtfertigt, indem trotz der Verschiedenheit der Jahrescurven die Phytoisotherme dieselbe bleibt.

Da diese Sätze der Boussaingault'schen Theorie zu widersprechen scheinen, so könnte man einwenden, dass es nicht denkbar sei, dass eine einzelne Art eine andere Abhängigkeit vom Clima habe, als eine Flora, die nur ein Aggregat von Arten sei: aber in der That widersprechen sich beide Theorien nicht, sondern die eben vorgetragene ist nur ein eingeschränkter Fall, die Temperatursphäre der Floren hat nur engere Gränzen. Der wesentliche Unterschied zwischen der Lebenssphäre der einzelnen Art und der einer Flora besteht in dem *natürlichen* Zusammenleben der Individuen, in ihrer physiognomisch-characterisirten Gruppirung, deren climatische Bedingung enger begränzt sein kann, als die des aus dieser wechselseitigen Beziehung losgerissenen Individuums. Die Boussaingault'sche Theorie enthielt gleichfalls ein Schwanken in der Vegetationsdauer der Art, und wich nur dadurch ab, dass es zugleich die möglichen Schwankungen der Temperatursphäre, als die möglichen Gränzen der Cultur, bestimmte, während solche Abweichungen beim natürlichen Vorkommen der Pflanzen zu verschwinden scheinen.

Dass die Phytoisothermen eine weit genauere Bestimmung der Beziehung zwischen Vegetation und Temperatur enthalten, als früher dazu angewendete Werthe, geht aus einer Vergleichung derselben hervor. Die Isothermen schwanken im Gebiete der mitteleuropäischen Flora um 8° (Kasan = 2°,4; Paris = 10°,8), während die Isotherme von Paris von der von Marseille (= 12°,3) nur um 1°,5 abweicht. Dagegen scheint die Phytoisotherme von Süd-Europa über 17° zu liegen, also die von Mittel-Europa um mehr als 4° zu übersteigen. Wäre hingegen die mittlere Jahreswärme das die Florengränzen bestimmende climatische Moment, so müsste man nach den eben angeführten Daten die willkührliche Annahme machen, dass 8 Wärmegrade unterhalb 11° einen geringern Einfluss auf die Vegetation zu äussern bestimmt seien, als 2 Grade über jenem Puncte. Die Isotheren schwanken in Mittel-Europa um 6° (Edinburg = 14°,1; Wien = 20°,3) und differiren von süd-europäischen nur um 2° (Rom = 22°,7). Es wird eine schöne Bestätigung der Theorie sein, wenn an einem Gränzorte der südeuropäischen Flora Beobachtungen über die Perioden des Pflanzenlebens in dem mehrfach erörtertem Sinne angestellt werden und daraus ein entschiedener Gegensatz gegen nahe gelegene Orte hervortritt, deren Vegetationsverhältnisse noch den nordischen Character tragen. Da die vorhandenen Zeitbestimmungen, z. B. die über Nord-America (Silliman's Journal Vol. 1.), zur Untersuchung der Phytoisothermen anderer Floren ungenügend sind und nicht auf die beiden angenommenen fixen Puncte bezogen werden können, so würde man eine weitere Anwendung der obigen Sätze für jetzt nicht zu machen im Stande sein, so wünschenswerth auch die Beantwortung der Frage ist, ob man auf diesem Wege die Ursache der Verschiedenheit natürlicher Floren unter gleichen Breiten und Isothermen finden werde, ob z. B. die Phytoisotherme der Flora von Nord-America eine Abweichung von der unsrigen zeige. Geht man von der Annahme aus, dass der Endpunct des Winterschlafs, so weit er von Temperatur-Differenzen abhängt, in der Jahrescurve durch ein schnelleres Ansteigen der Ordinaten, als dies in irgend einer andern Jahreszeit vorkommt, bezeichnet werde: so könnte man unmittelbar die Phytoisotherme aus jeder Jahrescurve ableiten. Ein Criterium dafür würde die Uebereinstimmung der Resultate an verschiedenen Orten derselben Flora sein.

Wenn Beobachtungen den Beweis liefern, dass die Bedingungen der extratropischen Floren durch ihre Phytoisothermen dargestellt werden, und dass ihre Gränzen sich ebensowohl durch Temperatur-Beobachtungen, wie durch botanische Untersuchungen bestimmen lassen, so wird auch hierin ein wesentlicher Gegensatz gegen die tropischen Floren nachgewiesen werden können, deren Winterschlaf, wie wir sehen, von der Vertheilung der Feuchtigkeit auf das Jahr abhing. Für diese letzteren werden die thermischen Bestimmungen, da die Differenzen in den Ordinaten ihrer Jahrescurven gering sind oder doch nie die Vegetation unterbrechen, durch die Isotherme mit hinreichender Genauigkeit ausgedrückt werden können. Dazu kommen ferner die Temperatur-Maxima und Minima, die auch in den Floren höherer Breiten neben der Phytoisotherme Berücksichtigung verdienen. Da die Temperatur-Maxima auf der Erde nirgends so gross sind, um das Pflanzenleben aufzuheben, da sie in ihren Extremen nur im Stande sind, mittelbar durch Entziehung der Feuchtigkeit ein Schlafen der Vegetation zu veranlassen: so wird man dagegen in den Temperatur-Minimis absolute Gränzen des vegetabilischen Daseins erkennen, und diese Gränzen des Pflanzenreichs in verticaler und horizontaler Richtung nach ihrer climatischen Gleichartigkeit untersuchen müssen. Hier genügt es, neben den Florengränzen unter einander auch auf ihre äussern Gränzen und deren Abhängigkeit vom Clima hinzudeuten. Spätere Untersuchungen haben zunächst die Aufgabe, die Phytoisotherme anderer Floren kennen zu lernen, wozu es erlaubt sein mag, das Interesse, das neue Beobachtungen haben würden, nochmals hervorzuheben.

Zum Schlusse stelle ich die Hauptergebnisse der bisherigen Untersuchung in folgenden Sätzen zusammen:

1) Die Vegetation der Erde zerfällt in scharf begränzte natürliche Floren, die gemeisame botanische und climatische Charactere haben.

2) Die Floren zerfallen in zwei Hauptclassen, je nachdem sie eine dauernde oder eine durch Winterschlaf unterbrochene Vegetation haben.

3) Floren mit dauernder Vegetation finden sich nur in der Nähe des Aequators.

4) Der Winterschlaf der Floren hängt entweder von Trockenheit oder von gesunkener Temperatur ab. Hierdurch unterscheiden sich die tropischen von den extratropischen Floren.

5) Das Clima einer tropischen Flora mit dauernder Vegetation wird durch die mittlere Jahrestemperatur gemessen.

6) Das Clima einer Passatflora wird durch die Dauer der Regenzeit und durch die mittlere Temperatur während derselben bestimmt.

7) Das Clima einer extratropischen Flora wird durch die mittlere Temperatur der Vegetationszeit gemessen.

8) Andere climatische Momente haben auf die Gränzbestimmung der natürlichen Floren keinen nachweisbaren Einfluss.

9) Die mittlere Temperatur der Vegetationszeit ist im ganzen Gebiete der mittel-europäischen Flora identisch, ebenso diejenige Ordinate der Jahrescurve, die den Endpuncten des Winterschlafs entspricht.

10) Die Endpuncte des Winterschlafs treten mit dem Aufsteigen des Frühlingssaftes und der herbstlichen Blattentfärbung ein.

11) Ob die climatischen Gesetze der mittel-europäischen Flora für alle extratropischen Floren Gültigkeit haben, kann aus Mangel an Beobachtungen über die Dauer der Vegetationszeit noch nicht nachgewiesen werden: ebenso we-

nig, ob es eine climatologische Diagnostik sämmtlicher Floren gebe.

12) Die Nordwestküste von Europa gehört zum Gebiete der mittel-europäischen Flora und man kann in Europa nur drei Floren unterscheiden: die Flora mediterranea, europaea media und alpina.

Göttingen, den 16. Februar 1838.

REPORT ON THE CONTRIBUTIONS TO BOTANICAL GEOGRAPHY DURING THE YEARS 1842 AND 1843

Heinrich Rudolf August Grisebach

REPORT

ON THE CONTRIBUTIONS TO

BOTANICAL GEOGRAPHY,

DURING THE YEAR 1842,

BY PROFESSOR GRISEBACH.

TRANSLATED BY W. B. MACDONALD, B.A.

BOTANICAL GEOGRAPHY.

MARTIUS has published comparative observations on the zones, within which the northern trees which occur also on the Alps are met with. These observations were made partly on the coasts of Scandinavia and partly in Switzerland, on the northern declivity of the Grimsel (Comptes rendus, 1842, and Ann. des Scien. Nat. 18, p. 193). He opposes the opinion of Wahlenberg, that the trees do not disappear towards the North in the same succession as they do on the Alps towards the snow-line. The relations which obtain in this respect, in the Bernese Oberland, probably more resemble what occurs in Scandinavia than those of the north of Switzerland, where Wahlenberg made his observations. On the Grimsel, Martius met with but one anomaly in comparing the boundaries of the trees there indigenous with those of Norway.

It might be supposed that, on account of the unequal distribution during the year of an equal mean temperature, the parallelism of the polar and altitude boundaries, as regards individual plants, would not be perfect; and this would be more evident in the case of a forest tree. The single exception, however, which is adduced by Martius is one only in appearance.

It consists in the fact that on the northern slope of the Grimsel, the oak ceases at 800 m., and the beech at 985m.; while in Sweden and Norway, on the contrary, the oak occurs further north than the beech. This fact is correct, and generally admitted.

But the oak which in the Swiss Alps grows at that height is *Q. Robur*, while in Norway, as well as Russia, *Q. pedunculata* only is met with. Martius is mistaken in asserting that *Q. Robur* occurs so far north as 61° N. In Sweden this tree is indigenous only on the S.W. coast in the beech forest district, as Wahlenberg had already remarked in the Flora Suecica. There may be some cultivated trees at Stockholm; but Martius himself mentions that there is a beech even in the Botanic Garden of Upsal, and that he has also seen one 20′ high, and bearing fruit even at Elfkarleby, 43″ north of Upsal. *Q. Robur* retains the same relative position, it remains below the beech; whilst *Q. pedunculata*, which in the middle of Europe occurs only on the plains, lives in the open air at Trondjem, in Norway, up to the 64°. In the distribution of the birch, which extends northward so far beyond the Coniferæ, the anomaly, as in the case of the oak, is also one only in appearance. According to Martius, *Betula alba*, on the Grimsel, attains an elevation of 1975m., leaving the fir and pine behind, and would there form the arboreal limit, as in Scandinavia, did not *Pinus Cembra* occur along with it, a tree which is not met with in the north of Europe. But such a vertical extension of the *birch* is quite unusual on the Alps. It is there a rare tree, and it is well known that the forest boundary on those mountains is almost everywhere formed by *Pinus Abies*, which in Norway is left so far behind the birch. It seems as if the birch, which in the north, both above and together with the Coniferæ, forms such extensive forests, did not, in the woody regions of the Alps in general, meet with the climatic conditions necessary for its existence. This contrariety, however, depends on a difference of species, for the birch of the high northern latitudes is *Betula pubescens*, of which Blasius, in his travels in Russia, expressly remarks, that it retains the white, smooth bark much longer than

B. alba, Aut. Now, as the Linnæan *B. alba* is recognized in the northern birch, this may be called *B. corticifraga*, from the bark on the older stems cracking off, and by its broken portions giving the tree its different habit.

Martius's original observations, which are related in his memoir, are the following:—In Sweden, the highest point north at which he observed the oak was at Laeby (60° 6′), and a planted one at Hudickswall (61° 44′). The apple and pear were observed farthest north at Sundswall (62° 23′); further, *Abies* lastly, south of Karasuando (formerly Enontekis, at 68° 15′; *Pinus sylvestris* and *Sorbus aucuparia*, near Bohekop (70°); *Betula* and *Juniperus* as far as Hammerfest (70° 40′).

On the northern declivity of the Grimsel, the limits of trees, measured by Martius, are:—

Quercus Robur	(800m.)	=	2460′
Fagus sylvatica	(985m.)	=	3030
Prunus Cerasus, Corylus Avellana	(1060m.)	=	3260
Pinus Abies	(1545m.)	=	4760
Sorbus aucuparia	(1620m.)	=	4990
Pinus Mughus	(1810m.)	=	5570
Betula alba	(1975m.)	=	6080
Pinus Cembra	(2100m.)	=	6460

Hinds has given some critical remarks on the division of the earth's surface into "natural floras" (Journ. of Bot. 1842, pp. 312-18). This essay, as well as a second one by the same author, "on the regions of Alpine vegetation" (*Ibid.* pp. 128-33), without being either complete or precise, contains only the most well-known facts, whilst on the other hand, erroneous generalizations prove that, without sufficient literary knowledge, the author has yielded to the impressions received in a long sea voyage. In a more detailed manner, though uncertain in particulars, and hurrying over important points of view with superficial brevity, he has treated of the first-mentioned subjects in Sir E. Belcher's Voyage (A Narrative of a Voyage round the World during the years 1836-42, 2 vols. 8vo, London), in an appendix to which (The Regions of Vegetation, by Hinds, vol. ii, pp. 325, 460) he distinguishes and characterises 48 natural Floras. As far as his views are novel, and derived from actual observation, they will be noticed among the special works.

Another Essay by Hinds, on the 'Connexion of Climate with Botanical Geography' (Ann. Nat. History, vol. ix), is much more correct, and at the same time very copious, embracing known meteorological relations. It contains no new points, but presents to the reader a summary collection of correct views on the subject. The influence, however, of the seasons, in certain particulars, is too little attended to.

In a treatise by Langethal, on the 'Dependence of Plants upon certain kinds of Soil' (Cotta's Anleitung zum Studium der Geognosie, 1842, Hft. iv, pp. 545-60), along with a classification of plants according to the substratum upon which they grow, appears the new but quite unproved hypothesis, that the occurrence of the mountain plants of central Germany, e. g. *Trientalis, Veronica montana, Circæa alpina, Arnica montana,* in the plains of Pomerania and Mecklenburg, does not probably depend on temperature, but on the proportion of moisture.

Were this fact limited to the coast, it might probably be thus explained, but the same plants grow also at Hanover and Brunswick in the plains 20 or 30 miles from the sea. The author asserts that *Ilex Aquifolium* perishes by the frost in Thuringia and Saxony, while it supports in Rügen a winter of 12°—18°. I doubt the fact, as this plant grows in all the beech forests of Hanover, where I have experienced a cold in winter of 24° R.

He supposes that the wet summer in Rügen may alter the capacity of the plant for heat; but were it the fact that it does not thrive properly in central Germany, on how many different circumstances might not this depend?

According to some statements of the author, which are to be farther followed up, certain plants will be found in different districts to be always distributed in entirely different localities; for instance, *Myosotis sylvatica* occurs in leafy woods in central Germany, and on the sunny plains of the peninsula of Mönkgut; *Vicia Sepium* in Germany among shady bushes; in England in meadows; *Alchemilla vulgaris* in Thuringia in the forests; in Switzerland in the meadows. I made a similar observation in reference to *Vaccinium uliginosum*, where scarcely any doubt can arise as to species, which grows in Norway everywhere in the woods, and often on the dryest soils under Coniferæ.

Poech has published 'New Observations on the Vegetation of the Calcareous and Slaty Alps' (Regensb. Flora, 1842, pp. 359-67).

Fries mentioned some problems in Botanical Geography at the meeting of the Scandinavian naturalists at Stockholm (1842): he particularly explained the destructive influence of cultivation on the natural character of the vegetation of the earth. As Swedish plants now extinct he enumerated *Trapa natans* and *Xanthium strumarium*. Also the so-called ruderal plants he holds to be originally indigenous; but to the question of Schouw as to their original locality, he could assign them no other habitat except mountain precipices and the sea-shore, where they may have been easily choked up or washed away, so that now they are only to be seen like the domestic animals in the neighbourhood of man.

St. Hilaire expatiates on those plants which have followed man, spontaneously, to other parts of the earth, in an essay 'De la dispersion des Plantes sur la Globe' (Nouv. Ann. d. Voyages, 1842, pp. 54, 62); he mentions, e. g. that in Brazil and St. Paulo *Marrubium vulgare* and *Conium*, in Porto Allegre *Rumex pulcher*, in Monte Video *Echium italicum*, in Minos Geraes *Verbascum Blattaria* and *Poa annua* are to be found.

Hinds has undertaken the useful task of summing up the different species of plants in each division of the earth, in the first four volumes of De Candolle's 'Prodromus' (Ann. Nat. Hist. ix, p. 415). Of 20094 species, 3210 are European, 5004 Asiatic, 3731 African, 2111 North American, 5742 South American, and 922 Australian. It is to be desired that the author would communicate the results of this laborious enumeration in greater detail.

I.—ARCTIC ZONE.

Ruprecht and Savelieff have explored Arctic Russia in Europe between the White Sea and Petschora (Bullet. Petersb. x, p. 29). The peninsula of Kanin is flat; what is designated in charts as a chain of mountains consists only of low hills. On the island of Kolgnieff the soil is constantly to the depth of an *Arschine,* as in Northern Siberia. The plants collected are not yet published.

These regions, though poor in species, are yet covered with a pretty thick vegetation, which begins to decrease towards the north coast of Kolgnieff; it differs considerably from the Flora of Lapland. The forests have receded from the coast of the Icy Sea. Undoubted signs have been discovered that large trees formerly grew near it, but now the nearest wood is four or five miles distant.

Baer has published the corrected results of the meteorological observations in Boothia Felix (70° 2′ N.L.) Bullet. Petersb. ix, p. 10.

RANGE OF THE TEMPERATURE IN BOOTHIA.

January	=	− 26° 7 F	July	=	+ 41° 3 F
February	=	− 32 1	August	=	+ 38 5
March	=	− 29 1	September	=	+ 25 4
April	=	− 2 4	October	=	+ 9 4
May	=	+ 15 4	November	=	− 6 3
June	=	+ 34 3	December	=	− 22 6

II.—EUROPE.

Beilschmied has compared Wirzen's 'Flora of Kasan' with Weinmann's 'Flora of Petersburg' (Regensb. Flora, 1842, pp. 561-70).

The sketch of the vegetation at Hochland, in the Gulf of Finland, by Schrenk (v. Baer's Beiträge zur Kenntniss des russischen. Reichs, bd. iv, pp. 143-62), contains a view of the indigenous vegetation, with a catalogue of all the species.

The small rocky island formed of porphyry, rising to a height of 350′, is surrounded by barren cliffs, whose covering of lichens is but rarely alternated with a green sward of *Arctostaphylos*, *Linnæa*, and other less characteristic forms. Thin woods of fir and pine of low growth, cover the slopes, here and there intermixed with birches. Where the wood is more dense, it shades an undergrowth of *Rubus idæus*, *Ribes rubrum* and *alpinum*, *Viburnum Opulus*, *Daphne Mezereum*, *Vaccinium Myrtillus*, and *uliginosum*, or a mossy turf of *Polytrichum commune* and *Hypnum splendens*. The plants in the shade are *Fragaria vesca*, *Pyrola uniflora* and *secunda*, *Oxalis acetosella*, *Anemone nemorosa*, *Convallaria majalis* and *Polygonatum*, *Majanthemum*, *Linnæa;* the Glumaceæ, *Carex panicea* and *Melica nutans*. Characteristic forms of the morasses composed of *Sphagnum*, in the valleys:—*Myrica*, *Ledum*, *Betula fruticosa*, *Vaccinium Oxycoccos* and *uliginosum*, *Salix rosmarinifolia*, *Rubus chamæmorus*, *Cornus suecica*, *Neottia cordata*, *Drosera*, *Eriophorum*, *Carices*. The banks of small lakes present besides, *Scheuchzeria* and *Rhynchospora fusca* and also *Lobelia*.

These include all the formations, which, with the same elements are so widely spread over the continent. The most remarkable plant of this island is certainly *Lychnis alpina*, found on a sterile rocky summit of a wall of porphyry, called Hauka-Wuori, 448′ high and almost perpendicular; this and *Sedum annuum* are the only mountain-plants.

A third Mantissa of Fries's Scandinavian Flora has appeared which is again very important in regard to the systematic arrangement of that Flora; it bears, with the earlier divisions the common title: 'Novitiæ Floræ suecicæ. Continuatio sistens Mantiss.' I. II. III. uno volumine, comprehensas. Accedunt de stirpibus in Norvegia recentius detectis prænotiones e maximâ parte communicatæ à Blytt. Lund. 1842, 8vo.

Oersted laid before the meeting of the Scandinavian naturalists observations on the distribution of the *Algæ* in the Sound. On the Danish and Swedish coasts, three Algan regions may be distinguished. The highest corresponds to the stratum of the shore within reach of the waves, on which the *Algæ* are loosened by storms, and become accumulated on the coast; in the middle region in particular grow the Fucoideæ and *Zostera*, and in the lower the Laminariæ and Florideæ are found adhering to the solid rocks. Agardh, who was present, remarked upon this statement that, the green Algæ flourished always nearest the light in shallow water, but that the Fucoideæ and Florideæ grew independent of it in deep water.

A treatise of considerable importance on the vegetation of the Scottish Highlands has appeared, by Watson, who has measured with the sympiezometer the boundaries of the altitude of 400 plants, which are there indigenous. (The plants of the Grampians, viewed in their relations to altitude, in 'Journal of Bot.' 1842, pp. 50, 72, 241-54.)

His account expressed in English feet, relates only to the mountains between Clova and Ben Nevis, near latitude 57°. The plants of the alpine region in Scotland, extend into a much lower level than on the mountains of Norway; however, I find in this catalogue only four species, which belonging to the range of alpine forms descend to the sea coast. This is the case with *Saxifraga aizoides, Rhodiola rosea, Alchemilla alpina, Polygonum viviparum.* I have met with the last three of these also, on the sea-shore of Norway. The facts communicated by Watson on the altitudinal limits of the alpine and ligneous plants, are all contained in the following list:

ALTITUDINAL LIMITS OF THE ALPINE PLANTS IN THE GRAMPIANS.

Thalictrum alpinum, L. . . .	1050′		3900′
Draba rupestris, Hook. (*hirta* Sm.)	3700		3900
Draba incana, L.	2000		3200
Arabis petræa, Hook.	2000		3000
Silene acaulis, L. . . .	1250		4300
Lychnis alpina, L. . . .	3000		3200
Alsine rubella, W. . . .	2550		3300
Cherleria sedoides, L. . . .	2500		4000
Spergula saginoides, L. . .	1950		2700
Stellaria cerastoides, L. . .	2700		3800
Cerastium alpinum, L. . .	2300		4000
latifolium, L. . .	3100		3600
Astragalus alpinus, L. . .		2500′	
Oxytropis campestris, D. C. . .		2000	
Dryas octopetala, L. . . .	2500		2700
Potentilla salisburgensis, Jacq. .	1500		2700
Rubus chamæmorus, L. . .	1750		3300
Sibbaldia procumbens, L. . .	1500		4100
Alchemilla alpina, L. . .	0		4000
Epilobium alpinum, L. . .	1400		3900
alsinifolium, Vill. .	800		2900
Rhodiola rosea, L. . . .	0		3900
Saxifraga cernua, L. . . .	3750		3900
rivularis, L. . .	2700		3600
nivalis, L. . .	2000		4000
hypnoides, Sm. . .	1200		3900
oppositifolia, L. . .	950		4000
stellaris, L. . .	1400		4100
aizoides, L. . .	0		3200
Cornus suecica, L. . . .	1750		2850
Menziesia cœrulea, Sm. . .		2700	
Azalea procumbens, L. . .	1850		3550
Arbutus alpina, L. . . .	1850		2700
Gnaphalium supinum, L. . .	1400		4250
Erigeron alpinum, L. . . .		2500	
Saussurea alpina, D. C. . .	2000		4000
Apargia Taraxaci, W. . .	2300		3000
Hieracium alpinum, L. . .	1850		3000
Veronica alpina, L. . . .	2500		3700

Veronica saxatilis, L. . . .	2100′		2700′
Myosotis alpestris, Schm. . . .	3100		3900
Oxyria reniformis, Hook. . . .	800		4000
Salix limosa, Wahl. (arenaria Sm.)	1050		2500
lanata, L.		2500	
reticulata, L.	2500		3300
herbacea, L.	1850		4300
Betula nana, L.	1600		2750
Luzula arcuata, Hook. . .		4300	
spicata, D. C. . .	1600		4300
Juncus trifidus, L.	2000		4250
castaneus, Sm. . .	2400		3000
biglumis, L.	2800		3000
triglumis, L.	1750		3000
Carex Vahlii, Schk.		2500	
rigida, Good . .	1850		1000
aquatilis, Hook. .		2700	
pulla, Good	2500		3100
atrata, L.	2500		2700
capillaris, L.	1700		2700
rariflora, Sm.	2300		2700
Poa alpina, L.	2500		4000
Alopecurus, alpinus, L. . .	2400		2700
Phleum alpinum, L.	2400		3500
Aira alpina, L.	2200		4100

According to the first appearance of the alpine Glumaceæ, the lower boundary of this region in the Grampians may be fixed at about 2500′, and Watson's statement of the arboreal limits leads to the same result, as the Pine ceases at 2230′, and the Birch at 2000′, though on Ben Nevis it may reach as high as 2700′.

ALTITUDINAL LIMITS OF THE LIGNEOUS PLANTS IN THE FOREST REGION.

	0′	2500′
Pyrus Malus, L. . .		
Prunus spinosa, L. . .	0	400
Viburnum Opulus, L. . .		
Ribes rubrum, β. Hook. . . .	0	600
Fraxinus excelsior, L. . .		
Salix caprea, L. . . .	0	800
Sambucus nigra, L. . .		
Ilex Aquifolium, L. . .	0	1000
Ulmus campestris, L. . .	0	1050
Ribes nigrum, L. . .		
Quercus pedunculata? . .		
Cratægus Oxyacantha, L. .		
Prunus Padus, L. . .	0	1100
Ribes Grossularia, L. . .		
Ulex europæus, L. . .		

Rosa canina, L. *Lonicera periclymenum*, L.	0′	1200′
Alnus glutinosa, G. *Corylus Avellana*, L. *Rosa villosa*, L. Sm.	0	1500
Populus tremula, L.	0	1600
Myrica Gale, L.	0	1700
Rubus idæus, L. *Salix fusca*, L. *Sarothamnus scoparius*, W. G.	0	1950
Genista anglica, L. *Erica cinerea*, L.	0	2150
Pinus sylvestris, L.	0	2230
Erica Tetralix, L.	0	2370
Sorbus aucuparia, L.	0	2500
Betula alba, E. Bot.	0	2000 2700

The following ligneous plants of the Scottish coast extend into the alpine regions.

Arbutus Uva ursi, L.	0′	2800′
Calluna vulgaris, Salisb.	0	3150
Vaccinium Myrtillus, L.	0	4200
uliginosum, L.	0	3500
Vitis idæa, L.	0	3300
Oxycoccos, L.	0	2700
Empetrum nigrum, L.	0	4000

Thus, only Ericeæ, and forms allied to them, thrive equally on the coast, and in the regions where no trees grow. In this list *Juniperus communis* alone is omitted, (0′—2750′) as the British botanists make no distinction between it and *J. nana*, W. Those shrubs peculiarly belonging to the alpine regions, which are contained in the first list, are certainly for the most part found distributed also in the forest region, but not as far as the sea coast; they are as follows: *Menziesia cœrulea*, Sm., *Arbutus alpina*, L., *Azalea procumbens*, L., *Betula nana*, L., and the *Salices*.

The works published on the British flora in 1842, are a fifth edition of Sir W. Hooker's 'British Flora' arranged according to the natural system (London, 8vo.) The 'Phytologist', a journal relating to the localities of British plants, which has appeared since June of that year.

Deakin's 'Florigraphia Britannica' (Sheffield, 1835-1842); a popular work with plates, on British plants.—Lee's 'Remarks on the Flora of the Malvern Hills in Worcestershire,' &c. (read in the Botanical Society of London).—Balfour has made a report of an excursion in the district of Braemar in the Scottish Highlands (Ann. Nat. Hist. x, p. 117.)—Edmonstone has sent to the Edinburgh Botanical Society, a list of his Flora of the Shetland Islands, increased by fifty phanerogamic species (Ann. Nat. Hist. ix, p. 69.)

Van der Bosch has published an appendix to his Flora of Zealand. (Van der Hoeven's Tijdschr. 1842, pp. 245-65.)

Four Decades of the sixth century of Reichenbach's 'Icones Floræ Germanicæ' have appeared in 1842, containing the Caryophylleæ; twenty-three Centuries of the 'Flora Germanicæ exsiccata' have been published. Eighty-eight numbers of the first part of Sturm's 'Flora Deutschlands' are now completed. An illustrated work on the 'Flora Deutschlands' by V. Schlechtendal and Schenk, has reached the first number of the fourth volume. A similar one on the plants of Thuringia by the same authors has reached the thirty-eighth number. Lincke has also published figures to the German Flora, of which twenty-one have appeared; and also, the wild plants of Prussia, of which seventeen numbers have appeared. The third volume of a 'Flora Deutschlands,' by Meigen, has come out. The fourth Century of the instructive collection of dried *fungi*, begun by Klotzsch and continued by Rabenhorst, has been published. Of German provincial Floras are to be mentioned: Klinsmann 'Novitiæ atque defectus floræ Gedanensis (in the 'Schriften der Danziger naturf. Gesellsch,' Bd. iv); Reichenbach 'Flora Saxonica', (Dresden, 1842, 8vo. Also, under the title of 'Der deutsche Botaniker,' Bd. 11), which besides the kingdom of Saxony comprehends also the Saxon dukedoms, the adjacent Prussian provinces, and the principalities included in this area; Heinhold and Holl 'Flora von

Sachsen,' (Dresden, 1842, 8vo,) which comprises the same district; Hampe's appendix to his catalogues of the plants found on the Harz, (Linnæa, 1842, pp. 380-3, and in the pamphlet: 'Vier Verzeichnisse zur Kenntniss des Harzes.,' Nordh. 1842, 8vo); Dolliner 'Enumeratio plantarum phan. in Austria inferiori crescentium' (Vind. 8vo); Maly 'Nachträge zur Steirischen Flor' (Regensburg Flora 1842, pp. 251-6.)

In the professional journies of Ratzeburg through the German forests, are included the character of the great forests of northern Germany, with reference especially to matters of forestry, though constant regard is paid to the nature of the soil as dependent upon the substratum, and to other relations connected with the botanico-geographical characteristics of the forest formation. (Berlin, 1842, 8vo.)

The author attempts to prove the influence of the geognostic formations on the intimate nature of the soil, and gives many hints which should be further followed by the botanist. The investigation of the soil has reference to the following geognostic formations, and is particularly directed to—(*a*) its power of retention of water; (*b*) its contents in parts which can be washed away (clay and fine sand); (*c*) in combustible parts; (*d*) in water; (*e*) in oxydes soluble in hydrochloric acid; (*f*) insoluble earthy components. In this way the sorts of soil of the following mountain formations are compared. What in this table is wanting in 100 parts, always corresponds to the insoluble earthy ingredients.

	a	*b*	*c*	*d*	*e*	*f*
1. Speckled sandstone of Solling and from Grubenhagen.	45·57	39·45				24·0 per cent.
	54·6		2·78	1·6	2·8	
2. Porphyry free from quartz from Ilefeld	22·25	8·65	0·31	0·06	0·46	
	33·8	18·79				28·65
3. Basalt from Solling	89·24	21·6			0·65	35
4. Clay slate from the Harz at Lauterburg	45·2	18·1	0·2	0·27	0·29	
5. Trachytes from the Rhenish Siebengebirge	38·58	20·08			0·62	30·54
6. Porous lava of the Laacher See	81·1	5·2	3·01	2·41	1·66	

The separated oxydes under (*e*) occurred in the following proportions:—

	No. 1.	No. 2.	No. 3.	No. 4.	No. 5.
Calcareous earth	0·83	0·126	0·07	0·04	0·07
Aluminous " .	0·605	0·288	0·25	0·18	0·28
Talcose . " .	0·224	0·038	0·01	0·065	(in the C. E.)
Oxide of iron .	0·908	(in the Al. E.)	0·31	(as No. 2)	0·22
Alkalies . . .	0·23	0·007	0·01	0·005	0·05
	2·797	0·459	0·65	0·29	0·62

The fertility of volcanic hills is therefore principally owing to their power of retaining water, as well in the basalt as in the lava soil of Eifel, without its being rich in aluminous constituents, capable of being washed away. This quality appears therefore dependent on the minerals constituting the sand of these soils. The author characterizes the sylvan vegetation on this and other geographic formations, and also completes this description by lists of the shade plants which occur in the greatest numbers. I here arrange from this catalogue the results for twenty of the most widely distributed wood plants, the occurrence of which the author compared in great masses in the ten following different localities:

I. In beech woods	*a*	On the Porphyry at Ilefeld.
	b	Clay slate of the Rhenish Schiefergebirges.
	c	Trachyte of the Siebengebirges.
	d	Granite of the Spessarts.
Mixed with oaks	*e*	Speckled sandstone of the Spessarts.
II. In oak woods	*f*	Alluvium of Upper Silesia.
III. In woods of *P. Abies*	*g*	Granite of the Riesengebirges.
IV. " *Picea*	*h*	Alluvium of Upper Silesia.
Mixed with *Abies*	*i*	Humus soil of the Upper Silesian alluvium.
	k	Granite of the Riesengebirges.

PREVAILING PLANTS.

Hypnum crista castrensis, a. g. h.
Schreberi, a. g. h.
splendens, a. g. h.
Aspidium filix fœmina, c. g. i.
Polypodium Dryopteris, a. e. g.
Agrostis vulgaris, b. c. k.
Aira flexuosa, b. c. g.
Deschampsia cæspitosa, a. b. g.
Luzula albida, a. b. c. e.
Urtica dioica, a. f. g.
Atropa belladonna, b. e. h.
Hieracium sylvaticum, c. e. g. h. k.
Prenanthes muralis, a. b. h. k.
Senecio Jacobæa, b. c. d.
Cirsium lanceolatum, a. f. h.
Fragaria vesca, a. b. h.
Epilobium montanum, a. b. c.
angustifolium, b. g. h.
Hypericum perforatum, a. b. h.
Oxalis acetosella, a. g. h. i.

Ratzeburg's descriptions of the German forests refer to the Harz, Solling, the Rhenish Schiefergebirge, the Eifel, the pit-coal formation between Trier and Südweiler, the beech and oak woods of which, he declares to be the most beautiful in Prussia, the Spessart, French Jura, the Riesengebirge, and the alluvium of the Oder in Upper Silesia, containing the largest forests, whose soil composed of humus is like bog with a mixture of loam, but free from acid and iron, and bears very splendid mixed woods of *P. picea* and *P. abies.* I may mention, from isolated observations of the author on geographic botany, what he says of the boundaries of the height of the beech. This tree forms in the Siegen a lofty forest, even at 2065′, and extends in these mountains higher than the oak, while in the Harz it does not reach to much above 2000′. In Hundsrück at Tronecken, the beech reaches to 2500′, but in the Riesengebirge as high as 4000′.

Wallroth draws out the following list of calcareous plants of the southern Harz in Ratzeburg's journeys (p. 15): *Calamagrostis montana*, *Stipa capillata*, *Sesleria cærulea*, *Festuca glauca*, *Carex humilis*, *Allium montanum*, Schm., *Anthericum ramosum*, *Orchis ustulata*, *Ophrys myodes*, *Epipactis atrorubens*, *Cephalanthera rubra*, *Betula pubescens*, *Thesium intermedium* and *montanum*, *Lithospermum purpuro-cæruleum*, *Alectorolophus angustifolius*, *Gentiana ciliata*, *Scabiosa suaveolens*, *Inula hirta*, *Cineraria campestris* and *spathulifolia*, *Scorzonera purpurea*, *Asperula tinctoria*, *Cornus mascula*, *Libanotis vulgaris*, *Potentilla supina*, *Rubus saxatilis*, *Rosa cinnamomea*, *Viola arenaria*, *Gypsophila fastigiata* and *repens*, *Reseda lutea*, *Biscutella lævigata*, *Hutchinsia petræa*, *Arabis auriculata*, *Erysimum virgatum* and *odoratum*, *Thalictrum montanum*. Some names are here altered, and plants which occur certainly in general on limestone, but are likewise distributed on other sub-soils, omitted, as also, calcareous forms, which I do not hold as specifically different from others.

List of the calcareous Cryptogamia of the southern Harz: *Gymnostomum curvirostrum*, *pusillum*, *trichodes*;

Jungermannia incisa and *gypsophila*, W.; *Marchantia commutata*, *hemisphærica*, and *umbonata*, W.; *Grimaldia inodora*, W., *punicea*, W., *ventricosa*, W., *Verrucaria mutabilis*, W.; *Patellaria cæsia*, W., *epipolia*, *saxicola*, W., *variabilis*, W., *intermedia*, W., *candida*, *lentigera*, *teicholyta*, W., *fulgens*, W., *decipiens*, W., *testacea*, W.; *Parmelia gypsophila*, *versicolor*, W.

Poech has described the vegetation of the calcareous formation of St. Iwan in Bohemia (Regensb. Flora, 1842, pp. 410-14.) Characteristic plants of the rocks there: *Dracocephalum austriacum*, L., *Hieracium echioides*, Kit., *Artemisia scoparia*, Kit., *Alsine setacea*, K., *Saxifraga cæspitosa*, L., *Potentilla cinerea*, Ch., *Sempervivum hirtum*, L., *Seseli glaucum*, Jacq., *Allium strictum*, Schr. and *fallax*, Don, *Gagea bohemica*, R. S., *Iris bohemica*, Schm. Growing among the oak bushes: *Carex Michelii*, Host., *Cytisus biflorus*, L'Hér., *Hieracium Nestleri*, Vill., *Silene nemoralis*, Kit. In the leaf-woods: *Adenophora suaveolens*, Fisch. Botanical travels through the Julian Alps by Sentner, and in Frioli by Tommasini, are of local interest. (Ibid. pp. 442-79, and (609-35.)

Schärer has communicated some facts on the distribution of the lichens on the highest summits of the Swiss Alps, (Linnæa, 1842, p. 65.) Saussure, senior, had brought from the highest point of Mont Blanc, *Parmelia polytropa*, Schär., and *Lecidea confluens*, Ach. In Sept. 1841, Agassiz found on the top of the Jungfrau (12,850′) the following: *Lecidea conglomerata*, Ach., and *Parmelia elegans* var. *miniata*, both with undeveloped fruit; of *Lec. confluens* var. *Steriza*, Ach., fruits without thallus; and two Umbilicariæ *U. atropruinosa* var. *reticulata*, Sch., and a new species *U. virginis*, Sch., allied to *U. hirsuta*.

Wierzbicki has made a report of botanical travels in the Banat. (Regensb. Flora, 1842, pp. 257-80.) We find in them more exact information on the Hungarian oak (*Q. conferta*, Kit.), which may be compared with *Q. apennina*, Lam. Besides this, *Q. robur* and *Q. cerris* grow in the woods of the Banat.

Of French local floras have appeared: Holandre (Nouvelle Flore de la Moselle, ed. ii,) and Delastre (Flore de la Vienne, Paris, 1842.) The works of Desmazières during the past year are confined to French Fungi. (Ann. Sc. Nat. 17, pp. 91-128.)

W. P. Schimper has described the moss and lichen-vegetation of the sandstone of the Vosges. (Regenb. Flora, 1842, pp. 337-59.) He remarks that, the Vosges sandstone, through weathering, forms a sandy soil; but the speckled sandstone of Alsace forms a clayey marl, which produces some moss and lichens, e. g. *Barbula aloides, brevirostris* and *rigida, Funaria hibernica, Grimmia ovalis* and *leucophæa, Lecidea reticularis,* which are not found on the Vosges sandstone. The granite group of the Ballons is covered with firs, while the Vosges sandstone mostly bears beech, among which single birches or oaks are found. Coniferæ have only lately sprung up in artificial plantations. At Offweiler, *Castanea* also forms a wood, but on the south side of the mountain it only reaches to a height of 600′. The Vosges sandstone is very rich in mosses and lichens, the localities have been specially given by Schimper.

In a review of Boissier's 'Flora of Granada,' I have drawn attention to the varied distribution of the S. European mountain-plants (Gött. gel. Anz., 1842, p. 599.) Thus the plants which inhabit the alpine region of the Sierra Nevada, fall into six classes, according to their distribution.

1. Endemic plants of the region which are found nowhere on the earth but above the tree-boundary of Sierra Nevada.

2. Endemic plants of Spain, which, in consequence of their indifference to climate, ascend from the lower regions to the high mountains.

3. Arctic-alpine plants, which occur also upon the high mountains of central Europe, on the Pyrenees and Alps, as far as the Caucasus, partly without difference of altitude; whose natural habitat everywhere is only above

the tree boundary, and some of which again appear in Lapland on the low hills. These form in many families about the half of the species observed in the Sierra Nevada, above the level of 5000′; but, on the other hand, in the Umbelliferæ only one quarter, in the Synanthereæ and Scrophularineæ only one fifth.

4. South European mountain-plants, which first appear beyond the Alps, and there grow on most mountains above the tree-boundary. Of these, however, the greater number are found only in the lowest part of the alpine zone, and descend abundantly into the forest region.

5. Central European plants, of which certain species, only extend to the mountain and alpine region of S. Europe, while a greater number grow on the Mediterranean, on an equally low level as in the plains on this side.

6. Lastly, there are a number of Mediterranean plants which ascend from the coast into the alpine region. It is evident that these different component parts of the alpine vegetation, although found here, united under the same conditions of climate, by no means possess the same receptivity in regard to the external influences to which they are exposed. In a systematic point of view, they are thus in part widely separated, and we consequently find that, the most characteristic families of the Alpine Flora, are richest in plants of the third class, and that forms of a similar type can be distinguished in the 1st and 4th classes in a still higher degree than in the rest.

I have remarked (ibid. 606) on the 'Distribution of the Leguminosæ' in the south of Europe, that the maximum of the Genistæ is in Spain, of the Trifolieæ in Italy, and that the Viciæ increase in Greece, while the Astragaleæ first preponderate in Asia Minor.

Schouw has made researches on the plants dug out in Pompeii, and believes he can prove from these remains that, *Agave-americana* and *Opuntia* were known in ancient Italy. He read a report on this subject at the meeting of naturalists in Stockholm, but I am not in a condition to be able to express an opinion on so remarkable and unexpected a result.

Tenore has written a treatise on the species of Cotton cultivated in the Neapolitan States, (in the 'Atti del real Istituto di Napoli,' 1840, pp. 175-206.) From the *Gossypium herbaceum*, L. (Cav. diss. t. 164, f. 2), cultivated most abundantly in the south of Europe, he distinguishes a new species *G. siamense*, Ten (l. c., t. 2), which bears flowers without spots. In Naples it is called *Cotone siamese* or *turcheso*, in commerce also *C. di Castellamare*. It has been generally considered as *G. religiosum*, but it is not a shrub, but rather an herbaceous biennial like *G. herbaceum;* it does not, however, seem different from *G. hirsutum*.

Of Italian local floras, and other contributions to the knowledge of the plants of that country, have appeared (Gr. Trevisan) 'Prospetto della Flora Euganea' (Padova, 1842, 8vo,) which contains a catalogue of Paduan plants, as a precursor of a complete Flora; Parlatore (Plantæ novæ s. minus notæ, Paris, 1842), 8vo, not unimportant systematically, on the subject of the Italian Gramineæ and some other plants; Meneghini (Alge italiane e dalmatiche illustrate, Fasc. 111, Padova, 1842, 8vo); J. G. Agardh (Algæ maris Mediterranei et Adriatici, Paris, 1842, 8vo); Zanardini (Synopsis Algarum in mari Adriatico hucusque collectarum, in the Memorie dell'Academia di Torino, 1842, pp. 105-256,) embraces 245 species of Algæ in 79 genera; De Notaris (Algologiæ maris Ligustici Specimen,—ibid. pp. 273-316) embraces 127 species in 56 genera; Meneghini (Monographia Nostochinearum Italicarum addito specimine de Rivulariis, Turin, 1824, 4to.) The work of Bertoloni on the Apennines of Bologna, mentioned in last year's report, has been published in the 'Novi Commentarii Bononienses,' the fifth volume of which appeared in 1842.

Hogg has published a catalogue of Sicilian plants, collected from former sources. (Ann. of Nat. Hist., x, pp. 287-335.) This useful collection is on the Linnæan system. The critical observations of the author, in comparison with Philippi are unimportant, except the assertion, that the cultivation of the vine on Ætna ceases at a height of 2600′, and not 3300′.

In Biasoletto's Travels in Dalmatia, no view is given of the relations of vegetation, except an appended catalogue of the plants collected; and although this contains the greatest portion of the Dalmatian flora, yet it has been rendered superfluous by the excellent work of Visiani, which has appeared in the meanwhile, (Flora Dalmatica, Lips., 1842, 4to, vol. 1;) in the introduction, some contributions are made to the botanico-geographical characteristics of the country. The annual course of vegetation on the coasts of Dalmatia is as follows: after a short mild winter, several spring plants come out, even at the end of December or beginning of January; *Corylus Avellana, Colchicum montanum, Helleborus multifidus, Erodium pimpinellifolium* flower at that time, and *Amygdalus* somewhat later; in February the whole vegetation awakes, and the blossoming of almost all the Dalmatian plants occurs in this and the following months up to May, when the dry hot summer begins which lasts till the end of September, a season during which vegetation is almost wholly interrupted, as the annual plants have quite disappeared, the bushes are no longer green, and even in the trees and shrubs the formative impulse stagnates; lastly follows the rainy autumn in October and November, in which many plants blossom for the second time. Visiani, quite in accordance with nature, divides the Flora of Dalmatia into three regions, the upper of which in that country, on account of its want of wood, probably descends unusually low. He characterizes these regions by the following plants:—

1. Coast-region, 0′—1410′. *Olea, Arbutus Unedo, Laurus nobilis, Nerium, Pinus Pinea* and *halepensis, Pistacia lentiscus, Phillyrea, Rosmarinus, Rhamnus Alaternus, Cistus villosus* and *Monspeliensis.—Trichonema Bulbocodium, Andrachne telephioides, Crozophora tinctoria, Arum tenuifolium.*

2. Hill-region, 1400′—3000′. *Fagus, Acer pseudoplatanus* and *obtusatum, Quercus Cerris.—Cytisus Weldeni, Rubus idæus.—Gentiana lutea, Valeriana montana* and

tripteris, *Teucrium Arduini*, *Prenanthes purpurea*, *Centaurea montana* and *tuberosa*, *Helleborus multifidus*, *Hypochaeris maculata*, *Cerastium grandiflorum*, *Primula suaveolens*. A singular mixture of plants of the central European mountains, and of Roumelia, in which in this locality, as on the Scottish coasts, some alpine forms occur at a lower level.

3. Subalpine region, 3000′—6000′. *Juniperus nana*, *Rosa alpina*, *Lonicera alpigena*.—*Dryas octopetala*, *Arabis alpina*, *Androsace villosa*, *Pæonia Russi*, *Silene graminea*, *pusilla* and *Tommasinii*, *Campanula pumilio* and *serpyllifolia*, *Arenaria Arduini* and *gracilis*. This region is rich in endemic species.

The author characterizes also most of the groups of plants of Dalmatia, in doing which, however, he pays too much attention to the rare species, to afford a distinct representation. He advances as a striking peculiarity, that many vicarious species of the same genus may be remarked, if the Dalmatian Flora be compared with that of the neighbouring countries, or its own regions with each other. As for example, in the wood or hill-region, *Thymus Serpyllum*, *Salvia pratensis*, *Phleum pratense*, *Thesium Linophyllum*, *Genista sylvestris*, *Onopordon Acanthium*, *Anthriscus trichosperma*, appear in the place of those inhabiting the coast-region, as *Thymus acicularis*, *Salvia verbenaca*, *Phleum echinatum*, *Thesium divaricatum*, *Genista dalmatica*, *Onopordon illyricum*, *Anthriscus Cerefolium*.

The portion of this flora which has as yet appeared, contains about a fourth of the whole in detailed systematic order, particularly the monocotyledons and apetalous dicotyledons. When the work is completed, the statistical numerical relations will be brought together.

Margot and Reuter have completed the catalogue of a Flora of Zante, begun in 1839, and which appeared in 1841, in the 'Mémoires de la Société de physique, &c. de Genève.' This list contains 621 phanerogamic and 42 cryptogamic plants, and is an important contribution to the Greek Flora. I do not notice any new species in the second division. Beside the localities, there is always one distinguishing synonyme for the species; the authors have but rarely found occasion for critical remarks.

Spruner in Athens, who for some years has distributed collections of Greek plants, has made a report on a botanic journey to the southern Pindus. (Regensb. Flora, 1842, p. 636.) He ascended the Velugo near Carpenitzi, which rises to 7000′ the highest top of Pindus in Greek Roumelia. This mountain which consists of limestone is clothed with wood to an altitude 5500′. The oak woods reach on the south to 3000′, the rest of the space is occupied by Coniferous forests. Above the tree-boundary succeeds a zone of *Traganthastragalus*, and on the topmost summit, as upon Athos, *Prunus prostrata*, Lab. grows. C. H. Schultz has described several Greek Synanthereæ collected by Fraas. (Regensb. Flora, 1824, Beibl. p. 158.)

III.—ASIA.

The Flora of the East has also been systematically studied last year in every direction. Boissier has continued his examination of Aucher-Eloy's collection, and published the results, partly in the 'Annales des Sciences Nat.' for 1842, and partly in a separate synoptical work, under the title 'Diagnoses Plantarum Orientalium' (Fasc. 1, 1842, Genev. 8vo), of which the first number appeared in 1843. The former work embraces only the Cruciferæ, of which Aucher has collected no fewer than 343 species. In this family the East seems inexhaustible in the multitude of fugacious annual species, and the narrow bounds of their distribution. A review of remarkable forms with numerous new species is given, particularly of *Hesperis, Alyssum, Cochlearia:* 2 new *Morettiæ* from Muscat; *Diceratium* B.: Persian subfrutex; *Parlatoria, Zerdana,* and *Strophades,* Sisymbreæ from Kurdistan and Persia; several *Drabæ* from the Armenian and Persian mountains; *Moriera, Brossardia,* and *Heldreichia,* new Persian Thlaspideæ; several species from the allied group of *Æthionema.* In Boissier's separate work, new species only are described, among which, besides those of Aucher, there are also some Greek and Roumelian ones from the following families: 3 Capparideæ, 5 Resedaceæ, 3 Violaceæ, 4 Polygaleæ, 101 Caryophylleæ, 4 Lineæ, 4 Hypericineæ, 6 Geraniaceæ, 4 Zygophylleæ, 2 Rutaceæ, 9 Ranunculaceæ, 1 Fumariacea, 15 Cruciferæ, 5 Rhamneæ, 1 Terebinthacea, with 190 Leguminosæ, and 14 Dipsaceæ. It is to be desired that the complete view of Aucher's collection should not be interrupted by this publication.

In Russegger's Travels in Greece, Lower Egypt, Syria, &c. (Bd. I, Stuttgart, 1841, 8vo), some information is to be found on the Flora of the Taurus, it has, however, been borrowed from the 'Researches' of Ainsworth, which appeared in 1838. The work contains a botanical appendix by Fenzl (pp. 883-990), in which the new species of plants collected by Kotschy in Syria and the Taurus are described at length. In the introduction, Fenzl reports on the considerable extent of the collections of Kotschy, which are preserved in the Vienna Museum. The first parcel sent home contained 700 species, mostly from the Western Taurus, the rest from the Orontes and from Karamania and Lebanon; another collection consists of about 300 species from Aleppo. The Synanthereæ, Labiatæ, Leguminosæ, and Umbelliferæ prevail in these herbaria; next follow the Caryophylleæ, Cruciferæ, and Scrophularinæ; the character of Mediterranean vegetation displays itself throughout with a mixture of the Caucasian forms. Even most of the new genera and species, although they certainly form the eighth part of the collection, bear the stamp of forms complementary to the European; a few only can be adduced as characteristic of the vegetation of those parts, e. g. *Pelargonium Endlicherianum* from the Taurus; *Heldreichia Kotschyi* from the alpine regions of the mountain; *Silene pharnacefolia* and *stentoria*, *Viola pentadactyla*, *Elæochytris meifolia*, and *Actinolema eryngioides* from Syria. The diagnoses of the new species have been printed separately by Fenzl in his 'Pugillus Plantarum Novarum Syriæ et Tauri Occidentalis primus,' (Vindob. 1842, 8vo,) and are also given by Walpers in his 'Repertorium.' In the year 1843, the first livraison of a book of plates, which Fenzl has grounded on the materials collected by Kotschy, has also appeared.

Forbes has given some account of the time of flowering of the plants, which bloom during the winter months in Lycia (Ann. Nat. Hist., ix, p. 251).

The first seven livraisons of Gr. Jaubert's and Spach's 'Illustrationes Plantarum Orientalium,' have appeared (1842), to which it is more convenient to recur afterwards.

Bertoloni has begun to labour completely at the plants collected by Chesney on the Euphrates expedition, and divided by Lindley, of which many cursory notices have been given. The species belonging to the first eight Linnæan classes have been described by him in the 'Commentarien des Instituts von Bologna,' (1842), but he does not seem to have made sufficient use of the more recent literature on the Flora of the East. There are figures of several new species.

C. Koch, who lately published at Cotta a description of his first journey in the Caucasus, with which I am not yet acquainted, has continued his catalogue of Caucasian and Armenian plants. During last year the following families have been discussed, (Linnæa, 1842, pp. 347-373): 71 Rosaceæ (new, 1 *Rubus*, 1 *Potentilla*,) 1 Myrtaceæ, 4 Crassulaceæ, 2 Saxifrageæ, 1 Philadelpheæ, 1 Lythraræe, *Trapa* from Mingrelia, 6 Onograriæ, 3 Ribesiaceæ (1 new *Ribes*), 60 Umbelliferæ (12 new species, among which are the newly characterized genera *Sympodium*, *Fuernrohria*, *Froriepia*, and *Eleutherospermum*), 1 Araliaceæ, 2 Corneæ.

Godet has made a catalogue of the plants found in the environs of the Caucasus baths of Petigorsk, which forms an appendix to the fourth part of Dubois de Montpéreux's 'Voyage autour du Caucase,' (Paris, 1840, 8vo,) and refers partly to his collections made on the Beschtau, one of the mountains of the Caucasus, about 4100′ high, composed of trachytic porphyry, completely covered with woods of *Fagus*, *Carpinus*, *Quercus*, and *Acer tataricum*.

Some notices on the course of the vegetation on the Caucasus are contained in the introduction to Eichwald's 'Fauna Caspiocaucasia,' (Petrop. 1841, 4to.) At the northern foot of the Caucasus, e. g. towards Derbend, and also at Kuban, the plants are already all withered in

July and by the end of September the country around is found quite barren from the drought. In the valleys of Imaretia, on the other hand, the summer is moist, for here all the aqueous vapour which rises from the Black Sea, attracted by the large woods, is changed into mist.

Two important and larger works on the Flora of Asiatic Russia, founded on accurate and very recent researches, are contained in the 'Bulletin de la Soc. imp. des Natur. de Moscou,' (Mosc., 1842), of which the first three numbers have been received. The one is a copious Flora of the Baikal region, by Turczaninow, the other a description of the plants collected in the year 1841 by Karelin and Kirilow, in the steppes of eastern Soongaria, and on the Alps of Alatau. Both treatises are arranged according to De Candolle, the former, still incomplete, characterizing 277 species, reaches to the termination of the Rutaceæ; the other comprehending 932 species, is already finished.

Turczaninow's Flora commences with an interesting introduction on the peculiarities of the Baikal Flora, comprising the districts of Nertschinsk, Werkné-Oudinsk, and Irkutsk; the first he usually calls the *Daurian*, the second, the *Transbaikal*, and the third, the *Cisbaikal* district. The complicated mountain tracts there often rise above the forest-boundary, but never reach the snow-line. Adjoining the hills and their forests are large steppes, partly stony, partly covered by a soil containing a bitter salt, rarely sandy; they often inclose lakes and marshes. The Baikal itself is nearly everywhere encircled by rocky mountains. The whole materials collected by Turczaninow appear to consist of about 1400 phanerogamic plants; he remarks that, 452 of these are also found in the 'Flora Suecica' of Wahlenberg, and 756 are also indigenous in the Altai. Three families represented in the Altai are wanting in the Baikal, the Frankeniacæ, Paronychieæ, and Apocyneæ; while the Menispermaceæ of the Baikal are wanting in the Altai. The lists of the plant-limits observed in the Baikal region are more important; they show that many plants do not pass beyond the Baikal lake; others are confined to Dauria; but the plants mentioned are too numerous, and later researches will doubtless shorten these lists considerably. I find also several errors in the catalogue of European plants, which indigenous in Dauria are wanting in Siberia, which lies between. As such statements are most to be depended upon with regard to ligneous plants, I shall here only mention them. 1. Among about ninety plants found in Irkutsk, but distributed eastward as far as the Baikal, are *Cotoneaster multiflora, Arctostaphylos alpina, Andromeda calyculata* and *polifolia, Daphne Mezereum, Salix microstachya*, T., *Betula fruticosa*. 2. Among about 160 plants found in the Transbaikal district, extending westwards also as far as the Baikal, but partly again appearing in the region of the Altai are: *Rhamnus daurica* and *Erythroxylon! Caragana microphylla! pygmæa* and *spinosa! Amygdalus pedunculata! Spiræa thalictroides* and *hypericifolia! Ribes diacantha, pulchellum*, T.! and *Dikuscha*, Fich.! those marked with an (!) are wanting eastwards in Dauria. 3. Among about 140, found only in Dauria are: *Rhamnus polymorpha*, T., *Armeniaca sibirica, Spiræa sericea*, T., *sorbifolia* and *lobata, Viburnum dauricum, Lonicera chrysantha*, T., *Betula daurica* and *Gmelini*, Bg., *Alnus sibirica*, Fisch., *Quercus mongolica*, Fisch., *Pinus daurica*, Fisch.

Of the families hitherto treated there are: 81 Ranunculaceæ, 1 Menispermeæ, 1 Berberideæ, 5 Nymphæaceæ, 3 Papaveraceæ, 6 Fumariaceæ, 78 Cruciferæ, 19 Violaceæ, 2 Droseraceæ, 2 Parnassiæ, 3 Polygaleæ, 22 Sileneæ, 1 Spergularia, 36 Alsineæ, 1 Lineæ, 1 Malvaceæ, 2 Hypericineæ, 8 Geranieæ, 1 Balsamineæ, 1 Oxalideæ, 1 Zygophylleæ, 2 Rutaceæ. The new species, as well as those described by Karelin and Kirilow, have already been given in the appendix to Lebedour's 'Flora Rossica.'

A more detailed analysis of the plants found in the Soongarian steppes, might have been given from the two catalogues of Karelin and Kirilow, in conjunction with

Schrenck's discoveries, if these collectors had more exactly distinguished the localities on the Alatau and Altai. In this year's publication I find about 300 distinct steppe-plants, while more than two thirds of those collected grow in the Alatau. According to that list, the steppe Flora of eastern Soongaria, contains the following series of prevailing families: (1) the Leguminosæ with 30-40 species, particularly Astragaleæ; and (2) Cruciferæ; (3) Synanthereæ, especially represented by *Artemisia* and *Cousinia*, about 30 species; (4) Cruciferæ with 20-25 species; and (5) Gramineæ; (6) Boragineæ with 10-15 species; (7) Liliaceæ and Umbelliferæ; then follow Labiatæ, Caryophylleæ, and Cyperaceæ.

Several plants lately discovered by Schrenck in Soongaria, particularly on the Tabaryatai and Alatau, are published in the 'Bulletin scient. de S. Petersbourg, (vol. x, pp. 253 and 353.) They belong to the following genera: *Picea*, *Populus*, *Stellera*, *Rheum*, *Rosa*, *Oxytropis* (3 species); *Astragalus*, *Swertia*, *Solenanthus*, *Calamintha*, *Chamæpeuce*, *Saussurea* (2 species); *Allium* (3 species); *Carex*, *Bromus*, *Triticum*. The new genera of Synanthereæ, *Cancrinia*, *Waldheimia*, *Richteria*, and *Acanthocephalus*, all from the Alatau, are described by Karelin and Kirilow in the 'Bullet. de Moscou (1842, pp. 124-8); in their larger work are two new genera of Cruciferæ, *Spirorhynchus*, and *Cryptospora*, as well as two new genera of Liliaceæ, *Ammolirion*, and *Henningia*.

Siebold's 'Flora Japonica,' which has been elaborated by Zuccarini, has advanced to the third number of the second volume (Lugd. Bat. 1842, 4to.) The last three numbers contain the Coniferæ. The following species of this family are Japanese: *Sciadopitys verticillata*; *Abies leptolepis*; *Tsuga firma*, *homolepis*, *bifida*, *jezoensis*, *polita*; *Pinus densiflora*, *Massoniana*. The species distributed most abundantly from the coast to 3500′ are *parviflora*, *koraiensis*.

We are indebted to Cantor for some information about the Island of Chusan on the Chinese coast (30° 0′ N. L.), who, in the months from July to September, collected there an herbarium of 150 species; the generic names of which have been revised by Griffith (Ann. Nat. Hist. ix, p. 265.) Chusan is about four geographic miles long, two broad, rises to the height of 1800′, and is covered by rocky hills, composed of clay-stone porphyry, and green stone. All the water is artificially collected for agricultural purposes; the woods have all been rooted up; and every spot of land is cultivated for the greatest possible amount of produce. In the cultivation of Chusan, rice holds the first rank, and yields, as it appears, two crops. Of other kinds of grain, Maize, *Coix Lacryma, Sorghum, Polygonum*, are mentioned. Among other vegetables, *Convolvolus batatas* (sweet potato) is the most important. The culture of tea is inconsiderable; the shrub flowered in July, had ripe fruit at the end of September, and flowered again at the commencement of November. *Stillingia sebifera* is cultivated to a considerable extent; it flowers in July and August, and the fruit arrives at maturity in November. The seeds, after having been taken out of the capsules, are thrown into large vessels of boiling water, upon the surface of which, after it is allowed to cool, the white tallow-like substance is left. Small plantations of *Elæococcus Vernicia* Juss. are cultivated for the purpose of making varnish. Oaks and pines are there planted for timber. The extremes of temperature must be considerable. In the year 1840, after a hot summer, heavy showers fell from the end of September till the end of November; snow fell in the end of December, and in January the thermometer sunk to 22° F. In the following summer, Cantor observed the range of temperature as follows:

	Highest.	Lowest.
July . . .	86°	79°
August . . .	93	76
September . . .	100	71
October . . .	84	58

The vegetation of Chusan has a true European character; the few Indian forms which occur are not characteristic, and many of them, such as the Palms and the Musaceæ, which are cultivated, do not bear ripe fruit. *Humulus Lupulus* is extraordinarily abundant, and Gützlaf's inquiries have established it beyond doubt to be indigenous. Cantor saw in a tea plantation a grouping of plants, highly characteristic of the Chusan Flora. A *Tea tree*, around which was entwined a *hop vine*, afforded shade to *Artemisia vulgaris, Hypericum perforatum, Viola canina*, and in its neighbourhood were *Rubus idæus* and *Fragaria*, together with *Pinus* and *Quercus*, a *Musa* and *Raphis flabelliformis*. The list of genera is of less interest, in consequence of the cultivated and ruderal plants not being distinguished from the rest. I here place together the forms which are not European, and which compose only the fifth part of the genera enumerated, and add to these the cultivated plants, or species otherwise remarkable: *Nelumbium, Sterculia, Gossypium, Citrus, Thea chinensis, Camellia, Vitis vinifera, Zanthoxylon, Phaseolus, Persica, Prunus, Pyrus, Cydonia, Eriobotrys japonica, Lagerstroemia indica, Myrtus, Punica, Cucumis Melo, Cucurbita maxima* and *lagenaria, Actinostemma* Gr. (new Cucurbitacea), *Hamamelis, Panax* (the Cinchonaceæ), *Pæderia* and *Gardenia, Olea fragrans, Convolvulus Batatas, Ipomœa cœrulea, Capsicum, Rosmarinus, Clerodendron, Sesamum, Celosia, Begonia, Polygonum fagopyrum, Rheum. Stillingia* and *Elæococcus* (v. sup.), *Phyllanthus, Chloranthus, Salix babylonica, Cannabis sativa, Morus, Ficus, Juglans, Quercus, Salisburia, Pinus, Juniperus, Cupressus, Zingiber, Musa, Pardanthus, Commelina, Rhaphis, Aseca, Catechu, Triticum, Zea, Saccharum, Bambusa, Oryza, Coix, Sorghum, Lygodium, Nephrodium, Pleopeltis.*

Hinds has given an account of the vegetation of Hong Kong, an island at the mouth of the Canton river, and his plants collected there have been described by Bentham. (Journ. of Bot. 1842, p. 476, 494.) Hong Kong is a bare rocky island, the valleys of which are covered by a fertile soil, resting upon granite, and are well cultivated. The climate is subject to a variation of temperature from 26° to 94° F. The hottest months, June, July, and August, have a mean temperature of 89°, 94°, 90° F.; the coldest, December, January, and February, 57 5, 51°, 51° 5. The atmospheric deposits, although irregular, as might be expected at the edge of the monsoons, are still very considerable; they are estimated at 70″, 6 of rain. The summer months, May to August, are the wettest, December and January the driest, yet rain falls every month. The vegetation treeless, except in a few instances of *Pinus sinensis*, wants all the southern richness; there are, however, many evergreen shrubs, and the forms belong to tropical families. Hinds remarks, what has already often been mentioned of the Southern Chinese Flora, that a mixture of European and Indian genera distinguishes it. The presence of the American Orchidean genus *Broughtonia* is very remarkable. The herbarium collected by Hinds contains only about 130 species, but amongst them no fewer than fifty-one families are represented. List of the genera: 17 ferns (*Davallia, Lindsæa, Adiantum, Blechnum, Meniscium, Pteris, Aspidium, Polypodium, Niphobolus, Lygodium, Mertensia, Osmunda*); 16 Synanthereæ, (*Vernonia, Diplopappus, Amphiraphis, Siegesbeckia, Wollastonia, Gnaphalium, Gynura, Emilia, Senecio, Barkhausia, Sonchus, Brachyrhamphus*); 8 Leguminosæ (*Crotalaria, Indigofera, Desmodium, Cantharospermum, Milletia, Cæsalpinia, Bauhinia*); 6 Rubiaceæ, (*Mussaenda, Ixora, Psychotria, Hedyotis*); 5 Myrsineæ (*Choripetalum, Embelia, Mæsa, Ardisia*); 5 Rosaceæ, (*Rubus, Rosa, Raphiolepis*); 4 Melastomaceæ, (*Osbeckia, Melastoma, Allomorphia*); 4 Smilaceæ, (*Smilax, Dianella, Asparagus*); 3 Ericeæ, (*Enkianthus*); 3 Verbenaceæ, (*Callicarpa, Vitex*); 3 Euphorbiaceæ, (*Glochidion, Melanthera*, Decaisn., *Ricinus*); 3 Urticeæ, (*Ficus, Sponia*); 3 Cyperaceæ, (*Lepidosperma, Scleria, Carex*); 3 Gramineæ, (*Rottboellia, Bambusa, Erianthus*); 2 Violaceæ, (*Viola*); 2 Celastrineæ, (*Evonymus, Catha*); 2 Apocyneæ, (*Cerbera, Hollarrhena*); 2 Labiatæ, (*Scutellaria, Leucas*); 2 Solaneæ, (*Solanum*); 2 Schrophularineæ, (*Pterostigma*,

Bonnaya); 2 Lentibulariæ, (*Utricularia*); 2 Acanthaceæ, (*Barleria, Rostellaria*); 2 Polygoneæ, (*Polygonum*); 2 Thymelaæ, (*Daphne, Cansiera.*) Single representatives of the following families: Anonaceæ (*Unona*), Bixineæ (*Phoberos*), Droseraceæ (*Drosera*), Caryophylleæ, (*Stellaria*), Malvaceæ, (*Sida*), Byttneriaceæ (*Waltheria*), Sterculiaceæ (*Helicteres*), Ternstroemiaceæ (*Polyspora*), Aurantiaceæ (*Atalantia*), Oxalideæ (*Oxalis*), Connaraceæ (*Connarus*), Zanthoxyleæ (*Zanthoxylum*), Rhamneæ (*Berchemia*), Myrtaceæ (*Myrtus*), Onagrariæ (*Jussieua*), Lythrariæ (*Ammannia*), Araliaceæ (*Paratropia*), Crassulaceæ (*Bryophyllum*), Lobeliaceæ (*Lobelia*), Ebenaceæ (*Diospyros*), Sapoteæ (*Sideroxylon*), Mitrasacmeæ (*Mitrasacme*), Amaranthaceæ (*Amaranthus*), Laurineæ (*Cassytha*), Coniferæ (*Pinus*), Orchideæ (*Broughtonia chinensis*, Lindl.), Restiaceæ (*Eriocaulon*), Lycopodiaceæ (*Lycopodium*), lastly, the genus *Blackwellia*. Hinds names the potato as the most important among the cultivated plants. Besides this two other tuberose plants, Yams and *Cocoes* (*Dioscora* and *Arum*) are also cultivated.

He points out the mixed character of the vegetation of Canton (the Regions of Veg. p. 427) by this, namely, that *Viola* flowers there in the shade of *Melastoma*; *Bambusæ* grow on the same heights with Coniferæ, and potatoes and the sugar-cane are cultivated in the same field; the woods contain some species of *Quercus*, besides *Pinus*.

Some account of the vegetation of the plateaus of Thibet are contained in 'Moorcroft's Travels in the Himalayan Provinces,' &c. (London, 1841, 2 vols. 8vo). Ladak forms a plateau of 11,000′ elevation, composed of narrow valleys and hills rising a little above them. Some of the northern valleys lie even 13,000′ high; the southern mountain passes 16,000′. The uncultivated soil, even in the valleys, is quite bare and sterile; the only trees belong to the genera *Populus* and *Salix*. The scanty vegetation consists of some steppe plants, grasses, and furze, as it is called by the Himalayah surveyors (according to Royle, *Astragalus Moorcroftianus*, *Gerardianus*, and *spi-*

nosissimus). Frost and snow begin early in September, and continue with little interruption till the commencement of May. The greatest cold on the 1st February, was 9° 5 F.; from the middle of December, the thermometer during the night rarely stood above 15° F. The rays of the sun are very powerful in summer. On the 4th July, the thermometer stood in the sun at 134° F., another time at 144°, and during the night at 74°. Even in winter the rays of the sun are warm for an hour or two about mid-day; on the 30th January, they raised the mercury to 83°. The great effect of the sun compensates for the shortness of the summer, and quickly ripens the grain. The barley in Pituk, 800′ below Lè, was ripe for the sickle two months after being sown. The wheat required four months, it must consequently be sown immediately after the melting of the snow. The climate is notoriously one of the driest in the globe; in summer it rains very seldom, and always in small quantity, the greatest portion of the moisture falls as snow The agriculture, which besides the two species of grain mentioned, also includes buckwheat, is however, very prolific by the Chinese mode of culture and watering. The only cultivated fruits are Apricots, Apples, and *Sarsin* (*Elæagnus Moorcroftii*). The highest spot where corn is cultivated is the village of Kiwar, the elevation of which, according to Trebeck Moorcroft's companion, is at least 13,000′. (i, p. 78.)

Moorcroft's reports also upon Cashmere, grounded upon a long residence, contain much that is important, compared with the results of other travellers. There is also here as in Ladak, a principal valley, but at half the altitude 5800′—5000′, besides a series of smaller valleys or glens; the mountains covered with snow are wooded below. The upper woods, according to V. Hügel, (Kaschmir and das Reich der Sick, 1840, 8vo, Bd. ii, s. 245,) consist of *Pinus Deodara*, and six other Coniferæ, these extend to 7000′. Below this, according to Jacquemont, a thicket of wild fruit trees is extended, formed of *Pyrus*, *Persica*, *Ar-*

meniaca, *Punica*, *Prunus*, *Morus*, *Juglans*, and entwined by vines; Moorcroft remarks that the vegetation, on account of the greater moisture, is, in fact, more luxuriant than in the other valleys of the Himalayah. Snow falls in the lower regions from December to March; and from the end of March till May, frequent heavy rains and violent gales succeed; the rest of the summer months are warm, and cultivated plants quickly ripen; the atmosphere, also, is much more protected from drought by lakes and rivers than in Thibet. The uncultivated soil forms meadows in the valleys, but most of the arable land is cultivated. A productive alluvial soil, rich in humus, produces very rich crops, particularly of rice, which is sown in the beginning of May and reaped in the end of August; an uncommonly short season of vegetation! The other esculent plants are wheat, barley, buckwheat, maize, millet, and pulse of various kinds; he describes circumstantially the floating gardens for cucurbitaceæ, the use of the water nut (*Singhara*), and of the *Nelumbium;* and is, in all respects, more correct than V. Hügel, who, for example, describes and figures the fruit of the *Trapa* as its roots. (Bd. ii, s. 278.)

The most important botanico-geographical results have been communicated by Griffith, in letters to Von Martius on his herbarium collected in Affghanistan, consisting of 1400 species. (Münch. gel. Anz. 1842, No. 87.)

The character of the vegetation resembles that of Asia Minor; the collection comprised 270 Gramineæ, 230 Synanthereæ, many Cruciferæ and Chenopodeæ. Thorny Staticeæ were very abundant, and amongst the Leguminosæ sixty Astragali. Ferns and Orchideæ, are perhaps wholly wanting in western Affghanistan, only some single plants occur near Hindostan.

Falconer read before the Linnean Society the description of two new genera from Peschawur, of the Myrsineæ *Edgeworthia*, and of the Asclepiadeæ *Campelepis*. (Ann. Nat. Hist. x, p. 362.)

Forty-nine livraisons of Jacquemont's 'Voyage dans

l'Inde' have now reached us, which will be reported on hereafter. Montagne has described the cryptogamous plants, collected on the Neilgherries by Perrottet. (Ann. Sc. Nat. xvii, pp. 243-56, and 18, p. ix, 23.) The number of leaved mosses extends to sixty-six species; they are principally found in the forest region and are distinguished by the manifold character of their forms, as no fewer than thirty-one genera are represented by them. *Hypnum* numbers fourteen, *Neckera* five, *Fissidens*, *Macromitrium*, and *Brachymenium*, each four species. The only new genus *Symphyodon* has been already described. As so many tropical mosses have been made known of late years, and as their area usually extends over several Floras, it is not surprising that only about a sixth part of them are new; but it appears more remarkable that so few European forms are contained in the collection. These also form about a sixth part of the species, of which again a third belongs to *Hypnum*. For example, we may mention some of the species which are remarkable for their general dispersion, and are even found on the Neilgherries; *Grimmia ovata*, H., *Mnium rostratum*, H., *Hypnum alopecurum*, L., *cupressiforme*, L., and *tamariscinum* H., in the forest region; moreover, *Ceratodon purpureus*, Brid., *Bryum argenteum*, L., *Funaria hygrometrica*, H., *Bartramia fontana*, *β.*, *falcata*, H., also upon the plateau. Perrottet has collected thirty-four species of *Hepaticæ*, of which five only are new; most of them also occur in Java. The European forms are *Lophocolea bidentata*, N., *Trichocolea Tomentella*, N., *Metzgeria furcata*, N. With these are ranked thirty-six lichens, almost only cortical ones, of widely-distributed forms, generally *Parmeliæ* or genera allied to *Sticta* and *Usnea*, finally, five species of *Collema*. Several Fungi (twenty-three species, a fourth part new), conclude this memoir.

The Fungi collected by Cuming on the Philippines have been determined by Berkeley (Journ. of Botany, 1842, pp. 142-57.) This collection contains about thirty-five species, of which only a fifth, according to

Junghuhn and Montagne, occur also in Java, and to this fifth belong also four *Polypori*, found in all tropical countries. Although, however, the species of the Javanese equatorial Flora are different from the monsoon Flora of the Philippines, yet the character of the Hymenomycetæ remains the same, in so far as in both places *Polypori annui* of the section *Apus* prevail. Berkeley has described two thirds of Cuming's species as new. The collection consisted for the most part of Hymenomycetæ.

Junghuhn has worked at the Javanese Balanophoreæ (Art. Leop. Cæs. 19, Suppl.); Hasskarl has described some new species from Java, and published systematic remarks on all the plants of that island. (Regensb. Flora, 1842. Bd. 2. Beiblätt, s. i, 114.) He had an opportunity of making observations on the living plants while gardener to the Botanic Garden at Bogor, in Java. The greatest part of his labour has been expended on the Leguminosæ, but he has only few new species, and their descriptions appear very incomplete.

Botta has published his travels in Southern Arabia. (Rélation d'un Voyage dans l'Yémen par Botta; Paris, 1841, 8vo.) I am only acquainted with this work, which probably also gives a view of the character of the vegetation of this country, from the analysis of it contained in the 'Annales des Voyages.'

IV.—AFRICA.

Seventy livraisons have now appeared of the work of Webb and Berthelot on the Canary Islands; the systematic account of the phanerogamic plants, according to De Candolle's arrangement, has proceeded as far as the conclusion of the Rubiaceæ.

Schnitzlein has made a report on the plants sent by Kotschy from Nubia and Kordofan (Regensb. Flora 1842, Beibl. 1, s. 129-49). This collection, which is to be disposed of by the Würtemberg Society of Travellers, contains about 400 species of 56 families. The richest are the Leguminosæ (61), Gramineæ (48), Synanthereæ (28), Euphorbiaceæ (22), Malvaceæ (20), Convolvulaceæ (18), Cyperaceæ (15), Acanthaceæ (14). More than half the Leguminosæ consist of Loteæ, the most of which belong to the division of the Galegeæ, particularly *Indigofera* (13 species), *Tephrosia* (6), *Sesbania* (4). The Trifolieæ are represented almost only by *Lotus*, the Genisteæ by *Crotalaria* (7 species). To the Loteæ succeed the Phaseoleæ with nine species, then the Cæsalpiniæ, with eight species of *Cassia* and *Bauhinia*, finally, five Mimoseæ, four Hedysareæ, and one *Moringa*. Among the Gramineæ the Paniceæ preponderate; the Chlorideæ are also numerous, and the Stipaceæ contain seven species of *Aristida*. The Synanthereæ belong to many different genera, and exhibit, so far as they are presented to us, no definite character. Of the Euphorbiaceæ, perhaps about a third part belongs to *Euphorbia*, whilst *Acalypha*, *Croton*, and *Phyllanthus* also number several species. The Malvaceæ consist, for the most part of species of *Pavonia, Sida, Hibiscus*, and *Abutilon;* the Cyperaceæ of *Cyperus.* The Acanthaceæ are rich in generic types, and the Convolvulaceæ so much the poorer.

The following families are not so numerously represented in these Nubian herbaria; 11 Boragineæ mostly belonging to *Heliotropium;* 10 Amaranthaceæ; 10 Scrophulariaceæ; 10 Cucurbitaceæ; 9 Portulaceæ, particularly *Mollugo* and *Trianthema;* 8 Rubiaceæ without Stellatæ; 7 Tiliaceæ; 7 Lythariæ, principally *Bergia*; the Capparideæ, Cruciferæ, Labiatæ, and Solaneæ number six representatives. The characteristic genera of the remaining families are: *Nymphæa* (3), *Boerhaavia* (3); among the Phytolacceæ, *Gieseckia* and *Limeum;* among the Combretaceæ, *Poivrea* and *Guiera;* Saxifrageæ, *Vahlia* (2); Polygoneæ, *Ceratogonon;* Gentianeæ, *Hippion;* Palms, *Crucifera thebaica.* Schnitzlein has very properly compared this collection with the neighbouring Floras, and has found that about the sixth part of Kotschy's Nubian plants are also met with in Egypt. It should, however, here be remarked that these species, almost without exception, are indigenous only to Upper Egypt, in the neighbourhood of the tropic, where the Nubian vegetation also begins. This is completely different from the Egyptian Flora, except some species washed down by the Nile, and others spread by cultivation.

One result, confirmed anew by this collection, is much more interesting, viz.: that the tropical Floras of the old world are not so strongly defined from each other as the American. Thus, of Kotschy's Nubian plants, on the one side, about the fifth part also grow in the East Indies; on the other side, according to Buchinger's correction of the accounts of Schnitzlein (Regensb. Flora 1842, 2, s. 479), of 400 about 120 species grow likewise on the west coast of Africa, from Senegambia to Guinea. But of this collection there remain not less than 140 new plants endemic to Nubia, and thus an independent Nubian Flora seems to be sufficiently established. It is desirable that the botanic collections of German travellers in the east of Africa should give occasion to a comprehensive systematic work; for although these herbaria are in many hands, yet we have only fragmentary communications upon them. To these belongs a memoir by C. H. Schultz on the Compositæ of Rüppel's and Schimper's Travels in Abyssinia, as well as of Kotschy's Nubian Travels. (Regensb. Flora 1842, s. 417-24, and s. 433-42).

The author possesses 163 Synanthereæ, mostly undescribed, from Abyssinia; of these 6 only are also contained in the Nubian collection. He mentions that he has received from Abyssinia and Nubia 20 Vernoniaceæ, 2 Eupatoriaceæ, 48 Asteroideæ, 72 Senecionideæ, 17 Cynareæ, 25 Cichoraceæ, and 2 Mutisiaceæ. He promises to work at these systematically, and publishes, by way of commencement, as new Abyssinian genera: *Wirtgenia*, *Dipterotheca*, *Schnittspahnia*. Steudel has described the new Cyperaceæ from Schimper's Abyssinian collections (ibid. s. 577, 599). Of 16 species of *Cyperus* collected by him 12 were new; of 9 *Kyllingiæ*, 7; 3 new species of *Mariscus* have also been characterized. Hochstetter has instituted the following new Abyssinian genera (ibid. s. 225): *Cyclonema* (Verbenaceæ) of which 3 other species have been collected by Krauss at Port Natal, *Lophostylis* (Polygaleæ), *Kurria* (Rubiaceæ), *Valoradia* (Plumbagineæ). The genus *Raphidophyllum*, mentioned in last year's report, has not been separated by Bentham from *Gerardia*.

Lindley has described several new Orchidaceæ from the Cape of Good Hope (Journ. of Botan. 1842, p. 14-18). They belong partly to *Disa*, *Penthea*, and *Disperis*, partly to the genus *Brownleea*, founded by Harvey. Harvey has, besides this, lately characterized the following genera of the Cape Flora (ibid. pp. 18-29); *Choristylis* (Escalloniæ), *Pentanisia* (Rubiaceæ), *Raphionacme* and *Chymocormus* (Asclepiadeæ), *Toxicophlæa* and *Piptolæna* (Apocyneæ), *Brehmia* (Strychneæ), and *Acanthopsis*

(Acanthaceæ). Hochstetter (l. c.) separates *Eurylohium* from *Stilba*, and describes the new Campanulacea *Rhigiophyllum*.

Several new genera, from Port Natal, contained in the collection of Krauss, have been described by Harvey (l. c.) in January, and by Hochstetter in April 1842; No. 348 is *Diplesthes*, Harv. (Hippocrateaceæ); No. 186, *Ctenomeria*, Harv. (Euphorbiaceæ), Rubiaceæ; No. 178, *Mitrastigma*, Harv. (*Phallaria lucida*, Hochst.); No. 121, *Kraussia floribunda*, Harv. (*Coffea Kraussiana*, Hochst.); No. 131, *Pachystigma*, Hochst.; No. 144. *Mitriostigma*, Hochst.; No. 129, *Lachnosiphonium*, Hochst.; and No. 427, *Cyathodiscus*, Hochst. (Daphnoideæ). Harvey has borrowed from other collections from Port Natal the new genera of Acanthaceæ, *Crabbea*, *Ruttia*, and *Sclerochiton*.

Meissner has begun to give contributions to the South African Flora from Krauss's collections (Journ. of Bot. 1842, pp. 459-76). The first part is confined to the Ranunculaceæ, Nymphæaceæ, Cruciferæ, Violarieæ, Droseraceæ, and Polygaleæ.

Banbury has made a report on his botanical travels in South Africa (Journ. of Bot. 1842, pp. 540-70). His description of the character of the vegetation in the environs of Cape Town is so much the more interesting, as being accompanied by Harvey, who is intimately acquainted with the Cape Flora, he was enabled to acquire an exact knowledge of the species. The first impression of the Cape vegetation was very unfavorable, as he landed in the dry season; the bushes, although varied in forms, yet everywhere appeared dried up, stinted in growth, and destitute of blossom; herbaceous and bulbous plants were entirely burnt up by the scorching heat; the valleys and lower declivities of the hills were thinly clad with *Ericæ* and similar shrubs, between which grew sapless Restiaceæ, grasses, and low thornbushes (*Cliffortia*); there are, however, some pretty considerable woods of *Leucadendron argenteum* in this steppe; it is the only tree of moderate height which appears to be indigenous to the peninsula of Cape Town, and is 30-40′ high; its branches are directed obliquely upwards, and the glittering white silver colour of the leaves gives it an uncommonly beautiful appearance, particularly when the thick foliage is set in motion by the wind. The bark of the stem is also gray, and always remains smooth, and even on the highest trees moss or lichens never grow. It is known that this tree, which is found so abundantly in the neighbourhood of Cape Town, is entirely confined, geographically, to that locality. Several of the Cape Proteaceæ, to which this silver tree or Witteboom also belongs, are social, and cover large surfaces with their unmixed vegetation. This is particularly the case on the peninsula, with two bushes, the Kreupelboom (*Leucospermum conocarpum*), which forms a large bush about 8-10′ high, with gray-coloured foliage, at the foot of the Devil's and Table Mountain, and the Sugarbush (*Protea mellifera*), one of the most beautiful, and, at the same time, most abundant Proteaceæ of the Cape, with bright green leaves and large variegated blossoms of green, red, and white. These flowers when they burst open secrete such a quantity of sugary fluid, that when reversed it runs out as if from a cup, by which swarms of insects are allured, and are continually passing in and out. The environs of Cape Town are very rich in Proteaceæ; they spring up readily in the driest soils, in loose arid sand, or among sharp fragments of rock. Though the Silver tree occupies such narrow limits, yet other forms have a wider range; *Protea cynaroides*, though usually only a foot high, has the largest blossoms of all the Cape Proteaceæ, and is distributed from the flats or the sandy isthmus of the peninsula, as far as the point of the Table Mountain, and from Cape Town to the eastern bounds of the colony; its flowers are of a pale red colour, and form a head nearly as large as the crown of a man's hat. *Cliffortia ruscifolia*, one of the most abundant shrubs of the peninsula, is a tough, low bush, the thorny foliage of which is very annoying to the traveller, more so than that of the broom, for its leaves, which are as pointed as needles, very readily break off, and remain sticking in the skin and clothes. This bush formation has the type of the European heaths, and put Banbury particularly in mind of the coasts of Provence, where the Ericæ compose the principal part of the vegetation.

The Ericæ of the Cape which, in their own country, as Lindley supposes, are not less beautiful than in our hot-houses, fall into three divisions according to their locality. Some grow in company in great masses like the European, and cover large surfaces, as *Erica corifolia, ramentacea, racemifera, flexuosa, baccans, Blaeria muscoides,* and *ericoides.* Others are certainly also very abundant, but vegetate in a scattered manner among other plants, as *E. mammosa, cerinthoides, Plukenetii, Sebana.* Finally, there are species which are only found singly here and there in a cleft of rock. *E. cerinthoides* has the widest area, and is even found eastward of Graham's Town.

The Pelargonia are of less importance to the character of the vegetation of the peninsula, although a considerable number of them are indigenous. Few plants have gained so much by culture as these; the most abundant species in the mountains is *Pelargonium cucullatum,* a shrub, with large purple-coloured flowers, which grows generally with *Leonotis Leonurus* in all the defiles and valley tracks; when both are in flower, the glittering red of the one and the burning orange of the other produce a very rich effect.

When Banbury returned to Cape Town from a journey into the interior, at the end of the wet and cold season, which corresponds to our summer months, vegetation had in the meanwhile unfolded itself, and was now brilliant in all its beauty, in the peculiar richness of its forms. During August, September, and the first half of October, the Ixiæ everywhere put forth their brilliant blossoms, and cast a lively glow over the Erica steppe. The Irideæ are not confined to one definite kind of soil or degree of moisture, as they are so rich in species at the Cape that every locality numbers its separate forms. Some grow in loose sand, others on hard clay or soil containing iron. The geognostic substratum of the peninsula is generally granite to a height of 1500′, sandstone in horizontal layers succeeds as far as the summit of Table Mountain, 3582′ high. Whilst on the outer coast, at Greenpoint, *Sparaxis grandiflora* and several species of *Babiana* blossom, the rose-coloured *Ixia scillaris*, the yellow *Ixia conica*, and the extremely varied forms of *Gladiolus* unfold themselves on the Devil's and Lion's Mountains. The splendid *Antholyza æthiopica* raises its slender ears of yellowish red blossoms above the turf of the Restiaceæ and grasses, on moist spots, by the banks of mountain brooks. *Babiana ringens* unfolds itself on the moist sand of the Muysenberg, with its singular scarlet flowers stuck in the ground. On the flats *Aristea* and *Watsonia* thrive, and in the open spaces of the city different species of *Trichonema*.

Together with the Irideæ in this season, the Orchideæ are next to be mentioned. *Disperis Capensis* is one of the most abundant, and grows everywhere among the bushes, it is called Hottentot caps, from the peculiar form of the purple and green spotted blossoms. The most beautiful, but also the rarest of all the Orchideæ of the peninsula, is *Disa grandiflora*, the ornament of Table Mountain, where it is only found in a single locality at the top, among the bushes, on a black morass; its colours are nearest those of *Tigridia*, but the scarlet is much brighter. Many other Orchideæ, particularly species of *Disa* and *Satyrium*, are native to this district, but none of the group of the Epiphytæ, for which the climate seems neither sufficiently moist nor warm; it is, however, worthy of remark that several parasitic Orchideæ are found at Graham's Town, where considerably less rain falls than in Cape Town.

The Amaryllideæ are less abundant than the Irideæ, yet species of *Brunsvigia* and *Hæmanthus* grow on sandy soil.

The dependence of the plant formation on the nature

of the soil is everywhere apparent; the sandy bottom of the flats is an inexhaustible mine of peculiar plants. Here, according to the degree of moisture, two formations of shrubs are to be distinguished, of which the one follows the streams, and consists of *Cliffortia strobilifera, Erica concinna, Psoralea pinnata, Leucadendron floridum, Brunia* and *Rhus*, the other clothes the flats with the social Ericæ, *Cliffortia ternata* and *juniperina, Chironia, Borbonia, Struthiola, Mimetes*, and Restiaceæ; this latter formation is only 2-3 feet high, and under it are numerous bulbous plants, *Lobeliæ*, Synanthereæ, and other herbs; *Erica concinna*, on the contrary, grows to the height of a man.

Banbury has also communicated the botanical relations of several other localities, but without describing their vegetation minutely. He twice ascended to the summit of Table Mountain in company with Harvey, where, on the highest platform, mostly veiled in clouds (table-cloth), vegetation is composed of the following forms: The Ericoid bushes are here formed of 7 *Ericæ;* 2 *Penææ, Grubbia rosmarinifolia*, the Bruniaceæ, *Stavia glutinosa*, and the Scrophularineous *Teedia*. Among these appear numerous Restiaceæ, seven bulbs of the family of the Orchideæ, Irideæ, and Amaryllideæ, also several Synanthereæ, the Umbelliferous *Hermas*, the Gentianaceous *Villarsia ovata*, the parasitic Scrophularineæ *Aulaya, Crassula*, and two Ferns, *Todea* and *Schizæa*. The summit of Table Mountain is also rich in mosses and lichens; the rocks in the waterfalls are clothed with *Andreæa subulata*, Harvey, and *Jungermannia Hymenophyllum*, Hook, *Racomitrium lanuginosum, Dicranum flexuosum* and *Sticta crocata* are also common here. Banbury is of the same opinion with those who believe that the richness of the Cape Flora, notwithstanding all the numerous and successful researches, is not yet nearly exhausted. Many plants are limited to such narrow boundaries, that the region where they grow may be long examined without finding their locality. Some have a very short period of vegetation, and many of them, particularly those of the 'Karroo,' require circumstances for their complete development which can only occur after periods of several years. In concluding his report with the remark that South Africa, as respects its natural flora, belongs to the most distinctly marked regions of the earth, Banbury takes a similar view to that which I expressed in last year's report on New Guinea, namely, that the distribution of animals and plants is not subject to exactly the same laws. Swainson and others have shown that many birds are indigenous in both Senegal and the Cape, whilst not a single plant is common to the two districts, and not even in Congo can a specimen be found of the group of plants characteristic of the Cape Flora.*

Bojer has described several new plants from the Comoro Islands, Madagascar, and Mauritius. (Ann. Sc. Nat. 18, pp. 184-92.) They belong to *Calpidia* and *Boerhaavia* (Nyctagineæ), *Hilsenbergia* and *Dombeya* with *Melhania* (Byttneriaceæ), and to *Erythroxylon.*

* The striking difference between the distribution of plants and that of animals in Africa is highly worthy of attention, and shows how both can be distributed independently of each other. South Africa has not only a considerable number of species of all classes of animals common in central Africa, but also even in their characteristic forms there is no distinction between the two Faunæ (dies. Arch. 1843, i, bd. s. 201). And although the Flora of New Holland is especially comparable with that of South Africa, this analogy in the Fauna is perceptible only in a very subordinate degree. (Arch. d. N. 1842, i, bd. s. 89.)

V.—AMERICA.

As the 'Travels' of the Prince v. Wied, in North America have been completed in the last year, we have now to report in detail on the botanical descriptions contained in them. They are accompanied, as formerly mentioned, by very characteristic sketches of the country, and, in regard to system, rely principally on the labours of Nees v. Esenbeck, on the plants collected by the Prince himself, which forms a supplement to the 2d vol. (pp. 429-54), under the title of 'Systematische Uebersicht der von der Reise auf dem Missouri zurückgebrachten Pflanzen.' A great part of this collection has been unfortunately lost; and, although the increase of material conducive to knowledge of prairie vegetation is, on this account, not considerable, yet the information on the relations of botanical geography is so much the richer and more valuable.

The prairies are as definitely separated from the vast primitive forests, only here and there opened up by the hand of man, which cover the greatest part of North America, as the bed of a large inland sea. Enclosed by the Rocky Mountains and the course of the Mississippi, extending from Texas to the Hudson, this undulating hilly plain, covered with Gramineæ, touches on the primitive forests in a long meridian line situated on the average twenty geographical miles to the west of the Great River, and twice in Illinois and Alabama stretches over this boundary into the wooded district like a bay of the sea. The cause of this great division of the United States into two zones of vegetation, independent of level, is altogether obscure. Under the tropics it is easier to conceive the connexion between the steppe formation and the prevailing winds, than in the temperate zone. The climatic relations also of the prairies are too little known, but we are indebted to the Prince v. Wied for important contributions. The very great contrast between the climate of the east coast of North America and of the prairies is shown not only in the atmospheric deposits, dependent in a great measure on the form of vegetation itself, but particularly in the variations of temperature.

Mädler, who has arranged the climatic data of the Prince, remarks, in this respect, that the annual as well as daily variations of the thermometer exceed all that has hitherto been known from these latitudes, even in the interior of Russia. There prevails here a continental climate in the strictest sense, presenting the greatest extremes, while beyond the Rocky Mountains it is exactly the contrary. (See last year's Report.)

The meteorological observations of the Prince v. Wied, made in the district of the Mandan Indians on the Upper Missouri do not, unfortunately, extend over the whole year. His own thermometrical measurements, instituted at Fort Clarke, include only the winter months, and consequently are of less interest in Botanical Geography. On the other hand, the results obtained by Mackenzie, from observations made at Fort Union, at the mouth of the Yellowstone river, on the Missouri, in about 48° N. 86° W. (Ferrol), relate to the whole season of vegetation, and are deficient only from October to December. Mackenzie has calculated the mean temperature according to the method of the least quadrate, and fixed it at nearly 5°·72 R. The observed mean-temperature of each month is—

		1833.	1832.
January . .	=	— 4°·75 R.	
February . .	=	— 6·46	
March . .	=	+ 0·21	
April . .	=	+ 7·03	+ 8°·91 R.
May . . .	=	+ 8·94	+ 6·90

		1833.	1832.
June	. . =	+ 15°·03	+ 15°·20
July	. . =	+ 18 ·41	+ 18 ·58
August	. . =	+ 17 ·20	
September	. =	+ 11 ·75	

The difference, therefore, between the mean temperature of the warmest and coldest months amounts to 23° ·2 R. The daily fluctuations are also extremely great. They are greatest in March, when they amount, at an average, to 11°·57 R. Once the thermometer rose, on the 14th March, in six hours 20°·9 R. It fell, on the 21st February, 1833, from 12h. till the following morning, 37°·3 R. The prevailing currents of air blow from the west. The principal direction is W., 13½° S. The north-west is the coldest wind, the south the warmest. According to the experience of the Prince v. Wied, the climate of Fort Union is dry and stormy. To the severe and uninterrupted winter succeeds, in spring, the wettest season of the year, during which the prairies are in blossom, which in the remaining months bear only parched-up, or snow-covered grass. There begins about the middle of July an absolutely dry season, which lasts until the end of autumn, with scarcely any rain. Here, as in the Russian steppes, vegetation is, as it were, interrupted by a double winter sleep. Even during April violent snow-storms sometimes occur; near the Mandan villages the leaves do not come out till May, although something earlier in the bank Salicetæ; it has even been known, that in the end of that month the trees were not yet green. (ii, s. 74.) The flowers of the prairie also unfold themselves in May, but in the end of June the hills about Fort Union were still almost without blossom. (i, s. 435.) At that time the whole country was covered with short dry grass, among which, in roundish patches, the stunted bushes of *Opuntia Missouriensis* (D. C.) were scattered in plenty, unfolding their yellow blossoms at the same time with *Artemisia Gnaphalodes* (Nutt.) Thus the season of vegetation in the prairies lasts from

May to July. July is the only month in which there is really no night frost. In the forests the leaves remain till October. In November the Missouri first freezes; then the snow remains lying, and does not begin to disappear till March. The soil in the prairies is composed of a sandy clay, often containing a mixture of saline particles. It would, however, be sufficiently fertile for agricultural purposes, were it not parched by a constant wind which blows from the plateau of the Rocky Mountains. The Indians, indeed, cultivate maize, but with success only on the lowlands on the banks of the streams, where the crops are sheltered from the west wind.

The botanical productions of the great plain of the prairies appear to consist of very few species, and present great uniformity. Even the Gramineæ, with which the whole steppe is more or less luxuriantly covered, are not numerous.

To the strongest grasses belong *Uniola spicata*, L., *Spartina patens*, Mühl., and *Atheropogon olygostachys*, Nutt., which last attains more than a man's height in the flat prairie near Fort Clarke. (ii, p. 81.) Other prairie grasses are *Hierochloa fragrans*, Kth., in many places in these Travels, by a singular mistake, called "Ribes aureum," (i, pp. 319-26; ii, p. 325); *Agropyrum repens*, P. B., *Sesleria dactyloides*, Nutt. Among the plants growing intermixed with these Gramineæ, several are social, but which, for the most part, change at wide distances, so that by the less number of these forms, the traveller may, to a certain extent, ascertain his position in this wide desert according to certain plants. Thus, on the long journey from the mouth of the Missouri to its falls near Fort Mackenzie, under the Rocky Mountains, there are successively enumerated as such characteristic forms: *Oxytropis Lamberti*, P., *Cristaria coccinea*, P., *Allium reticulatum*, Fras., *Amorpha nana*, Nutt., *Rudbeckia columnaris*, P., *Solidago fragrans*, W. But all these, and other plants, are by far surpassed by *Artemisia gnaphalodes*, Nutt., which is spread over nearly the whole district, and often, together with the prairie grasses, covers wide tracts almost exclusively. Next to these, may be ranked the two Cacteæ of the prairie, *Opuntia Missouriensis*, and one of the Mamillariæ described by Asa Gray, which, together with *Yucca angustifolia*, P., especially characterize the steppes of the Missouri. Finally, there are also met with universally some low shrubs, which cover large tracts of country, and are then suddenly replaced by another as abundant species.

On ascending the Missouri, the Buffalo-Beng-Shrub (*Shepherdia argentea*) makes its first appearance towards the south as a prairie shrub, at about 42° N., and thence onwards becomes more and more frequent. In the country of the Mandan Indians, in 47° N., *Juniperus repens* commences, by which, together with *I. communis*, L., the hills on the Missouri near Fort Clarke are covered. Above the mouth of the Yellowstone appears the "Pulpy Thorn" of Clarke, (*Sarcobatus Maximiliani*, Nees.,) (a new genus, doubtfully placed among the *Urticeæ*); which, from this place to the Rocky Mountains, grows everywhere, mixed with the *Artemisia*. The characteristic forms of the prairie plants are, according to the collection of the Prince v. Wied, the following: Leguminosæ—*Astragalus, Oxytropis, Thermopsis, Amorpha, Sophora;* Synanthereæ; e. g. *Iva, Artemisia, Senecio, Solidago, Helianthus, Rudbeckia, Chrysopsis, Sideranthus, Aster, Erigeron, Stenactis, Achillæa, Cirsium, Jamesia,* (*Prenanthes*, Torr.); Cruciferæ—e. g. *Vesicaria, Erysimum asperum*, D.C.; Boragineæ—*Batschia, Myosotis, Lithospermum, Echinospermum*; certain genera of the Malvaceæ—*Cristaria;* Onagrariæ—*Ænothera*, with several species, *Linum, Galium;* of the Hydrophylleæ—*Ellisia;* Scrophularineæ—*Penstemon;* Santaleæ—*Comandra;* Chenopodeæ—*Kochia diæca*, Nutt.; and of the Polygoneæ—*Eriogonum sericeum*, P.

Woods occur but rarely, and, as it appears, principally in the border districts of the prairies, but they do occur in the low grounds on the banks of the rivers, and afford the only timber to the inhabitants. In consequence of the prairies being traversed by the Missouri and its numerous tributary streams, they are much more accessible to culture than any other steppes, and for the same reason, at the present time they also afford shelter to innumerable animals. These woods on the banks of the rivers consist, for the most part, of *Populus angulata*, W., or of willows, particularly *S. lucida*, W., and *longifolia*, Torr. However, there also occur even in central Missouri, and consequently in the midst of the prairie, of larger trees, two *Oaks*, one *Ash*, and *Negundo.*

Descending the stream, in the border district of the prairie of Osage, the number of trees gradually increases, until at St. Louis the sylvan character of Indiana is already attained. From this point onwards, even the so-called "red cedars" of the Mississippi spread themselves outwards pretty nearly to the Missouri. At 43° N. there is an island in the stream called "Cedar Island." On this is situate a clump of Coniferæ (*Juniperus barbadensis*, L.), mixed, however, with leaf-trees which rises fifty feet above the poplars and willows by which they are surrounded, and under which is an undergrowth of *Shepherdia* and *Cornus sericea.*

The same appearance recurs on the N.W. border of the prairie. At the foot of the Rocky Mountains, near Fort Mackenzie, the woods of *Populus angulata*, however, always predominate, although there are also groves of Pines (*Fichten*), consisting, in this part, of *Pinus flexilis*, Jau., which species disappears eastwards, at the mouth of the Yellow-stone.

Between Fort Union and the "Cedar Island" there is, consequently, scarcely anything but *Poplars* and *Willows.* The jungle in these poplar woods consists of several roses, particularly *Rosa Maximiliani*, Nees., also of *Amelanchier sanguinea*, D. C., *Prunus serotina*, Ehrb., *Symphoria*, *Cornus*, *Ribes*, especially *R. aureum*, P., and *Shepherdia ;* the climbers are *Clematis cordata*, P., *Vitis cordifolia*, Mich., *Celastrus scandens* and *Humulus.* On the Mississippi also, near St. Louis, are still found these woods of *Populus*, which are highly characteristic of the whole of this country. In many places the Poplars form dense woods, surrounded towards the river by Salicetæ. In the neighbourhood of rock, however, the poplars on the Mississippi and on the Lower Ohio are replaced by the North American cedar (*Juniperus Virginiana.*) The communications of the Prince v. Wied on the character of the North American forests are of greater importance. During a prolonged residence in the Alleghany Mountains of Pennsylvania, and afterwards at New Harmony, in Indiana, he had an opportunity of comparing the difference of the forest formation, in the Northern and Western States. In Pennsylvania, the woods on the Lecha, e. g., near Bethlehem, consist of *Quercus coccinea, rubra, tinctoria, alba, Juglans nigra, Castanea, Prinos, Laurus sassafras*, with a thick underwood of *Rhododendron maximum*, and the climbing *Vitis Labrusca.* In other leaf forests, e. g., in the passes, 1050′ high, between the Delaware and Susquehannah, are found—besides the ever predominating Oak, Chesnut, and also Walnut, *Fagus, Carpinus, Betula, Ulmus, Nyssa, Acer,* and *Liriodendron.* In the Alleghany, above these forests, follows a region of Coniferæ, which is formed partly of pines, (*Pinus rigida*, Mill.,) partly of firs, (*P. Canadensis*, Ait.) On the lower boundaries of the Coniferous woods, in the district of the Delaware, grow oak bushes, e. g., *Q. Banister*, Mich., from among which single trees of *Pinus rigida* arise. In the Western Alleghany, on the contrary, the forests at the mean altitude, e. g., in the passes, 2400′-3000′ high, between the Susquehannah and Ohio, are, for the most part, a mixture of leaf-trees and Coniferæ, to which, towards Pittsburg, oaks, either alone or with chesnut trees and *Robiniæ*, succeed. The climbers consist of *Vitis, Ampelopsis*, and *Smilax*

There is an accurate account of the vegetation of the Coniferous region in a catalogue appended to the Travels of v. Schweinitz, which contains the plants found on the Pokono, a summit of the "Blue Ridge," or Eastern Alleghany, not far from Bethlehem. This mountain was also ascended by the Prince. The underwood here, likewise consists of *Rhododendron maximum*, and several other Rhodoreæ, as *Kalmia latifolia*, *Andromeda racemosa*, some Vaccinieæ, *Comptonia asplenifolia*, and the before-mentioned oak. There are still other shrubs and subfrutices, namely, *Gaultheria procumbens*, *Rhodora canadensis*, species of *Oxycoccus*, *Cornus*, *Prinos laevigata*, and *Dronia glabra*. From the burnt down forests of the Alleghany, arise *Rhus typhinum*, *Phytolacca*, and *Verbascum*. We have received accounts of the vegetation of the Southern Alleghany, from Asa Gray. (Jour. of Bot. 1842, pp. 217-37 ; and 1843, pp. 113-25.) The extensive Virginian longitudinal valley, between the "Alleghany Proper" and the "Blue Ridge," extending in a breadth of from 4-6 geographical miles, nearly from the 36th to the 42d deg. N., is everywhere covered with a rich fertile soil, mostly containing lime, but it has already, in a great measure, lost its primitive vegetation. European weeds, particularly *Echium vulgare*, have here established themselves over extensive tracts. This plant has taken complete possession of a space more than twenty geographical miles wide of uncultivated ground; often even on the cultivated fields, since wherever limestone exists the whole plain becomes blue with it. *Bupleurum rotundifolium*, *Marrubium vulgare*, *Euphorbia Lathyris*, and *Melissa Nepeta*, are also found here now very plentifully. In the circuit of this valley there are several tree-boundaries, *Robinia pseudacacia*, which is not found on the east of the mountains, first appears where the Potomac crosses the Blue Ridge, and is abundant from this place towards the south. To the south of the Lexington, *Gleditschia triacanthos* first grows, and also *Negundo*. Here also the Papaw tree begins, (*Uvaria triloba*,) which, in the Western States, acquires so much importance; and towards the New River, for the first time, *Æsculus flavus* is seen. Asa Gray's researches extended as far as the mountain-chains in North Carolina, where, in the Black Mountain, they attain almost their greatest elevation, rising to 6476′, by Michaux's barometrical measurement. The dicotyledonous woods about Jefferson are similar to those of Pennsylvania. They consist of *Castaneæ*, *Quercus alba*, *Liriodendron*, *Magnolia auriculata*, and sometimes *Acer saccharinum*. The mountains are usually wooded to the tops, as the "Grandfather" in the Blue Ridge, 5556′ high, which the traveller ascended. The Coniferous region of this mountain consists solely of Firs, more particularly of *Pinus Fraseri*, P., together with *P. nigra*, Ait. The blocks of mica slate, and fallen trees of this region are thickly covered with moss and lichens. Here is repeated exactly the character of the dark solitary forests on the river St. Lawrence, only that the trees in the mountains of Carolina are smaller than in the northern plains of New York and Vermont. This resemblance extends over the whole vegetation. The shrubs and herbage, also, are, for the most part, Canadian. With these are mixed some endemic mountain forms, as *Astilbe decandra*, *Chelone Lyoni*, *Aconitum reclinatum*, Gr., *Saxifragra Careyana*, G. Among the shrubs of the Southern Alleghany, Gray mentions particularly *Rhododendron Catawbiense*, *Menziesia globularis*, *Leiophyllum serpyllifolium*, D. G., *Vaccinium erythrocarpum*, *Sorbus Americana*, *Pyrus melanocarpa*, thus, almost only Rhodoreæ and Rosaceæ. If we compare the woods of the Eastern and Western States, there will be the same general result, that on this side of the Alleghany the underwood is chiefly formed of Rhodoreæ, on the other side, of an Anonacea, of *Uvaria triloba;* on the Lower Mississippi we find, instead of thick strong underwoods, the tall reed of the woods, the *Miegia macrosperma*, and lastly, in the primitive forests, on the boundaries of the prairies, only *Equisetum hyemale*. The Prince v. Wied first became aware of these differences in the formation of the underwood when travelling to the west of the Alleghany, he traversed the primitive forests on the Ohio below Pittsburg. Here *Rhododendron maximum* had already disappeared, and was represented by the Papaw, a tree 20′-30′ high, with violet-brownish flowers, beautiful, large, smooth, bright green leaves, and edible fruit. Of tall forest trees, the plane and the beech here begin to prevail, intermixed with *Liriodendron*, *Acer*, *Tilia*, *Juglans*, *Fraxinus* and *Ulmus*, while the oak and chesnut disappear. The common climbers still continue to be *Vitis* and *Ampelopsis*. From this place the Papaw accompanied the traveller through Ohio, Indiana, Illinois, and Missouri, till in the prairies of Osage it is seen for the last time. But towards the west it gradually decreases. Already, on the Lower Ohio, there are tracts, in which the soil of the primitive forests, instead of underwood, is covered with *Miegia*. However, this wood-reed does not grow so high here as in Louisiana; it is only 8′-10′ high, but forms a dense coppice, which in winter remains green, while under the trees there no evergreen forms occur. On the Lower Missouri the wood-reed is entirely wanting, but the already-mentioned *Equisetum*, about 2′ high, here covers the soil in the natural forests so densely that a walking-stick can scarcely reach the ground between its stalks. The woods of Indiana are composed of very many kinds of trees, of which the Prince v. Wied counted about sixty. In the lofty forest the following genera are found: *Platanus occidentalis* (Buttonwood), *Liriodendron* (Poplar), *Acer*, 6 species, *Quercus*, 9 species, particularly *A. macrocarpa*, *Gymnocladus canadensis* (Coffee-tree), *Juglans*, 10 species, *Gleditschia*, 2 species, (Locust,) *Liquidambar Styraciflua*, (Sweet-gum,) *Catalpa*, *Tilia*, *Ulmus*, 3 species, *Fraxinus*, 2 species, *Nyssa sylvatica*, (Black-gum,) *Fagus Americana*, *Robinia pseudacacia*, *Diospyros*. In these forests the underwood is usually 15′-30′ high, and into its formation, besides the Uvariæ, *Laurus Benzoin*, (Spice-wood,) and *Cercis canadensis*, (Red-bud,) enter. The remaining low arboreal forms belong to *Populus*, *Carpinus*, *Ostrya*, *Morus*, *Celtis*, *Laurus Sassafras*, *Cornus florida*, *Pyrus coronaria*, *Mespilus arborea*, and *Prunus Virginiana*.

Smaller thickets are formed principally of *Evonymus*, *Cratægus*, *Spiræa*, *Rubus*, *Corylus*, and *Salix*. Other forms found there are *Symphoria glomerata*, *Hydrangea arborescens*, *Ceanothus Americanus*, *Staphylea trifolia*, *Amorpha fruticosa*, *Hamamelis Virginica*, and *Gonolobium hirsutum*. The climbers are here more numerous than in Pennsylvania, several *Bignoniæ* make their appearance, especially *B. radicans*, and besides several species of *Vitis* and *Smilax*, *Celastrus scandens*, and *Clematis Virginiana*, also occur. But the most remarkable climber of Indiana is *Rhus radicans* "the Poison vine," the tendrils of which lie close to the stem, and are fixed by innumerable aerial roots. From this felt-like mass of roots the branches project in a rectangular direction, and then curve upwards, with their pinnate leaves. In travelling through those forests from the Alleghany onwards, what is most striking, besides the difference in the underwood here, is that the Coniferæ are entirely wanting, and that the Chesnut and Magnolia have likewise disappeared.

The soil in the primitive woods of Indiana consists mostly of a black *humus-mould;* where this is not the case, or where, for example, in some places between Harmony and Vincennes, a looser sand alternates with the stiff covering of humus, the character of the woods also changes suddenly, and instead of luxuriant mixed woods, the low-growing *Quercus nigra* alone is met with, from 30-40′ high. The climate of Indiana is rude. Cotton here no longer thrives. The chief product of husbandry is maize, which grows to the height of 12-15′. The potato is also cultivated, and the four kinds of grain of the north of Europe.

Hinds has given a report of the character of vegetation on the north-west coast of America from his own observation. (The Regions of Vegetation, pp. 331-35.) A continuous natural forest covers the coast, from 68°-46° N.; but on the Columbia River there is a sudden change in the vegetation, the mouth of that river forming a distinct boundary line to the Flora of California. The dense

8

forest that stretches so far to the north from the Columbia, contains but few species. All the large trees are Coniferæ—three kinds of fir (*Abies*) and *Cupressus thyoides*, L. The smaller arboreal forms belong to *Cratægus*, *Prunus*, *Betula*, *Salix*, and in the south also *Diospyros*. The situation of this coast, open and exposed to frequent storms from the west, in conjunction with the constant wetness of the climate, which on the Columbia produces 53″6 rain, occasions in these natural forests, a very early decay of the trees. Hinds remarks, that here the ground is so thickly covered with fallen stems as to remind one of the period of the coal formation. Trees scarcely full-grown, clothed with cryptogamic vegetation, lie everywhere, horizontally, in the forest. The underwood is in these north-western pine-forests developed in great luxuriance. There are numerous forms of *Vaccinium*, *Menziesia*, *Rubus*, and *Ribes*. Farther south there are added to these, *Lonicera*, *Mahonia*, *Gaultheria*. A social fern, *Aspidium munitum*, by itself alone covers extensive tracts in the forest. Various species of the same genus appear in different latitudes, particularly of *Ribes*, *Lupinus*, and the Rosaceæ. Two plants with very large leaves, one an Aroideous the other an Araliaceous plant, are here very widely spread. *Dracontium Kamtschaticum* and *Panax horridum* both occur from the Columbia, as far as 61° N. For the rest the genera are perhaps, almost without exception, as is known, European, and nearly the half of the species are also found in Europe or Siberia; only once, in the vicinity of their northern boundary, are these forests broken in upon on the coast, towards the Aleutian Islands, by a tract of country without trees, which in summer bears a luxuriant vegetation of *Rosa*, *Salix*, and *Lupinus*. These together form a thick vegetation with which many herbaceous plants, as *Mimulus luteus*, *Geranium eriostemon*, *Lupinus Nootkaensis*, *Epilobium latifolium*, *Polemonium humile*, together with some *Ferns*, and European grasses, are mixed.

South of the Columbia, where the Californian Flora begins, the woods of Abies suddenly cease, and give place in the forests almost entirely to Pines and Oaks. The country here, however, is for the most part open. The formations of the Californian Flora are not sufficiently known, nor are they, with the exception of the forests, happily characterized by Hinds. (pp. 345-48.) According to that traveller, the oak woods of California contain two deciduous species and two evergreens. The latter are found along the coast, only however, between 34° and 38° N. The other trees are not numerous: *Acer, Æsculus*, two Laurineæ (*Tetranthera Californica*, and *Laurus* species), and on the banks of rivers, *Platanus, Fraxinus, Juglans, Salix*. The jungle in the woods consists of *Rubus, Ribes, Rhus, Vaccinium, Cornus, Lonicera*, and ligneous Synanthereæ. In the Californian peninsula, the most important forms of vegetation belong to the Cacteæ, but they extend only to 34° N. Hinds considers the Californian Flora to extend on the south-east as far as the Colorado.

Martius has published some diagnoses of newly obtained plants from Illinois, and Missouri. (Bullét. de l'Académie de Bruxelles, 1841, i, pp. 65-8.) Schouw has read to the Copenhagen Academy, some of Liebmann's 'Botanical Letters' from Mexico, which lie before me only in the still unfinished translation by Hornschuch. (Regensb. Flora, 1843, s. 109-18.) In "Tierra caliente" between Vera Cruz and Xicaltepec, the Palmaceæ are much more plentiful than has hitherto been believed. The most plentiful Palm is *Acrocomia spinosa*, Mart., but at Laguna verde, *Sabal Mexicanum*, M., forms close, thick, 40' high, palm groves, which are not intermixed with any other kind of tree.

Liebmann describes more accurately the forests of the "Tierra fria," near Turutlan. The soil of the elevated plains consists of sandy clay, which usually rests on sandstone, and is very fertile when it does not remain too long dry. The temperature of the earth here, amounted to 13° R. Ridges of hills rise above the plateau, and on these alone do the forests occur, while the plain is bare and naked.

In the forests at Turutlan, nine species of Pine are met with, particularly *P. Montezumæ*, *P. Teocote*, and *P. Ayacahuite*, C. Ehrenb., of which, the last mentioned is the most remarkable. It attains a height of about 120′, and is the most resinous of all, and its cones reach the astonishing length of 15-16″. Intermixed with the Coniferæ, grow five species of *Quercus* and *Alnus*. There are very few shrubs, or little undergrowth in these woods. To the former belong (generally diffused over the plateau) *Myrica jalapensis* and *Helianthemum glomeratum*; to the latter, *Fragaria Mexicana*, and a variety of *Pteris aquilina*. On the Pines is found as a parasite *Viscum vaginatum*.

Kickx has described some new Mexican *fungi*. (Bullét, de l'Académie de Bruxelles, 1841, ii, pp. 72-81.) Breutel has published his 'Botanical Observations on St. Kitts and St. Thomas,' (Regensb. Flora, 1842, s. 549-60.) Of Bentham's elaboration of Schomburgh's Plants, the Ferns (79 species), and the Lycopodiaceæ (6) have appeared. J. Smith has undertaken this division. (Journal of Botany, 1842, pp. 193-203.)

Of the important collection of plants which Hostmann has begun to send from Surinam, and which is to be published by Sir W. Hooker, only some *fungi* have as yet been described by Berkeley. (Journ. of Bot. 1842, pp. 138-42.) Splitgerber has published a series of new plants from Surinam (v. d. Hoeven n. Vriese Tijdschr. 1842, pp. 5-16 and 95-114.) These belong to the Bignoniaceæ, (among which is the new genus *Couralia*), Dilleniaceæ, Anonaceæ, Tiliaceæ, Ternstrœmiaceæ, Guttiferæ, Sapindaceæ, Leguminosæ.

Miguel has also written on the Bignoniaceæ of Surinam. (Regensb. Flora, 1842, pp. 224-341.)

In the third to the fifth numbers of the 'Flora Brasiliensis,' by Endlicher and V. Martius, another series of plant formations has been illustrated by plates and copious descriptions. The bank vegetation of the Amazon is developed between the river and the primitive forest on the Laa-Ygapo, and especially on the sandy islands. It consists partly of low trees growing in clumps, of a Salicete formation, here consisting of *Salix Humboldtiana*, and of a Euphorbiaceous plant, *Alchornea castaneæfolia*, and partly of thickets of *Cecropia peltata* physiognomically characterized by its white bark, squarrose branches, and large-lobed, green glittering leaves with white hairs beneath. The Caa-ygapo is also included in this description. The following trees are here mentioned as a supplement to the earlier description of this primitive forest: Euphorbiaceæ (*Hura, Sapium, Pera*); *Hippocratea*; Laurineæ (*Nectandra*); Myrtaceæ (*Psidium, Eugenia*); Bombaceæ (*Pachira*); Leguminosæ (*Phellocarpus, Pterocarpus*); Palmæ (*Bactris Maraja, Astrocaryum vulgare* and *acaule*); also climbers Nandhirobeæ (*Feuillea*); Cucurbitaceæ (*Elaterium, Melothria, Anguria, Convolvulus, Bignonia*. The water plants of the Amazon present nothing peculiar; they are almost all distributed through the whole of tropical America, as *Pistia, Limnanthemum Humboldtianum, Cabomba, Azolla microphylla*. Plate 18 is a supplement to plates 1 and 11 It represents the vegetation of the banks of the Itahype, in Bahia, which is distinguished by the greatest variety of character in its forms. A social Aroideous plant (*Arum liniferum*), a sedge-like Gramineous plant (*Gynerium*), a *Rapatea*, forming a large-leaved turf, and the Marantaceous *Thalia* are next figured in their usual localities. Then follow the trees of the primitive forest with their climbers, in the first place a *Sterculia*, a *Zanthoxylon*, and the Palm, *Euterpe edulis*.

The "Catinga" (Tab. 10) is a forest formation of the trade-wind Flora of Bahia, which loses its leaves during the dry season. The trees here are generally only 20-40′ high, and stand more apart than in the primitive forest. Should the rainy season fail, which frequently occurs in the interior of Brazil, there is no vegetation to be seen, except the Cacteæ, during the whole time. Compared with the woods of Europe, those of the Catinga, with much physiognomical similarity, yet present a much greater number of forms. When the fall of the leaf commences, succulent plants, and others better protected by the texture of down on the leaves, remain green, such as the Bromeliaceæ, the Capparideæ, *Colicodendron*, and species of *Croton*. Some trees regain their leaves much more easily than others when the moisture is greater, as the Euphorbiacea *Cnidoscolus*. The woods of the Catinga are also distinguished from those of temperate zones by the multitude of the parasites and climbers. To the first belong here particularly the Bromeliaceæ, Cacteæ, and Loranthaceæ, rarely the Aroideæ, Orchideæ, or Ferns. In no formation of Brazil are the Cacteæ so numerous and various as here; they thrive well also on the thin poor soil of these woods. The Bombaceæ, *Cavanillesia*, also *Bursera*, *Spondias*, *Cnidoscolus*, and the Palm *Cocos coronata*, are examples of the trees constituting the Catinga.

The other plates represent landscapes in the provinces of Rio and S. Paulo, they are principally intended to show the forms of plants, as that of the Rhizophoræ (Tab. 12), *Rhizophora Mangle*, which, with two Avicenniæ and several Combretaceæ, forms the mangrove woods of Brazil; the Tree-ferns (Tab. 14) are represented by *Alsophila paleolata;* the arborescent grasses (Tab. 15), by *Guadua Tacoara*, and in Tab. 13 are the parasites of the primitive forest, which have been treated of most copiously by Martius, and the results published partly here and partly in the 'Münch. gelehrt. Anzeigen' (1842, Nr. 44-9.) This work is of much physiological interest. The true parasites of Brazil belong mostly to the following families: Fungi, Balanophoreæ, Cytineæ, Rafflessiaceæ, Burmanniaceæ (*Gonyanthus*), Orchideæ, Aroideæ (*Philodendron*, *Anthurium*), Laurineæ (*Cassytha*), Convolvulaceæ (*Cuscuta*), Orobancheæ, Ericeæ, Loranthaceæ, Marcgraaviaceæ, Guttiferæ. *Voyra*, however, should not be in the list of these plants, as Martius does not hold it to be a true parasite, since it resembles Orobanche.

The systematic portion of the Cyperaceæ by Nees v. Esenbeck, and the Smilaceæ and Dioscoreæ by myself, are contained in the three Nos. of the 'Flora Brasiliensis,' which have appeared in 1842.

St Hilaire, Tulasne, and Naudin, have begun to publish supplements to the 'Flora Brasili meridionalis,' of the former. (Ann. Sc. Nat. 17, pp. 129-43, and 18, pp. 24-54 and 209-13.) This work extends from the Ranunculaceæ to the conclusion of the Malvaceæ, and contains a considerable number of new species.

Gardner, upon whose recent collections from Minas Geraes, Sir W. Hooker has made a report (Journ. of Bot. 1842, p. 295), has begun to work at a catalogue of the plants found by him. (Ibid. pp. 158-93 and 528-48.) A summary of his herbarium collected from 1836-41 :— 400 species from Rio; 600 species from the first journey to the Organ Mountains; 500 species from Pernambuco (Oct. 1837—Jan. 1838); 200 species from Alagoas (Feb.—April, 1838); 600 species from Crato, in Ciara (Sept. 1838 –Jan. 1839); 400 species from Oeiras, in Piauhy (April—July, 1839); 500 species from Piauhy and Goyaz (Aug.—Sept.); 1400 species from Goyaz (Oct. 1839—April, 1840); a considerable collection from Minas (May—Oct.); and from a second journey to the Organ Mountains in the spring of 1841. Three or four hundred species have been already enumerated in the catalogue in geographical order.

An interesting paper on the Paraguay tea by Sir W. Hooker, illustrated by plates, is to be found in his Journal (Journ. of Bot. 1842, pp. 30-42.) The author confirms the opinion of St. Hilaire, that the *Ilex Paraguayensis* of Paraguay is identical with that indigenous to Brazil.

Hinds remarks on the equatorial limits of the Peruvian Flora (l. c.) that the woods of Guayaquil commence at 4° S. L., and are here distinctly separated from the coast vegetation of Peru; in this neighbourhood also lie the equatorial limits of the Peruvian Mist, the "Garuas," for whilst at Guayaquil heavy showers fall, half a degree south at Tumbez there is no rain during the year. The "Garuas" extend southwards to 36° S. L., at Valparaiso they cease to be regular.

Bridges has made a report on an excursion in the Andes near Valparaiso (Journ of Bot. 1842, pp. 258-63.) He remarks, that a third of the whole vegetation in the upper region consists of Synanthereæ.

Steudel has published the Cyperaceæ collected by Bertero in Chili and Juan Fernandez. (Regensb. Flora, 1842, pp. 599-605). Miers has described the new genus of Irideæ *Solenomelus* from Chili at a meeting of the Linnæan Society. (Ann. Nat. Hist. ix, p. 244.)

VI.—AUSTRALIA.

Hinds describes only about 500 plants as indigenous to the Society Islands. (The Regions, &c. p. 382.) The Marquesas, and Harvey's and Gambier's Archipelago have the same vegetation. On the Pomotu's, or Dangerous Islands, Hinds has collected 47 species, and terms their Flora very poor.

He (ibid. p. 384) includes in the Flora of New Guinea that of the Archipelago situated to the east as far as Tonga, thus including the Navigator and Friendly Islands. In this region are observed anomalies in the division of the seasons; most distinctly at New Guinea, but much less so at a greater distance in the Pacific Ocean; whilst in the Indian Ocean, for example at Celebes, where, agreeably to theory, the S. E. monsoon blowing from May to October, brings the dry season, and the N. W. wind prevailing in the other months, the rainy season, we observe the reverse in the Pacific.

During the S. E. monsoon frequent heavy rains occur. This wind begins in March or April, and lasts six months; the heat and moisture are much greater than in the central regions of the Great Ocean. The N. W. monsoon succeeds the rainy season, which blows till the following March, and here produces a dry season. (Hinds in Jour. of Bot., 1842, p. 670.) The farther eastward we proceed in the Pacific, the less are the seasons developed. Hinds has collected Herbaria on the Fidji islands, Tanna, New Ireland, and New Guinea, which Bentham has begun to work at. The vegetation of the Fidji Islands is, according to Hinds's observation, much more copious than that of the Society Islands; Leguminosæ are abundant, Mangrove woods appear, and also a leafless *Acacia, Chamærops*, and *Passiflora*.

New Ireland, like New Guinea, is densely wooded, and its climate very moist. The forest trees are very high, but almost without underwood, or other shade plants. The Palms here become more various, Hinds mentions the genera *Areca* and *Caryota;* a Cycadaceous plant (?) *Pandanus, Myristica*, and *Ficus*, are also characteristic; Ferns and Orchideæ are likewise numerous.

Drummond has continued the communication of his researches at Swan River. (Journ. of Bot.) Preiss has made known some information on the extent of the Herbaria collected by him in the same country. (Linnæa, 1842, p. 384.) Colenso has made a report on several New Zealand plants. (Journ. of Bot. 1842, pp. 298-305.)

REPORT

ON THE CONTRIBUTIONS TO

BOTANICAL GEOGRAPHY,

DURING THE YEAR 1843,

BY PROFESSOR GRISEBACH.

TRANSLATED BY GEORGE BUSK, F.R.C.S.

BOTANICAL GEOGRAPHY.

The most important work of the past year, in the department of general climatology, is Humboldt's 'Central Asia,' (Asie centrale. Recherches sur les chaines de montagnes et la climatologie comparée. Paris, 1843, 3 vols. 8vo.) In the first two volumes the relations of that part of Asia lying between the Altai and Himalayas, as to position and elevation, are deduced from a renewed analysis of all known sources. Particularly is it proved that the hitherto received notions with regard to the elevation and extent of the high lands of central Asia have been very much exaggerated. It had been already shown that the Chinese province of Thian-schan-pelu, or the land between the Altai and Thian-schan, belongs to the depression of the Caspio-Siberian steppes. But in the same way, also, the province of Thian-schan-nanlu, between Thian-schan and Kuenlün, has been excluded from the high land, because here, under the latitude of Italy, the cotton-tree flourishes, whilst in Jarkand the grape-vine thrives, and in Khotan the breeding of the silkworm is carried on successfully. (iii, p. 20.) The desert of Gobi, according to the measurements of Fuss and Bunge, in their journey to Pekin, has a mean elevation of 4000′ and is therefore on a level with the plateau of Persia. (i, p. 9.) The celebrated table land of Lesser Thibet alone, attains the level of the lake of Titicaca (12,000′), and its mean elevation is probably lower. (Vid. last Report, p. 403.) In the third volume, some of Humboldt's most important treatises on general climatology, have been again discussed, and enriched by new measurements, partly there published for the first time. Among these are included the researches on the causes of isothermal curves, and on the snow-line.

Extract from a table of all the measurements of the snow-limit in *toises :*

I. NORTHERN HEMISPHERE.

Mageröe . .		71¼°	=	370 T.	
Norway . .	70°	70¼°	=	550	(v. Buch)
	67°	67½°	=	650	(Wahlenb.)
	60°	62°	=	800	
Iceland . .		65°	=	480	(Morcks and Olafsen)
The Aldan Chain in Siberia . .	60° 55′		=	700	
Ural . . .	59° 40′		=	750 ?	
Kamschatka .	56° 40′		=	820	(A. Erman.)
Unalaschka . .	53° 44′		=	550	(Lütke)
Altai . . .	49¼°	51°	=	1100	(Ledeb. and Bunge)
Alps . . .	45¾°	46°	=	1390	
Caucasus . .	43° 21′		=	1730	(Kupfer)
	42° 42′		=	1660	(Dubois)
Ararat . .	39° 42′		=	2216 ?	(Parrot)
Argaeus . .	38° 33′		=	1674	(Hamilton)
Bolor . .		37½°	=	2660	(Wood)
Hindu-Kho .		34½°	=	2030	(Burnes)
Himalayas					
Northern decliv.	30¾°	31°	=	2600	
Southern decliv.			=	2030	
Pyrenees . .	42½°	43°	=	1400	
Sierra Nevada .	37° 10′		=	1750 ?	
Etna . .	37½°		=	1490	
Abyssinia . .	13° 10′		=	2200	(Rüppel)
Mexico . .	19°	19¼°	=	2310	(Humboldt)
South America .	8° 5′		=	2335	(Codazzi)
	4° 46′		=	2397	(Humboldt)
	2° 18′		=	2405	(Boussingault)

II. EQUATOR.

Quito = 2475 T. (Humboldt)

III. SOUTHERN HEMISPHERE.

Quito . .	0° —	1½°	=	2470 T.	(Humboldt)
Chili					
Eastern Cord.	14½°		=	2490 T.	(Pentland)

Western Cord. .	18°		=	2897	(Pentland)
Chili . .	33°		=	2300	(Gillies)
	41°	44°	=	940	(Darwin)
Straits of Magellan	53°	54°	=	580	(King)

The tables which accompanied Humboldt's celebrated 'Treatise on the Isothermal Lines,' have also been filled up with all the more recent measurements, and have been arranged by Mahlmann for Humboldt's work. They embrace 315 places, the mean temperature of which, the temperature of the four seasons, and of the warmest and coldest months, is given. In the latter particulars therefore, these tables are more copious than the immediately preceding work of Mahlmann, (Dove's Repertorium, Bd. iv, 1841,) in which of 700 to 800 places the mean temperature only is given and, where it is possible, the summer and winter temperature. Humboldt, resting upon data which, in comparison with his Treatise of 1817, are now increased fivefold, divides the surface of the earth into eight zones of temperature, the extent of each of which is determined by the following limits of mean temperature:

I.—18° to 0° C. e. g. Melville Island,—18° 7 (74° 8′ N. L.), Nain in Labrador,—3° 6 (57° 2 N. L.)

II.+0° 1 to 5° C. Ulcaborg, + 0° 7′ (65° N. L.), Quebec, 3° 1 (46° 8′ N. L., and 300′ alt.)

III.—5° 1 to 7° 5 C. Upsala, 5° 3 (59° 9 N. L.), Utica 7° 4 (43° 1 N. L., and 450′ alt.)

IV.—7° 6 to 10° C. Orkney Isles 8° (58° 9′ N. L.), Berlin 8° 6 (52° 5′ N. L. and 108′ alt.), Fort Providence 8° 5 (41° 8′ N. L.)

V.—10° 1 to 15° C. Metz 10° 3 (49° N. L.), St. Louis 12° 9 (38° 6 N. L.)

VI.—15° 1 to 20° C. Florence 15° 2 (43° 8′ N. L. and 200′ alt.), New Orleans 19° 4 (30° N. L.)

VII.—20° 1 to 25° C. Cairo 22° 3 (30° N. L.), Macao 22° 5 (22° 2′ N. L.)

VIII.—25° 1 to 31° C. Calcutta 25° 7 (22° 6′ (N. L.), Guayaquil 26° (2° 2′ N. L.), Poudicherry 29° 6 (11° 9′ N. L.), Massahua 31° 5 (15° 6′ N. L.)

Observations on the periodical phenomena of vegetation are now arranged under the direction of Quetelet, according to a connected plan in England, France, Germany, Italy, Switzerland, Belgium, and Holland, and published since 1843 in the 'Mémoires de l'Acad. de Bruxelles.'

E. Meyer has proposed a simple method of signs, in order to distinguish, in a list of the plants of any floral district, those which reach their areal limits somewhere in it. (Bot. Zeit. 1843, p. 209.)

The signs selected are the following :—

$|\overline{\underline{\ast}}|$ The endemic plants of the Flora ; $\overline{\ast}$ Plants which reach their northern limit therein ; $|_{\ast}$, $_{\ast}|$, $\underline{\ast}$, in the same way stand for the western, eastern, and southern limits respectively.

With respect to the numerical proportion of Monocotyledons to Dicotyledons, E. Meyer asserts (Drege's Dokumente, s. u. s. 28), that the law developed by Schouw of the decrease of Monocotyledons towards mean latitudes (35°-45° N. L.) does not hold good as regards mountains, where the Dicotyledons increase in the neighbourhood of the snow-limit. The hygrometrical conditions of the atmosphere may perhaps explain these phenomena, and the alpine region which lies above the clouds, in the heat of the summer corresponds with the Mediterranean basin, where the Monocotyledons decrease in the most marked degree.

The geographical relations of several families of plants have been monographically treated in the past year, by Watson as respects the Ranunculaceæ, Nymphæaceæ, and Papaveraceæ (the Geographical Distribution of British Plants); the Malpighiaceæ, by A. Jussieu (Monographie des Malpighiacées, Paris, 1843); the Rosaceæ, by Frankenheim ; the Piperaceæ, by Miguel (Systema Piperacearum, Roterod. 1843, 8vo.) As the numerical relations obtained by such researches are extremely variable, I quote only some general results.

Ranunculaceæ. In Steudel's 'Nomenclator' 830 species are enumerated. Met with in all the polar expeditions, the number of species diminishes in the temperate zone towards the tropic, or they retreat into the higher regions of the mountains. Compared with the sum of the phanerogamia, they are most numerous in the polar circle, but the absolute number of species is greatest in the north temperate zone; 22 species are found in arctic America; Hooker enumerates 74 in British North America; Pursch 73 in the United States; Wahlenberg 44 in Sweden; Koch 109 in Germany; Sibthorp 60 in Greece; Desfontaines 30 in North Africa; and Humboldt 20 species on the Andes.

Nymphæaceæ. By Steudel 57 species, of which Asia possesses 20, North America 14, South America 9, Europe 8, Africa 7, the West Indies 2, Madagascar and Java 1 or 2. But to this distribution Watson opposes, that Torrey and Gray are only acquainted with 5 species in the United States, and Hooker with only as many in British America.

Papaveraceæ. There are distinguished of this family, including the Fumariaceæ, about 170 species. They are distributed in the arctic zone, and are found also within the tropics, though rare. They are most numerous in the warmer parts of the north temperate zone.

Malpighiaceæ. Of this family America possesses 528 species, viz., Brazil 290, Mexico 61, the West Indies 56, Columbia 45, Guiana 42, Peru 31; the Old World, on the contrary, presents only 55 species, of which 14 occur in India, 11 Madagascar, 9 West Africa, 9 in the Island of Sunda, 5 East Africa, 3 Australia, 2 Arabia, 2 China. There are but few instances of the Malpighiaceæ passing beyond the tropics; in North America *Hiræa septentrionalis* does not occur beyond 26° N. L.; in Nepal, *Hiptage* not beyond 28°, but in the southern hemisphere one *Acridocarpus* is found near Port Natal (30°), and *Higmaphyllon litorale* extends to Buenos Ayres. In the Mexican Andes the family does not ascend above 6000′, or scarcely passes beyond the limit of tropical vegetation. It is also met with at an equal elevation near the equator. In New Holland it is at present entirely unknown.

Rosaceæ. The author enumerates about 1100 species.

Of these 175 occur in central Europe, and about an equal number in North America, 92 in southern Europe, 74 in the Himalayas, 61 in the Alps, 85 in the tropical Andes; including, however, the Chrysobalaneæ.

Piperaceæ. This family is richest in species in tropical America; about a fourth part of that number are found in Asia, and isolated species in the South Sea Islands, and only a few in Africa. In the northern hemisphere, with few exceptions, scarcely any species pass beyond the tropic. In Africa only to 14° N. L. on the Senegal; from Arabia one species only, *Peperomia arabica* (22°) is known; on the Himalayas there are some at 30·5°; in China at 22·5°; in America, the only one, *Enckea Californica*, grows at Monterey up to 38°.

In Quito, *Piper peploides* extends to the altitude of 1590 *toises*. In the southern hemisphere, the Piperaceæ pass beyond the tropic to the greatest distance; they flourish at the Cape at 35° S. L., and one *Macropiper* as far as 45° S. L. in New Zealand.

I.—EUROPE.

THE new data concerning the climate of European Russia contained in Humboldt's work on Central Asia, differ materially from the earlier less accurate accounts. A comprehensive idea of the climatic relations of eastern Europe is given, founded on observations of temperature at Petersburg, Moscow, and Casan.

Petersburg. (As. Centr. 3, p. 56). The measurements are by Wisniewsky, and were already known, though not accurately computed.

		MEAN TEMP.			MEAN TEMP.
December	=	− 5·2° C.	June	=	+ 15° C.
January	=	− 9·5	July	=	+ 17·3
February	=	− 7·5	August	=	+ 15·8
Winter	=	− 7·4° C.	Summer	=	16° C.
March	=	− 3·7°	September	=	+ 10′·5°
April	+	+ 2·6	October	=	+ 5·1
May	=	+ 15·4	November	=	− 0·8
Spring	=	+ 2·5° C.	Autumn	=	+ 4·8° C.

An. Temp. = 3·9° C.

Moscow. (Ib. iii, p. 554.) The observations are by Spaski, and contained in the Bullet. Mosc. 1842.

Sea level	=	400′			
Winter	=	− 9·5° C.	Summer	=	+ 17·4° C.
Spring	=	+ 4·5	Autumn	=	+ 4·1

An. Temp. = 1·9° C.

Casan. (Ib. iii, p. 555.) The observations are by Knorre, and published in the same place.

Height above the Black Sea = 240′

Winter	= — 14·3° C.	Summer	= + 16·2° C.
Spring	= + 3·2	Autumn	= + 2·7

An. Temp. = 1·9° C.

Blasius has given an excellent exposition of the distribution of organic nature in European Russia, which, with regard to botany, contains a general determination and characterization of the provinces proposed by Ledebour, and mentioned in the Report for 1841. (Reise im Europ. Russland in den Jahren 1840 und 1841; 2 vols. 8vo; Brunswick.) In Northern Russia the author has especially investigated the districts on Lake Onega, and the southern part of the government of Wologda. In the Central province he has explored an extensive region from Jareslaw on the Volga, across the districts on the Oka as far as the Dwina and the Dnieper; and in the south he travelled across the Ukraine as far as the Steppes.

Northern Russia is chiefly distinguished from the Central province by its dense forests, in which *Pinus sylvestris*, L., and *P. Abies*, L., are the predominant species, and whose vast extent is only broken by swamps, or where, in the neighbourhood of the fluvial valleys, the trees have been thinned and destroyed by man. Amongst the pines and firs are intermingled here and there, *Alnus incana*, L., and *Betula pubescens*, Ehrb., which in some parts constitute by themselves large forests. The limits between cultivation and the wilderness are everywhere indicated, especially by alder bushes. Besides which, the only forms of leaf-trees are *Populus tremula*, L., *Sorbus Aucuparia*, L., and *Prunus Padus*, L. The pines and firs form two distinct forest formations, differing in the proportion of the argillaceous constituent of the soil. The clayey, often marshy low lands of the old red sand-stone are covered by thick fir woods, among which occur the *aspen* and *alder;* the sandy diluvial hillocks bear *Pinus sylvestris*, L., and *Betula pubescens*,

Ehrb., and represent the forest character of the North German plain, the soil of which has been formed at the same time. On this diluvium, where the soil is deficient in clay, are met with also heaths of *Calluna* (Bd. 1, p. 102), which do not occur in the Silurian plains and trap formations. However, the diluvium is not altogether free from bogs, where *Ledum* and *Andromeda calyculata*, L., flourish, but even here also, the fir (*Tanne*) does not grow, but only the pine (*Kiefer*), which does not shun the water, and requires only a light sandy soil. (p. 161.)

The characteristic plants of the coniferous woods of North Russia, are *Rubus arcticus*, L., *saxatilis*, L., *Chamæmorus*, L., *Vaccinium Myrtillus* L., *uliginosum*, L., *Oxycoccus*, L., *Rubus idæus*, L., *Rosa canina*, L., *cinnamomea*, L., *Linnaea borealis*, L.

In the pine and birch forests, principally *Cetrariæ* or *Antennaria dioica*, Br. The forest meadows are filled with *Ranunculus reptans*, L. On the mountain limestone grow *Peristylus albidus*, Bl. and *viridis*, Bl., and on the Lake Onega, *Aconitum septentrionale*, Mart. (*A. Napellus*, Blas.), most luxuriantly.

In the North Russian bogs of the argillaceous lowlands, Blasius distinguishes two plant formations.

1. "The dwarf-birch formation." The uncertain depth is covered by a dense shaking turf of *Sphagnum* with *Vaccinium Oxycoccus*, L., from which bushes of *Betula nana*, L., and *fruticosa*, Pall., from three to five feet high, spring up everywhere. In common with these grow several Ericeæ, northern *Rubi* and *Salices*: *Ledum palustre*, L., *Andromeda polifolia*, L., and *calyculata*, L., *Arctostaphylos Uva ursi*, Spr., *Vaccinium Vitis Idæa*, L., and *uliginosum*, L., *Rubus arcticus*, L., *Chamæmorus*, L., and *saxatilis*, L., *Salix bicolor*, Ehrb., *limosa*, Wahl., *glauca*, L., *myrtilloides*, L., and *rosmarinifolia*, L.

2. "Formation of the reed grasses and Carices." The soil is covered with water, but the bottom is firmer and more clayey than it is under the birch bushes, and without any covering of *Sphagnum*. Tufts of reed grasses are placed

on the surface close together. About twenty species of *Carex* are enumerated, and above these project the crowded white heads of *Eriophorum*. (Bd. 1, p, 43.) Ligneous plants are wanting, but *Calla* and *Pedicularis* partly supply their place. The open pools and lakes which occur in these swamps, present almost the same forms as in Germany: *Nymphaea alba*, L., *Nuphar luteum*, Sm., and *pumilum*, Sm., *Stratiotes aloides*, L., *Hydrocharis*, white flowered *Ranunculi* and *Caltha*. (p. 252.)

The cultivated spaces form only oases in these boundless plains, which, from the White Sea, as far as the watershed towards the district of the Volga, are everywhere covered with these four formations. The country is intersected only by the river valleys, in a peculiar way. These, with their broad irregular water-courses, form deep ravines in the great plain, which is elsewhere only slightly undulating. Thus, Ustjug-weliki on the Dwina, is 330′ above the sea, and the highest plateau of the forest plain, in its neighbourhood, in general, 600′. It is on the broad ridges of land forming the declivities especially, that the swamps extend for many miles. Towards the rivers the plain is suddenly depressed, and below the forest forms two terraces, which constitute the spacious valley. The inferior terrace is quite horizontal, and is reached by the overflowing of the stream. It is uninhabited, and contains fertile meadows or wastes, banks bare of vegetation, and islands. The water-course lies throughout, on the right hand, close at the base of the steep upper terrace. (Bd. 1, p. 238.) On the desolate sandy banks throughout all Russia, even to the southern Steppes, *Salix acutifolia*, W., grows, whose roots, from 40′-60′ long, are closely interlaced in the loose soil. The meadows are formed in the first instance from clay and marl deposited from the river, and being annually irrigated and supplied with fresh marl from it, possess the most luxuriant turf. The dunes on Lake Onega bear, on the contrary, *Calluna* with *Empetrum*. The upper terrace lies about 40′-60′ above the bottom of the valley. It is undulating, and extends as far as the base of the wooded diluvium. It is inhabited, and the greater part of it cultivated, and contains dry sloping meadows, blooming with Orchideæ, Labiatæ, and Synanthereæ, which give place lower down to bogs; all the hollows also of the ground, especially along the margin of the forest, are occupied by marshy meadows.

As far as the condition of the soil is concerned, the land is everywhere adapted to the cultivation of all the central European Cerealia, but the climate is opposed to agriculture. The destruction of the forests, which has been so ruinous in Central Russia, has, in these regions, effected but little alteration in the character of the country, and that only in the neighbourhood of the river valleys; nevertheless, since the memory of man, two of the noblest and most useful trees have almost entirely disappeared from these districts. In regions where Pallas still saw large forest ranges of *Pinus Larix*, L., Blasius counted scarcely half-a-dozen trees in a distance of 60-80 miles. In the same way *P. Cembra*, L., the Russian cedar, which, at a former period extended further to the west, at present is first met with in central Witschegda, east of the Dwina. Blasius met with the finest forests along the course of the Suchona, in the government of Wologda. Here the stems of the fir and aspen rise to a height of 100′ to 150′; and the birch not unfrequently attains more than 100 feet. (Bd. 1, p. 164.)

Blasius has indicated more accurately the natural boundaries of North and Central Russia. They are accurately defined by the ridge of the Waldai, that is, the line of the watershed between the northern and southern streams. Its level lies only 200′ above the highest elevations of the north; it may be assumed to have an average elevation of 800′ (for example, near Grjansowez, between Wologda and Jareslaw, the mean altitude above the sea is 760′); and yet this low range everywhere distinctly divides two extensive botanical districts. It is the southern limit of *Alnus incana*, D.C., and the northern limit of orchards and of many leaf-trees, for instance, of *Betula corticifraga*, which, however, at first occurs mixed with *B. pubescens*, Ehrb., but further south it composes the birch woods exclusively. The Coniferæ decrease, *Populus tremula*, L., becomes more abundant, and forms dense forests. The birch and aspen contend for the mastery with the pine until the oak appears, and from this point the forests of leaf-trees predominate. *Fraxinus excelsior*, L., *Tilia*, and *Quercus pedunculata*, Ehrb., first make their appearance near Jareslaw. *Q. Robur*, L., on the contrary, is exotic to Central Russia, and eastward does not appear to reach even the Dnieper. The underwood consists of *Corylus Avellana*, L., mixed occasionally with *Evonymus europæus*, L., and *verrucosus*, Scop., *Rhamnus frangula*, L., and *cathartica*, L. Jareslaw is, besides these, the northern boundary for the following plants: *Berteroa incana*, D. C., *Lunaria rediviva*, L., *Lavatera thuringiaca*, L., *Chærophyllum aromaticum*, L., *Eryngium planum*, L., *Scrophularia vernalis*, L., &c.

The northern marsh willows are replaced by *Salix fusca*, L., *cinerea*, L., *Caprea*, L., and *Alnus incana*, D. C., is represented by *Alnus glutinosa*, G. Thus almost all the plant formations assume another character, but the physiognomy of the whole country is much more strikingly altered by the increased extent of cultivation. The cultivated land and forest in Central Russia are in equal proportions; in this district, or that constituting Great Russia, the forests have been cleared. On the Oka, where the woods are formed of oaks mixed with the aspen and birch, they are for the most part limited to the neighbourhood of the rivers, and the adjacent valleys and ravines, indicating the gradual commencement of the treeless steppes. Here is already seen on dry elevations a thick vegetation of Artemisieæ (*A. scoparia*, Kit., *vulgaris*, L., *campestris*, L., and *Absinthium*, L.), which extends to the willows on the bank of the river, where now *Salix acutifolia* grows mixed with other species, such as *S. alba*, L., *fragilis*, L., *viminalis*, L., &c.

Central Russia is geognostically defined, on the north, by the predominance of dolomite on the old red sandstone; further on its natural character is marked by the marly soil of the new red sandstone and mountain limestone, or by the chalk marl, which present themselves tolerably free on the surface in long tracts. Northern Russia, on the contrary, presents the sandy and argillaceous strata of the old red sandstone, and thicker diluvial formations. On the Osero, the central region of vegetation encroaches to some extent upon the northern, owing to the chalky soil; on the other hand, between the Dwina and the Dnieper the northern botanical region extends further to the south, in consequence of the opposite geological conditions which there obtain.

South Russia commences where extensive diluvial deposits cover the chalk, and tertiary formations, and are again themselves covered by the humose soil termed *black earth*, or "Tschernon Sem." On the Dnieper, its northern limit lies near Tschernigoff, whence it passes through the southern part of the government of Kursk, and reaches the Volga in the neighbourhood of Simbirsk; at which point the arenaceous covering of the chalk is immediately contiguous to the new red sandstone of the north. From these geognostic relations, it is evident that the vegetation of the steppes is as distinctly defined from the district of the leaf-trees, as those are from the northern Coniferæ. On the Desna, which falls into the Dnieper near Kiew, first appear the wild fruit trees, *Pyrus communis*, L., and *Malus*, L., together with *Prunus Cerasus*, L., and with these commences the Southern region of vegetation. These trees are distinguished, even at a distance, from the other leaf-trees, by their crooked, crowded branches, and dark-coloured bark; the apple trees have a stem about the height of a man, above which they spread out into equal-sized branches. (Bd. 2, p. 221.) But the whole surface of the country is entirely treeless. It is only in the swampy hollows, and in the depths of the river valleys, the only places which in the north have been cleared, that an arboreal vegetation flourishes; but even here there are nowhere continuous woods as far as the diluvial deposit on the surface extends. The Coniferæ have long altogether disappeared, and of the leaf-trees, the birch soon retires. The oak is the most abundant tree, and is always associated with the fruit trees; in this way narrow strips of wood are formed which, in proportion to the size of the steppe, are very inconsiderable. Cultivation is confined to the "black earth," on the border of the steppe. This narrow strip of land scarcely reaches to Krementschug, on the Dnieper, where Blasius found the northern limit of the culture of the vine. Close to this point the steppe commences with lofty shrubs, species of *Artemisia*, *Verbascum*, *Achillea*, *Euphorbia*, and *Cynara*, which are mixed with the tall dry grass, and as they are used for firing, have received the name of "Burian," fuel. In the spring these plains are covered with a flowery carpet, but which, after a few months, is again scorched up and withered by the burning sun. In the short autumn the atmosphere is again foggy, causing a renewed verdure, but snow-storms soon succeed, and the waste surface remains covered with deep snow through the long winter.

Ukraine Proper, or the government of Charkow, forms a peculiar transition between the Steppes and Central Russia. It is a hilly country, in consequence of the chalk projecting from the diluvial sand. Hence forests are produced, which cover a considerable part of this fertile country. On passing from the plain of Poltawa towards Charkow, the "black earth" is observed to diminish in thickness on the watershed of the district of the Dnieper and the Don near Walki, and here the forests first appear. They consist of oak, lime, aspen, poplar, ash, and *Acer Tataricum*, L., but always mixed with the wild pear. The underwood consists principally of *Corylus*. The unwooded surface is here thickly covered with steppe-shrubs, two to three feet high, viz. *Cytisus supinus*, *Caragana*, and the dwarf cherry (*Prunus Chamæcerasus*,

Jacq.) The Flora of this province is distinctly that of South Russia, and this renders it probable that the climate exercises a more general influence than the soil, which in the Ukraine is calcareous, like that of Central Russia.

On the southern declivity of the Taurian mountain ridge, M. Wagner found the forests from Alupka as far as Ajuga-Dagh composed of *P. Laricio*, M.B., the region occupied by which extends from 600′ to 3000′. On the northern side, where the winter cold is much greater, they are replaced by the beech. *Arbutus Andrachne*, L., occurs only on the south side from the coast, as high as 1200′, though usually solitary, and its seeds appear to have been carried thither by birds of passage from Anatolia. (Augsburg Zeitung, 1843, Nos. 47-8.)

Of Ledebour's 'Flora Rossica' (vide Report for 1841, p. 416), the third and fourth parts appeared in 1843, and the fifth in 1844. (Vol. i, fasc. iii, vol. ii, fasc. iv, v.) The statistical relations of those families which have been treated of since the former Report are Balsamineæ 3; Oxalideæ 2; Zygophylleæ 10—in the European steppes, however, only *Zyg. Tabago*, L., and at the mouth of the Ural, *Zyg. Eichwaldii*, C. A. M.; Biebersteineæ 2; Rutaceæ 14—under which are included two species of *Tetradictis*, a genus probably belonging to the Crassulaceæ; Diosmeæ 1; Celastrineæ 6, and 1 Staphylea; 10 Rhamneæ, and 1 *Nitraria;* 2 Juglandeæ, both indigenous to the Caucasus; Anacardiaceæ 3; Papilionaceæ 568, among which is *Astragalus* with 168, *Oxytropis* with 61 species—genera confined to Asia are only *Thermopsis, Leobordea, Güldenstadtia, Halimodendron, Sphærophysa, Eremosparton, Lespedeza, Ammodendron, Gleditschia,* all with a single or few species; Mimoseæ 2, viz. *Lagonychium stephanianum*, M.B., and *Acacia fulibrissia*, W.—both only in the Caucasian provinces; Amygdaleæ 18; Rosaceæ 155—among which are *Spiræa* with 18 species, *Potentilla* with 60 species, 16 distinct species of *Rubus*, and 17 of *Rosa*—the Asiatic forms are *Coluria, Drya-*

danthe, Chamærhodos, Hulthemia; Pomaceæ 42, chiefly 19 species of *Pyrus* and *Sorbus, Punica* 1; Onagrariæ 23; Halorageæ 2; Hippurideæ 3; Callitrichineæ 5; Cratophylleæ 3: Lythrareæ 15, viz. two species of *Peplis* and *Middendorfia,* e. g. on the Dnieper, 2 *Ammanniæ* and *Ameletia* in Caucasia—the rest are *Lythra;* Tamariscineæ 15, for the most part Asiatic, although there are five species in the steppes of South Russia; Reaumuriaceæ 3, viz. *Reaumuria,* from the Caucasus to the Sea of Azof, *Eichwaldia* on the east side of the Caspian, and *Hololachna* in Zungaria; Philadelpheæ 1; Cucurbitaceæ 9, viz. on the Caucasus single representatives of *Lagenaria, Cucumis, Cucurbita,* and *Sicyas angulatus,* L., from thence westward as far as Podolia; Portulaceæ 16, of which 11 species of *Claytonia* occur in Eastern Siberia; Sclerantheæ 2; Paronychieæ 17; Crassulaceæ 59, e. g. 12 species *Umbilicus,* for the most part from the Caucasus and Ural; Grossulariæ 13, mostly Siberian; Saxifrageæ 70, besides 57 *Saxifragæ,* and 6 *Chrysosplenia*—in East Asia there are single species of *Leptarhena, Mitella, Tellina, Tiarella,* and *Heuchera;* Umbelliferæ 331, most numerous in Caucasia, almost disappearing in Eastern Siberia, yet there are 92 species in the Altai. The genera with most numerous species in Russia are *Heracleum* 23, *Peucedanum* 21, *Seseli* 18, *Bupleurum* 18, and *Ferula* 15; Araliaceæ 2, viz. *Hedera* and *Panax horridus* in the Kodjak Islands; Hamamelideæ 1—*Parrotia* in Talüsch; Corneæ 5; Loranthaceæ 3; Caprifoliaceæ 23; Rubiaceæ 77—among which, in the Caucasus, is the Hedyotideous, *Karamyschewia,* and the Spermacoceous *Guillonia,* both with a single species; Valerianeæ 41—among which in Siberia were 4 *Patrineæ,* in Armenia 1 *Dufresnea;* Dipsaceæ 36, with *Morina parviflora,* Kar., in the Alatau.

Works on the Flora of Finland have been commenced by Nylander (Spicilegium plantarum Fennicarum; Helsingf. 1843. Centur. I. 31 pp. 8vo, 1844. Cent. II. 38 pp. 8vo.) Further (Stirpes cotyledoneæ parocciæ Pojo—ib. 1844, 22 pp. 8vo,) and Wirzén (Prodromus Floræ Fennicæ—ib. 1843,

32 pp. 8vo.) The 'Spicilegium' contains critical remarks upon doubtful species, particularly on the *Carices*. Wirzén's work follows the sexual system, and extends at present only to the grasses.

Nylander, in 1842, travelled over Russian Lapland, from Uleaborg to Kola, on the Arctic Sea, and in 1843, East Finland and the governments between the Ladoga and the White Sea. Catalogues of the curiosities collected on the first journey are given in Lindblom's Zeitschrift. (Botaniske Notiser, 1842, 1844.)

Lund has described his botanical travels in Nordland and Finmark. (Reise ig jennem Nordlandene og Vestfinmarken i Sommaren, 1841; Christiania, 1842, 8vo.) He visited Tromsöe, where the birch was in flower at the end of August, and also Alten, Hammerfest, Mageröe, as far as the North Cape, and some other points. His review of the Finmark Flora contains 402 phanerogamia, in 50 families, whilst, in the whole of Norway, following Blytt's statement, he enumerates 84 families, with about 1100 phanerogamia. The families most rich in species in Finmark, are the following: Cyperaceæ 51; Gramineæ 42; Synanthereæ 33; Caryophylleæ 27; Cruciferæ 29; Rosaceæ 18; Junceæ 17; Ranunculaceæ 16; Ericeæ 15; Scrophularineæ 15; Saliceæ 15. Then follow 12 Leguminosæ, and 12 Orchideæ. The more interesting plants are: *Viola epipsila*, Led., nearly to the North Cape, *Lychnis affinis*, Vahl., *Potentilla nivea*, L., near Tromsöe, *Conioselinum tataricum*, Blytt., (Fisch?) near Alta, *Galium triflorum*, Mich.

The observed polar limits of the ligneous plants are, (1) near Alta, *Rubus idæus*, L., *Ribes rubrum*, L., *Myricaria germanica*, Desv., *Menziesia cærulea*, Sm., *Andromeda tetragona*, L., *Arctostaphylos Uva ursi*, Spr., *Rhododendron lapponicum*, Wahl., *Ledum palustre*, L., *Salix pentandra*, L., *arbuscula*, L., *hastato-herbacea*, Laestad., *Populus tremula*, L., *Alnus incana*, D. C. (2) Near Hammerfest, *Prunus Padus*, L. (3) *Pinus sylvestris*, L., as far north as 70°, that is, within one mile and a half south-east of Kistrand, on the Persanger Fjord. (4.) At Mageröe itself are still found *Sorbus Aucuparia*, L., *Calluna*, *Andromeda hypnoides*, L., *A. polifolia*, L., *Arctostaphylos alpina*, Spr., *Azalea procumbens*, L., *Vaccinium Myrtillus*, L., *V. uliginosum*, L., *Vitis idæa*, L., *Empetrum nigrum*, L., *Diapensia lapponica*, L., *Salix glauca*, L., *S. lapponum*, Wahl., *S. Myrsinites*, L., *S. reticulata*, L., *S. herbacea*, L., *Betula pubescens*, Ehrb., (*glutinosa*, Ld.,) *B. nana*, L., *Juniperus communis*, L.

Beurling, who, at the meeting of Scandinavian naturalists, in the year 1842, described the physiognomy of the neighbourhood of Stockholm, has travelled through Sweden in 1843, and the botanical results of his journey have been given in the 'Memoirs of the Stockholm Academy.' (Verhandelungen der Stockholmer Akademie.)

Zetterstedt's botanical tour through Jemtland, in the year 1840, has been translated in the 'Botan. Zeitung for 1844.' This Report contains lists of the localities, although without a more general characterizing of the vegetation of this province of Sweden. V. Düben has described an excursion in Bohuslan, made in 1841. (Lindblom's Botaniske Aviser, 1843. s. 75.) The first livraisons of Gaymard's 'Voyages en Scandinavie' have appeared. The plates give a graphic representation of the natural character of the north, but the explanatory text connected with them has not yet been published.

The fourth edition of Hartmann's Scandinavian Flora had been already published. (Handbok i Skandinavien's Flora, innefattende Sveriges och Norrige's Vexter, till och med Messorna; Stock. 1843.) Högberg's 'Svensk Flora, (Oerebro, 1843) is an inconsiderable compilation. Anderson's 'Observationes stirpium circa Christinehamn provenientium' (Upsala, 1842, 4to) contains some new localities for plants. Kröningsvärd has written a 'Flora dalekarlica.' (Fahlun, 1843, 8vo, 66 pp.) Torssel has published a catalogue of the Scandinavian Lichenes (343), and Byssaceæ (43). (Enumeratio Lichenum et Byssacearum Scandinaviæ hucusque cognitorum; Upsal. 1843, 12mo.) The eighth Century of Fries' 'Normalherbarium (vide Report for 1841) has appeared, and the fortieth part of the 'Flora Danica.'

The statistical relations of the British Flora have afforded a subject for new labours to Watson. The first part of a great work on this subject (The Geographical Distribution of British Plants, London, 1843, 8vo), though extending only to the Ranunculaceæ, Nympheaceæ, and Papaveraceæ, contains not fewer than 259 pages. This is the most copious collection of localities which has perhaps ever been made.

The horizontal and vertical distribution of each individual species is displayed in a table, which is even repeated forty times in this volume. Bielschmied has given a summary view of these special results in the Regensburg Flora. (1843, s. 641.) The only observations of more general interest are those illustrating the distribution of the three above-named families over the whole earth, of which we have spoken before.

The vegetation of the rocky island of St. Kilda, lying westward of the Hebrides in the Atlantic (58° N.L.), has been described by M'Gillivray. (Edinb. N. Philos. Journ. 1842, pp. 47-70, and 178-80. Also extracted by Bielschmied in Regensb. Flora, 1843, s. 455.) This island, about half a German mile in length, and scarcely half as wide, constitutes a rock 1380 feet high, consisting of trap and syenite, and presents in parts pasture land with Scottish vegetation; there are, however, only fifty indigenous phanerogamia. The characteristic species are *Cochlearia danica*, L., *Silene maritima*, Wilh., *Sedum anglicum*, Huds., *Rhodiola rosea*, L., *Ligusticum scoticum*, L., *Anagallis tenella*, L., *Salix herbacea*, L., *Carex rigida*, Good.; *Salix herbacea*, L., occurs here at a lower level than in Scotland, where Watson has not met with it below 1850 feet. The winter is very mild. Barley and oats are cultivated.

Dickie has investigated the geographical relations of the vegetation in Aberdeenshire. (Notes on the distribution of the plants of Aberdeenshire, in Hooker's London Journal of Botany, ii, pp. 131-35, and 355-58.) This is a supplement to Watson's work on the Grampians, mentioned in the last year's Report, and from it are derived the following corrections and additions to the altitudinal limits of the ligneous plants:

Quercus Robur, L. . .	0′ — 1500′
Lonicera Periclymenum, L. .	0 — 1500
Rosa canina, L. . . .	0 — 1860
spinosissima, L. . .	0 — 2000

Besides these, the extreme limit in altitude of a considerable number of plants belonging to a lower region is determined.

The author makes the following corrections in the list of alpine plants:

Arabis petræa, Hook. . .	1740′ (also washed down to 800′.)
Cerastium latifolium, L. . .	1740
Rubus Chamæmorus, L. . .	1000
Saxifraga oppositifolia, L. . .	On the shore near Aberdour.
Cornus suecica, L. . . .	1200
Veronica alpina, L. . . .	2300
Salix reticulata, L. . . .	2000
Juncus castaneus, Sm. . .	2300
triglumis, L. . .	1200
Carex rupestris, All. . .	2000— ?
lagopina, Wahl. (*leporina*, Ant.)	3560

Babington has published a British Flora on the plan of Koch's Synopsis. (Manual of British Botany, containing the flowering plants and ferns, arranged according to the natural orders; London, 1843, 8vo.) Of Withering's British Plants (corrected and condensed by M'Gillivray. Aberdeen, 1843), the fifth edition has appeared. Of dried collections of British plants, we have 'Salicetum Britannicum exsiccatum,' containing dried specimens of the British willows, edited by Leefe (Fasc. i, 1842, fol. with thirty-two forms), and Berkeley's 'British Fungi' (four fasc. of dried specimens; London, 1843).

The 'Flora Batava' (vide Report for 1841) has advanced in 1843 to the 130th part (aflevering). Dozy has given a supplement to his Catalogue (mentioned ib.) of the Jungermanniæ and Marchantiæ, found near Leyden. (In v. d. Hoeven's Tijdschrift, 1843, s. 108-14.) Kickx has given the first Century of the Flemish Cryptogamic Flora, in the 13th vol. of the 'Mémoires de l'Acad. de Bruxelles,' the greatest part of which consists of the Fungi. (Recherches pour servir à la Flore cryptogamique des Flandres; Bruxelles, 1840, p. 46, 4to.)

De la Fons has published some remarks upon the plants of the upper valley of the Maas, which possess only a local interest. (Ann. d. Sc. Nat. 19, pp. 317-19.)

The six last decades of the sixth Century of Reichenbach's 'Icones Floræ Germanicæ,' have appeared, which conclude the Caryophylleæ, and contain the Celastrineæ Liliaceæ, and part of the Lineæ. The 'Flora Germaniæ exsiccata' contains at present twenty-five Centuries. Of Sturm's 'Flora Deutschlands' the 21st and 22d parts of the third division have appeared, containing the Fungi by Rostkovius. The work of Schlechtendal and Schenk with figures, mentioned in the last year's Report, has advanced in 1843 to the tenth part of the fourth volume; and that upon Thuringia to the forty-seventh part; and a new edition of the former has also been commenced. The publications by Lincke, mentioned (ib.) have both advanced to the thirty-third part. D. Dietrich has begun a work with plates on the German Cryptogamia, of which the first part includes twenty-six illuminated plates of Ferns. (Deutschland's Kryptogamische Gewächse; Jena, 1843, 8vo.)

Koch has published a second edition of his celebrated Synopsis Floræ Germanicæ, (Frankfort, 1843,) which has been augmented with numerous special researches and additions. An abridged edition of this work appeared in 1844. (Taschenbuch der Deutschen und Schweizerischen Flora von Koch; Leipsig, 12mo.) A second edition also of Kittel's 'German Flora' has been prepared. Scheele has made critical remarks upon certain German plants, but without sufficient literary aid. (In Regensb. Flora, 1843, pp. 296, 421, 557.) Of Rabenhoorst's collection of dried Fungi of the German Flora, the fifth and sixth Centuries have appeared.

German provincial Floras and similar works :—Langethal, on the north of Germany (die Gewächse des. n. D. für Landwirthe, &c., Jena, 1843, 8vo.) Schmidt on the Prussian province (Preussens Pflanzen; Danzig, 1843, 8vo. Roeper on Mecklenburg (Zur Flora M.'s Th. 1, Rostock, 1843, 8vo), containing the vascular Cryptogamia, and valuable with respect to Morphology. Scholtz, Flora of the environs of Breslau; (Breslau, 1843, 8vo.) Döll, Rhenish Flora; (Frank. 1843, 8vo), including the vegetation of the Rhine district, from the lake of Boden as far as the Moselle and Lahn, and important as regards systematic Botany. Hackl, List of Plants in the southern division of the Leitmeritzer circle in Bohemia (in the Medic. Jahrb. des österr. Staats, 1843, p. 105, &c.) More special treatises: by John, on some Plants of the neighbourhood of Berlin—in the (Bot. Zeitung, 1843, pp. 689-92); by Preuss, upon some localities for Plants in the Oberlausitz,—in the (Regensb. Flora, 1843, pp. 671-72); by Wimmer, on the Silesian Hieracia in the (Uebersicht der Arbeit. der schles. Gesellsch. für 1843); by Hampe, the latest supplement to the Harz Flora—in (Linnæa, 1843, pp. 671-74); by Traunsteiner, on the *Salices* of the Tyrol—in the (Zeitschr. des Ferdinandeums, 1842.)

Among these works the Flora of Upper Silesia, by Grabowski, is distinguished by its giving the altitudinal limits .In the Gesenke (compare Report for 1840), according to Grabowski's measurements, the extreme limits in altitude of the ligneous plants are as follows:

1. In the Fir (*Tannen*) region (1500′-3600′, accordinp to Wimmer), *Pinus Abies*, L., and *Picea*, L. reach 4000′; *Juniperus nana*, W. 4500′; *Betula pubescens*, Ehrb. and *Sorbus Aucuparia*, L. 3900′; *Populus tremula*, L. 3800′; *Pinus Larix*, L. 3000′; *Juniperus communis*, L., 2600′; *Betula alba*, L., 2500′; *Acer Pseudoplatanus*, L., 2400′; *Prunus Padus*, L., 2300′; *Pyrus communis*, L., 2200′; *Fagus sylvatica*, L., 2000′; *Alnus glutinosa*, G., 1800′; *Prunus avium*, L., 1700.′

2. In the Oak and Pine (*Kiefer*) region, *Quercus robur*,

G., 1500′; *Fraxinus excelsior*, L., 1480′; *Ulmus campestris*, L., and *Pinus sylvestris*, L., 1300′; *Taxus baccata*, L., 1200′; *Populus alba*, L., 1000′. Wheat and barley are grown up to 1000′; rye to 1800′; and oats on the average up to 2000′.

I am not as yet acquainted with Reichenbach's Memoir on the Botanical relations of Saxony (contained in the Gäa von Sachsen.) A botanical sketch of the Kylfhäuser in Thuringia, by Ekart, is merely a collection of lists of plants in those localities, which are known from Wallroth's communications. (Regensb. Flora, 1843, pp. 169-82.) Kirschleger has compared the vegetation of the Black Forest, of the Jura and of the Vosges. (Congrés scientif, 1842, and translated in the Regensb. Flora, 1843, pp. 186-94.)

Since the more general influence of climate upon the vegetation of these three mountains is the same in each, and the more so as that portion of the Jura which lies to the south of Neufchatel is excluded, the author correctly attributes the varieties of the vegetation, described by him, to the nature of the soil. The mountain-region from 2400′ to 4800′ presents these contrasts in the most marked degree. The Jura at this elevation affords 116 Phanerogamia, which are not met with in the Black Forest, nor the Vosges, which, on the other hand, present fifty-two species wanting on the Jura. So much richer in plants is the calcareous Jura, but to this abundance the nearer proximity of the Alps also contributes. The following, together with many alpine plants, are characteristic of that region: *Erysimum ochroleucum*, D. C., *Thlaspi montanum*, L., *Saponaria acymoides*, L., *Arenaria grandiflora*, All., *Linum montanum*, Schl., *Hypericum Richeri*, Vill., *Acer opulifolium*, Vill., *Genista Halleri*, Regn., *Heracleum alpinum*, L., *Centranthus angustifolius*, D. C., *Hieracium rupestre*, All., *Prenanthes tenuifolia*, L., *Sideritis hyssopifolia*, L., *Fritillaria Meleagris*, L. The Vosges, again, present a much more peculiar vegetation than the Black Forest.

The following are characteristic forms of these two mountains, which are absent from the Jura, and are also not among plants widely distributed elsewhere; *Nasturtium pyrenaicum!* Br., *Brassica Cheiranthus*, Vill., *Hypericum elodes*, L., in Lotharingia, *Angelica pyrenæa*, Spr., *Galium tenerum*, Schl., *Carlina longifolia*, Rchb., *Hieracium longifolium*, Schl., *Sonchus Plumieri*, L., *Campanula hederacea*, L., *Pyrola media*, Scr., *Digitalis purpurea*, L., and its hybrid *Epipogium aphyllum!* Rich.; of these, however, only the two marked (!) occur in the Black Forest, the rest only on the Vosges. The vegetation of the lower region also presents a variety of contrasts, according to the geological formation. The Jura limestone, together with the basalt and trachyte of the "Kaiserstuhl" are, in this respect, exactly opposed to the sandstone and granite. The following are the plants of the Jura limestone in the valley of the Rhine and on the Vorberg, 2400′: e. g. *Thalictrum montanum*, Wallr., *Hutchinsia petræa*, Br., *Althæa hirsuta*, L., *Alsine fasciculata*, M. K., *Trinia vulgaris*, D. C., *Bunium Bulbocastanum*, L., *Artemisia camphorata*, Vill., *Crepis pulchra*, L., *Melittis melissophyllum*, L., *Euphorbia verrucosa*, Lam., *E. falcata*, L., *Gymnadenia odoratissima*, Rich., *Himantoglossum hircinum*, Spr., *Orchis simia*, Lam., *Ophrys aranifera*, Huds., *apifera*, Huds., *Aceras anthropophora*, Br., *Allium rotundum*, L. Plants of the granite and sandstone are, e. g. *Sisymbrium pannonicum*, Jaq., *Mænchia erecta*, G., *Potentilla recta*, L., *P. inclinata*, Vill., *Lactuca virosa*, L.

Heufler has endeavoured to characterize the plant regions of the Tyrol. (Tiroler Bote, 1842, Nos. 19-27.) The botanical part of the subject, however, has been treated too generally, and the altitudinal limits can only be considered as approximate estimates. An evergreen vegetation of *Quercus Ilex* and *Phillyrea media* occurs only in the Sarcathal. The vegetation of the Reichenauer and Flatnitzer Alps, on the borders of Styria and Carinthia, has been described by Pacher. (Regensburg Flora, 1843, s. 803-11); this paper is only of local interest.

In the geological work on the Venetian Alps, by Fuchs, (Solothurn, 1843, fol.) of which I have no further knowledge, a section treats of the limits of vegetation in the southern Alps. Mohl has communicated observations on the arboreal vegetation in the Swiss Alps. (Bot. Zeit. 1843, p. 409 et seq.) They form a sequel to the observations of Martius, mentioned in the last year's Report. The author corrects some statements of Wahlenberg, which apply indeed to northern Switzerland, though not to the central chain, which was imperfectly examined by him. In this country *P. Abies* decreases in the higher forest region, and is replaced by an abundance of *P. Larix* and *P. Cembra*.

Near Zermatten, where the red pine does not attain the altitude of 5000′, the arboreal limit, formed by the two last-named Coniferæ, is placed as high as 7000′. The Beech and Oak also disappear on the central chain at a lower level than in northern Switzerland; the former in Oberhaslithal at 3000′, the latter at 2460′. Allowing that these and similar differences in the arboreal vegetation of the calcareous and slaty Alps, are to be referred to the geological substratum, the same explanation does not apply to the cultivated plants, in which similar relations are pointed out by V. Martius.

NORTH SWITZ.—ACCORDING TO WAHL:		CENTRAL CHAIN:
Cherry tree up to	2900′	4480′ in the Matterthal.
Apple	3000	3400
Walnut	2000	3600 in the Lauterbrunnenthal (Kastof)
Vine	1700	2500 near Stalden.
Cerealia	2700	Wheat 5400′ near Zermatten (Gaudin); Barley 6100 near Zermatten (Martius)

Mohl is inclined to refer these differences to climatic causes. He believes, that although the mean temperature of the seasons might be expected to produce a diametrically opposite effect, yet that, on the other hand, with reference to the amount of atmospheric deposit, at least to its increased quantity in the summer? as well as with regard to the hygrometrical condition of the air?

the greater elevation of the country in southern Switzerland possesses a more continental climate than the regions explored by Wahlenberg. It is certainly true that the central chain of the Alps, in its climatic relations, more nearly approaches the character of a *plateau* than do the steep, lesser, calcareous Alps; but it appears to me that the greater number of the phenomena adduced so prominently by V. Martius are explicable by the different conformation of the valleys in the slaty mountains, whilst the cultivation of the soil is limited by the form of the surface in the calcareous Alps and conglomerates.

The remarkable local differences in the altitudinal limits of the trees are also indicated in a paper by Heer, 'On the Forest Cultivation in the Swiss Alps.' (Schweiz. Zeitsch. für Land und Gartenbau, 1843.) The extremes are collected in the following table:

	NORTH SWITZ:	SOUTH SWITZ:
Fagus sylvatica . .	4250′	4660′ in Tessin.
On the North declivity to	3900	
South	4550	
Acer Pseudoplatanus . .	4800	
On the North declivity to	4700	
South	5000	
Pinus Picea, L. . .	5000	
Abies, L. . .	5500	5100 near Airolo.

In the Ober Engadin it ascends, on the other hand, as high as 6100′, in the Unter Engadin to 6600′.

Pinus Larix, L. . . .	to 6000′	6500′ in Graubundten.

In the Engadin it also attains a greater elevation, and the greatest on the south side of the pass between Scarl and Münster; that near the Wormser Jochs as high as 7150′.

Pinus Cembra, L.		6500′

In the Engadin higher, and highest near Stelvio, where it attains 7280′

Pinus sylvestris, L. .	to 5500′	6000′
Pumilio, H. K.	6200	6750 in Graubündten.
Betula		5000 in the Engadin. 6000 in the Albignathal.

These facts afford a scale of the influence of locality on the distribution of plants in Switzerland, an influence which causes the close approximation of so many various climates, determined by the position, inclination, and surface-formation of the valleys and heights. It is only through the complete analysis of all these relations that an isolated abnormal phenomenon can be here explained. But on the whole, these local conditions equalize each other, and the peculiarities of the Oberland and Valaise, described by Mohl, lose in general importance, when they are compared with the Engadin, a valley also appertaining to the system of the central chain, and running north-east.

Systematic works on the Flora of Switzerland, are: Hagenbach (Supplementum Floræ basileensis; Basel, 1843, 8vo.) J. B. Brown (Catalogue des Plantes qui croissent naturellement dans les environs de Thoune et dans la partie de l'Oberland Bernois qui est le plus souvent visitée par les voyageurs; Thun. 1843, 8vo: Catalogue of the Phanerogamia and Mosses, with their habitats.) Rapin (Le Guide du Botaniste dans le canton de Vaud, comprenant les descriptions de toutes les plantes vasculaires qui croissent spontanément dans ce Canton; Laus. 1843, 8vo.) Blanchet (Essai sur l'histoire naturelle des environs de Vevey, 1843, 8vo): a work with which I am unacquainted. Reuter (Supplément au Catalogue des Plantes vasculaires qui croissent naturellement aux environs de Genève; Genève, 1841, 8vo): fifty-one pages, with a figure of *Arabis hybrida*, R. The rarer plants of the neighbourhood of Pfäfers are enumerated by Kaiser. (Die Heilquelle zu. Pf. St. Gallen, 1843.) Schærer's 'Lichenes Helvetici exsiccati' have reached the 18th part, and contain 450 species. The last part is accompanied with the conclusion of the 'Lichenum helveticorum Spicilegium.'

Kirschlager gives a review of the botanical relations of the environs of Strasburg. (Congres scientif. l. c.) He enumerates on this, for the most part cultivated alluvium, 960 sp., which he divides into the following formations:— "in arvis" 290 sp.; "in pratis" 300 sp.; "in campis" (incultis, &c.) 120 sp.; "in sylvis" 280 sp.; "in paludibus" 80 sp.; "in aquis" 110 sp.; "in ripâ Rheni" 20 sp.

The meadow vegetation on the Orne, from the village of Louvigny (south of Caen) to the sea, has been described by several botanists in Normandy—Hardouin, Leclerc, Fourneaux, and Eudes-Deslongchamps. (Mém. de la Soc. Linnéene de Normandie, vol. 7.) This paper shows the influence of the soil upon the distribution of meadow plants. Where inundations occur regularly, *Agrostis vulgaris* replaces *Hordeum secalinum* and *Cynosurus*, which elsewhere are the chief constituents of the grass turf, or in places where the marine tides overflow twice a month, the *Agrostis* gives way to *Glyceria maritima* and *Festuca rubra*, var. *maritima*. Of Schultz's 'Flora Galliæ et Germ. exsiccata,' six Centuries have appeared up to the present time. The following works relate to the Flora of France: Cosson, Germain, and Weddel (Introduction à une Flore analytique et descriptive des environs de Paris; Paris, 1842, 12mo.) By the same authors, (Supplément au Catalogue raisonné des plantes de Paris; Paris, 1843, 12mo.) A new edition has appeared of Bautier (Tableau Analytique de la Flore Parisienne; Paris, 1843;) as also of Mérat entitled (Revue de la Flore Parisienne; Paris, 1843); the latter in opposition to the more exact work of Cosson, &c. Godron (Flore de Lorraine—includes the departments Meurthe, Moselle, Meuse, and Vosges; Nancy, 1843. 3 vols. 12mo.) By the same author (Monographie des Rubus, qui croissent naturellement aux environs de Nancy—Ibid. 1843, 8vo.) Desmaziéres (Dixieme notice sur quelques plantes cryptogames recemment découvertes en France—Ann. Sc. Nat. 19, pp. 335-73), contains new Fungi, especially *Pyrenomycetes*, and some *Pezizæ*.

Tulasne has described the subterranean Lycoperdaceæ of the neighbourhood of Paris, with several new species, and the new genera *Hydnobolites* and *Delastria*. (Ann. Sc. Nat. 19, pp. 373-81.)

Massot has published a table on the plant limits on the Canigou in the Pyrenees. (Comptes Rendus, v. 17; also printed in the Regensb. Flora, 1844, p. 84, and in the

Bot. Zeitung, 1844, p. 427.) These measurements are important as regards the alpine plants.

Descending from the summit, which rises to a height of 2785 met., the ligneous plants appear in the following order:

Rhododendron ferrugineum, L. (1322 m.)	2540 m.
Genista purgans, L.	—
Pinus Abies, L. (1500 m.)	2415
Sambucus racemosa, L.	2063
Betula alba	1987
Pinus Picea, L.	1950
Sorbus Aucuparia, L.	1838
Populus tremula, L.	1640
Amelanchier vulgaris, Mch.	—
Limits of cultivation of Potato and Rye, harvest in beginning of September	—
Fagus sylvatica, L.	1623 m.
Corylus Avellana, L.	—
Lonicera Xylosteum	—
Sorbus Aria, Cr.	1566
Rubus fruticosus, L.	1322
Cratægus Oxyacantha. L.	1250
Prunus spinosa, L.	1050
Ilex Aquifolium, L.	987
Cornus sanguinea, L.	—
Rye harvest middle of July	—
Castanea vesca, G.	800
Alnus glutinosa, G.	800
Sarothamnus scoparius, W. G.	—
Attempted cultivation of the vine	750
Acer monspessulanum	700
Euonymus Europæus, L.	—
Abundant cultivation of vine	550
Cultivation of olive	420

According to Bory, the indigenous oak of the mountains of Andalusia is the *Quercus bætica*, Webb., identical with *Q. Robur*, Desf., and widely distributed in Algeria. The former has named it *Q. Mirbeckii*. (Comptes Rendus, v. 17.)

Systematic remarks upon some South European Gramineæ have been communicated by Link. (Linnæa, 1843, pp. 385-407).

An interesting memoir on the character of the vegetation of New Castille has been read by Reuter before the Geneva Natural History Society (Essai sur la végétation de la Nouvelle Castille; Geneva, 1843, iv, 34 pp.) The plateau of Madrid, which is elevated more than 2000′, is bounded on the north by the Sierra of Guadarrama, (the mountains Carpétano-Vétoniques of Boissier), which are covered with snow for eight months in the year. The mean temperature of Madrid (2050′) amounts to 15° C. according to Humboldt, that of the summer to 24°·8, and of the winter to 6°·1 (Schouw); but the thermometer always falls in the winter below the freezing point, so that there is almost every year skating on the pond of the Retiro; it seldom sinks below 6°, although in 1830 it fell to 10°, and in 1802 to 11°·25 C. In summer, the thermometer occasionally rises in still air in the shade to from 37° to 41°. Rain falls only in winter and spring, with the prevailing north winds, which are cooled by passing over the mountains. In the spring these winds alternate with westerly and southerly breezes, which characterize the summer, and are accompanied with greater heat and drought. The autumn also is throughout warmer, till December. The epochs of vegetation appear to occur a month earlier than in Geneva. At the end of March, the trees were already in leaf, and the Cherry and Syringa in blossom. The vegetation of the herbaceous plants commences in the beginning of March, and is quite over by the end of June, excepting some shrubs which withstand the drought. (p. 12.)

The plateau, which presents undulating ridges of low hills, is, in the neighbourhood of the metropolis, covered for the most part with wheat and barley fields, and being destitute of wood or even of shrubs, presents the most uniform aspect, and is everywhere bounded by the same confined horizon. The plant formations are throughout dependent on the nature of the soil, and consequently fall under four classes: that of the clay, of the gypsum, of the sand, and of the granite. The argillaceous substratum stretches southwards from Madrid over the greater part of La Mancha. The hills, for example, from Aranjuez to Alcala, consist of saliferous gypsum, in the springs from which common salt effloresces and *Halophyta* flourish. To the northward and westward of Madrid as far as the mountains, the surface is composed of close-grained sand without stones, which, in consequence of the drought, acquires almost the same degree of cohesion as the clay. Lastly, the granite soil constitutes the Sierra of Guadarrama itself, and blocks of it are also scattered over the sandy surface. These mountains attain an altitude of 7-8000′, and the passes into Old Castille are about 4500′ to 5500′. Limestone is not found in the neighbourhood of Madrid; it first appears on the east towards Cuença, and together with it the extensive shrub formations of Catalonia commence, which are not found in the plateau of New Castille.

The corn on the sandy soil is but poor, on the clay it is perhaps 4′ high; "Garbanzos" (*Cicer arietinum* and "Algarrobas" (*Ervum monanthos*) are especially cultivated for food. The vine and olive occur only in sheltered places, but the latter is always small and poor. Meadows are altogether wanting; even the pastures (Kräuterwiesen) at Manzanares consist only of annual grasses and Leguminosæ, which at the approach of summer are soon choked by thorny plants; e. g. *Centaurea calcitrapa, Eryngium campestre, Ononis spinosa, Xanthium spinosum,* or where the ground is more marshy are replaced by large tufts of *Juncus acutus* and *Scirpus holoschænus.* According to ancient chronicles, it would appear the elevated plain of Madrid was formerly wooded (p. 13), and remains of these woods, in the form of stunted, widely scattered *oaks*, especially *Q. Ilex*, are still visible on the sandy hills of the Casa de Campo, and of the Prado, mixed with leafless Genisteæ (*Retama sphærocarpa, Sarothamnus scoparius*), but these, together with the trees growing on the banks of the river (*Salix, Populus, Ulmus, Fraxinus angustifolia*, Vahl.), and some shrubs (*Tamarix gallica, Cratægus, Rosa, Rubus, Rhamnus, Osiris*), are the only ligneous plants of the plateau. It is evident that the want of wood is only the consequence of the aridity;

this is proved by the lofty growth of the plantations in the valley of the Tagus near Aranjuez, as well as of the more recently planted avenues in Madrid, which are maintained by irrigation.

Summary of the Plant formations:—(1) Argillaceous soil. In the fields the first plants which appear are *Brassica orientalis*, *Lathyrus erectus*, Læg., *Turgenia*, *Glaucium corniculatum*, *Polygonum Bellardi;* these soon become choked by thorny Synantheræ; *Picnemon*, *Scolymus*, *Xanthium*, *Onopordon nervosum*, Boiss.; at the end of summer the only surviving plant is *Ecballion*, which at last develops its fruit, *Crozophora* also is abundant. The uncultivated plains and hills (campi) are covered with aromatic plants, a class which in Spanish is termed "Tomillares" from *tomillo* (Thymus). Here the vegetation consists of *Thymus tenuifolius*, *Teucrium capitatum*, and *Sideritis hirsuta*, with which various plants, characteristic of the country, are mixed, e. g. *Queria*, *Minuartia*, *Astragalus macrorhizus*, and *narbonensis*, *Echinops strigosus*, *Cynosurus Lima*, *Stipa barbata.*—Riparial plants are: *Althæa officinalis*, *Lavatera triloba*, *Cochlearia glastifolia*, *Gypsophila perfoliata*, *Sonchus crassifolius.*—Halophyta flourish most luxuriantly in the pool of Ontigola, near Aranjuez: *Spergularia marina*, *Frankenia pulverulenta*, *Erythræa spicata*, *Atriplex*, *Suæda setigera*, *fruticosa*, and *maritima*, *Salicornia*, *Hordeum maritimum*, and (cultivated) *Salsola Soda*.

(2) Gypsum. The vegetation peculiar to this substratum is, together with the soil, extended through the whole of Arragon. The steep declivities are covered with plots of *Frankenia thymifolia*, associated with *Peganum*, *Lepidium subulatum*, and *Cardamine*, *Helianthemum squamatum*, *Gypsophila struthium*, *Zollikoferia*, *Salsola vermiculata.*—Other characteristic plants are, *Vella Pseudocytisus*, *Iberis subvelutina*, Gass., *Herniaria fruticosa*, *Centaurea hyssopifolia*, *Statice dichotoma*, Cav. Extending from the south of Spain to Aranjuez, and clothing

the ridges of the downs, grows the social *Stipa tenacissima*, which is used for a variety of purposes; with this are associated several Cisteæ, *Pimpinella dichotoma*, *Rosmarinus*, *Fritillaria*. Many isolated thickets are formed of *Quercus coccifera*, with *Rhamnus lycioides*, *Retama sphærocarpa*, and *Bupleurum frutescens*.

(3) The sandy soil is characterized by numerous Cruciferæ, which perhaps nowhere present such a variety of species, nor grow individually so much associated together as here; and in spring they give a yellow colour to the cultivated plains. With this prevailing colour are also mingled blue Boragineæ and white Anthemideæ; *Diplotaxis catholica* and *virgata*, *Sisymbrium contortum* and *hirsutum*, Lag.; *Brassica lævigata* and *valentina*, *Sinapis heterophylla*, Lag.; *Anchusa undulata* and *italica*, *Echium violaceum; Anthemis mixta*, *pubescens*, and *arvensis;* and also *Malcolmia patula*, *Hypecoum grandiflorum* and *pendulum*, *Ræmeria hybrida*, *Cerastium dichotomum*, *Veronica digitata*, *Aphanes cornucopioides*, and several species of *Linaria*, particularly the extremely social *L. ramosissima*, Boiss., and *L. hirta* and *spartea*. When this rich vegetation has disappeared, the fields are overgrown with *Tanacetum microphyllum*, D.C. The "tomillares" occupy extensive tracts, and here consist of *Thymus tenuifolius* and *Mastichina*, *Santolina rosmarinifolia*, and *Lavandula pedunculata*. Among these, in the spring, is seen a multifarious vegetation of annual herbs and grasses; several Cisteæ, particularly *Helianth. sanguineum*, Lag., and *Ægyptiacum*, *Astrolobium durum*, *Campanula Loeflingii*, *Myosotis lutea*, *Pyrethrum pulverulentum*, *Prolongoa pectinata*, *Aira involucrata*, *minuta*, *lendigera*, and *articulata*, *Holcus setiglumis*, *Bromus ovatus*, *Psilurus anistatus*, *Hordeum crinitum*. After the disappearance of these, larger herbaceous plants spring up, especially Umbelliferæ: *Thapsia villosa*, *Margotia laserpitioides*, *Daucus crinitus*, *Magydaris panacina*, *Pimpinella villosa*, *Verbascum sinuatum* and *pulverulentum*, *Ruta montana*, *Onopordon illyricum*, *Centaurea ornata*.

(4) On the granite of the Sierra of Guadarrama, these "tomillares" extend up to about 4000′, becoming gradually mixed with other plants. The greater moisture of the soil produces here several central European plants. Extensive pasture grounds for horned cattle, which are protected by fences from the flocks of sheep (and termed "Dahesa"), are covered with bushes of *Quercus Toza* and *faginea;* on the rocks grow *Jasminum fruticans, Lonicera etrusca, Daphne Gnidium, Juniperus Oxycedrus.* Here also Rose-Cistuses first appear; *C. ladaniferus* and *laurifolius.* Several new species of plants were found by Reuter in this region, which for the rest differs but little from the plateau, e. g. *Ranunculus carpetanus, Pæonia Broteri, Silene Agrostemma, Hispidella;* and also are here found, *Caucalis hispanica,* Lam., *Digitalis thapsi, Dianthus lusitanicus, Antirrhinum hispanicum,* Chav., *Macrochloa arenaria;* several Orchideæ, Irideæ; under the shade of the oak bushes, *Arenaria montana, Bunium denudatum, Valeriana tuberosa, Scilla nutans.* The higher mountain region, above 4000′, is that of the Genistæ, since it is almost entirely covered with *Genista purgans.* Solitary shrubs of *Juniperus* and *Adenocarpus hispanicus* occur; upon the latter of which lives the true "Cantharis." In this shrub region are found *Arabis Boryi,* Bois., *Linaria delphinoides,* Lag., *saxatilis,* Chav., and *nivea,* Boiss., *Senecio Tournefortii* and *Duriæi,* Gay, *Narcissus apodanthos.* Some higher points rise above this Genista region, and bear a thick firm turf of *Festuca curvifolia,* Lag., mixed with *Armeria juniperifolia,* W. Of alpine plants there are only a few traces, such as *Saxifraga nervosa* and *hypnoides, Ledum hirsutum* and *brevifolium,* the annuals of the sandy plain of Madrid, however, flourish even here. In the neighbourhood of the mountain-rivulets the turf is composed of *Nardus stricta* with *Pedicularis sylvatica, Jasione carpetana* and *Veronica serpyllifolia.*

It is only on the northern declivity of the Sierra that forests of a bifoliate fir (*P. sylvestris*) occur, and here large tracts are covered with *Pteris.* The Sierra de

Gredos, the most westerly and highest elevation of this ridge, differs but little in its vegetation, and is in a still higher degree poor in plants, and of a uniform aspect.

Much more interesting appear to be the mountains south of Toledo, which were explored by Reuter at too late a season of the year. These extensive rounded hills belong to the vegetation form of the Monte Baxo, under which the Spaniard understands the oak bushes which grow in clumps. But the contrast presented by the Sierra Nevada is much greater, since all the plants which are common to this and that of Guadarrama, grow, without exception, also in Asturia and on the Pyrenees.

Reuter collected above 1250 species of plants in New Castille. The new species (about 50) are published by him, in concert with Boissier, in the 'Bibliothèque Universelle de Genève' (1840). The families most rich in species of this collection are the following: 143 Synanthereæ, 123 Gramineæ, 110 Leguminosæ, 76 Cruciferæ, 61 Caryophylleæ, 54 Labiatæ, 52 Scrophularineæ, 38 Rosaceæ, 33 Ranunculaceæ, 38 Boragineæ. The recurrence of a series of Castilian plants in the Crimea, without their being found in any of the intermediate countries is remarkable. Reuter explains this extraordinary fact by the analogy of the extreme climates, and of the geological substratum, which is especially evident in the heavy clayey soil and saliferous gypsum. The plants to which this explanation applies, are: *Lepidium perfoliatum*, *Menioeus linifolius*, *Mollugo cerviana*, *Minuartia dichotoma*, *Queria hispanica*, *Callipeltis*, *Campanula fastigiata*, *Veronica digitata*, *Acinos graveolens*, *Rochelia stellulata*, *Plantago Loefflingii*.

Contributions to the Flora of Italy. Of Bertoloni's 'Flora Italica' the fifth volume has appeared, containing the 11, 12, and 13 classes (Bologna, 8vo); and also the second volume of Moxis' 'Flora Lardoa,' an original work, indispensable in the systematic study of the plants of South Europe; this volume, in which De Candolle's arrangement of the families is followed, includes the Rosaceæ and the whole of the Ericeæ, from No. 411-779, together with Pl. 73 to 93—(Turin, 1840-43, 4to.) Puccinelli (Synopsis plantarum in agro Luccensi sponte nascentium Lucca, 1842.) Id. (Additamentum ad Synops. Lucc.—Giornale Bot. Ital. fasc. 1.) Gussone (Synopsis Floræ Siculæ, 1842:—a new working out of his Prodromus) Todaro (Orchideæ Siculæ, 1842.) Gasparrini (Nonnullarum plantarum descriptiones—Rendiconto, accad. Nap. 1842, extracted in the Botan. Zeitung, 1843, s. 643); 1 *Geranium*, and 1 *Fumaria* from Calabria, 1 *Cerinthe* from Naples, 1 *Sedum* from the Nebrodes.

Ball has published some remarks on his botanical tour in Sicily, and has taken occasion to give a very complete list of the Sicilian Gramineæ (240 sp.). (Ann. Nat. Hist. 11, pp. 338-51.)

The statement in the last year's Report, that Schouw had indicated *Opuntia* and *Agave* as occurring in Pompeii, appears, according to the Bot. Zeit. (1844, s. 581), to have originated only in an erroneous translation of his account.

On the vegetation around Pola, in Istria, are some remarks containing only what is known, by Von Heufler, (in the Regensb. Flora, 1843, s. 767.)

Zanardini, in a new systematic memoir, has completed his catalogue of the Dalmatian Algæ up to 272 species (Saggio di classificazione della Ficee; Venezia, 1843, 64 pp. 4.)

In Davy's work on the Ionian Islands (Notes on the Ionian Islands and Malta. London, 1842, 2 vols. 8vo,) are contained two years' observations on the climate of Constantinople, from which I extract what is most important as regards vegetation. (ii, p. 400.)

MEAN TEMPERATURE.

	1839.	1840.
January . . .	= + 2·2°	= + 4·8° C.
February . . .	= + 5·6	= + 4·1
March . . .	= + 4·4	= + 4·6
April . . .	= + 6·1	= + 7·7
May . . .	= + 11·1	= + 15·5
June . . .	= + 21·1	= + 20·6
Max. of temp. . .	+ 31·7°	= + 32·7° C.
Min. . .	— 1·7	= — 4·4

	1839.	1840.
July . . .	= + 22·2°	= + 24·5° C.
August . . .	= + 26·7	= + 22·9
September . . .	= + 20·	= + 20·6
October . . .	= + 17·2	= + 15·6
November . . .	= + 13·9	= + 12·7
December . . .	= + 7·8	= + 3·2
Mean temp. . . .	= + 13·3	= + 14·7 C.

Prevailing winds, N.E. (215 and 199 days), S.W. (99 and 113 days). Rain fell (102 and 122 days). Amount of rain, 1840=31·65″; May, June, July, and August, almost without rain; between 1″ and 2″ in November; between 2″ and 3″ in February and April; between 3″ and 5″ in May, September, October, December; above 6″ in January.

According to Davy's observations, the temperature of the springs in the Ionian Islands, at the level of the sea, fluctuates between 16° and 18° C.

The mean temperature of Malta (i, p. 261), equals 17·8° C.; the max. temp.=31·1°, and the min. temp.= +5° C.

The same work contains an important series of observations on the saline contents and temperature of the Mediterranean Sea. The common opinion that it is specifically lighter and warmer than the Atlantic, is in no way supported by these observations.

In Forbes's researches on the distribution of the lower animals in the Ægean Sea, the Algæ are also considered, though only in a general way. (Report on the Mollusca and Radiata of the Ægean Sea; from the Report of the

British Association, &c. for 1843.) In the eight regions admitted by Forbes, from 0° to 1380′ deep, the prevailing Algæ are distributed in the following relations:

(1) 0—12′.

a. Above low-water mark, *Dictyota dichotoma* and *Corallina officinalis.*

b. Below low-water mark. The characteristic Fucoid is *Padina pavonia.*

(2) 12′—60′. The mud is usually green from the presence of *Caulerpa prolifera.* The sandy bottom is rich in *Zostera oceanica.*

(3) 60′—120′. *Caulerpa* and *Zostera* gradually diminish in quantity.

(4) 120′—210′. Fucoids are abundant, especially *Dictyomenia volubilis, Sargassum satirifolium, Codium Bursa* and *flabelliforme, Cystosira.* Corallines increase. Nulliporæ and Spongiæ frequent.

(5) 210′—330′. Fucoids diminish in number; *Dictyomenia volubilis* is rare, *Rytiphlœa tinctoria* and *Chrysimenia uvaria,* more abundant. The bottom is constituted for the most part of Nulliporæ and shells.

(6) 330′—474′. Fucoids are extremely rare. The sea bottom consists of Nulliporæ. Although at this depth, the higher Algæ scarcely any longer exist, yet many phytophagous Testacea are met with, from which circumstance the opinion that the Nulliporæ are plants receives new and very important support.

(7) 474′—630′. The Algæ, except the Nulliporæ, which still usually constitute the sea bottom, entirely disappear.

(8) 630′—1380′. Here the Nulliporæ also appear to be wanting; as the sea bottom from this point onwards consists of yellow mud, with remains of Foraminifera.

Of my 'Spicilegium Floræ rumelicæ et bithynicæ,' in which about 2000 plants have been systematically studied, the first volume containing the Polypetala (almost the half of the whole), has appeared. (Brunswick, 8vo.) On the conclusion of this work I shall recur to it.

The Report on Koch's journey by the Danube to Constantinople, seems to have appeared without the knowledge of the author (Bot. Zeitung, 1843, s. 605), and must be passed over on account of the uncertainty of the names of plants (e. g. *Pinus Cembra* and *Ammodendron* on the Bosphorus).

Tenore has published remarks upon Sibthorp's 'Flora Græca,' which should not be overlooked in the comparison of the Italian and Greek Flora. (Rendiconto accad. Nap. 1842, extracted in Bot. Zeitung, 1843, s. 877.)

Schultz has proposed a considerable number of new Greek *Orobanchæ* (Reg. Flora, 1843, s. 125); but the descriptions are defective, and the species, without doubt, for the greater part untenable.

II.—ASIA.

Aucher-Eloy's oriental journals have been edited by Count Jaubert. (Relations de Voyages en Orient de 1830-38, par Aucher-Eloy, revues par le Cte. Jaubert; Paris, 1843, 2 vols. 8vo.) The scientific contents of this work are not considerable, but the importance of the author's collections, which have already for the most part been arranged, invests even a simple Itinerary, from which the locality and time of flowering of most of the plants can be determined, with a great degree of interest. Before entering upon the review of these travels, and as Aucher-Eloy has made no estimates of altitude, I will premise an observation of Ainsworth, which occurs in his last book of travels (Travels and Researches in Asia Minor, Mesopotamia, Chaldea, and Armenia. London, 1842, 2 vols. 8vo., ib. ii, p. 374), and in which the altitudinal relations of a part of the regions explored by Aucher-Eloy have been strikingly characterized from his own measurements. Asia Minor is a highland, encircled by a level or hilly littoral tract, to which on the north side, a second terrace of less elevated plains succeeds, e. g. that of Duzcha, E. of Nicomedia, 250′; of Boli, 570; of Vezir Köpri, above the mouth of the Kizil-Irmak, 800′, &c. Thence commences on the south, the elevated plateau, which, descending gradually from Persia towards the Ægean Sea, is near Angora, 2700′, and even near Kastamuni, close to the Black Sea, S.W. of Sinope, still 2400′ high, but rising near Erzerum to a height of 6000′. This plateau, with its scattered conical hills, the highest of which, that of Argäus, near Cæsarea, according to Hamilton, rises to 12809′, contains numerous basins, having no exit for the water; viz., Ak-Scher, 2300′; Konia, 2200′; the great salt lake, Koch-Hissar, south of Angora, 2800′; Erekli, on the northern base of the Taurus, 2600′; Kara-Hissar, near Cæsarea, 3420′; the lake Van, 5460′; and Urmia, 4300′. The Taurus, or southern border chain of this great highland, descends on the south abruptly, partly to the littoral tract and partly to the plains of Assyria and Mesopotamia, the latter of which are nowhere more than 700′ above the sea.

The first journey of Aucher-Eloy occupied from November 1830 to October 1831. It included Egypt, where he remained from December to March, and Syria, where he passed the months from April to July; August he devoted to a visit to Cyprus.

The imperfect journal of 1832 shows that Aucher-Eloy in this year explored Smyrna and Rhodes, whence he returned by way of Moylah and Guzel-Hissar. The third journey is comprehended in the year 1834. In May he arrived at Cæsarea from Constantinople, by way of Nicomedia and Angora; in April, at Scanderoon and Antioch, by way of Tarsus; in May he explored the country about Aleppo and Aintab: between Antioch and Aleppo he remarked the sudden transition from the Mediterranean to the Syrian vegetation (vol. i, p. 84); in June he crossed the passes of the Taurus to Malatia, on the Euphrates, and followed that stream downwards to the vicinity of Arabkir, and proceeded, in July, by Erzingan to Erzerum. Fourth journey in 1835. February, Constantinople, Brussa, Kutaja, Ophium, Kara-Hissar, Ak-Scher; March, Konia, Adana: *Crocus*, *Hyacinthus*, *Anemone coronaria*, and others, in flower on the 9th March on the southern declivity of the Taurus. Skanderoon: *Phœnix* abundant on the coast; groves of *Myrtus*, *Laurus*, *Styrax*, and *Arbutus Andrachne* towards Antioch. Aleppo: the period of vegetation of the steppe lasts from the end of February to June. (Ib. p. 177.) April: Bir, Mardin, Mossul: as prevailing steppe-plants between the two latter towns, are mentioned *Serratula cerinthefolia*, D. C., *Sinapis oliveriana*, *Avenæ* sp. (Ib. p. 191.) May: along the Tigris to Bagdad; banks of the river covered with *Tamarix gallica*, *Populus euphratica*, Oliv., *Capparis leucophylla*, *Sinapis lævigata;* below Dor (34° N.L.) the Date-palm becomes abundant; considerable palm groves near Hilla and Kerbela; prevailing plants of the saline steppes: *Tamarix pycnocarpa*, Decaisn., *gallica*, *Chenopodium fruticosum*, *Zygophyllum simplex*, *Peganum*, *Fagonia Bruguieri*, *Cucumis* sp., *Ajuga elongata*, M. B., *Savignya ægyptiaca*. (ib. p. 227). June: Kermanschah; limits of *Phœnix* towards Persia, near Hadschi-Kara-Khani, S.W. of Elluan (ib. p. 231), Hamadan, ascent of the Elvend. July: Scheschnau, Ispahan. August: excursion to Zerdaka, a mountain lying to the S.W. (32° N.L.), the height of which Aucher Eloy estimates at more than 10,000′; journey continued by Cashan to Teheran. September: excursion to Demavend, Kasbin, Tauris. The chain of the Elbruz presents no Coniferous region. The forests consist of *Quercus*, *Fagus*, *Ulmus*, *Celtis*, *Diospyros*, *Gleditschia caspica*, *Acacia Julibrissin*, *Platanus*, succeeded by bushes of *Paliurus* and *Juniperus hispanica*, A. E, in the alpine region, another *Juniperus*, *Rosa*, and *Berberis*. (Ib. p. 335.)

Fifth journey in 1836.—Smyrna, Chios, Syra, Athens, Parnassus, Eubœa, Thessaly, Olympus, Hajion-Oros, Skyros, Lemnos, Imbros, Hellespont, Brussa.

Sixth and last journey, 1837 and 1838.—March: Nicomedia, Angora. April: Tokat, Baibut. May: Erzerum, Koi, Lake Urmia. June: Tauris, Ardebil, coast of Ghilan, Rescht. July: Erzevil on the southern declivity of the Elbruz. August, September: exploration of that mountain, second ascent of the Demavend. September to December: remains in Teheran. January: Ispahan, Schiraz; spring vegetation commences in the middle of January with a *Bulbocodium* (*Colchicum crocifolium*, Boiss.), and in February the country is covered with flowers; the only rainy season is from the 15th January, to the 15th March; Bushire on the Persian Gulf. February: Dscharun, Lar., *Mimosæ* become frequent; Bender-Abassi. March: passage to Muscat, the coast near Sohar is covered with palm groves (vol. ii, p. 545), excursion inland as far as the mountain Akadar (about 5000′ high). April: passage to Bender-Said, and back to Bender-Abassi. May: Forg, Darap, limits of palm vegetation between Darap and Fasa (ib. p. 600), Shiraz. June: returns to Ispahan, where he dies in October.

Ainsworth (l. c. et sup. vol. ii, p. 131) describes the annual course of vegetation in the environs of Mossul. During the moist February, the mean temperature of which = 10° C., the vernal plants, which constitute the only adornment of the steppe, bud forth. In the beginning of March, *Anemone* and *Narcissus* flower; in the second week of the same month, species of *Ranunculus*, the Fig, and Apricot shoot; in the third week, flowering Cruciferæ and Orchideæ, *Ranunculus Asiaticus* and *Traganth-Astragalus*. Towards April, about twenty Phanerogamia were in flower, viz., *Gladiolus, Sternbergia, Trollius asiaticus*, and a small Anthemideous plant; the almond trees blossomed, and the water-melon put forth buds. In the latter half of this month, the temperature = 15° C. With May commenced the dry season; to the spring grasses now succeeded other species of *Chrysurus, Dactylactenium*, &c.; the prevailing Phanerogamia were Euphorbiæ and Synanthereæ; the corn-harvest lasted from the middle to the end of the month, at which time the mean temperature = 30° C., now all the Phanerogamia began to wither, and only a white *Trifolium* and *Nigella damascena* continued in flower. Finally, no plant remained except the prevailing steppe-plants, species of *Artemisia* and *Mimosa*. In July the heat reaches 40° C., and from this time the hibernation of the vegetation continues till the following year.

As the most abundant plants of the Mesopotamian steppes, the light red soil of which, according to Aucher Eloy, rests upon a calcareous formation, with the rolled fragments of which it is mixed, Ainsworth enumerates (ib. p. 177) *Artemisia fragrans* and *Absinthium*; here and there are found other social plants, e. g. *Allium*, *Rœmeria*, *Silene*, *Erigeron*, (*Aster pulchellus*, Ains.), Anthemideæ, &c. Where the ground is less arid, an *Avena* prevails for miles together, together with some other Gramineæ and Synanthereæ, *Chrysanthemum*, *Gnaphalium*, *Crepis*, *Centaurea*. The steppe is nowhere altogether barren, but bare tracts are often only covered with lichens, especially with a gray *Lecidea* with black apothecia, a *Cetraria*, and some *Verrucariæ*.

The region of oak forests in the high mountains of Kurdistan, near Amadia, extends from 1500′-2500′, according to Ainsworth's measurements. (Ib. p. 194.)

On the Lake Urmia, the steppe-vegetation consists of almost the same plants as in the low-lying country of Mesopotamia and Babylon, although its level is almost 4000′ higher. (Ib. p. 301.) But the Artemiseæ are for the most part represented by species of *Traganth-Astragalus*, *A. verus*, and *tragacanthoides*. Where the steppe is free from salt, *Nigella damascena* grows, together with *Capparis spinosa* and *ovata*, as near Mossul, or the surface is covered with *Ononis* and a *Mesembryanthemum*, which flourishes as near Hilla. The vegetation of the saline steppe on the Lake Urmia consists of Chenopodeæ: *Salsola*, *Salicornia*.

M. Wagner ascended the Great Ararat, and found the arboreal limits to be constituted by some tufts of birch, at an elevation, estimated by Parrot, to be 7800′. The mountain declivities of Armenia, however, are almost as bare of trees as the elevated plains. The traveller was assured by the natives, that forests formerly existed in districts at present entirely bare. (Augsb. Zeit., 1843, No. 214.)

The systematic works connected with the Flora of Hither Asia, have made considerable progress. Of Boissier's 'Diagnoses Plantarum orientalium' (vid. last Report), the second and third parts have appeared, and this important work will be concluded in 1844 with the fourth and fifth parts. The new species described in it belong to the following families.—5 Rhamneæ from Persia, Kurdistan, and Cilicia; 1 *Rhus* from Muscat; about 180 Leguminosæ, of which 54 are Persian, for the most part *Astragali* (39), two species of *Taverniera*, 1 *Crotolaria* from Bender-Abassi, and 1 *Tephrosia*, the rest for the greater part from Asiatic Turkey; also about 10 *Astragali*, then follow *Trifolium* (11), *Trigonella* (10), *Onobrychis* (9); but many species are not as yet sufficiently determined, for comparison with those of Sibthorp and Willdenow; 10 Rosaceæ, mostly *Potentillæ* from Anatolia, 1 *Cotoneaster* from the Bithynian Olympus, found by Boissier, 2 *Amelanchier* species; 5 Paronychieæ with a new genus *Sclerocephalus* (*Paron. sclerocarpa*, Decaisn.), indigenous on Sinai and near Muscat (Mascate); 1 Reaumuriaceous plant; *Eichwaldia Persica* from the plateau of the Persian steppes. 6 Crassulaceæ, among which 3 *Umbilici* from Persia and Babylon; 6 *Saxifragæ* from Cadmus, Bithynian Olympus, Taygetus and Parnassus; the numerous Umbelliferæ have been published in the 'Ann. d. Sci. Nat.' for 1844; about 45 Rubiaceæ, many of which, however, must be reduced, the most remarkable are the *Wendlandia*, found by Kotschy in Kurdistan, and the new genus *Mericarpæa* from Mesopotamia; 8 Valerianeæ; 13 Dipsaceæ. Above 40 Synanthereæ, of which the most numerous are *Anthemis* (9) and *Centaurea* (8), though with several of the species untenable; a newly instituted genus *Cephalorrhynchus* with the habit of *Crepis pulchra*, found by Boissier in Lydia; 6 Campanulaceæ from Anatolia; 1 Primulaceæ; 2 Asclepiadeæ; 1 Convolvulaceæ; 17 Borragineæ; 47 Scrophularineæ, amongst the most numerous of which is *Verbascum* (18) mostly from Anatolia, *Scrophularia* (9), *Veronica* (9); what is remarkable, 1 *Gymnandra* near Erzerum, 1 *Wulferia* near Seleucis; 1 Acanthaceæ from Caria; 65 Labiatæ, among which are *Salvia* (7), characteristic of Persia, *Nepeta* (5), *Phlomis*, 2 *Otostegiæ*, 1 *Lagochilus*, and the two new genera *Zataria* and *Sestinia*; 3 Polygoneæ; 3 Santaleæ. 3 Aristolochiæ. 7 Euphorbiaceæ; 1 *Orchis*. 15

Liliaceæ, with the new genus *Chionodona* found by Boissier in the alpine region of the Tmolus, near Sardis; 4 Colchicaceæ: 13 Gramineæ, with the new genera *Rhizocephalus* from Mesopotamia, and *Nephelochloa* from Caria. In the appendix are contained: 1 Fumariaceæ from Spain (*Aplectrocapnos*), 5 Cruciferæ, 4 Caryophylleæ, 1 Lineæ, 2 Rutaceæ, 1 Leguminosæ, 1 Dipsaceæ, 3 Gentianeæ, among which is a *Swertia* from Persia, also published by me in De Candolle's 'Prodromus.'

Henzel's illustrated work, mentioned in the former annual Report, is entitled 'Illustrationes et descriptiones plantarum novarum Syriæ et Tauri occidentalis' (Stuttgard, 1843, fasc. i, with 14 lithographic plates, 4to.) This livraison contains, besides complete descriptions of the species published in the 'pugillus:' 12 Leguminosæ, with the new genus *Hammatolobium* from the Taurus; 2 Rosaceæ (*Potentilla*); 1 Geraniaceæ; 1 *Euphorbia*; 4 Hypericineæ; 18 Caryophylleæ, chiefly species of *Silene* and *Dianthus*; 4 Violariceæ; 7 Cruciferæ; 1 Ranunculaceæ; 3 Crassulaceæ; 10 Umbelliferæ.

The 'Illustrationes plantarum orientalium' of Comte Jaubert and Spach (vide last Report) make rapid progress. The first volume of 100 plates was completed in 1843, and the second is already commenced with the 12th livraison. The following genera have been completed: *Argyrolobium, Cicer, Hypericum, Gaillonia, Statice, Quercus.* I shall afterwards refer to this work more in detail. In the 'Ann. d. Sc. Nat.' Spach has at the same time discussed several oriental genera, particularly *Spartium, Leobordea, Argyrolobium, Ebenus, Amygdalus, Gaillonia,* and the section Armeriastrum of *Statice.*

Schlechtendal has described some plants collected by Kotschy in Kurdistan (Linnæa, 1843, pp. 124-28): 3 Umbelliferæ, with the new genus *Polycyrtus*, 1 *Fedia*, 1 *Althæa*, 1 *Hyosciamus*. 7 new Umbelliferæ from the same source have been described by Fenzel (Regensb. Flora, 1843, s. 457-63); among which are the new genera *Callistroma, Elæosticta, Anisopleura, Uloptera.*

The Flora of Cyprus has been composed by Pöch, and the work has been founded principally on an herbarium collected in that island by Kotschy in 1840. (Enumeratio plantarum, hucusque cognitarum ins. Cypri; Vindob. 1842, 8vo, pp. 42.)

In the whole 310 species are enumerated, of which four are given as new: *Pterocephalus multiflorus*, *Teucrium Kotschyanum*, *Quercus alnifolia*, *Crocus veneris*. The diagnoses of these have been given in the 'Regensb. Flora,' 1844, s. 454. Flotow has determined some lichens collected in Cyprus. (Linnæa, 1843, s. 18-20.)

I have, unfortunately, not yet obtained C. Koch's 'Travels in the Caucasus.' His list of Caucasian and Armenian plants, however, has been previously published. (Linnæa, 1843, s. 31-50, and s. 273-314.) The following families were treated of in the past year:—5 Caprifoliaceæ, 21 Rubiaceæ (new, 1 *Galium*); 7 Valerianeæ (new, 1 *Dufresnea*, 1 *Valerianella*); 16 Dipsaceæ (new, 2 *Scabiosæ*); 178 Synanthereæ (new, 1 *Centaurea*, 3 *Cirsia*, 1 *Carduus*, 1 *Anthemis*, 1 *Pyrethrum*, 2 *Senecio*, 1 *Antennaria*, 2 *Podosperma*, 1 *Scorzonera*, 1 *Lactuca*, 2 *Crepis* species, 2 *Mulgedia*); 16 Campanulaceæ, 2 Cucurbitaceæ, 7 Ericeæ, 1 *Diospyros*, 1 *Ilex*, 2 Oleineæ, 2 Asclepiadeæ, 2 Apocyneæ, 10 Gentianeæ, 3 Convolvulaceæ, 10 Solaneæ, 67 Scrophularineæ (new, 2 *Verbascum*, 1 *Celsia*, 2 *Scrophularia*, 1 *Linaria*, 3 *Veronica*, 1 *Gymnandra*, 1 *Odontites*, 1 *Pedicularis*); 10 Orobancheæ (new, 1 *Phelipæa*, 2 *Orobanche*-species); 1 *Sesamum*, 1 *Globularia*, 1 *Verbena*, 81 Labiatæ (new, 2 *Ziziphora*, 1 *Satureja*, 1 *Micromeria*, 1 *Lamium*); 43 Boragineæ (new, 1 *Omphalodes*, 1 *Caccinia*, 1 *Onosma*); 16 Primulaceæ, 5 Plantagineæ, 1 Laurineæ, 2 Thymeleæ, 3 Eleagneæ, 2 Santaleæ, 24 Chenopodeæ (new, 1 *Spinacia*, 1 *Halimocnemis*), and the new genus *Halanthium* from the Araxes.

Trigonometrical measurements of the altitude of the Caucasus above the level of the Black Sea, by Fuss, Sabler, and Sawitsch differ materially from the previous statements. For three of the most known mountains they have been communicated by Humboldt. (Asie Centr. ii, p. 57.) The western peak of the Elbruz measures 2882 toises (18493 English feet); the eastern 2880 toises; the Kasbeck 2585 toises; the Beschtau 710 toises.

Basiner gives a concise description of the autumnal vegetation on the sea of Aral, in his 'Journey from Orenburg to Khiva.' (Bull. Petersb. ii, pp. 199-204.) The steppe between the Caspian Sea and that of Aral is called "Ust-Jurt;" it was traversed by Basiner, and, according to the report of Tschihatscheff, forms a plateau lying 500′ above the plain of Orenburg. (Humb. Asie Cent. iii, p. 558.) The last-named officer, who accompanied the unfortunate expedition of the Russians against Khiva, has given an account of the extreme climate of this region, where the cold in winter descends to — 43·7° C., whilst in summer it may be observed to reach + 42·2° C. When Basiner travelled with an embassy from Orenburg to Khiva, by the same route, the steppes were already burnt up by the summer heats.

Between Orenburg and the Aral he observed in many places the plain covered for miles together with *Salsola arbuscula*, and *Atraphaxis spinosa*. On the rocky declivities of the "Ust Jurt," above the Aral, these were mixed with other Chenopodeæ. On the sand hills *Pterococcus aphyllus* was particularly abundant, easily distinguished by the slender leafless branches, and the fruit dependent on filiform peduncles.

Two shrubs vegetated among the testaceous tertiary rocks on the Aral, *Tamarix ramosissima*, Led., and the frequently mentioned "Saxaul" (*Anabasis Ammodendron*, C.A.M.), which resembles a bundle of green painted twigs. Further to the south near Aiburgir, N.W. of Kunä-Urgendsch, Basiner met with a large and moderately thick grove of "Saxaul," in which stems 15′ high occurred. This was the first wood after passing the Ileck, but a wood without leaves of any kind, although verdant and blooming.

The other characteristic plants of the "Ust-Jurt" coincide with the usual forms of the steppes of South Russia, and they extend also as far as Khiva. In Khiva *Karelinia caspia*, Led., and *Alhagi camelorum*, Fisch., are generally distributed, and *Salsola subaphylla*, C.A.M., and *Halimocnemis sclerosperma*, C.A.M., are not unfrequent. The saliferous loamy soil, however, of Khiva was in many places completely bare of plants. The meadows of the Kanat, so celebrated in the East, are indebted for their existence only to artificial irrigation, and in them *Poa pilosa*, *Setaria glauca*, *Melilotus*, and *Plantago*, contend with the Chenopodeæ, *Kochia hyssopifolia*, and *Atriplex Hermanni*. The embassy returned, in the middle of winter, to the western bank of the Amu-Deria, which presents a shrub formation of *Elæagnus angustifolia*, L., *Halimodendron argenteum*, D.C., *Tamarix ramosissima*, Led., and *Populus diversifolia*, Schrk. At the end 3 new species are described: 1 *Asperula*, 1 *Lepidium*, and the fruit of *Sium cyminosma*, which is cultivated in Khiva, supplies the place of *S. Sisarum*.

The remarkable journey of Middendorf in northernmost Siberia, almost to the promontory of Taimoor, may be said to have approached the limits of the next world. (Erman's Archiv für Russland, 1843. H. 3.) Descending the river of that name, the traveller did not turn back until he had attained the 76° N.L., when he had nearly reached the open Arctic Sea, under unspeakable difficulties. He then lay sick eighteen days, and forsaken by his companions during the month of September, buried in the snow on the Lake Taimoor, and was with difficulty saved. The last traveller in this region had been Laptiew (1739-1743), who advanced to 77° 29′, and traces of whose expedition were discovered by Middendorf. The whole peninsula on the Lake Taimoor is inhabited by only two Samoiede families, who pasture their herds of reindeer there in the summer, and drive them southwards in the winter. The collection of objects of natural history made under 74° N.L. is not yet arranged. Middendorf found arboreal vegetation even beyond 70° N.

At the meeting of Scandinavian naturalists at Stockholm (1842), Eichwald gave an account of an Alga of the Aleutian Isles, which is used for food, *Bromicolla aleutica*. At Unimah it forms a layer two feet thick, of a nostoc-like substance, which is covered with a gramineous vegetation. Whenever the provision of fish of the natives fails, these Algæ are collected and eaten.

Systematic works on the Flora of Northern Asia.

Schrenk has explored the regions on the Tschu, a river of the Zungarian steppes. The new species which he found have been already published. (Bullet. Petersb. 2. Nos. 32, 37.) They belong to the following genera: *Lepidium, Diplotaxis,* 3 species, *Silene, Zygophyllum, Euphorbia,* 5 species, *Astragalus, Oxytropis, Rosa,* 2 species, *Lythrum, Rubia, Microphysa* (nov. gen. Stellat.) *Cousinia, Apocynum, Pedicularis, Diploloma,* (nov. gen. Boragin.), *Solenanthus, Echinospermum, Plantago, Brachylepis, Rheum, Allium, Juncus.*

To these, besides, are added 9 Chenopodeæ (with both the new genera *Pterocalyx,* Schr., and *Halostachys,* C.A.M.), and 2 Staticeæ, which have been described in the Bull. de l'Acad. de Mosc. (Mars, 1843.) A monograph on the Siberian Rosacean genus *Chamærrhodos,* by Bunge, is given in the Ann. Sc. Nat. (Vol. xix. p. 176-178). A monograph by Besser on the *Artemisiæ,* which is only just printed, (Mém. Péters. Divers savans, v. 4, 1843), is highly important towards a knowledge of the steppe vegetation.

Kützing has given the characters of the Fucoideæ collected by Tilesius on the coast of Japan, which had been already admitted in his Phycologia generalis. (Bot. Zeit. 1843. s. 53-57.)

In the 'Souvenirs d'un Voyage dans l'Inde par Delessert' (Paris, 1843-4), are reports upon the climate of the Neilgherries, derived partly from his own, and partly from Baikie's measurements. The two English stations are named Kotagherry and Ootacamund, the former is situated 1983·5m. above the sea, the latter 2255m., both in. 11°-12° N.L.

	Mean Temperature.	Difference between Max. and Min.
Kotagherry	= 16·1	8°
Ootacamund	= 14·4	

An important systematic work has been commenced by Bentham, which is intended to embrace all the East Indian Leguminosæ, as well as those of tropical and Southern Africa. (Hook. Lond. Journ. of Bot. 1843. pp. 423-81, and 559-613.) This monograph relates principally to the collections of Wallich, Royle, Wight, Jaquemont, Griffith, Helfer, &c., from India; of Kotschy, Heudelot, and Vogel, from tropical Africa; of Burchell and other travellers at the Cape. Up to the present time, the Podalirieæ, the Liparieæ, and part of the Genisteæ, particularly the Crotalariæ, have been published, already about 300 species. Among these there are about 100 Genisteæ, 37 diadelphian Genisteæ (Liparieæ), and 27 Podalirieæ; about 80 Genisteæ, and 3 Podalirieæ, Indian, from the Himalayah; about 40 Genisteæ belong to tropical Africa.

Griffith has described the following new genera: *Jenkinsia* (Thymeleæ) from Assam, *Enkleia* (Thymeleæ) from Malacca, *Leptonium* from Assam, and *Champereia* from Malacca (both transition forms from the Santaleæ to the Olacineæ), *Plagiopteron* from Silhet (Euphorbiaceæ?), *Siphonodon* (Ilicineæ) from Malacca. (Calcutta, Journ. of Nat. Hist. vol. iv, 1843; also in the Regensb. Flora 1844, p. 432.) In the same Journal, which has not yet come to hand, Jack's botanical labours on Sumatra, &c., are also said to be collected.

The great illustrated work on the Flora of Java, which has been edited by Blume under the title of 'Rumphia' (Lugd. Batav. fol.), has advanced in 1843 to the end of the second volume, which relates principally to the Palms. A second work by Hasskarl, which has very little connexion with that mentioned in the last annual Report, is published in v. d. Hoeven's Tijdschrift (1843, pp. 115-50). It contains systematic observations upon Javanese plants, and descriptions of new species from the following families: 1 *Fern*, 2 Cyperaceæ of the new genus *Pandanophyllum* (near *Chrysitrix*), 1 Xyrideæ, 1 Commelineæ, 1 Melanthaceæ, 1 Amaryllideæ, 1 *Canna*, 1 *Artocarpus*, 4 Labiatæ, 1 *Begonia*, 1 Malvaceæ, 1 Meliaceæ, 4 Euphorbiaceæ, 1 *Connarus*, 1 *Rubus*, 2 Leguminosæ.

Junghuhn's 'Travels in Java' (vide Report for 1841), have been edited in more detail with the assistance of Nees v. Esenbeck. (Lüdde's Zeitschr. für vergleich. Erdkunde, Bd. 2, 3.) In the investigation of the mountain vegetation of Java, the traveller enjoyed great advantages from his residence at Djocjokarta, at the southern base of the volcano of Merapi, which is more than 8000′ high. He ascended this mountain repeatedly, and has described its vegetation. The forests of the lower region (vol. ii, p. 457), consist of hundreds of species of trees, but species of *Ficus* and other Urticeæ prevail, and after them Magnoliaceæ, with an undergrowth of Melastomaceæ and Scitamineæ.

To these succeed the oaks, especially *Quercus pruinosa*, Bl., with stems 100′ high, covered to the summit with Orchideæ and other parasites, and with *Usneæ* a foot long, and mosses. In these woods the Palmaceous forms are represented by *Areca humilis*, W., and the tree-ferns by *Chnoophora glauca*, Bl. A coniferous belt is wanting on the Merapi. Above the oak-region the forest consists of *Celtis*, which is called "Angring," and this affords shade to species of *Rubus* which, as in other mountains, are joined with *Podocarpus* (*Rub. javanicus*, Bl., *moluccanus* L., *lineatus*, Reinw.) In this region even, volcanic blocks occur, which throughout Java, above the altitude of 5000′, are clothed with *Polypodium vulcanicum*. On other declivities of the Merapi the *Celtis* is replaced by *Acacia montana* (vol. iii, p. 68), or the *Rubus* by *Gaultheria* and *Thibaudia*. These Ericeæ, together with other shrubs, here constitute an alpine region above the arboreal limits, a formation which reaches to the upper trachyte declivities.

The most abundant species is *Gaultheria punctata*, Bl.,

and a ligneous *Gnaphalium* (*G. javanicum*, Bl.) Mixed with these, grow *Thibaudia varingifolia*, Bl., *Rhododendron tubiflorum*, Bl., and other Ericeæ, besides *Hypericum javanicum*, *Polygonum paniculatum*. Junghuhn found: *Gualtheria repens*, with *Lycopodia*, some mosses, and *Polypodium vulcanicum*, up to the margin of the crater.

The views of the south coast of Java are not sufficiently definite to convey any precise ideas; but few travellers have as yet displayed the talent of being able to exhibit a luxuriant tropical vegetation. Junghuhn is especially acquainted with the *fungi*, of which family he has been the first to describe many Javanese forms. He treats of them in the description of the Zuider Hills, a wooded range which runs along the Bay of Pashitan, on the south coast. At this part, thin groves of *Tectonia grandis*, *Emblica officinalis*, and low leguminose trees, alternate with the moist primitive forests; the open spaces in these groves are occupied by a dense growth of high grass, the "Allang-Allang" of the Javanese. It is, however, in the shades of the lofty primitive forests that the fungi are to be met with (ii, p. 358). In this equatorial Flora their appearance is not limited to any particular season of the year. The rains continue, at least in the mountains, all the year round, and the rich "humus" soil is constantly moist and spongy. Each particular species even is not limited to any special season, and the same species is continually making its reappearance. On the other hand, the larger *fungi* are not here so much associated together as in the temperate zone. They everywhere occur only solitary, a condition consequent upon their habitat on decaying trees; for in these forests the *Agarici* of the North are replaced by parasitic *Polypori*.

In the forestlike plantations of Batavia and Weltevreden the most abundant trees are the following (ii, p. 89): *Garcinia Mangostana*, *Mangifera indica*, *Artocarpus*, *Nephelium lappaceum*, *Citrus*, *Averrhoa*, *Morinda*, *Eugenia*, *Anona*, *Persea*, *Durio zibethinus*, *Carica*, *Cocos*, *Areca*, *Tamarindus*, *Canarium*, *Morus*, *Hibiscus tiliaceus*,

Musa paradisaica, *Bambusa arundinacea*, *Bixa*. On the plain of Batavia there are no longer any primitive forests, and only thickets of *Psidium*, *Mussænda glabra*, and *Melastoma malabaricum*. The soil here consists of a reddish brown, rich clay, which, towards the sea-shore, acquires an admixture of "humus," until it finally passes into the pure "humus," or swamp of the Rhizophorous region. In these morasses the river water is mingled with that of the sea. True Rhizophori do not occur here, but *Bruguiera caryophylloides*, Bl., *B. Rhedii*, Bl., and species of *Ægiceras* with climbers of *Ipomæa maritima*, Br., *Verbesina*, and *Borassus*, with Loranthaceæ, or a jungle composed of *Nipa fruticosa* and *Acanthus ilicifolius*. (Ib. ii, p. 141.)

Botta (vide last Report) has given an account of his Travels in Arabia Felix, chiefly as an introduction to the description of the plants collected by him. (Archives du Muséum d'histoire nat. v. ii, pp. 63-88.)

The traveller started from Hodeida (15° N.L.) to Zebid, and explored the mountains in the district of Taas, particularly Mount Saber, celebrated from the time of Forskäl, on account of its botanical riches. It is a lofty precipitous mass of trachyte, at the northern base of which lies the city of Taas. The plain of Taas, which is situated at an elevation about half that of the mountain, is at present a waste, in consequence of the political disturbances of Yemen, and is covered with fleshy *Euphorbiæ*. On Mount Saber, on the other hand, which is easily defended, together with the cultivation of wheat and oats, the culture of *Celastrus edulis* (Cât.) flourishes; the buds of this plant and the unexpanded shoots are eaten without any other preparation, and produce a slight and pleasant excitement of the nerves. The trade in this production is in Yemen even more considerable than that in coffee. A single person may consume five francs' worth in the course of a day. Coffee plantations occur only on the south side of Mount Saber. In the country, the pulp only of the coffee-fruit is used; the infusion of the beans being but little esteemed. The hill is, besides these, rich in all varieties of fruits, both of the tropical and temperate zones, which are here cultivated: Banana, Anona, Vine, Almond, Pomaceæ. On ascending the summit, Botta, lastly, observed also European plants (*Rubus*, *Geranium*) succeeding the tropical region, which he states to be characterized by thorny Solaneæ and Orchideæ. At a considerable elevation woods occur of a tree-like *Juniperus*, under the shade of which, however, tropical Aroideæ (*Arisæma*) and Labiatæ (*Coleus*) are still met with. From the summit of the Saber, which appears to rise far above the other mountains of Yemen, Botta saw both the Red Sea and the Gulf of Aden. From thence he returned to the coast, and lay sick for a long time at Mecca. The plants collected are, in great part, different from those of Forskäl. As they were collected in the same region, but at another period of the year (October and November), and in part at higher altitudes, the author conceives that they so far complete the Flora of South Arabia that little more remains to be explored in that country. (p. 81.) We only wish that the materials may be completely worked out. In the sketch of the botanical regions, which Botta has appended to his account, the absence of systematic knowledge of the forms collected is to be regretted; the description, however, of the relations of cultivation is interesting. The low strip of country along the coast of Western Arabia, called by the natives "Téhama," and which is occasionally some miles in width, and sometimes very narrow, is not everywhere capable of cultivation, being sandy and barren; but the country is capable of being rendered extremely fertile by artificial irrigation. The usual crops are Maize, Doura and Indigo. Extensive plantations of the date-palm are also here met with. The woods consist entirely of various *Acaciæ*, and present many forms of plants common also to Sennaar: *Indigofera*, *Aristolochia indica*, *Capparis*, *Amyris*, *Cissus*, *Cadaba*, Asclepiadeæ and thorny Solaneæ. The Halophyta of the coast consist of *Salsola* and *Suæda*. The altitude of the mountains, according to Botta, is considerable, he estimates that of Saber as much above 8000′. The rainy season there occurs in the months of May, or July to October, whilst in "Téhama," where the heat is greater, the rainy season is limited to the winter months after December, and depends on the monsoon of the Gulf of Arabia.

Decaisne has begun to work upon the plants collected by Botta in Yemen, but at present only the Cryptogamia have appeared. (Loc. cit. p. 89-199, with seven plates.) The whole collection, however, according to the prefatory observations, contains only 500 sp. and more perhaps could not be obtained in two months. Whether the Algæ of the Red Sea are included in this or not, can only be learnt from the continuation of Decaisne's work, which at present is confined almost entirely to the Algæ, of which Botta has sent 53 sp., e. g. 7 sp. *Caulerpa,* 3 sp. *Dictyota,* 12 sp. *Sargassum,* &c., most perfectly figured by the author. To these succeed 13 *Ferns,* among which 5 are European and 2 new; lastly, 2 Lycopodiaceæ, edited by Spring.

III.—AFRICA.

Bory de St. Vincent has made a report to the French Academy on the Flora of Algiers, upon the return, and as an earnest of the results of the scientific expedition sent there in 1840-42. (Comptes rendus, t. 16.) The herbarium collected contained about 3000 species, of which only 60 are undescribed; for the most part they correspond with the productions of Spain and Portugal. The Cryptogamia collected amount to 400 species. The forests of the Lesser Atlas are laid waste. The chesnut, evergreen oak, and laurel, have become rare. The Syrian cedar occurs only on isolated declivities, and is said to be very abundant on the Great Atlas.

Bory distinguishes three botanical regions, the Numidian, Mauritanian, Tingitanian. The eastern, or Numidian, extends from Biserta to Collo; La Cala is situated in the centre of it; the country is well wooded, and contains many plants of central Europe. Algiers lies in the middle of the Mauritanian, or central region, and in its environs plants of the south of Europe prevail, and the Banana ripens its fruit. The Tingitanian, or western region, stretches from Cape Tanes to Morocco; it contains various African forms, among which, however, the *Stapelia* from Oran, adduced as an example, ought not to be included.

The 'Characteristik' of the Flora of Kordofan, mentioned in the last annual Report, is again enlarged and corrected by Brunner (Regensb. Flora, 1843, s. 473), and A. Braun (ib. s. 498). The former paper is unimportant, the latter relates only to two Alismaceæ, and contains valuable remarks upon that family.

A. Braun has also communicated a numerical review of the plants sent by Schimper from Abyssinia. (Ib. pp. 749-52.) These herbaria consist of 1250 species, two thirds of which are new. The first two parcels sent by Schimper are here included. The families are arranged in consecutive order, according to the number of species they respectively contain:—Gramineæ (141), Synanthereæ (140), Leguminosæ (116), Cyperaceæ (60), Acanthaceæ (46), Malvaceæ, including the Tiliaceæ (42), Labiatæ (40), Scrophularineæ (33), Rubiaceæ (26), Urticeæ (25), Euphorbiaceæ (24). Added to which are 18 species Cruciferæ and Boragineæ, 16 species Terebinthaceæ and Orchideæ, 15 species Amarantaceæ, 14 species Ranunculaceæ, 13 species Convolvulaceæ and Asclepiadeæ, 11 species Combretaceæ, Crassulaceæ, and Solaneæ, and 10 species Capparideæ and Verbenaceæ.

Included in the above number (1250 species) are about 100 Cryptogamia, viz. 27 Ferns, 50 Mosses, 17 Lichens, &c.

New Abyssinian plants collected by Feret and Galinier have been described by Raffeneau-Delile. (Ann. Sc. Nat. xx, pp. 88-95.) At present the number of species is only 16, but among them are several new genera: *Teclea* (Zanthoxyleæ), *Laneoma* and *Ozoroa* (Terebinthaceæ), *Feretia* and *Galiniera* (Rubiaceæ). Flotow has defined the lichens collected by Schimper (Linnæa, 1843, H. 1.); few of the species are new. Hochstetter has proposed the following new African genera (Regensb. Flora, 1843, ss. 69-83): *Xylotheca* from Port Natal (Bixineæ), *Candelabria* ib. (Samydeæ), *Diotocarpus* ib. (Rubiaceæ): *Kurria* of the former report. = *Hymenodictyon*, Wall., *Haplanthera*, *Monothecium*, and *Tyloglossa*, from Abyssinia and Nubia (Acanthaceæ), *Lachnopylis* from Abyssinia (Loganiaceæ?) *Pterygocarpus* ib. and *Apoxyanthera* from Natal (Asclepiadeæ).

E. Meyer has published a very important memoir on

the Flora of the Cape, which is based upon the complete list of localities given by Drège, and which are printed together with it. (Zwei pflanzengeographische Dokumente von Drège nebst einer Einleitung von E. Meyer: als besondere Beigabe zur Regensburg Flora, 1843.) It contains 230 species and a chart. Drège's herbarium contains about 7000 different species (6595 Phanerogamia and 497 Cryptogamia), and was collected in a district of scarcely 4000 square miles. E. Meyer estimates the number of the Phanerogamia of the Cape, hitherto known, at 9000 species, and their whole number indigenous to the colony, chiefly in the districts explored by Drège, at 11500. He considers a very limited range of distribution for each species to be a fundamental feature of the Cape Flora, the areal range of most of them being five times more contracted than is the case in the Flora of Europe, where, according to Schouw, the average area extends over from 10° to 15° of latitude. Social plants form but a very small proportion of the mass of vegetation, and even those that are met with, are for the most part, much less numerously associated than the meadow grasses or forest trees of Europe. Among the social forms, Drège enumerates some Proteaceæ, small-flowered Ericeæ, *Elytropappus rhinocerotis* (Stœbe, Th.), which covers large tracts on the 'Karroo,' *Galenia*, and in the eastern plain of the 'Karroo' is found *Mesembryanthemum spinosum*, one of the most social plants of the country. Besides these, the *Cliffortiæ*, *Prosopis elephantina*, *Acacia horrida*, are tolerably abundant on the streams of the 'Karroo,' and some Bruniaceæ, Oxalideæ, Asclepiadeæ, and Aloes species; on the "Giftberge," on the west coast, *Toxicodendron capense;* near Port Natal, the Rhizophori and *Hyphæne coriacea ;* lastly, the social water plant *Prionium serratum* (*Juncus*, Th.)

The Cape genera, as at present defined, contain on the average 6 to 8 species; on which account, the unusual number of species in some of the genera, characteristic of the Flora, is the more remarkable. Thus Drège collected

151 species of *Senecio*, 148 species *Pelargonium*, of *Erica* proportionately few, only 139 species; of *Helichrysum* 108 species; of *Aspalathus*, *Hermannia*, *Oxalis*, and *Restio*, between 80 and 90; of *Mesembryanthemum*, *Crassula*, *Euphorbia*, and *Indigofera*, between 60 and 70; of *Polygala*, *Muraltia*, *Rhus*, *Cliffortia*, *Anthericum*, and *Heliophila*, between 50 and 60, &c.

The statistical relations of the families are exhibited by E. Meyer with great precision, assuming Endlicher's genera as the basis of his conclusions. The Monocotyledons constitute 21, and the Dicotyledons 125 families. Of these, 38 appear to be wanting in New Holland, whilst, on the other hand, that continent presents 18 species not met with at the Cape. The families most rich in species form in Drège's collection the following series: 1110 Synanthereæ, i. e. almost 17 per cent., 510 Leguminosæ, 312 Gramineæ, 286 Irideæ, 264 Liliaceæ, 191 Restiaceæ, 184 Cyperaceæ, 170 Scrophularineæ, 169 Geraniaceæ, 167 Ericeæ, 157 Proteaceæ, 135 Euphorbiaceæ, 122 Orchideæ, 112 Polygaleæ, 108 Crassulaceæ, 104 Asclepiadeæ, 104 Umbelliferæ, 99 Byttneriaceæ, 88 Rubiaceæ, 87 Cruciferæ, 83 Oxalideæ, 79 Labiatæ, 75 Thymeleæ, 75 Campanulaceæ, 71 Rosaceæ, 69 Mesembryanthemeæ, 69 Selagineæ, 68 Malvaceæ, 66 Acanthaceæ, 65 Anacardiaceæ. Besides these, E. Meyer regards the following less numerous families as characteristic: Lobeliaceæ (56), Rhamneæ (56), Smilaceæ (51), Zygophylleæ, (44), Celastrineæ (40), Bruniaceæ (38), Hypoxideæ, (37), Cucurbitaceæ (35), Ebenaceæ (27), Penæaceæ (11), Cycadeæ (9), Stilbeæ (4). Some families range over only a portion of the colony: the Ericeæ, of which Bentham has already described 455 species from the Cape, were collected in smaller numbers by Drège, since the majority of that family are confined to the mountains of the extreme south; no Proteaceæ are found above thirty miles from the coast, but even within those limits the family does not extend to the borders of the colony. The Crassulaceæ occur principally in the great plain of the 'Karroo;' the *Her-*

mannia are found in the northern districts; the species of *Heliophila* on the west coast, between 30° and 34° S. L., and the *Rhus* species on the east, as also the Hypoxideæ. Four families are endemic in the Cape Flora: the Selagineæ, Bruniaceæ, Penæaceæ, and Stilbeæ.

The Cape Flora is physiognomically characterized by the abundance of large flowered Monocotyledons with coloured perigon, by the succulent plants, and Erica-forms. Of the latter, at least single genera occur in most of the great families, e. g. *Stæbe* among the Synanthereæ, *Aspalathus* among the Leguminosæ, some *Proteæ*, and among the Rhamneæ *Phytica;* and to the same class, besides the Ericeæ themselves, belong most of the Diosmeæ, Bruniaceæ, Stilbeæ, Penæaceæ, Thymeleæ. Among the succulent forms, may be instanced the Crassulaceæ, Mesembryanthemeæ, the Stapeliæ, many Euphorbiaceæ, several Portulaceæ, and Aloes.

Lofty arboreal forms are notoriously wanting, and dense forests entirely so. Drège adduces the following list of ligneous plants above twenty feet high, some of which attain a height of fifty feet: 3 Coniferæ, (3 species *Podocarpus* = Geelhout); of the Urticeæ, *Ficus Lichtensteinii;* the Laurineæ, *Ocotea bullata;* 3 species of *Olea* (Yserhout, among which *O. exasperata* is the thickest tree of the colony, but only about 30′ in height); of the Araliaceæ *Cassonia paniculata;* certain Meliaceæ (*Trichilia*); 1 Tiliacea (*Grewia*); of the Celastrineæ, *Curtisia faginea; Ilex crocea*, the Rhamneæ, *Olinia acuminata;* the Diosmeæ, *Calodendron capense;* the Myrtaceæ, *Jambosa cymifera;* and Leguminosæ, *Virgilia grandis.* Of parasitic Dicotyledons, Drège has collected 42 species, among which are 17 Lorantheæ, 5 *Cassytæ,* 1 Cactus, 12 Orobancheæ, 3 Cytineæ, 1 Balanophoreæ, 3 Cuscuteæ.

The Monocotyledons of Drège's collection, are in proportion to the Dicotyledons as 1 to 3·2, as is the case also in the same latitude in New Holland. From the coast towards the elevated plains of the interior, the Monocotyledons at first increase, then decidedly diminish, and again increase on the most elevated terraces; but the proportionate number of these latter does not correspond with the above given statement of E. Meyer. To the lower terrace of the country he assigns a mean altitude of 500′, to the central, one of 2000′, and to the upper, one of 3500′, above which then, the mountains, where the Monocotyledons again increase in number, still rise to an altitude of more than 8000′ above the sea. Meyer endeavours to explain these differences in the distribution of the Monocotyledons, by the quantity of atmospheric deposit, an increase in which induces a corresponding increase in the Monocotyledons.

According to Drège, it nowhere rains more frequently or more copiously than on the south-west coast of the colony. From stage to stage in ascent the quantity of rain diminishes, in the same proportion that the Dicotyledons increase. Similar differences are also observable on the coast line. At the mouth of the Gariep, the winter rains of the Cape are said to be almost entirely wanting, and the summer rains seldom fall; on the east coast on the other hand the influence of the trade-wind is felt in the opposite condition of a dry winter, and a tropically moist summer, by which the peculiar vegetation of Port Natal is explained. The Verbenaceæ and Acanthaceæ begin to increase even at Algoa Bay. The following forms consequently are characteristic at Natal: tropical Leguminose genera, Myrtaceæ, Rubiaceæ, two Palmæ, and other plants of the torrid zone; although this settlement lies at 30° S. L., or more to the south than the mouth of the Gariep.

Bunbury has continued the reports of his botanical wanderings in the Cape country (vid. last Report.) (Hook. Lond. Journ. of Bot., ii, pp. 15-41.) He has described his journey from Cape Town to Graham's Town. In the coast region he found a distinct limit of vegetation at the mouth of the Gamtos river; from this point towards Algoa Bay commences a district which is characterized by fleshy *Euphorbiæ* and other succulent plants, as also by *Schotia speciosa* (Boerboontje.)

Rather further to the west as far as the River Kromme, the *Zamiæ* make their first appearance, with respect to which Meyer observes erroneously that they are first seen in Albany. For the rest, the whole journey occupied only seventeen days, and consequently little opportunity for observation was afforded, but a further supplement is to appear in 1844.

The systematic contributions to the Cape Flora, drawn principally from the herbaria of Krauss, have been continued by Meissner in the same journal (pp. 53-105 and 527-59.) This second part contains the following families: 2 Tiliaceæ, 1 *Aitonia,* 30 Oxalideæ, 7 Zygophylleæ, 1 Ochnaceæ from Natal, 1 Rhamneæ, 1 Bruniaceæ, 166 Leguminosæ, 10 Rosaceæ, 1 Portulaceæ, 1 Cunoniaceæ, 28 Umbelliferæ, 1 Hamamelideæ, 1 Corneæ, 3 Loranthaceæ, 5 Rubiaceæ, 1 Lobeliaceæ, 1 Jasmineæ, 1 Apocyneæ, 25 Asclepiadeæ, 1 Scrophularineæ, 1 Orobancheæ, 10 Amarantaceæ, 6 Chenopodeæ, 12 Polygoneæ, 28 Thymeleæ, 3 Penæaceæ, 4 Euphorbiaceæ. Bartling has given detailed descriptions of twenty-two new Diosmeæ, or which have been made known only by Ecklow; among which is the new genus *Gymnonychium.* (Linnæa., 1843, pp. 353-82.) Some new Cape plants have been published by Fenzel (ib. pp. 323-34); the new Amarantacean genus *Sericocoma* with three species; the newly proposed Asclepiadea, *Anisotoma;* and one *Veronica,* all from Drège's collection. Flotow has given definitions of fifty-five Cape Lichens, and descriptions of the new ones (ib. pp. 20-30); Berkeley has contributed thirty-one fungi from Zeyher's collections. (Journ. of Botan., pp. 507-24.)

Bojer has again, as in the last year, described new plants from the islands lying off the south-east of Africa. (Ann. d. Sc. Nat. xx, pp. 53-61, and 95-106.) Among which are 1 Anonacea, 2 Menispermeæ, 8 Capparideæ, 4 Polygaleæ, 2 Pittosporeæ, 1 Lineæ, 6 Tiliaceæ, 2 Leguminosæ, with the new Dalbergiea *Chadsia.*

IV.—ISLANDS OF THE ATLANTIC.

I HAVE here merely to refer to the interesting memoir on the botanical characteristics of the Azores, by Seubert and Hochstetter, with which the present annual volume of these Reports commenced. Watson has, at the same time, given a report upon his botanical voyage to the Azores. (Hook. Lond. Journ. of Bot. 2. pp. 1-9, 125-31, and 394-408). The endemic vegetation was observed by Watson, beyond the cultivated district of Fayal, first between Horta and Flamingos, where the low hills on the strand are covered with *Myrica Faya* and *Myrsine retusa*; together with these grows *Erica azorica*, Hochst. (*E. arborea*, S. H. p. 21), which, however, according to Watson, is probably only a variety of *E. scoparia*, L. Near Flamingos, there are added to these, two European Ericeæ: *Menziesia Dabœci*, D.C., and *Calluna*. It appears to be probable, from this description, that the region of the Laurel Woods 1500′-2500′, which consists almost entirely of the same ligneous plants as the coast formation near Flamingos, originally extended everywhere to the sea. The forest above Flamingos is composed of *Erica scoparia*, *Myrica Faya*, *Myrsine retusa*, and *Juniperus Oxycedrus*, S.H., which is regarded by Watson as distinctly an endemic species; with these are mixed *Vaccinium maderense*, L.K. (a small flowered variety of which probably is *V. cylindraceum*, Sm., *V. longiflorum*, Wickstr., and *V. padifolium*, S.H.), besides *Rubus Hochstetterorum*, S., *Ilex Perado*, *Viburnum Tinus*, *Persea azorica*, S., *Laurus canariensis*, S.H.), and *Euphorbia stygiana*, W. (*E. mellifera*, S.) The margin of the crater of Fayal is situated 3170′ above the sea, and sinks down internally to an inclosed lake, the altitude of which is only 1670′. This moist hollow, the diameter of which is about a mile, is thickly covered with Ferns and the endemic evergreen shrubs. The Phanerogamia are for the most part the same as on the outside of the crater, but the endemic species are here much more densely crowded. The water-plants on the lake, however, are again European. The description of the Peak of Pico, agrees in all respects with that by Seubert and Hochstetter. The altitude of the peak, by barometrical measurement, amounted to 7616 feet Eng., and it would appear that the upper limits of some plants are placed higher than was supposed by Seubert and Hochstetter. At the summit the only plants are *Thymus micans*, and an undetermined species of *Agrostis*, with some mosses and lichens. Highest limit for *Calluna* 7000′, for *Erica scoparia* 6000′.

The younger Hooker, on his Antarctic voyage, explored the Cape de Verd Islands. (Journ. of Bot. p. 250.) The interior of St. Jago, of which island the coasts are completely barren, presents a luxuriant vegetation; on the mountains are forms of the Atlas and south of Europe, and in the valleys tropical genera. It is only now that a little is known of this Flora; in the opinion of the traveller, the mountains, immediately after the rainy season, would afford the richest booty. Any one who is desirous of exploring them must proceed directly from Porto Praya to St. Domingo, because, at that time, for several miles around the capital no plant is any longer visible. Foyo, the volcano of which is said to be about 7000′ feet high, might be more interesting than St. Jago. St. Antonio also is covered with wood, and Sal, a saliferous plain.

Hooker did not land on the island of St. Paul (0° 58′ N.L.), but Darwin, who explored it, remarks (Journ. of Research, p. 10), that although there are several indigenous insects and spiders, not a single plant, not even a lichen, is to be found, and only Algæ in the greatest variety.

St. Helena has already in great measure lost its endemic Flora, (ib. p. 582). The great forest of the elevated plain which existed at the commencement of the last century has been destroyed, and together with it, doubtless, many plants which, like the productions of a former world, have now disappeared from the earth. Darwin agrees with Beatson in ascribing this change to the introduction of the goat, which prevents the growth of the seedling trees. Instead of the endemic Flora, European plants have in great measure now become distributed over the soil of St. Helena. The most abundant tree at present is the Scotch Fir, but Hooker also observed (l. c. p. 252) *P. Dammara*, *Casuarina*, *Acaciæ*, and Pittosporeæ from New Zealand, *Eucalyptus* from New Holland, and Scitamineæ and Aroideæ from the East Indies.

V.—AMERICA.

Nuttall has continued the description of the plants collected on his journey through North America to the Sandwich Islands, (vid. Report for 1841, in the Transactions of the American Philosophical Society, 1843, p. 251.) This memoir contains the Campanulaceæ, Lobeliaceæ, Ericeæ, and allied families; several extensive genera have been divided by the author: e. g. *Vaccinium* and *Andromeda*. Engelman has published a distinguished monograph on the North American Cuscuteæ. (Silliman's American Journal of Science, vol. 43, pp. 333-45, 1842. Extracted in London Journ. of Bot. 1843, pp. 184-99.) In the same American Journal, has also appeared a continuation of Dewey's 'Caricographia.' (Vol. 43, pp. 90-2, with five figures.) Bruch and Schimper have examined Drummond's collection of Canadian mosses, and the results have been published by Shuttleworth. (Journal of Bot. pp. 663-70.)

An account of the botanical geography of the Mexican volcano Orizaba, was given by Liebmann at the meeting of Scandinavian Naturalists, at Stockholm, in 1840. (Also translated in the Botan. Zeitung, 1844, s. 688 et seq.)

(1) Hot Region. (0′ to 3000′.) The sloping savannahs to the west of Vera Cruz, beyond Santa Fé, and at an altitude of 200′, are interrupted by a forest, in which the characteristic trees are *Mimosa, Bombax, Citrus, Combretum.* Then follows an extremely fertile marly soil, derived from the disintegration of the porphyritic blocks of the Orizaba, the forests on which present magnificent groups

of the undescribed "*Palma real.*" Hence the sloping grass savannah again extends to an elevation of 3000′, with *Mimosa* bushes, and the Ternstræmiacea *Wittelsbachia*, a *Convolvulus* and a *Bignonia*.

(2) THE WARM, MOIST REGION. (3000′ to 6000′.) At 3000′ commence the moist mountain forests, in which the oak makes its appearance in numerous forms, and with it grow six species of *Chamædorea*, sometimes erect, sometimes creeping palms. This is the richest botanical region of Mexico, and in which with a mean temperature of 21° C., and a rainy season of eight or nine months, about 200 Orchideæ are indigenous. Here commences a ferruginous, hard clayey soil, which overlies the volcanic rock to a height of 11000′. The oak is most luxuriant between 4000′ and 5000′; about twenty species are met with, and several are limited to those altitudes. As is the case in Java, the oaks grow in a thick tropical forest of Laurineæ, Myrtaceæ, Terebinthaceæ, Malpighiaceæ, and Anonaceæ. The underwood is formed of *Melastoma*, *Treeferns*, *Mimosa*, the Monimiea *Citrosma*, *Bambusa*, *Yucca*, *Jatropha*, and *Croton*, *Triumfetta*, *Magnolia*, arboreal Synanthereæ, *Symplocos*, *Æsculus*, Araliaceæ, &c. The climbers are Smilaceæ, Sapindaceæ, *Cissus*, Apocyneæ, Asclepiadeæ, Bignoniaceæ, Passifloræ, Leguminosæ, and Cucurbitaceæ. In these forests the upper limit of the cultivation of coffee and cotton is attained at from 4000′ to 5000′, and that of sugar and the plantain at 5500′.

(3) REGION OF THE OAK FORESTS. (6000′ to 7800′.) At 6000′ the foot of the Cordilleras is reached, where another climate and other forms of plants commence. In the neighbourhood of the city of Cosco-matepec, where, together with maize, the orchard trees of Europe and fruits of the south are cultivated, and where the fertile plains of the plateau begin, the most abundant trees are the following: *Yucca gloriosa*, *Cratægus pubescens*, *Sambucus bipinnata*, *Clethra tinifolia*, *Persea gratissima*, *Cornus*; the climbers are here composed of *Convolvulus*, *Vitis* and *Rubus*. The Palms have

already ceased at 5000′, but in the interior highland other species reappear up to 8000′. Tree-ferns are also wanting on the Cordillera, flourishing only between 2500′ and 5000′; and the last fruticose myrtles are seen at 4800′. The Orizaba itself is a more lofty peak, rising 1700′ above the border of the plateau. The traveller explored it in the month of September, in the middle of the rainy season. The lower forest belt (6000′ to 7800′), consists chiefly of species of *Quercus;* the other trees are *Lacepedea pinnata, Ulmus, Alnus, Clethra,* a Verbenacea, and an Araliacea; an underwood of *Cornus tolucensis, Viburnum, Triumfetta, Rubus,* with climbers of *Vitis, Ipomæa Purga,* a *Bidens,* and *Cuscuta,* Alstræmeriæ; parasitic Ferns, *Viscum,* Orchideæ, *Piper,* in three small forms, *Cereus flagelliformis.* The open spaces are covered with *Cassia* and *Mimosa* bushes, the herbaceous plants and grasses are rich in forms, and among the characteristic ones are *Ranunculus, Thalictrum, Anada, Hypericum, Drymaria, Oxalis, Geranium, Euphorbia, Desmodium, Rexia, Lopezia, Cuphea, Georgina, Lobelia, Salvia, Erythræa, Iresine, Cyperus, Panicum, Paspalus, Festuca, Vilfa, Lycopodium;* Ferns, Mosses, and Lichens are also tolerably abundant. Even at an elevation of 7000′ the vegetation is remarkably changed. *Vaccinium, Gualtheria, Andromeda,* become frequent, and particularly a new, tree-like *Arbutus, Fuchsia microphylla;* among the herbaceous plants: *Chimaphila, Dracocephalum, Tagetes, Carduus,* a Gentianea, several Orchideæ, *Ferraria,* and Commelineæ.

(4) Coniferous Region. (7800′ to 11000′.) The first Coniferæ appear at 6800′, *Pinus leiophylla,* but the oaks are not supplanted by the Coniferæ below 7800′. At this altitude vigorous trees of *Pinus Montezuma* predominate, with parasitic Tillandsiæ and Usneæ. At 9000′ commence the forests of the Oyamel pine (*Abies religiosa*), but *P. Montezuma* again constitutes the upper belt of the continuous pine forest at 11000′, solitary or situated individuals occur as high as 14,000′ on the N.W. side of

13

the peak. The traveller staid a fortnight in the Coniferous region, in a herdsman's hut in the 'Vaqueria del Jacal.' (10000′.) The mean temperature at that season was 11° C. At the end of autumn instead of rain, snow falls, which lies from November to March. The uniformity of the north does not prevail in these Coniferous forests. Leaf-trees are everywhere intermixed; such as the oak and alder; shade plants continue numerous; the ravines (barrancas) nourish a luxuriant vegetation; whole mountain sides are bare of trees, and covered with a high grass, together with alpine plants. The plants of the Coniferous region especially, present the greatest variety of forms, of which Liebmann furnishes a copious list. The following belong to the characteristic families: Leguminosæ (*Lupinus*), Umbelliferæ, Ericeæ (*Clethra, Vaccinium, Pyrola,* &c.), Synanthereæ (*Eupatorium, Stevia, Bidens, Baccharis, Aster,* &c.), Scrophularineæ (*Chelone, Lamouzouxia, Gerardia, Castilleja*), Labiatæ (*Salvia Stachys*), *Verbena*, Orchideæ (*Spiranthes, Serapias*), *Veratrum*, Irideæ (*Sisyrinchium*). Ferns, &c. Bushes of Laurineæ, Rhamneæ, *Tilia, Viburnum, Cornus,* Synanthereæ, *Salix* occur together with the Ericeæ. In one "barranca" Liebmann observed a thicket of bamboo, a form of plant that elsewhere disappeared at 3000′.

(5) Region of the Steviæ. (11000′ to 13600′.) Low Synantherean shrubs (*Stevia purpurea, arbutifolia,* &c.) represent the sub-alpine Erica-forms, as they do on the Cordilleras of South America, where instead of *Stevia*, the genus *Baccharis* makes its appearance. Still they do not, as here, extend to the uppermost limits of vegetation. A more abundant shrub in the lower part of the region is *Spiræa argentea*. Besides these, alpine genera for the most part grow upon masses of volcanic rock, which now appears instead of the clayey soil. Characteristic forms are: Cruciferæ (*Draba, Nasturtium*), Alsineæ, *Viola, Lupinus,* Rosaceæ (*Alchemilla, Potentilla*), Umbelliferæ (*Eryngium, Seseli, Œnanthe*), *Tiarella, Pedicularis, Lithospermum, Stachys,* Synanthe-

reæ (*Erigeron, Hieracium, Hypochæris*), *Veratrum, Sisyrinchium, Serapias*, Junceæ, *Carex.*

(6) Alpine Mountain Plains. (13600′ to 14800′.) The soil of the highest parts of the surface below the crater consists of a mixture of volcanic sand with ashes; it bears a gramineous vegetation, the species of which correspond with those found by Humboldt on the Nevado of Tolucca: *Festuca toluccensis, Bromus lividus, Avena elongata, Deyeuxia recta, Crypsis stricta, Agrostis,* and other Festuca species. Instead of the *Steviæ,* are here seen thick-leaved, hoary shrubs of *Senecio.* The smaller craters are clothed with *Mahonia ilicina* and *Juniperus mexicana.* Under the Gramineæ grow several Synanthereæ (*Conyza, Helichrysum, Carduus, Saussurea*), *Gualtheria ciliata,* and of the other alpine genera, *Cerastium, Viola, Draba,* extend up to this height. The thawed snow affords growth to a formation of *Ranunculus* and *Potentilla,* associated with which are some Glumaceæ and *Veronica* (*Luzula, Carex, Phleum, Agrostis.*) Mosses and Lichens abound, and among the latter also, especially the northern Umbilicariæ (*U. pustulata,* 10000 to 14000′, *U. vellea,* 13000 to 14000′, *U. cylindrica* and *proboscidea,* 14000 to 14800′). The base of the great crater, which slopes at an angle of 30°, is situated at an altitude of 14300′, and is covered with broken rocks difficult of ascent. Here grow the last phanerogamous plants, which have, for the most part, not been observed by Schiede: 1 Hydrophyllea (*Phacelia lactea,* Liebm.), 1 *Castilleja, Saussurea, Carduus nivalis, Arenaria, Cherleria, Draba vulcanica,* Liebm., one of the fruticose Senecios, and most of the above-mentioned Gramineæ. The traveller observed the last Phanerogamia at the elevation of 14600′. The large rocks continue to be covered with Cryptogamia from this point up to 14800′; these are, besides the Umbilicariæ, *Tortula ruralis, Parmelia Ehrharti, Lecidea atro-alba, citrina geographica, Cenomyce pixidata,* and, as the last of all, 50′ higher than the rest, *Parmelia elegans,* a lichen which Agassiz also

found among the highest on the Jungfrau in Switzerland. (Vide last Report, p. 389.) The snow-line on the Orizaba may probably be taken at 15000′.

Much may be anticipated for systematic botany, as relates to the Flora of Mexico, when Liebmann's collections are fully studied, and these expectations have been raised to a greater height, among other things, by the essays already read by that traveller at the meeting of Scandinavian naturalists, held at Christiania in 1843, after his prosperous return. In the meanwhile other contributions to the knowledge of the Flora of Central America have been simultaneously made public in England and Belgium. Bentham has published a second and last part of his 'Plantæ Hartwegeanæ.' (London, 1842, 8vo.) This extends from Nos. 518-631, the greater part of which have been collected not in Mexico, but in Guatemala. The new genera are: *Hemichæna* (Scrophularineæ), *Lindenia* (Rubiaceæ), *Oxylepis* (Heliantheæ), *Caloseris* (Trixideæ), *Lampra* (Commelineæ). Very copious is the 'Enumeratio synoptica plantarum a Galeotti in Mexico lectarum,' which was commenced by Martens and Galeotti, and which is already tolerably far advanced. This work is contained in the 'Bullet. de l'Acad. de Bruxelles' (1843, vol. x. P. i, pp. 110, 208, 341; P. ii, pp. 31, 178, 302. 1844, vol. xi, P. i, pp. 121, 227, 355.) The families at present discussed, with numerous new and copiously described species, are the following: 7 Irideæ, 1 Hæmodoraceæ, 2 Hypoxideæ, 14 Amaryllideæ, 10 Bromeliaceæ, 4 Scitamineæ, 1 *Najada*, 4 Aroideæ, 1 Typhacea, 3 Palmæ, 11 Coniferæ, 24 Piperaceæ, 2 Myriaceæ, 35 Cupuliferæ, 2 Betulineæ, 2 Plataneæ, 8 Saliceæ, 5 Chenopodiaceæ, 14 Amarantaceæ, 12 Polygoneæ, 10 Nyctagineæ, with the new genera *Tinantia*, 4 Laurineæ, 2 Thymeleæ, 13 Valerianeæ, 83 Rubiaceæ, 5 Caprifoliaceæ, with the new genus *Vetalea*, 15 Apocyneæ, 40 Asclepiadeæ, 17 Gentianeæ, with the genus *Arembergia*, considered new, and for the most part new species, 8 Spigeliaceæ, and 205 Leguminosæ, with the new genera

Mikelertia and *Robynsia*. Besides these, there had previously appeared the Ericeæ and Vaccineæ, under the title of 'Notice sur les plantes des familles des Vacciniées et des Ericacées, recueillies au Mexique par Galeotti, et publiées par Martens et Galeotti.' (Ib. 1842. p. 526.) Schlechtendal's recent communications on the Mexican Flora (vide Report for 1840) relate to the Burseraceæ, especially *Elaphrium* (Linnæa, 1842. H. 6, and 1843, s. 245), and also to the Dioscoreæ. (Ib. s. 602.)

E. Otto has now collected the results of his American travels (vide Report for 1840). (Reis erinnerungen an Cuba, Nord und Südamerika, 1838-41; Berlin, 1843, 8vo.) Starting from Caraccas, he explored the district of the Orinoco.

Jameson's observations on the Flora of the Ecuador (Lond. Journ. of Bot. 2, pp. 643-61) are at present too fragmentary to allow of their being more closely considered; they will not certainly, however, be unimportant when continued, as is promised by the author. Of Bentham's work on the plants collected by Schomburgh in Guiana, the following families have appeared: Euphorbiaceæ, by Klotzsch (32 species, with the new genera *Schismatopera*, *Dactylostemon*, *Tragantlius*, *Brachystachys*, *Geiseleria*, *Discocarpus*, the two penultimate separated from *Croton*); by Bentham himself, the Dilleniaceæ (1 sp.), Nymphæaceæ (1 sp.), Cabombeæ (1 sp.), Sarraceniaceæ (1 sp.), Ternstræmiaceæ (10 sp., with the two anomalous new genera *Catostemma* and *Ochthocosmos*), Guttiferæ (8 sp.), Marcgraviaceæ (1 sp.), Hypericineæ (3 sp.), Erythroxyleæ (6 sp.), Trigoniaceæ (3 sp.), Humiriaceæ (4 sp.), Olacineæ (3 sp.), with which Bentham places the new genus *Ptychopetalum* from Cayenne, Rhizoboleæ, (1 sp.); by Lindley, the Orchideæ (66 sp.) The number of species at present published amounts to 846. (Hooker, London Journ. of Bot. 1843, pp. 42-52, 359-78, 670-74.)

Miquel has published in several journals concerning new plants from Surinam, especially from the herbarium of Focke; in the first part of the 'Annals of Nat. Hist.

for 1843,' which has accidentally not reached us from the bookseller; and further in the 'Linnæa,' and in b. d. Hoeven's 'Tijdschrift.' The continuation in the 'Linnæa' (1843, s. 58-74) contains species from various families, and little that is new; viz. 1 Cyperaceæ, 1 Xyrideæ, 2 Pontedereæ, 1 Smilaceæ, 1 Hæmodoreæ, 1 Aroideæ, 4 Synanthereæ, 4 Rubiaceæ (among which is the new genus *Bruinsmania*). The 'Animadversiones in herbarium Surinamense, quod in colonia Surin. legit H. C. Focke, auct. Miquel,' (Tijdschr. voor natuurlijke Geschiedenis, 1843, pp. 75-93,) include the following families: Cacteæ (without any new sp.), 2 Portulaceæ (1 n. sp.), 1 Phytolacceæ, 7 Malvaceæ (2 n. sp.), 1 new Byttneriaceæ, 3 Guttiferæ (1 n. sp.), 1 new Marcgraaviaceæ, 1 Hypericineæ, 6 Malpighiaceæ (1 n. sp.), 1 new Erythroxyleæ, 2 new Sapindaceæ, 1 Polygaleæ, 1 Euphorbiaceæ, 1 Anacardiaceæ, 1 Myrtaceæ, 16 Leguminosæ.

Focke, in Paramaribo, has caused to be published by Miquel, a systematic catalogue of all the plants cultivated in Surinam. (Hoeven's Tijdschrift, l. c. pp. 373-85.) The following is an abstract of the Dutch names of the most important tropical products: *Anona muricata*, L. (Zuurzak); *A. squamosa*, L. (Kaneelappel); *Terminalia latifolia* (Amandelboom, Tafelboom); *Eugenia pimenta*, D. C. (Bayberry tree); *Jambosa vulgaris*, D. C. (Pomme de Rose); *J. malaccensis*, D. C. (Schambo); *Passiflora quadrangularis*, L. (Marquisade, Grenadille); *Mammea americana*, L. (Mammi); *Caryocar tomentosum*, W. (Bokkenoot); *Hibiscus esculentus*, L. (Okro); *H. Rosa sinensis*, L. (Engelsche Roos); *H. Sabdariffa*, L. (Roode Zuring); *Spondias dulcis*, Forst. (Pomme de Cythère); *Ricinus communis*, L. (Krapatta); *Janipha Loefflingii*, Kth. (Zoete Cassave); *J. Manihot*, Kth. (Bittere Cassave); *Malpighia glabra*, L. (Sure Kers); *Averrhoa Bilimbi*, L. (Bilambi); *Abrus precatorius*, L. (Weesboontje); *Erythrina corallodendron*, L. (Koffij-mama); *Arachis hypogæa*, L. (Pienda); *Poinciana pulcherrima*, L. (Sabinabloem); *Parkinsonia aculeata*, L. (Jerusalemsdoren); *Anacardium occidentale*, L. (Cachou); *Mangifera indica*, L. (Manja); *Artocarpus incisa*, L. (Broodboom); *Persea gratissima*, G. (Advocaat); *Coccoloba uvifera*, L. (Zeedruif); *Achras Sapota*, L. (Sapotilla); *Chrysophyllum Cainito*, L. (Star-appel); *Sesamum orientale*, L. (Abonjera); *Crescentia Cujete*, L. (Kalebasboom); *Justicia picta*, L. (Portret-boom); *Lycopersicum esculentum*, Dun. (Tomati); *Solanum ovigerum*, Dun. (Antroeri); *Plumeria rubra*, L. (Frangipane); *Cycas revoluta*, Th. (Sayo); *Amomum granum paradisi*, L. (Malaguetsche Peper); *Musa paradisiaca*, L. (Banane); *M. humilis*, L. (Dwerg-Banane); *M. sapientum*, L. (Bakove, Bakoeba); *Agave fœtida*, L. (Ingi-sopo); *Yucca stricta*, Ker. (Bajonet); *Bambusa arundinacea*, W. (Guinea-Gras).

Steudel has commenced the determination of the plants in Hoffman's herbarium from Surinam, mentioned in last year's Report. (Regensb. Flora, 1843, pp. 753-65.) The species described as new belong to the following families: Anonaceæ (3 sp.), Sterculiaceæ (2), Tiliaceæ (2), Sapindaceæ (1), Homalineæ (1), Leguminosæ (21), Rosaceæ (5), Combretaceæ (2), Myrtaceæ (3), Paronychieæ (1), Umbelliferæ (1), Rubiaceæ (2), Solaneæ (2); Verbenaceæ (1); Spigeliaceæ (2), Gentianeæ (2). The diagnoses are short, and descriptions are not added.

Of Pöppig's illustrated work upon the plants collected by him on his South American travels, (Nova genera, &c., Lips. 1843-4), the 5th and 6th Decades of the third vol. have appeared. Orbigny's 'Travels' have appeared regularly up to the 74th livraison. Casaretto has published 8 Decades of Brazilian plants, a work which has not yet come under my notice. (Novarum Stirpium Braziliensium Decades. Genuæ, 1842-44, 8vo, 72 pp.)

Meyen's collection of dried plants, made on his voyage round the world, has been subjected to the united labours of several botanists, and been published in the Transactions of the Leopoldine Academy. (Nov. Act. Nat. Curioser. vol. xvi. Supplem. secund.; Vratisl. 1843.) This collection contains about 1500 species, but the number of new ones is proportionately not great. The Leguminosæ were submitted to the late Vogel, Nees v. Esenbeck undertook the Glumaceæ, Philydreæ, Acanthaceæ, Solaneæ, and in conjunction with Lindenberg and Gottsche, the Lycopodiaceæ; Meyen himself, with Flotow, the Lichens; Klotzsch the Euphorbiaceæ and Fungi; Schauer the Myrtaceæ, Apocyneæ, Asclepiadeæ, and the rest of the Monocotyledons; I took the Gentianeæ; Walpers the rest of the Dicotyledons, and Goldmann the Ferns.

The greater number of the plants collected are from South America, especially from Chili and Peru, but more interest attaches to the herbaria from Manilla (about 200 species) and Macao (about 220 species).

Gardner's publication on the Flora of Brazil (vid. last Report) has been continued. He has described four new genera from the Organ Mountains at Rio: *Bowmania* (Nassauviæ), *Leucopholis* (id.), *Hockinia* (Gentianeæ), *Napeanthus* (Cyrtandreæ). (Hooker, Lond. Journ. of Bot. ii, pp. 9-15.) The continuation of the list of his collection, which is arranged geographically, contains 125 species from the Organ Mountains, among which are many new species, and the new genera, *Isodesmia* (Hedysareæ). (Ib. pp. 329-55.) 39 *Fungi*, collected by Gardner, have been determined and described by Berkeley. (Ib. pp. 629-43.)

Darwin, whose spirited descriptions of the natural relations of South America and the South Sea Islands, presents such a variety of interest, has occupied himself with the problem of determining the reason of the absence of forests from Monte Video to Patagonia. (Journ. of Researches, p. 53.) On the banks of the large streams in Monte Video willows occur, and report speaks of a palm forest near Arroyo-Tapes. The traveller observed a solitary palm at 35° S.L. But these are the only exceptions to the treelessness of a country in which the orchard trees of Europe flourish extremely well. Plains, such as the Pampas of Buenos Ayres, are entirely treeless, a circumstance that appears to depend upon the prevailing winds, and the consequent hygrometrical conditions of the air. But these conditions do not obtain in Monte Video, where the hilly rocky surface presents the utmost variety of soils, including clay, and where there is no want of water. Here there is, in winter, a regular rainy season, and even the summer is not inordinately dry. New Holland, south of the tropic, is much drier, and yet it is always well wooded on the coasts. Consequently, the author is of opinion that the absence of trees at Monte Video can only be referred to geological causes, to an original peculiarity in this centre of creation. If on that occasion the larger ligneous plants were omitted, they could not readily extend themselves to this region from other centres of creation, for the trees of Brazil require a tropical climate, and there is no other forest country adjacent to this coast. South America generally presents forests, only in much moister regions, as on the west coast, southwards from 38° S.L., where the west winds of the Pacific prevail, and in Brazil as far as the trade wind extends. Thus, the districts on either side of the Cordilleras, which interrupt the course of the winds and deprive them of moisture, as well within as without the tropical zone, are in opposite conditions. Opposite to the primitive forests of Brazil lies the west coast, which from 4° to 30° S.L. is barren and treeless, and the wooded coasts from Chiloe to Tierra del Fuego are in the same way opposed to the scanty vegetation of Patagonia. Thence it might be concluded that Monte Video, as regards the arboreal vegetation of South America, possesses a too dry climate although not so in comparison with other wooded parts of the globe. The Falkland Islands also are barren of trees, although, with respect to climatic and geognostic relations, they are placed under precisely the same conditions as the forests of Tierra del Fuego.

The Rio Colorado constitutes the southern limits of the Pampas vegetation; its mouth lies at 40° S.L. (Ib. p. 87.) Here the nature of the soil of the steppe is changed, and with it the character of the vegetation. Between the Rio Negro and the Colorado, the gravelly surface bears a turf of grass with low thorny bushes, and this continues to be the type of the vegetation along the whole of the shores of Patagonia. The entire surface of the country, in like manner, from the Straits of Magellan to the Colorado, consists of a gravelly soil, which is for the most part composed of porphyry, derived from the Cordilleras. North of the Colorado the gravel becomes gradually smaller, till it finally passes into the calcareous argillaceous soil of the Pampas, which sort of soil, without stones, occupies a large basin, extending to the granite of Monte Video. The climate north of the Colorado continues to be no less arid and sterile, but the soil produces a variety of herbaceous plants and grasses, whilst the thorny shrubs are lost, together with the gravel. The vegetation of the Pampas of Bahia Blanca was still in its winter sleep at the beginning of September, around the White Bay (p. 115), but in the middle of this month the plains were covered with bloom; just as in all steppes, both of the Old World and New, the abundant blossoms of spring usually burst out suddenly. Previously to the bursting of the flowers the mean diurnal temperature had been 10°·6 C., but it then rose to 14°·4 C., that is to say, to a height at which the hibernation still continues at Monte Video. Hence it might be supposed that a different degree of irritability existed in the two floras separated by the Rio de la Plata; but probably in this problem the moisture of the atmosphere is concerned, which at Monte Video, longer than in the hot Pampas, prevents the evaporation of the plants, the cause of the rise of their sap in the spring. The steppe vegetation of the Pampas, from the Rio Salado to Buenos Ayres, is much more luxuriant than it is in these southern border districts, but probably only in consequence of the greater use of pasture. (p. 137.) In common with the wild horses and other domestic animals, which, since the first colonization of these countries in the year 1535, have spread themselves widely over the steppes European plants have also been introduced, and having completely supplanted the endemic vegetation over extensive tracts, have given the country, in many districts, from the Plata to the Cordilleras, its present natural character, in the same manner as the Opuntiæ and Agave have become characteristic on the shores of the Mediterranean. In this region, where at the present time horses of European origin only exist, Darwin has discovered the remains of a fossil indigenous horse of the latest geological period, and exactly in the same way, together with an endemic thistle which covers extensive tracts on the Rio de la Plata, has the European *Cynara Carduncalus* obtained possession of the soil over much wider districts.

This lofty growth of thistles is, on account of its extreme density, quite impenetrable by man or beast. Darwin is acquainted with no instance of an introduced plant occurring in such enormous quantity, and found on prolonged land journeys the same growth frequently recurring; he even observed it beyond the Plata, and saw many square miles in Monte Video thickly covered with the same thistle.

In South Patagonia, after having already explored a series of points on the coast, Darwin ascended the St. Cruz (50° S. L.) up to the foot of the Cordillera. The whole of Patagonia consists of a tertiary plain, gradually rising in a succession of terraces up to the Andes, and precipitous towards the sea from a height of 1200 feet. The rounded gravel which covers this plain, reposes upon a whitish soil, the argillaceous porphyritic detritus, in which the plants are rooted. Amongst the few productions of this soil, *Opuntia Darwinii*, Hensl., is characteristic. The terraces are frequently cut across by level valleys, but without water, and in these, the thorny shrubs abound. The climate is so dry that one may travel many days together without meeting with a drop of water.

The most striking contrast to these steppes is afforded by the clay-slate mountains of Tierra del Fuego, which are, however, very closely contiguous to them; these mountains are everywhere, down to the sea-shore, covered with a continuous gloomy forest. (p. 227.) The valleys of this mountainous region, like those in Norway, are lower than the sea-level, and constitute "Fjorde." The forest consists principally of *Fagus betuloides*, Mirb., (*Betula antarctica*, Forst.,) the other species of *Beech* and *Drimys* occurring only in inconsiderable numbers. The forest extends on the steep declivities, where there is scarcely anywhere a speck of earth, up to the altitude of 1000′ to 1500′, to this succeeds the region of alpine plants, which grow on a boggy soil, and reach to the line of perpetual snow (3500′.) A turf formation is also frequent in the forest region, in a wilderness of fallen and still vegetating trees. On this account, and owing to the yellowish, brown-green of the beech leaves, which do not fall in winter, the landscape acquires a gloomy character, which is also not often enlivened by sunshine. The peat formation (p. 349), which extends on the north as far as the Chonos Archipelago (45°), and is no longer met with in Chiloe, is constituted, in open situations, principally of the social *Astelia pumila*, Br., (*Anthericum trifarium*, Sol.,) a genus allied to the Junceæ, and which consequently here represents the *Narthecium* of the "Emsmoore," which is systematically placed immediately next to it. Together with the *Astelia*, grow *Myrtus nummularia*, *Empetrum rubrum*, and *Juncus grandiflorus*, which take a part in the formation of the peat. In the Falkland Islands, in a corresponding soil, all the plants are converted into peat, especially the Gramineæ. At the eastern entrance of the Straits of Magellan, the Patagonian steppes pass over into the shores of Tierra del Fuego; in the interior, the strait separates the one Flora from the other rather abruptly (p. 263), a condition with which may be compared the analogous contrast between Jutland and Norway. The cause of this remarkable contrast in the southern extremity of America, is to be sought, according to Darwin, in the amount of atmospheric deposit. At first sight such a difference in the character of the landscape, within a distance of four geographical miles, appears almost wonderful, but the climate itself presents the same degree of contrast; on the one side are the rounded mountains of Port Famine, continually drenched with showers and mists, which are collected by the stormy movement of the atmosphere, and at a distance of twelve miles, on "Gregory Bay," are dry barren plains, under a hot and cloudless azure sky. The mean temperature of Port Famine is probably = 5°·3 C., that of summer = 10°, in winter = + 0°·6. (King and Darwin.)

Dr. Hooker has described his stay in the Falkland Islands during the winter. (Journ. of Bot. ii, pp. 280-305.)

Urville enumerates, in his 'Flora' of these islands, 217 species. Hooker has added to the list, especially in the Cryptogamia. The only shrubs are *Chiliotrichum amelloides*, *Empetrum rubrum* and *Pernettia empetrifolia*, and on the western island, *Veronica decussata*. The famous "Tussac grass" (*Dactylis cæspitosa*, Forst., = *Festuca flabellata*, Lam.,) which grows to a height of six feet, and forms extensive patches of turf upon the peaty substratum, and the naturalization of which in Ireland, on account of its great nutritious properties, has been thought of, is limited, however, only to peculiar localities, and yields in importance as a food for cattle to the much more generally distributed and also very nutritious *Festuca Alopecurus*, Urv., with which every peat bog is covered.

"Hermit Island" lying to the west of Cape Horn, is the southernmost point at which Hooker, on his antarctic voyage, observed any arboreal vegetation. (Ib. p. 365.) An herbarium of eighty-four phanerogamous plants, corresponds with the forms of Tierra del Fuego, and the Falkland Islands; the tree is Darwin's evergreen *Beech*, to which, besides the synonyms given above, Hooker has applied as probable ones, those of *Fagus Forsteri*, Hook., and *F. dubia*, Mirb.

Darwin compares the forests of Chiloe for their luxuriant vegetation with those of the tropics. Various evergreen species, particularly Laurineæ and *Drimys*, are intermixed, and loaded with parasitic Monocotyledons, and under their shade grow various large *ferns* and tree-grasses. This vegetation, on the west coast of the continent at 45° S. L., borders on the uniform forest which extends from hence on the west side of the Andes as far as Tierra del Fuego. Its existence in such a high latitude, is owing to the extraordinary moisture of the climate. In Chiloe it rains both winter and summer, and Darwin believes that no other country in either temperate zone is subject to so much atmospheric deposit. The winds are usually violent, and the sky almost constantly covered with clouds. Even in Valdivia the forest character becomes remarkably changed (46°), the evergreen trees decrease, and at Valparaiso (33°), where, during the summer, rainless south winds prevail, and the atmospheric deposits take place almost only in the three winter months, there is scarcely a tree.

VI.—AUSTRALIA AND SOUTH SEA ISLANDS.

Darwin has given a general description of the botanical characters of the Gallapagos, the endemic Flora of which is still almost entirely unknown. (Ib. p. 453.) Covered with innumerable craters, these islands rise to the altitude of 3000′—4000′, and possess, on account of the low temperature proper to the surrounding ocean, a not very hot climate. On the coasts it seldom rains, but on the mountains the clouds hang low, and consequently at the altitude of about 1000′ a tolerably luxuriant vegetation makes its appearance, instead of the barren aspect presented by the littoral region. The modern lavas, however, spread over the declivities are entirely bare. Both animals and plants indicate for the most part an endemic creation. The plants are characterized by the sparing development of the leaves, and cannot be at all referred to the equatorial position of the islands. Ligneous plants are rare: amongst the most abundant in the lower region is a fruticose Euphorbiaceous plant, with small brownish leaves, together with an *Acacia* and the treelike *Opuntia galopagea*, with large, oval, compressed joints springing from the cylindrical stem; in the mountain region is an arboreal Synantherous plant, as well as ferns and grasses, but no tree-fern, and no palms.

Bentham has continued his investigation of the Flora of the Fidji Islands, &c. (vide last Report), from the collection of Hinds and Barclay. (Journal of Bot. pp. 211-40.) This list of not quite 200 species, appears now to be closed; it contains the new genera *Vavæa* from the Friendly Isles (allied to the doubtful Cedrelaceous plant, *Ixio-*

nanthes), *Cardiophora* from New Ireland (Terebinthiaceæ), *Lasiostoma* from New Guinea (Rubiaceæ), *Chætosus*, ib. (Apocyneæ), *Leucosmia* from the Fidji Islands (Aquilarineæ).

Lhotsky has endeavoured to characterize certain districts on the east coast of Australia by their productions. (Some data towards the Botanical Geography of New Holland, ib. pp. 135-41.) He distinguishes the following formations:

(1) "The coast vegetation" from Sydney southwards to the Illawarra. Quicksands or sandstone rocks with a scanty soil; numerous lagunes with salt or brackish water. The only tree, a *Eucalyptus;* thick bush of Epacrideæ, Proteaceæ, Podaliriæ, *Boronia* and *Comesperma*, social Xanthorrhoeæ and *Xerotes*. These shrubs constitute an almost impenetrable growth, and are of no economical use.

(2) "Watered rocky valleys along the coast." This appears to be the only locality for both the palms of New South Wales: *Corypha australis* and *Seaforthia elegans*. Here grow also the arboreal Amaryllideous plant, *Doryanthes*, a tree-fern (*Alsophila*), the Magnoliaceous *Tasmania*, besides some Malvaceæ, Rubiaceæ, and *Callicoma*.

(3) "Vegetation of the clayey soil." Extensive tracts of this formation are covered by the thin Eucalyptus forest, well known from R. Brown's celebrated sketch; it contains little underwood, but excellent pasture ground, with the greatest variety of herbaceous plants.

(4) "The vegetation of the Minero-Downs" includes the large tracts of pasture along the foot of the Blue Mountains. Excepting *Hakea* and *Brunonia*, ligneous plants are entirely wanting on this plain. In November it is clothed with a luxuriant vernal vegetation, which withers up under the heats of summer, and after April it presents only the aspect of a yellow-coloured steppe; the plants, however, consist chiefly of Gramineæ and Cyperaceæ. On these downs, between which and the argillaceous district there is no very marked line of demarcation,

the breeding of cattle is carried on, the wealth of the colony.

(5) "Vegetation of the Blue Mountains." The upper declivities of "Mount William the Fourth," which was ascended by Lhotsky, and on the summit of which the boiling point=196° F. were found by him to be covered with *Eucalyptus* trees 12′ to 20′ in height.

The Australian grasses in Lindley's collection have been determined by Nees v. Esenbeck, and besides several new species, the genera *Gamelythrum* and *Amphibromus* have on this occasion been instituted by him. (Journ. of Bot. ii, pp. 409-20.) Schauer has given a report on the Myrtaceæ collected by Preiss on the Swan River (178 species.) (Regensburg Flora, 1843, s. 405-10.) A. Braun has described 8 New Holland *Charæ*, for the most part collected by Preiss; they are all without an external cellular layer. (Linnæa, 1843, pp. 113-19.)

In Van Diemen's Land, a journal has appeared since 1842, (The Tasmanian Journal of Natural Science, Agriculture, &c.,) with botanical contributions by Gunn and Colenso. According to an extract from the first volume (Bot. Zeit. 1844, p. 140), the former has communicated remarks upon the Flora of Geelong, Port Philip; and the latter described some ferns from New Zealand.

Dieffenbach in his 'New Zealand Travels' speaks of the statistical relations of the indigenous Flora. (Travels in New Zealand, London, 1843, i, pp. 419-31.) Up to the present time, only about 630 species from New Zealand have been made known, and this paucity is to be referred, in the traveller's opinion, not to an incomplete exploration, but to the actual poverty of the Flora, the greater part of which appears to him to be already known. Principal families are: ninety-four Ferns, which constitute the distinguishing characteristic of the Flora, not simply in consequence of the variety of forms, but chiefly from the quantity of the individual plants, since enormous tracts of open country are covered with them, as representatives of the grasses in other Floras; three Tree-ferns (*Cyathea medul-*

laris and *dealbata, Dicsonia squarrosa*) attain a height of from 30′ to 40′, and grow in even greater numbers together, deep in the forest; 24 Gramineæ; 20 Cyperaceæ; representatives of the Junceæ, among which the European social *Juncus filiformis,* widely distributed, is said to indicate a thin layer of fertile soil, above the sterile argillaceous deposit; of Palms only *Areca sapida,* but which is seldom wanting in the denser forests; certain Liliaceous forms, characteristic of the open regions; *Phormium* almost everywhere; *Dracæna australis* forming a jungle on the river banks; of the Smilaceæ, *Ripogonum parviflorum,* Br., together with a Pandanea (*Freycinetia Banksii*), which is the most abundant climber of the forest; Orchideæ rare, however 3 Epiphyta are met with; *Typha angustifolia* usually covers the swamps, as in Europe; 2 Piperaceæ, common; 11 Coniferæ, among which the most important the Kawri Pine (*Dammara australis*) is confined to the northern extremity of the north island, and the others (*Dacrydium, Podocarpus, Phyllocladus*) do not occur associated, but dispersed about in the forests; 9 Epacrideæ; some Araliaceæ of curious forms (*Panax, Aralia Schefleri,* &c.); several Cunoniaceæ, amongst which *Leiospermum racemosum* constitutes large forests in all parts of New Zealand; 20 Onagrariæ; 13 Myrtaceæ, including some widely-distributed forest trees (2 species *Leptospermum,* 9 *Metrosideros, Eugenia,* and *Myrtus bullata,* indigenous also in Chili); 6 Pimelea-species, but only 2 Proteaceæ (*Persoonia tora* and *Knightia excelsa*); 3 Laurineæ, 2 of which are extensively distributed, and constitute a riparial growth; *Laurus tawa* covers the upper region of the mountains on Cook's Strait; of the Atherospermeæ, *Laurelia,* a bulky tree; 12 Scrophularineæ, with 9 Veronica-species, some of which are fruticose; of the Cyrtandraceæ, only *Rhabdothamnus Solandri;* of the Myoporineæ, *Avicennia tomentosa,* which forms the mangrove woods of New Zealand; of the Verbenaceæ, the important tree *Vitex littoralis,* the "New Zealand oak" of the settlers.

"Lord Auckland's Islands," lying to the south of New Zealand (51° S. L.), were explored by Dr. Hooker, during several weeks in the most favorable season (Nov. Dec.), and afforded him, without reckoning the Algæ, an herbarium of 120 species. The Flora, as stated by Dieffenbach, is probably not endemic, but introduced from New Zealand. This opinion is supported by the circumstance, that even in this high latitude true tree-ferns are met with, though here forming only a low stem. The surface of this mountainous Archipelago, is pretty equally covered with forest—bush and open pasture ground. From the sea-shore to the forest, European genera especially are found, together with the prevailing ferns which are also very numerous in the woods. The forest trees cover a thick undergrowth; the more lofty trees consist of a *Veronica*, an Araliaceous plant, some Myrtaceæ, and Epacrideæ, and these trees are often placed so thick, that they completely shade the ground. The tree-ferns belong to *Aspidium*. To this forest succeeds a region of shrubs, in which the Veronica tree is wanting, and the constituents of the undergrowth become more and more dwarfish. Above the fruticose region is one, consisting of the Gramineæ, where the mountain-meadows consist of *Bromus*, and one *Hierochloa*, together with some herbaceous plants, e. g. two Umbelliferæ; and strongly defined by a different vegetation, an alpine region occupies the summit of the mountain, with European genera and *Acæna*. Characteristic forms: a very social *Asphodelus*, with gold yellow flowers, *Veronica*, *Gentiana*, *Coprosma*, *Dracophyllum*, *Astelia*, &c. With the Flora of the "Lord Auckland's Archipelago," that of the neighbouring Campbell Island corresponds, even in the two arboreal ferns (52½° S. L.), only the south-west or windward side of the island is wholly without ligneous plants.

On "Keelings Islands," which have become celebrated through Darwin's 'Researches on the Coral Islands' (12° S. L.), and which, from the introduced Cocoa Palms, have also received the name of Cocoa Islands, that traveller collected about 20 plants, to which the indigenous Flora is confined. (Journal, p. 541.) According to Henslow's examination, 20 species belong to 19 different genera, and 16 natural families; and as these have all been washed from Java or New Holland, there is not a single endemic species produced by the coral reefs.

Kerguelen's Land (50° S. L. in the Indian Ocean) was a long winter station on Dr. Hooker's voyage. (Journ. of Bot. ii, pp. 257-63.) He there collected many Cryptogamia, and obtained an herbarium of 130 species, among which were 30 lichens, which cover the 2000′ high mountains in large quantities. The peculiar nature of the climate appears to have enabled Hooker to collect also the few Phanerogamia in a condition admitting of their determination. The following genera are met with: *Agrostis* and 4 other grasses, 1 *Juncus*, 1 *Ranunculus*, 1 *Callitriche*, 1 large cabbage-like Cruciferous plant, 1 social Umbellifer (probably *Bolax*,) 1 *Acæna*, 1 *Silene?* 1 Portulaceæ, Rubiaceæ, and Synanthereæ, and 3 Phanerogamia of uncertain relationship. Moreover of Cryptogamia, still 1 fern, 2 *Lycopodia*, 23 mosses, mostly corresponding to the Arctic forms, 10 *Jungermanniæ*, 1 *Marchantia*, 10 *Confervæ*, and 39 other Algæ, 1 Fungus. This great poverty of the Flora cannot be considered to be caused by the climate, which though indeed stormy, is otherwise not so very inclement, but is to be explained geologically.

The French works with figures which have been published, as a sequel to the Antarctic Voyage of Dumont d'Urville, as well as to the 'Expedition of the Venus,' contain botanical sections, but are as yet far from completion.

REPORT ON THE PROGRESS OF GEOGRAPHICAL BOTANY DURING THE YEARS 1844 AND 1845

Heinrich Rudolf August Grisebach

REPORT

ON THE PROGRESS OF

GEOGRAPHICAL BOTANY,

DURING THE YEAR 1844.

BY DR. A. GRISEBACH,

EXTRAORDINARY PROFESSOR IN THE UNIVERSITY OF GOTTINGEN.

GEOGRAPHICAL BOTANY.

In the first volume of the 'Physical Atlas' of Berghaus, which is now completed, six sheets are devoted to graphic representations in the department of Botanical Geography. The first sheet, entitled 'Outlines,' forms a sequel to the works of Humboldt and Schouw, and refers principally to the geographic subdivision of vegetable formations; in the vertical direction it illustrates the serial gradations of the regions, whilst in the horizontal direction it shows the areal boundaries of the natural Floras. This representation, however, appeared as early as 1838, and on future revision would require considerable improvements. The second sheet, which treats of the Districts in which the most important products of culture are distributed, is of greater interest. Its design consists in an attempt to subdivide the province of agriculture throughout the entire inhabited surface of the earth, according to the kinds of Cerealia which predominate, whence general relations are found between the climate and the productive power of different countries. In the Old World the author distinguishes the following zones between the polar limits to agriculture and the equator.

1. Zone of barley and rye. It might with propriety be called the zone of the summer Cerealia, inasmuch as the duration of the winter is the most important condition which prevents the culture of the more productive

and certain winter corn. In this more comprehensive point of view, the separate denomination of the South of Scandinavia as the district of the exclusive cultivation of rye, and of Scotland as that of barley, disappears, as circumstances not founded upon climatic conditions.

2. Zone of rye and wheat. This is considered as extending southwards to about the fiftieth degree of latitude, or as far as the polar limits of the cultivation of the vine.

3. Zone of wheat. To this, those parts of Europe and Western Asia belong which lie south of the fiftieth degree. In several districts maize is cultivated as well as wheat.

4. Zone of rice and wheat in those provinces which are subject to the influence of tropical seasons. In tropical Western Africa rice and maize occupy the place of the former.

In America, where these relations are modified by the greater extent to which maize is cultivated, Berghaus distinguishes the following zones: rye, wheat, and barley (i. e. summer Cerealia); rye and maize; wheat and maize; wheat; in the tropical zone maize is the principal cereal grain. With these sketches the author has combined indications of the distribution of other nutritive plants, and has illustrated, in separate charts, the districts in which the most important plants of commerce are produced. The two following sheets contain the statistical numerical proportions of the Flora of Europe, which, not being susceptible of tabular arrangement, and being subject to very important differences in the views taken of the definition of the species and botanical groups, were not adapted, in the present state of botanical geography, to graphic representation. Although the same applies still more to the last sheet upon Germany, which appeared in 1841, nevertheless the review of the polar and equatorial limits of numerous woody and cultivated plants in Europe, deserves great praise, inasmuch as the observations made use of in it have appeared to us, on the frequent use of this chart, very numerous. Moreover, many of the sheets intended to illustrate meteorological relations appear indispensable also to the botanist.

M. Römer has commenced the publication of a Memoir entitled 'Botanical Geography and Geographical Botany,' which treats of the subdivision of the surface of the earth into natural Floras. (Lüdde Zeitschr. für vergl. Erdkunde, Bd. iii, pp. 527-534.)

A paper by E. Fries, entitled "The Native Land of Plants," in his peculiar style, the special interest of which is confined to the Swedish public, but also frequently touches acutely more general questions, treats of different botanico-geographical subjects, especially of the native country of the so-called ruderal plants. (Botaniska Utflygta, Bd. i, pp. 229-328, translated in Hornschuch's Archiv Skandinav. Beiträge zur Naturgesch. Bd. i, H. 3.) The *original* native country of many cultivated plants cannot now be determined by empirical proof, but only by rational investigation. Thus rape is no longer met with in its wild state, but when we adduce proof from all extra-European countries that it is not indigenous to them, we must conclude that it is of European origin, although its wild state has disappeared through cultivation. Many plants have been extirpated by use; this is now gradually taking place with *Gentiana lutea*, in the Alps, and *Inula Helenium* in the west of Sweden. The contact of Nature with man exerts no less a modifying influence upon the vegetable kingdom than upon the animal creation. The original vegetation of a country must in general, therefore, be regarded as more rich in species, and in this manner in Sweden and Germany, even under our own eyes, the localities of rare plants are disappearing one after the other, as e. g. of *Trapa*, *Xanthium*, and *Stipa*.

The excellent work of A. Wagner, on the 'Geographical Distribution of the Mammalia,' (Abhandlungen der mathem. physik. Klasse der Bairischen Akad. Bd. iv), which belongs to an allied province, but was not designed without regard to the geographical relations of other organisms, must not be passed over here without notice. The question of the original native country of the various organisms is acutely investigated by the author, and it is found that the distribution of animals, as of plants, cannot be satisfactorily explained by the climatic and local conditions of their existence, but that the most rigid facts, together with the physical relations at present in existence, point to other, perhaps historical causes, with which we are at present unacquainted, and which the author considers as the effects of a general order of the creation, which ought, however, rather to be kept in view *by us* as objects worthy of future investigation. From the observations, in Belgium, upon the Periodical Phenomena of Vegetation, published by Quetelet, and mentioned in the previous Yearly Report, the following brief extract, containing the period of the appearance and fall of the leaves, in the year 1841, of some generally diffused woody plants, may be of use in the determination of the Phyto-isotherms of Northern Europe; and for this purpose it will be conjoined with some observations simultaneously made by Hartmann, in Gefle (60° N. l.) (Bot. Notis. 1842.)

	Appearance of the Leaves.			1841.	Fall of the Leaves.	
	Gefle.	Brussels.	Lyons.	Ghent.	Brussels.	Ghent.
Æsculus Hippocastanum	15 May	...	29 March	...	25-30 Oct.	24 Oct.
Acer pseudo-platanus	...	23 April	...	27 March	25-30 Oct.	...
Vitis vinifera	...	23 April	...	...	10-15 Nov.	...
Tilia europæa	21 May	26 March	...	24 March	20-25 Oct.	12 Sept.
Juglans regia	...	27 April	...	25 April	...	3 Oct.
Prunus Cerasus	...	27 March	...	...	...	27 Oct.
Pyrus Malus	...	24 March	...	17 March	1-5 Nov.	29 Oct.
Sorbus aucuparia	12 May	...	...	...	...	...
Ribes Grossularia	...	12 March	17 March	14 March	...	...
Ribes rubrum	...	18 March	20 March	17 March	...	...
Sambucus nigra	...	18 March	15 March	14 March	5-10 Nov.	24 Oct.
Syringa vulgaris	...	12 March	15 March	17 March	5-10 Nov.	24 Oct.
Fraxinus excelsior	25 May	...	...	...	...	...
Daphne Mezereum	3 May	16 March	24 March	...	...	...
Ulmus campestris	22 May	29 March	...	26 March	1-5 Nov.	31 Oct.
Salix babylonica	...	17 March	24 March	17 March	15-20 Nov.	...
Populus fastigiata	...	1 April	...	...	20-25 Oct.	24 Sept.
Populus tremula	19 May	...	...	...	...	...
Corylus Avellana	16 May	24 March	25 March	18 March	...	27 Oct.
Quercus Robur	...	28 April	...	...	10-15 Nov.	...
Betula alba	14 May	27 March	...	...	1-5 Nov.	...
Alnus glutinosa	20 May	...	...	...	...	...

Of monographs upon individual groups of plants, in which attention is paid to their geographical distribution, published during the past year, the following require mention: Parlatore on the Fumariaceæ (Giornale Botan. Ital., i, p. 97 et seq.); v. Martius, on the Erythroxylaceæ (Bairische Abhandl., iii, pp. 325-32); Lomler, on the Distribution of the Coniferæ (Ratisbon Flora, 1844, pp. 440-3).

Fumariaceæ.—Only 13 species; these are distributed throughout both temperate zones, for the most part, indeed, secondarily transferred from one region to the other. With the exception of the Cape *Discocapnus*, they all grow in the South of Europe, between the 34th and 40th degrees of latitude, and diminish so rapidly from this zone in both meridional directions, that beyond the 50th degree, 3 species only are met with; a statement which, however, is not correct as regards Germany. Spain contains several endemic forms.

Erythroxylaceæ.—Of 58 species of the genus *Erythroxylon*, Brazil contains 29; the West Indies, 8; Guiana, 7; Columbia, 4; and Mexico and Peru, one each; hence tropical America contains 50 altogether: 5 species grow in Madagascar and the Mauritius, single representatives at the Cape, in the East Indies, and on the north coast of New Holland. In America the district of their distribution extends from the tropic of Cancer to that of Capricorn, in the Old World, from 15° N. lat. to 30° S. lat.

Coniferæ.—Lomler enumerates only 208 species. Of these, he calculated that 165 exist in the northern and 51 in the southern hemisphere; moreover, there are 22 in Europe, 87 in Asia, 16 in Africa, 83 in America, and 35 in Australia; lastly, 24 in the tropic zone, 159 in the north temperate, and 33 in the south temperate zone. These statements can only be regarded as preliminary steps to our knowledge on this point.

I.—EUROPE.

A work, containing copper-plate engravings of the plants of Russia, has been begun by Trautvetter (Plantarum Imagines et Descriptiones. Monachii, 1844, 4 fasc., 1-4; at present 20 plates). Also a continuation of the old 'Bieberstein Centuries' (M. de Bieberstein Centuria Plantarum Rossiæ Meridionalis Iconibus illustrata; Pt. ii, Dec. 1-3. Petropoli, 1844,) has been commenced in St. Petersburg. Engelmann has published a paper upon the Genera of Plants found in the Russian provinces of the Baltic (Genera Plantarum, or the Genera of Plants growing wild in Esthonia, Livonia, and Courland; Mitau, 1844-8).

A. F. C. v. Fischer has written upon the botanical relations of Southern and Central Lithuania, especially in the circle of Sluzk (Mittheilungen d. Natur. Gesellschaft zu Bern, for the years 1843-4; Bern, 8). In the immediate neighbourhood of Sluzk, in the district of the source of the Niemen and several tributary streams of the Dnieper, the author only found about 600 Phanerogamia, a catalogue of which he gives, with remarks upon their statistics. In these districts, heathy plains, overgrown with *Calluna* (together with *Juniperus* and *Genista tinctoria*) are still common. Dwarf underwood, consisting of the oak (*Quercus pedunculata*) covers large spaces, and stamps the physiognomy of Lithuania towards the western districts of the Baltic plain. In moist low grounds *Salix angustifolia* and *livida* predominate. The large forests consist of pines or fir trees; the truly foliaceous trees, which are less common, are mostly birch, and in Polesia, the oak, which grows mixed with the birch, poplar, mountain-ash, &c. The following may be mentioned as geographically characteristic species: *Thalictrum aquilegifolium* L., *simplex* L., and *angustifolium* Jacq., *Anemone patens* L.,

Viola stricta Horn., *Dianthus arenarius* L., *Euonymus verrucosus* Scop., *Trifolium lupinaster* L. (in pinetis siccioribus raro), *Spiræa Aruncus* L., *Geum strictum* Ait., *Potentilla norvegica* L., *Agrimonia pilosa* Led., *Saxifraga Hirculus* L., *Cnidium venosum* Kch., *Chærophyllum aromaticum* L., *Inula Helenium* L. (in sylvis udis), *I. hirta* L., *Cirsium rivulare* Kch., *Andromeda calyculata* L., *Pyrola media* Sw., *Polemonium cæruleum* L., *Pulmonaria azurea* Bess., *Pedicularis Sceptrum* L., *Dracocephalum Ruyschiana* L., *Melittis Melissophyllum* L., *Amaranthus sylvestris* Desf., *Thesium ebracteatum* Hayn., *Euphorbia virgata* Kit., *Salix nigricans* Fr., *livida* Wahlb. (*depressa* Fr.), *myrtilloides* L., *versifolia* Wahlb., *lapponum* L., *Betula fruticosa* Pall., *Typha pendula* nov. sp.,* *Malaxis monophyllos* Sw., *Cypripedium Calceolus* L., *Gladiolus imbricatus* L., *Fritillaria* sp., *Veratrum Lobelianum* Bernh., *Tofieldia calyculata* Wahlb., *Carex divulsa* Good., *pilosa* Scop:, *Hierochloa odorata* Wahlb., *Calamagrostis stricta* Spr.

Wahlberg has published some remarks upon the plants of Quickjock in Swedish Lapland (Öfversigt af Kongl. Vetenskabs-Akademicens Förhandl., 1844, p. 23). *Rubus castoreus* Laestad. is a bastard of *R. articus* and *saxatilis* occurring in two forms.

Lindblom has published some observations upon the Botanical Relations of Norway (Bot. Notiser., 1842-3.) In the outset, we meet with the unfounded assertion, that in most of the regions of the coast of Norway, Alpine plants extend as low down as the level of the sea; an occurrence which is limited to individual species only, and may be compared to the growth of Alpine plants on the Isarkies, near Munich. This statement, made by Lindblom, is one of those erroneous generalizations, borrowed by one person from another. Alpine plants do not occur in Norway below the limit of trees, any more

* T. spicis cylindricis, masc. et fœm. contiguis, foliis planis linearibus culmo longioribus pendulis. (An. T. Shuttleworthii, Kch.?)

than in the Alps. Then follow observations upon the limits at which plants occur in the direction from west to east, to which scientific value must be attributed, on account of the climatic contrasts between the internal districts and the western coast of the south of Norway.

a. Plants of the western coast, which, according to Lindblom, are not found in the inner district. (The polar limit of their distribution is expressed numerically according to the degree of latitude, their occurrence in Sweden is inclosed in parentheses).—

Fumaria capreolata. 59°.
Hypericum pulchrum. 63½°.
— *montanum.*
Vaerdal in Trondjem.
Vicia orobus.—62½°, i. e. the limit of the occurrence of the oak.
Sanguisorba officinalis. 60°.
(Isld. Gottland.)
Bunium flexuosum. 63°.
Myrrhis odorata. 63°.
Chryosplenium oppositifolium.—62½°.
Rosa pimpinellifolia. 60°.
Ilex aquifolium. 62½°.
(Bohuslän.)
Galium saxatile. 62½°.
(South of Sweden.)
Centaurea nigra.—
Snaasen in Trondjem.
Hypochæris radicata. 62½°.
(— Bohuslän.)
Erica cinerea. 62½°.
Pyrola media. 61°.
(— Bohuslän.)
Lysimachia nemorum.—63°.
(Schonen.)
Primula acaulis. 63°(=).
Digitalis purpurea. 63°.
(— Bohuslän.)
Lamium intermedium. 61°.
Teucrium Scorodonia. 59°.
Luzula maxima. 68°.
Carex binervis. 63°.
— *salina.* 70°.
— *maritima.* 70°.
Aira præcox. 62½°.
(— Bohuslän.)
Bromus tectorum. 61°.
Brachypodium gracile. 62½°.

b. Plants belonging to the western coast, which occur only on the southern coast, e. g. at Christiania, or in the valleys of the Fjeldplateaux, but not in the true inner districts of the south of Norway.—

Arabis petræa.—62°.
Rosa pomifera.—63°.
(South of Sweden.)
Sorbus aria.—63½°.
(— Bohuslän.)
Sorbus hybrida.—62° (?)
(Gottland.)
Hedera helix.—60½°.
Lonicera periclymenum.—
Valderhong in Trondjem.
(— Bohuslän.)
Sambucus nigra.—
Valderhong. (— Bohuslän.)
Gentiana purpurea.—62½°

Mentha sativa.—63°. (South of Sweden.)

Fagus sylvatica.—61°. (— Bohuslän.)

Quercus Robur.—62°. According to Blom, 63°. (South of Sweden.)

Allium ursinum.—63°. (— Bohuslän.)

c. Inland plants of the eastern districts of southern Norway, which are absent on the western coast. (Excluding those of the Fjeld.)—

Pulsatilla vernalis.
Trollius Europæus.
Berberis vulgaris.
Astragalus glycyphyllus.
Ledum rupestre.
Galium trifidum.
Hieracium cymosum.
Pyrola chlorantha.
Dracocephalum Ruyschiana.
Thymus Chamædrys.
Pedicularis Sceptrum.
Salix daphnoides.
— *amygdalina.*
Carex capitata.
— *parallela.*

d. Plants of the Eastern Fjeld, principally observed on the Dovre-fjeld, but not found on the western coast (some species which I myself found at Hardanger, and which are therefore more widely distributed, are omitted in this list, viz. *Aconitum septentrionale*, *Draba hirta*, *Gentiana nivalis*, and *Salix arbuscula*).—

Ranunculus hyperboreus.
Lychnis apetala.
Alsine hirta.
Oxytropis lapponica.
Phaca oroboides.
— *frigida.*
Potentilla nivea.
Saxifraga cernua.
Saxifraga controversa.
Primula stricta.
Gentiana tenella.
Kœnigia Islandica.
Junctus arcticus.
Kobresia caricina.
Elyna spicata.
Carex microglochin.

A remarkable peculiarity of the highlands of Norway, and which is not merely indicated, but satisfactorily established by this catalogue, yet cannot be explained by means of the variations of climate pointed out above, consists in the fact that the Alpine vegetation appears to attain its maximum, as regards the number of species, on the Dovre mountains, and that it diminishes from this locality both towards the west and the south. Moreover in these directions the individual numbers of many characteristic species also become less, the Fjeldplateau gradually assuming the condition of a steppe. In this respect, Lindblom's observations on the desert of the Bygle- and Hekle-Fjelds, or the most southern part of the high-lands, which were made many years ago, but are again brought forward in the present memoir, are instructive. The predominating plants of some tracts in this part, e. g. between Siredal and Lysefjord, are *Molinia cœrulea* and *Solidago virgaurea*, and these displace all others. The alpine plants of this region, as shown by the following list of them, also grow in Hardanger, and do not resemble those of the Brocken or the Sudeten, to which, among the whole of the Scandinavian mountains they are most nearly situated.

Ranunculus pigmæus.
Arabis alpina; Cardamine bellidifolia.
Silene acaulis; Lychnis alpina; Stellaria alpestris; Cerastium trigynum alpinum; Sagina Linnæi.
Epilobium alpinum, alsinifolium.
Dryas octopetala; Potentilla maculata; Sibbaldia procumbens; Alchemilla alpina.
Rhodiola rosea.
Saxifraga Cotyledon, stellaris, aizoides, rivularis, oppositifolia, nivalis.
Saussurea alpina; Hieracium aurantiacum, alpinum.
Phyllodoce taxifolia; Cassiope hypnoides; Arctostaphylos alpina; Loiseleuria procumbens.
Gentiana purpurea.
Veronica alpina, saxatilis; Bartsia alpina.
Oxyria reniformis.
Salix glauca, Myrsinites, Lapponum, retusa, herbacea.
Betula nana.
Tofieldia borealis.
Juncus biglumis, trifidus; Luzula arcuata and *spicata.*
Aira alpina and *atropurpurea; Poa alpina; Phleum alpinum.*
Carex rariflora, pulla, lagopina, rigida, vaginata, atrata, rotundata, capillaris and *alpina; Eriophorum capitatum.*
Lycopodium alpinum.
Polypodium alpestre.

The second section of Lindblom's memoir treats of the distribution of the Norwegian Ferns, which, according to theory, ought to be more common on the western coast than in the inland districts, but which, in fact, do not correspond to this view. The author is certainly of an opposite opinion, and states that the number of individuals increases towards the west, which I should much doubt; but it is certain that *Hymenophyllum Wilsoni* can alone be considered as an evidence of the marine climate, whilst the inland country contains five more of the thirty-three ferns which are here enumerated than the west, viz. *Polypodium calcareum; Aspidium Thelypteris, cristatum, montanum*, and *crenatum* Sommf. On the western coast, *Aspidium aculeatum* and *Asplenium adiantum nigrum*, which are not found in the east, extend as far as Trondjem, but they must be considered as forms belonging to the south, not to the coast.

I can only refer here to my paper upon Hardanger (see Wiegm. Arch., p. 1-28); still I cannot omit this opportunity of replying to the editor of the 'Botaniska Notiser' (see that journal, 1844, appendix, p. 64), that the beech is certainly cultivated beyond Christiansund. Blom, whose authority is Blytt, makes this statement. (Das Königreich Norwegen. Leipz., 1843, p. 48.) I did not say that it grew wild there, as Lindblom has erroneously stated in his translation, and the only object I then had in view, was to show how far north the climate was suitable to the growth of that tree. I found single specimens of *Helianthemum alpestre* on rocks near the herdsmen's hut Oppedals-Stölen, and have given specimens of *Phippsia* from the same region to several botanists. However, I place little value upon these new localities, of which I had several, and I should consider it as the best recompense for the labour of my memoir, if Lindblom and other able Scandinavian naturalists, instead of filling their Journal with unsatisfactory lists of the results of their excursions, and critical minutiæ regarding the distinction of species and their nomenclature, were also induced by it to direct their scientific attention more and more to the conditions of the distribution of plants in the North of Europe.

Blytt, whose Flora of Norway has long been in preparation, but is still looked for in vain, has published a Catalogue of the Plants growing wild at Christiania (Enumeratio Plantarum, quæ circa Christianiam sponte nascuntur, Christiania, 1844, p. 4). It contains 790 vascular plants. Fries has continued the publication of his Critical Remarks upon Swedish plants and their stations (Bot. Notis. 1844, p. 1, 49, 75 et seq.) Parts ix and x of his Normal Herbarium have appeared. Anderson and Lindblom have worked at the Alpine *Epilobia* of Sweden (id.) Ångerström has issued some contributions to our knowledge of the Scandinavian Mosses (Nov. Act. soc. Upsal. 12, pp. 345-80).

Lindblom's Botaniska Notiser also contains the following Memoirs upon the Topography of Swedish plants: Borgström, Contributions to the Flora of Wärmeland (1842); Lindgren and Torssell, Mosses of Upsal (1842-3); Forssell, Catalogue of the more rare Plants which occur in Norrtelge (north-east of Stockholm) (id.); Hofberg, Localities at Strengnäs on the Lake Malern (1842-3); Von Post, Botanical Conditions of the Western Bank of Lake Malern (1844), of some interest, on account of the careful observation of the localities in which 480 phanerogamous plants are distributed; Hamnström, New Localities in Nerike (1842); Lindgren, Localities at Lake Wener, with critical remarks (1842-3); Holmgren, Kalén, and Hamnström, Localities in East Gothland (1841-3); Lagerheim, the same in West Gothland (1844); Sieurin, Diary of Travels in North Holland, containing habitats (id.); Lindblom and Borgström, Habitats at Schonen (1843-4). Nyman has published Contributions to the Flora of Gothland, by which the number of vascular plants found upon this island is increased to more than 800 (Vetenskaps Akademicens Handlingar för år 1840, pp. 123-51). The results of Beurling's voyage now communicated in these Memoirs, are confined to lists of localities, principally in Jemtland; they are copiously detailed as regards the mountain Åreskuten.

On the death of C. E. Sowerby, the proprietor of the

'English Botany,' his successor, J. D. C. Sowerby commenced a new series of the parts of this illustrated work, of which, with the aid of Wilson, Berkeley, Babington, and Borrer, up to 1844, the first three parts have appeared (Supplement to English Botany, second series, Nos. 1-3. London). The Botanical Society of London, following the example of that of Edinburgh, have published a catalogue of British plants (The London Catalogue of British Plants, published under the direction of the Botanical Society of London. London). In consequence of critical elaboration, this catalogue contains considerably fewer species (1305 indigenous, and 132 acclimated phanerogamia) than the Edinburgh one, and is ascribed to the pen of Watson. The 'Phytologist,' a journal which was noticed in the yearly report for 1842, is still continued. I may refer to the list of contents given in the 'Botanische Zeitung.'

Watson has made some critical remarks upon individual British plants (London Journal of Botany, iii, pp. 63-81). Newman has issued a description of British Ferns (A History of British Ferns and allied Plants. London, 1844). 'The Annals of Natural History' (vol. xiii, xiv) contain the following contributions to the British Flora: Ball, on *Œnanthe;* Taylor, contributions to our knowledge of the *Jungermanniæ;* Harvey, description of the new Irish genus of Algæ, *Rhododermis;* Berkeley, contributions to Mycology; Dickie, critical catalogue of the Marine Algæ existing at Aberdeen; Spruce, catalogue of the Mosses and Hepaticæ of Teesdale, in Yorkshire; Salwey, of the Lichens of Wales; Graham, on the results of his journey through Wales; Babington, on the Irish Saxifrages.

Babington has shown that *Neottia gemmipara* Lm., the rarest of all the European Orchidaceæ, which was discovered by Drummond near Cork, in 1810, and has only recently been again found, is identical with the *Spiranthes cernua* Rich. of North America (Proceed. of the Linnæan Society, 1844).

Sande Lacoste, and Dozy have been occupied in the study of the cryptogamic plants of the Netherlands. The former has made known localities of the Mosses; the latter, in conjunction with Molkenboer, has published a catalogue of the Fungi indigenous to that country, and some newly-discovered Mosses (both in v. d. Hoeven's Tidjschrift, f. 1844, p. 165 and 377).

The general works upon the Flora of Germany, mentioned in the previous Annual Reports, have been continued. Four decades of the seventh volume of Reichenbach's 'Icones,' containing the Aroideæ and the allied groups, have appeared. A cheaper edition, containing a more copious text, was commenced at the same time, with the title of 'Deutschland's Flora;' Parts 23 and 24 of the third section of 'Sturm's Flora;' the fifth volume of Schlechtendal and Schenk's illustrated work; and Parts 48-56 of that upon Thüringia; Parts 34-49 of Link's Publication; and Parts 2-4 of D. Dietrich's 'Cryptogamia.'

Rabenhorst has published the first volume of a 'Cryptogamic Flora,' containing the Fungi (Deutschland's Kryptogamen—Flora. Bd. i. Leipzig, 1844-8). This compilation is adapted to the present time, but does not entirely come up to our expectations. Of the author's valuable collection of dried Fungi, the seventh, and in the following year the eighth, "Centurie" have appeared. Hampe is preparing a similar herbarium of the 'Cryptogamia of the North of Germany,' which comprises at present 230 Mosses, 80 Hepaticæ, and 80 Lichens. (By the author at Blankenberg, on the Hartz.)

In Wallroth's 'Contributions to Botany,' two parts of which are before us, individual genera of the flora of Germany are treated monographically; especially *Agrimonia*, *Armeria* (with two well-marked Hartz mountain-plants, *Agrim. odorata* D. C., Syn. *A. procera* Wallr., and *Armeria humilis* Lk., Syn. *A. filicaulis* Boiss.! *A. Halleri* Wallr.), *Lampsana*, and *Xanthium*. Then follow critical remarks; as, e. g., upon *Senecio paludosus*,

Salix hastata, from which Wallroth distinguishes the form which he discovered on the gypsum-chain of the southern Hartz mountains, as *S. surculosa*. Scheele has continued his work upon German and individual Exotic Plants, which was noticed in the last Annual Report (Ratisbon Flora, 1844, and Linnæa, 1844); and Petermann has followed him in attempts of the same kind, to contribute to our knowledge of native species (Ratisbon Flora, id.).

Provincial Topographies and Sketches of the Vegetation in the Province of the German and Prussian Flora:—Kamp, Catalogue of Plants growing wild around Memel (Preuss. Provinzialblätter, 1844, p. 451-569); Leo Meier, On the Flora of Gerdana in Eastern Prussia (Bot. Zeit., 1844); Roeper, Contributions to the Flora of Mecklenberg (Part 2, Rostock, 1844), representing the Graminaceæ in the manner pointed out above; Fiedler, Synopsis of the Mosses of Mecklenberg (Schwerin, 1844-8); Häcker, Flora of Lübeck (1844-8); K. Müller, Contributions to a Cryptogamic Flora of Oldenburg (Bot. Zeitung, 1844), with additions and corrections by H. Koch (id.); Wimmer's Flora of Silesia, which was mentioned in the Annual Report for 1840, has appeared in a second and enlarged edition (Breslau, 1844); Reichenbach, Upon the Botanical Conditions of the Flora of Saxony (i. e. Gaea of Saxony, 1843-8), contains nothing more than a catalogue of rare plants from the separate districts, in the form of extracts from the author's 'Flora Saxonica;' Pfeiffer, Sketch of the Plants hitherto found in Kur-Hesse (Cassel, 1844-8); this is to be regarded as preliminary to a critical Flora of Hesse, and contains a large number of new localities, especially on the basaltic mountains of Cassel; by the same author, A few words upon the Subalpine Flora of Meissner (loc. cit., 1844); Wirtgen, Supplements to the Flora of the Prussia Rhine Provinces (Verhandlungen des naturhistorischen Vereins der Preussischen Rheinlande, Jahrg. 1); Thieme, Catalogue of the Plants growing at

Hainsberg, in the Territory of Aix (Ratisbon Flora, 1844, p. 209-21); Löhr, Manual of the Flora of Trêves and Luxembourg, with a notice of the surrounding districts (Trêves, 1844-8); Lechler, Supplement to the Flora of Würtemberg (Stuttgard, 1844-8); Sailer, Flora of Linz (Linz, 1844-8), an extract from the Flora of Upper Austria, mentioned in the Annual Report for 1841; Sauter, Report upon a Journey to Lungau (Ratisbon Flora, 1844, p. 813-16).

E. v. Berg, at Lauterberg, on the Hartz mountains, endeavoured to prove that the Coniferæ are gradually becoming more widely distributed in the north of Germany (Das Verdrängen der Laubwälder durch die Fichte und Kiefer. Darmstadt, 1844-8). The fact, in the case of the Hartz mountains, rests upon authentic testimony; but how far this change, which in many places has been completed in the space of twenty years, has been produced by external natural conditions, or merely by the economic management of the forests, is difficult to ascertain. In Lüneburg also, where, e. g., in the struggle between the two methods of culture, it was not decided in favour of the pine until after the lapse of a century, as also in Solling, on the Upper Weser, where deciduous forests are still very extensive, the same conditions have prevailed as on the upper Hartz mountains. On the western Hartz mountains, the red pine generally succeeds the beech; but in some parts, on the removal of the latter, the remains of oaks have been found as high as a level of 2000′, i. e. an elevation at which they have long since ceased to grow. When we consider that the tree-limits on the Hartz mountains lie extremely low, in comparison with those of the north of Europe, and that even the Coniferæ do not ascend higher on the Brocken mountains than at 9—10° further north in Norway, the fact of the culture of the oak and beech at a former period, would render it, at any rate, tolerably probable that secular changes had taken place in the climate, by means of which the distribution of the forest trees had been produced, and by which Steenstrup's succession of forest growth in maritime countries would be brought into connexion with the extermination of the Coniferæ in the elevated regions of the upper Hartz mountains.

The work of Fuchs, on the Venetian Alps, contains an account of the limits of vegetation in the southern dolomitic Alps, especially the district of Agordo; it fills up an important gap in the observations upon the vertical distribution of the Alpine plants (Vienna, 1844, fol.) Unfortunately, however, in the case of most of the plants, the lower limits of altitude only are given; and of these, a local value only can be attributed to many measurements. The results, expressed in French feet, are as follows:—

a. Upper limits.

Ficus Carica, and limit of the cultivation of the *vine*, 1500′. (At Agordo, *Vitis* grows very luxuriantly at a level of 2000′, but no wine is made.

Castanea vesca, 2000′ at Agordo.

Juglans regia, 3500′ at Frassene.

Zea Mays, 2500′ in the valley of Cordevolethal.

Cerealia, excluding wheat, 4400′ at the Col di S. Lucia; 4600′ at Buchenstein.

Dense Forest of Coniferæ, 5500′. In the regions of the mountain-pine, individual larches and fig-trees; 6309′ at Sasso di Palma.

Fagus sylvatica, 5000′; e. g. at Monte Luna, 4915′, still higher at Bosco Medona and in the Val Pegolera.

Pinus Cembra, 6665′ at the Col di Lana.

Upper limit of the Phanerogamia, = 9000′; *Aretia Vitaliana*, and some Saxifrages.

b. Lower limits.

Ranunculus aconitifolius, 3500′.
montanus, 7000′.
glacialis, 8000′.
Pyrenaicus, 8000′.
Anemone baldensis, 4500′.
Aconitum Anthora, 4500′.
Napellus, 6500′.
Stoerkianum, 6500′.
Arabis cærulea, 7000′.
Hutchinsia alpina, 7000′.
Hutchinsia rotundifolia, 7000′.
Papaver pyrenaicum, 5500′.
Viola biflora, 3500′.
Silene acaulis, 5500′.
pumilio, 7000′.
Cerastium latifolium, 6500′.
Cytisus alpinus, 1300′.
purpureus, 2000′.
Trifolium alpinum, 5500′.
Phaca astragalina, 6500′.

Phaca alpina, 6500′.
Hedysarum obscurum, 7000′.
Dryas octopetala, 2000′.
Potentilla caulescens, 1300′.
nitida, 6500′.
Geum montanum, 5500′.
reptans, 8000′.
Sibbaldia procumbens, 5500′.
Rosa alpina, 5500′.
Sedum atratum, 7000′.
Rhodiola rosea, 7000′.
Saxifraga Aizoon, 1300′.
aizoides, 1500′.
cæsia, 1500′.
rotundifolia, 2000′.
mutata, 2500′.
Burseriana, 2500′.
cuneifolia, 3500′.
stellaris, 5500′.
aspera, 5500′.
controversa, 6500′.
muscoides, 6500′.
planifolia, 7000′.
androsacea, 7000′.
sedoides, 7000′.
bryoides, 7000.
oppositifolia, 8000′.
Bupleurum graminifolium, 6500′.
Lonicera nigra, 4500′.
'*alpigena*,' 4500′.
Valeriana saxatilis, 1300′.
Aster alpinus, 1500′.
Tussilago alpina, 2000′.
Cacalia alpina, 4500′.
Arnica montana, 2000′.
Bellidiastrum, 1300′.
Gnaphalium Leontopodium, 1500′.
Chrysanthemum alpinum, 7000′.
Anthemis alpina, 6500′.
Achillæa Clavennae, 4500′.
moschata, 7000′.
Doronicum scorpioides, 7000′.
Aronicum Clusii, 7000′.
Senecio abrotanifolius, 5500′.
carniolicus, 7000′.
Cirsium ochroleucum, 2500′.
spinosissimum, 5500′.
Carduus defloratus, 5500′.
Saussurea alpina, 7000′.
Sonchus alpinus, 4500′.
Phyteuma comosum, 1300′.
Scheuchzeri, 1300′.
hemisphæricum, 5500′.
orbiculare, 5500′.
Sieberi, 7000′.
pauciflorum, 7000′.
Campanula barbata. 4500′.
Morettiana, 4500′.
Rhododendron hirsutum, 1300′.
Chamæcistus, 1300′.
Arbutus uva ursi, 2500′.
alpina, 5500′.
Azalea procumbens, 7000′.
Vaccinium Myrtillus, 2000′.
Vitis idæa, 2000′.
Primula Allionii, 2500′.
glutinosa, 7000′.
minima, 7000′.
longiflora, 6500′.
Auricula, 6500′.
Soldanella alpina, 2500′.
minima, 2500′.
Cortusa Matthioli, 7000′.
Androsace alpina, 7000′.
obtusifolia, 7000′.
Aretia Vitaliana, 8000′.
Pinguicula alpina, 2000′.
grandiflora, 2000′.
Gentiana acaulis, 1300′.
germanica, 1300′.
utriculosa, 2000′.
cruciata, 3500′.
asclepiadea, 3500′.
ciliata, 3500′.
punctata, 5500′.
bavarica, 5500′.

Gentiana nivalis, 5500′.
pumila, 5500′.
Linaria alpina, 1300′.
Euphrasia tricuspidata, 1300′.
Salisburgensis, 1500′.
Pedicularis tuberosa, 4500′.
rostrata, 6500′.
verticillata, 6500′.
rosea, 6500′.
Bartsia alpina, 6500′.
Pæderota Bonarota, 1500′.
Veronica alpina, 4500′.
aphylla, 4500′.
Horminum pyrenaicum, 1300′.
Betonica Alopecuros, 1300′.
Myosotis nana, 8000′.
Globularia nudicaulis, 1300′.
cordifolia, 1300′.
Daphne striata, 1500′.
Pinus Pumilio, 1400′. Between Agordo and Peron.
Nigritella angustifolia, 4500′.
Himantoglossum viride, 4500′.
Crocus vernus, 2000′.
Czackia Liliastrum, 2500′.
Luzula nivea, 5500′.
Carex atrata, 5500′.
firma, 5500′.

Giacich has enumerated the rare plants of Monte Maggiore, in Istria (Ratisbon Flora, 1844, pp. 274-6). Hauffel gives a sketch of the Carices of Hungary, Croatia, Slavonia, and Siebenbürgen (Id. pp. 527-36). The author here again refers his *C. rhynchocarpa* to *C. brevicollis* Lam., and regards *C. saxatilis* Baumg. as *C. dacica*.

Moritzi has written a new Manual of the Swiss Flora (Die Flora der Schweiz. Zurich, 1844-8). Trog has published a Catalogue of Swiss Fungi (Berner Mittheilungen, pp. 17-92), in which 1121 species are mentioned.

The upper limit at which the larch occurs on the south side of the Mont Blanc chain at Cramont, in Courmayeur, was found by the measurement of Forbes, to be 7200′ Engl.; and on the north side, on the rocks les Echellets which belong to the Mer de Glace, 6800′. (Travels through the Alps of Savoy. Edinb., 1843, pp. 68 and 215.)

The seventh and eighth centuries of F. Schultz's 'Flora Galliæ et Germaniæ exsiccata' have been issued, and are accompanied by critical remarks upon individual plants (see Bot. Zeitung, 1845). By the same author, four French plants are proposed as new in the 'Ratisbon Flora' (1844, pp. 806-9); *Orobanche brachysepala* Sch. according to the description given, and comparison with the original plants, is identical with *O. apiculata* Wallr. Rchb. (Spicil. rum., 2, p. 58); *O. macrosepala* is probably my *O. Bartlingii*, a name which obtained priority by several months.

French local Floras: J. Lloyd, 'Flora de la Loire inférieure' (Nantes, 1844, 12); Guépin, 'Supplément à la Flore de Maine et Loire' (Angers, 1842).

Martins has worked out an exposition of the climatal contrasts which occur within the boundaries of France (les Régions Climatoriales de la France), in the 'Bibliothèque de Genève, 1844, pp. 138-60, and pp. 347-50. The author distinguishes the five following climates in France.

1. *Climate of the Vosges.* This comprises a district in the north-east of France, which is bounded by the cities of Basle, Dijon, Auxerre, and Mezières. Mean temperature = 9° 6 C. The relatively most intense winters predominate in this region, the difference between the mean summer and winter heat amounts to 18° C. (18° 6 and 0° 6 C.); the greatest cold observed in Strasburg and Metz amounted to about 23° C. The mean quantity of rain (from meteorological observations made at Strasburg, Mühlhausen, Nancy, Metz, and Geneva) = 669 mm.; of this, 19 p. c. fall in the winter, 23 in the spring, 31 in the summer, and 27 in the autumn. Average number of rainy days = 137. Predominant winds, those from the south-west and north-east.

2. *Climate of the Seine,* or north-west of France, as far as the Loire and Cher. Mean temperature 10° 9 C. Difference between the mean summer- and winter-temperature = 13° 6 C.; diminishing in the direction from Brussels (=14° 3 C.) to Brest (= 10° 8 C.); the former average value is the arithmetical mean of observations made at Dunkirk, Arras, Abbeville, Paris, Cherbourg, Angers, and Denainvilliers. Average amount of rain = 548mm.; in Finisterre, however, it amounts to 900mm. (from observations made at Paris, Brussels, and Denainvilliers, 21 p. c. of the rain falls in winter, 22 in spring, 30

in summer, and 27 in the autumn. Average number of rainy days = 140. The prevailing wind is the south-west, the next is the north-east.

3. *Climate of Garonne,* or south-west of France, as far as the Pyrenees. The eastern boundary is situated in Auvergne, but cannot at present be accurately defined; it probably includes the plateau of Auvergne, and follows the course of the Rhone and Saône. Mean temperature = 12° 7 C. Difference between summer- and winter-heat = 160° C.; on account of the smaller extent of the coast-line, the marine climate is less developed here than in the north-west; mean summer-temperature = 20° 6 C., winter-temperature = 5° C. Greatest intensity of cold at Poitiers, La Rochelle, Toulouse, and Agen, where it attains to —12° C. Average amount of rain = 586 mm.; of which, 25 p. c. fall in winter, 21 in spring, 23 in summer, and 34 in autumn. Average number of rainy days = 130. The prevailing wind is the south-west, which in the neighbourhood of the Pyrenees passes into the west.

4. *Climate of the Rhone,* comprises the valley of the Rhone, from Dijon and Besançon to Viviers, and the mountainous regions of the Higher Alps; the boundary in the department of the Lower Alps is at present undefined. Mean temperature = 11° C. Difference between summer and winter = 18° 6 C. Mean temperature of summer = 21° 3 C., of winter 2° 5 C. Average quantity of rain = 946 mm., i. e. the greatest amount precipitated throughout the whole of France; of which 20 p. c. falls in the winter, 24 in the spring, 23 in the summer, and 34 in the autumn. Number of rainy days in the valley of the Saône = 120-30, in the valley of the Rhone = 100-15. Prevailing winds, north and south.

5. *Mediterranean climate.* The northern boundary runs through the Rhone at Viviers, near Montélimart, thence follows a line drawn on one side to Montpellier, on the other to Marseilles, and, lastly, comprises the coast-districts of Provence and the regions of Aude as far as

the Pyrenees. Mean temperature = 14° 8 C. Mean summer temperature = 22° 6 C., and winter temperature = 6° 5C. Greatest intensity of cold observed — 11° 5C. Average amount of rain = 651 mm.; of which 25 p. c. fall in the winter, 24 in the spring, 11 in the summer, and 41 in the autumn. Prevailing wind, north-west. —(Mistral.)

A work by Grenier, relating to the botanical conditions of the French Jura, appears of importance; at present, however, I am only acquainted with it from Von Schlechtendal's review (Thèse de Géographie Botanique du Dép. de Doubs, Strasbourg, 1844-8). According to this work, the upper limit of the oak here occurs at an altitude of 6-700 metres, that of the beech at 8-900 metres; above these deciduous trees comes the Coniferous region, covered with both kinds of fir-trees.

Lloyd's Flora of the Mouth of the Loire also notices the local conditions of vegetation. The diffusion of several plants belonging to the south of Europe, along the sea-beach, as far as the 47th degree of latitude, is characteristic: e. g. on the lagunes, *Inula crithmoides*, *Sonchus maritimus*, several *Statices*, *Salicornia fruticosa*, *Scirpus Savii*, *Spartina stricta*; on the downs, *Matthiola sinuata*, *Silene portensis*, *Tribulus terrestris*, *Otanthus maritimus*, *Ephedra distachya*, *Pancratium maritimum*, &c. But on the heaths of Bretagne are also found *Erica ciliaris*, *vagans*, and *scoparia*, *Simethis bicolor* Kth. (*Phalangium* D. C.), *Asphodelus albus*, *Pinguicula Lusitanica*, *Serapias triloba*, in conjunction with northern plants, as *Ulex Europæus*, *Narthecium ossifragum*, *Anagallis tenella*, *Hypericum elodes*, *Myrica Gale*, and *Alisma ranunculoides*.

To this place belong, on the French coast of the Mediterranean, the investigations of Duchartre upon the vegetation of the district around Béziers in the Dép. Hérault (Comptes rendus, 1844, v, 18, pp. 254-9). This work gives an accurate and complete survey of the vegetable formations which occur there. The author divides them into two principal classes, according as their growth is consequent on proximity of the sea or not.

1. The following formations belong to the coast-plants: *a. Formation* of the *Dunes.*—Herbs or low shrubs, which are either of very pubescent or of glaucescent tint. To the former belong, e. g. *Matthiola sinuata, Medicago marina, Orlaya maritima, Mercurialis tomentosa. Diotis candidissima;* to the latter, *Eryngium maritimum, Echinophora, Euphorbia Paralias,* and *Crucianella maritima.* The shrubby plants consist of *Astragalus massiliensis* and *Ephedra distachya.* As regards the number of forms, the Grasses predominate (12 species are known), the Cruciferæ come next, with 4 species, the Leguminosæ and Euphorbiaceæ number 3 species, and the Chenopodeaceæ, Polygonaceæ, and Synantheraceæ: altogether more than 40 species grow there. Among the dunes 2 allied species of *Juncus* are found (*J. acutus* and *maritimus*), a formation peculiar to the humid soil, which in the Landes is denominated Joncasses, and forms the transition to the following formation.

b. Formation of the Salt-water Marshes.—Shrubs and herbaceous perennials, with succulent leaves. Chenopodeæ and Statices predominate here, both as regards the number of individuals and of species: among woody plants, *Tamarix Gallica* is found arborescent. Characteristic forms among the Chenopodeaceæ (11 sp.): *Chenopodium fruticosum, Ch. setigerum, Salicornia,* 3 sp., *Salsola,* 2 sp., *Atriplex,* 3 sp.; of the Statices (5 and more species) *St. oleifolia, bellidifolia,* and *ferulacea;* of other plants (15 sp.), *Frankenia,* 2 sp., *Spergularia,* 2 sp., and *Artemisia Gallica.* The Graminaceæ are here represented by *Crypsis schœnoides* only.

2. The plants independent of the influence of the sea, resolve themselves into formations of a moist and dry soil; the latter are either independent of the cultivation of the land or not so.

A. Water-plants.

a. Fresh-water Formation.—Amongst numerous Graminaceæ, Cyperaceæ and Naiadeæ, with Nymphæaceæ and Typheæ, it possesses but few forms which are characteristic of the climate; as, e. g. *Vallisneria spiralis* and *Marsilea pubescens*, Ten. (*M. Fabri*, Dun.)

b. Formation of those soils which are occasionally overflowed, are characterised, e. g. by *Mentha cervina*. These are the localities of *Cicendia Candollei* and *Conyza sicula*.

B. Plants of the uncultivated part of the district. The author believes that three or four formations may be distinguished, of which the first, that of the Cistaceæ, is more distinct than the rest are from each other.

a. Garrigues, i. e. *Formation of the Cisti*. A stony soil is densely covered with shrubs of *Cistus* or some other firmly interlaced and frequently thorny bush. These shrubby forms are the following: *Cistus crispus, salvifolius, albidus*, and *monspeliensis; Ulex provincialis* and *Europæus; Daphne Gnidium; Quercus coccifera; Erica scoparia* and *cinerea, Calluna vulgaris; Phillyrea angustifolia* and *latifolia; Lavandula Stœchas; Osiris alba; Juniperus Oxycedrus* and *communis*, and *Rosmarinus officinalis*. Among other plants, the following are characteristic: a number of species of *Helianthemum*, growing with the *Cisti*, some *Euphorbias, Santolina, Helichrysum Stœchas, Aphyllanthes*, &c. From this formation also more than 40 species are enumerated.

b. Duchartre has not been able to distinguish the peculiarities of those surfaces and hills (campi) which are covered with shrubs and annuals, and to illustrate them according to their characteristic vegetable forms. We shall, therefore, pass over this part of his memoir, and merely mention some of the more rare species which belong here: *Biscutella coronopifolia, Linum salsoloides, Centaurea Pouzini*, and *Echium Pyrenaicum*.

c. Plants of the cultivated soil.

a. Formation of the ruderal plants. The species are all widely diffused.

b. Plants accompanying those which are cultivated. The author makes several divisions of these, which it is not necessary to detail. The number of species enumerated is very considerable, but they are not characteristic of the south of France, as distinguished from other countries on the Mediterranean.

c. Formation of the meadow-lands. The same remark applies to this: *Euphorbia pilosa* and *Iris spuria*, however, deserve to be mentioned.

d. Formation of the forests. The evergreen forests consist of *Quercus Ilex:* there are no others. Underwood: *Pistacia Lentiscus* and *Terebinthus*, *Erica arborea* and *Calluna*, *Sarothamnus scoparius*, *Cytisus capitatus*, *Genista Scorpius*, *Spartium junceum*, &c.

From the appended sketch of the cultivated plants, it is seen that the preparation of soda from Halophytes has entirely ceased in that district, that the cultivation of the olive is very much on the decrease, in consequence of several cold winters having destroyed the plantations, and that latterly attempts have been made to cultivate *Ricinus* on a large scale. The principal production of Béziers is wine; the cerealia do not suffice for home consumption.

Desmoulins has given a description of his botanical journey in the Pyrenees, during which he made some observations upon the vertical limits of the Alpine flora of the Pic du Midi (Etat de la Végétation sur le Pic du Midi de Bigorre. Bordeaux, 1844, 8vo.) We extract from it the following additions to the earlier statements of De Candolle and Ramond:

Cochlearia pyrenaica, 5500′—6000′.
Herniaria pyrenaica, 3000′—7500′.
Paronychia polygonifolia, 6000′—7500′.
— *serpyllifolia*, 7500′—8400′.
Astragalus depressus, 6000′—7500′.
Vicia pyrenaica, —8500′.
Carduus carlinoides, 6000′—8100′.
— *carlinifolius*, 3000′—6900′.
Cirsium eriophorum, 0′—6600′.
Scabiosa pyrenaica, —8400′.

Pedicularis pyrenaica, —9000′.
Crocus nudiflorus, —7500′.
Anictangium ciliatum, —8400′.
Parmelia chrysoleuca, 5400′—9000′.
— *cartilaginea, elegans, cinerea, badia*, —9000′.
Lecidea vesicularis biformis, 6000′—7500′.
— *polycarpa, atrobrunnea, morio, geographica, umbilicata*, —9000′.
Umbilicaria cylindrica, 6000′—9000′.

Some interesting letters, written by M. Willkomm during a journey in Spain, have been published in the 'Botanische Zeitung' (1844-5). They commence in May 1844, bearing date from Valencia, where the author remained until the middle of June. He then went to Madrid, botanised at Aranjuez in the beginning of July, passed over the Sierra Morena, reached Granada, and during the latter part of the summer and the autumn, explored the Sierra Nevada and the Alpuxarras. In the present report, we shall confine ourselves to the first part of the journey, intending to recur to the notices regarding the south of Spain next year, when the conclusion of Boissier's illustrated work, together with Willkomm's observations of 1845, will conjointly furnish a more copious illustration of that part of the subject. In the Huerta of Valencia, the original vegetation is greatly displaced by cultivation: wheat, rice, and hemp are principally cultivated; mulberry trees, olives, and fruits belonging to the south are common, date-palms, from 40 to 60 feet high, are frequently met with. On the Lagune Albufera is a wood of *Pinus Halepensis*, containing abundance of the original plants of this district: the underwood here consists of *Quercus coccifera*, *Myrtus*, and *Chamærops*, and growing with them we find *Pistacia Lentiscus*, *Rhamnus lycioides*, *Erica arborea*, *Rosmarinus*, *Juniperus Oxycedrus*, and *Ruscus aculeatus*. The adjacent sandy hills contain *Cistus albidus* and *salvifolius*; *Passerina hirsuta* and *Solanum sodomeum*, the stems of which are of the thickness of the arm.

The Sierra de Chiva, 12 miles north of Valencia, belongs to the limestone mountains, which separating from the

Spanish plateau, between the Ebro and Xucar, traverse the province from west to east as far as the sea. This broad mountain-range, which is about 6000′ in height, and intersected with deep Barrancos, was once covered with forests of Coniferæ, the only remains of which at the present time are isolated stems of *Pinus Halepensis*. The dry slopes, which are almost entirely free from springs, are now overgrown with a low bush (Montebaxo), the extreme summits only being bare. Willkomm admits the following stages in the Mediterranean vegetation of this region, which attains an unusual elevation, ascending to 4000′.

0—500′. To about this height the Opuntias and Agaves extend, together with the culture of *Ceratonia*. The Montebaxo consists of *Chamærops*, *Erica arborea*, *Daphne Gnidium*, *Retama sphærocarpa*, *Ulex*, *Rosmarinus*, and some oaks.

500′—2000′, i. e. as far as the upper limits of *Chamærops* (also of *Retama*, *Juniperus Oxycedrus*, and *Pistacia Lentiscus*). *Rosmarinus* and *Chamærops* predominate; in addition to those already mentioned, *Erica arborea* from among those of the first stage, and *Rhamnus lycioides*, *Pistachia Terebinthus*, and some *Cisti* are here first met with. Characteristic Grasses: *Macrochloa tenacissima* and *Stipa juncea*.

2000′—4000′ up to the limits of the cultivation of the olive and wheat. The greater part, however, of the slopes at this level consists of uncultivated mountain-land. In a Montebaxo, the principal plants associated here with *Rhamnus*, *Rosmarinus*, *Erica*, and *Cisti*, are *Juniperus Phœnicea*, *Fraxinus* sp., *Arbutus unedo*, and *Quercus Ilex*.

Isolated pine-trees and a Montebaxo formed of *Ulex Australis* and *Juniperus Phœnicea* characterise the region extending from 4000′—5500′, which may be distinguished from the Mediterranean by the occurrence of the plants of the north of Europe. On the summit of the Monte de la S. Maria (5500′—6000′), of woody plants, *Arctostaphylos uva ursi*, *Taxus*, and some Cotoneasters are also found; and with them few shrubs only and a single species of Saxifrage.

V. Martens, in a general work, has described the botanical geography of Italy from literary sources (Italy, Stuttgard, 1844, 8vo, 3 vols.)

Works upon the Flora of Italy. The first two parts of the sixth vol. of Bertolini's Flora Italica, which treat of the 14th class, have appeared (Bologna, 8vo.) The Flora of Nice, by A. Risso (Nice, 1844, 8vo), is of no scientific value. We have not yet received Cesati's paper upon that of Lombardy (Saggio sulla Geographica botanica e sulla Flora della Lombardia. Milano, 1844, 8vo, p. 74). Purcinelli Additamentum ad synopsin plantarum in agro Luccensi sponte nascentium (in the Giornale Botanico Italiano. 1844, pp. 118-123). Savi Florula Gorgonica (id. pp. 243-283), a catalogue enumerating 290 sp. of vascular plants observed in Gorgona, a small island opposite Leghorn, and covered with *Cisti*, *Ericas*, and Leguminous shrubs, may be considered as a companion to the Flora of Capraja, published some years ago by Moris and Notaris. De Notaris, Appendix to his Specimen Algologiæ Ligusticæ (id. pp. 191, 311). Meneghini, Algarum species novæ vel minus notæ (id. pp. 296-306), 33 species from the coasts of Italy and Dalmatia. The fourth part of Alghe Italiane e Dalmatiche, by the same author, has appeared (Padova, 1843, 8vo). Tenore has shown that the Dalmatian *Arenaria Arduini* is identical with his former *A. Rosani* (Rendic. Acad. 1842, p. 266). V. Heldreich describes four new Sicilian plants (Ratisbon Flora, 1844, p. 65): 1 *Helianthemum*, 1 *Elichrysum*, 1 *Centaurea*, and 1 *Lithospermum*. Nyman's Observationes in Floram Siculam (Linnæa, 1844, pp. 625-665) contain a catalogue of his collection which is for sale in Sweden, with descriptive remarks. The only new plant is *Parietaria populifolia*, N. from Malta.

Link distinguishes a new *Erica anthura*, obtained from Spalatro (Sitz. des Ges. naturf. Freunde, 1844, in the Ratisbon Flora, 1845). Visiani raises *Turinea Neu-*

mayerina Vis., which was figured, in the Flora Dalmatica, into a separate genus as *Amphoricarpos* (Giorn. Bot. It., vol. i, p. 196).

Ebel's essay on Montenegro (Zwölf Tage auf Montenegro, Hft. 2. Königsberg, 1844, 8vo), contains a catalogue of all the Phanerogamous plants hitherto observed in Dalmatia (2003 sp.), with a statement of the frequency of their occurrence, expressed in a manner peculiar to the author, but the localities are not given. It contains preliminary observations upon the statistical relations of the flora of Dalmatia, in which the most abundant families form the following series, according to the number of species contained in them: Synantheraceæ (225 sp.), Leguminosæ (220 sp.), Graminaceæ (142 sp.), Cruciferæ (107 sp.), Umbelliferæ (103 sp.), Labiatæ (91 sp.), Caryophyllaceæ (85 sp.), Scrophulariaceæ (82 sp.), Liliaceæ (61 sp.), Rosaceæ (59 sp.), Ranunculaceæ (54 sp.), Orchidaceæ (46 sp.), Cyperaceæ (43 sp.), Boraginaceæ (42 sp.) The reports upon the vegetation of Montenegro itself, the productions of which, according to the author, entirely agree with those of Dalmatia, belong here. This small tract of land, which is covered with arid, rocky mountain-pastures, and elevated into limestone summits, which are either barren or slightly surrounded with forests of pines, and from which narrow fluviatile valleys descend to the sea of Scutari, is extremely unfruitful from a deficiency of soil and water. Nevertheless, the plants appear, as in Dalmatia, to be various, 450 sp. having already been mentioned by the author: there are no new ones among them, the two which are proposed as new are untenable.

In my work upon Rumelia and Bithynia (Spicilegium Floræ Rumelicæ et Bithynicæ, exhibens synopsin plantarum, quas anno 1839 legi: accedunt species, quas in iisdem terris lectas communicarunt Friedrichsthal, Friwaldzki, Pestalozza vel plene descriptas reliquerunt Buxbaum, Forskal, Sibthorp, alii; vols. i, ii, Brunsnigæ, 1843-4, 8vo), 2300 Phanerogamous plants are treated of systematically, and in regard to their geographical distribution. The families containing most species form the following series: Synantheraceæ (264 sp.), Leguminosæ (203 sp.), Graminaceæ (156 sp.), Labiatæ (134 sp.), Caryophyllaceæ (130 sp.), Cruciferæ (121 sp.), Umbelliferæ (114 sp.), Scrophulariaceæ (90 sp.), Ranunculaceæ (78 sp.), Rosaceæ (68 sp.), Boraginaceæ (55 sp.), Liliaceæ (53 sp.), Rubiaceæ (48 sp.), Campanulaceæ (41 sp.), Orchidaceæ (41 sp.), Cyperaceæ (41 sp.) When this series is compared with that given above for Dalmatia, the increase in the Labiatæ and Caryophyllaceæ becomes one of the characteristic peculiarities of Rumelia. The former family does not reach the centre of its distribution through the south of Europe until we arrive at Greece; but the Silenaceæ, which abound in endemic forms of *Dianthus* and *Silene*, do not appear to be anywhere more numerous than in Rumelia. The increase in the Ranunculaceæ, Boraginaceæ, and Campanulaceæ is also worthy of consideration; but I must confine the deductions to these few facts, since if carried further than the extent of our present knowledge admits, they would lose in truth. The extent of our knowledge of the flora of Rumelia is much better shown by the examination of those vegetable forms which are endemic to that country, than by sketches of the entire vegetation, in which so many constituents are still wanting. Of these 2300 species of plants, about the seventh part are peculiar to the peninsula of Europe: from these about 80, which have only been found in Bithynia, are excluded, a great part of which, however, will probably be found also on this side of the Bosphorus. Moreover, if we take into account the distribution of Greek plants over the south, and of Dalmatian over the west of Rumelia, we may consider more than two thirds of the endemic plants of the south-east of Europe as known. *Summary of the endemic plants of Rumelia:* 23 Leguminosæ, principally species of *Trifolium* (5), and *Astragalus* (9), mostly belonging to the evergreen region; 5 Rosaceæ, of these, 3 Dryadeæ to the mountainous region; 2 Rutaceæ (*Haplophyllum*); 4 Euphorbias, of which

2 belong to the alpine region; 2 Geraniaceæ to the alpine region; 25 Caryophyllaceæ, especially species of *Silene* (6), and *Dianthus* (10), only 5 Alsineæ: species from all three regions, but the pinks mostly indigenous to the central European and alpine; 5 Hypericineæ (*Hypericum*) from the evergreen region; 14 Cruciferæ, one half of which consist of alpine species of *Arabis, Cardamine, Koniga, Thlaspi,* and *Eunomia;* 15 Ranunculaceæ, with 7 species of *Ranunculus,* mostly from the evergreen region; 2 Crassulaceæ; 3 Saxifrages from the alpine region; 21 Umbelliferæ, increasing towards the coast; 2 Ericaceæ: *Erica verticillata* and *Arbutus Andrachnus;* 3 Primulaceæ; 26 Scrophulariaceæ, principally alpine *Pediculares* (3), species of *Veronica* (4), *Digitalis* (3), *Scrophularia* (4), and *Verbasca* of the evergreen region (8); 2 *Orobanches;* 9 Boraginaceæ, among these 4 species of *Alkanna,* 2 of *Borago;* 20 Labiatæ, of these 6 species of *Stachys* in both the lower regions; 9 Rubiaceæ in the evergreen and alpine region (instead of the term *Galium trichophorum,* which was elsewhere proposed at the same time, I prefer that of *G. Trichodes*); 2 Valerianaceæ; 9 Dipsaceæ; 40 Synantheraceæ, principally Anthemideæ and Cynareæ, mostly from the genera *Anthemis* (6: mostly in the evergreen region), *Achillea* (5: mostly in the alpine region), *Senecio* (4), *Centaurea* (5), *Cirsium* (5); 13 Campanulaceæ, of which 10 were *Campanulæ,* most of which belonged to the evergreen region; 2 Amentaceæ: *Quercus Ægilops* and *infectoria;* 3 Coniferæ: *Pinus maritima* in the lower, *Juniperus sabinioides* in the middle, and *Pinus Peuce* on the boundary of the alpine region; 3 Orchidaceæ; 4 Iridaceæ, species of *Crocus* in the evergreen region; 12 Liliaceæ, e. g. *Ornithogalum* (3); 2 Cyperaceæ; 11 Graminaceæ from all three regions. The remaining endemic plants are as yet single members of their families: the Bithynian, &c. Of the Cryptogamia, we are not yet acquainted with 200 species.

Heldreich observed at Athens a form of *Arbutus,* which was probably *A. hybrida* Ker., but is regarded by him as a distinct species, intermediate between *A. Unedo* and *Andrachne* (Ratisbon Flora, 1844, p. 13). He denies its hybrid origin, because *A. Unedo* flowers in October and November, *A. Andrachne* in February and March; I have, however, met with both plants in flower at the same time in Bythynia.

II.—ASIA.

Among the endemic plants of Bithynia described in the 'Spicileg. Rumelic.,' part of which belong to the evergreen coast-region, part to the high mountains of Olympus and Bolu, the following are the principal families represented: 5 Leguminosæ (mostly *Trifolia*); 2 Geraniaceæ; 5 Caryophyllaceæ (consisting of 3 Sileneæ and 2 *Dianthi*, all from Olympus); 4 *Hyperica*, 9 Cruciferæ (all from Olympus, and consisting of 3 sp. of *Arabis*, 2 sp. *Eunomia*, &c.); 3 Papaveraceæ; 2 Ranunculaceæ; 5 Umbelliferæ (mostly from Olympus); 4 Scrophulariaceæ; 2 Boraginaceæ; 3 Labiatæ; 3 Rubiaceæ; 13 Synantheraceæ; 4 Campanulaceæ; 3 Liliaceæ; 3 Graminaceæ, &c.

The oriental Umbelliferæ have been worked out by Boissier (Ann. Sc. Nat. 1844), comprising 300 species. The number of species proposed as new is very large. The new genera distinguished are the following: *Lereschia* (*Cryptotænia Thomasii* D. C.); *Elwendia*, from Persia, near *Carum; Microsciadium* (*Cyminum minutum* Urv.); *Muretia* (*Bunium* sect. *Chryseis* D. C.); *Diplotænia* from Persia, near *Peucedanum; Stenotænia* from the same place, near *Pastinaca; Ducrosia* (*Zozimiæ*, sp. D. C.); *Ainsworthia* (*Hasselquistia cordata* L.); *Trigonosciadium* from Mesopotamia, near *Heracleum; Synelcosciadium* (*Heracl. Carmeli* Lab.); *Polylophium*, (Thapsieæ) from Persia; *Smyrniopsis* near *Smyrnium; Meliocarpus* near *Prangos; Turgeniopsis* (*Turgenia fœniculacea* Fzl.); *Lisæa*

(*Turgeniæ*, sp. D. C.); *Rhabdosciadium*, one of the Scandicineæ from Persia; *Thecocarpus*, from the same place and from the same division; *Osmosciadium*, one of the Coriandreæ, from Cappadocia.

C. Koch's travels to the Caucasus (Reise durch Russland nach dem Kaukasischen Isthmus in den J. 1836-8. Bd. i, ii. Stuttgard, 1842-3) contains reports upon the autumnal vegetation of Ossetia and Imiretia, as also upon the vernal flora of Russian Armenia; the author's investigations were subsequently interrupted by protracted illness, but he finally resumed them in a second journey. In the military road of the Caucasus, Koch represents the prairies of Kabarda, near Uruch, as very luxuriant and abundant in plants; herbs and the Grasses grow here in such luxuriance, that a man can readily conceal himself without lying down (i, p. 250). The Graminaceæ are mostly the same as the meadow-grasses of central Europe, whilst among the shrubs many Caucasian species are met with; they are diffused by the rivers over these surfaces which are situated opposite to the high mountain chain. By this circumstance and the development of the vegetation in the height of summer, when the Russian heaths are burnt up, the meadows of Kabarda differ essentially from the steppes, with which C. Koch has compared them. In fact, judging from certain kinds of plants, the steppe climate still prevails here; this is shown by the Artemisiæ, Cynaraceæ, and Astragalaceæ; but the influence of the neighbouring mountains modifies the character of the vegetation as determined by the climate. The plants of the steppes are destroyed in the summer by the drought, whilst Kabarda is well watered from the Caucasus.

C. Koch remained during October in Ossetia, in the middle of the high Caucasus, and the offsets connecting it towards the south with the Armenian highlands, and then travelled in Imiretia until the end of the year, certainly too late a period to allow of the botanical character of the country being completely ascertained. The reports are partly limited to lists of the localities of the autumnal plants which he was then able to collect. The alpine flora, even at elevations of 7—8000′ was but slightly represented by its characteristic forms (ii, p. 69); these high mountains are altogether more sterile than the Alps, which the author attributes primarily to the rarity of glaciers in the Caucasus, as if the alpine meadows of the Tyrol were only fertilized by melting ice. He then goes so far as to assert (p. 91), that the disintegrated soil of Ossetia, the steep rocks and precipitous defiles of this alpine district, are not adapted to the production of humus, and that this is the cause of the total absence of a luxuriant vegetation there. But the author is not clear upon this point, and does not separate general from local conditions; for he speaks at the same time of clay-slate plateaux, but little supplied with water and destitute of woods, extending between the defiles and valleys to the ridge and lateral offsets of the Caucasus. Upon this form of mountain and peculiarity of soil the alpine poverty of Ossetia appears to depend; that it also prevails over the well-wooded slopes of the northern Caucasus is not probable. But Ossetia shares this deficiency of alpine vegetable forms with the mountains of the south of Europe, where alpine pastures abounding in species are but rarely developed, and where this phenomenon is occasioned by the deficiency of water upon narrow crests and summits. Ossetia does not possess the fine forests of the northern promontories of the Caucasus. Even in the true forest region there is a perceptible deficiency of wood, and frequently the soil is scarcely covered with a scanty underwood: e. g. at Zrchinwall (p. 55) consisting of *Corylus, Cornus mascula, Paliurus, Cratægus, Prunus insititia*, and *Juniperus*. The traveller only met with wooded slopes at Dschedschora, in the district of Gudaro, (p. 82); here deciduous trees predominated, and the Coniferæ present were *Pinus abies, picea*, and *orientalis, Taxus*, and *Juniperus communis*. The deciduous forests consisted of the oak, beech, maple, lime, and alder (*Quercus iberica* Stev. and *Robur* (?), *Carpinus orientalis, Fagus, Acer platanoides, Tilia parvifolia, Alnus denticulata* C. A. M.); the underwood of *Euonymus latifolius, Rhamnus frangula* and *cathartica, Staphylea pinnata, Viburnum orientale, Argyrolobium lotoides,* and *Lonicera cærulea.*

The Imiretian slopes of the Caucasus in the upper valley of Rion (p. 129) are more abundantly wooded; above the vine mountains of the latter, mixed forests of deciduous trees ascend to a considerable height; in addition to the trees above mentioned, the chesnut, various fruit trees, and poplars were found at Oni, as also among the shrubs, *Ilex, Azalea pontica,* and *Rhododendron Caucasicum, Rhus Cotinus,* together with *Smilax excelsa.* At Glola, wild fruit-trees, especially *Pyrus communis,* and *Prunus avium,* extended to beyond 5000′. A region of subalpine shrubs, of which *Arctostaphylos* and *Azalea pontica* ascend together high up at the glacier of Rion, immediately succeeds the deciduous forests; and with them, subalpine herbaceous perennials, as *Aconitum nasutum* Fisch., *Pyrethrum macrophyllum, Doronicum Caucasicum,* &c. Lower down in the valley of Rion, Koch describes a fine primitive forest at Kutais (p. 166), consisting of magnificent trunks of *Carpinus orientalis,* oaks, and high tops of the chesnut and plane-trees projected singly; in thickets, luxuriant lianes of the grape-vine, *Smilax,* and ivy, upon the branches of which the mistletoe grew, and from which *Usneæ* were suspended.

The journey from Tiflis to Eriwan through Georgia and Russian Armenia was made during the months of April and May, in 1837, and yielded a rich booty. The forests of Somchetien differ from those of Imiretia, in the more regular growth of the trees, and in the absence of evergreen shrubs and lianes (p. 350). They consist of *Quercus iberica* and *pedunculata, Carpinus Betulus* and *orientalis, Acer platanoides* and *pseudoplatanus;* with isolated examples of *Ulmus excelsa* Bork., *Fagus,* and *Acer tartaricum.* The soil of these forests consists of a thick layer of humus, and this black earth produces the beautiful mountain pastures which alternate with the former. The traveller soon ascended the high mountain chain between Kur and Araxes, forming the boundary between Georgia and Armenia (this he calls the lower Caucasus, Güldenstedt and Klaproth call it the promontory of Ararat), which, according to Parrot's measurement, ascends to an elevation of 12,780′. But vegetation was still backward in this region, for even in the Armenian highlands, few of the herbaceous plants and Grasses which, with a thorny underwood of Tragacanth, cover the bare heights, were in flower (p. 386). However, on going from Alagäs, near the valley of Araxes, to Eriwan, Koch was amply compensated by the banks of the Kasach (p. 397). The climate is so dry, that even in May the soil is parched and barren, whilst in the more elevated regions vegetation has scarcely commenced. But by artificial irrigation, the cultivation of the soil may be effected even during the hot and dry months, the fields and orchards surrounding the villages then resemble oases in a stony desert. Fruit trees were planted generally, especially peach trees and apricots; besides these there was a natural arboreal vegetation along the Kasach valley, consisting of *Elæagnus* and *Populus*, with *Prunus incana*, and *Tamarix*. In Eriwan, especially, the greatest attention is paid to the cultivation of fruits and the vine, and the traveller had never met with more beautiful gardens than he saw there.

Schrenck continued his travels in Soongarei in 1843, and has already made known the plants he found in that year (Bulletin Pétersbourg, iii, pp. 106-10, 209-12, 305-9). They belong to the following genera: *Ranunculus* (2 sp.), *Stubendorfia*, nov. gen. Crucifer., *Isatis*, *Geranium*, *Zygophyllum*, *Haplophyllum*, *Euphorbia* (2 sp.), *Sophora*, *Oxytropis*, *Astragalus*, *Seseli*, *Lomatopodium*, nov. gen. Umbelliferæ, *Carum*, *Artemisia* (2 sp.), *Chamægeron* near *Henricea*, *Saussurea*, *Cousinea* (4 sp.), *Plagiobasis*, nov. gen. near the preceding, *Jurinea*, *Serratula*, *Echinops*, *Echinospermum*, *Eremostachys*, *Arthrophytum*, nov. gen.

Chenopod., *Pterococcus*, *Statice*, *Populus*, *Ephedra*, *Allium*, *Typha*, and *Triticum* (2 sp.)

Middendorf has commenced the arrangement of the results of his journey through the north of Siberia, which was mentioned in the preceding annual report (Bulletin Pétersbourg, iii, pp. 150 et seq.) The Tundres of the Taimyr country, i. e. the peninsula situated between the lower Jenisei and the Katanga, contain in their diluvial loam, in addition to the mammalia of the diluvium, large masses of wood either in a bituminous state, such as is found in the peat moors, or converted into peat. In such of these tracts, however, as were beyond the tree limit, the stems were only met with lying horizontally, and were compared by Middendorf to the floating timber of the arctic coast, and from which, by the rising of the land, they may have gradually attained the interior. The trees appear to be of the same kinds as those in the forests of New Siberia and the fluviatile valleys of Siberia, consisting principally of the beech and larch; they have not yet, however, been examined microscopically, hence these statements require confirmation. The climate of the Taimyr country appeared to be less cold than might have been expected: from the 6th of June to the 8th of August there was no frost there; constant fogs and storms (especially in summer, so that in May, June, and half of July, the altitude of the sun could only be taken three times,) indicated great irregularities in the distribution of heat in the atmosphere. The high surface of the country, which rises to an elevation of 1000′, was perfectly free from snow in the summer; even in the winter, storms sweep the snow into the lowest parts, frequently leaving the heights bare. In the middle of July, Middendorf saw at Taimyr 1500 square miles (Eng.) free from snow, in a few narrow valleys only was any still remaining. The lakes only freeze to a depth of eight feet; the layer of snow then protects them from any deeper penetration of the frost. As regards the botanical results, we must wait for further reports, on account of the want of accurate determinations of the plants.

The thermometric observations made by Stchoukine, from 1830-1844 at Irkutsk (1330 English feet above the level of the sea), give (the months being reckoned as lunar) the following average of temperature:

January . .	— 19°·9 c.	July . .	18°·5 c.
February . .	— 13°·6 c.	August . .	+ 13°·75 c.
March . .	— 3°·25 c.	September .	+ 6°·75 c.
April . .	+ 5°·75 c.	October . .	— 3°·75 c.
May . .	+ 12°·25 c.	November. .	— 14°·25 c.
June . .	+ 17°·6 c.	December. .	— 19°·9 c.

Mean temp. = + 0°·01 c.
Maximum = + 35° (once in 1843, 39°·5 c.)
Minimum = — 35°

Turczaninow's "Flora of the Baikal Regions" (see the Annual Report for 1842) has been continued, and has now reached the end of the Umbelliferæ (Bulletin de la Soc. de Moscou, 1843-44). The following families have thus far been treated of: 3 Rhamneæ, 94 Leguminosæ, 69 Rosaceæ, 5 Onagrariæ, 6 Halorageæ, 1 of the Ceratophylleæ, 1 of the Lythrariæ, 2 Tamariscineæ, 1 of the Portulaceæ, 8 Crassulaceæ, 1 *Nitraria*, 9 Grossularieæ, 19 Saxifrageæ, 48 Umbelliferæ, with the recently separated genera *Physolophium* (*Angelica saxatilis* Turcz.) and *Czernævia* (*Conioselinum Czernævia* F. M.) Altogether 542 Polypetalous plants have now been fully treated of.

Kittlitz's work contains some very interesting illustrations of the characters of the vegetation of Kamtschatka; his botanical sketches of the countries, which were made during the well-known voyage of the younger Mertens round the world, and described in the text with a perfect comprehension of the physiognomical characteristics, form one of the most valuable contributions to botanical geography made during the past year (Vierundzwanzig Vegetations-Ansichten von Küstenländern und Inseln des stillen Oceans, aufgenommen in den Jahren 1827-9 durch F. H. v. Kittlitz. Siegen und Wiesbaden, 1844-5, 4to). As we cannot omit making a full report upon this work, we shall preserve as far as possible the excellent language of the text which accompanies the copper-plates; they afford a sample of the author's power of observation.

The physiognomy of Central and Northern Europe agrees with that of Kamtschatka much more completely than we should anticipate, considering the great difference between their longitudes: the number of European plants is very considerable (p. 53). The peninsula is divided into an eastern and western half by its mountain-chain. In the former, rise the conical volcanic mountains, of which the Kliutschewsk, according to Erman, is 14,800′ in height, or as Kittlitz expresses himself, they rival the Peak of Teneriffe in height, and excel all other volcanoes in the perfection of their conical form. They alternate with long mountain-chains whose rugged tops are covered with snow, whilst the remainder of the district is adorned with the growth of noble forests and pasture. On the west side, however, the coast is low and marshy, passing towards the interior into a broad plain of fertile land, the soil of which is watered by numerous streams, and is covered partly with woods, partly with luxuriant grassy plains in their original and natural state. For the purpose of carrying out this sketch completely in detail, the author has given five tables, which indicate the botanical character of the forests and grassy plains in the summer months (July to September).

Grass Plain at Awatscha, therefore in the neighbourhood of Peter-Paul's harbour (plate XVII). This picture represents a luxuriant woody prairie, abounding in plants, and containing scattered groups of shrubs, and the open surface of which is inclosed by a wood of birch (*Betula Ermani*). This birch is the principal forest tree of the country; it somewhat resembles the oak in the knotty and flexuous growth of its stem, and differs moreover from *Betula alba* in its bark, which is gray and much torn, whilst the leaves agree with those of the common birch. A thicket of alders and willows denotes the vicinity of the stream; some of these are shrubby, others tall in growth, resembling that of the poplar, and with these woody plants the gregarious *Spiræa Kamtschatika* (Schalameynik), a plant which throughout the summer characterises Kamtschatka above all other countries, and here repeats the Panax-form of the north-west of America in a physiognomical point of view: "A plant of wonderfully rapid growth, which in a few weeks acquires a height of more than 10 feet, whilst in the autumn it disappears still more quickly, without leaving a trace, for a single frosty night is sufficient to cut it off to the ground." Above the large, crenate leaves, the stems display in July their white bunches of flowers, which subsequently acquire a gray tint. Single plants of a very tall *Heracleum* (*H. Panaces?*) grow among the Spiræas, from the juice of which the natives prepare sugar. The grass covering these prairies attains an astonishing height; at first, indeed, it is overshadowed by shrubs of *Cratægus* and *Salix*, with thick stems, which project here and there, but these at a later period scarcely extend above the rapidly developed culms of the grass. The same applies to the herbaceous perennials, which are mixed in large numbers with the Grasses, and of which the following are mentioned: 2 *Sanguisorbæ*, *Angelica*, *Epilobium angustifolium*, *Senecio cannabifolius*, *Cacalia hastata*, 2 lilies with large orange flowers (one with stems of the height of a man, probably *L. Kamtschatkense* Lour.), and *Fritillaria Kamtschatkensis*, the latter under the name of Sarannah. Of these, *Senecio* and *Epilobium* are the principal ones which contribute to the physiognomy of the land. The former, although as high as a man, is laden with flowers, and frequently colours the surface of the meadows of a pure yellow colour, whilst the latter produces a splendid red. The Sarannah, which is everywhere met with in short grass, yields in its tubers an excellent article of food, which, although difficult to dig up, often supplies the place of bread.

Plate XVIII leads us to the *Forest on the Upper Kamtschatka river*, which, lying in a valley running longitudinally towards the east, traverses plains that are extensive towards the north, and almost everywhere wooded. Here, but here only, a different kind of birch constitutes the predominant forest tree, which the author regards as one of the European species, and denominates *Betula alba* (*B. pubescens* of Erman). It is so distinctly separated geographically in the neighbourhood of the river from *B. Ermani*, that on the road from Ganal to Puschtschina, whilst from the coast to this place the latter only is met with, the white birch suddenly begins to form the forests as soon as the upper course of the Kamtschatka river is reached. Together with the birch, we here find drawn a group of tall balsam poplars as straight as a line; this tree by itself forms large woods in the middle of Kamtschatka. The underwood and shrubs consist principally of *Spiræa*, next of *Lonicera*, *Cratægus*, *Prunus*, and *Salix*. In the glades, in the midst of scanty grass, a dark blue *Iris* grows; it is everywhere common, forming an incomparable ornament to the country, and is succeeded at a later period by several Synantheraceous perennials with beautiful flowers, as *Aster*, *Achillea*, and *Sonchus Sibiricus*.

Forests of Central Kamtschatka (pl. XIX, XX). A strip of land extends across the middle of the peninsula, from the west towards Cape Kronotzkoi; it is wooded with Coniferæ, no trace of which exists in the other districts. The forests consist of two kinds of fir-trees, the larger of which resembles the Canadian larch, the other has the growth of our red pine, with which it is probably identical: here also the birch and the aspen are associated with them. As the pine-forests of Kamtschatka differ from those of the north-west of America in their dryness, so also the underwood merely consists of a thicket, 3 feet high, composed of Roses and *Loniceræ*, and beneath them again a large number of bacciferous plants are concealed, *Vaccinia*, *Rubi*, and *Empetrum*, exactly as occurs under similar conditions in Scandinavia, so much so, that even the species mentioned of these genera are identical with them. Among the edible fruits, *Rubus arcticus* has the most agreeable taste; the elongated dark blue berries of a *Lonicera* come next, their taste is not inferior to that of the finest cherries, and they are prepared with milk or Sarannah to form a favorite article of food with the natives. The Kamtschatka river is constantly changing the course of its valley, and hence, like the rivers of Russia, its banks are steep (Jar) on the side excavated by the current, whilst sandbanks (Pessok) are deposited by the water on the opposite side. On the former, the old pine-forest extends down to the river, and by the falling in of the banks is carried away as floating timber; on the latter, different woody plants have settled, the period of formation of which is later than that of the former: first, thickets of willows, then larger deciduous trees, willows, alders, and poplars appear to follow. The difference in the foliage commonly expresses a difference not of the age of the trees, but of the period at which the district became wooded.

Mountain-Forests of the Eastern Coast (pl. XXI), extending over its steep declivities. These forests, which are also composed of *Betula Ermani*, and sometimes contain tall trees of *Salix*, appear far lighter than those in the fluviatile valleys; but the thickets of underwood and shrubs extending between the trees are proportionately thicker, and contain a larger number of plants. This character is evident, even at a level of 500′, and extends high up the mountain. But at a greater elevation, the birch trees gradually diminish in number, preserving the same state of growth, until at last they disappear, and give place entirely to the shrubs, just as the latter are displaced by the alpine flora, according to the same law. These thickets of shrubs are in general impenetrable to man, and represent the pine-region in Kamtschatka. They consist of *Pyrus sambucifolia* Cham., *Alnus incana*, and a pine which is probably a variety of Pinus Cembra, and is called Kedrownik. The former of these shrubs predominates in the lower regions, and disappears at an elevation of 1000′. The Kedrownik grows even in the vicinity of the coast, but appears to be most widely diffused between 1000′ and 2000′. Its nuts are nutritious, and are eaten, as is also the fruit of *Pyrus sambucifolia.* The most extensive thickets consist of the northern alder, which also grows in the lower regions, in common with two others, but between 2000′ and 3000′ exists alone, limited by the alpine flora, bare stones, and eternal snow. On all the high mountains of Kamtschatka there exists a region in which it exclusively covers the soil. Its upper limit had been previously determined correctly by Erman to be 2890′, i. e. more than 2000′ beneath the snow-line (5000′); but *Salix arctica* (4974′), *Parrya Ermani*, and *Saxifragi Merkii* ascend as high as the latter.

Grass Plain in the West of Kamtschatka, on the Bolschaja-Reka (pl. XXII). The south-western slope of the peninsula is comparatively poor in pictorial beauty and botanical variety; the forest-growth is less than in the east, the morasses are more extensive, and bushes of willow predominate almost everywhere with the peat-moors. The landscape, which was taken in September, is remarkable from the astonishing height of two withering Umbelliferous plants, which give a most peculiar character to the grass plains of the west. They probably belong to the genera *Angelica* and *Heracleum;* their strong stems appear more than fifteen feet high; thus, growing in numbers, they project far beyond the Grasses and other herbaceous plants. We next have a tall gregarious *Urtica*, 10′ high, and from which the natives prepare a valuable yarn. The remaining plants agree generally with those of the grass plains in the east.

The Algæ of Kamtschatka are described and figured in the splendid illustrated work of Postels and Ruprecht; they were also collected in the expedition of the younger Mertens (Illustrationes algarum in itinere nauarchi Lütke collectarum. Petropoli, 1840, fol.)

Zuccarini has published a very valuable sketch of the Flora of Japan (Notizen über die Flora von Japan und die bisher hierüber vorliegenden wissenschaftlichen Leistungen: in the Münchener gelehrte Anzeigen für 1841 and 1844, id. pp. 430 et seq.)

It must first be remarked, in regard to the notice given in the Annual Report for 1843, of the progress of the author's Flora Japonica, that this work has indeed experienced an interruption, but that by the completion of the part which treats of the *Coniferæ*, the number of this order found in Japan has been increased, far beyond that previously given, i. e. to 30, which are distributed through 14 genera. Zuccarini's present work contains a catalogue of all the genera as yet known in Japan, with the number of species in each family. The latter amount in all to about 1650 species; but as Zuccarini estimates the number of the Japanese plants contained in the herbaria of the Netherlands at 2400 species, the statistics must ultimately be altered in proportion as the still remaining families in V. Siebold's work are worked out. With the proportional numbers of the genera and families this will not be so much the case; hence Zuccarini's sketch acquires a permanent value. He enumerates the following as the most remarkable general results of his investigation: 1. The large number of families of plants represented in Japan, of which, according to Endlicher's system, there are 172. 2. The large number of genera in proportion to the species, for 621 are already mentioned in the catalogue, and probably 700 are contained in the herbaria (it must, however, be remarked, that Zuccarini has included the Chinese genera found in Beechey's voyage, as also those from the Bonin Islands). 3. The limitation of endemic genera to a single species, corresponding to the monotypes of the Canary Isles; a condition which applies to the greater part of the new genera from Japan, whilst the remainder contain at present only two, or, at the most, four or five species, and some monotypes also of North America and India, and the European Humulus, possess in Japan a second, but only a second species. 4. The very large number of woody plants in so high a latitude, both from woody families belonging to the temperate as well as the tropic zone, from the latter of which representatives of the Palms, Pandaneæ, Lauraceæ, Ternstrœmiaceæ, &c., together with numerous bamboos, are here in part diffused further towards the north than in other meridians of the northern hemisphere. 5. The endemic character of the flora of Japan, which is not connected, like Siberia, with that of Europe, having but very few species in common with Europe. We have not space, unfortunately, to enter more minutely upon the consideration of the components of the catalogue of the genera; we shall, therefore, merely confine ourselves to the mention of those families which are remarkable from the number of species they contain, and to the enumeration of some of the characteristic botanical forms of Japan. The predominating families are: Synantheraceæ (124), Graminaceæ (90), Rosaceæ (90), Leguminosæ (72), Liliaceæ, in the extended sense (60, of these 25 are Smilaceæ), Cyperaceæ (48), Labiatæ (47), Ranunculaceæ (42), Umbelliferæ (40), Amentaceæ (38), Orchidaceæ (35, principally of North American and European genera), Ericaceæ (36, of North American genera), Coniferæ (30), Urticaceæ in the extended sense (about 30), Cruciferæ (30). Characteristic forms (excluding several tropical representatives): Melastomas (4), Zanthoxyleæ (6), Aurantiaceæ (10), Ternstrœmiaceæ (19), an *Opuntia*, the source of which is, however, doubtful, Magnoliaceæ (10), one of the Proteaceæ (*Helicia* Z.), Lauraceæ (18,) Palmeæ (4), Musaceæ (4 *Musæ*), Scitamineæ (7), the Hæmadorous *Aletris*, Dioscoreæ (5), 1 *Philhydrum*, Commelyneæ (5), *Eriocaulon* (4), and *Cycas* (1).

Zuccarini and v. Siebold have described some new genera of Japanese plants in the memoirs of the Academy of Bavaria (Plantarum quas in Japonia collegit de Siebold genera nova. Fasc. 1, l. c. iii, pp. 719-49). List of these genera: *Pityrosperma* (a Ranunculaceous plant with 3 species, one of which is *Actæa Japonica* Thunb.); *Pteridophyllum* (connecting link between *Hypecoum* and *Fumaria*); *Eucapnos* (*Diclytra spectabilis* D. C.); *Trochostigma*, with 5 sp. (probably the type of a new family allied to the Ternstrœmiaceæ); *Corchoropsis* (one of the Tiliaceæ); *Tripetaleia* (doubtfully placed among the Oleaceæ); *Stephanandra* (affinity also doubtful, probably belonging to the Rosaceæ); *Cerascidos* (one of the apetalous Amygdalaceæ plants); *Platycaria* (one of the Juglandeæ); *Schizocodon* (Polemoniaceæ); *Conandron* (allied to *Ramondia*); *Phyllostachys* (Bambuseæ). According to the author, the bamboo-stems of commerce, as also the pepper-canes as they are called, are obtained from the Bambuseæ, which are common in Japan, and of which there are 15; they seldom, however, flower, and therefore the species are but imperfectly known.

Royle has drawn up some remarks upon the vegetation of Afghanistan, Cashmere, and Thibet, from the truly very inconsiderable collections of Vigne (Travels in Cashmere, Ladak, Iskardo, &c., by G. T. Vigne, 2d edit. London, 1844, 8vo, Appendix). However, these fragmentary reports are of interest, on account of Royle's intimate acquaintance with the botanical character of the Himalayan mountains, the use he has made of other sources, and the general plan of his investigation. Thus he starts with the question of what constitutes the northern and western boundaries of the indigenous plants of the Indian plains. He considers it as an established fact, that the western boundary of the Indian flora along the Indus is formed by the Soliman mountains, and, in fact, the influence of the monsoon and summer-rains, upon which the vegetation of the tropical plains is dependent, disappears entirely in the district of this meridional chain, on the line from Kelat to Peshawar. Royle is especially indebted for the observations upon the western localities of Indian plants to the traveller Falconer, who is now his successor in the Botanic Garden at Saharunpore. The latter found *Butea frondosa* even on the Jhelum, the most westerly of the Punjab rivers; the Chenopodeæ of northern India accompanied it as far as Peshawar. Above Attock, on the Indus, the characteristic plants of the British Himalaya again recurred.

Even from Attock, according to Elphinstone (Cabul, p. 130), the tropical rains extend northwards as far as Hindu-Rusch, without the high flats of Afghanistan being moistened by them; for Surat would there form its western boundary, at which place, in summer, whilst e. g. rain still falls in Pukkely, the sky is overcast for a month only, and merely occasional showers fall. Thus the double harvests of the Indian year, which are occasioned by the rainy season, cannot be obtained west of Jellalabad (Irvine, Journal of As. Soc. of Bengal). Hence between Jellalabad and Gundamuc, on the road to Cabul, the periods of development of the vegetation are suddenly changed. "In Gundamuc," writes Burnes, "the willows flowered at the end of February. On the 11th of March the first spring flower appeared; it was a sweet-smelling Iris. The apricots began to unfold their buds on the 1st of April; the wheat here was three inches above the ground, whilst in Jellalabad it was already cut." But when we take into consideration the elevation of the soil above the Indus and its tributary streams, it appears clear that the tropical conditions of the vegetation only extend so far west in the valleys. In fact, Royle does not allude to the important question, to what elevation the mountain-slopes which limit these fluviatile valleys on every side are reached by tropical rains; but as regards Cashmere, a valley lying far to the east of Peshawar, we know that the atmospheric precipitations of the spring cease to occur at that period at which the rainy season commences in the Indian plains and the low valleys of the Himalaya. Thus it appears, from all the descriptions, that the more elevated regions in the neighbourhood of Attock and Peshawar are not subjected to the monsoon. This explains Elphinstone's statement, that a number of English plants thrive in the gardens at Cohaut, where plum and peach trees were in flower at the end of February, and where weeping willows, plane, and apple trees were thriving upon European meadow plains. From these reports, it is probable that the entire district west and north of the

Jhelum, or of the salt-chain, which is intersected by innumerable offsets of the Himalaya and Soliman mountains, with the exception of the lower fluviatile valleys, is free from all those Indian vegetable forms which, up to the foot of these mountains, are extended in an uninterrupted distribution over the Punjab.

But Royle's investigation passes over unnoticed a still more important aspect of the question regarding the boundaries of the Indian flora. Hitherto we have only treated of tropical forms of vegetation, to the growth of which the rainy season is unfavorable; but in addition to these, India possesses in the Himalaya and the monsoon region that mountain vegetation also, in which the European type is repeated. Here the question arises whether the areal limits of the latter are the same as those of the former, with which, in fact, they partly grow in common on the western chain of the British Himalaya, without, however, being favoured in the same degree, during their period of vegetation, by the tropical rain. The knowledge of this remarkable coexistence of the productions of two climates, for which we are also principally indebted to Royle's former investigations, has not induced him to devote his attention to the question of whether there are not forests of Himalayan trees in other regions which do not shade tropical plants in the rainy season. However, the simultaneous publication of Jacquemont's Journal at Cashmere has thrown some light upon this obscure point (Voyage dans l'Inde, vol. iii, p. 169). The traveller describes his journey from the Punjab to Cashmere over the Pirpanjol, the Himalayan Pass, which Royle, relying upon Bernier's descriptions, had formerly marked as a sharply-defined limit of the vegetation of the Indian flora, which assertion he now himself withdraws pretty openly. During the ascent, the pomegranate and olive trees disappeared at an elevation of 16—1700m., and soon after, *Pinus longifolia* also. A region of oaks, *Pinus attenuata*, and firs was next met with, which, on the northern slope of the chain, extended above the level of the Pass (2681 m.), whilst on this side it terminated below the alpine meadows. The alpine vegetation presented merely local differences from that of the British Himalaya; its spring-plants were in flower at the commencement of May. On the north side, therefore, Jacquemont first met with the same trees he had left on the southern slope, and further down, in the district of the valley of Cashmere, he arrived at forests of an *Æsculus* of the same species as that indigenous to the British Himalaya. The older opinions concerning the Pirpanjol, which Royle disseminated, are contradicted by these Reports. But as there is no tropical region in the valley of Cashmere, we have here also a proof that the diffusion of the Himalayan plants is not limited by the boundaries of the monsoon. The tropical forms of India may be wanting in Cashmere —and there is no evidence to show that they exist there —and yet the forest trees may appear the very same, and the character of the vegetation for the most part identical with that of the British Himalaya; in fact, the greater number of species may be common to both of them. The natives of the Pirpanjol say that it is always raining there (p. 225); hence this Pass may form one of the points of the boundary, as far as which tropical forms accompany the wooded slopes of the Himalaya. When thus considered, all the known facts are connected under a common point of view, but they are by no means sufficient for determining the absolute sphere of diffusion of all the Indian plants. Although Royle has rendered it probable that this area does not extend west of the monsoon-limit, yet the line at which the Himalayan plants cease, towards the north, is either totally unknown to us, as is the case beyond Cashmere, or merely indicated by uncertain evidence.

Royle's statements regarding the flora of the elevated plains of Afghanistan are very general; but where Griffith is his authority, the fragmentary notices derived from his letters are substantiated by the catalogue of a series of Afghanistan genera, the seeds of which were also transmitted by Griffith. They are nearly all European forms, and principally the following: *Aconitum* and *Papaver;* 5 European Cruciferæ and *Tauscheria; Silene* and *Arenaria; Ruta* and *Peganum; Euphorbia* and *Phyllanthus;* several Astragalaceæ and *Caragana; Rosa* and *Cratægus; Epilobium; Prangos pabularia;* several *Carduaceæ* with *Centaurea* and *Cicorium; Campanula; Heliotropium* and *Onosma; Pedicularis, Linaria, Veronica,* and *Verbascum; Hyoscyamus, Samolus, Plantago, Hippophäe, Rumex,* and *Polygonum; Blitum; Iris* and *Tulipa.* Irwine treats fully of the cultivated plants of Cabul (loc. cit.) Wheat, barley, lentils, and peas are sown; they are protected during the winter by a layer of snow, and harvested in June. To the summer crops, which usually require irrigation, belong *Phaseolus radiatus, Cicer arietinum, Panicum miliaceum* and *Italicum,* maize, and rice; these are sown in May, and harvested in the months of August and September. Besides the European vegetables, *Solanum melongena* and several Cucurbitaceæ are cultivated, which require much manure and water. The meadows yield abundant crops of hay, and contain some excellent species of trefoil: one of these is denominated *Trifolium giganteum; Medicago sativa* is also widely diffused. The fruit trees of Cabul are celebrated: in addition to the fruits of central and eastern Europe, those of *Elæagnus* (Sinjet and Sinjilla) and *Edgeworthia buxifolia,* one of the Theophrasteæ, are mentioned.

Falconer discovered, in Cashmere, the plant yielding the Costus of the ancients, a substance which still forms an article of commerce in India, under the name of Koost or Koot. It is obtained from the aromatic root of a new alpine species of *Carlina* (*Aucklandia*), which Falconer has accurately described (Linnæan Trans., xix, p. 23). He has also proposed there (p. 101) the Asclepiadaceous genus *Campelepes,* from Peshawar. Falconer's so-called *Fothergilla,* which forms large bushes in Cashmere, and the wood of which, according to Vigne, is called Chob-i-pan, is a new type of the Persian *Parrotia* (*P. Jacquemontiana* Decs.)

From the elevated valley of Astore, between Cashmere and Thibet, Vigne brought the following plants: *Aconitum heterophyllum*, *Anemone discolor*, *Podophyllum*, *Dianthus*, *Geranium*, *Epilobium*, several Gentians, *Swertia* and *Ophelia Chirata*, *Polemonium cæruleum*, and *Dracocephalum Royleanum*. Here, far above the tree-limit, we find the elevated plain Deosuh, at an altitude of 13,000′, the soil of which is rendered verdant by dwarf-willows and alpine herbs, whilst the valley in which the Indus runs in Thibet is bare, a few plants occurring only at the snow-line. Falconer found here a new *Rheum* and two species of *Pyrola*, which, as Royle remarks, are the only Ericaceous plants in Thibet. Vigne's plants from Iskardo agree pretty accurately with the older collections from Kunawar: *Actæa*, some Cruciferæ, *Silene Moorcroftiana*, *Acer microphyllum*, *Myricaria*, *Biebersteinia odora*, *Astragalaceæ*, several *Potentillæ*, *Saxifraga stenophylla*, *Hippophäe* and *Salsola*.

Jacquemont's work on his travels, which has been mentioned above, is now complete, and affords extensive contributions to our knowledge of India in a botanico-geographical point of view, especially the flora of the British Himalayas and those of Thibet (Journal, vol. i-iii. Paris, 1841. Vol. v, Descriptions des Collections. Ib. 1844, 4to. 2 vols. plates). The admirably-kept journal of this traveller, which is printed unaltered, contains, of course in a fragmentary form only, the impressions produced by the character of the vegetation of the Himalayas, and separate regions of India; but in the last section of the work, the more rare and new plants of Jacquemont's herbarium are treated in systematic detail by Cambessèdes and Decaisne, and illustrated with 180 plates.

In Lesser Thibet, J. travelled on the road to Ladak, in the valley of Spiti, as far as Danker, where at an elevation of 17,000′, at the limit to vegetable life, he found the new Anthemideous genus *Allardia*, a *Nepeta*, and an *Urtica*. The villages in the valley of Spiti, according to Jacquemont, are situated on a higher level than that formerly stated by Royle, e. g. Nako at 3658 m.; and the cultivation of the Cerealia, which is limited to *Hordeum hexastichon* and *cœleste*, and a *Panicum*, extends here to 3962 m., whilst in the southern Himalaya it only extends to 3048 m. Woody plants are not entirely absent from this elevated valley; even low trees, an indigenous *Juniperus*, and cultivated poplars and willows, are met with. The character of the vegetation, however, lies in the bushes, which was also noticed by Moorcroft These consist not only of thorny *Astragali*, but also of *Genista*, *Rosa*, *Ephedra*, and *Juniperus*. The absolute limit of elevation of the Phanerogamia west of Bekar was most accurately determined by Jacquemont. Here, in two passes leading from Thibet, Gantong (5486 m.) and Kimbrong (5581 m. according to Gerard's measurement), he left these plants below him. The leguminous shrubs of the valleys of Kunawar and Lesser Thibet were not found on the slopes of this pass, only a few alpine plants, the last of which was met with at Gantong, about 2—300 m. beneath the summit, hence at a level of 5200 m. Here he found two *Potentillæ*, *Corydalis physocarpa*, the new Caryophyllaceous plant *Periandra cæspitosa*, which resembles in appearance *Silene Acaulis*, with *Allardia* and *Eritrichium Jacquemontii* (Decs. ii, p. 309). Much lower down, the traveller met with a rose, forming the last shrub, and considerably lower still a *Juniperus*. At Kinbrong the vegetation also disappeared 300 m. below the pass, with a *Ranunculus*, *Corydalis*, and *Ligularia nana*; but at a level of 5400 m. Jacquemont saw an isolated green spot in the stony desert-waste. This was the highest evidence of vegetable life which he perceived (ii, p. 298). He estimated the snow-limit here at little less than 6000 m., so that between the last plants and perpetual snow there is an intermediate bare region extending through about 2000′.

As regards Kunawar, that remarkable transition-district between the British Himalaya and Thibet, on the central Sutlej, where the influence of the monsoon on the seasons ceases, and the dryness of Thibet commences, Jacquemont's botanical observations agree with the more copious reports of Royle. The forests are very inconsiderable, the growth of grass poor, and kept down by Tragacanth-shrubs (*Astragali*), which are distributed as far as here; the alpine flora is also very scanty (ii, p. 269). Jacquemont devotes particular attention to the cultivation of the grape-vine, which is confined to this part of the Himalaya, not extending beyond the limits of the tropical rain (ii, pp. 416 et seq.) Although the grape-vine is cultivated at an elevation of 10,000′, this is only the case in the bottom of the valley, not on the mountain-slopes, for it only there receives the reflected rays of the sun, which are necessary to ripen the grapes, and there it is also protected from that radiation of heat which exerts too powerful an effect in cooling the earth on mountains. Moreover, even in the valley of the Sutlej, irrigation is indispensable to this branch of culture; but although the grapes under these circumstances mostly ripen well, they are usually dried in the sun, and used to make raisins, as the wine does not keep long, and even when new was found almost undrinkable by the Frenchman. We find the grape-vine as far upwards as Nako, in the valley of Spiti, and downwards as far as the mouth of the Buspa, where the climatic line above mentioned lies, and where the Sutlej intersects the high southern chains of the Himalayas.

The chains of the Southern Himalayas, which are situated immediately opposite to the plain of the north of India, do not possess any of that variation of soil, by means of which their vegetation might equal the flora of the Alps in variety, notwithstanding the mixture of forms of Tropical and European plants. Plane surfaces are scarcely anywhere found; and as we have already remarked, the broad valleys of Cashmere and Nepaul form exceptions to the mountain-character. Perpendicular precipices are also absent. We find everywhere vast inclined plains, and the mountain-stream usually entirely fills up the bottom of the valley. Jacquemont says (ii, p. 130), "the vegetation which covers the inclined soil is as uniform as this conformation. Variety of localities causes a region to abound in plants, but here all the localities are alike." In the upper regions the forests are generally thin, and belong principally to the valleys. On viewing from a distance one of these immense declivities, on which there are scarcely any forests, we perceive lines of a darker green accompanying the few rivulets which water the mountain-slopes, at great distances apart. Between them the green is uniformly pale, for neither meadows nor mountain-pastures thrive there; but, with the exception of the summits of the rocks, an irregular and unfruitful growth of plants prevails among the blocks and the crumbled portions of rocks. High mountains occur, which, from the valleys to the crests, are covered with this mixture of rocks and plants only. More commonly a thin forest is distributed over a soil of this kind, between 6000′ and 7500′, consisting either of pine trees on the southern declivities, or the oak, with *Rhododendron arboreum*, on those which are colder. It is only at the foot of the mountains that dense forests, such as those on the Alps, flourish. The elevated forests of the Coniferous region of the Alps are not met with on the Himalayas.

At Massuri, Jacquemont measured the lower limit of the oak forest containing *Rhododendron arboreum*, and estimated it at 1768 m. (ii, p. 52). This measurement is tolerably near that given by Royle, who, in this district, determined the elevation of 5000′ to be the level at which the forms of the European forests appear in the place of tropical trees. In his ascent of the Kedarkanta, in the district of the source of the Jumna, Jacquemont also estimated the upper tree-limit at 3500 m. (ii, p. 127). The pine forests (species of *Abies*) terminated here, and were succeeded by a shrubby formation of *Rhododendron* (probably *Rh. lepidotum* Wall.); where this also disappears, the alpine soil is covered with turf, consisting of Grasses and *Carex,* among which Ranunculaceæ most commonly spring up, with *Iris, Corydalis,* and *Phalangium.* The above measurement of the tree-limit appears to deserve the more confidence and to form an indication of climatic conditions, inasmuch as on the Kedarkanta the soil and inclination of the summit were favorable to forest growth.

Towards the end of his extended tours through the East Indian peninsula, Jacquemont's attention was drawn to an important peculiarity in the progress of the vegetation on the eastern coast of the district of the Ganges (iii, p. 550). In Bengal the soil remains green throughout the year, because the water flows off these plains so slowly, that it is retained deep in the soil during the dry season ; also because in the winter dense fogs, and in the hot and dry months of spring, transient thunder-showers occur. Thus, when the traveller landed, on the 5th of May, at Calcutta (therefore on the coast), the turf was just as green as at the period of the heaviest precipitations in August. The treeless country of Puna, in the western Ghauts, however, in 1832 remained perfectly arid and parched, even in the latter third of June, just like the soil of the steppes; the surface of earth was without a trace of moisture, and, as it were, glowing in the sun's rays. Yet on the 1st of July the whole country was green, even the barest rocks had become covered with turf with wonderful rapidity. Hence the character of the monsoon flora is much more distinctly stamped here than at Calcutta: but the Bengal coast is anomalous in this respect. In the greater part of India, the vegetation of most of the plants is interrupted for a longer period by the dry season, than in Europe by the winter. The large shrubs, the sugar-cane plantations, and the turf of Paniceæ wither and dry up in November, and their vegetative life is not again aroused until June or July of the following year. At Puna the rainy season then lasted but little more than three months, and ceased at the beginning of November; but that year threatened to be unproductive, in consequence of too small an amount of rain having fallen.

In the descriptive part of Jacquemont's work, which, arranged in accordance with De Candolle's system, is worked out by Cambessèdes as far as the conclusion of the Rosaceæ, and the remainder by Decaisne, in addition to a large number of new species, the following genera, mostly from the Himalayas are proposed: *Christolea* and *Donepea* (Cruciferæ), *Oligomeris* (Resedaceæ), *Periandra* (vid. sup.), *Anquetilla* (Xanthoxylaceæ) *Leptopus* (near *Phyllanthus*), *Allardia* (v. s.), *Melanoseris* (Cichoraceæ), *Belenia* (Solaneæ), *Dargeria* (Scrophulariaceæ), *Lasiosiphon* (*Gnidiæ* sp. plures) *Girardinia* (*Urticæ* sp.), and *Diplosiphon* (a remarkable Epigynous and Monocotyledonous water-plant, the natural affinity of which is not determined).

The continuation of Bentham's work upon the Indian and African Leguminosæ, which was noticed in last year's Report, includes about a hundred Genistas, most of them from the Cape (London Journal of Bot. iii, p. 338-65).

The new parts of Kortbal's Monographs on the Flora of the Indian Archipelago (Annual Report for 1841), contain the Melastomaceæ, Oaks, and the following genera: *Cratoxylon* and *Tridesmis*, *Hippocratea* and *Salacia*, and *Maranthes; Boschia*, nov. gen. (Sterculiaceæ), *Omphocarpus*, n. g. (near *Grewia*), *Paravinia*, n. g., and *Cleisocratera*, n. g. (Rubiaceæ). De Vriese has described a *Casuarina* (*C. Sumatrana* J.) found by Junghuhn in Sumatra in v. d. Hoeven's Tijdschrift (1844, p. 113), also some Javanese plants (id. p. 336-47); the only new plant is an *Æschynanthus*. New contributions by Hasskarl on various families of the Javanese flora are published partly in the same Journal (p. 49, iii; pp. 178-228), and partly in the Ratisbon Flora (1844, pp. 583 et seq.) Montagne has described some new Javanese Mosses (London Journal of Botany, 1844, pp. 632-4). Dozy and Molkenboer have commenced an illustrated work on the Mosses of the Indian Archipelago (Musci frondosi inediti Archipelagi Indici, Fasc. I. Lugdun. Batav. 1844). The preliminary diagnostic characters of about 75 new species have been published by them in the 'Annales des Sciences Naturelles' (1844, ii, p. 297-316); among these are the new genera *Cryptocarpon*, *Endotrichon*, and *Symphysodon*.

An extremely important systematic and illustrated work, on the Flora of Java, which is now concluded, is that published by Bennett and R. Brown from Horsfield's herbarium (Plantæ Javanicæ rariores descriptæ iconibusque illustratæ. Descriptiones et characteres plurimarum elaboravit I. Bennett, observationes structuram, et affinitates præsertim respicientes passim adjicit, Rob. Brown, pt. i, Londini, 1838; pt. ii, 1840; pt. iii, 1844). This work contains 45 plates, and the following new genera: *Scleracline* and *Polytoca* (Graminaceæ), *Hexameria* (Orchidaceæ), *Cyrtoceras* (Asclepiadaceæ), *Stylodiscus* (*Andrachne trifoliata* Roxb.), *Euchresta* (*Andira Horsfieldii* Lesch.), *Mecopus* and *Phylacium* (Leguminosæ), *Saccopetalum* (Anonaceæ), *Lasiolepis* (near *Harrisonia* Br.), *Pterocymbium* (Sterculiaceæ), and several types from other countries, which are elucidated in these copious disquisitions.

Junghuhn's diaries of his travels in Java, which have been already alluded to, were indeed first published, with additions, in 1845 (Topographische und naturwissenschaftliche Reisen durch Java, von F. Junghuhn, herausgegeben von Nees v. Essenbeck. Magdeburg, 8vo), but for the sake of connecting them with the preceding Annual Report, we think it better to report upon them now. In the western portion of the island, as at Gedé, the traveller found the mountain-ridges covered far and wide on both slopes with Rosamala-forests, i. e. with *Liquid-ambar Altingiana* Bl., the stems of which are recognised even at a distance by their tall straight growth and white colour, and which overshadow a thicket of Scitamineæ, *Melastomæ*, *Rubus*, and other shrubs (p. 165). A rich red soil here covers almost the whole of the trachyte of Gedé. According to several measurements, the region of the Rosamala-forests is situated at a level between 2000′ and 4000′ (p. 436): this tree, which is confined to the west of Java, occurs singly as high as 4500′, and as low as 1500′. It is one of the most gigantic formations of the vegetable world, and attains on an average a height of 150′: the stems when cut down measure 15′ in circumference, 12′ above the root; their length below the point at which they branch amounts to 90′—100′, and the crowns extend to a height of from 50′ to 80′ beyond this. Cocoa palms would scarcely reach as high as these crowns. Above the Rosamala-forest on Pang-Gerango, came forests of Laurineæ, *Castaneæ*, Oaks, *Schima*, and *Fagræa*, which were far more abundantly filled with climbing plants (e.g. Freycinetias and *Calamus*) and parasites (Orchidaceæ and Ferns); and these again were succeeded by the Podocarpeæ. But even beyond the limits of *Podocarpus* the arboreal form is not wanting here, as is the case on other mountains. On the summit of Pang-Gerango itself, at a level of 9200′, *Thibaudia vulgaris* J. and an undetermined diœcious plant, 30 feet in height, with various other trees, form a wood abounding in Mosses, which, however, from its manner of growth, appears to belong to a vigorous growth of mountain pines (p. 452), although, even as far as this, a slender tree-fern, *Cyathœa oligocarpa*, from 15′ to 20′ high (extending from 5500′—9200′), is met with (see Annual Report for 1841, p. 449). "But," says Junghuhn, "we search in vain throughout the island for another example of such a wood on a mountain-top: all the mountains, far below this altitude, are either bare, being covered with lava and crumbled rocks, or overgrown with grass meadows of *Festuca nubigena* J. or with social Casuarinas." Junghuhn estimated the upper forest-limit on the Tjernai volcano (p. 235) at 7000′; it is formed by *Podocarpus imbricata* Bl., and is immediately succeeded by the subalpine shrubs (see preceding Ann. Report), and this appears to be the general manner in which the forests are distributed throughout the island. The true climatic arboreal limit of Java, which is only attained on the Pang-Gerango, and is here indicated by the mountain-pine formation of the wood on the mountain-top, is thus situated several thousands of feet higher than the apparent one, which is merely produced by local conditions of soil, and thus Junghuhn, by his ascent of this mountain, has thrown some light upon an anomaly which has hitherto been almost inexplicable, viz. that the tree-limit in Java is so much lower than in the Himalaya, and that in general subalpine Ericaceous shrubs, with the northern alpine genera (e. g. *Ranunculus, Viola,* and *Gentiana*), descend there to an equally low level of 7—8000′. Yet the difficulty in explaining these deviations is not, in fact, completely removed by these observations, but merely confined within narrower limits; for although Pang-Gerango teaches us, that at 9200′ the most luxuriant woods still imitate the crooked stems of the mountain-pine, yet we find in India forests of tall fir trees at a level of more than 10,000′.

At the foot of the mud-volcano Galungung, Junghuhn describes the occurrence of almost impenetrable rush-formations, the marshy surface being thickly covered with *Saccharum Klaga*, 15′ in height, around which an *Equisetum* and *Epidendra* are coiled. Above these marshes, on the slope of the mountain, the forest of Urticeæ and Magnoliaceæ commences, including all those accessory components which render the attempt to describe tropical forests apparently impossible, even although we should not aspire to represent its copiousness by words and expressions, but merely to seize the distinctions in its mode of development and the conditions under which it occurs.

Just as the Rosamala-forests in the west of Java determine the physiognomy of the mountains, when covered by them, so in the eastern portion of the island do the forest-regions of *Casuarina equisetifolia*, which, however, are not met with below a level of 4000′; hence, although they ascend higher than other forms of trees, they are confined to the more limited space on individual elevated points. No trace of *Casuarinæ* is found west of Merapi, a mountain from which they are almost extirpated, whilst they do not appear to be absent from any of the mountain-tops which ascend on the east of it (p. 372).

Junghuhn gives the following statements regarding the altitudinal limits of some branches of cultivation in Java. Coffee might probably be cultivated as far as a level of 5000′, but at present the plantations do not usually extend beyond 3000′ or 4000′ (p. 234). *Artocarpus integrifolia* and *Arenga saccharifera*—3000′, *Duris zibethinus*—2000′ (p. 419).

Kittlitz gives two landscapes of districts in Manilla (Plates XXIII, XXIV), which, like all the others, are extremely characteristic, but deficient in sufficient botanical elucidation. Montagne has described the *Algæ* of the Philippine Islands from Cumming's collections (Lond. Journ. of Bot. 1844, pp. 658-62).

III.—AFRICA.

Of the botanical investigations of the French in Algeria but few notices have yet been published. Durieu met with extensive forests of cedar at Blidah, on the Lesser Atlas (Comptes rendus, vol. v, p. 18). As far as a level of from 7—800m. the mountain-slope was inhabited, and the soil cultivated; the oak then began to be intermixed with the fruit trees, and soon after single majestic cedars, 40 meters in height, were seen. But it was only on the southern declivity that the traveller met with connected forests of this tree, which are cut down annually by the inhabitants; they do not, however, appear to be destroyed as at Mount Lebanon, but are apparently readily reproduced. At Mascara, Durieu found *Callitris quadrivalvis* common, and increasing in frequency thence towards the south (Comptes rendus, vol. v, p. 19). Bory de St. Vincent has described some new species of *Isoëtes*, partly living upon a dry soil, from Algeria (l. c. vol. xviii).

We may now recur to Russeger's travels (Annual Report for 1842), since his work has proceeded to a considerable extent, and commenced the illustration of the conditions of the climate and soil in a more tangible and definite style than was the case in the first volume, which treated of the East (Reisen in Europa, Asien, und Africa, Bd. 2. Stuttgard, 1843-45. In 1844, appeared the first part of this volume, including Egypt and Nubia, and the first number of the second part, containing Eastern Soudan). The climate of Lower Egypt, as far as Cairo, is that of the Mediterranean—a wet winter (ii, p. 263) and a serene summer. In Cairo we find the rainless zone of the north of Africa. At Cairo, according to the quinquennial average of Clot Bey, there are twelve rainy days in the year, with 0·034m. of rain. The absence of rain, both in Upper Egypt (Cairo to Nubia) and in the Sahara, depends upon constant north winds; hence Egypt is climatically a part of the Sahara.

The swollen state of the Nile, produced by the tropical rainy season, lasts from June to the end of September (i, p. 229). The months of October and November form the period at which the Cerealia are sown in those tracts of ground which are artificially flooded by canals; the harvests occur in February and March. Here, according to the kind of grain, a second crop may be sown in April, and reaped immediately before the irrigation. In other fields, the crop cannot be sown until December or January, and only once reaped, in May.

Sketch of the most important Branches of Cultivation, arranged according to the usual period at which the Crops are Sown and Reaped.

	SOWN.	HARVEST.
January.	Beans (Cerealia).	Sugar-cane.
February.	Rice, maize.	Barley, melons.
March.	Cotton.	Cerealia, maize.
April.	(Cerealia.)	
May.		Figs, dates, grapes (Cerealia).
June.		Beans (Cerealia).
July.		Cotton.
September.		Oranges, olives, rice.
October.	Cerealia.	Rice.
November.	Cerealia.	Maize.
December.	(Cerealia.)	

In the rainless zone of the north of Africa, in consequence of the duration of the polar currents, the great diurnal differences of temperature allow of the formation of dew to a slight extent; it occurs very copiously in the lower valley of the Nile, is formed in Upper Egypt, and appears also to fertilize the oases. Russeger, however, did not meet with any dew in the Desert of Nubia, but in that of Lybia it is common (ii, p. 253). The oases lying to the west of Egypt, according to Russeger, generally obtain their soil-water from the Nile, which flows laterally to them over beds of clay (p. 271). Thus they form a valley which is filled with springs, excavated below the level of the Nile, and parallel to this river. The other oases of the Sahara appear to be produced merely by the formation of dews. Borgu, Darfur. and Kordorfan, however, in this sense are not oases, but savannahs, situated within the rainy climate (p. 283).

The tropical rains extend in most years to, at the most, 18° N. lat. (i, p. 224), i. e. two degrees north of Chartum, the point at which the two arms of the Nile become confluent. The heavy rains fall there in the summer, and correspond to the south winds which blow at this time, and which prevail below 15° N. lat. from April to September, and alternate every six months with the north winds. The northern border of this monsoon-zone, which in the south of the Desert or Soudan produces savannahs, is not accurately determined. A short rainy season may occasionally occur beyond 18° N. lat., when the south winds blow as far as this part. However, the dry chamsin of the Desert, which blows from the same direction, and which Russeger regards as a local and electrical phenomenon, must not be confounded with these general south winds which bring rain. Even between 16° and 18° N. lat. the rainy period is irregular, and in many seasons abbreviated: at Chartum it lasts five months. Russeger assumes the following mean values as marking the north border of the tropical rainy zone throughout Africa:—21° N. lat. at the Red Sea,

18° at the Nile, 16° north of Tschad (according to Denham), and 20° in Senegambia (ii, p. 546). He forms a law on the great diurnal differences of temperature between the night and day, even within the rainy zone, which, if generally confirmed, would constitute a characteristic peculiarity of tropical Africa.

The whole of Nubia, as far south as 18° N. lat., except the valley of the Nile and the coast, consists, like Egypt, of rocky and sandy deserts. The heights here extend scarcely 1000′ above the plain, on the coast only ascending to 4000′, and in Dschebel Olba, according to Wellsted, to 8000′. The coast of the Red Sea is not free from rain; but on the Nubian side, the summer rains produced by the south-west monsoon extend almost as far as the latitude at which the tropical winter rains (as in Lower Egypt) commence. Suakim is situated on the northern border of the full rainy season (19° N. lat.); here, however, it occurs six weeks later (middle of July) than below 17°, and the summer rains are proportionately retarded and abbreviated as far as 21° N. lat., from which latitude northwards the winter rains commence. Although the upper portion of the sea north of Suakim is set in motion throughout the entire year by north winds, yet the African coast of the Arabian Sea is never free throughout the entire year from humid currents of air. This explains how it is that the entire coast line of Nubia is furnished with willows and other trees, whilst the inner country does not contain even oases. During the journey through the Desert, from Korosco to El Muchaireff, which occupied fifty hours, and is usually made to avoid the great bend of the Nile, Russeger only once met with brackish water, and that was in the middle of the journey.

The Nile leaves the zone of tropical rain at the influx of the Atbara, and again comes in contact with it by its bend at Dongola for a short distance. South of the mouth of the Atbara, savannahs begin to alternate with tropical forests, and this is the case throughout the whole of Soudan: no more deserts are met with, except where the soil is rocky; they gradually pass into savannahs (ii, p. 525). The savannahs, during the rainy season, are overgrown with thick grass; in the other months they resemble a dry stubble-field. The forests consist of *Mimosæ*, and are crowded along the banks of the stream, as in Guiana. Near the rivers the rain district also extends further north; hence, at a considerable distance from them, even beyond the 18th degree, the creeks of the desert encroach upon the savannah.

Throughout the entire district of the Nile, at least as far as the 10th degree south, there are no terrace-like elevations of the soil west of Abyssinia, only immense plains. The terraces of Sennaar, Fazokl, &c., are geographical over-estimates (ii, p. 539). According to Russeger's barometrical measurements, the following places are situated at the annexed altitudes above the Mediterranean: Assuan (Syene), 342′, Par.; Korosko, 450′; Abuhammed, 963′; El Muchaireff, 1331′; Chartum, 1431′; Torra, on the White Nile, 1595; Eleis (13°), 1667′; and the capital of Kordofan, El Obeehd, 2018′. Russeger found the northern limit to the occurrence of *Adansonia*, in the savannahs of Kordofan, to exist below 14° N. lat.

On the coast of Adel, on the road from Tajura to the foot of the mountains of South Abyssinia, according to Harris's report of his travels (The Highlands of Ethiopia. London, 1844, vol. i, p. 412), the entire country was desert, and almost dried up in June, i. e. before the commencement of the rainy season, and the soil entirely uncultivated. When the heavy rains commenced it was stormy and unhealthy; one of the most uninhabitable parts of Africa. The flora was uncommonly poor; the woody plants consisted of shrubs of *Mimosæ* and *Cadaba Indica*, one of the Capparidaceæ; subsequently isolated Palms, *Cucifera Thebaica*, and below 11° N. lat. *Phœnix* were met with. The only other plants found at the end of the dry season were a few Capparidaceæ and Malvaceæ; and of other botanical groups of the steppes, single forms only, as *Stapelia*, *Pergularia*, and some succulent Euphorbias; but at the river Hawasch the vegetation became social, by the formation of thickets of *Tamarix* or *Balsamodendron Myrrha*, with single Capparidaceous trees (i, p. 416). At the foot of the high mountains of Abyssinia, *Aloe Socotrina* was also met with, and soon after *Tamarix Indica*, with which the desert steppe was overgrown.

Harris read a paper on *Balsamodendron Myrrha* before the Linnæan Society (Ann. Nat. Hist. xiii, p. 220). This important shrub is called by the Danakil tribes, who inhabit the coast of Adel, Kurbeta. Myrrh (Hofali) is the milky juice which escapes from wounds made in it, dried in the air; it is usually collected in January, the period at which the buds unfold, and in March, when the seeds are ripe. *Balsamodendron Opobalsamum* grows on the opposite Arabian coast, at Cape Adem. The Frankincense trees of the mountains of Cape Guardafui, have not been botanically determined.

Harris's botanical reports upon Shoa are very unsatisfactory (The Highlands, &c. ii, pp. 395 et seq.) The pine of North Abyssinia is replaced in Shoa by the Det, a *Juniperus*, 160′ in height, with a stem from 4′ to 5′ in diameter, and with the growth of a cypress. Forest trees are also mentioned: *Taxus* (Sigba), *Ficus* (Schoala), and *F. Sycomorus* (Worka); moreover, Rüppel's Lobeliaceous tree, *Rhynchopetalum montanum* (Jibera), is common at Aukober, the stem of which is 15′ in height, and bears a crown of large leaves. Shrubs: an *Erica* (Asta) and *Polygonum frutescens* (Umboatoo) distributed generally. *Celastrus edulis* (Choat) is cultivated commonly, and resembles tea in its action and taste (ii, p. 423).

Harris's meteorological observations, which were made from August to December in 1841, and from January to July in 1842, in Aukober, the capital of Shoa, are of more importance. This place is situated below 9° 35′

N. lat., 8200′ above the level of the sea, and upon an open, cultivated flat. The climatic values are as follow:

	Mean Temperature.	Number of Rainy Days.	Quarter of the Wind.
January . .	11° 1 C.	—	East
February . .	12° 5 „	7	East and South
March . .	14° „	4	East
April . .	12° 9 „	14 (storms ?)	East
May . . .	15° 4 „	4	East
June . . .	16° 7 „	8	East
July . . .	14° 5 „	28 } rainy season	Changeable
August . .	13° 2 „	26 } rainy season	Changeable
September .	13° „	13 } rainy season	North and East
October . .	11° 2 „	4	North and East
November .	11° „	4	North and East
December .	11° „	—	East

Mean temperature, 13° 1 C.; Maximum, 20° 6; Minimum, 5° C.

In Koolo (4° N. lat.), south of Enarea, on the confines of the pigmy Doco-negro tribes, according to the reports of the natives, the rainy season lasts from May to February with but slight intermission (iii, p. 64). To the north-west of this part, below the 5th degree of north latitude, the country of Susa is situated, high up on the prolongation of the Abyssinian mountains, and there, as at Shoa, the rainy season lasts only three months; but it must be colder there, for the mountains appeared to reach the sky, and were covered with perpetual snow. This is the district in which Bruce supposed the White Nile to arise.

Hochstetter has described some new Grasses found in Nubia and Abyssinia, from Kotschy and Schimper's herbarium, and after making some critical remarks upon the results obtained by Raffeneau, Endlicher, and himself in this branch of the Flora, proposes the following new African genera (Ratisbon flora, 1844): *Chasmanthera*, a Menispermaceous plant from Abyssinia; *Paulo-Wilhelmia* (Dombeyaceæ), from Nubia; also from Abyssinia, the Umbelliferæ, *Agrochoris*, *Haplosciadium*, and *Gymnosciadium*; *Discopodium* (Solanaceæ); *Hymenostigma*, and

Acidanthera (Iridaceæ); and *Clinostylis* (Liliaceæ). Fresenius has written notices of some Abyssinian plants (Bot. Zeitung, 1844, pp. 353-7). Fenzl has announced a work upon Kotschy's collections from Africa, and in it has enumerated a series of new forms, but without giving any descriptions (Ratisbon Flora, 1844, pp. 309-12).

A valuable notice of the plants collected by Krauss in the most southern regions of the Cape Colony, and in Natal, with a report on his travels, and a botanico-geographical introduction, has been published by the author (Ratisbon Flora, 1844, p. 46). He accurately describes the large elevated forests, which are limited to a small area, considering the extent of the whole colony, and extend along the south coast between Gauritz and the Kromme river and the foot of the Onteniqua mountains. According to his account, Drège's representation, contained in E. Meyer's work, of the generally poor character of the forests of the Cape, is not perfectly correct. At least in this district there exists a quantity of timber collected into woods, which Krauss characterises as impenetrable thickets. He mentions some gigantic stems of *Podocarpus*, which four men cannot span; moreover *Crocoxylon excelsum* (Safranhout), *Ocotea bullata* (Stinkhout), *Curtisea faginea* (Hassagaihout), and *Elædendron Capense*—trees, the large, densely leafy crowns of which are elevated far above the thicket beneath, and covered with numerous climbing plants. Underwood, e. g. *Burchellia*, *Gardenia*, *Canthium*, *Plectronia*, *Tecoma*, *Grewia*, *Sparmannia*, and *Rubus*. Lianes: *Cissus*, *Clematis*, *Cynoctonum*, and *Secamone;* Ferns in the deeply-shaded parts. After a tedious ascent, and laborious struggle through the chaos of bushes, is finally reached a thin wood, the trees become smaller and smaller, and their limit is soon attained. They are succeeded by shrubs of Synantheraceæ, Thymeleaceæ, Bruniaceæ, Proteaceæ, and Ericaceæ.

Krauss confirms the statement that the river Camtos constitutes a distinct limit of vegetation. This stream might form the boundary between the flora of the Cape, and of the Caffre country; for here certain types of tropical Natal commence, whilst the Proteaceæ, Ericaceæ, Selagineæ, &c., diminish. The shrub-formations in Algoa Bay are taller and thicker than in the western districts; they serve as places of concealment for large Pachydermata. Characteristic forms of plants: Celastrineæ, *Euphorbia Canariensis*, *Strelitza*, *Zamia*, *Tamus*, *Pelargonium*, &c. This remarkable difference between the eastern and western provinces of the Cape Colony, which Bunbury also (London Journal of Botany, 1844, pp. 230-63) mentions and enlarges upon more in detail, is by no means to be so simply explained as the tropical peculiarities of the flora of Natal. At Graham's Town in Albany, Bunbury only found 13 plants in the extensive surrounding country, and these occur also at the Cape. Ericaceæ and Proteaceæ are rare, arborescent Euphorbias common, and the Restiaceæ replaced by Grasses. Extending along the Great Fish River, we find the wildest thickets of shrubs with arborescent Euphorbias, *Strelitzia*, and *Zamia horrida*; these are more impenetrable, and from the presence of spinous trees, more inaccessible than the natural Brazilian forests: they merely form the abode of large Pachydermata and border-robbers of the Caffre race. Tropical families of plants, single species of which only occur at the Cape, become numerous in Albany, especially Acanthaceæ, Apocyneæ, Bignoniaceæ, Rubiaceæ, and Capparidaceæ. These and other similar facts evidently indicate an approximation to the flora of Natal, although by no means to the extent these two authors suppose, viz. that the vegetation of Albany and Natal gradually run into each other. As long as the intermediate districts of the Caffre country are so little known, this must remain hypothetical, but is rendered extremely improbable by climatic laws. A resemblance of certain families and forms is by no means a resemblance of species and their combination into formations. But the increase of tropical forms in Albany is even considerably more mysterious than the contrast between Albany and the west of the colony. In explaining the latter, we must bear in mind the narrow district throughout which the Cape plants are distributed, the tropical forms appear to indicate climatic influences which do not exist; for Albany is, if anything, unusually dry in comparison with the other regions of the colony, except the district of the Gareep. Rain, says Bunbury (p. 247), is rare and uncertain; when precipitations do occur, it is only the case during south or south-east sea winds. The climate is indeed considered very healthy, yet it is subjected to great and sudden changes of temperature, with stormy and dry winds from the west and north. Hence Albany does not exhibit any trace of that periodical rainy season, which at Port Natal, as the most southern point (30° S. lat.) of the regular tropical seasons, gives rise to the trade-wind character of the flora, and yet, in so dry a climate, the mode of formation of the plants is more similar to that of the trade-wind flora, than at the Cape, where in the winter regular precipitations occur almost as in the south of Europe. We must, therefore, in Albany, admit the occurrence of one of those botanico-geographical facts, where even a tropical constituent of the vegetation appears to be dependent not only upon climatic conditions, but historical or geological events.

According to Krauss, Natal is well watered by numerous rivers, which arise in the coast-chain Quathlamba; these mountains are nearly 10,000′ high, and run through the coast-country of the new colony in every direction. The vegetation springs up in September, and during the months of October, November, and December, corresponding with the atmospheric precipitations, attains the greatest splendour. During this moist season, the thermometer varies between 19° and 31° c. Vegetable life is suddenly arrested as early as January, the grass plains appearing dark yellow, and the forests flowerless and uniformly green. Rain seldom falls from January to March; the air during this period is hot and oppressive, and the temperature between 26° and 32° 5 c. The same appears to be the case with the two following months, which Krauss did not spend in Natal; July and August are fine, the days hot (as high as 31°), but cool in the morning and evening; the thermometer, however, seldom falling so low as 15° c. A changeable, windy, and disagreeable period begins in September—the precursor of the rain. From these statements, the course of the seasons is the same as in the East Indies, except that the rainy season of three months occurs during the spring in the southern hemisphere, i. e. three months later than in the former.

Sketch of the predominating botanical formations:

1. Coast or forest region.

a. Forests of Rhizophoreæ in the mud between the ebb and flow of the tide (Mengerhout of the colonists). *Bruguiera Gymnorrhiza, Rhizophora mucronata, Avicennia tomentosa.*

b. Dense, tropical, mixed forests, which can only be traversed by the paths formed by the elephants and buffaloes. Among the trees several belong to the new genera published by Hochstetter, with *Ficus, Tabernæmontana, Zygia, Milletia, Phœnix reclinata,* &c. Underwood, Lianes, and the other components of tropical forests are copiously developed.

c. Grass plains with various shrubs.—*Musa.*

2. Hilly region, with beautiful pasture land, constituting the flower of the colony. The woods consist of Acacias. The Aloe and tall-stemmed Euphorbias resemble those in Karro. The highly nutritious grass, which consists principally of Andropogineæ, contains numerous shrubs, especially tropical forms of Leguminosæ, Scrophulariaceæ, Labiatæ, Acanthaceæ, and *Gnidia Kraussiana.*

3. Mountainous region. The above-mentioned extensive grass plains are succeeded upwards by a woody belt of *Podocarpus,* with numerous Ferns, and above this mountain-meadows of Cyperaceæ, with Orchidaceæ, *Ixia, Hypoxis,* and *Watsonia,* are distributed. The largest

number of the representatives of the botanical forms of the Cape occur in this region; but hitherto only two Proteaceæ, one *Aspalathus*, two Geraniaceæ, one *Muraltia*, one *Maternia*, and one *Barosma*, have been found in Natal, and not a single species of *Erica*, *Phylica*, *Selago*. *Oxalis*, Zygophylleæ, &c.

The summary of Krauss's Herbaria contains the diagnoses of several new species from Natal, and some from the Cape colony, published under the authority of those naturalists who have worked out the collections for the traveller. Among them the following new genera are proposed: By Bischoff, *Sphærothylax* (Podostomeæ); by Meissner, *Bunburya* (Rubiaceæ); by C. H. Schultz, *Monopàppus* (Helichryseæ); and *Antrospermum* (Arctotideæ). Kunze has described some new Ferns from the Cape and Natal (Linnæa, 1844, pp. 113-24).

Bojer has continued his descriptions of new species of plants from the Mauritius and Madagascar (Troisième Rapport de la Soc. de St. Maurice); on this occasion they refer to the Anonaceæ, Menispermaceæ, Capparidaceæ, and Leguminosæ. Gardner has made a brief report upon some excursions in the Mauritius (London Journal of Bot., 1844, pp. 481-85).

IV.—ISLANDS OF THE ATLANTIC OCEAN.

Seubert has published a copious Flora of the Azores, in which his former memoir, which has been noticed in this work, is satisfactorily carried out, and brought to systematic perfection (Flora Azorica, Bonnæ, 1844, 4to). Of about 400 plants from the Azores, upon which his observations are made, fifty sp. are endemic, twenty-three sp. belong also to the Canarian Archipelago, five sp. to the continent of Africa, and six sp. to that of America; the remainder occur also in Europe. Of the endemic species, seven are Synantheraceæ, as many Cyperaceæ, and five Graminaceæ. Immediately after the appearance of the Flora Azorica, Watson published a list of the plants which he collected in the Azores (Lond. Journ. of Botany, 1844, pp. 582-617); and thus increased the number of the Phanerogamia of these islands, which have as yet been made known, to about sixty species. As the plants belonging to the south of Europe found there are of less interest, we shall confine ourselves to his contributions to our knowledge of the endemic flora. He has admitted the following of Seubert's species into this category: *Plantago Azorica* Hochst. as a variety of *P. lanceolata*, and *Juncus lucidus* Hochst. as a synonyme of *J. tenuis* W.; also *Luzula purpureo-splendens* S., according to an older syn. *L. purpurea* Watson; and *Bellis Azorica* as a distinct genus, denominated *Seubertia*. Lastly, he has described five new endemic forms: *Hypericum decipiens* (*H. perforatum* S.?), *Petroselinum trifoliatum*, *Campanula Vidalii*, *Myosotis Azorica*, and *Euphrasia Azorica* (*E. grandiflora* Hochst.?) *Vaccinium cylindraceum* Sm. appears different to him from *V. Maderense* Lk.; but *Erica Azorica* Hochst. only a var. of *E. Scoparia*. The following may be mentioned as interesting discoveries of the plants of Madeira, and other adjacent floras in the Azores: *Melanoselinum decipiens* Hoffm., *Tolpis macrorrhiza* D. C., *Mirabilis divaricata* Lour., and *Persea Indica* Spr.

Seventy-five parts of the work of Webb and Berthollet on the Canary Islands are out. They carry the systematic part as far as the Synantheraceæ.

Reid has communicated some reports upon the cedar of the Bermuda Archipelago (Lond. Journ. of Bot., 1844, p. 266, and 1843, p. 1). The inhabitants erroneously consider this Coniferous plant (*Juniperus Bermudiana*) to be the same as the Virginian cedar (*Juniperus Virginiana*). Even the climate of these islands is very different from that of the opposite coasts of the American continent, as water never freezes in the Bermudas. The most magnificent oranges are produced there, being protected from the winds of the Atlantic by the large forests of these cedars, which cover all the uncultivated regions. This tree is also called the pencil cedar, although the wood does not appear to be used at present in the manufacture of lead-pencils in England. It is much prized for ship-building. Reid thinks that the Bermuda cedar does not occur in the hot climate of the West Indies, but it is very common on the mountains of Jamaica.

V.—AMERICA.

The plants collected by Simpson and Dease, in their voyage of discovery on the arctic coast of America, have been named by Sir W. Hooker (Narrative of the Discoveries on the North Coast of America, by T. Simpson. London, 1843, 8vo, Appendix). These plants had, however, been previously found in Franklin's travels in the same region, and admitted into Hooker's Flora of British America, with the single exception of *Salix nivalis* Hooker, which was discovered by Drummond on the Rocky Mountains, and occurs also on the coast below 71° N. lat., west of Mackenzie.

A coast-landscape of Unalaschka, by Kittlitz (pl. IV), represents luxuriant meadows, in which various subalpine shrubs, forming most luxuriant thickets of plants, are intermixed with strong turf composed of Cyperaceæ: amongst the former are *Aconitum*, *Heracleum*, *Epilobium*, and especially *Lupinus*. The dwarf shrubs, also, of the alpine region, *Salices* and *Rhododendron Kamtschaticum*, extend on these islands, which are situated beyond the tree-limit, into the vicinity of the sea. Two views of the island of Sitcha, the forests of which they represent (pl. II and III), may serve as contrasts. They give a distinct representation of the mixed foliage of the Canadian larch (*Pinus Canadensis*) and a species of pine (*P. Mertensiana*), the growth of *Panax horridum*, the palmate auricled leaves of which are sometimes crowded together upon a turfy kind of brushwood, at others upon shrubby stems, consisting of the *Vaccinia* and *Rubi*, forming the underwood, and other botanical forms, with which Bongard's sketch has made us acquainted.

After an interval of two years, the third part of the second volume of Torrey and Asa Gray's 'Flora of North America,' containing the completion of the Synantheraceæ, has appeared. A. Braun has described the *Equiseta* and *Charæ* of North America (Silliman's Journal of Science, vol. xlvi). Mac Nab read before the Edinburgh Botanical Society a botanical journal, which he had kept at Hudson (Ann. Nat. Hist., xiv, pp. 223-25).

Asa Gray has continued the report upon his botanical journey in the south of Alleghany (London Journ. of Bot. 1844, pp. 230-42). On the summit of the Roan, in Tennessee, the altitude of which is 6000′, *Rhododendron Catawbiense* forms a fertile subalpine shrubby formation, the turf of which consists of *Carex Pennsylvanica* and other species of this genus, with *Aira flexuosa* and *Juncus tenuis*. Beneath the shrubs, *Lilium, Veratrum, Potentilla, Geum*, some Ranunculaceæ, Umbelliferæ, Saxifrageæ, and *Solidago*, with *Rudbeckia, Liatris*, &c., are mentioned. The remaining woody plants, in addition to the Rhododoreæ and Rosaceæ, mentioned in the Annual Report for 1842, consisted of *Pyrus arbutifolia, Cratægus punctata, Ribes rotundifolium, Diervilla trifida, Vaccinium Constablæi* n. sp., and *Alnus crispa*. *Pinus Fraseri* is the tree which extends to the greatest altitude; it occurs near the summit in a dwarf and crooked form. At the end, A. Gray describes the new genus *Shortia* (*galacifolia*) from specimens in fruit in the herbarium of Michaux, who discovered it on the mountains of Carolina. It has not since been found, and its flowers are unknown. This remarkable plant unites the habit of *Pyrola uniflora* with the leaves of *Galax*. Nuttall described another genus (*Simmondia*) from S. Diego, in Upper California, as a new type of the Garryaceæ (l. c. p. 400, t. 16). The collections of Hinds (Ann. Rep. for 1842) have afforded the matter for an important systematic illustrated work, which Bentham is working out, and the traveller elucidating by botanico-geographical remarks (The Botany of the Voyage of H.M.S. Sulphur. Edited and superintended by R. Brinsley Hinds. The botanical descriptions by G. Bentham. London, 1844, 4to). Five parts are now published. The representation of the character of the vegetation of California given in this work has a decided preference over the earlier ones, upon which we have previously reported. The flora of California resolves itself into two districts, a northern one, extending from the Columbia river to S. Diego (33° N. lat.), and a southern one, from here to the vicinity of the tropic, where the tropical forms of plants commence: the former corresponds to about the limits of Upper, and the latter to those of Lower California. South of Columbia (46°), where the forests of *Abies* terminate, the deciduous forests gradually continue to disappear: above S. Francisco (38°), there are no large forests, and altogether but few trees. In sailing, in Upper California, up the S. Francisco from the coast, a broad alluvial plain presents itself; it is open, and here and there sparingly wooded with oak trees like a natural park: the river flows through it, and floods it in the moist season. Bentham determined the trees to be, *Quercus agrifolia* and *Hindsii* and *Oreodaphne californica; Fraxinus latifolia* and *Æsculus californica* are also found there, and *Salices*, with *Platanus californica*, grow upon the banks of the river. At S. Pedro, the flora of Lower California prevails, and extends as far as the Magdalene Bay (24° 38′), where the most northern mangrove forests are found. Between these two points, the soil at different landing-places was either covered with low shrubs, which frequently fill the air with agreeable odours, or (in October and November) bare, like the steppes, and ornamented between the isolated portions of underwood with herbaceous plants with very beautiful flowers. Here the Synantheraceæ predominate, in the most varied forms and colours; in fact, they constitute more than the fourth part of Hinds's collection. Next to these, the Euphorbiaceæ, Polygonaceæ, and Onagrariæ are more abundant than the remaining families: the entire Californian herbarium, however, only includes about 200 sp. The arid and frequently sandy soil is physiognomically characterised by different *Cacti*, 2 of which are distributed exactly as far as S. Pedro, and accurately define the extent of the flora. At Magdalene Bay, other tropical forms also begin to appear with the mangrove forests in considerable numbers, which are mixed in the text with the steppe-plants of Lower California, but which must of course be distinguished geographically from them. The Euphorbia-shrubs only are common to both the districts of the peninsula; nevertheless, the species which predominate within and without the tropic are different. Magdalene Bay appears to form a well-defined floral limit northwards. Together with Cape Lucas, it yielded one half of the plants contained in Hinds's Californian herbarium. But whether this tropical southern point of the peninsula forms a distinct and third botanical district, or should be considered as belonging to that of the western coast of Mexico, remains at present undecided, inasmuch as most of the plants collected here have not yet been described. The following are the families of the latter collection which contain the largest number of species: Synantheraceæ ($\frac{1}{5}$), Euphorbiaceæ ($\frac{1}{9}$), Leguminosæ ($\frac{1}{10}$), Graminaceæ, Solanaceæ, Malvaceæ, and Nyctagineæ. New genera from California, by Bentham: *Stegnosperma* (Phytolaccaceæ), *Serophytum* and *Eremocarpus* (Euphorbiaceæ), *Helogyne, Perityle, Coreocarpus, Acoma, Amauria* (*Synantheraceæ*), *Eriodictyum* (Hydroleaceæ). F. D. Bennett in a short time collected some 70 sp. on the tropical south point of California; they have not yet been made known (Narrative of a Whaling Voyage. London, 1840, vol. ii, p. 18). He saw there columnar *Cacti*, from 15′—20′ in height, and speaks of the luxuriance of the forests and of numerous succulent and bulbous plants.

Martens and Galeotti have continued their papers on

the flora of Mexico (Bullet. de l'Acad. de Bruxelles, 1844, vol. xi, part 2, pp. 61, 185, 319; 1845, vol. xii, p. 129): they contain 74 Labiatæ, with the new genus *Dekinia*, 39 Verbenaceæ, 9 Cordiaceæ, 30 Boraginaceæ, and 63 Solanaceæ. The Ferns (170 sp.) and Lycopodiaceæ (12 sp.) are treated of by them in detail (Mémoires de l'Acad. de Bruxelles, 1842), and are illustrated by copper-plates. Kunze has described the Ferns and the allied families collected by Leibold in Mexico (128 sp.) (Linnæa, 1844, pp. 303-52). V. Schlechtendal's continuation of his 'Contributions to the Flora of Mexico' contain the Sapindaceæ, a new Dioscoreaceous plant, and *Hydrotænia* (l. c. pp. 48, 112, 224). Bateman has published a splendid illustrated work on the Orchidaceæ of Guatemala and Mexico, with 40 plates (Orchidaceæ of Guatemala and Mexico. London, 1843, imp. fol.)

Galeotti, in his 'Memoir on the Mexican Ferns,' has also investigated their distribution in the regions which he has assumed, and commenced a similar work in connexion with Richard, in which the Orchidaceæ of Mexico, where, according to Richard's judgment, the forms of this family are more abundant than in any other country in the world, are treated monographically, from a collection of 500 species (i. e. $\frac{1}{8}$ of all that are known), and the geographical distribution of which is given (Comptes rendus, vol. xviii, pp. 497-503) in a preliminary paper. The regions assumed by Galeotti in these two papers include the greater part of Mexico, without, however, as was the case with Liebmann's 'Characteristics of Oribaza,' their being supported by a sufficient number of special investigations. We cannot judge of the value of Galeotti's botanical division of the country until, as it is undoubtedly his intention to do, a special work is published on the botanico-geographical relations of all the Mexican families of plants. The altitudes given do not always agree with those of Liebmann, nay, in some cases, they do not agree with each other: how far this is due to inaccurate observation, and how far to local variation in the limits of the plants, cannot be satisfactorily determined. In the following sketch of Galeotti's regions, the local displacements are added within brackets to the altitudes given.

1. Hot regions. 0′—3000′ (2500′). Vegetation from December to May (end of October to June) languid: most of the trees leafless.

a. Eastern coast with forests of Rhizophoras. Mean temperature = 95° c.

b. Moist mixed forests, not, however, containing many Ferns (R. chaude tempérée des ravins). Mean temperature, 25°—19° c.

c. Coast forest of the Pacific, 25°—19°.

2. Temperate regions.

a. Eastern slope. 3000′—6000′ (5500′, 7000′). This region differs from the coast in its great humidity and evergreen foliage. It contains tree-ferns, *Liquid-ambar*, evergreen oaks (à feuilles luisantes), and numerous Orchidaceæ. Mean temperature = 19° to 15°. In Oaxaca, this region is less distinctly separated from the others: the pine trees here descend as low as 3000′, whilst Myrtaceæ, *Melastomæ*, &c., are found even at an elevation of 7000′. The soil is calcareous, and Galeotti found there only 21 Ferns, whilst on the volcanic soil of Vera Cruz, he found 77 species at the same level.

b. Western slope. 3000′ (1000′) to 6500′. Mean temperature = 20° to 15°. To it a large part of Oaxaca, of Mechoracan, and Xalisco belong. Tree-ferns are not found there, and in fact few Ferns, but a large number of oaks, with numerous Orchidaceæ growing parasitically upon the bark, and some Palms.

c. Plateau, and the slopes adjacent to it, mean temperature = 20° to 15° (21° to 18°). The internal slopes of Mexico differ in every case, botanically speaking, from those situated externally, and inclined towards the two oceans. Their dry climate, for the most part, excludes the vegetation of Ferns and Orchidaceæ. These elevated surfaces are characterised by the large number of *Cacti*:

spiny *Mimosæ* and unparasitic Bromeliaceæ are common. The latter, with *Agave,* are frequently the only plants occurring on the calcareous soil; or on other kinds of mountains, the surface is extensively covered with *Prosopis dulcis* and *Mimosæ. Bronnia spinosa* is also characteristic.

3. Régions froides.

a. Eastern slope. The determinations of the altitudes of the upper stages of vegetation, e. g. at Oribaza, are in part inaccurate; thus, according to Liebmann's investigations, the statement that vegetation ceases at 12,500′ or 13,000′ is incorrect; this portion of the sketch is therefore passed over. This region has yielded 52 Ferns, most of which grow upon limestone, and also numerous Orchidaceæ (especially between 7500′ and 8000′).

b. Western slope and high mountains of the plateau. Botanical characteristics wanting. The upper limit to vegetation is situated, according to Galeotti, on Popocatapetl, at 11,500′, on the Pic of Toluca, at 13,000′.

c. Most elevated surfaces of the plateau. No botanical characteristics.

The second and larger section of Hinds's and Bentham's work (v. s.), which is, however, not yet perfectly completed, includes the west coast of America from S. Blas (21° 32′ N. lat.), to Guayaquil (2° 30′ S. lat.) On this long coast-line the flora is adapted to a moist tropical climate, and the shore covered with a dense forest; but the plants north and south of Panama are not the same. Nor are the seasons contemporaneous; the tropical rains commence at Guayaquil in the beginning of the year; towards the north, they gradually occur later, so that at S. Blas they commence at the end of June. They divide the year into two periods of vegetation; the Bay of Choco alone forms an exception, for there atmospheric precipitations last from ten to eleven months, producing a vegetation which is constantly green, and abounding in flowers. The forests of Guayaquil appear to be comparatively poor in forms, because the rainy season there, and with it the luxuriant growth of the plants, in the vicinity of the Garua, only lasts for a short period. Of the characteristic tropical forms, some are absent, or rarely found; as Epiphytes, all the Monocotyledons, and the Ferns. North of Guayaquil the desert tracts again recur, in which the coast-stream at Salango (2° S. lat.) clothes a spot of land, like an island, with tropical trees; but as soon as the equator is passed northwards on this coast the vegetation acquires variety and strength. The Orchidaceæ and other Epiphytes then become more common; the number of forest forms rapidly increases in the same proportion as the duration of the rainy season augments, as far as the Bay of Choco (3°—7° N. lat.), where the vegetation of the western coast is most copiously developed; but the solstitial point is also reached at the same time. In this climate, the boundary of which is on this side of the equator, but which is still equatorial, the western coast contains its only Tree-ferns, and even here the *Cacti*, the characteristic plants of the trade-wind flora of America, are absent. At Panama (9° N. lat.) we again find a proportionate change of the two tropical seasons, hence no Tree-ferns nor Scitamineæ are met with there, but arborescent Cacti and other succulent plants. Most of the new species of the collection described by Bentham are from this south region of the western trade-wind coast (9°N. lat. to 3°S. lat.) North of Panama the influx of Mexican types is perceptible; Helianthea become numerous; the forests of mahogany at Realejo are also succeeded above by a region of *Pinus occidentalis*, and the oak is found even 1500′ above Acapulco. 654 species of the rich collection have already been described in the parts at present published, which extend from the Polypetalæ to the Scrophulariaceæ. Families containing most species: Capparidaceæ (10), Malvaceæ (31), Byttneriaceæ (11), Sapindaceæ (12), Leguminosæ (125), Melastomaceæ (23), Rubiaceæ (39), Synantheraceæ (95), Apocynaceæ (13), Bignoniaceæ (17), Convolvulaceæ (39), Boraginaceæ (23), Solaneæ (25), and Scrophulariaceæ (at present 17). Considering the large number of new species, the number of undescribed genera is not great: *Triplandron* (Guttiferæ), *Pentagonia* (Rubiaceæ), *Oxypappus* (Synantheraceæ), *Stemmadenia*, 3 sp. (Apocynaceæ), *Diastema* (Gesneriaceæ), *Thinogetum* (Solaneæ), and *Leptoglossis* (Scrophulariaceæ).

Purdie, a collector for the Kew Gardens, has reported upon his travels in the West Indies (Lond. Journ. of Bot., 1844, pp. 501-33). Among others he ascended the peak of the Blue Mountains, in Jamaica, where the forests of the summit consist of *Podocarpus coriacea* (Yacca). In other respects these, as also Moritz's Botanical Letters from Cumana and Caracas (Bot. Zeit., 1844, pp. 173, 195, 431), merely give notices of the plants collected.

Miquel has continued his Contributions to the Flora of Guiana (Linnæa, 1844): some new Capparidaceæ, Sapindaceæ, Malpighiaceæ, Dilleniaceæ, Leguminosæ, Melastomaceæ (*Hartigia*, n. gen.), Memecyleæ, Passifloreæ, Onagrariæ, Cucurbitaceæ, Loranthaceæ, Rubiaceæ, Convolvulaceæ, Cuscutaceæ, Bignoniaceæ (*Callichlamys = Bign. latifolia* Rich.), *Avicenniæ*, Nyctaginaceæ, Polygonaceæ, Piperaceæ (*Nematanthera*, n. gen.), Bromeliaceæ, Musaceæ, Scitamineæ, Hydrocharidaceæ, Commelynaceæ, Xyrideæ, and Aroideæ. Steudel (Ratisbon Flora, 1844): on the Melastomaceæ from Surinam, and various plants in the collections of Hoffmann and Thappler, which are for sale. Robert Schomburgk (Lond. Journ. of Bot., 1844, pp. 621-31): a new Rubiaceous plant, and two Lauraceæ, from British Guiana. Berkeley on *Stereum hydrophorum* (Ann. Nat. Hist., xiv, p. 327).

Richard Schomburgk, who accompanied his brother during his last travels in British Guiana, has described in his letters the botanical characters of the explored regions (Bot. Zeit., 1844-5). We thus obtain an interesting supplement to the previous work of Robert Schomburgk on his travels, in which the botanical determination of the plants was omitted, and which, now that a great part of the previous herbaria have been described, may be added to the descriptions of the country. The forest at Essequibo, from which *Mora excelsa* projects to an altitude of 160′, formed the first opportunity for the traveller to develope his descriptive talent. After having vividly delineated the crowded growth of the trees, the climbing plants and the creeping shrubs, which connect the stems in impenetrable meshes, and the parasites of the fallen trunks, he dwells upon a point with which we are less familiar—the *light of tropical forests.* On the ground the eye would miss the splendour of the flowers of other regions, and detect only Fungi, Ferns, and decaying vegetable structures; for even at noon a subdued light prevails in the forest, since scarcely anywhere is a portion of the sky visible through the closely-interlaced branches; but although the light is subdued beneath so dense a covering of foliage, there is more light than in dark pine forests. V. Kittlitz comes to the same conclusion as to the remarkable and as yet but little studied question, of how plants still thrive so well, and their green organs are able to respire, in shaded parts of the most dense vegetation which the crust of the earth anywhere produces (Vegetations-Ansichten, p. 6). "I was astonished," writes he, "to find so much light beneath the noblest trees, the widely-spread foliage of which scarcely anywhere allowed the sky to be seen. Remaining the same at the most varied times of the day, it could not be ascribed to the perpendicular light of noon, but only to those innumerable undulations of light which, falling from above through the crowded masses of leaves in every direction, being reflected from stem to stem and from branch to branch, finally reach the lower space in the thicket, and there produce a tone of dull lustre peculiar to tropical nature. In fact, what would become of that whole world of plants destined to live in this shade, if nature had not given the huge masses of foliage, which produce it, a *structure and distribution,* which permits it, although reflected a thousand times, still to reach in sufficient power the plants living beneath." This problem may be expressed more definitely as follows. We have to explain why the shadow of obscure deciduous forests in the temperate zone are principally illuminated by transmitted, and in the tropics by reflected light, and why the Coniferous forests are poorer in these two luminous sources, and therefore so frequently deprived of plants growing in the shade. We first think of the *Mimosæ* and forms of Palms, of the compound, and therefore imperfectly shading forms of leaves, which thus contribute powerfully to the light tone of the tropical forests. But trees possessing this character form a part only, not the whole; for those forms with simple leaves, as the laurel- and Bombax-type, preponderate, in variety of form or size of the leaf. And even the form of the leaves of the Lauraceæ, which recurs in so many tropical families, is wanting in that transparent texture to which the light of the half-shaded parts of the northern deciduous forests is owing. But Kittlitz has pointed out another more universal character of the trees of tropical forests, in the arrangement of the leaves, which appears intended to complete the former. In climates where cold or aridity cause the winter-sleep of woody plants, they develope a very much larger number of small branches, which usually form a more connected, although on the whole poorer, stratum of leaves than in the tropics. This, therefore, throws a deeper shadow upon the ground, although it is more transparent; not so deep, however, as in the Coniferous forests, the crowded leaves of which are opaque. On the other hand, it is evident that the uninterrupted heat and moisture of the equatorial climate also ensure a longer duration of the first-formed branches, many of which, in the temperate zone, fall off or remain undeveloped, and must therefore produce fresh ramifications to allow of the necessary number of leaves being formed; these first branches attracting the currents of sap, continue to grow excentrically, and hence leave between their uppermost tufts of leaves, i. e.

the youngest and softest part, more or less broad intervals. Under this double condition of the formation and distribution of the foliage, we may perceive universally in the latter climate "a certain and wholly peculiar permeability"—seen only in its simplest and most developed state in the Palms—even in woody plants, which in other respects but little resemble the latter, and in which the more copious development of the ramifications of the stem produces this prevailing character, inasmuch as they imitate and replace the natural growth of the summit of Palms. "Large masses of very delicate foliage in this manner obtain so light an aspect, that they appear as it were to float in the air; but, even down to the smallest Fern upon the soil, everything exhibits a tendency to an excentric distribution, which does not permit the separate organs to press upon one another, but by the constant crossing of lines in every direction, produces spaces for the transmission of air and light." Here Nature addresses man like the noblest works of mediæval architecture, the pointed arches of which, of Arabian origin, have, it is supposed, borrowed that openness conjoined with gigantic masses and infinite variety of form, from two palm-stems with their penniform leaves in contact.

As the second principal formation of Guiana, R. Schomburgk describes the vegetation on the banks of the streams, at the border of the forest, as made generally known by V. Martius and Pöppig, from the north of Brazil. The underwood surpasses the retreating gigantic stems; a belt of Cecropias and bamboos forms the foreground; herbaceous lianes wind around the trees and bushes as in a most luxuriant hedge, on the borders of which beautiful flowering plants augment still more the most abundant variety.

From Essequibo the travellers went to the tributary stream, Rupununi, to arrive at the savannahs on the sea of Amuku, which in these regions cover the ridges of the land almost down to the water-line, and are only separated from the rivers by seams of woods from 100′

to 200 in breadth. The main mass of the vegetation in the savannah consists of scabrous, straggling Graminaceæ and Cyperaceæ, from 3′ to 4′ in height, as *Pariana campestris, Chætospora capitata, Elionurus ciliaris, Sataria composita, Mariscus lævis,* intermixed with prickly or arborescent underwood of various kinds, as *Curatella Americana, Byrsonima, Plumieria,* Leguminosæ, Myrtaceæ, some Synantheraceæ, and Malvaceæ. The marshy places are denoted by *Mauritia flexuosa,* with Melastomas, Scitamineæ, Polygaleæ, and *Byttneria scabra;* the surface of the water itself, *Pontederia* and Nymphæaceæ.

Pöppig's illustrated work upon Tropical America is now completed by the 7—10 decades of the third volume (Lipsiæ, 1844, 4to). The 75th to 78th parts of Orbigny's Travels have appeared. Klotzsch has commenced the publication of 'Contributions to the Flora of Tropical America,' from the Museum of Berlin (Linnæa, 1844), comprising at present the vascular Cryptogamic plants and the Jungermanniæ, by C. Müller.

V. Tschudi's zoological work upon Peru contains, in the introduction, an interesting division of the Peruvian Andes, according to their climatal conditions and botanical characters (Untersuchungen über die Fauna Peruana, Lief i. St. Gallen, 1844, 4to). The climatal regions of Peru, the elevated surfaces and valleys of which are produced by the structure of the two Cordilleras, and are not dependent upon the polar altitude, are, according to Tschudi, as follows:

1. Western slope (contains no woods).

a. Coast region (0′—1500′). Mean temperature in the hot season = 27° c.; during the garua = 19°·75. A band of sand, 1620 miles in length, and from 18 to 60 miles in breadth, extending across the rivers, which intersect it many times, subdivides it into two principal formations; for the banks of the river form oases of cultivation in the Peruvian coast-steppe, the barren hilly surfaces of which are covered with fine quicksand, and are devoid of springs and, during the dry seasons, of vegetation. This hot and dry season lasts from November until the end of April. The garua, a thin mist, which is thickest in August and September, reanimates the steppes from May to October. It only extends 1400′ high vertically in the atmosphere. As long as this prevails the steppe is verdant, and sends forth numerous Liliaceous forms into flower. The south winds generally last throughout the entire year; and V. Tschudi considers the formation of the garua as still unexplained. May they not arise, as winterly precipitations, from an admixture of the lower trade-wind with the east winds descending from the Andes, and which, during the summer, are not in a condition to separate the moisture from the coast trade-wind?

b. Internal coast-region (1500′—4000′). This comprises the fan-shaped expansion of the west valleys of the Cordilleras, which, at the time of the garua, is affected with a true rainy season. Mean temp. in the dry season, $= 29°{\cdot}25$; in the rainy season $= 22°{\cdot}75$. The vegetation is not very luxuriant, but the cultivated tracts are extraordinarily productive. The sugar-cane thrives well even at 3600′. Of fruits, *Anona tripetala* (Chirimoya) and *Passiflora quadrangularis* (Granadilla) are peculiar to this region.

c. Western Sierra (4000′—11,500′), or that slope of the Cordilleras which is gently inclined below, and steep above, with its narrow transverse valleys. The air is dry; the nights are very cold in summer; the prevailing wind is the east. In summer the mean temperature at noon is $= 22°{\cdot}4$, at night $= 10°$; in winter, the mean diurnal temperature is $= 19°$. This is the region of the tropical Cerealia, and that in which the potato thrives readily and in profusion. *Oxalis tuberosa* (Oca) commences in it. The *Cacti* are among the characteristic plants of this slope, which contains but little wood.

d. Western Cordillera, comprises the west slope of the Andes above 11,000′, and the east declivity of this western crest as far downwards as 14,000′. It forms a wild mountainous tract, containing steep rocky declivities, valleys expanded into small plains, and numerous alpine lakes, and is bounded by glaciers and perpetual snow. Cutting, ice-cold winds from the east and south-east prevail constantly. Mean temp. of the day in summer $= + 11°\cdot25$, of the night $= - 7°\cdot1$; in winter, i. e. during the rainy season, of the day $= + 7°\cdot5$, of the night $= + 2°\cdot6$. The vegetation extends to 15,500′, and consists of low *Cacti* and alpine plants.

2. Eastern declivity (two regions containing no forests, two wooded).

a. Puna region (11,000′—14,000′), or the large undulating plateau between the two Cordilleras, and which attains a mean altitude of 12,000′. Sparingly overgrown flats alternate with extensive marshes, lakes, and alpine rivulets. Cold west and south-west winds blow throughout the year, most violently from September to May, with fearful thunderstorms, which occur almost daily during these months. Thus the rainy season commences in the opposite half of the year to any one travelling from the coast to the interior. The sky is clear from May to September, and the nights very cold; the temperature is altogether very variable; it frequently varies within twenty-four hours 22 or 25 degrees; not unfrequently, on these cold heights, we suddenly encounter warm currents of air from the S.S.E., which at times are only from two to three paces in breadth, while in other cases as much as several hundred feet, and which rapidly recur (p. xxiv). Tschudi gives, as approximative mean values of the temperature: Summer (November to April, which period is there incorrectly called the winter), of the night $=+1°\cdot5$, of noon $=8°\cdot75$; winter (May to October, there incorrectly called summer), of the night $=-6°\cdot25$, of noon $=12°\cdot1$. The vegetation of Puna is poor; *Stipa Ichu* is abundant; Synantheraceæ, Malpighaceæ, Leguminosæ, Verbenaceæ, Scrophulariacæ, and Solaneæ are mentioned. Barley does not ripen at 13,050′.

b. Eastern Sierra (11,000′—8000′), consists of broad,

open, fluviatile valleys, the most thickly populated in Peru, and is separated from that of the Puna by rocky declivities; rainy season, with frequent hail, from October to February. During the winter months (also called summer in the text) dry east winds prevail; night frosts set in after the end of the rainy season, and the Cerealia are harvested. Mean temp. during the rainy season: of the night = + 5°·1, of the day = + 14°·1; during the winter (March to September), of the night = – 4°·25, at noon + 17°·1. But great local differences occur in the hot bottoms of those valleys which are sheltered from the winds, where fruits of the south of Europe, as peach trees, thrive, sometimes even at an altitude of more than 10,000′; the principal cereal grain appears to be maize. The slope of this region, which, like the former, is destitute of woods, contains a profusion of *Cacti*, and on the banks of the streams only we find woods of *Salix Humboldtiana*, 20′ in height; even European fruit trees do not thrive when cultivated. In the valleys, however, this region extends directly into the forest region, from which it is also separated by a second Puna, i. e. by the crest of the central Cordillera.

c. Upper forest region or Ceja-region (from Ceja de la Montagne, i. e. the brow of the mountain) (8000′—5500′), comprises the eastern slope of the internal Cordillera, and its western slope in the north of Peru, with the longitudinal valley of Huallaga. It consists of steep rocks and narrow, wooded mountain-ridges. The climate is humid, cold, and rough, with prevailing south winds. Towards evening dense mists are formed, which during the night rest upon the forests, and which the wind carries away with it from the morning until the serene evening. These mists extend downwards as far as 6500′, and often resolve themselves into very heavy showers. The differences in the seasons are not mentioned; the observations upon the temperature are also incomplete. Low trees and shrubs covered with mosses commence even at 9500′, and increase in size and strength as we ascend. Cerealia cannot be cultivated in this region, which is not exposed to the direct rays of the sun; potatoes grow abundantly.

d. Lower forest region (5500′—2000′), is composed of immense forests, savannahs, and marshes. Its humidity is great throughout the year; for even in the dry season (May to September) thunderstorms are common. The true rainy season begins in October, and lasts until March or April. Mean temp. = 30°; when the wind is in the east, the nocturnal temperature sinks to 18°·75. This region forms the commencement of the primæval forests of the Amazon.

Contributions to the Flora of Brazil: Moricand, Plantes nouvelles ou rares d'Amérique, livr. 8, Tab. 71-84 (Genève, 1844, 4to); Naudin, Description de Genres nouveaux de Mélastomacées (Ann. d. Sc. Nat., 1844, ii, pp. 140-56): *Tulasnea, Brachyandra, Eriocnema, Augustinea, Stenodon,* and *Miocarpus;* Fischer and C. A. Meyer, *Asterostigma,* n. g. Aroideæ (Bull. Pétersb. iii, p. 148). Miers, *Triuris* and *Peltophyllum,* forming the new family Triurideæ, allied to the Juncagineæ (Transact. Linn. Soc. xix, pp. 77, 155); Sir W. Hooker and Wilson, Enumeration of the Mosses and Hepaticæ collected in Brazil, by G. Gardner (Lond. Journ. of Bot., 1844, pp. 149-67). K. Müller, Enumeration of the Mosses collected by Gardner in Brazil (Bot. Zeit. 1844, p. 708), gives no description of the new species, so that the preceding publication, which is founded on more complete materials, acquires priority as regards the nomenclature.

Tenore has described a new *Aristolochia,* from Buenos Ayres, which he obtained from Bonpland's collection of seeds, and has taken this opportunity of republishing the diagnosis of some plants derived from the same source, and described in his catalogue of seeds (Rendiconto di Napoli, 1842, pp. 345-8).

Kittlitz's first plate represents the botanical character of the coast of Valparaiso. It gives a view of one of the steppes during the dry season, the bare soil of which appears only to produce *Cacti* and shrubs with stellate prickles, but in which, during August and September, the most luxuriant grass plains, with their bulbous plants, are found. The following are some of the physiognomically important forms of plants represented in this drawing: the Caves (*Mimosa Cavenia*), the dwarf-pine-like Lithi (*Rhus caustica*), *Cereus Peruvianus*, *Puretia courctata*, Synantheraceous shrubs, bamboos, &c.

Miers has proposed two genera of Iridaceæ from Chili —*Solenomelus* (*Cruckshankia* ej. ol.) and *Symphostemon* (*Sisyrinchium odoratissimum* Cav.) (Transact. Linn. Soc. xix, p. 95). Sir W. Hooker has determined the Alerse tree of the south of Chili, which is used as timber for building, to be *Thuja tetragona* (Lond. Journ. Bot., 1844, p. 144). Berkeley has described an edible Fungus from Terra del Fuego; *Cyttaria* n. gen., near *Bulgaria*, also containing a species from Chili (Transact. Linn. Soc., xix, p. 37).

VI.—AUSTRALIA AND SOUTH-SEA ISLANDS.

F. D. Bennett remarks that westerly winds, corresponding to the monsoon, not infrequently extend eastwards over the Pacific Ocean towards the Society Islands, and especially in February and March are not infrequently taken advantage of, for making voyages in a south-easterly direction; consequently in regions which, in other respects, are completely under the influence of the south-east trade-wind (Whaling Voyage, i, p. 159). The botanical communications, which form an appendix to the account of his voyage, and which treat especially of the cultivated plants of the South Sea Islands, contain, in addition to numerous well-known facts, many names of Polynesian plants.

The illustrations of the Caroline and Marian Islands, and the archipelago of Bonin, are among the most excellent and richest views in Kittlitz's work; but the systematic determination of the plants figured is entirely wanting—a deficiency caused by the early death of Mertens. The vegetation of tropical forests has, in fact, never been more distinctly represented than in these landscapes, except by Rugendas. The characteristic types of the most important physiognomical forms of tropical foliage are principally found in the following plates: Bamboo form, indicated by *Artocarpus* (pl. X); Banana, form expressed by the *Rhizophoræ* of the mangrove forests (pl. V), and stems of *Ficus* supported by aerial roots (pl. VI); Cycadeæ (pl. XI), Palmeæ (pl. IX, XVI), Musaceæ (pl. VII), *Pandanus* (pl. X, XI, XII, XV), and that of the Tree-ferns (pl. XVI). Of other physiognomical forms, Lianes (pl. VIII, XV), *Freycinetiæ* (pl. VI), parasitical Ferns (pl. V, VI), Aroideæ (pl. VII), *Agave,* imitated by stemless species of *Pandanus* (pl. XI, XII), herbaceous Ferns (pl. VIII). Pl. XIII represents the savannahs on the Marian Islands; grass plains with *Casuarina, Cycas,* and *Pandanus.*

Suttor's paper, read before the Linnæan Society, upon the Forest Trees of New Holland contains, according to the extracts before us, only well-known facts (Ann. Nat. Hist., xiii, p. 217). Drummond's Letters from the Swan River have been continued (Lond. Bot. Journ., 1844, pp. 263, 300). They contain, for the most part, notices of individual plants which were transmitted to Hooker. The extensive herbaria brought by Preiss from the Swan River have been systematically described in detail in a separate work, edited by Lehmann, by a number of scientific men, mostly Germans (Plantæ Preissianæ sive enumeratio plantarum, quas in Australasia occidentali et meridionali—occidentali collegit L. Preiss. Ed. Chr. Lehmann. Vol. i. Hamburghi, 1844-1845, 8vo). The coadjutors were—Bartling, Bunge, Klotzsch, Lehmann, Meissner, Miquel, Nees v. Essenbeck, Putterlick, Schauer, Sonder, Steetz, Steudel, and De Vriese. Summary of the families treated of, with the enumeration of new genera, and those containing most species: 247 Leguminose (Meissn.); 63 *Acaciæ*, 10 *Chorozema*, 15 *Gompholobium*, 11 *Jacksonia*, 23 *Daviesia*, 15 *Gastrolobium*, 10 *Bossiæa*, Rosaceæ 1 (N.), Chrysobalaneæ 1 (N.); 161 Myrtaceæ (Sch.), 15 *Verticordia*, 14 *Calycothrix*, *Symphomyrtus* n. gen., 15 *Eucalyptus*, 33 *Melaleuca*, 10 *Beaufortia*, 15 *Calothamnus;* 3 Halorageæ (N.); Onagrariæ 1 (N.); 2 Oxalideæ (Steud.); Lineæ 1 (Bartl.); 6 Geraniaceæ (N.); 2 Zygophyllaceæ (Miq.); 25 Diosmeæ (Bartl.); *Boronia* 15; 12 Euphorbiaceæ (Kl.); *Trachycaryon* (*Croton* sp. Lab.); *Calyptrostigma* (*Croton* sp. Lab.); *Lopadocalyx* n. gen.; 3 Stackhousiaceæ (Bg.); 22 Rhamneæ (Steud.); *Pomaderris* 10; *Cryptandra* 10; 13 Pittosporaceæ (Putterl.); 17 Polygalaceæ; *Comesperma* (Steud.); 15 Tremandraceæ (Steetz); 11 *Tetratheca; Platytheca* n. gen.; 10 Sapindaceæ (Miq.); Olacineæ 1 (Miq.); Hypericineæ 1 (N.); 32 Byttneriaceæ (Steud.); *Thomasia* 19; *Fleischeria* n. gen.; 11 Malvaceæ (Miq.); Phytolaccaceæ 1 (Lehm.); 5 Caryophyllaceæ (Bartl.); 5 Portulaceæ (Miq.); *Tetragonella* n. gen.; 2 Mesembryanthemeæ (Lehm.); Frankeniaceæ 1 (N.); 20 Droseraceæ (Lehm.); *Drosera* 17; 8 Cruciferæ (Bg.); *Monoploca* (*Lepidii* sp. D. C.); 6 Ranunculaceæ (Steud.); 44 Dilleniaceæ (Steud.); *Hibbertia* 26; *Candollea* 11; 3 Crassulaceæ (N.); Cephaloteæ 1 (Lehm.); 8 Loranthaceæ (Miq.); 31 Umbelliferæ (Bg.); *Platysace* n. gen.; *Schœnolæna* n. gen.; 99 Epacridaceæ (Sond.); *Astroloma*, *Brachyloma* n. gen.; *Leucopogon* 47; *Andersonia*, 14; 3 Primulaceæ (N.); 8 Lentibulariæ (Lehm.); 6 Scrophulariaceæ (Bartl.); 5 Solanaceæ (N.); 5 Convolvulaceæ (De V.); 5 Boraginaceæ (Lehm.); 8 Myoporinaceæ (Bartl.); 2 Verbenaceæ (Bartl.); Avicennieæ 1 (Miq.); 25 Labiatæ (Bartl.); *Colobandra* 6, n. gen.; *Anisandra* n. gen.; 6 Gentianaceæ (N.); Apocynaceæ 1 (Lehm.); 5 Loganiaceæ (N.); 4 Rubiaceæ (Bartl.); 69 Stylidieæ (Sond.); *Stylidium* 64; *Coleostylis* (*Stylidii* sp. Benth.) *Forsteropsis* n. gen.; 18 Lobeliaceæ (De V.); *Lobelia* 17; *Vlamingia* n. gen.; 59 Goodenovieæ (De V.); *Dampiera* 15; *Scævola* 27; 101 Synantheraceæ (Steetz);

Eurybia 11; *Gymnogyne* n. gen.; *Silphiosperma* n. gen.; *Pogonolepis* n. gen.; *Pachysurus* n. gen.; *Chrysodiscus* n. gen.; *Chthonacephalus* n. gen.; *Anisolepis* n. gen.; *Pterochæta* n. gen.; *Siemssenia* n. gen.; *Hyalosperma* n. gen.; *Schœnia* (*Helichrysi* sp.); 2 Plantaginaceæ (N.); 208 Proteaceæ (Meissn.), *Petrophila* 21, *Isopogon* 15, *Adenanthos* 10, *Conospermum* 17, *Grevillea* 29, *Hakea* 46, *Banksia* 19, *Dryandra* 22; 16 Thymeleaceæ; *Pimelea* (Meissn.); 7 Laurinaceæ, *Cassyta* (N.); Nyctagineæ 1 (N.); 6 Polygonaceæ (Meissn.); 14 Amarantaceæ (N.); *Trichinium* 19; 14 Chenopodeaceæ (N.): Urticaceæ 1 (N.); 9 Casuarineæ (Miq.); 2 Coniferæ (Miq.); *Actinostrobus* n. gen.; Cycadaceæ 1; *Macrozamia* (Lehm.) Hence at present about 1450 Dicotyledons.

Gunn has addressed some botanical letters from Van Diemen's Land to the editor of the 'London Journal of Botany' (1844, pp. 485-96). He describes an excursion to the western highlands of the island, and gives statements of the localities of rare plants, with a more detailed report upon a new species of *Eucalyptus* (*E. Gunnii*) (Hooker, *fil.*), which in December and January contains a large quantity of a saccharine and fermentable juice, whence it is called by the colonists the cider-tree. As it forms extensive mountain-forests, it will probably hereafter become an important product of Tasmania. Harvey has described some new *Algæ* from Van Diemen's Land (id. pp. 407, 428); amongst them the Rhodomeleaceous plant, *Pollexfenia*, which is also indigenous to the Cape. Contributions to the Flora of New Zealand: Catalogue of a Collection of Plants from New Zealand, by Stephenson, determined by J. D. Hooker (Lond. Journ. Bot., 1844, pp. 411-18)—it contains but few new species; Hepaticæ Novæ Zeelandiæ et Tasmaniæ, by J. D. Hooker and Taylor (id. pp. 556-82); Diagnoses of New Zealand Plants, by Raoul, preliminary to his illustrated work, which was published in 1846 (Ann. Sc. Nat., 1844, ii, pp. 113-23), with the new genera: *Ileodictyon* (Fungi), *Pukateria* (Corneaceæ ?), and *Tetrapathea* (Passifloreæ).

Colenso's Botanical Diary of his travels during several months through the little known interior of the northern islands of New Zealand (Lond. Journ. Bot. 1844, pp. 4-62) contains numerous localities of, and reports upon, newly-discovered plants which have not yet been made public, but will be described in Dr. Hooker's illustrated work.

The first three parts of the latter work have appeared; they contain a general introduction upon the botanical characters of high latitudes of the southern hemisphere, and also the commencement of a flora of the Auckland Archipelago (The Botany of the Antarctic Voyage of H. M. Discovery Ships Erebus and Terror, under the command of Sir J. Ross; by Jos. Dalt. Hooker. Parts i-iii, London, 1844, 4to.) Being, during the summer, almost always either in high barren latitudes, or on the open sea, Hooker had but little opportunity of collecting other than such plants of the antarctic flora as flowered in the winter or spring. But he considers this defect, which concerns the copiousness of the materials which he collected, as of little consequence, as he was in the favorable position of being able to make use of the botanical results obtained in all the earlier British voyages to the South Pole, but of still less, in consequence of a climatal peculiarity which he developes in the introduction, and regards as the most characteristic feature of the antarctic vegetation. He was surprised on finding at Kerguelen's Land, the same plants in flower which Cook had met with at other seasons, and this result he subsequently found to occur generally. The vast preponderance of water in the high southern latitudes produces an uniformity in the distribution of heat throughout the year, and the more we approach the pole the more distinctly does this appear to increase. The seasons there are not distinguished, as in the north, by their temperature, but by scarcely anything more than the variation in the amount of light; all the months are cold, but the thermometer varies, as in the tropics, within narrow limits. In the region of floating icebergs, between 55° and 65° S. lat., seldom a day occurred during the summer in which the temperature rose or sunk beyond the limits of 0° c. and — 6°·6 c. South winds, with much snow, alternate there with aerial currents from the north, which being loaded with aqueous vapour, incessantly diffuse white fogs of indescribable density over the surface of the ocean. These precipitations are also formed on islands situated in the vicinity of this region, throughout the year, by the admixture of the winds from the land and sea depriving them of their solar climate, and for the most part preventing the change of temperature dependent upon the position of the sun. A climate so inhospitable and uniform excludes any variety in the forms of plants, but confers a luxuriance of growth upon the indigenous plants, of which the arctic regions must necessarily be deprived, because their vegetation is subjected to a prolonged winter-sleep. This is so remarkably the case, that notwithstanding the differences in the climatic conditions, most of the genera and forms of the antarctic agree with those of the arctic flora in the most important points, excluding only the Auckland Islands, which appear to belong to the same primitiveformation with New Zealand. But notwithstanding this similarity of the types, the species of the southern district are peculiar; which could not have been expected to be otherwise in the case of islands, not only separated climatally to such an extent, but are also situated beyond the reach of all continents, the oceanic currents of which usually plant the waste shores. Many antarctic species indicate their endemic origin by the limited district through which they are distributed in the region itself. However, the special botanical results of Hooker's voyage, the description of which far excels his former communications in fulness and arrangement of the matter, are reserved for the next Annual Report. The Cryptogamia have also been partly described in the 'London Journal of Botany' for 1844, including 72 Hepaticæ from the Auckland Islands, by

Hooker and Taylor, (p. 366;) 66 sp. more from the Falklands, Cape Horn, and Kerguelen's Land, by the same author, (p. 454;) 73 antarctic Jungermanniæ, by Hooker and Wilson; with the new genera *Lophiodon* and *Hymenodon* (p. 533), and 151 antarctic Lichens, by Hooker and Taylor, (p. 632.)

Dr. Hooker paid particular attention to the distribution of the *Algæ* floating in the high latitudes of the Southern Ocean. (Antarct. Voy., Introduct.) *Macrocystis* and *Urvillea* were found common as far as the northern limit to the floating ice, in one instance they extended to 64° S. lat.; but they usually disappeared much sooner, e. g. south-east of America, below 55° S. lat. But in the latter meridian a new form of *Alga* appeared below 63° S. lat., which although previously found in D'Urville's expedition, has since been described as *Scytothalia Jacquinotii.* Here, near the coast of Palmer's Land, on Cockburn's Island, (64° S. lat.), no Phanerogamous plants were met with, only 20 Cryptogamia. These appear to be the last forms of plants in the direction of the antarctic pole; for even the Algæ are absent on that continental coast upon which the active volcano Erebus and the extinct volcano Terror are situated, and where the soil at the level of the sea appeared for the first time entirely deprived of vegetation,—a sight never before witnessed, and from which nature appears to have preserved even the highest latitudes of the north.

REPORT

ON THE PROGRESS OF

GEOGRAPHICAL AND SYSTEMATIC BOTANY,

DURING THE YEAR 1845.

BY DR. A. GRISEBACH,

EXTRAORDINARY PROFESSOR IN THE UNIVERSITY OF GOTTINGEN.

GEOGRAPHICAL AND SYSTEMATIC BOTANY.

The consideration that the greater number of literary productions in the province of Systematic Botany are connected with the working out of individual local floras, and ought therefore to have been noticed in the preceding botanico-geographical Annual Reports, has convinced the author of them, that by altering the arrangement of the matter, with an appropriate limitation of the descriptive notices, Systematic botany also may be made to enter into these notices, without the allotted space being exceeded. The year 1845 has moreover proved comparatively poor in botanico-geographical results, so that the present period appears appropriate for a first attempt at enlarging the botanical Annual Reports in correspondence with this view. Hence by comprising, in combination with those upon Vegetable Physiology, the entire domain of Botany, they will now obtain a completeness corresponding to that of the Zoological Reports, and, as I hope, will acquire more practical utility. An important limitation will, however, still exist in the Botanical Reports, viz. that both the descriptions of the plants, as also the notices of individual species of genera already known, are excluded from the sketch given of the systematic works; not only, however, does want of space compel us to omit these considerations, but it would also be superfluous to repeat here what is annually done in so admirable a manner in several botanical periodicals and repertories,

A.—BOTANICAL GEOGRAPHY.

R. B. Hinds has continued his general observations upon botanical geography (see Annual Report for 1842) during the past year (Memoirs on Geographic Botany, Ann. Nat. Hist. vol. xv); like the former, however, they contain little more than already known facts and views, not unfrequently mingled with errors, both as regards the facts and deductions.

The present papers contain, e. g. estimates of the whole number of plants known;* remarks upon centres of creation, the existence of which the author denies; on the distribution of certain families; on the mean area of the extension of each species; elements for the comparison of two floras; on physiognomy, &c. I shall only enter upon the consideration of one of these views, and that because it places a simultaneous work by Forbes, remarkable for its originality, in a proper light. The old hypothesis of a single centre of creation, from which all plants upon the earth were distributed, as also the later supposition, that this migration of organisms originated in one or a few centres, Hinds opposes by the general law, that wherever plants met with the conditions necessary for their existence, the present vegetation was originally produced. He does not admit, in opposing every migration of plants, even such variations in the original state as would cause the extirpation of individual species, and their disappearance from among the number of living organisms; whilst such an event, e. g. in the case of the endemic plants of St. Helena, has been as satisfactorily determined as in that of *Didus ineptus*. The historical change in the constituents of forests, and the migration of individual plants which is in progress before our eyes, and is not merely produced by man, are incompatible with a law expressed so generally. The fact that certain islands in the Indian Ocean, as e. g. Darwin showed, contain only plants which have been washed on shore, by which they are thickly covered, whilst the islands in their vicinity possess an endemic vegetation, contradicts the supposition of the existence of a generally diffused productive force, or at least limits it to distinct creative epochs. When we

---

* Hinds estimates the number of known plants at 89,170; those existing on the earth at 134,000 species. His statements are founded upon the number of species contained in the first four volumes of De Candolle's 'Prodromus.' These amount to 20,100; of which 3875 are Leguminosæ, 1631 Rubiaceæ, 1009 Umbelliferæ, 990 Cruciferæ, 759 Caryophyllaceæ, 715 Myrtaceæ, &c.

reflect on the well-known facts which Hinds brings forward, without either certainty or accuracy of detail, as the foundation of his opinions, they leave room for the formation of other hypotheses besides his. His laws are as follows: 1. In proportion as the districts of vegetation are further separated from each other by the sea, so is the number of species of plants common to them less. Hence the large number of species common to the three large portions of the earth in the arctic zone, and hence the greater the contrast, in comparing the floras of corresponding climates, the further south we proceed, the various portions of the earth in the southern hemisphere being further separated from each other. 2. If we divide the whole earth into six floral districts—which would certainly be arbitrary enough—we obtain for each almost exclusively endemic species, and we may add, that the same result would hold good were we to admit more than thirty districts. 3. In different natural floras under corresponding climates, we certainly find similar forms, but not the same species. 4. In some islands the vegetation is entirely endemic; these cannot, therefore, have obtained their plants by migration from external sources, &c. All these and similar facts certainly are opposed to the migration of plants from a single point of the earth's surface to all the rest, and which scarcely any naturalist now believes; but there is a broad interval in the argument, which has not yet been filled up with facts, between these considerations and the assertion, that centres of creation do not exist, but that every point has produced the plants it possesses. We know that some regions of the earth contain many more endemic species than others, without the soil or climate being sufficient to explain this increase. The number of endemic forms diminish in the direction of any climatal boundary, as it were, like the radii of a circle, in the centre of which a centre of creation is situated ; hence, e. g. in Europe, we may speak of western, eastern, or southern forms of plants, which gradually disappear, one after the other, towards the east, the west, or the north. There does not appear to be any other difference between an island which possessed endemic plants only, as St. Helena, and a continental district, which, like Spain or Illyria, abounds in endemic plants, than that other plants which have migrated from external sources are associated with the latter, which could not readily occur in the former case, on account of the distance of any continent. On reviewing all the known facts, and seeking for the simplest theory by which their relation may be explained, we are compelled to adopt the supposition of the existence of as many centres of creation as there are districts of endemic plants upon the globe. The difficulty of determining in each case the original centre of creation, on account of the mixture of the groups in the wide and connected districts of continents, is so great, that it must always remain the main object of botanical geography. It is only the problem of the groups of creation that gives this science a peculiar importance, and raises it above the imputation of being an aggregate of heterogeneous laws belonging to different sciences; for in this light only does it present a definite and independent method of investigation, a progressive course of development. Commencing with observations upon the geographical area of each individual species of plant, the object of botanical geography is first to determine what limits to this distribution have been placed by the composition of the soil or the subdivision of the continent. It then points out the climatal sphere of the species, and if it ascertains, after this twofold limitation, that the natural area is more contracted than is accounted for, the geological problem presents itself—what has not been the result of soil and climate must depend upon historical causes, the history of the earth. If the same soil and the same climate have produced only similar but not identical forms, this refers us to a creative act of a different kind, therefore to a geological epoch.

In connexion with this combination of geological and botanico-geographical investigation, Forbes has made an attempt of a different kind, viz. the application of the distribution of plants to geological deductions (Report of the Meeting of the British Association, held at Cambridge, Ann. Nat. History, xvi, p. 126). On comparing the specific centres of the endemic plants of Great Britain, i. e. the centres of their geographical areas, it is evident that the flora of the greater part of the surface of the country belongs to that of Germany. The specific centres of the few species peculiar to the British Isles occur in the same region. In addition to this principal area, four smaller districts of vegetation may be distinguished according to the same law: 1. The mountainous districts of the west of Ireland contain a number of plants in common with the north-west of Spain and the Pyrenees. 2. The south country, Devonshire, Cornwall, and the Channel Islands, in common with the west of France. 3. The south-east of England, especially its chalky districts, with the north of France. 4. The mountains of Wales, the north of England and Scotland, with the plains of Norway. Forbes does not consider this connexion as explicable by soil and climate, and therefore seeks for geological causes, in conformity with the above law. He believes that these are to be found in a former connexion of Great Britain with the continent, probably existing in earlier geological periods, especially during the tertiary period: not that this connexion, made use of for his explanation, has been geologically determined; but he endeavours to support his geological hypotheses by these botanico-geographical relations. Following up this design, which is certainly not free from objection, Forbes then not only maintains generally these connexions of land, but by supposing former elevations and subsidence of the soil, arrives at certain views regarding the series of changes which have occurred; in fact, he distinguishes the floras according to the periods at which they were formed. I should have no hesitation myself in granting, that when two different floras really belong to the same soil and climate, by far the simplest hypothesis is to attribute their origin to different geological epochs; but if, as I believe to be the case, climatal conditions sufficient to account for the above distribution of British plants were present, the error would not lie in the method, but its application, which has led Forbes to the following results. According to his view, the districts of vegetation distinguished above correspond to as many geological eras, so that the flora of the west of Ireland would be the oldest, that of the mountains the fourth, and that related to Germany the youngest. The first mentioned descend from a time when a chain of mountains, running across the Atlantic Ocean, connected Ireland with Spain; this would explain its difference from the vegetation of the mountains, although it would still correspond with the mountain character. Moreover, in the second and third periods the English Channel was closed, first towards the west, then towards the east also, by the connexion of land, and thus the distribution of French plants in England was occasioned. Forbes explains the alpine flora of the mountains by means of Agassiz's glacial period; the mountain summits of Britain were then low islands, extending to Norway, and were clothed with an arctic vegetation, which, after the gradual upheaval and consequent change of climate, became limited to the summits of the newly-formed and still existing mountains. Lastly, the bed of the North Sea itself was upheaved, and extensive plains laid dry between England and Germany, upon which the elk and other extinct quadrupeds lived, and over which the plants of Germany migrated; until at last the sea, in consequence of fresh depression, flowed back, after the important object of transplanting Roses and *Rubi* beyond the ocean had been accomplished. Further than this, hypothetical views could not easily be carried, and I have translated them almost entirely here, only because Forbes appears desirous by this paper of opening a new path in botanical geography, for this first lecture has since been followed by others. The criticism of his undertaking lies simply in the denial of one of the first statements with which he commences: actual natural forces, the sea, rivers, currents of air, by which seeds are diffused, or animals, and even man, are, in the majority of cases, insufficient means for effecting the migration of plants across the British seas. I maintain that these forces are quite sufficient, provided the imported seeds meet with a suitable climate and proper soil. Those western-Europe plants,—which, being produced through the agency of the coast-climate of the Atlantic, and, according to the degree of this dependence, becoming distributed sometimes to a greater, sometimes a less distance within the continent, the author refers in the latter case to Spain, in the former to France,—are not met with equally on the coast-line of the continent, but are often absent from wide tracts, the soil of which is not favorable to their growth; when e. g. we do not find *Erica cinerea* anywhere from the Rhine to the Fjord of Bergen, who would, in this case, suppose that former connexions of land had disappeared, when for the most part the connexion still exists, without contributing to the distribution of this shrub?

Since the Alps contain so many alpine species in common with the arctic regions, it is still more easily seen how little the continent situated between these two terminal points serves to explain these agreements. The plains which, without this alpine attire, extend e. g. from Cola to the Carpathians, are, however, less adapted to the transport of foreign plants than a sea which rapidly carries over the seeds in its current. Or when Forbes has recourse to the glacial period in explaining the diffusion of the arctic plants: how will he account for so many central European species of Sierra Nevada or Pindus traversing the extensive tracts of land which separate them from their centre of creation? How, by the most complex dislocations, will he bring the *Minuartiæ* and *Queriæ* into geological connexion, which do not flourish anywhere between Castile and the Crimea? There is no reason why water should form a greater obstacle to the distribution of plants than a soil which does not conduct them; extensive oceans certainly form barriers, when there is no current to carry them across, or when the climates of the two coasts are dissimilar.

A. Erman has written a paper on the periods of vegetation in different climates (Arch. für Russland, Bd. 5, pp. 617-40).

He examines the question, of what relation the stages of development of vegetation hold to the temperature, at which in different latitudes they appear in the same species. His investigations lead to the negative result, that a law communicated to him conjecturally by Quetelet is unfounded. It consisted in the assumption that similar stages of development occur in two different places, when the sum of the squares of the diurnal temperature from the commencement of the period of vegetation is the same for both. He shows, also, that the stages of development and the sum of the temperature acting upon them in different places, are by no means in direct proportion.

We must mention a remark made by J. D. Hooker in regard to botanico-geographical physiognomy (On *Fitchia*, in Lond. Journ. of Bot., 1845, p. 640).

On many remote islands possessing endemic floras, we find woody plants belonging to the family of Synantheraceæ, which contribute essentially to the character of the districts, and belong to peculiar genera, of which representatives do not occur on the continents. The following sketch will serve to illustrate this:

St. Helena contains 4 gen. 10 sp. of Synantheraceæ, all woody plants;
Juan Fernandez „ 8 „ 17 „ of „ of which 3 gen. and 12 sp. are woody plants;
Gallapagos contains 13 gen. 21 sp. of Synantheraceæ, of which 3 gen. and 8 sp. are woody plants;
New Zealand contains 30 gen. 60 sp. of Synantheraceæ, of which 8 gen. and 14 sp. are woody plants.

Elizabeth Island, which belongs to the botanical district of the South Sea Islands, in the southern hemisphere, but is situated nearer to the island of Juan Fernandez and the continent of America than the others, also contains the new arborescent Cichoraceous genus, *Fitchia*, whilst the other islands of this archipelago do not contain similar forms of plants.

I.—EUROPE.

Of Ledebour's Flora Rossica (see the Annual Report for 1841 and 1843), the 6th part appeared in 1845, and the 7th in 1846 (vol. ii, part 2).

The statistical proportions of the families treated of in them are as follows: Synantheraceæ, 890 sp. The Vernoniaceæ are only represented by the Caucasian *Gundelia;* of the Eupatoriæ, in addition to the West-European genera, *Nardosmia* and 7 arctic species; the Asteroideæ comprise the exclusively Asiatic genera *Turczaninowia, Calimeris, Arctogeron, Diplopappus, Rhinactina, Myriactis, Brachyactis, Dicrocephala, Karelinia, Eclipta,* and *Siegesbeckia,* which extends as far as the Crimea; among the Senecionidæ, including the Heleniæ, *Richteria* and *Cancrinia, Brachanthemum,* one of the Chrysanthemeæ from Siberia, *Waldheimia* from Altai, *Cladochæta* and *Amblyocarpum* from the Caucasus, and *Senecillis* from Podopolien. The genera most abundant in species are *Artemisia* (83 sp.), *Senecio* (52 sp.), *Achillea* (31 sp.), *Pyrethrum* (29 sp.); among the Cynaraceæ, including *Acanthocephalus* and *Haplotaxis* (3 sp.), and *Ancathia* from Altai, *Alfredia* (4 sp.) from Siberia, and *Cousinia* (20 sp.) and *Acroptilon* from the Steppes, and *Acantholepis, Chordinia,* and *Oligochæta* from Armenia. Those containing most species are *Centaurea* (61 sp.), *Cirsium* (51 sp.), *Serratula* with *Jurinea* (36 sp.), *Saussuria* (32 sp.); the Cichoraceæ contain *Heteracia* and *Microrhynchus* from the Steppes, *Asterothrix* from the Caucasus, *Intybellia* from the Crimea, *Youngia* (5 sp.) from Armenia and Siberia, *Ixeris* and *Nabalus* from Siberia, *Apargidium* from Sitcha; and of larger genera, *Hieracium* (25 sp.), *Crepis* (23 sp.), *Scorzonera* (19 sp.), *Lactuca* (17 sp.), *Tragopogon* (17 sp.); Lobeliaceæ, 2 sp.; *L. dortmanna*, and in Eastern Siberia *L. sessilifolia;* Campanulaceæ, 66 sp., containing *Michauxia* and *Symphyandra* from the Caucasus, *Platycodon* from Darien; the genera containing most species being *Campanula* (36 sp.) and *Adenophora* (10 sp.); Vacciniæ, 11 sp., including 4 sp. from Sitcha, 1 sp. from the Aleutian Islands, and *V. arctostaphylos* from the Caucasus; Ericeæ, 36 sp.; 2 *Rhododendra* and *Azalea Pontica,* confined to the Caucasus, extending hence to Dombrowitza in Lithuania, 4 sp. of *Cassiope, Bryanthus,* 2 sp. of *Amothamnus,* 5 sp. of *Rhododendron* confined to Siberia, 2 sp. of *Cassiope, Menziesia,* 1 sp. of

Phyllodoce, *Kalmia*, and *Cladothamnus* confined to Sitcha; Pyroleæ, 7 sp., corresponding to the German species; Monotropeæ, 1 sp.

The 5th and 6th parts of Trautvetter's illustrated work 'Plantarum imagines Floram Rossicam illustrantes.' Monachii, 1845, 4to, (see the preceding Annual Report) have appeared, containing pl. XXI—XXX.

The Academy of St. Petersburg has commenced the publication of 'Contributions to the Botany of the Empire of Russia' (Part 1, Petersburg, 1844, 30 pages 8vo.; part 2, 67 pages and 6 plates; part 3, 56 pages; part 4, 93 pages; ib. 1845).

The first part contains a local flora of the province of Tambow (incomplete, containing 312 sp.), forming Ruprecht's fourth contribution to the flora of St. Petersburg. The same author has published an account of the Ferns and Charæ of the empire of Russia in the third part; it also describes some new Ferns from Siberia, Mongolia, and American Russia, as also of the Charæ from the Soongari.

The second part, in which Ruprecht has described his botanical travels in the extreme north of European Russia, is of more general interest. In the unfavorable summer of the year 1841, he collected in the eastern part of the province of Archangel, especially Mezen, the peninsula of Kanin, and the island of Kalujew. The natural character of the country differs from that of Scandinavian Lapland in the circumstance that the forest-limit recedes almost to the vicinity of the arctic circle; hence extensive, low, treeless plains are spread along the arctic ocean. Thus the pine forests are entirely absent at Kanin (excepting a wood composed of *Abies*, situated below 67¼° N. lat., and which is now dying); they cease at the Indega river, about fifteen English miles from the sea, and scarcely extend over the arctic circle beyond Petschora. The cultivation of barley and potatoes also is only carried on as far as the city of Mezen. The forests are succeeded northwards, first by a zone of low birches and willow-shrubs, the dwarf birch with the arctic Ericaceæ follow next, and lastly, with the latter the continuous turf of the alpine regions terminates. A few Ranunculaceæ, Saxifrageæ, and Grasses, which only partially cover the soil, are all that are subsequently met with. In these travels 342 Phanerogamic plants altogether were collected. They also differ from those of the flora of Scandinavian Lapland, in containing a large proportion of species which are not Scandinavian. Eleven new species, which are illustrated by figures, belong to the genera *Ranunculus*, *Viola*, *Parnassia*, *Salix*, and *Poa* (7 sp., 1 sp. of each of the others).

Czerniaïew has published some scattered remarks on

the influence of climate upon the vegetation of Ukraine, at the same time introducing the description of some new Fungi (Bulletin des Naturalistes de Moscou, tom. xviii, part 2, pp. 132-57).

Many plants are excluded by the low Isochimena, whilst the high summer temperature appears favorable to the culture of Maize and several Cucurbitaceæ, by which also the author endeavours to explain the remarkable fact, that the berries of *Solanum nigrum* in Ukraine lose their narcotic principle, and when ripe become saccharine and eatable. The forests and fields there are protected from the persistent aridity of the summer, which acts to so great a degree upon the adjacent steppes, by the soil of humus, which is from 10 to 15 feet in depth (Tscherno Sem; compare Ann. Rep. for 1843.) Hence the principal forest trees which thrive there are such as send out deep roots, as the oak, the lime, elms, and pear trees; the red fir (*P. Abies*), which predominates on the shallow soil of Scandinavia, is unknown in Ukraine, and the ash is frequently killed during the dry season. The deep soil of humus causes several herbaceous plants to grow there to an unusual height; *Cephalaria Tartarica* is found 9 feet, *Delphinium elatum* 5—6 feet in height; Thistles and Umbelliferæ are usually twice the size of those of other regions; of the Fungi, the pileus of *Polyporus* and *Lenzites* is found three feet in breadth; and the new *Morchella alba* a foot in height. But the most remarkable object presented by this luxuriant development is the new Bovista, *Lycoperdon horrendum*, a spherical Fungus, 3 feet in diameter. This Fungus, says the author, might really produce no slight amount of terror; when in the dark forest it suddenly comes into sight, it makes one imagine a phantom in white or brown garments and in a stooping attitude. This black earth of the south of Russia, which produces this luxuriant growth, must indeed contain a large store of nutritive matter for the vegetable world; for barley grows there as in the best districts of England or Germany, without ever requiring manure. As regards the Fungi of Ukraine, Czerniaïew attributes the very numerous varieties of their forms to the paucity of the species of Mosses, Lichens, and Ferns. According to his observations, Ukraine alone contains more than 1000 Hymenomycetæ, whilst the abundance of Gasteromycetes is still more characteristic. Weinmann, in his 'Prodromus,' which was published in 1836, enumerates 300 Gasteromycetes as existing in the whole of Russia, whilst Czerniaïew has found almost twice this number of species in Ukraine alone: among them there are several new forms, and a few new genera.

Weinmann has studied the Mosses of Russia (Bulletin Moscou, tom. xviii, pt. i, pp. 409-89, and pt. ii, pp. 417-503); his new species belong to *Funaria* (1 sp.) and *Hypnum* (4 sp.) Kaleniczenko describes ten new plants from the south of Russia and the Caucasus (Ibid. pt. i, pp. 229-40);

2 Umbelliferæ (*Pimpinella* and *Pastinaca*), 2 Leguminosæ (*Arthrolobium*), 6 Synantheraceæ (*Inula*, 2 sp., *Centaurea*, 3 sp., and *Jurinea*).

The travels of A. Bravais and Ch. Martins (Bibliothèque univers. de Genève, 1845, vol. ii, pp. 147-73) traverse the north of Scandinavia almost exactly in the same way as that described by V. Buch in his celebrated work on the extreme regions of the north, when he returned from Alten-Fjord, in Finmark, to Tourneå, on the Gulf of Bothnia. But the French travellers believe that they have measured the limits of vegetation under more favorable circumstances; hence their results must find a place here. They completed the difficult journey from the 6th to the 26th of September, 1839: remarking at the same time, that partly on account of the water which they had to pass over, partly on account of the swarms of gnats met with in Lapland during the summer, which have to be avoided, September is the only month fit for travelling.

In the forests of Alten (70° N. lat.), the pines were 60′ in height, the birch on an average 45′. On the third day they passed over the upper terrace of the plateau of Kjölen. Under the name of Nuppivara, it ascends here to an elevation of only 600 metres; but it is of the same conformation as the far more elevated, undulating plateaux of the Langfjelde, which abound in lakes. The bare soil of the rocks contains only a scanty underwood of *Betula nana* with *Empetrum*, and *Andromeda tetragona*, or *Salix Lapponum* with *Juniperus communis*. On the south side, birch forests are again next met with, but they do not extend above Kautokeino further than to Karesuando (68° 36′); for from this point the whole country, as far as the Gulf of Bothnia, is covered with a single continuous pine forest.

Limits of vegetation measured:

Northern slope of Kjölen, in the valley of Alten.

Pinus sylvestris, dense forest	. . . .	249 metres
„ isolated dwarf	. . . . . .	500 „
Betula pubescens, dense forest	. . . . . .	380 „
„ in the form of a stunted tree	. .	432 „
„ „	local .	534 „

Southern slope of the Nuppivara.

Betula pubescens	. . . . . .	477 m., 480 m.

(The figures in apposition denote the results of different measurements.)

Sorbus Aucuparia	. . . .	477 m.

Dividing line between the rivers of the North Sea and of the Baltic:
Region extending from Kalanito to Suvajervi.

Pinus sylvestris	341 m., 374 m.
Betula pubescens	493 m., 498 m., 520 m., 530 m.
Sorbus Aucuparia	474 m.

At Karesuando.

Pinus sylvestris	410 m.

A list of the Phanerogamous plants occurring at Karesuando, by Læstadius, is inserted in the report of these travels.

A report by Blytt on his botanical journey through the valley of Valders in Norway, contains, for the most part, merely copious lists of localities (Bot. Notiser, 1845, Nr. 1-3). The author, however, in his account of the calcareous vegetation of Torpe, adds some remarks upon the influence of lime on the distribution of Norwegian plants.

Very few peculiarly calcareous plants are found there, and several species which in other countries are confined to a calcareous soil, grow upon the gneiss-formations of Norway. Blytt only enumerates the following Phanerogamia as belonging to the chalk in Norway: *Anemone ranunculoides, Trifolium montanum*, Libanotis*, Monotropa, Stachys arvensis, Carduus acanthoides*, Ophrys myodes*, Neottia nidus avis,* and *Malaxis Loeselii;* those species only marked with an * are, according to my knowledge, peculiar to calcareous soils in other regions, nor are the Lichens and Mosses enumerated, so in every case. Blytt then criticises Unger's well-known catalogue of lime-plants, and in so doing, he separates the following species, which grow in Norway on the gneiss-formation, and part of them on it only: *Hepatica triloba, Corydalis fabacea, Astragalus glycyphyllus, Dryas, Rubus saxatilis, Sorbus aria, Cotoneaster vulgaris, Saxifraga oppositifolia, Asperula odorata, Pyrola rotundifolia, Arctostaphylos alpina, Fagus, Taxus, Convallaria majalis, verticillata, Polygonatum, Calamogrostis sylvatica,* and *Brachypodium gracile.—Grimmia apocarpa, Hypnum Halleri, Lecidea vesicularis* and *candida,* and *Gyalecta cupularis.*

Blytt points out similar differences between Norway and Tyrol, in regard to those plants which, according to Unger, are more common on a calcareous soil on the Alps than upon other substrata.

W. P. Schimper has given a description of the Dovrefjeld, especially the Mosses found on it, several new species even of which he has discovered on this soil, which has been so frequently described (Ratisbon Flora, 1845, pp. 113-28.)

Swedish works upon the botanical topography of Scandinavia: Anderson, Plantæ vasculares circa Quickjock Lapponiæ lulensis (Upsal, 1845, 8vo, 36 pages); contains 356 sp.; Lagerheim and Sjögren, Botanical Remarks made during a journey from Stockholm to Snaasahög in Jemtland, in 1844 (Bot. Notiser, 1845, Nr. 11); Schagerström, Conspectus vegetationis Uplandicæ (Upsal, 1845, 8vo, 83 pages): contains 870 sp.; Lindeberg, an Excursion to the Malar Lake (Bot. Notiser, 1845, Nr. 12); Lindgren, Notices upon the Vegetation of the Wener Lake (ibid.), with a description of some new pileate Fungi; Lindeberg, on the Country around Grenna on the Wetter Lake (ibid. Nr. 4).—Systematic contributions to the Flora of Sweden: Anderson, Salices Lapponiæ, cum figuris 28 specierum (Upsal, 1845, 8vo, 90 pages): described according to Fries's views; Lund, Conspectus Hymenomycetum circa Holmiam crescentium (Christiania, 1845, 8vo, 118 pages).

On the botanical topography of Denmark: Petit, Remarks on the Vegetation of the south-west of Zealand (Kröyer's naturhistor. Tidskr., second series, vol. i); J. Lange, on the Vegetation of Laaland and Falster (ibid.) In the case of a tolerably large number of the plants mentioned here, the north limit to their distribution is situated on these islands.

Watson is preparing some new works on the botanical geography of Great Britain, and has made a report upon the plan adopted in them (Lond. Journ. of Bot. 1845, pp. 199-208). He very properly takes care to separate the topographical details of the localities from the more general investigations, which are of real scientific interest. The two botanical regions which he distinguishes in Great Britain, he denominates the Agrarian and the Arctic; the area of the region of the Cerealia coincides with the distribution of *Pteris aquilina.*

Further contributions to the botanical topography of Great Britain: Balfour, on some Excursions on the Scottish Peninsula of Cantire, and the Island of Isla, one of the Hebrides (Ann. Nat. Hist. xv, pp. 425-6); Gardiner, on the Highlands of Braemar (Botanical Rambles in Braemar, Dundee, 1844), described in a picturesque style; Moore, on the rare Plants of Yorkshire (Report of the British Association held at York, pp. 70-1); Andrews, on the Island of Arran and the West Coast of Ireland (Lond. Journ. of Bot. 1845, pp. 569-70).

Local British Floras: On the county of Cork in Ireland (The Botanist's Guide for the County of Cork: Contributions towards a Fauna and Flora of the county of Cork, London, 1845, 8vo), contains 885 Phanerogamia, and 936 Cryptogamia; Jenner, on the country around Tunbridge Wells, in Kent (A Flora of Tunbridge Wells; Tunbr., 1845, 8vo), includes also the Cryptogamia.

Systematic works on British plants: Bell Salter, three new species of *Rubus* (Ann. Nat. Hist. xv, p. 305); Babington on *Cuscuta* (ibid. xvi, pp. 1-3), containing figures of *C. trifolii* and *C. approximata* Bab., the latter brought over from the East Indies, with seeds of *Melilotus;* Parnell, on the Grasses (Descriptions of the Grasses of Great Britain, illustrated by 210 figures); Spruce, on some newly-discovered Mosses (Lond. Journ. of Bot., 1845, pp. 169-95); 23 Jungermanniæ with 4 new species; Taylor, on 6 Jungermanniæ new to great Britain, (ibid. pp. 276-8,) amongst them one new species; Salwey, the rare Lichens of Wales (Ann. Nat. Hist., xvi, pp. 90-9); Hassall, a History of the British Fresh-water Algæ, including the Desmideæ and Diatomaceæ, with upwards of 100 plates (London, 1845, 2 vols. 8vo). The Phytologist is continued. The following collections of dried plants must be mentioned: Salicetum Britannicum auct. Leefe (see Ann. Rep. for 1843), fasc. 2, see the critical remarks by Sonder (in Ann. Nat. Hist. xv, p. 275); M'Calla, Algæ Hibernicæ, vol. i (Dublin, 1845, 4to), with 50 species; Ayres, Mycologia Britannica (London, 1844), containing 50 sp., may be regarded as a continuation of Berkeley's work.

Van den Bosch has published the third part of his Flora of Zealand (see Ann. Rep. for 1842), containing the Lichens and some additions (V. d. Hoeven, Tydschrift, vol. xii, pp. 1-22), e. g. in the coast districts of the Netherlands: *Cerastium tetrandrum*, *Trifolium subterraneum*, and *Centaurea nigra*, are found common, also *Salix holosericea*, *Carex trinervis* (*C. rigida* Fl. Leydens), and *Zygodon viridissimus*. The contributions to the Cryptogamic Flora of the Netherlands, by Dozy and Molkenboer, have been continued (ibid. pp. 257-88): on the Fungi, containing among them some new species which are illustrated by figures. General works on the Flora of Germany: Reichenbach's Icones, vol. vii, Dec. 5-10, with the Naiadeæ, Alismaceæ, Hydrocharidaceæ, Nymphæaceæ, and a supplement to the Grasses; Sturm's Flora, sect. 1, parts 89, 90, principally containing species of *Viola* and Labiatæ; V. Schlechtendal and Schenk's illustrated work, vol. vi; Lincke's Publication, parts 50-9; Koch's synopsis, ed. 2, fasc. 3 (Lips. 1845), containing the Ferns, with appendices and the Index: an abridgment of this work appeared as a pirated edition under the false name of Herold; Nees v. Esenbeck's Genera Plantarum Floræ Germanicæ, continued by Putterlick and Endlicher, fasc. 24 (Bonn, 1845, 8vo). Special systematic works on the Flora of Germany: Sauter's new Contributions to the Flora of Germany (Ratisbon Flora, 1845, pp. 129-32): unimportant notices, with the diagnosis of a new *Riccia*; Perktold, the *Hypna* of Tyrol (Neue Zeitschr. des Ferdinandeums, Bd. 11); Rabenhorst's Cryptogamic Flora of Germany (see the preceding Ann. Rep.) vol. ii. pt. 1, containing the Lichens; Roemer, the Algæ of Germany (Hanover, 1845, 4to; with 11 plates), confined to the fresh-water Algæ, principally the forms found by the author on the upper Hartz, and imperfectly illustrated by bad lithographic drawings; arranged according to Kützing's system, but paying no attention to the history of development; Kützing's Phycologia Germanica (Nordhausen, 1845, 8vo). It includes the entire flora, and only appeared a few weeks after the preceding work, Roemer's materials having been made use of; yet it is carried out perfectly independently of these, and although subject to known criticisms in a systematic point of view, is indispensable to the understanding of the author's larger work on the Algæ.

Local Floras of Germany: F. Wimmer, Flora of Silesia. Concluding volume (Breslau, 1845, 12mo); J. C. Metsch, Flora Hennebergica, a contribution to the Flora of the Prussian portion of the Forest of Thuringia (Schleusing, 1845, 8vo); F. Schultz, Flora of Palatinate (Speier, 1846, 8vo); it appeared, however, in 1845.

In Metsch's memoir upon the plants of the mouth of the Swine (Ratisbon Flora, 1845, pp. 705-8) is contained a sketch of the vegetable formations on the island of Usedom.

The sandy soil is in some places extended into plains, at others depressed, so as to receive deposits of peat or salt-water lakes; sometimes it forms elevations, which are partly covered with the pines, or even considerable forests of beech. The dunes along the coast are consolidated by roots of Glumaceæ or *Salix*. A few of the characteristic plants only can be mentioned here, as the author only enumerates the more rare species:

1. Formation of plants on the dunes, e. g. *Ammophila arenaria* and *Baltica*, *Elymus arenarius*, *Carex arenaria*, *Kochia hirsuta*, *Halimus portulacoides*, *Petasites spurius*, and *Anthyllis maritima*.

2. Formation of the sea plants, e. g. *Aster salignus*, *Erythræa linariifolia*, *Zannichellia pedicellata*, *Juncus balticus*, *Scirpus Rothii*, and *Hierochloa borealis*.

3. Formation of the marsh plants, e. g. *Thalictrum aquilegifolium*, *Barbarea stricta*, *Helosciadium inundatum*, *Lysimachia thyrsiflora*, *Euphorbia palustris*, *Salix daphnoides* and *rosmarinifolia*, *Stratiotes*, *Carex filiformis*, and *Calamagrostis stricta*.

4. Formation of the peat plants, e. g. *Ledum palustre*, *Betula fruticosa*, *Empetrum*, and *Myosotis sparsiflora*.

5. Formation of herbaceous plants on sunny hills, e. g. *Thalictrum minus* and *simplex*, *Silene viscosa*, and *Ononis hircina*.

6. Formation of the forests, e. g. *Arabis arenosa*, *Vicia villosa*, *Peucedanum Oreoselinum*, *Arctostaphylos officinalis*, *Pyrola chlorantha*, *media*, and *umbellata*, and *Goodyera repens*.

V. Mohl has written a memoir on the Flora of Wurtemberg (Würtemb. naturwissenschaftliche Jahreshefte. Jahrg. i, pp. 69-109. Stutt., 1845, 8vo).

He commences with general remarks on the scientific importance of local floras. He states their object to consist in the investigation of the limits of the distribution of the species within a larger district, and for this purpose to compare them, as e. g. the flora of Wurtemberg with that of the adjacent countries. In this manner he shows, that when a natural subdivision of Germany is made, the flora of Wurtemberg belongs to those of the adjacent districts, and does not contain distinct centres of vegetation of its own. Mohl distinguishes four separate regions : the fluviatile system of the Neckar and the Tauber, the Black Forest, the rugged Alp, and the tertiary plain of Upper Swabia.

1. The Neckar region, lying between the Swabian Jura and the Black Forest, in regard to the distribution of its plants, may be considered as a portion of the Rhine district. The eastern limit to the occurrence of a tolerable number of plants exists at the Jura; but with the single exception of *Orobus albus* (at Tübingen), no species is found from the Neckar to the Tauber, which does not also exist in the valley of the Rhine. But the district on this side is poor in comparison to the latter; for "in correspondence with a general phenomenon," as the bed of a river becomes narrowed, many plants disappear which are common lower down the stream. The author also gives a list of more than fifty species which prove this connexion, and from which we shall select the following as characteristic forms of the Rhine district: *Helianthemum ælandicum* (*vineale*), *Myagrum perfoliatum*, *Isatis tinctoria*, *Diplotaxis tenuifolia* and *muralis*, *Althæa hirsuta*, *Lathyrus hirsutus*, *Rosa gallica*, *Helosciadium nodiflorum*, *Œnanthe peucedanifolia*, *Carum bulbocastanum*, *Crepis pulchra*, *Lactuca saligna*, *Artemisia pontica*, *Centaurea nigra*, *Heliotropium Europæum*, *Calamintha officinalis*, *Mentha rotundifolia*, *Parietaria diffusa*, *Spiranthes æstivalis*, and *Scirpus mucronatus*. Geologically considered, the district of the Neckar and the Tauber belongs to the muschel-kalk, lias, and the keuper formations. Of these, the muschel-kalk, as in Thuringia, exerts the most important influence upon the distribution of the plants, whilst the lias and keuper, being less homogeneous formations, favour a greater chemical variation in the soil. A list of about 20 sand-plants contrasted with 100 plants belonging to the muschel-kalk, shows how the latter augments the number of indigenous species to a greater extent than the other formations.

2. The Black Forest, the soil of which is derived from *bunter* sandstone or Plutonic rocks, contains in the district of Wurtemberg but few Phanerogamia peculiar to it, whilst the higher elevations of these mountains, which are also poor in plants, belong to Baden. Among the plants of the Black Forest of Wurtemberg, excepting those which also occur in other regions of the kingdom, there is not one which is not distributed over the greater part of the mountains of Germany, so that e. g. all those mentioned, excluding *Crocus vernus*, occur also on the Hartz. If we bring into relation with these facts Kirschleger's general remarks upon the whole of the Black Forest (see

Ann. Rep. for 1843), we cannot enumerate these mountains among the independent centres of vegetation of the flora of Germany, because the whole of their Phanerogamia may be regarded as having migrated from the Alps, the Vosges, the Jura, or the Rhenish mountains.

3. The rugged Alp (Swabian Jura) possesses the characteristic vegetation of the Jura-limestone, which is uniformly distributed from Switzerland to Franconia. However, although the mean level of the elevated surface amounts to more than 2000′, and individual summits ascend to more than 3000′, the alpine forms of plants, which are common on the higher Jura of Switzerland, are for the most part absent here, and even the few species (7 sp.) which belong to this category, are in most cases found at single spots only; whilst on the other hand, many calcareous plants from the valleys of the Bavarian Alps are common here. About 50 species are found in Wurtemberg on the rugged Alp only, 34 calcareous plants occur in common with the region of the Neckar, 18 species with Upper Swabia only, 16 others with both these districts, and 5 with the Black Forest. Perhaps the following from the list of plants peculiar to the Swabian Jura might be mentioned as characteristic forms, omitting those which are diffused over the calcareous Alps: *Thalictrum galioides, Thlaspi montanum, Sisymbrium austriacum, Erysimum crepidifolium* and *odoratum, Dianthus cæsius, Linum flavum* (at Ulm), *Coronilla montana* and *vaginalis, Sorbus latifolia, Leontodon incanum, Doronicum Pardalianches, Jasione perennis, Specularia hybrida, Digitalis lutea, Nepeta nuda, Orchis pallens, Aceras anthropophora,* and *Iris Germanica.*

4. The tertiary plain of Upper Swabia lying between the Jura and the Alps, 1250′ to 2000′ above the level of the sea, is geographically a portion of the plateau of Upper Bavaria, and also contains its vegetation, whilst the Jura agrees with it far less than might have been expected. Even the peat-moor formation is the same here as in the bogs of Bavaria. Upper Swabia, although it has been least explored botanically, probably contains more plants than any other part of Wurtemberg, on account of its fertile calcareous Molasse-soil, the considerable variations of its elevation, the abundance of water, and the proximity of the Alps from which, as in Bavaria, many plants are washed down.—The list of those plants of Upper Swabia which have not hitherto been observed in other parts of Wurtemberg, includes about 90 species.

The following from among them, excluding the alpine plants, may be mentioned as characteristic: *Ceratocephalus falcatus, Viola lactea, Linum viscosum, Alsine stricta, Potentilla norvegica, Saxifraga Hirculus, Helosciadium repens, Gentiana utricularis, Pedicularis sceptrum, Primula acaulis, Betula humilis, Stratiotes, Iris graminea, Allium suaveolens, Juncus tenuis, Carex capitata, microglochin, chordorrhiza, cyperoides,* and *Heleonastes.*

Mohl makes some ingenious observations on the distribution of alpine plants towards the Bavarian plateau of Upper Swabia. He distinguishes several kinds of distribution: 1. The seeds are constantly being carried down anew by the waters, and the individuals which germinate are, therefore, only accidental inhabitants of the drift on the banks, not having any fixed locality, e. g. on the Iller. *Campanula cæspitosa, Hutchinsia alpina,* &c. 2. Other alpine plants, which also grow upon the Alps themselves, on the drift of the rivers, again meet with the conditions necessary for their existence in the elevated plains: hence they constitute a permanent formation there, e. g. *Myricaria, Salix daphnoides,* and *Epilobium rosmarinifolium.* 3. Other plants of the alpine flora occur in the plain of the peat-moors, far distant from the present alpine rivers, distributed socially, e. g. *Bartsia alpina, Primula auricula, Gentiana acaulis,* in large masses on the bogs of Upper Bavaria, and *Veratrum album* also in Upper Swabia. On the Alps, part of these plants grow in totally different localities; yet, according to the opinion of the author, there is no doubt that, like those above mentioned, they emanated from the Alps, although the conditions under which these depositions occurred cannot now be ascertained. In this respect, he declares that Zuccarini's view is a very hazardous hypothesis, who supposes that the first seeds were carried down in remote ages by the same rivers which filled up the whole of the tertiary plain with alpine Molasse, and gave rise to the continent. This view is inadmissible, because the phenomenon of the occurrence of the alpine plants in the peat-moors is evidently the same as that which is now going on in the north of Germany, where e. g. *Primula farinosa, Swertia perennis,* and *Salix daphnoides* are met with under the same conditions. The humous pasture-soil of the Alps is not so entirely different from peat, nor the climate of Upper Bavaria so very different from that of Mecklenberg, as regards many plants, as to render inadmissible every explanation of this simultaneous growth of individual species in remote plains and on the mountains, by means of the soil and climate: we need not then hypothetically devise any geological causes. Are not aerial currents sufficient to convey the minute seeds of the Gentianeæ, or the cotton of a willow, to all those parts of Germany, nay, even of Europe, where the climate and soil permit their germination and growth? What their original locality was, whether a plain or a mountain, appears to me an idle question, because it is incapable of scientific solution. 4. The same applies to all those alpine plants which have attained any extent of diffusion in the south-east corner of Upper Swabia, e. g. *Rhododendron ferrugineum, Campanula barbata, Streptopus amplexifolius,* &c. Mohl considers that these are the original plants of Upper Swabia, and that their origin is not to be looked for in the Alps; this view will appear totally untenable to every one who is acquainted with the botanical relations of the Alps, from personal observation, although we are not in a condition to ascertain how they arrived at their present locality. The latter point appears simple to me, when we recollect that the greater number of these plants thrive also on the Sudeten and other remote mountains, and thus probably possess a wide climatic sphere, and also means of more easy diffusion by the air; but how the first question can be decided by personal observation, I do not understand, inasmuch as a plant may be diffused equally as luxuriantly and generally on a secondary as on a primary locality, as e. g. the thistle of the Pampas of Buenos Ayres teaches us, which in the Old World, where it is indigenous, is found at individual spots only, whilst in the former place it covers the plains in the most intimate community. The conclusion of this important work is formed by a catalogue of the names of all the Phanerogamia hitherto found in Wurtemberg, without the localities; it contains only 1287 species, i. e. more than 100 species less than are known (according to my manuscript) in the kingdom of Hanover: hence Mohl appears to be justified in the opinion, that many remain to be discovered in Wurtemberg.

The Topography of the Upper Pinzgau (see Ann. Rep. for 1840), contains a work by A. Sauter, with which I am only at present acquainted from Beilschmied's Abstract (Ratisbon Flora, 1845, pp. 501-7), upon the botanico-geographical relations of this district, which includes the longitudinal valley of the upper Salzach, between the Tauern and the clay-slate alps of Kitzbühl.

It contains, in addition to lists of the more rare species, a sketch of the botanical regions, but the source of the altitudes given is not stated. 1. Region of the cultivated country. 2400′—4000′ on the south side, —3000′ on the north side of the mountain. Pasture land alternates there with forests; meadows and fields are more rare. Most of the deciduous trees, and *Alnus incana* is very common, do not ascend higher. 2. Forest region.—On the average 3500′—5500′. *Pinus abies*, however, which constitutes the main part, appears only to thrive as high as 5000′, *P. Picea* at 4000′, whilst *P. Cembra* here and there covers the upper slopes, and in the Tauern chain ascends as high as 6000′, the same height as *P. Larix*. 3. Alpine Region.—On the average, 5503′—8000′. It also contains but few meadow surfaces, mostly naked rocks and detritus. The sub-alpine forests do not form a dense zone there; *Rhododendron ferrugineum* occurs in groups as far as 6000′; dwarf willows, *Empetrum*, *Arctostaphylos*, and *Azalea procumbens*, as far as 7000′.

Perini read a paper on the Botanical Regions of Trient, in the south of Tyrol, before the Association of Italian Naturalists (Atti di VI riunione, p. 460).

L. v. Heufler has described a botanical excursion in the north of Istria (Die Golazberge in der Tschitscherei. Triest, 1845, 4to).

On the 16th of June the author collected 300 species of plants on a mountain-ridge of the Karst, only 3410′ in height, to the south of the Fiummean road; they are enumerated in the order of their occurrence in this luxuriously-printed work. In the map which is appended, the following regions of the coast-country of Illyria, in the direction from the Adriatic Sea to the coast of Terglou, are distinguished, but how the altitudes were determined is not stated: 1, 0′—500′; region of Olives. 2, —2000′; oak-region (with regard to which it is incorrect to state that the region of the species of the north of Europe is not different from that of the Mediterranean species). 3, 2000′—4800′; beech region. 4, —6500′; pine region. 5, —8500′; region of the alpine plants. 6, —9036′; region of snow. The vegetation of the Golaz mountains resolves itself into the oak forests (1500′—2000′: *Quercus Robur, pedunculata, Cerris,* and *pubescens*), beech forests (2000′—3410′), alpine pastures, and the rocky formation. In addition to this main subdivision, separate arrangements in groups are also mentioned, e. g. shrubs of *Ornus* in the lower, and of *Corylus Avellana* in the upper part of the oak-region, herbaceous meadows of *Cytisus* and *Genista*, &c.

Wierzbicki has published a Catalogue of the Plants found in the Banat (Ratisbon Flora, 1845, pp. 321-25), subsequent to the appearance of the most recent work on the flora of this province (Rochel's Travels in the Banat, 1838); also, Prof. Fuss, of Hermannstadt, a list of 319 plants of Siebenburg, with their localities (Archiv des Vereins für Siebenbürg. Landeskunde, (Bd. 2, Hft. 3).

O. Heer's memoir upon the Upper Limits of Animal and Vegetable Life in the Swiss Alps (Zürich, 1845, 4to), is more important in a geological than a botanical point of view, on account of the descriptions and illustrations of new insects belonging to the snow-region contained in it; however, it also contains some valuable observations on the forms of plants, which, under certain conditions, vegetate above the snow limit (8500′).

Lichens extend far above the Phanerogamia and Mosses; they exist even on the summit of Mont Blanc. The highest of the Phanerogamia was *Androsace glacialis* (*pennina* Gaud.), and occurred at 10,700′ on Pic Linard; and from this altitude down to 10,000′, the following were found in succession on different glaciers, i. e. on account of the position or inclination of the surfaces free from snow on the Rhetian Alps: *Gentiana bavarica*, var. *imbricata, Silene acaulis, Chrysanthemum alpinum, Ranunculus glacialis, Cerastium latifolium*, var. *glaciale, Saxifraga oppositifolia* and *bryoides, Cher-*

leria and *Poa laxa*. Associated with these between 10,000′ and 9000′ we find 50 more, and as far as 8500′, i. e. at the snow-line, 46 others; so that the entire flora of the Rhetian Alps consists of 106 Phanerogamia, belonging to 23 families. All these plants are perennials, most of them cæspitose, hence they are propagated without the seed arriving at maturity; all of them are depressed and small, and are thus less influenced by the heat of the air than of the soil: in fact, the only two woody plants are dwarf willows, their stems being almost entirely inserted into the earth. Yet the temperature of the soil at these altitudes is probably for a short time only above the freezing point. The author very correctly explains their being enabled to grow, notwithstanding this circumstance, by the short period of their vegetation, as when transplanted into the low country, they become, without exception, spring plants, which, in a few weeks after budding, ripen their fruit, their winter sleep being proportionately prolonged. Moreover, in the low country they all exhibit but little susceptibility to cold, so that even at the period of their flowering, although exposed to frost, they are not at all injured. Even if, in their elevated position, the spring season should not occur, they would endure a state of hybernation for several years, without being destroyed. The conditions of vegetation being so different from those of the level country, explains the fact that the Phanerogamia of the snow-region never become spontaneously distributed in the valleys. The case is different with the Cryptogamia; for the lower the degree of organization, says Heer, the less does the form require to be modified, to adapt it to a different climate.

Mougeot and Nestler have published, in connexion with W. P. Schimper, the twelfth century of their well-known collection of dried Cryptogamia from the Vosges (Stirpes Cryptogamæ Vogeso-Rhenanæ, fasc. 12. Bruyère, 1844, 4to).

French local floras and systematic contributions upon French plants: Observations sur quelques Plantes Lorraines par Godron (Nancy, 1835, 8vo, pp. 31), containing a supplement to his Flora of Lorraine; Choulette, Synopsis de la Flore de Lorraine et d'Alsace, Partie i, Tableau Analytiques (Strasb., 1845, 16mo); Cosson et Germain, Flore descriptive et analytique des Environs de Paris (Paris, 1845, 8vo, 2 vols.), remarkable for its accuracy, and the systematic investigations contained in it; it gives satisfactory elucidations, e. g. of *Astrocarpus Clusii*, *Trifolium Parisiense*, *Euphrasia Jaubertiana*, *Potamogeton tuberculatus*, and *Carex Mairii*; Puol, Catalogue des Plantes qui croissent dans le Département de Lot (in the Annuaire du Département, pp. 1845, 1846), extending to Hexandria; F. Schultz, Continuation of the communications on the Orobancheæ of France (Ratisbon Flora, 1845, p. 738); Desmazières, Eleventh Contribution to the knowledge of the Cryptogamia of France, containing Fungi (Ann. Sc. nat., 1845, 3, pp. 357-70).

The excellent observations of Ch. Martins on the climate of France, which were mentioned in the preceding Annual Report, are now published in greater detail, and have been augmented by a description of the botanico-geographical relations (Essai sur la Météorologie et la Géographie botanique de la France; separate division of the encyclopædic work, Patria; La France ancienne et moderne. Paris, 8vo).

The French botanical geography, however, is merely based upon the facts given in Duby's 'Botanicon Gallicum.' The distribution of about 3700 Phanerogamous plants through France is shown by a series of lists. 1. 1250 species are diffused over the entire country; i. e. they occur in the local floras of Boreau, Godron, Cosson and Germain, and Dumortier, and in Bentham's 'Catalogue of the Flora of the Pyrenees.' 2. About 30 species, most of which are distributed over Central Europe, correspond, in France, to the climate of the Vosges and Seine. (See the preceding Ann. Rep.) 3. About 30 species belonging to the valley of the Rhine, are confined to the Vosges climate: the mountain plants of the Vosges, which, however, appear also to occur on other mountains of France, are separated from these, as also the southern forms of the valley of the Rhine (10 species), but which according to their distribution ought rather to be arranged in the third list. 4. About 30 species of plants belonging to the north-west, corresponding to the climate of the Seine. 5. The centre of France forms a transition-district from the north to the south, and has only 3 species peculiar to it. 6. 750 species belonging to the south of France, correspond to the climate of the Garonne and Rhone, but they also occur in the Mediterranean district. 7. 800 species are confined to the Mediterranean climate. 8. 500 species belong to the subalpine region of the mountains of France, the level of which, between the 46° and 49° of north latitude, Martins estimates at from 600 m. to 1600 m.; south of 45°, from 1000 m. to 1800 m. 9. 300 species grow beyond these limits in the alpine region. The conclusion consists of plants arranged according to the localities.

In the same paper Martins also publishes the following measurements of the limits of vegetation in Dauphiny:

Cultivation of barley. Upper limit.
Col de la Vachère. North side, 1745 m. South side, 2110 m.
Fagus sylvatica. Upper limit.
Grande Chartreuse, 1465 m. Col des Sept Lacs, 1475 m.
Pinus abies. Upper limit.
Grande Chartreuse, 1631 m. In a shrubby form, 1900 m.
P. picea. Upper limit.
Col des Sept Lacs. North side, 1770 m. South side, 2045 m.
Alnus viridis. Upper limit.
Col des Sept Lacs. North side, 1910 m.
Sorbus aucuparia. Upper limit.
Col de la Vachère. North side, 2000 m.
Rhododendron. Lower limit.
Col des Sept Lacs. North side, 1160 m. Grande Chartreuse, 1660 m.
Col de la Vachère. North side, 2125 m.
Upper limit.
Col de la Vachère, 2410 m.
Pinus Cembra. Lower limit.
Col de la Vachère, 1740 m.
Upper limit.
Col Longet, 2515 m.
P. larix. Upper limit.
Col Longet, 2515 m.

Daum has described two barren, almost desert regions on the south coast of France (Bemerkungen über die Landwirthschaft in Südfrankreich. Charlottenb. 1844, 8vo).

The plain of Crau, which lies to the south of Arles, covers nearly seventy-five square miles; it consists of a gravelly surface, containing scattered, but nutritive herbs and grasses, upon which no kind of agriculture can be carried on, but 30,000 fine sheep find pasture from late in the autumn until the spring, when they are driven upon the pastures of the Maritime Alps; and which it is now being commenced to convert into meadow-land by artificial irrigation; next the plain of Camargue, in the delta of the Rhone, a boggy salt-marsh, nearly half of which consists of flooded land and bog, the remainder of pasture-land and a few fields, and which, by means of a large capital, it is also being attempted to turn to advantage by canal drainage. As regards the agriculture of the province, the traveller remarks that, on account of the grape-vine requiring a large quantity of manure, without affording nutriment to cattle, the principal object is the cultivation of fodder, for as there are no meadows, this is necessarily obtained from lucerne. These facts show the natural character of the country.

Cuynat's Topography of Catalonia (Mémoires de

l'Académie de Dijon, 1845), contains a Catalogue of the Plants found in this Spanish province (2, pp. 91-100).

Six hundred species are enumerated, but most of them are so much more extensively distributed on the Mediterranean coasts, as to prevent our obtaining an accurate idea of the peculiar nature of the vegetation of Catalonia, which has not yet been described.

Willkomm's sketch of Monserrat, with which he concludes his Botanical Reports upon Spain, may be mentioned as a contribution towards filling up this deficiency (Bot. Zeit., 1846).

This isolated conglomerate mountain, which the traveller visited in April, is scarcely more than 3000′ in height; but the summit is only accessible by a deep rocky valley, which runs in a north-westerly direction, whilst the outer sides rise so steeply that they cannot be ascended. In Catalonia, the "warm region," which undoubtedly corresponds to the region of *Chamærops* (see below), extends scarcely more than 1000′; hence the greater part of the Monserrat belongs to "the mountainous region" (the region of evergreen oaks), and the summit reaches the subalpine region (region of the pines). The Mediterranean, as also the Central European plants, are mixed on this mountain with a number of Pyrenean plants. In the lower region, the heights at Bruch are covered with forests of *Pinus halepensis* and *pinea*; the other parts are covered with freely-vegetating "montebaxo," consisting of evergreen oaks, *Pistacia lentiscus*, *Erica arborea*, and other shrubs. Characteristic plants: *Genista hispanica*, *Euphorbia oleifolia* G., *Globularia Alypum*, *Coris monspeliensis*, and *Passerina tinctoria* Pourr. At the central elevation: *Polygala saxatilis* Lag.; *Erodium supracanum*, *Sarcocapnos enneaphylla*, *Carduus tenuiflorus* Salzm.; *Ramondia pyrenaica*, and *Convolvulus saxatilis*. The upper region was not developed at that season; however, *Arctostaphylos uva ursi*, *Globularia nana* Lam., and *Narcissus Jonquilla*, were in flower.

The families containing most species in the flora of Castile form the following series, according to Reuter's collection, which contains 1232 sp. (Boissier, Voy. en Espagne, i, p. 207): Graminaceæ (161 sp.), Leguminosæ (130), Synantheraceæ (125), Cruciferæ (74), Caryophyllaceæ (52), Rosaceæ (38), Ranunculaceæ (33), Boragineæ (31), and Chenopodeaceæ (26).

According to Willkomm (loc. cit.), the Sierra Morena contains an uncommonly uniform vegetation, notwithstanding its great length and breadth. With an average breadth of 8 geogr. miles, it extends from Murcia to Algarve, only forming, however, an intermediate mountain chain, the crest of which, for the most part, only ascends to 2—3000′, and the greatest elevations of which are hardly 5000′. By the density of its forests or tall shadowing shrubs, and by this connected green and fresh vegetation, the Sierra Morena differs from all other mountains of Andalusia, which only contain isolated spots of wood, and a low, barren "montebaxo." Geologically considered, the principal rocks of the Sierra Morena consist of sandstone, which, in the form of grauwacke, forms the greater part of the mountain-chain, occurring from 4 to 6 geog. miles broad, in gently-rounded mountains and undulating summits, alternating at Almaden with clay slate, in the province of Huelva with gneiss, and southwards, near the low plain of the Guadalquiver, inclosed by other sandstone formations. The central portion of the mountains is interrupted by the immense granitic formation of Cordova, which, in the form of an elevated plain, inclines from Hinojosa towards the north, and becomes connected there with white quartzose rocks, which, between Almaden and Fuencaliente, appear to form the highest chain of the whole Sierra. In the opinion of the traveller, the character of the vegetation is associated with these geological variations, which are lithologically unimportant. The predominating shrub on the grauwacke, as far as Portugal, is *Cistus ladiniferus*, which extends over the Sierra Morena for a length of more than 50 geog. miles, and "frequently covers whole square leagues exclusively." Next to this, the most common are *Phillyrea angustifolia*, *Rosmarinus*, and a *Helianthemum*. The forests on the grauwacke consist of evergreen oaks, of *Quercus Ilex*, *Ballota*, and *Suber*, but the first is mostly only a shrub; the dry, arid, but densely-populated granite plateau is very sterile, but still possesses extensive woods of *Quercus Ilex* and *Ballota*, only a dwarf growth, however, of *Quercus Ilex*, mixed with *Cistus ladaniferus*, *Phillyrea angustifolia*, and *Arbutus unedo*. The southern sandstone chains are furnished with an extremely luxuriant and varied "montebaxo," which, near the city of Cordova, alternates with woods of pines and cork-oaks. The quartzose rocks of Mancha are also covered by a "montebaxo" abounding in forms, among the shrubs of which *Cistus populifolius* is distinguished. Finally, in Huelva, Portuguese forms are associated with the other shrubs, as *Genista tridentata*, *Ulex genistoides*, and with these *Erica umbellata*, *Teucrium fruticans*, and *Helianthemum halimifolium*. Unfortunately, Willkomm has not made us ac-acquainted with the vernal vegetation of the Sierra Morena, which would be the most interesting. But the dryness of the summer commences here in July, and from that time until the autumn no more flowering plants are met with. Different kinds of bulbous plants appear very uniformly distributed in the autumn, as *Scilla maritima*, *Scilla autumnalis*, *Leucojum autumnale*, *Merendera Bulbocodium*, &c.

Boissier's large work upon the southern border of Granada and the Sierra Nevada forms the most valuable contribution to botanical geography which has been made during the past year (Voyage botanique dans le Midi de de l'Espagne, t. i. Narration and Géographie botanique, pp. 241. Paris, 1845, 4to); as regards the earlier systematic parts of this illustrated work (t. ii), see the Ann. Rep. for 1840-1.

Boissier's excellent account includes the coast-terraces between Gibraltar and Almeria, towards the centre of the country as far as the elevated surfaces of Andalusia, and thus entirely includes the highest mountain-chains of the south of Spain. Along the entire coast-line a series of isolated mountain-chains, consisting of marmoraceous limestone, arise directly, in almost every case, without any intervening land, the western extremity always ascending highest, whilst towards the east the ridge gradually falls: to this system belong the Serrania de Ronda (6000′), Sierra Tejeda (6600′), and Sierra Gador (7000′). These chains, which run parallel with the coast, may be considered as forming the southern mountain-border of the Spanish plateau; for its northern foot, at an elevation of from 2000′ to 2500′, passes directly into the elevated plain of Ronda, the Vega in Granada, or the great plains of Guadix and Baza. In a line from Ducal to Almeria, not far from Granada, the chain of the Sierra Nevada, twenty-two leagues only in length, is inserted between the boundary chain and the elevated plain; it is nearly twice their altitude, but narrow, its highest summits ascending to an elevation of 11,000′. In fact, the passes in the western portion are not situated below 9500′, whilst toward the east the mean height of the crest appears to diminish to 6000′. The Sierra Nevada is mainly composed of mica-slate, but on its flanks secondary and tertiary formations have been carried up with it as far as an altitude of 6000′. The district of Alpujarra forms an important constituent of these mountains; it includes the longitudinal valley running between the coast-chain S. Contraviesa and the Sierra Nevada, together with the southern intersecting valleys of the latter. The following are some of the heights which were measured by Boissier with the barometer: The city of Ronda, on the plateau, 2300′; Granada, 2200′; Sierra Nevada, the farm of San Geronimo, 5064′; Col de Vacares, 9472′; Picacho de Veleta, 10,728′; Mulahacen, 10,980′.

Four botanical districts, which Boissier distinguished in South Granada, yielded him vascular 1900 plants, which he is inclined to regard as forming three fourths of all the indigenous plants of this district. The author considers the following as among the general characters of the flora—viz. that many forms cover the soil in gregarious condition, and that the south of Spain contains more thorny plants than any other country in Europe, and hence resembles the steppes of Western Asia, although the families which develope the thorns are not the same. The hot region (région chaude) comprises the coast declivity up to a level of 2000′. Intense atmospheric precipitations fall during October and November; the vernal rainy season, which is less regular, lasts from February to March, sometimes until April; uninterrupted drought prevails from April to the end of September. Thus the dry season there, is probably of longer duration than in any other point of the Mediterranean flora. Observations are given, made at Malaga, during nearly three years (1836—1839), by Haenseler, on the distribution of heat; the extreme temperatures of which, as also the monthly average, calculated from the corresponding months of the year in which the observations were made,* yields the following values:

	Med.	Max.	Min.
January . . .	12°·25	17°·22	6°·2
February . . .	14 ·3	18 ·25	6 ·1
March . . .	15 ·8	21 ·62	10 ·0
April . . .	17 ·8	25 ·0	11 ·25
May . . .	21 ·2	24 ·5	15 ·72
June . . .	23 ·4	26 ·87	20 ·12
July . . .	26 ·2	31 ·87	23 ·5
August . . .	26 ·8	30 ·6	23 ·75
September . .	24 ·4	29 ·87	19 ·37
October . .	22 ·25	25 ·5	19 ·25
November . .	18 ·15	22 ·75	11 ·2
December . .	15 ·75	21 ·0	8 ·5

Mean annual temperature, 17°·3

The vegetation passes through phases corresponding to this climate. After the dry season, Liliaceæ are developed during the first rain of October or November; these are succeeded by the annuals, which flower throughout the entire winter. The flowering season of most of the plants is in April and May; in June and July, when all the annual plants have withered, herbaceous plants belonging to the families of the Synantheracæ, Umbelliferæ, and Labiatæ flower; lastly, from August to September, the most profound repose of vegetable life prevails, so that two or three Liliaceæ, *Mandragora*, and *Atractylis gummifera*, are all that remain. The hot region is principally characterised botanically by *Chamærops*, which covers large tracts and prevents cultivation; as in Valencia, it only ascends to 2000′. Among the cultivated plants, the orange also corresponds accurately to the extent of this region. The soil of other parts is principally devoted to the cultivation of the grape-vine, the fruit of which ripens at the end of August. The Cerealia require artificial irrigation: on those parts which are reached by the water from the mountains, either in its natural descent or by aqueducts, we sometimes find most luxuriant fields of maize and wheat, shaded by orange and mulberry trees. But such oases are rare on these bare and arid slopes, on which wheat is reaped as early as the latter half of June, and barley in May. However, on a narrow coast-district which surrounds the coast-chains,

* The average which I have calculated refers, for June, July, and August, to two, and for the other months to three years.

sometimes in the form of a surface containing salt-water lagunes, at others as a line of hills, and which at Malaga alone is covered with a more extensive alluvial plain, the cultivated plants are confined to the hot zone (0′—600′), as the sugar-cane, the cotton-plant, and sweet potato, the Date-palm and *Ceratonia*, as also the migrated Agaves and Opuntias, with several indigenous plants, as *Aloe perfoliata* and *Withania;* excepting the white poplar, indigenous trees are absent from this littoral district. Boissier enumerates altogether 19 trees as belonging to the hot region, part of which, however, like the Agrumæ, are of foreign origin. The following only can be regarded as indigenous: *Ceratonia*, which ascends 2000′, *Zizyphus*, *Punica*, *Celtis australis*, and *Populus alba*, with those which extend into the following region, where they become more common—viz. *Ficus carica* (0′—3000′, on the southern slope 4000′), *Olea europæa* (vid. supra), *Quercus Ballota* and *lusitanica* (3000′), *Q. Suber* (4000′), *Q. Ilex* (4500′), and *Pinus Pinaster* (vid. supra), The following are the most important formations of the hot regions: *a. Maquis* (Montebaxo). Shrubs from 3′—6′ in height cover the greater part of the sloping soil, consisting of *Chamærops*, several *Cisti*, viz. *C. ladaniferus*, *albidus*, and *Clusii; Pistacia lentiscus*, *Rhamnus lycioides*, and *Phillyrea*, numerous *Genistæ*, most commonly *Genista umbellata* and *Retama sphærocarpa*, and some oaks, beneath which numerous annuals and Grasses flower in the winter and spring, and, more rarely, herbs which are developed at a later period. Shrubs of *Nerium* denote the humid soil of the banks of rivers. *b. Campi.* On bare wastes are found predominating *Thymbra capitata*, *Lavandula multifida*, *Teucrium Polium*, and numerous herbs, among which *Centrophyllum arborescens* is pre-eminent. In other places, these are replaced by the social *Macrochloa tenacissima*. In addition to these two principal formations, there are the Halophytes of the littoral district, those plants which are indigenous to the marshes of Malaga, and lastly, the plants of the cultivated part of the country, with its hedges of Agaves and Opuntias. The following plants belong to the endemic forms of the hot region of Granada: *Caltha europæa* (*Celastrus Voy.*), *Genista umbellata* and *gibraltarica*, *Sarothamnus bæticus* and *malacitanus*, *Ulex bæticus*, *Leobordea lupinifolia*, *Ononis Gibraltarica* and *filicaulis*, *Elæoselinum Lagascæ* and *fœtidum*, *Lonicera canescens*, *Withania frutescens*, *Triguera ambrosiaca*, *Lycium intricatum*, *Lafuentea rotundifolia* (according to Willkomm, but it is absent according to Boissier), *Digitalis laciniata*, *Sideritis lasiantha* and *arborescens*, *Salsola Webbii*, *Passerina canescens* and *villosa*, *Osiris quadrifida*, *Euphorbia medicaginea* and *trinervia*, *Quercus Mesto*, *Salix pedicellata*, and *Ephedra altissima*.

The second region (région montagneuse), or the region of the Spanish plateau, is peculiar to Spain, and cannot be compared with the mountain-vegetation of other European countries. By way of introduction to Boissier's description, I shall premise here a remark upon the climatal cause of this peculiar condition. In Italy, Dalmatia, and in Turkey, we find immediately above the evergreen region, slopes abounding in forests of Central European forms of trees, and other plants indigenous to this side of the Alps: angiospermous trees, which lose their leaves in winter, even at an altitude of 1200′ to 1500′, frequently begin to denote this second Central European region. In Spain, Boissier, with other authors, distinguishes two evergreen regions: a lower one, which in the character of its vegetation appears to agree with the Italian or Dalmatian, and reaches to 1500′ in Catalonia, and to 2000′ in Granada; and an upper one, which, extending from 2000′ to above 5000′, includes the greater part of Spain, and has no analogy with any other throughout the entire south of Europe. It has been shown by Schouw's investigations, that the climatal cause of the evergreen vegetation of the Mediterranean lies in the aridity of the summer, to which the plants of the north of Europe are not subjected. Out of Spain, the latter again meet with conditions necessary for their existence on the mountain-chains of the south of Europe, in the vicinity of the region of clouds, where, even in the summer, the air contains mists formed from its watery vapour, and where the low scale of temperature is the same as in the climate of the north. The elevated plains of Spain are, however, in summer even more arid than the coast-regions; the humid and mild spring, which stimulates all the plants to flower, is there succeeded by a hot and dry summer and a cold winter; the three seasons of the steppes of Russia are distinguished. Although this explains the fact that some plants of the plateau of Spain recur in the Crimea, or on the elevated surfaces of Asia Minor, yet their number is small; for the contrast of the insular and the extreme climate of the interior of the continent exerts such influence here, that the greater part of the plants of Spain are not exposed to the great winter-cold of the eastern elevated surfaces and steppes. Hence the greater part of the flora of the plateau of Spain must consist of endemic plants, because these climatal conditions do not exist elsewhere in Europe. This is still more strikingly perceptible in central Spain (see the Ann. Rep. for 1843) than in Granada, where the plateau-character is less developed on the mountain-slopes, and the vegetation contains fewer forms. But it is clear that more plants of the evergreen coast-region may occur in a climate of this kind than in that of Northern and Central Europe. Returning to the observations of Boissier, he estimates the region which corresponds in Granada to the plateau of Spain as extending from 2000′ to 4500′ on the northern, and to 5000′ on the southern slopes. Within this region, but not far from its lower boundary, the cities of Granada and Ronda are situated, where, in the winter, the thermometer regularly falls 6—8° below 32° F. for a few days. At its upper limit, e. g. in the village of Trevelez, in the Alpujarras, the snow lies on the ground for four months, from December to April. The summer-heat is frequently greater at Granada than on the coast, but the nocturnal cooling is very considerable. The distribution of the atmospheric precipitations is the same as in the lower region, except that in the summer thunderstorms are common on the Sierra Nevada, and hence the soil rarely dries up so completely as lower down. The agriculture consists principally in the cultivation of wheat and maize, the upper limit to which coincides with the boundary of the region. The wheat is reaped in July, or, in more elevated localities, at the commencement of August. The cultivation of fruit trees extends to the same level as that of wheat; the chesnut, mulberry, and walnut to 5000′; pears and cherries somewhat higher (the latter, in parts, to 6500′). But the most remarkable phenomenon is, that here, quite independently of their horizontal area, the olive and grape-vine extend to nearly the same level (*Olea* on the northern slope to 3000′, on the southern slope to 4200′; *Vitis* to 3500′ and 4200′). The formations of the second region are nearly the same as in Castile: *a.* " Maquis" of the same aspect as in the lower region, but composed of mostly different species. Genisteæ and Cisteæ are more common here; those most so are—*Cistus populifolius, Genista hirsuta,* with *Sarothamnus arboreus, Ulex provincialis, Daphne Gnidium, Rosmarinus,* &c. *b.* Thin forests of *Pinus Pinaster* (1200′ to 4000′) and *P. halepensis* (2000′ to 3000′), or of evergreen oaks, as *Quercus Ilex, Ballota* and *Suber* (vid. sup.) The underwood here also consists of shrubs of *Cistus,* the density in the growth of which increases in proportion as the intervals between the trees augment. Characteristic forms of the forest-vegetation: *Cistus laurifolius, populifolius,* and *salvifolius, Lithospermum prostratum, Herniaria incana, Scabiosa tomentosa,* &c. In the Serrania de Ronda, this thicket is replaced by a mixed kind of forest, consisting of *Abies Pinsapo* B. (3500′ to 6000′), and *Quercus alpestris* B. (3000 to 6000′). In addition to those above mentioned, the following are the only other trees which occur in this region: *Fraxinus excelsior* (3000′ to 5000′), *Ulmus campestris* (2000′ to 4000′), *Populus nigra* (2000′ to 5000′), and *Pinus pinea* (3000′). *c.* " Tomillares." Low shrubs and herbaceous plants belonging to the families of the Labiatæ, Synantheraceæ, and Cistineæ form a dense expanse of vegetation, among which stellate patches of high turf, consisting of *Stipa,* are distinguished. Characteristic forms: *Thymus Mastichina, zygis,* and *hirtus, Salvia Hispanorum, Teucrium capitatum, Sideritis hirsuta, Helianthemum hirtum, Stipa Lagascæ, Linum suffruticosum, Artemisia Barretieri* and *campestris, Lavandula Spica* and *Stœchas, Helichrysum serotinum, Santolina rosmarinifolia. d.* Meadows of rigid, tall grasses, which are but little touched by cattle, and consist of *Avena filifolia* and *bromoides, Festuca granatensis* and *Macrochloa tenacissima,* cover particular slopes. *e.* Vegetation, consisting of Cynaraceæ, on the untilled fields, on the clay. *f.* Gypsum-vegetation with Halophytes (See Reuter's description of Castile, in the Ann. Rep. for 1843), principally distributed over the elevated surfaces of Guadix and Baza. Characteristic plants, mostly glaucous, and part furnished with fleshy leaves: *Peganum, Frankenia thymifolia* and *corymbosa, Lepidium subulatum, Ononis crassifolia, Helianthemum squamatum, Statice, Atriplex,*

Salsola, and *Juncus acutus*. The following are among the endemic forms of the second region of Granada, in addition to those already mentioned, e. g. *Aplectrocapnos bætica*, *Crambe filiformis*, *Hypericum bæticum* and *caprifolium*, *Rhamnus velutinus*, *Ulex bæticus*, *Genista biflora* and *Haenseleri*, *Sarothamnus affinis*, *Ononis speciosa*, *Anthyllis tejedensis*, *Saxifraga gemmulosa*, *Elæoselinum millefolium*, *Lonicera splendida*, *Santolina canescens*, *viscosa*, and *pectinata*, *Centaurea acaulis*, *Clementei*, *prolongi*, and *granatensis*, *Cynara alba*, *Chamæpeuce hispanica*, *Digitalis laciniata*, *Salvia candelabrum*, *Thymus longiflorus*, *Teucrium fragile* and *Haenseleri*, *Salsola Webbii* and *genistoides*, *Euphorbia Clementei* and *leucotricha*, and *Oligomeris glaucescens*.

Third region (Bossier's alpine region), from 4500′ (5000′) to 8000′. The name applied to this region is not fortunately selected, because at the most it could only be compared with that of the subalpine vegetation of the Alps. But considering that, in addition to many endemic plants, at least a large proportion (two fifths) consist of the plants of Central Europe, it would have been more appropriate to have named it after them. But the question of whether the region possesses natural boundaries is of more importance than that regarding the name. On this point it is at once evident, that the tree-limit, which in most mountains so sharply defines the Central European from the alpine region, coincides in the present instance with the latter (6000′ to 7000′) according to Boissier's estimation. Among the trees which vegetate here we have *Pinus sylvestris*, *Taxus*, *Salix caprea*, and *Sorbus Aria*; hence, in fact, forms belonging to the forests of Central Europe. Now if, as in the Sierra Nevada, the forests had diminished to such an extent, or in the course of time had disappeared, so as at the present day not to have any considerable influence upon the natural character of the mountains, still, to allow of comparison with other mountains, it is requisite to determine the regions according to the sphere of distribution of those species which inhabit a large portion of Europe, and thus afford the most certain standard for the climate of any particular region. In the distribution of a mountain into regions, the height at which the vegetation undergoes a decided change is not the only point to be taken into consideration, but also where the climates corresponding to those of other latitudes approximatively occur; a determination which can only be made by the comparison of the vertical distribution of the same plants. There is another decided reason, which renders it essential to form the regions of the Sierra Nevada beyond the tree-limit (part of Boissier's alpine region and his snow-region) into one. Boissier does not give any other decided difference between the two, than that in the snow-region flakes of snow remain during the summer, and that the taller shrubs are absent. That as the altitude increases, the alpine plants themselves alter considerably, is always found to be the case in the upper mountainous regions; hence the latter might be subdivided, without the accuracy and distinctness of the representation being increased. But the region of

Genista aspalathoides evidently corresponds to the *Rhododendra* of the Alps, the dwarf birch and the shrubby willows of the north; formations which are always considered as belonging to the alpine region, or have been separated from it as subalpine. Hence I propose the following regions for Granada, which are comparable with those of other mountainous countries in the south of Europe, and whereby the most remarkable circumstance, that the alpine region has a very wide, and the central European a very narrow altitudinal extent, vanishes when brought into relation with the general fact which I have elsewhere determined, that in Europe the tree-limit does not ascend southward of the Alps.

A. Evergreen region. 0′—5000′ (4500′).

a. Region of *Chamærops.* 0′—2000′. (Boissier's hot region.)

b. Region of the *Cisti.* 2′—5000′ (4500′). (Boissier's mountainous region.)

B. Central European region, or pine-region. 5000′ (4500′) to 6500′. (Part of Boissier's alpine region.)

C. Alpine region. 6500′—11,000′.

a. Region of alpine shrubs. 6500′—8000′. (Part of Alpine region.)

b. Region of alpine shrubs and grasses. 8000′—11′000. (Boissier's snow-region).

But we must now follow Boissier's further description, and hence take in the two regions which are contained in his alpine region. In the upper part of it the snow settles even at the end of September, and the last masses of snow do not melt until the commencement of June (hence corresponding with the alpine region of the Alps), whilst in the lower part, the soil is only covered with snow for four months (hence the climate of the Coniferous region). The distribution of heat corresponds generally to the position of the coasts; the winter is not cold, and the temperature of the summer never exceeds 77° F. The atmospheric precipitations are distributed over the whole year; in spring, and even throughout the summer, mists and thunder-storms keep the soil moist, and this to a greater extent on the northern than on the southern slope, which explains the greater abundance of plants on the north side of the mountains; consequently all the vegetative conditions of that portion of Europe on this side of the Alps are present. Agriculture is carried on there only as in gardens, at the chalets (Hatos): potatoes and barley, the latter usually only as high as 6300′; at a single spot on the southern slope as far as 7600′. There are no fixed dwellings; the land is only used for pasturage, without, however, affording the same nourishment to cattle as other mountains, for connected portions of meadow-turf are rare, and even here shrubs and thorny plants cover the greater portion of the slope.—Formations of the Central European region: *a.* Shrubs of *Sarothamnus scoparius*, *Genista ramosissima*, and *Quercus Toza*, ascending to 6000′; at the chalets, these are replaced by thickets of *Rosa canina* and *Berberis vul-*

garis.—*b*. Thin forests of *Pinus sylvestris* (5000′—6500′), of small extent on the Sierra Nevada, the trees being only from 20′ to 30′ in height; on the Serrania de Ronda, the Pinsapo forest previously mentioned with isolated trees of *Taxus* (5000′—6000′). The Sierra de las Almijarras, south of the city of Granada, is also partly covered with fir trees up to the summit (vid. sup.), hence the present forests appear to be only the remains of the zone of Coniferæ, which once covered the whole of these mountains, and is now destroyed. In the fluviatile valleys of the Sierra Nevada, isolated groups of trees, forming the remains of large forests, among which are the following, part of which only occur as single trees: *Sorbus Aria* (5000′—6500′), *Cotoneaster granatensis* (5000′—6000′), *Adenocarpus decorticans* (4500′—5500′), *Acer opulifolium* (5000′—6000′), *Fraxinus excelsior* (3000′—5000′), *Salix Caprea* (6000′—6500′), and *Lonicera arborea* (6000′—7000′), which are the tallest trees growing there.—*c*. Thorny, low shrubs of a stunted-growth form, an isohypsilous formation with *a*, which is principally found on a calcareous soil: *Erinacea hispanica*, *Genista horrida*, *Astragalus creticus*, *Vella spinosa*, and *Ptilotrichum spinosum*.—This region also contains numerous rock plants, particularly those belonging to the limestone; lastly, boggy springs, with limited meadows, exist in the valleys, and these are the localities where most of the Central European species exist.—Formations which in my opinion should be enumerated as belonging to the alpine region: *a*. Formation of the Piorno (*Genista aspalathoides*). This shrub, which is sometimes locally replaced by *Juniperus nana* and *Sabina*, forms a broad, connected zone of vegetation (—8000′), and is distributed downwards, like the Rhododendrons, over a tract which extends for a considerable distance into the forests (—5500′). *b*. Isolated pastures of rigid Grasses occur on the sloping ground between the Piorno-thickets. They consist of *Avena filifolia*, *Festuca granatensis*, and *duriuscula*, and *Agrostis nevadensis*.—Among the endemic forms of Boissier's third region, in addition to those already mentioned, we have the following, e. g. *Sarcocapnos crassifolia*, *Silene Boryi*, *tejedensis*, and *nevadensis*, *Arenaria pungens* and *armeriastrum*, *Erodium trichomanefolium*, and 3 other species, *Anthyllis tejedensis* and *Ramburei*, *Astragalus nevadensis*, *Prunus Ramburei*, *Saxifraga Haenseleri*, *Reuteriana*, *arundana*, *biternata*, and *spathulata*, *Reutera gracilis* and *procumbens*, *Butinia bunioides*, *Scabiosa pulsatilloides*, *Pyrethrum radicans* and *arundanum*, *Senecio Boissieri* and *elodes*, *Haenselera granatensis*, *Odontites granatensis*, *Thymus granatensis* and *membranaceus*, *Teucrium fragile*, *compactum*, and other species, and *Passerina elliptica* and *nitida*.

Fourth region (region of snow), 8000′—11,000′. Isolated patches of snow never entirely disappear: a connected layer of snow covers the ground for at least eight months. The soil is constantly kept moist, even in the summer, by the melting snow. Chalets are no longer met with, although cattle are driven up to this elevation. The vegetation consists of alpine herbs and grasses. Only four species of low shrubs belong decidedly to this region, but of these *Ptilotrichum spinosum* and *Salix hastata* are extremely rare; the two others, *Vaccinium uliginosum* and *Reseda complicata* do not raise their woody stem from the ground. The alpine meadows are called "Borreguiles," and they form here a fine sward of *Nardus stricta, Agrostis nevadensis, Festuca Halleri,* and *duriuscula:* and upon this turf, *Leontodon, Ranunculi, Gentianæ,* and other alpine plants grow. In other cases, herbaceous plants, growing in the form of tufts, preponderate, and displace the grass-plat; as *Silene rupestris, Arenaria tetraquetra, Potentilla nevadensis, Artemisia granatensis,* and *Plantago nivalis.* The alpine rivulets arise from small lakes, as also from a single spot of glacier-ice, near which, in moist defiles, we meet with taller herbaceous plants, as *Eryngium glaciale, Carduus carlinoides,* and *Digitalis purpurea.* Lastly, come the plants of the loose drift, as *Papaver pyrenaicum, Ptilotrichum purpureum, Viola nevadensis,* &c.; as also of neighbouring rocks, e. g. *Androsace imbricata, Draba hispanica, Arabis Boryi,* and *Saxifraga mixta.* The following plants are endemic, in addition to those already mentioned: *Ranunculus acetosellifolius* and *demissus, Lepidium stylatum, Silene Boryi, Arenaria pungens, Bunium nivale, Meum nevadense, Erigeron frigidus, Leontodon Boryi* and *microcephalus, Crepis oporinoides, Jasione amethystina, Gentiana Boryi, Echium flavum, Linaria glacialis, Holcus cæspitosus, Trisetum glaciale, Festuca pseudoeskia* and *Clementei.* Boissier gives the following proportions of the flora of Granada, which are valuable in a statistical point of view. The following are the families containing most species in the systematic section of Boissier's work: 239 Synantheraceæ (containing 80 Cynaraceæ, 65 Cichoraceæ, 64 Senecionideæ, 29 Asteroideæ, 1 of the Eupatorineæ), 202 Leguminosæ, 164 Graminaceæ, 105 Cruciferæ, 97 Umbelliferæ, 95 Labiatæ, 90 Caryophyllaceæ (comprising 39 Silenaceæ, 31 Alsinaceæ, and 20 Paronychieæ), 63 Scrophulariaceæ, 38 Cistineæ, 38 Ranunculaceæ, 37 Rubiaceæ, 36 Boraginaceæ, 34 Chenopodiaceæ, 33 Rosaceæ, 33 Liliaceæ, 32 Cyperaceæ, and 30 Orchidaceæ. Statistical sketch of the four regions assumed by Boissier. In the *first* region, 1070 species were observed, of which one sixth only occurred also in the second region, and a few plants only in sunny places in the upper region. Of this total number, 542 species are ⊙, 442 ♃ and 46 ⊝; as far as I am acquainted, this is the only region at present known, in which the number of annuals is as large or larger than that of the perennials. Of the 442 ♃, 19 are trees (vid. sup.), 58 are shrubs less, and 68 more than 3′ in height, the remainder are herbaceous plants. Of the shrubs, 22 are Leguminosæ (14 Genisteæ), 14 Cistineæ, 13 Labiatæ (low under-shrubs), 6 Chenopodiaceæ, 4 Asparageæ (2 Smilax), 4 Amentaceæ, 4 Solaneæ, &c. The region contains 860 Dicotyledons, 200 Monocotyledons, 10 vascular Cryptogamia, distributed through 82 families, of which the following contain most species: Leguminosæ (147), Synantheraceæ (124), Graminaceæ (106), Cruciferæ (47), Umbelliferæ (47), Labiatæ (46), Caryophyllaceæ (46), Chenopodiaceæ (33),

Scrophulariaceæ (26), Cistineæ (21), and Boragineæ (20). Cryptogamic plants are very rare, because the dryness does not suit the Mosses, nor the limestone the Lichens. According to its geographical distribution, the flora of the first region resolves itself into the constituents: *a.* About 200 are endemic to Spain, or consist of species distributed as far as Barbary or Provence; 12 species only are also found in the East. Characteristic families: 12 Cruciferæ (half of which are Brassicaceæ), 20 Leguminosæ (13 of which are Genisteæ), 24 Synantheraceæ (11 consisting of Cynaraceæ), 12 Scrophulariaceæ (8 *Linariæ*), and 13 Labiatæ. *b.* About 770 Mediterranean plants, the distribution of which is more extensive, but confined to the Mediterranean Sea. *c.* 200 Central European plants, most of which are either ruderal or marsh plants. Boissier observed 698 species in the *second region*, of which one seventh ascend into the third region, and some still higher. They consist of 202 ⊙, 465 ♃, and 31 ⊙. To the ♃ belong 21 trees, 43 tall and 68 low shrubs: of the tall shrubs, 11 consist of Leguminosæ (10 Genisteæ), 4 Cistineæ, 4 Caprifoliaceæ, and 4 Rosaceæ; of the undershrubs of the Tomillares, 13 consist of Labiatæ (4 species of *Thymus*), 12 Synantheraceæ (5 species of *Santolina*), 7 Cistineæ, 7 Leguminosæ (Genisteæ and *Astragalus tumidus*), 4 Ericaceæ, 4 Chenopodiaceæ, and 3 Thymeleaceæ. The region contains 597 Dicotyledons, 93 Monocotyledons, and 8 Ferns, distributed through 65 families, the following of which contain most species: Synantheraceæ (97), Leguminosæ (50), Labiatæ (44), Cruciferæ (41), Umbelliferæ (40), Graminaceæ (36), Scrophulariaceæ (27), and Cistineæ (23). Cryptogamic plants are common, tree-lichens commence at 3300′ in the Serrania de Ronda. The second region contains the following components, arranged according to their geographic distribution: *a.* 220 Spanish plants, of which 22 are distributed as far as Provence; 9 species are also found in the East. Characteristic families: 15 Cruciferæ, 15 Leguminosæ (11 Genisteæ), 15 Umbelliferæ, 33 Synantheraceæ (19 Cynaraceæ), 15 Scrophulariaceæ, and 17 Labiatæ. *b.* About 220 Mediterranean plants. Boissier collected 422 species in the *third region:* of these 333 are ♃, 78 ⊙, and 11 ⊙. To the ♃, belong 14 trees, 44 mostly low shrubs: among these 9 Labiatæ (*Thymus* 5 species), 8 Leguminosæ (6 Genisteæ, 2 Astragaleæ), 5 Rosaceæ, and 4 Thymeleaceæ. The region contains 358 Dicotyledons, 54 Monocotyledons, and 10 Ferns, distributed through 52 families, among which the following contain most species: Synantheraceæ (55), Leguminosæ (29), Graminaceæ (29), Cruciferæ (29), Caryophyllaceæ (29), Labiatæ (27), Scrophulariaceæ (24), and Umbelliferæ (20). Arranged according to their geographical distribution, the third region contains: *a.* 182 Spanish plants, of which 101 species appear at present to be confined to Granada. Characteristic families: 12 Cruciferæ, 14 Caryophyllaceæ, 15 Leguminosæ, 21 Synantheraceæ, 12 Scrophulariaceæ, and 16 Labiatæ. *b.* 185 Central European plants. *c.* 55 Mediterranean plants. In the *fourth region*, Boissier found 117 species, of which one third also occur in the third. Of these, 5 only are ⊙, 3 ⊙, and 109 ♃; moreover, 97 Dicotyledons, 16 Monocotyledons, and 4 Ferns, distributed through 34 families, of which the following contain most species: Synantheraceæ (16), Graminaceæ (11), Cruciferæ (11), Caryophyllaceæ (8), Scrophulariaceæ (8), Ranunculaceæ (5), and Gentianaceæ (5). Of Cryptogamic plants, Lichens growing upon rocks are common. Arranged according to geographic distribution, the fourth region contains: *a.* 45 Spanish plants, of which 30 species are at present peculiar to the Sierra Nevada; 13 species are also endemic to the Pyrenees. *b.* 66 alpine plants, part of which also occur in plains of the north of Europe. *c.* 6 species, which the Sierra Nevada contains in common with other mountains of the south of Europe.

Wilkomm's botanical letters on his travels (vid. supra), relate to the whole of Andalusia, and serve both to confirm and complete Boissier's systematically developed description.

The German traveller at once remarked, that the Sierra Nevada was much more bare and poorer in shrubs than the other mountains of Spain. He found, with Boissier, that the northern slope was richer in plants and more humid than the Alpujarras. In the Serrania de Ronda, he saw the forests of Pinsapo, a tree which unites the growth of the pine with the bark and arrangement of the branches of the red fir, but which differs greatly in the thick and short leaves. At a remote period, a great portion of the Serrania was covered with Pinsapo forests, but the trees have gradually been so cut down, that the Pinsapo is only now seen as a tree on elevated spots; it is, however, found as a shrub from 3000′ downwards. The Sierra Tejada, also a dolomitic mountain between Granada and Velez Malaga, was formerly covered by forests of *Taxus,* from the Spanish name of which, Tajo, the mountain derives its name. Now isolated trees only occur there at the source of the Tagus (Fuente del Tejo). The low eastern prolongation of the Sierra Tejada, the above-mentioned Sierra de las Almijarras is still partly wooded, between Motril and Granada, with pine trees and oaks, consisting of *Pinus pinea, halepensis* and *Pinaster, Quercus Ilex* and *lusitanica.* The slope of this mountain-chain, towards the coast, is also the native place of *Catha Europæa,* a shrub, which is found common between Nerja and Motril. The mountains on the eastern portion of Granada appear to agree in their vegetation most with the Sierra Nevada; as does also the barren Sierra de Alfacar (7000′), which separates the fertile Vega of Granada from the waste and arid elevated plain around Guadix: *Lavandula spica,* with some *Cisti,* there forms the covering of the bare slope. The vegetation of the plains of Guadix and Baza is that of a haloid and gypseous soil. On the boundary near Murcia, we next meet with the high limestone mountain of Sagra, at

Huescar (almost 8000′), where there exist large numbers of pines (*Pinus sylvestris*): the vegetation here also appears to resemble that of the Sierra Nevada, inasmuch as *Lonicera arborea*, e. g. is found there. The same applies to the Sierra de Filabres at Almeria (7000′), where Willkomm met with a number of plants endemic to the Sierra Nevada.

The low plain of Andalusia, or the district of the valley of the Guadalquivir, is inferior to the highlands of the south as regards the abundance of plants it contains, but is at the same time more carefully cultivated, especially around Seville. Willkomm found the tracts lying waste between Seville and Huelva, covered with dwarf-palms, also woods of pines and cork; oaks were common. In the autumn, numerous Liliaceous plants generally flowered there, with *Carenoa lutea* B. (*Pancratium humile* Cav.), one of the Amaryllidaceæ. Along the sandy coast, within the lagunes and salt-marshes, which at Huelva e. g., extend inwards to a considerable distance, pine forests extend from the Straits of Gibraltar to the mouth of the Guadiana, the underwood of which, opposite the coast-branches of the Sierra Morena, consists of *Cistus ladaniferus*, *Ulex Boivini*, &c. On the eastern terminal point of this line of coast, on the Sierra of Algesiras, the traveller met with a splendid forest of tall cork-oaks and olive trees, such as does not exist elsewhere in Spain, where *Rhododendron ponticum* is also mentioned as occurring by Boissier.

Willkomm visited Algarve in February 1846. The inhabitants distinguish three regions in it. *a.* The sandy line of coast (Cousta), which scarcely extends two leagues into the country, was originally a desert ravaged by the sea, but by industry has been converted, especially near Tavira, into a paradise-like district of gardens, containing plantations of southern fruits, vineyards, and corn-fields. Between Faro and Albufeira, this cultivated surface is interrupted by an extensive pine forest, containing *Erica umbellata*. Characteristic plants: *Empetrum album*, *Ulex Boivini* and *genistoides*, *Myagrum iberioides*, *Arenaria emarginata*, *Linaria præcox* and *linogrisea*, *Aristolochia bætica* (*glauca* Brot.) and *Scilla odorata* and *pumila*. *b.* The hilly country (Barrocál), extending to 1000′, is very much divided, and consists of various calcareous conglomerates; it is also fertile and well watered; still a considerable extent of the good soil lies waste, and is covered with Montebaxo. The vegetation was still backward; at Loulé, e. g. *Erica australis* and *lusitanica*, *Osyris quadripartita*, and several *Narcissi*, were found. *c.* The mountainous region (Serra), a terminal, undulating prolongation of the Sierra Morena, like the latter consisting of grauwacke and clay-slate, the western portion only, the Sierra de Monchique, being composed of granite and basalt, appeared dark green, yet not susceptible of cultivation. It is a remarkable fact, and tends to show the great influence of the geological substratum, that even here the shrubs of the Spanish Sierra Morena predominate, *Cistus ladaniferus* being very common, but mixed with the two *Ericæ* of the Barrocál. The valleys of the Sierra de Monchique are, however, wooded with chesnut trees and cork-oaks, with which *Rhododendron ponticum* grows in common. This mountain does not ascend more than 3800′ according to Portuguese measurements, but the upper vegetation is subalpine, and corresponds with the altitudes of 5000′—6000′ of Andalusia, which, however, I should not ascribe, as the author does, so much to the influence of the capacity of the granite for heat, as to the cessation of the influence of the plateau. The depression of the climate below the natural standard is not occasioned by storms, but the adjacent sea and low country cause a normal fall of the temperature in a vertical direction; whence Andalusia and the whole of Spain are abnormal in this respect, and the limits of vegetation extend to a disproportionate height.

Schouw has now published his work upon the Coniferous trees of Italy (see Ann. Rep. for 1844), in greater detail (Ann. Sc. nat. 1845, tom. i, p. 230).

Districts of the distribution of the species: 1. *Pinus sylvestris* L. (including *P. uncinata* D. C.) South slope of the Alps, 6000′—below 1000′ at Tagliamento, an Apennine of Montserrat. 2. *Pinus Pumilio*, Hk. South slope of the Alps, 4000′—7500′. 3. *P. magellensis* Sch. (*P. Pumilio* Ten. and *Mughus* Guss.) appears to hold the same relation to 4 as 2 to 1. The Abruzzi at Majella, 5000′—8300′. 4. *Pinus Laricio* Poir. (*P. sylvestris* and *nigrescens* Ten.) forms the forest of Ætna, 4′—6000′. Calabria and Abruzzi at Majella. *Pinus nigricans* Host. and *P. Pallasiana* are probably the same, as I have also assumed. 5. *Pinus Pinaster* Ait. Apennines. 0′—2800′ on Monte Pisano. 6. *Pinus Pinea* L. Apennines as far as Genoa, with 5. 0′—1500′ in the Western Apennines, 2000′ in the south of Italy. 7. *Pinus halepensis* Lamb. The whole of Italy, as far as the Apennines; 0′—2000′ on the Somma at Spoleto. 8. *P. brutia* Ten. Calabria, on the Aspromonte at Reggio, 2400′—3600′. It does not, however, appear to me sufficiently distinct from *Pinus Laricio*. 9. *Pinus Cembra* L. Alps, 4—6500′. 10. *Abies excelsa* D. C. (*P. Abies* L.), Alps; 1000′ (Tolmezzo)—7000′ (Stilfser Joch). 11. *Abies pectinata* D. C. (*Pinus Picea* L.) From the Alps to Madonia; 1000′—4500′ in the Alps, 5500′ in the Apennines. 12. *Larix europæa*. Alps; 1500′ (at Piave)—7000′. 13. *Cupressus sempervirens* L. Alps to Sicily, 0′—2500′. 14. *Juniperus communis* L. South to 40°; Alps, 0′—5000′. 15. *Juniperus nana*. W. Alps, 5′—7500′; Apennines. 16. *J. hemisphærica* Prl. Ætna, 5′—7000′; Calabria. 17. *J. Oxycedrus* L. Apennines, from 1000′—3000′. 18. *J. macrocarpa* Sibth. Along both the seas from Pisa to Sicily. 19. *J. Sabina* L., Alps; Apennines. 20. *J. phœnicea* L. Along both the seas from Nice to Sicily. 21. *Taxus baccata* L. Alps; Apennines.

Parlatore has commenced a Flora of Palermo (Flora Palermitana, vol. i. Firenze, 1845, 8vo). The first volume contains only the Grasses (130 species), with very full descriptions.

New species: *Avena Heldreichii, Melica nebrodensis, Vulpia panormitana* and *attenuata* (*Festuca sicula* Mor.)

Regarding the periods of the vegetation of wheat (*Triticum vulgare hybernum*), we find in Daum's work (l. c. p. 347), the following observations made in the year 1842:

	Average period of Sowing.		Time of Harvest.	
Malta	Dec. 1		May 13	164 days
Sicily	Dec. 1	(Palermo)	May 20	171 „
Naples	Nov. 16		June 2	195 „
Rome	Nov. 1		July 2	242 „
Berlin				299 „

Link read some observations upon the vegetation of Dalmatia, before the Geographical Society of Berlin (Monatsberichte, Bd. 2).

Visiani has had individual plants of Greece and Asia Minor drawn, from Parolini's collections (Memorie dell' Instituto Veneto, vol. i, 1843): 6 of the Labiatæ (*Thymus, Stachys*), 2 Boragineæ (*Anchusa, Lycopsis*), *Dianthus Webbianus* and *Sedum Listoniæ*. The new species which he has there proposed had in part, however, been previously described elsewhere.

II.—ASIA.

Parts 11—18 of the 'Illustrationes plantarum orientalium' of Gr. Jaubert and Spach (see Ann. Rep. for 1843) have appeared (Paris, 1844-45). The following families and genera are described in detail: Polygoneæ, Asarineæ, Chenopodeæ, Leguminosæ, principally Genisteæ (including *G. gracilis*, t. 143 = *G. carinalis*, m.), *Cousinia*.—In Lorent's oriental travels (Wanderungen im Morgenlande. Mannheim, 1845, 8vo), 35 species are described by Hochstetter as new, most of them from Syria and Armenia.—Endlicher and Diesing have described 6 Algæ which were collected by Kotschy in the Persian Gulf (Bot. Zeit. 1845, p. 268).

Mahlmann has elaborated Chanykoff's observations on

the climate of Bokhara (Berliner Monatsber. für Erdkunde, Bd. 2, pp. 132-40).

North winds are always prevalent in Chanat, hence they are in the direction from the steppe to Hindu-Kusch, which explains its freedom from rain and its continental distribution of heat. During eight months the wind was in the opposite direction ten times only. In the city of Bokhara (1116′ above the sea) Chanykoff however found the mean temperature during a severe winter = 30° F., i. e. lower, with the exception of Pekin, than has previously been anywhere observed in the same latitude. The trees bud between the 20th of March and the 10th of April. The vegetation of the steppes between Samarkand and Karschi lasts from the middle of March to the end of April; but the temperature remains high from the middle of March to the end of November, and is excessive in the summer.

Tschihatcheff's work upon his travels in eastern Altai, principally the district of the course of the Jenisi (Voyage dans l'Altai oriental. Paris, 1845, 4to), contains a list of the plants collected by the author in a portion of the district, part of which had not been previously examined; when determined by Turczaninow they were found to agree with those of adjacent countries.

The following were the trees: *Larix sibirica, Abies Pichta, Pinus sylvestris* and *Cembra, Alnus viridis, Betula alba, Salix Pontederana, pentandra,* and *stipularis* Turcz., *Populus alba, tremula,* and *laurifolia,* and *Sorbus aucuparia.* —The following families and genera have been described (Bull. Moscou, 1845), as forming an addition to Turczaninow's Flora of the Baikal Regions (see the preceding Ann. Rep.): 1 *Adona,* 1 *Cornus,* 6 Caprifoliaceæ, 7 Rubiaceæ, 6 Valerianeæ, and 2 species of *Scabiosa.*

G. Reichenbach has described some Orchidaceæ in Göring's collection from Japan (Bot. Zeit. 1845, p. 333).

The following divisions of R. Wight's illustrated work upon the flora of Hindostan (Ann. Rep. for 1840) have appeared according to the advertisement: Vol. ii, part 1 of the Illustrations of Indian botany, with 39 plates (Madras, 1841); of the Icones plantarum Indiæ orientalis, the conclusion of the first volume, consisting of 16 parts and 318 plates; vol. ii, with 318 plates (ib. 1840-42); and vol. iii, parts 1-3, with 409 plates (ib. 1843-46). Wight has also published a Spicilegium Neilgherense, with 50 plates (ib. 1846, 4to), in which particular plants of Nielgherry are figured: the latter appears, however, to be only an abridgment of his previous work (see Gardner's remarks in the Lond. Journ. of Bot. 1845, p. 565).—As stated in a letter, a memoir by Madden upon the Coniferæ of India is contained in the 'Quarterly Med. and Lit. Journal,' 1845, pp. 34-118, published at Delhi.—Gardner, the Brazilian traveller, who is now superintendent of the garden at Columbo in Ceylon, has reported upon his botanical excursions in Nielgherry (l. c. pp. 393-409 and 551-67); he enumerates the localities of the plants existing there.

De Vriese has commenced publishing an illustrated work upon select plants of the Dutch East Indian possessions (Nouvelles Recherches sur la Flore des Possessions Néerlandaises aux Indes Orientales. Fasc. 1, with 3 plates. Amsterdam, 1845, fol.): it contains a description of some new Styraceæ from Sumatra and Java, a figure of *Casuarina sumatrana*, as also of the new *Pinus Merkusii* from Sumatra.—Hasskarl has continued his remarks upon various points relating to the plants of Java both in the 'Ratisbon Flora' (1845, pp. 225 et seq., containing the Rubiaceæ), as also in V. d. Hoeven's Zeitschrift (Bd. 12, pp. 77 et seq., comprising the Malvaceæ and the allied families).—Montagne is describing the Lichens and Mosses of the Philippine Isles, from Cumming's collections (Lond. Journ. of Bot. 1845, pp. 3-11).

III.—AFRICA.

Fresenius has published Contributions to the Flora of Abyssinia from Rüppell's collections (Mus. Senckenbergian. vol. iii, 1845): containing copious descriptions of those Polygoneæ which have already been made known, and some new Synantheraceæ.

He gives, at the same time, a figure of the Lobeliaceous tree of Abyssinia, Gibarra (*Rhynchopetalum montanum* = Jibera of the preceding Ann. Rep.), and has represented its habit as follows: from 6'-7' in height, stem hollow, a crown of lanceolate leaves and tall bunches of flowers; hence the dimensions of the plants observed by Rüppell in Simen, between 11,000' and 12,000' were less in those found by Harris in Shoa. (See the preceding Ann. Rep. p. 382.) C. H. Schultz has described some of Rüppell's new Cichoraceæ from Abyssinia (loc. cit. p. 47).

Endlicher and Diesing are describing new Algæ from the Natal Colony (Bot. Zeit., 1845, pp. 288-90).

IV.—AMERICA.

Seller has made some isolated systematic remarks upon a collection of plants from the coasts of Davis's Straits and Baffin's Bay, in the 'Annals of Natural History' (vol. xvi, pp. 166-74).

Forry has compared the results obtained at the meteorological stations of the United States since 1819, and traced the distribution of heat in various points of view (Amer. Journal of Science, 1844, extracted into the Biblioth. de Genève, vol. lvii, pp. 140-50).

The unusual, nay, unparalleled accumulation of fresh water in the Canadian lakes, which, at a mean level of 1000′, include a surface of almost 4000 square geog. miles, procures for the northern states an insular climate for a very considerable distance into the eastern forest-region. Hence the difference between the summer and winter does not become excessive until we arrive just beyond the Mississipi, as also between the lakes and the Atlantic Ocean; in Lower Canada, e. g. the extremes of temperature are somewhat greater than in Michigan on the one hand, and the coast of Nova Scotia on the other. In the southern states, the annual curve resulting from the influence of the two oceans becomes still less arched than in the north, until, in Florida, it gives way to an almost tropical uniformity. The difference between the temperature of the summer and winter amounts there at Key West to only 43°F.; flowers bud there throughout the year without any general winter-sleep. During a space of six years the thermometer never rose, at this station, above 89°·6 F., and never sunk below 44°·6 F. The atmospheric precipitations are unequally distributed in Florida: in the central districts there are 309 fine days in the year, on the coast 250, and at the lakes, in the northern part of the state, only 117; but the air generally abounds in moisture, and the formation of dew is common.

Macnab has continued the Botanical Report of his travels (Ann. Nat. Hist., xv, pp. 65 and 351). Berkeley has published an account of some new Fungi from Ohio (Lond. Journ. of Bot., 1845, pp. 298-313).

Geyer's reports upon the characters of the vegetation of the prairies on this side of and beyond the Rocky Mountains (Lond. Journ. of Bot., 1845, pp. 479-92, and 653-62), in connexion with Frémont's investigations (vid. inf.), are immediately connected with the descriptions given by the Prince of Wied, to which the former are far inferior in regard to their too aphoristic style, but are as superior in systematic botanical knowledge.

The traveller ascended the Platte River from the State of Missouri, through the Osage district, as far as its source in the Rocky Mountains, traversed the mountains and the Colorado of California at about the forty-second parallel, and thus arrived at the Oregon territory. The western and southern limit of the prairies, which, to the south of Arkansas (according to De Mofras' map), are connected with the forests of New Mexico (37°) are not far from the Lower Kanza, in the district of Osages (39° N. lat.) Hence even here the forests of the valleys along the river become more numerous, the prairies abound to a greater extent in flowers, and the period of the summer drought is shortened. The most common species among the deciduous trees of Illinois, which are almost the same as those mentioned by the Prince of Wied (Ann. Rep. for 1842), and which also form the forests of the banks of the river on the Lower Missouri, gradually meet here with their western boundary, and they diminish in height the nearer they approach the sandy valley of the Platte. The herbaceous plants of this fertile prairie, however, become proportionately more numerous, and produce an uninterrupted succession of flowers throughout the spring and entire summer. In April isolated spring plants appear; in May and June the whole of the undulating surface for an immense distance is in flower, the plants consisting, e. g. of *Amorpha canescens*, *Batschia*, *Castilleja*, *Pentstemon*, *Cypripedium candidum*, &c.; taller herbaceous plants follow: *Petalostemon*, *Baptisia*, *Phlox aristata*, *Asclepias tuberosa*, *Lilium canadense*, and *Melanthium virginicum*; and finally, in the latter part of the summer, almost exclusively Synantheraceæ, from tall Helianthеæ down to the dwarf *Aster sericeus*.

The rich soil of the prairies terminates at the river Platte, with the limestone of the Missouri, which favours the vegetation detailed above. The lower terrace follows next; it is 900′-1000′ in height, and further up the stream is connected with the elevated surface of the upper steppe. The stony and sandy crust of the earth is formed of the detritus of granite, which is expanded over horizontally laminated sandstone and bituminous slate. The woods on the islands of the river then consist of *Populus canadensis*, *Ulmus americana* and *fulva*, *Negundo* and *Celtis occidentalis*; on the bank, thickets of *Salix longifolia*, with *Amorpha fruticosa*, *Rosa parvifolia*, *Rubus occiden-*

talis, and *Rhus glabrum*, predominate on the open prairie, which in May and June is rendered moist by the atmospheric precipitations: the vegetation, nevertheless, scarcely lasts longer than these short weeks of spring. The following forms may be mentioned as characteristic of the flora of the prairie; they are subdivided according to their localities, although the author has not arranged them in the form of a summary:—Of the Leguminosæ, *Astragalus*, e. g. *A. adsurgens* and *caryocarpus*, *Oxytropis*, *Phaca*, *Petalostemon*, *Psoralea*, *Glycyrrhiza*, and *Schrankia*; Malvaceæ, *Sida coccinea*; Cactaceæ, *Mamillaria simplex* and *Opuntia missurica*; Onagrariæ, *Œnothera* and *Gaura*; Synantheraceæ, principally Helianthcæ, e. g. *Echinacea*, *Rudbeckia*, *Heliopsis*; moreover, *Artemisia*, e. g. *A. caudata*, and *Lygodesmia*; Scrophulariaceæ, *Pentstemon* and *Castilleja*; Hydrophyllaceæ, *Ellisia*; Boragineæ, *Batschia*; Nyctagineæ, *Calymenia*; Liliaceæ, *Yucca*; Graminaceæ, e. g. *Sesleria dactyloides*, *Crypsis*, *Stipa*, *Agrostis*, *Eriocoma*, &c.

The remaining large district is denominated by Geyer the upper saline desert region, the area of which extends far inwards symmetrically, on both sides of the Rocky Mountains, from Missouri to Lower Oregon, a desert elevated surface resting upon sandy rocks, and gradually ascending from 1200′ to more than 4000′; so that the chains of the Rocky Mountains, in spite of their elevated central ridges, cannot by any means be regarded as forming a boundary of vegetation. The boundaries of this immense steppe, which everywhere affords pasture, Geyer considers as formed, in the north, by the Saskatchawan and Lake Winnipeg; in the east (the same as the Prince of Wied), by a line running longitudinally through Ioway, or the former district of the Sioux (Great Sioux river and Moine's river); in the south, by the Upper Arkansas; in the west, by the mouth of the Wallawalla, in the Oregon district (more distinctly by Frémont, the union of the two principal forks of this river, the Lewis river and the Upper Columbia); hence about 38°-54° N. lat. and 77°-101° W. long. from Ferro. With the exception of the pine and snow-clad central chain of the Rocky Mountains, this space contains no forests. The prevailing character of the flora is generally the same as that described by the Prince of Wied, that of the Upper Missouri. Beyond the Rocky Mountains also, as in the district of the source of the River Platte, the steppe is covered with two social shrubby Artemisias (*Art. tridentata* and *cana*). The Pulpy-thorn, *Sarcobatus vermicularis* (*S. Maximiliani* N.), also called the Salt-cedar, is found everywhere on the saline soil as low down as Oregon; it is a shrub, with numerous stems from 3′-8′ in height, with diverging thorny branches and dark-green succulent leaves. Considering the similarity of the climate and soil of the prairies and the Russian steppes, it is an interesting fact, that this genus, which was first recognised as distinct, from the examination of Wied's collections, according to both Lindley and Torrey (*Fremontia* ej., *Batis vermicularis*, Hook.), is a true member of the Chenopodiaceæ (Lond. Journ. of Bot., 1845, pp. 1 and 481), and grows in common with other Halophytes belonging to the same botanical group. The other most common thickets of the upper steppe consist of *Elæagnus argentea* and *Shepherdia argentea;* then *Amorpha frutescens, Rosa parvifolia,* and woody Synantheraceæ, e. g. *Iva* and *Bigelovia. Juniperus andina* (*J. repens* of Wied), with *Yucca angustifolia,* appear to be confined to the Missouri country below the mouth of the Yellowstone river. Geyer's further distinctions of several districts of vegetation in the region of the upper terraces has not been carried out sufficiently clearly. The following may be regarded as characteristic forms:—Of the Leguminosæ, *Astragalus, Homolobus, Psoralea, Glycyrrhiza, Hosackia, Schrankia,* and *Amorpha;* Cruciferæ, *Stanleya pinnatifida;* Loaseæ, *Bartonia ornata;* Onagrariæ, *Œnothera;* Cacteæ, *Opuntia missurica;* Umbelliferæ, *Cymopterus;* Synantheraceæ, in addition to the above-mentioned shrubs, several Chrysopsideæ, Cichoraceæ, *Achillea;* Scrophulariaceæ, the same genera as those of the lower terraces; Chenopodiaceæ, in addition to *Sarcobatus: Kochia, Salsola, Chenopodium,* and *Atriplex;* Liliaceæ, *Calochortus* and *Allium, Iris, Triglochin maritimum, Carex;* Graminaceæ, e. g. *Triticum missuricum, Hordeum jubatum,* and *Ceratochloa.*

Geyer's description is rendered geographically more clear by the excellent diary kept by FREMONT of his travels, who, being the chief of an expedition of discovery, and furnished with botanical knowledge, explored the whole of the steppes of the North American prairies, down to Lower Oregon and Upper California, in different directions, with the most fortunate results (Narrative of the Exploring Expedition to the Rocky Mountains in 1842, and to Oregon and North California in 1843-44. Washington, 1845. I am only acquainted with it from the English edition, London, 1846-8). On this side of the Rocky Mountains, Frémont followed the same course to the River Platte as Geyer; on the second occasion, he ascended the Kanza and its accessory streams, to the central chain. The country ascends very gradually from the bifurcation of the Kanza (79° W. long.), to the foot of the Rocky Mountains, and on the west side of the mountain the land sinks to the conflux of the Lewis and Oregon, as is evident from the following line of level which was determined barometrically by Frémont, and intersects the entire steppe from east to west. Bifurcation of the Kanza (79° W. long.) = 926′; River Platte (81°) = 2000′; River Platte (83°) = 2700′; Fort Laramie, on the Platte (87°) = 4470′; and almost in the same meridian, Fort Vrains (40° 16′ N. lat.) = 4930′; as also the River Arkansas (38° 15′ N. lat.) = 4880′; Artemisia-steppe, at the eastern foot of the Rocky Mountains (41° 36′ N. lat. and 90° W. long.) = 6820′; south pass through the Rocky Mountains, in a deep depression, which does not possess any mountain character (42° 27′) = 7490′; foot of the Rocky Mountains, at the upper part of the course of the Colorado of California (41° 46′) = 6230′; Fort Hall, on the Lewis (43° N. lat., 95° W. long.) = 4500′; River Lewis (43° 49′ and 99°) 2100′; River Lewis (44° 17′ and 100° W. long.) = 1880′.

The open prairie-steppe beyond the Rocky Mountains is covered generally with shrubs of *Artemisia*, between which, however, cattle everywhere find food in nutritive grasses. *Purshia tridentata*, one of the Spiræaceæ, which frequently accompanies the *Artemisiæ*, is a shrub peculiar to this part. The nutritive plants used there by the Indian hunters in cases of necessity, corresponding to the *Psoralea esculenta* on the Missouri, consist of *Valeriana edulis* (Tobacco-root), *Cirsium virginianum*, a species of *Anethum* (Yampeh), and *Kamassa* (Kamas), Fr. indescr. The bank-forests of the cotton-wood (*Populus*) are not met with until we arrive at the lower regions: they appear to be entirely wanting on the upper terrace. At the bifurcation of the Oregon, where the prairie terminates (101° W. long.), the wooded promontories of the western chain of high mountains commence, which may be compared to the Rocky Mountains in extent, and projecting everywhere above the snow-limit, probably exceed them in height. Being a continuation of the Californian Andes, it is called, in Upper California, Sierra Nevada, in Oregon, the Blue Mountains, and the Cascade-chain, where, on the south side of the united rivers, near Fort Vancouver, it rises into high snow-mountains, as at Mount Hood. At Oregon, the forests of this high mountain-chain (explored between 2700′ and 3800′), which are only interrupted by the most splendid meadow-slopes, consist of birch, but above all of various Coniferous trees remarkable for their enormous dimensions, which are such as are not met with in any other part of the globe. The larches were sometimes 200′ high (p. 182), the firs were of the same height, with stems 7′ in diameter; in the former, the unbranched stem beneath the crown was sometimes 100′ in length. White spruces which gave off branches down to the root appeared nevertheless to measure 180′, perhaps 200′.—The Cascade-chain separates the mild climate of the western coast of the Oregon district from the dry prairies equally as definitely, but in an inverse direction, as the Peruvian Andes do the west desert coast-region from the more humid highlands. This meridional mountain-chain, which intersects the River Columbia about twenty-five or thirty miles from its mouth, receives the mists and rain which are driven over to it from the Pacific Ocean, but which do not penetrate to the clear sky of the steppe. At the rapids of the Columbia, the "Dalles" within the mountain-line, the rainy season is already unknown, which on the coast denotes the winter, and this season is only recognisable there (45° N. lat.) by a slight fall of snow, which scarcely lasts on the ground for two months. The cause of the winter rainy season at the mouth of the Oregon, where west winds predominate, appears to me to depend simply upon the fact, that in the summer the sea, whilst in the winter the continent, is the coldest, so that during the latter period of the year, the humid winds from the sea must quickly lose their moisture in passing over the coast-district. But the steppe lying behind the mountains is a highland; as such it exceeds the coast in warmth and dryness, and cannot, therefore, readily precipitate the aqueous vapour from the westerly current of air. The same, however, holds good here, from whatever other points of the compass the wind may blow, so that instead of steppes, deserts would be expanded between the Rocky Mountains and the Californian Andes, if this internal country were not also so copiously watered by these mountains, and thereby also subject to local precipitations. The climatal relations of the Oregon-district also perfectly explain the drought of the prairies in the Missouri described by the Prince of Wied.

From Columbia, Frémont went to the eastern foot of the Sierra Nevada, as far as the 39th degree of south latitude, following the boundary-line between the steppe- and the forest-regions. Below the 42d degree, at the south water-boundary of the district of the Oregon-river region, the inland country is elevated into a mountain-chain running from east to west, and not void of forests, and this appears to form a connexion between the Californian Andes (S. Nevada) and the Rocky Mountains.

South of the chain, a desert highland is situated; it is probably for the most part uninhabitable, and, from the nature of its soil and its declivity, it may be compared with the uninhabitable regions of Persia; it ought to be called the Californian salt-desert (Frémont's great interior basin). An Indian guide pointed to it, saying at the same time, "There are the great llanos—no hay agua, no hay zacatá, nada"—i. e. Plains without water, without herbage: "Every animal that enters it must die." Entirely surrounded by mountains which form its borders, bounded to the north by the river-limit of the Oregon, to the south by a similar chain, covered with snow, towards the Colorado, and on both sides by the Sierra Nevada and the Rocky Mountains, it only contains internal streams, which lose themselves in the desert or in salt-water lakes, and is perhaps dry and destitute of springs for the space of many days' journey. As the greater part has not hitherto been explored by any traveller, we are confined, with regard to the altitudes, to the following measurements made by Frémont, which, in fact, only relate to the external margin: on the plateau of the great salt-water lake Utah (41° 30′ N. lat. and 95° W. long.) = 4200′; Lake Pyramid, at the foot of the Sierra Nevada (39° 51′) = 4890′; foot of the Sierra Nevada (38° 50′) = 5020′; on the boundary-mountains, Bear River, on the slope of the Rocky Mountains (42° and 93°) = 6400′; pass from the Bear River to Colorado (41° 39′) = 8230′; pass over the Sierra Nevada to the Bay of St. Francisco (38° 44′) = 9338′.—The salt-desert differs from the prairies of Missouri, as from the *Artemisia*-steppes of Oregon, by its excessive dryness, rocky soil with volcanic heaps, by the more general presence of salt in the soil, and, as a consequence of these conditions, by the absence of the growth of nutritive grasses; nevertheless, from the strength and number of the rivers which enter it from the boundary-mountains, we may conclude as to the existence of oases on its streams. The vegetation consists almost exclusively of shrubby Chenopodiaceæ, with which in tracts Artemisias are mixed, and along the Sierra Nevada, and to the south of the 41st degree of latitude, *Ephedra occidentalis,* forming an evergreen shrub. The most common of the Chenopodiaceæ found here is also *Sarcobatus vermicularis; Obione* is next mentioned, of which *Obione rigida* Torr. and Fr., with another new species, occurred at Utah; *Salicornia* also covered the banks of this lake. The woods of the boundary-mountains to the north of Utah consisted of deciduous trees: *Populus, Salix, Quercus, Cratægus, Alnus,* and *Cerasus.* Below the 39th degree of longitude, the Sierra Nevada was crossed with great difficulties in the depth of the winter, so as to reach the valley of Sacramento. The lowest forest-zone on the desert side of the mountains consisted of a pine, the seeds of which were edible, *Pinus monophylla* Torr., a tree from 12′ to 20′ in height, and with a stem at the most 8″ in diameter, which, with some roots and the salmon found in the waters, form the food of the Indians. Further upwards this pine (nut-pine) was found somewhat larger, its diameter amounting to 15″. But it was not until an altitude of 6000′ had been reached, that Coniferous forests of a taller growth and of a different species were met with, accompanied by a more luxuriant vegetation, in which the first indications of a fairer climate were met with, At 8000′ the trees were almost as gigantic as in Oregon: red pines as high as 140′ and 10′ in diameter (*Pinus Colorado* of the Mexicans) predominating, and with them tall cedars 130′ in height, and two species of fir of an equally tall growth (white spruce and hemlock spruce). Trap-rocks form the fertile soil of these splendid tall forests, to a considerable depth. On the west side of the mountains, below the Coniferous zone, Frémont arrived at a region of evergreen and other oaks, which corresponds with Hinds's representation of the character of the region of St. Francisco: here, after the impressions left by the deserts, the traveller was delighted with the most luxuriant spring flowers in the valleys of the Sacramento and St. Joachim.

On his return, Frémont crossed over the Californian Andes by a much lower pass, below the 36th degree, and travelling parallel with the Colorado, on the southern border of the salt-desert, returned to the Great Salt-lake and the Rocky Mountains. This road, which forms the course taken by the caravans in going from New Mexico to California, was rocky and mountainous (sloping off towards the Colorado from about 5000′ to 2000′): the vegetation was scanty, corresponding to the character of the flora of California. A tall Zygophyllaceous shrub (*Zygoph. californicum* Torr., Fr.), a *Yucca,* and numerous *Cacti* constitute the principal forms of plants over extensive tracts; and from the north towards this part, as far as the woods of the *Yucca,* the *Artemisia tridentata* of the steppes extends: the traveller would not, however, give the preference to the former, since the stiff and unsymmetrical form of the *Yucca* appeared to him the most repugnant formation in nature. Among the shrubs of this region, he mentions *Ephedra occidentalis, Garrya elliptica,* which forms dense thickets on the banks of the rivers, and *Spirolobium odoratum* Torr., one of the Mimoseæ, 20′ in height. The northern forms distributed thus far were (36° N. lat.): *Pinus monophylla, Purshia tridentata,* and *Populus* and *Salix* on the banks of the rivers.

The snow-line of the Rocky Mountains was estimated on the Snow-Peak (42°—43° N. lat.) at 11,800′ (i. e. 1800′ above the measured point 10,000′), This mountain, the summit of which, 13,570′ high, Frémont ascended, belongs to the accessory chain of the Wind River mountains, but is regarded as the most elevated of the entire system. Above the Coniferous region, the altitudinal limits of which were not determined there, it contains a copious alpine vegetation, which, according to the examples brought forward, are principally characterised by Hudsonian forms, just as those of the Alps are by arctic forms.

Frémont's observations upon the tree-limits of the continent of North America are extremely remarkable; they show that they are much higher than in corresponding latitudes of Europe. Not only in the Californian Andes were the pine forests found to extend above 8000′, but on the east side of the Rocky Mountains, in the district of the so-called Park, in the region of the source of the southern bifurcation of the River Platte and the Arkansas (39° 20′ N. lat.), Frémont found that even at an elevation of 10,430′, "the pine forest continued, and grass was good. We continued our road, occasionally through open pines, with a very gradual ascent; and having ascended perhaps 800 feet, we reached the summit of the dividing ridge, which would thus have an estimated height of 11,200′" (p. 314.) Hence the altitude of the tree-limit of the Rocky Mountains in the latitude of Valencia may be assumed as 11,000′; the most elevated tree-limits of the south of Europe, the isotherms of which are of such very different temperatures, scarcely ascend beyond 7000′. If the influence of the highlands of North America is so great in moderating the vertical diminution of the temperature of the summer, we are justified in anticipating similar phenomena in central Asia. There is especially, one observation which corresponds with this supposition, and it is the only one with which I am acquainted, viz. that relating to the valley of Spiti, in Lesser Thibet, where, according to Jacquemont, at the same altitude, but a more southern latitude (32° N. lat.), dwarf trees alone occur. But it is not only the heat which causes the dense tall forests to ascend to such considerable elevations in North America; the humidity of the air or of the soil must also be taken into account. In the south of Europe, the tree-limit does not ascend in proportion to the increase of heat, since it is frequently situated at a greater altitude on the south side of the Alps than at any more southern point of the continent. In Thibet, where the highlands even ascend to the level of the tree limit, the limitation of the growth of trees is not caused by cold, but by dryness. Now, it is a circumstance common to both the mountain-chains of North America, that under the latitudes of the south of Europe, they ascend far above the limit of perpetual snow. Here the drying influence of the plateau, which lies far below the forests, is removed by the masses of snow; not so, however, in Thibet, where the country corresponding to the plateau ascends to the snow-line. On the mountains of North America, as on the south side of the Alps, sufficient water is thawed in summer from the large fields of snow to irrigate the elevated forests: thus they are provided with a permanent source of moisture; even when the prairies are without rain for months, they never dry up, whilst upon Pindus and the Apennines the winter snow soon disappears; whilst in Thibet the melted snow upon the elevated surface evaporates again, without fertilizing the soil.

In a botanical appendix to Duflot de Mofras' work upon the western coast of North America (Exploration du Territoire de l'Orégon, &c., 2 vols. 8vo. Paris, 1844), a list of about 300 Californian plants is given. It is, however, derived from old sources, and is disfigured by errors of the press to such an extent as to be rendered almost useless.

In the work itself we find the following statements regarding the course of the seasons in California: 1. In Upper California, e. g. in the latitude of St. Francisco (38° N.), the rainy season, with prevailing south-east winds, lasts from October to March. From April to September north-west winds blow; it then never rains, although fogs are common on the coast; the soil also then loses its verdure (ii, p. 46). On account of the length of the drought, the total amount of the atmospheric precipitations is less than in the south of Europe. 2. The vegetating season of the arid western coast of Lower California, however (30° to 23° N. lat.), occurs with its atmospheric precipitations during the summer (i, p. 239). 3. On the eastern coast of this peninsula, at Cape Lucas, in the Gulf of California (mer vermeille), and on the north-west coast of Mexico we find an inversion of the trade-wind, (inversion d'alizé, i, p. 171), as south-westerly or westerly winds predominate there. In Mazatlan (23° 12′) the rainy season coincides with south-westerly and westerly, and the dry season with north-westerly winds (i, p. 172); the same is the case at St. Lucas, where these latter monsoons prevail from November to May (i, p. 229). Within the gulf, where, although it is beyond the tropic, the monsoons are the same, the amount of rain appears to diminish very considerably. We cannot imagine anything more miserable and neglected than these two coasts, which lie waste from a deficiency of water (i, p. 205).

I find a notice of Plantæ Lindheimerianæ, by Gray, which probably contains an account of Lindheimer's collection from Texas; but I am unacquainted further with the work.

A. Richard and Galeotti intend publishing a monograph of the Mexican Orchidaceæ, which will include 460 species; of these about a third part are new. They have published preliminary diagnoses of the new species (Ann. Sc. nat., 1845, t. iii, pp. 15-33).—V. Schlechtendal's contribution to the flora of Mexico, for the present year, refers to the Asphodeleæ (Bot. Zeit., 1845).

Purdie (Ann. Rep. for 1843) has continued his botanical reports from Jamaica (Lond. Journ. of Bot., 1845, pp. 14-27).

The Cactaceæ, which are common on the south coast of the island, are absent from the north side. In the former locality, at Bath, he found coast-mountains about 3000′ high, covered with a tall forest of *Podocarpus Purdiena* Hook., one of the largest forest-trees of Jamaica; one of them, which had been cut down, measured more than 100′, 40′ up to the crown, and at the height of a man above the root, it was $3\frac{1}{2}$′ in diameter. *Podocarpus coriacea* occurs above a level of 5000′ or 6000′. The coffee-plantations are situated on the south side of the island, e. g. at the pass from Kingston to Bath, between 3000′ and 6000′. *Coffea* does not thrive at a greater elevation.

Kunze has enumerated and described the new species among the Ferns collected by Moritz in Caracas (Bot. Zeit., 1845, pp. 281-8).—Of Bentham's descriptions of Schomburgk's plants from Guiana, the Polygonaceæ (14 sp.) and Thymeleaceæ (3 sp.) have appeared, as also those of the Acanthaceæ (17 sp.) from Nees v. Esenbeck (Lond. Journ. of Bot., 1845, pp. 622-37). Schomburgk has himself described individual species of his collection (ibid., pp. 12, 375).—Gardner has published the diagnoses of 100 new plants, discovered by himself in Brazil, as a continuation of his former work (ibid., pp. 97-136). —K. Müller has resumed the description of Gardner's Mosses (Bot. Zeit., 1845, pp. 89 et seq.)—The continuation of Naudin's contributions to the flora of Brazil (see the preceding Report) comprises the Melastomaceæ (Ann. Sc. nat., iii, pp. 169-92, and iv, pp. 48-57).—

Jameson has described a botanical excursion made at Chimborazo (Lond. Journ. of Bot., 1845, pp. 378-85).

The aqueous vapours of the winds from the sea are precipitated upon the west side of the western Cordillera of Ecuador, to which Chimborazo belongs. Hence, simultaneously with the rainy season of the coast of Guayaquil, wet weather prevails there from the end of December to the middle of May, whilst on the eastern slope, and on the elevated surface of Riobamba, the weather is fine. This contrast exerts an important influence upon the vegetation; hence the numerous Calceolarias and the Alstræmerias are confined to the western slope; hence also, in the upper regions, tall-stemmed woody plants are isohypsilous with the shrubs of the central Cordillera. Between 13,000′ and 14,000′ *Polylepis lanuginosa*, one of the Sanguisorbeæ, forms a distinct woody zone, regarding which Jameson remarks, that this tree will grow at a higher level than any other upon the globe. Lower down, on the road from Riobamba to the locality of Guaranda, which is situated on the western side of the Chimborazo chain, there exists a meadow-region of the same extent, until, at 12,000′, woods of *Aristotelia Maqui* and *Columellia sericea* are again met with, in which the underwood consists of shrubby Synantheraceæ, Rosaceæ, Melastomaceæ, and Scrophulariaceæ. The report concludes with a list of the families of plants which were found to occur between 12,000′ and 14,000′. Nearly 250 species observed there by Jameson belong to about 50 families. Those containing most species are: 29 Synantheraceæ, 15 Scrophulariaceæ, 11 Graminaceæ, 11 Rosaceæ, 8 Leguminosæ, 7 Gentianaceæ, 7 Umbelliferæ, and 7 Cruciferæ; 14 Ferns and 13 Mosses; also the following characteristic alpine forms: Ranunculaceæ (5), Caryophyllaceæ (4), Ericaceæ (4), Vacciniæ (3), Valerianaceæ (4), Orchidaceæ (5), and Cyperaceæ (3). South American forms: Loaseæ (2), Passifloreæ (1), *Escallonia* (1), *Columellia* (1), Solanaceæ (5), and Lobeliaceæ (2). The following tropical forms are also found at this level: Melastomaceæ (4), Homaliaceæ (1), Loranthaceæ (2), and Bromeliaceæ (2).

Bridges has reported upon the first-fruits of his botanical travels in Bolivia (ibid., p. 571).—The first part of the botanical division of a very important work upon Chili, by Cl. Gay, has reached us (Historia Fisica y Politica de Chile, por Cl. Gay. Botanica, tom. i, pp. 1-104. Paris, 1845, 8vo). The diagnoses are in Latin, the descriptions in Spanish. The prodromus, when completed, will contain all the plants of Chili, and a select number will be illustrated by copper-plate engravings; but it also includes garden plants.

The genera treated of in the first part, the indigenous ones, are the following: Ranunculaceæ: *Anemone*, 7; *Hamadryas*, 2; *Barneoudia*, *Ranunculus*, 18; *Psychrophila*, 4; and *Pæonia*. Magnoliaceæ: *Drymis*, 2. Anonaceæ: *Anona*, 1· Lardizabalcæ: *Lardizabala*, 2; and *Boquila*, 1. Berberideæ: *Berberis*, 23. Papaveraceæ: *Argemone*, 3; and *Papaver* and *Fumaria*, 1.

V.—AUSTRALIA AND THE SOUTH-SEA ISLANDS.

J. D. Hooker opposes the opinion that all or most of the South-Sea Islands belong to the same primitive formation (Lond. Journ. of Bot., 1845, p. 642).

The resemblance of their vegetation is rather apparent than real, and is principally evidenced in littoral plants, and in those which with man have migrated beyond their native country towards the East. However, that the original vegetation, with which those naturalized have become associated, is endemic to the larger groups of islands, at least, is shown by a comparison of the flora of the Sandwich and Society Islands, for instance, both of which are subject to the same climatal conditions; one being situated north, and the other south of the equator. Few only of the prominent genera are found in both groups. The Society Islands are the poorest, but tropical in their forms, and less peculiar: here the extensive families of the torrid zone predominate, as the Malvaceæ, Leguminosæ, Apocyneæ, Urticaceæ, Melastomaceæ, and the Myrtaceæ. Of the forms peculiar to the Sandwich Islands, the Synantheraceæ, Lobeliaceæ, Goodenoviaceæ, and Cyrtandraceæ, few or no representatives are found. Other families, as the Grasses, Euphorbiaceæ, Rubiaceæ, &c., which are numerous in both Archipelagos, occur for the most part in isolated species. The same view of the endemic character of the Flora of the Sandwich Islands is adopted by Hinds. (Ann. Nat. Hist. xv, pp. 91-3.) With other Floras and those of the most different kinds, isolated points of resemblance only can be shown. Of 165 species which the traveller collected upon the coast there, half are endemic. In a physiognomical point of view the amount of forests is small in comparison with that of other tropical countries, the trees are not tall and only crowded in moist sheltered valleys. Cinchonaceæ, Guttiferæ, Sapindaceæ, Euphorbiaceæ, are found there, mixed with Tree-Ferns, and a single Palm which was originally endemic.

The work of Strzelecki upon New Holland contains a number of valuable details on the conditions of the vegetation of this continent (Physical Description of New South Wales and Van Diemen's Land. London, 1845, 8vo).

The extra-tropical south-eastern coast is exposed pretty uniformly to variable winds, which are dependent upon the monsoons of the adjacent oceans, but are not the same in the different latitudes. At Port Jackson and Port Macquarie (32° S. lat.) equatorial winds prevail in the summer and polar currents in the winter; in Port Philip (south-eastern extremity of the continent) equatorial currents in the winter, and polar currents in the summer; in Van Diemen's Land the equatorial winds predominate throughout the whole year (p. 168). The amount of rain is far more considerable on the coast than we should expect: on the average it amounts to 48″ 6 in New South Wales, and to 41″ 3 in Van Diemen's Land (p. 192). The temperature is far more uniform than in corresponding latitudes of the northern hemisphere, as is shown by the following table (p. 229).

	Port Macquarie.	Port Jackson.	Port Philip.	Woolnorth in Van Diemen's Land.
Mean temperature	+20° c.	+19° 2 c.	+16° 3 c.	+14° 1 c.
Mean summer temperature	+23° 9″	+23° 2″	+20° 8″	+16°
Mean winter temperature	+16° 1″	+15° 1″	+11° 9″	+12° 3
Maximum of summer	+31° 3″*	+27° 8″†	+32° 5†	+20°4‡
" winter	+ 8° 2*	+ 7° 4	+ 2° 7†	+ 8°‡

The influence of the geological conditions upon the vegetation and the cultivation of the soil is exceedingly variable, according to Strzelecki, as shown by a comparison of New South Wales with Van Diemen's Land. In New South Wales, granite, sandstone, and conglomerate predominate; limestone is confined to but few localities; in Van Diemen's Land, porphyry, greenstone, basalt, and trachyte predominate; limestone is also common (p. 360). In the former, the silica existing in the soil favours the nocturnal diminution of temperature, and would act still more prejudicially if the vegetation, which is more dense, did not frequently give rise to the formation of clouds (p. 219). But the small quantity of soluble constituents in the soil renders it only adapted for indigenous plants, as for pasture-land, but not for agriculture.

The botanical letters from New Holland, by Leichardt (Lond. Journ. of Bot., 1845, pp. 278-291), whose great voyage of discovery through the interior of the continent, which has never been surpassed in its results, was described without a view to publication, excite the most sanguine hope that the botanical characteristics of Australia, taken up by such talent for observation, and described in an equally successful manner, will, at some

* The warmest month is November; the coldest, August.
† The warmest month is November; the coldest, July.
‡ The warmest month is January; the coldest, August.

future time, acquire important elucidations from this traveller.

Systematic contributions to the Flora of Australia: Sonder's diagnoses of 76 new Algæ, belonging to Preiss's collection from the Swan River (Bot. Zeit., 1845, pp. 49-57)—Berkeley's new Fungi (54 sp.), from the same locality, and belonging to Drummond's collection.

J. D. Hooker has written a memoir upon the distribution of the Coniferæ in the southern hemisphere (Lond. Journ. of Bot., 1845, pp. 137-157).

Van Diemen's Land contains ten different Coniferæ, which are endemic to the island. They occur in limited localities, and most of them were discovered by Gunn; they are *Callitris australis* Br. (Oyster-bay pine), a tree from 50'—70' in height; *C. Gunnii* J. D. Hook. (native cypress), 6'—10' in height; *Arthrotaxis*, 3 sp.; *Micocachrys tetragona* J. D. Hook., a tree from 15'—20' high; *Podocarpus alpina* Br., a shrub on Mount Wellington, at an altitude of from 3'—4000'; *Podocarpus Lawrencii* J. D. Hook.; *Phyllocladus asplenifolia* Rich. (celery-topped pine), 50'—60' high; *Dacrydium Franklinii* J. D. Hook. (Huon-pine), the most beautiful tree of them all, from 60' to 100' in height, with a diameter of from 2'—8', of limited occurrence, but used at the harbour of Macquarie as ship-timber.

Sketch of the distribution of the Coniferæ as yet discovered in the southern hemisphere: 16 species in New Holland (10 *Callitris*, 4 *Podocarpus* and 2 *Araucaria* at Moreton Bay), 10 species in Tasmania (vid. sup.), 13 species in New Zealand and the South Sea Islands (6 species of *Podocarpus*, of which the Kaikatia, *Podocarpus dacrydioides* Rich. is most common at the Bay of Islands, 3 *Dacrydium*, *Thuja Doniana* Hook., *Phyllocladus trichomanoides* Don., *Dammara Australis* = Kauri pine, *Araucaria excelsa* Ait. =Norfolk Island pine, probably confined to this island; 8 species in South America (4 *Podocarpus* in Chili and the Brazils, *Thuja chilensis* Hook.; *andina* Pöpp.; *Thuja tetragona* Hook.= Alerse of Chiloe, *Araucaria brasiliensis*=Brazilian pine, *Araucaria imbricata* = Chili pine, on the Andes, from 37° to 46° S. lat.; *Juniperus uvifera* Don., from Cape Horn, remains doubtful; about 6 species in the South of Africa and the Mauritius (2 *Podocarpus*, 3 *Pachylepis*, including *Pachylepis Commersoni* from the Mauritius, and *Juniperus capensis* Lam., doubtful.

We have now received 15 Parts of J. D. Hooker's illustrated work upon his Antarctic Voyage. (The Botany of the Antarctic Voyage. London, 1845, 4to.)

The character of the vegetation of Lord Auckland's islands is more clearly described than before (Ann. Rep. for 1843). It was previously mentioned that these islands, the volcanic soil of which ascends in the form of gentle hills to an elevation of 1500′, were uniformly covered with forest-, shrub-, and pasture-land. The forest on the abundant humous soil of the coast consists of *Metrosideros lucida*, mixed with an arborescent *Dracophyllum*, the underwood consisting of *Coprosma*, one of the Rubiaceæ, shrubs of *Veronica* and *Panax*. As in New Zealand, beneath the woody plants, social Ferns are abundant. One of them, *Aspidium venustum*, Hombr. Jacquin., the luxuriant foliage of which spreads out from the summit of a stem from 2′ to 4′ in height, and 6″ in diameter, reminding us, in its growth, of the climate of the Tree-ferns of New Zealand, just as the Dwarf-palm does of a tropical climate. Above the forest region, which is confined to the coast, the bush-land alone is found as far as an elevation of 800′, where it is gradually replaced by treeless pastures of herbaceous plants and grasses. The herbaceous plants display flowers equalling alpine plants in brilliancy of colour, and for the most part are vicarious species of alpine types of plants, as *Gentiana*, *Veronica*, *Cardamine*, and *Ranunculus*.

Campbell's Island is girdled with rocks, like St. Helena, and does not therefore contain a connected forest-region. Being covered internally by meadows, it only contains the Ferns found beneath the bushes in the Auckland Islands, in isolated sheltered localities. Of the antarctic forms, a large golden-yellow Liliaceous plant (*Chrysobactron*) is found on the rocky heights, in such luxuriance, that the colour of its flowers is perceived by those sailing by at the distance of a mile from the coast.

Summary of the Flora of the Lord Auckland's Islands and Campbell's Island: 3 Ranunculaceæ (*Ranunculus*), 4 Cruciferæ (*Cardamine*), 4 Caryophyllaceæ (*Stellaria*, 3 *Colobanthus*), 1 *Drosera*, 1 *Geranium*, 3 Rosaceæ (*Sieversia* and 2 *Acæna*), 3 *Epilobium*, 1 *Callitriche*, 1 *Metrosideros*, 1 *Montia*, 1 *Bulliardia*; 3 Umbelliferæ (1 *Pozoa* and 2 *Anistome*), 1 *Panax*, 1 *Aralia*; 7 Rubiaceæ (6 *Coprosma* and *Nertera*); 11 Synantheraceæ (*Trincuron*, *Ceratella*, 3 *Leptinella*, *Ozothamnus*, *Helichrysum*, 2 *Pleurophyllum*, *Celmisia*, and *Gnaphalium*), 3 Stylidiaceæ (2 *Dracophyllum* and *Forstera*), 1 of the Lobeliaceæ (*Pratia*), 1 of the Epacridaceæ (*Androstoma*), 1 of the Myrsinaceæ (*Suttonia*), 2 *Gentiana*, 2 *Myosotis*, 3 *Veronica*, 2 *Plantago*, *Rumex*, 1; *Urtica*, 2; 8 Orchidaceæ (2 *Thelimitra*, 2 *Caladenia*, *Chiloglottis*, *Acianthus*, and 2 undetermined), 2 Asphodeleæ (*Chrysobactron* and *Astelia*), 5 Junceæ (2 *Juncus*, 2 *Rostkovia*, and *Luzula*), 1 of the Restiaceæ (*Gaimardia*), 6 Cyperaceæ (3 *Carex*, *Uncinia*, *Isolepis*, and *Oreobolus*), 14 Graminaceæ (2 *Hierochloë*, 4 *Agrostis*, *Trisetum*, *Bromus*, 2 *Festuca*, 3 *Poa*, and *Catabrosa*), 17 Ferns (5 *Hymenophyllum*, *Aspidium*, 3 *Asplenium*, *Pteris*, 2 *Lomaria*, 2 *Polypodium*, *Phymatodes*, *Grammitis*, and *Schizæa*); 66 Mosses, described in connexion with Wilson; 85 Hepaticæ, described by J. D. Hooker and Taylor; 30 Lichens, by the same; 57 Algæ, by J. D. Hooker and Harvey; and 15 Fungi, by Berkeley. Several of the Cryptogamic species are European, but few only of the Phanerogamia, which have either been introduced, or, forming varieties, their determination appeared doubtful.

The Flora of the Antarctic countries commences with the eleventh part of the work, and includes all latitudes situated between 45° and 64° S. lat.; comprising the following points, which were explored by the traveller, Fuegia, the south-west of Patagonia, the Falkland Islands, Palmer's Land, and Kerguelen's Land. Summary of the families as yet treated of: Ranunculaceæ (*Anemone*, 8 *Ranunculus*, 3 *Hamadryas*, and 3 *Caltha*), 1 of the Magnoliaceæ (*Drimys*), 3 *Berberis*, 11 Cruciferæ (*Arabis*, 2 *Cardamine*, 3 *Draba*, *Pringlea antiscorbutica* = the cabbage of Kerguelen's Land, see Ann. Rep. for 1843, *Thlaspi*, *Senebiera*, 2 *Sisymbrium*), 1 of the Bixaceæ (*Azara*, in South Chili), 4 *Viola*, 1 *Drosera*, 13 Caryophyllaceæ (*Lychnis*, *Sagina*, 4 *Colobanthus*, 4 *Stellaria*, *Arenaria*, 2 *Cerastium*), 4 *Geranium*, 2 *Oxalis*, 2 Celastrineæ in Fuegia (*Maytenus* and *Myginda*), 1 of the Rhamneæ from the same place (*Colletia*), 8 Leguminosæ (2 *Adesmia*, 3 *Vicia*, and 3 *Lathyrus*), 15 Rosaceæ (2 *Geum*, *Rubus*, *Fragaria*, *Potentilla*, and 10 *Acæna*), 2 Onagrariæ (*Fuchsia* in Fuegia, and *Epilobium*), 6 Halorageæ (*Myriophyllum*, *Hippuris*, *Callitriche*, and 3 *Gunnera*), 5 Myrtaceæ (*Metrosideros* in the Chonos Islands, 2 *Myrtus*, and 2 *Eugenia*), 1 *Montia*, 1 *Bulliarda*, 1 *Ribes*, 8 Saxifrageæ (2 *Escallonia*, *Cornidia*, 2 *Saxifraga*, 2 *Chrysosplenium*, and *Donatia*). The Umbelliferæ are not yet completed.

The description of the Antarctic Cryptogamia in the 'London Journal of Botany' (see the preceding Ann. Rep.) has been continued; 38 new Hepaticæ have been published by J. D. Hooker and Taylor (1845, p. 79-97), 76 new Algæ by J. D. Hooker and Harvey (p. 249 to 276, and 293 to 298), who have also enumerated the Algæ of New Zealand, at present 124 species (p. 521 to 551).

The first volume of the botanical text of the illustrated work, founded upon Dumont d'Urville's antarctic voyage, has now appeared; it contains the Cellular Plants, by Montagne. (Voyage au Pole Sud et dans l'Océanie sur les corvettes *Astrolabe* and *Zélée*. Botanique, t. i, Plantes cellulaires. Paris, 1845, 8vo.)

They consist altogether of 138 Algæ, 42 Lichens, 48 Hepaticæ, and 40 Mosses. The preface contains lists of the Cryptogamia found in both hemispheres, between the pole and the 50th parallel (they consist of 9 Algæ, 66 Lichens, 11 Hepaticæ, and 14 Mosses); also a list of those species which occur both in high and tropical latitudes (171 species); and lastly, of cosmopolite species (8 Algæ, 6 Lichens, 5 Hepaticæ, and 10 Jungermanniæ). Montagne's new genera had been previously described in a preliminary work. The copper-plates belonging to the Cryptogamic section, by Hombron and Jacquinot, the text of which has not at present appeared, although exquisitely drawn, have been severely criticised by the younger Hooker (Lond. Journal of Botany, 1845, p. 28).

B.—SYSTEMATIC BOTANY.

In conformity with the character of the earlier systematic literature, the description of new forms still predominates, so that even those most capable of the task are still too much drawn away from the more profound establishment of the System of Plants. But as in this Report the latter direction will principally be kept in view, its brevity will not only be excusable in consequence of a defective knowledge of the literature, of which important papers often reach me too late, but will also be an intentional result of the plan of the work.

In January, 1845, the ninth volume of De Candolle's Prodromus Systematis Naturalis (Paris, 8vo) was published, the tenth followed it in April 1846. The families treated of will be mentioned presently. Of Walper's Repertorium of the diagnoses contained in recent botanical works, (Repertorium Botanices Systematicæ, Lips., 1845-6, 8vo,) the conclusion of the Labiatæ appeared in the last part of the third volume; in the fourth volume, which has not been continued beyond the first fasciculus, the Verbenaceæ, Myoporinaceæ, Selaginaceæ, Globulariaceæ, and Plantaginaceæ, and in the fifth volume, supplements to the Polypetalous families treated of in the first volume, principally consisting of a reprint of Jussieu's Monograph of the Malpighiaceæ; these extracts and reprints are, however, anything but accurate, as is well known.

A new part of Sir W. Hooker's 'Icones Plantarum,' containing 50 plates, has been published. (Part 15, vol. viii, p. 1, Nos. 701-750. London, 1845, 8vo.)

Leguminosæ. Bentham is describing the Mimoseæ, and has given a complete synopsis of the genera and species of this group (Lond. Journal of Botany, 1844-5); during the past year, only *Inga*, with 134 species. He reduces this genus within narrower limits (= *Euinga* Endl.), and remarks that either the Monadelphous Mimoseæ, i. e. one third of all which are known, must be arranged in a single genus, or the formation of the leaves must be recognised as a generic character. Thus he separates *Inga* from *Picetholobium* (which has doubly pinnate leaves) by the simple pinnation only; but in this manner also obtains habitual characters in the more elongated and pubescent flowers, and in the thicker legumes, which are tumid at the margin. It must undoubtedly be admitted as a correct maxim, that when the higher systematic sections, as the families, are limited by the characters of their vegetative parts, the inferior categories, such as the tribes and genera, may also be made to depend upon them, when a natural subdivision of the group is effected by that means.—*Alexandra*, the new genius of the Sophoreæ, a tree from British Guiana, with colossal flowers, has been described by Rob. Schomburgk (l. c. 1845, p. 12).—The revision of the genus *Genista*, by Spach (Ann. Sc. Nat., iii ser. vol. 2, 3), contains a considerable number of new species; but, like the previous systematic works by this author, it cannot be regarded as conclusive, or as a description of the matter contained in it corresponding with the genius of science, but merely an excessively prolix enumeration of descriptive details. Part of the new species consists of unimportant forms, as evidenced by the description of several belonging to *Genista tinctoria;* the diagnoses, of excessive, totally unnecessary length, do not by any means afford a synopsis of the distinctive characters, but rather defeat their object, and, mingled with the extended yet special and abbreviated descriptions, which do not facilitate the recognition of the species as such, since they combine variable with constant characters,—necessarily render this more difficult. The arrangement of the sections and subgenera is of more importance; they are also unnecessarily increased; but they contain analytical details and new observations, which will be of use for a future monograph. *Dendrosparton* Sp. (iii, p, 152) = *Spartium ætnense* Biv., and *Gonocytisus* Sp. (p. 153)= *Sp. angulatum* L., are separated from *Genista*, and placed in distinct genera.

Myrtaceæ. J. D. Hooker and Harvey have described *Backhousia*, a new genus from New South Wales (Bot. Mag. 1845, i, 4133).

Melastomaceæ. Naudin separates *Microlicia alsinefolia* D. C. and *variabilis* Mart. from *Microlicia*, on account of the somewhat different structure of their anthers, forming them into the new genus *Uranthera*, and retains *Chætostoma* D. C., the proposed character, however, does not contain any characters distinctive from *Microlicia* (Ann. Sc. Nat. iii, 3, p. 189, 190). He elevates *Arthrostemma* sect. *Monochætum* into a separate genus, under the name of the section (4, p. 48). New genera: *Octomeris* Naud., a shrub belonging to the Andes, to which *Mel. octona* Humb. Bonpl. also belongs (p. 52); *Stephanotrichum* Naud. (p. 54), and *Chiloporus* Naud. (p. 57), both from New Granada.

Lythrarieæ. Planchon (Lond. Journ. of Botany, 1845, p. 474) refers *Henslovia* Wall. (one of the Hensloviaceæ Lindl.) to this order. He ascribes to this genus a *Capsula loculicida, valvis medio septiferis basi et apice connexis*, and places it near *Abatia* R. P. From the figure given in the 'Flora Peruviana,' he also regards *Alzatea* R. P. (*Celastrinea dubia*) as belonging to the Lythrariceæ, and places *Crypteronia* Bl. (*Rhamnea dub.* Endl.) and *Quilanum* Blanc. (*dub. sedis* Endl.) as doubtful synonymes of *Henslovia*; he is, however, only guided by the descriptions of the plants.

Diosmeæ. To this order Planchon refers a diœcious genus of woody plants from the Malay Archipelago, which he describes as *Rabelaisia* n. gen., without, however, being acquainted with the structure of the ovary (l. c. p. 519). At the same time the author proposes some changes in the limitation of the Diosmeæ, with which he considers the Zanthoxylaceæ ought to be united, after having separated *Brucea* and *Ailanthus* from the latter group, as proposed by Bennet, and in conjunction with *Soulamea* (*Cardiophora* Benth. according to the author's dissections), which has hitherto been placed among the anomalous Polygaleæ, combined them with the Simarubaceæ. —Torrey and Frémont have described a new genus *Thamnosma*, from Upper California, nearly related to *Zanthoxylon* (Frém. Exploring Expedit. Americ. edit. from the Bot. Zeit. 1847, p. 141).

Ochnaceæ. Sir W. Hooker refers *Hostmannia* n. gen. (Hook. ic. t. 709) from Surinam, to this family, notwithstanding its bilocular ovary.

Euphorbiaceæ. Planchon has described 2 Australian genera (l. c. p. 471, t. 15, 16): *Stachystemon* Pl., which is nearly related to *Pseudanthus*, and *Bertya* Pl. to *Calyptostigma*.

Sapindaceæ. Snake-seed, which has lately been introduced into commerce, consists of the spiral embryos, divested of the testa, of *Ophiocaryon* Schomb., one of the Sapindaceæ, the snake-nut tree of Essequibo, formerly referred by its discoverer, Rob. Schomburgk, to the Anacardiaceæ, but which he has now described more completely, and placed in its proper family (l. c. p. 375-8).

Malvaceæ. Duchartre has published an important account of his researches upon the development of the flowers of the Malvaceæ (Ann. Sc. Nat. iii, p. 123-50), upon the merit of which Ad. Jussieu has expressed himself at length (Compt. rendus, 1845, Aug. p. 417-26). The outer calyx, at the period of its earliest formation, appears to represent a bracteal system. Duchartre considers that the synsepalous calyx is formed in the same manner as all monophyllous floral envelopes, not by the growing together of originally distinct organs, as Schleiden believes, but the tube of the calyx is first formed, from the upper margin of which five sepals spring up. According to my more recent investigations, which were principally made upon the calyx of the Onagrariæ, this view is essentially in accordance with nature; but the sequence of the phenomena is incorrectly described. The free apices of the organ are first formed, the basilar formative points then unite, in consequence of the lateral growth of each individually, and thus, *after* the formation of the lobes, a connected tube of the calyx springs up from the torus. The marginal union of floral organs of the same whorl, when occurring, must be considered merely as an exception, in opposition to the universality of this process. Duchartre's most important discovery relates to the position of the stamens, and serves to confirm the supposed affinity of the Malvaceæ to the Rhamnaceæ. After the formation of the calyx, the stamens are developed somewhat before the corolla (as in several families with the stamens opposite), as five rudiments of leaves (mamelons) which alternate with the segments of the calyx. These divide while their formation is yet scarcely completed, at first into two segments (dédoublement collatéral), in the same manner as a divided leaf (their development proceeding more rapidly at the two sides than in the median line, the five primitive eminences become five pairs of minute rounded tubercles). Almost simultaneously with the division of those stamens which are first formed, the petals appear; they are opposite the former, and a considerable distance apart. The polyandrous character is produced by the frequent repetition of the same formation in front of the above ten stamens, united in pairs, that is, on their inner side (parallel deduplication: five new pairs of tubercles are formed in a circle, which is situated more internal and opposite to the first). This multiplication of the stamens is not considered by Duchartre as arising from the formation of new and opposite whorls upon the torus; but he appears to regard them, and certainly with truth, as formed from the expansion of the substance of the primary leaf toward the interior. The Polyandrous character is often increased by a second collateral division of the individual stamens. In fact, in *Malope trifida*, and some other species, Duchartre has even finally observed a third collateral division, both of the anthers and of the stamens, so that here, and perhaps generally, the unilocular anthers ought to be considered as the halves of a truly dimidiated stamen. Five teeth to the tube of the filaments, which alternate with the petals, appear to be universal in the buds, and are considered as forming a second circle of stamens, without any convincing argument being brought forward. Regarding the pistil of the Malvaceæ, Duchartre assumes the existence of five primary forms, the two former of which agree in the circumstance, that at first a pentagonal collar-like protuberance (bourrelet pentagonal) arises from the torus at the circumference of the apex of the axis (mammelon central), the angles of which are opposite the petals (at least this position is mentioned as occurring in *Malope*); either numerous carpels then shoot out from the margin of this protuberance (Malopeæ), or five only from its angles (Hibisceæ). The formation of the carpels is also preceded in the Malveæ and Sideæ by a protuberance, which is not, however, pentagonal, but annular. The number of carpels growing from its margin is undetermined. Lastly, the greatest deviations occur in *Pavonia*, and some allied genera, in which, upon an annular protuberance, the rudiments of ten styles are said to appear first, and subsequently to be fused into five ovaries.

Hypericaceæ. Cosson and Germain (Flore de Paris) admit Spach's genus *Elodea* (*Hypericum elodes*), which differs from *Hypericum* in its parietal placentation, whilst the latter has a central placenta. To me, however, the difference appears merely to consist in the parietal placentæ of *Hypericum* meeting in the axis of the fruit, whilst in *Elodea* this is not the case: whether this is a generic character or not must be decided by a future monograph of the family, Spach's work not being sufficient for this purpose.

Caryophyllaceæ. J. Gay's monograph of *Holosteum* (Ann. Sc. Nat. iii, 4, p. 23-44) is characterised by the author's well-known accuracy; but is impaired by the prolixity which is unfortunately so frequently combined with such accuracy, particularly with endless quotations. The following new genera are proposed by Gay in this memoir: *Rhodalsine* G. (p. 25) = *Arenaria procumbens* V. It appears to differ from all the other Alsineæ, in the stamens being biserial, which is, however, a very relative character only: and *Greniera* (p. 27) = *Alsine Douglasii* Fzl. and *Arenaria tenella* Nutt.; characterised by its seeds, which are compressed discoidally.

Cactaceæ. We are indebted, for a scientific summary of the Cactaceæ, to the Prince Salm-Dyck, who possesses the largest collection in existence on the Continent (about 700 forms), and is also one of those best acquainted with this difficult botanical group (Cacteæ in horto Dyckensi cultæ, additis tribuum generumque characteribus emendatis a principe Jos. de Salm-Dyck Paris, 1845, 8vo). *Pfeiffera* S. (p. 40), is a new genus described in this work.

Cucurbitaceæ. Wight has written in the 'Madras Journal of Science,' and Gardner in the 'Lond. Journ. of Bot.' (1845, p. 401), in favour of Seringe—De Candolle's view, that the middle of the carpels is situated in the axis of the fruit, and that the cells of the fruit are formed by the revolute incurvation of their margins, and have endeavoured to support this paradoxical theory by the course of development of the ovary. According to Gardner, the external wall of the fruit is formed by the tube of the calyx only, with which, in *Coccinia indica*, the dissepiments are merely loosely in contact, without adhering to it. The course of the bundles of vascular tissue also, the principal trunks of which, both in this plant and in *Bryonia*, are situated in the axis, is in favour of Seringe's view; but the principal point in the solution of this question consists in distinguishing the placentæ from true carpels, which has not yet been accomplished. It is still extremely improbable that three leaves should grow out of the point of the axis.—Payer remarks (Ann. Sc. Nat., iii, 3, p. 163) that at the lower joints, where three vascular bundles enter the petiole, the stem of the Cucurbitaceæ is not furnished with tendrils, whilst on the other hand, in the case of the upper leaves, according as one or two tendrils are present, it receives two vascular bundles only, or the central one alone. He thus explains the oblique position of the axillary bud, which is always situated opposite the central vascular bundle, and thus, where as usual a single tendril only accompanies the leaf, obtains an oblique position. But he does not thus prove that the tendrils are segments of a leaf or stipules, whilst if they are regarded as entire leaves, this may be proved by the earlier stage of development, before the formation of any vessels (Wiegm. Archiv., 1846, p. 24).

CRUCIFERÆ. Barnéoud has described the small group of the Schizopetaleæ, to which, in addition to the principal genus (containing 2 species), *Perreymondia*, n. gen. Barn. from Chili (with 4 species) belongs (Ann. Sc. Nat. iii, 3, p. 165-8). The characters are limited to the divided petals and the branched hairs, *Perreymondia* not possessing the divided cotyledons, but an ordinary notorrhizal embryo, and as this is the only difference, it cannot constitute a distinct genus. Trautvetter separates *Matthiola deflexa*, Bg. from the genus *Matthiola*, as *Microstigma*, Tr. (Pl. Ross. imagines T. 25). New genera: *Lyrocarpa*, Hook. Harvey (Lond. Journ. of Bot. 1845, p. 76); it has a panduriform silicule, and was discovered by Coulter in California; *Dithyrea*, Harv. (l. c. p. 77), allied to *Biscutella*, from the same source; *Oxystylis*, Torr. Frém. (Explor. Exp. and l. c. p. 41), well characterised, approaching the Capparidaceæ, also from California; *Pringlea*, Anders. d. Hook. (Antarct. Voy. p. 238, pl. 90-1), the above-mentioned cabbage of Kerguelen's Land.

PAPAVERACEÆ. New genera from California: *Romneya*, Harv. (l. c. p. 73), principally distinguished from *Papaver* by the trimerism of the two outer whorls; *Arctomecon*, Torr. Frém. (l. c. p. 40), only differing from *Papaver*, according to the description, in the strophiolate seeds.

RANUNCULACEÆ. For a notice of Barnéoud's work, the whole of which has not yet been published (Compt. rend. 1845, ii. p. 352-4), see Link's 'Physiological Report' (p. 95). Cl. Gay has established two genera from Chili: *Psychrophila* (Hist. de Chile Bot. i. p. 47, t. 2), separated from *Caltha* and *Barneoudia* (ib. p. 29, t. 1, f. 2), allied to *Helleborus*.

SAXIFRAGACEÆ. Gardner describes a shrub which he discovered on the Organ mountains at Rio, as *Raleighia* (Lond. Journ. of Bot. 1845, p. 97), with the following essential characters: Calyx four-partite, valvular, corolla absent, stamens numerous and perigynous, ovary one-celled, with a single style and 3 (— 2) placentæ, supporting numerous ovules, which are subsequently situated on the median line of the valves of the capsule; seeds with an axile embryo; leaves opposite, connate at the base, and serrated. The author refers it to the Bixaceæ, but Bentham rightly places it in the Cunoniaceæ near *Belangera*, as it forms a transition link from this to the parietal families, by its truly parietal placentæ, but differs from the latter by their attachment. Planchon takes a different view of *Raleighia* (ib. p. 476); from the examination of dried specimens, he regards this genus as scarcely generally distinct from *Abatia* (Lythrareæ, vid. Sup.), which cannot be the case unless both Gardner and Bentham have described the fruit and seeds totally incorrectly.

UMBELLIFERÆ. New genus from Lord Auckland's Archipelago: *Anisotome* J. D. Hook. (Antarct. Voy. p. 76, t. 8-10). The Umbelliferæ, Crassulaceæ, &c. proposed in the 'Phytographia Canariensis,' will be passed over until the work is completed.

EPACRIDACEÆ. New genus: *Androstoma* J. D. Hook., from the Auckland Islands (Antarct. Voy. p. 44, t. 30).

MYRSINACEÆ. New genus: *Labisia*, Lindl. (Bot. Reg. 1845, t. 48), from Penang, differing in the induplicate æstivation of the corolla.

BIGNONIACEÆ. In the Prodromus, this family is treated of by De Candolle, jun., in conjunction with the Sesameæ (vol. ix), in accordance with the description prepared by De Candolle, sen. The Sesameæ, which in this work also include the Pedalineæ, appear only to be separated from the Bignoniaceæ, because the type of the fruit is assumed to be quinary. The African species are separated from *Sesamum*, as *Sesamopteris*. The following genera are distinguished from Bignonia: *Pachyptera, Macfadyena*, =*B. uncinata* Mey., *Anemopægma* Mart., *Distictis* Mart., *Pithecotenium* Mart., *Cybistax* Mart., *Adenocalymna* Mart., *Sparratosperma* Mart., *Heterophragma* = *B. quadrilocularis* Roxb., *Craterocoma* Mart. The Crescentieæ, which are arranged by Endlicher among the Gesneriaceæ, form, in this work, the second tribe of the Bignoniaceæ, being characterised by the indehiscent fruit and the wingless seeds, found principally in Madagascar: *Tanæcium pinnatum* W. is separated from *Tanæcium*, as *Kigelia*. *Parmentiera* is a new genus from Mexico. The position of *Bravaisa*= *Bignonia bibracteata* Bert. remains doubtful.

GESNERIACEÆ. The Gesneriaceæ having appeared earlier in the Prodromus, the Cyrtandraceæ form an independent family in the ninth volume, the elder De Candolle having left them prepared. *Ramondia* and *Haberlea*, with *Conandron* from Japan, are correctly brought into this family as a distinct group, on account of the septicidal dehiscence of the capsule.

ACANTHACEÆ. New genera: *Lankesteria* Lindl., from Sierra Leone (Bot. reg. Miscell., 1845, p. 86?; *Whitfieldia* Hooker, from the same source (Bot. Mag., p. 4155); *Salpixanthia* Hook., from Jamaica (ibid., p. 4158).

SCROPHULARIACEÆ. Bentham's Monograph fills the greater part of the tenth volume of the Prodromus. With the exception of the Salpiglossideæ, which, notwithstanding the stamens being anisomerous, would have been more properly excluded, and placed among the Solanaceæ, all the genera possess an imbricate æstivation of the corolla. The position of the fourth and fifth petals, which form the upper lip in those furnished with labiate flowers, separates the two principal tribes, being external during æstivation in the Antirrhineæ, whilst in the Rhinantheæ they are included. New genera. Salpiglossideæ: *Leptoglossis*, from Peru. Antirrhineæ: from the west of North America, *Chioniphila* and *Eunanus* = *Mimulus nanus* Hook. and others; from Chili, *Melosperma*. Rhinantheæ: *Tricholoma* (near *Limosella*), from

New Zealand; *Camptoloma*, from the south of Africa; *Bryodes*, from the Mauritius; *Synthyris* (to which *Wulfenia reniformis*, Benth. belongs), from the west of North America; *Radamæa* and *Raphispermum*, from Madagascar; *Micrargeria*, from the East Indies; *Synnema* = *Pedicularis avana* Wall., from Ava. *Otophylla*, *Silvia*, and *Graderia* are separated from *Gerardia*. As regards the special style of Bentham's Monograph, it is characterised by the excellence of the natural division of the genera, and the appropriate contraction of the forms; the new species are very numerous. Webb has published some remarks upon the affinity of the genus *Campylanthus* Rth., from the Canary Isles (Ann. Sc. nat. 3, iii, p. 33), the position of which Bentham had left doubtful. It differs from the Veroniceæ in the character of the stamens; in the latter two posterior, and in *Campylanthus* two anterior stamens are developed, as is also the case in *Anticharis* and *Achetaria*. Webb places them in a distinct group, in which we are more inclined to agree with him than when he wishes to place it near to the Salpiglossideæ and Solanaceæ, from which it differs in the æstivation.

SOLANACEÆ. *Cyphomandra* Mart. *Solani sp.* R. P., described in a monograph by Sendtner, differs from *Solanum* in its large connective (Ratisbon Flora, 1845, pp. 161-76). New genera: *Jochroma* Benth. = *Habrothamnus* Lindl. ol. (Bot. reg., 1845, tab. xx.), from Ecuador; *Salpichroa* and *Hebecladus* Miers = *Atropæ sp.* Amer. austr. (Lond. Journ. of Bot., 1845, p. 321); *Lycioplesium* and *Chænestes* Mrs. = *Lycii sp.* Amer. austr. (ib. pp. 330-36); *Dorystigma* Mrs. = *Jaborosæ sp. Chilensis*, Hook. ol. (ib. p. 347); *Trechonætes* Mrs. (ib. p. 350), from Chili; *Pionandra* Mrs. = *Witheringiæ sp.* Mart., &c. (ib. p. 353).

NOLANEÆ. In 1844, Lindley, in the 'Botanical Register,' divided *Nolana* into five natural genera, and characterised the species belonging to them. Miers has recently also studied the characters of this small group (l. c. pp. 365-469), and described a new type from Chili, *Alibrexia* (p. 505). Miers regards it as a connecting link between Boraginaceæ and the Convolvulaceæ: from the former it differs principally in its habit, and in the position of the embryo; from the latter in its distinct ovaries. *Grabowskya* (one of the Boraginaceæ according to Endlicher, of the Solanaceæ according to others) forms the transition to the former, the Dichondreæ to the latter. If it is desired to preserve a limit between the Boraginaceæ and the Convolvulaceæ, the Nolaneæ must either be regarded as a distinct family, and the Dichondreæ placed with them; or, keeping in view the inflorescence and the æstivation of the Boraginaceæ, both groups must be united with the Convolvulaceæ. However, Miers retains *Grabowskya* as a distinct tribe among the Nolanaceæ, and leaves the Dichondreæ with the Convolvulaceæ.

ERYCIBEÆ. Both the elder and the younger De Candolle, in the Prodomus (vol. ix), have separated *Erycibe* also, another connecting link between the Convolvulaceæ and the Boraginaceæ, as a distinct family, principally in consequence of the absence of the style, and the one-celled ovary.

Hydroleaceæ. They have been worked out in the Prodromus (vol. x) by Choisy. A. De Candolle remarks that in *Hydrolea* the dehiscence of the capsule is marginicidal, whilst in the other genera it is loculicidal; and he thinks that in the latter the ovary ought to be regarded as unilocular with placentæ projecting toward the axis, whence he prefers uniting them with the Hydrophyllaceæ. Choisy takes the opposite view, without, however, enfeebling this argument.

Hydrophyllaceæ. They are treated of by A. De Candolle in the Prodromus (vol. ix), who separates *Eutoca* into two types—*Microgenetes* from Chili, and *Miltitzia* from California.

Polemoniaceæ. Also described by Bentham in the above work.

Convolvulaceæ. Choisy's description in the Prodromus (ibid.) is inferior to the other parts of the work in a critical point of view. The new genera admitted are *Marcellia* Mart. from Brazil, and *Seddera* Hochst., Steud., from Abyssinia. Pfeiffer (Bot. Zeit. 1845, p. 673) separates *Cuscuta epilinum* from the genus *Cuscuta* as *Epilinella*, as this species possesses a pentasepalous calyx; also those species which are furnished with capitate stigmata, as *Engelmannia*, a name which will be at once given up.

Boraginaceæ. A. De Candolle has described them excellently well in the Prodromus (vol. ix, x), from the preparatory writings of his father, and divides them into four tribes: Cordieæ, Ehreticæ, Heliotropeæ, and Borragineæ. New types: *Gynaion* (a monstrosity of Cordia?), from the Himalaya; *Meratia*, allied to *Myosotis*, from Caracas. *Heliophytum* and *Pentacarya* are separated from *Heliotropium*; *Maharanga*, with a basilar corona, from the Himalaya, from *Onosmodium*; *Pentalophus*, from *Lithospermum*, from the prairies; *Gruvelia*, from Chili, from *Cynoglossum*; from *Echinospermum*, *Heterocarium*, several species from the steppes of Asia. Moris separates *Buglossites* = *B. laxiflora*, D. C. from *Borago* (Turin Cat. of Seeds, f. 1845).

Avicenneæ. Griffith read before the Linnæan Society the history of the development of the ovum of *Avicennia* (Proceedings of Linn. Soc., Nov. 1844, in the Ann. of Nat. Hist., xv, p. 197). *Avicennia* has a free central placenta, with suspended ovules, which do not appear to possess any integuments, and were considered by St. Hilaire as funiculi. The embryo of the fertile ovule, after impregnation, grows in the axis of the nucleus on both sides, passes out from its anterior surface, and there attains its principal development, and which does not appear to be accompanied by the deposition of albumen until it reaches the exterior of the original ovule. At a subsequent period, a furrow is formed upon the anterior surface of the albumen, and which corresponds to the points of the cotyledons of the embryo, whilst the embryo-sac passes back into the placenta, and ramifies within it. Finally the embryo itself passes through the groove mentioned above, so that in the ripe seeds the radicle is the only part inclosed by the albumen, whilst the cotyledons project, free, from it.

Gentianaceæ. I have described them in the 'Prodromus' (vol. ix). New types: *Gyrandra*, from Mexico, and *Pagæa*, from South America. I have separated *Lapithea* from *Sabbatia*, *Exochænium* from *Sebæa*, *Pladera* from *Canscora*, and *Petalostylis* from *Leianthus*.

Loganiaceæ. According to the limits of this group given in the 'Prodromus (vol. ix), in which it was described by P. De Candolle, and revised by his son, it contains the aberrant forms of several allied families; i. e. in addition to the types admitted by Endlicher, the Spigeliaceæ, with *Mitrasacme*, *Mitreola*, and *Polypremum;* moreover *Lachnopylis* Hochst., and *Gelsemium* Juss.

Jasminaceæ. According to Wight and Gardner's investigations (Calcutta Journ. of Nat. Hist., and Lond. Journ. of Bot. 1845, p. 398), the genus *Azima* Lam. (*Monetia* L'Hér.), which has been doubtfully referred to *Ilex*, is intermediate between this family and the Oleaceæ. It differs from the Oleaceæ principally in having four stamens, erect ovules, and the absence of albumen; from the Jasminaceæ in its distinct petals and diœcious stamens, i. e. in characters which occur individually among the Oleaceæ. In habit it resembles the climbing Jasminaceæ.

Caprifoliaceæ. C. A. Meyer has produced a monograph of those species of *Cornus* in which the involucrum is absent (Mém. de St. Pétersb. 1845, reprinted in the Ann. Sc. nat. iii, 4, pp. 58-74). It comprises thirteen species, four of which are new.

Synantheraceæ. New genera: Antarctic, by J. D. Hooker—*Trineuron*, *Ceratella*, and *Pleurophyllum*, from the Auckland Islands (Antarct. Voy. pt. ii); *Brachyactis* Led. (Fl. Ross. ii, p. 495) = *Conyza Altaica*, D. C.; *Leucopodum* Gardn. (Lond. Journ. of Bot. 1845, p. 124), a *Conyza* with opposite leaves, from Brazil; *Nicolletia* Gray (Frém. Explor. Exped., and l. c., p. 55), one of the Tagetineæ from California; *Ceradia* Lindl. (Bot. Reg. Misc. 1845, p. 11), a succulent member of the Erechthiteæ, from the West of Africa; *Fitchia* J. D. Hook. (Lond. Jour. of Bot. 1845, p. 640), a Cichoraceous tree; *Harpochæna* Bung. (Delect. sem. Dorpat. 1845) = *Acanthocephalus* Kar. Kir., transferred to the Hyoserideæ; *Heterachæna* Fres. (Mus. Senckenb. iii, p. 74), one of the Cichoraceæ, from Abyssinia. Spach has treated of *Microlonchus* monographically (Ann. Sc. nat. iii, 4, pp. 161-9); eight species are distinguished, part of them from Algeria.

Plantaginaceæ. Barnéoud's Monograph (Monographie générale de la Famille des Plantaginées; Paris, 1845, iv, p. 52) is merely an exposition of the species (114 sp., of which 14 are new), with short diagnoses, composed from the abundant materials existing in the museums of Paris and Geneva. Barnéoud has discovered that, in *Littorella*, before impregnation, the ovary is two-celled, and that one of the two ovules which arise from the base of the thin septum soon disappears. The author has passed over the character of *Plantago* existing in the structure of the anthers. De Candolle's *Bougueria*, from Bolivia, is figured by Hooker (Lond. Journ. of Bot. 1845, t. 19).

ARISTOLOCHIACEÆ. Griffith establishes a new genus, *Asiphonia*, from Malacca (Linn. Trans. xix, p. 333), and has fully described *Thottea* Rottb. (ib. p. 325.)

RAFFLESIACEÆ. An important memoir upon this group from Griffith was read before the Linnæan Society (Linn. Trans. xix, pp. 303-47, t. 34-9), shortly after that by R. Brown (see Link's Physiological Rep.) Griffith considers the Rhizantheæ as an artificially-formed class, and regards it as a retrograde step in systematic botany. The embryo which he calls homogeneous, does not differ from that of other parasites, e. g. Orchidaceæ, Orobancheæ, &c. as R. Brown has already pointed out; but the ovules of *Balanophora* and *Sarcophyte* consist of simple sacs, without any integument or definable punctum, probably analogous to the naked nucleus of the Loranthaceæ; they cannot therefore remain united with the Rafflesiaceæ, the organization of the ovule of which is perfect. A new genus of Rafflesiaceæ, *Sapria*, from the Himalaya, is fully described by Griffith. Investigations of the Cytineæ follow. Griffith considers the stamens of *Hydnora*, with Meyer, as indefinite, and united into a tripartite column. He also considers the anthers of *Cytinus* (*C. dioicus* Juss.) as probably unilocular. He does not regard the terminal teeth of the column as rudimentary stigmata, but as productions of the connective. He compares the structure of the pistil of *Hydnora* to that of *Papaver* and *Nymphæa*; the organic connexion of the stigma with the placentæ is such, that it might give rise to a new objection to Schleiden's axile placentation (*stigma discoideum, trilobum, e lamellis plurimis in placentas totidem pendulas undique ovuliferas productis*). Griffith's views agree in the most important points with those of R. Brown. The latter retains his former notion, that *Rafflesia* forms a link to the Cytineæ, which, as Griffith appears to admit, are intimately allied to the Asarineæ, but that the latter have no affinity with the Balanophoreæ. Brown's memoir (Linn. Trans. xix, pp. 221-39, pl. 22-30) consisting of the paper which was read as early as 1834, and then published in the form of an extract, is now given complete, and a supplement (ib. pp. 240-9) with a systematic summary of the Rafflesiaceæ added. They are subdivided into the following tribes: Rafflesieæ (*Rafflesia, Sapria*, and *Brugmansia*); Hydnoreæ (*Hydnora*); Cytineæ (*Cytinus*); Apodantheæ (*Apodanthes* and *Pilostyles*). The character of the family runs thus: *Perianthium monophyllum, regulare; corolla O (in Apodantheis 4 petala); stamina: antheræ numerosæ, simplici serie; ovarium: placentis pluribus polyspermis, ovulis orthotropis v. in quibusdam recurvatione apicis, penitus v. partim, liberi funiculi quasi anatropis* (thus *lycotropis m.*)*; pericarpium indehiscens, polyspermum; embryo indivisus, cum v. absque albumine; parasiticæ radicibus* (*v. Apodantheæ ramis*) *Dicotyledonearum.* The ovary of *Rafflesia*, when in flower, is almost wholly free from the perigone, and wholly so when in fruit. The structure of the pistil remains a morphological puzzle; its numerous irregular cavities, the walls of which are covered with ovules, if the processes terminating the disc be regarded as styles, would lead to the supposition that the ovary is composed of separate (but cohering) simple pistils, arranged concentrically in several rows around an imaginary axis. But this supposition is contradicted by the new *Rafflesia Cummingii* from Manilla, where the number of the ovaries is considerably larger than these processes of the disc. Nor is this difficulty cleared up by the placentæ of *Hydnora*, the structure of which is described in the same manner by both Griffith and Brown (the placentæ may be said to be continuations of the subdivisions of the stigmata). The ovary of *Hydnora* may be regarded as composed of three confluent pistils, having placentæ really parietal, but only produced at the top of the cavity. The testa of the seeds of *Rafflesia* is hard, and corresponds to the simple integument of the ovule; the embryo is inclosed in a loose cellular tissue (albumen) as a cylindrical body (*embryo indivisus*). In *Hydnora*, the spherical embryo is inclosed in a cartilaginous albumen; in the testa of the minute seeds of *Cytinus* a homogeneous nucleus only can be detected, as in the Orchidaceæ.

BALANOPHOREÆ. R. Brown (ibid.) does not at present speak positively regarding their affinity, but makes the following remarks upon their union with the Rafflesiaceæ into the class of Rhizantheæ: 1. That an embryo exactly similar to that of these parasites exists in the Orchidaceæ and Orobanchaceæ. 2. That the anatomical structure of tissue (paucity of vessels, and their limitation to the form of spiral vessels) cannot serve as a character of the Rhizantheæ: *a*, because the Coniferæ so nearly agree in structure with the Winteraceæ; *b*, on account of the peculiarity of the woody tissue of several tropical climbers, which do not recur in allied genera; *c*, because in many families great deviations of anatomical structure are limited to individual forms of plants, as in the Loranthaceæ, e. g. in which the woody tissue of *Myzodendron* Bks. (instead of *Misodendron* Auct.) consists entirely of scalariform vessels. According to Griffith (l. c.) the Balanophoreæ consist of the following genera: *Balanophora* (to which five new Indian species are added), *Langsdorffia*, *Phæocordylis* Gr. (*Sarcocordylis* Walt.?), *Helosis*, and *Scybalium*. As regards their systematic position, Griffith considers them as probably forms of Urticaceæ with homogeneous embryos; but on the other hand he remarks, that their pistil resembles that of the Mosses and Hepaticæ, and that the style is closed before impregnation and perforated after. In *Phæocordylis*, the hairs in which the fruits are imbedded resemble the paraphyses of *Neckera*. Griffith's view of the structure of the family is seen more distinctly from the character of *Balanophora*: *Flores diclines* (*rarissime monoclines*); ♂ *bracteati*, *perigonio* 3—5 *sepalo valvato, staminibus totidem monadelphis bilocularibus* (*in unica specie multilocularibus*); ♀ *ovariis nudis stipitatis, receptaculo apice incrassato-glanduloso affixis, stylo setaceo persistente, stigmate inconspicuo, fructu pistilliformi sicco.* From the Balanophoreæ, Griffith excludes *Sarcophyte,* the affinity of which is unknown, perhaps tending towards the Urticaceæ, and *Mystropetalon* Harv., which exhibits some affinity to *Cynomorium* alone, and must be regarded either as a distinct family (*planta sui ordinis*), or as a doubtful form of Loranthaceæ, with a homogeneous embryo. Both genera are accurately described from Harvey's specimens. The description of *Sarcophyte* differs very considerably from that usually given (thus : *columnæ stamineæ* 3 (— 4), *antheris indefinitis unilocularibus stipitatis*), but Endlicher has already made the same remarks regarding the structure of the stamens.

THYMELEACEÆ. New genera from Guiana, in Schomburgk's collection: *Lasiadenia* Benth. (Lond. Journ. of Bot. 1845, p. 632), and *Goodallia* Benth. (ib. p. 633).

SANTALACEÆ. Griffith states, that in *Osyris* the development of the ovule, which in other respects resembles that of *Santalum,* the embryo-sac grows out from the nucleus, and, as in *Avicennia,* the albumen is only deposited in this projecting portion (Proceed. of Linn. Soc., Nov. 1844, in Ann. Nat. Hist., xv, p. 197).

LORANTHACEÆ. R. Brown forms the Myzodendreæ from *Myzodendron,* and gives them the following characters: *Ovula* 3, *in apice placentæ centralis suspensa, unum fertile* (approximating the Santalaceæ in this character); *flos* ♂ *nudus ; appendices plumosæ in* ♀ *et embryo indivisus, radicula ex albumine exserta* (Linn. Trans., xix, p. 232.)

POLYGONACEÆ. New genera: *Pteropyrum* Jaub. Sp. (Illustr. or. t. 107-9), endemic in Persia and Arabia; *Thysanella* Gray (Pl. Lindheimer) = *Polygonum fimbriatum ; Symmeria* Benth. (Lond. Journ. of Bot., 1845, p. 630), a diœcious tree from Essequibo.

CHENOPODIACEÆ. *Sarcobatus* N., is figured in the American edition of Frémont's Exploring Expedition, as *Fremontia* (tab. iii). New genera: *Pterochiton* Torr., Fré. (ibid. &c. p. 57), from the west of North America; *Physogeton* and *Halothamnus* Jaub. Sp. (Ill. or. t. 135-6), from Persia.

URTICACEÆ. Gasparrini separates the following species generically from *Ficus* (Ann. Sc. nat., iii, p. 338-48): *Tenorea* = *F. stipulata ; Urostigma* = *F. religiosa* and six other species; *Visiania* = *F. elastica ; Cystogyne* = *F. leucosticta ; Galoglychia* = *F. Saussureana,* D. C. and *galactophora* Ten.; *Covellia* = *F. ulmifolia.* He also regards *F. Carica,* at the same time attributing too much weight to the circumstance of Cynips Psenes living only on the wild fig-tree (*Caprificus*), as consisting not merely of two different species, but two genera, which he distinguishes as *Ficus* = *F. Carica fœmina* L. and *Caprificus* = *F. Carica androgyna.*

SAURURACEÆ. Decaisne describes *Spathium chinense* Lour., and forms the genus *Gymnotheca* from it (Ann. Sc. nat., iii, pp. 100-2).

PIPERACEÆ. Miquel has published a very copious supplement to his monograph, which is composed from Hooker's herbarium (Lond. Journ. of Bot., 1845, pp. 410-70).

CONIFERÆ. Koch read a paper on the Diagnosis of the European Species of Pine, before the Association of German Naturalists (Ratisbon Flora, 1845, pp. 673-83).—The new genus, *Micocachrys* J. D. Hook. (Lond. Journ. of Bot., 1845, p. 142), has been mentioned above.

GNETACEÆ. We are indebted to the investigations of C. A. Meyer for a profound monograph of *Ephedra*, an extract from which, containing the diagnoses of nineteen new species, has appeared during the past year (Bullet. Pétersb., v. 33-36).

CYCADEÆ. Link endeavours to show that the position of the Cycadeæ beside the Coniferæ is untenable, and that they are more nearly related to the Palms (Ratisbon Flora, 1845, p. 289). Independently of the embryo, Schleiden's observation of the cambial layer beneath the bark (Grundzüge der Bot., 2, Ausg. ii, p. 152), also completely opposes these views. Link considers the leaves of the Cycadeæ, in accordance with Miquel's views, as axial organs. Miquel has explained his views upon the flowers, and especially the ovule and embryo of the Cycadeæ, more clearly than in his monograph, and has illustrated his investigations by plates (Ann. Sc. nat., iii, 3, pp. 193-206, pl. 8-9). He raises several objections to the view that the unilocular anthers are mere cells of anthers. They grow out like anthers, from the spadix; like them are surrounded by a layer of spiral cells, open by a fissure; are sometimes separated from each other by rows of hairs, and develope the pollen in the same manner as the unilocular anthers of other plants; but all these conditions apply also to dimidiate anthers, as in *Salvia*, so that the cells of the anthers of the Cycadeæ only differ from the latter in their greater number. Although the development of the embryo of the Cycadeæ has not yet been observed, nevertheless, from the comparison of the ovule with the seeds, and from the development of the former in the unimpregnated state, it appears certain that impregnation takes place according to the same law as in the Coniferæ. But since the main principle of the natural system of botany depends essentially upon the reproductive organs, we thus have a well-defined and peculiar character of the Gymnosperms in opposition to all other Phanerogamia, as also the Loranthaceæ with naked seeds—viz. that the embryo is not developed immediately from the pollen-tube, but from the terminal cell of a cellular cord, the funiculus (R. Br., embryo-blastanon Hart., Miq.), which, after impregnation, proceeds from peculiar receptacles of the embryonic vesicles (*corpuscula* R. Br.) into the endosperm. The unimpregnated ovule of the Cycadeæ differs from that of the Coniferæ in the separation of several enveloping layers of cells, the innermost layer of which contains spiral vessels, so that they appear to correspond rather with the idea of an arillus than an integumentary system, as Miquel supposes (*stratum externosum carnosum, secundum ligneum, tertium = textus cellularis laxus intus spiroideis vasis pertensus*). The view that these cellular layers are formed before the nucleus requires confirmation, and probably only depends upon the observation not being commenced sufficiently early. In the upper persistent portion of the ovule, or the nuclear tubercle (amnios R. Br., nuclear tubercle Schleid.), Miquel has found two or more embryoblastic vesicles (*cavitates* Miq., *corpuscula* R. Br.) arranged around the axis of the organ, passing unnoticed Schleiden's description of the Coniferous ovule, according to which these originate in the upper portion of the endosperm. He expressly states that the nuclear tubercle, or rather the embryoblastic vesicles, correspond to the embryo-sac, and not to the cavity in which the albumen is formed, which, according to his view, would therefore be albumen formed within the nucleus. He regards the nuclear tubercle as "a compound amnios" the "individual embryo-sacs" of which would be the embryoblastic vesicles. Moreover, Miquel has shown in the seeds the connexion of the coiled funiculus with the embryoblastic vesicles, as also the anastomoses of the former; and lastly, has found that the form of the embryo is different in all the four genera, by which circumstance they may be distinguished. In the supplement (iv, p. 79), although he has never been able to discover the embryoblastic vesicles in the unimpregnated ovule, he admits the truth of R. Brown's observations (Ann. Nat. Hist., 1844, May), according to which it may bé formed independently of impregnation. This observation is confirmed by the corresponding statements of Gottsche (Bot. Zeit. 1845, p. 402), who has written a detailed critical memoir upon the inflorescence of the Cycadeæ and the Coniferæ (ibid. Nos. 22-27), in which his observations were extended to living Cycadeæ. According to Gottsche, the embryoblastic vesicles, which in *Cupressus* are merely simple enlarged cells of the endosperm, in *Macrozamia* and *Encephalartos*, in which they are 1‴ in length and about ½‴ in breadth (pp. 399-400), possess a cellular wall, which, however, may constitute a later stage of development. Gottsche has not, however, succeeded in bringing into accordance the contradictory views upon the function of these sacs in the act of impregnation by new observations; although, in opposition to Hartig and in favour of Schleiden, he believes (p. 417) that the pollen-tubes penetrate the embryoblastic vesicles in the Cycadeæ also. Although I have no hesitation in admitting that this point has been satisfactorily determined by observation in the Coniferæ, yet there is a wide interval, which has not yet been filled up, before we can arrive at Schleiden's view, that the pollen-tube is prolonged inwards as far as the embryo-blaston itself, with which neither the illustrated descriptions of Brown or Miquel of the origin of the embryo-blaston, from a spherical cell inclosed in its vesicles, and resembling a pollen-granule, can be brought into connexion. These two figures, one of which is taken from the Coniferæ, the other from the Cycadeæ, agree so completely that they cannot be doubted. In my opinion they only admit of one interpretation, that in the pre-formed embryoblastic vesicles, the apex of the pollen-tube only produces a primary embryonal cell; and that at a subsequent period long after the pollen-tube has been destroyed, this grows into the embryoblaston in the same manner as the pollen-tube does at first from the pollen-cell. In accordance with this hypothesis, the simplest expression of the impregnation of the Gymnosperms would be found in their embryo not being formed in the pollen-tube itself, but in the apex of one of its secondary cells, the development of which does not commence until a long period after it had lost its parent-cell.

PALMÆ. The eighth part of V. Martius's large work on the Palms has appeared, (Monach. 1845, fol.,) containing the conclusion of the text, a Memoir upon Fossil Palms, by Unger, and the commencement of the morphology of the family from the pen of Martius himself. The text includes the completion of *Phœnix* and the Cocoineæ. The contents of the morphological section, in which at present the stem and the formation of the leaves are treated of, are more histological and physiological than systematic. The fibrous root formed during germination is soon succeeded by the formation of the rhizome from an axillary system of branches of the base of the stem (§ 23) with new radicles, which may protrude through any part of the cortical layer of the rhizome (§ 24) whilst the branches of a higher order arise only from the axillary buds of the rudiments of the leaves of the rhizome, and therefore, like the stem, they only possess the vascular bundles of the leaves, and not unfrequently grow into turiones. After the earlier radicles are dead, the older palm-stem rests upon adventitious roots, which arise laterally from the lower part of the stem, generally in the vicinity of the scars of leaves; which contradicts Schleiden's explanation of this phenomenon. (Grundzüg. i, Ausg. 2, p. 122.) The structure of the stem is very fully described. The results agree essentially with those of Mohl: the remark that the vascular bundles do not always return to the bark on the same side of the stem as that upon which the corresponding leaf is situated, but on the opposite side, so that they traverse the entire stem in an oblique direction, is new. The vascular tissue of the root appears to be different from that of the stem. The morphology of the leaf is not yet completed, and partly differs, in the *genesis*, from Mirbel. It appears clear, from the plates, that the segments are really formed by the subdivision of a simple lamina. The course of the lateral vascular bundles indicates the segments even when the lamina is simple.

TYPHACEÆ. Schnitzlein has described this group. (Die Natürl. Familie der Typhaceen mit besonderer Rücksicht auf die Deutschen Arten. Nördlingen, 1845, iv, p. 28.) The morphological considerations are based upon an accurate investigation of the structure of *Typha angustifolia* and *Sparganium natans*. The author considers the Typhaceæ as more closely allied to the Cyperaceæ than to the Aroideæ, which is, however, at variance with the structure of the seed. He regards the sterile stamens as perigones, which must be rigidly shown by the history of their development.

ORCHIDACEÆ. New genera: *Dialissa* Lindl., (a century of new genera and species of Orchidaceous plants, in Ann. of Nat. Hist., xv, p. 107,) allied to *Stelis*, from New Granada; *Helcia* Lindl. (Bot. Reg. 1845, Misc. p. 18), near *Trichopilia*, from Guyaquil; *Porpax* Lindl. (ib. p. 63), near *Eria*, from the East Indies; *Galeottia* Rich. Galeott. (Orchidographic Mexicaine, in Ann. Sc. Nat. iii, 3, p. 25), near *Maxillaria*; *Galeoglossum* and *Ocampoa*, Rich. Gal. (ib. p. 31), two Neottieæ. Lindley has given systematic notices of *Miltonia*, (Bot. Reg. t. 8), of *Odontoglossum* (ib. Misc. p. 49-59), and of several sections of *Epidendrum* (ib. pp. 22-9 and 65-79): the Monograph last mentioned, which was commenced some time since, is thus completed.

IRIDACEÆ. Herbert has continued his description of *Crocus*. (Bot. Reg. 1845, t. 37, and Misc. pp. 1-8, 31, 80-3.)

TACCACEÆ. This new genus, *Thismia*, Griff., from Tenasserim, borders closely upon this group, (having a remarkable affinity with *Burmannia*), it is a monocotyledonous representative of the Rhizantheæ. (Linn. Transact. xix, p. 343.)

AMARYLLIDACEÆ. Herbert separates *Phædranassa* Herb. = *H. dubius* Knt., from *Hæmanthus*. (Bot. Reg. 1845, Misc. p. 16.)

LILIACEÆ. New genus: *Chrysobactron* J. D. Hook. (Antarct. Voy., p. 72, pl. 44, 45), vide supra. Lindley has written a short Monograph of the New Holland genus, *Blandfordia*. (Bot. Reg. 1845, tab. xviii.)

JUNCACEÆ. Decaisne has described a diœcious plant *Goudotia*, new genus, growing upon the Andes of New Granada, at an elevation of 5000 meters, and forming a turf; he arranges it among the Juncaceæ, (Ann. Sc. Nat. 3, 4, p. 84), from which it differs in its coloured six-partite perigone, which is surrounded by a tripartite involucre; the position assigned to it above, especially as the structure of the seed is not known, can, therefore, only be regarded as provisional.

CYPERACEÆ.—Schlechtendal has published some remarks upon *Scleria* (Bot. Zeit., 1845, Nos. 28, 30).

GRAMINACEÆ.—Mohl considers the 2 paleæ as the products of different axes, and thus endeavours to overthrow R. Brown's theory of the flowers of Grasses, in which a viviparous monstrosity of *Poa alpina* formed the basis of the argument supporting his view (Bot. Zeit., 1845, pp. 33-7). I have endeavoured to defend the view that these organs are bracts (Gott. gel. Anz., pp. 683-7). Parlatore has formed the genus *Antinoria* of *Airopsis agrostidea* D. C., *Aira agrostidea* Guss. (Fl. palermit, i, p. 92).

FERNS.—The eighth part of the first volume of Kunze's illustrated work (Die Farnkräuter in colorirten Abbildungen, Leipsig, 1845, p. 4), has been published, containing pl. 71-80.—Presl has written a supplement to his Pteridographia, in which the number of genera and species appears to be considerably increased (Supplementum tentaminis Pteridographiæ, continens genera et species ordinum q. d. Marattiaceæ, Ophioglossaceæ, Osmundaceæ Schizæaceæ et Lygodiaceæ. Pragæ, 1845, 4to, pp. 119). The third part of Sir W. Hooker's Species filicum is issued from the press, and contains twenty plates. J. Smith has separated some species of *Oxygonium* from the Archipelago, as *Syngramma* (Lond. Journ. of Bot., 1845, p. 168).

MOSSES.—Nägeli has written a valuable memoir, entering fully into physiological details, upon the growth of the vegetative organs in Mosses and Hepaticæ (Zeitschr. für wissenschaftl. Bot., Hft. 2, pp. 131-209), from which he arrives at the systematic conclusion, that the formation of the leaf of Mosses is subject to a peculiar law; the point of the organ is formed last, the base first, by the formation of cells, whilst the growth of the individual cells ceases sooner at the point than at the base of the organ. Regarding their germination, Nägeli remarks (p. 175), that it is the same as in the Ferns; in both, the axis is formed from a single parent-cell of the pro-embryo, whence, "the earlier view that the pro-embryo forms a tissue, and that the stem is formed from this tissue by the growth of numerous cellular threads is contradicted." However, in both families these parent-cells are only capable of growing upwards, whence it follows, that all the roots have a lateral origin, but not, as Schleiden believes, that no roots are present. Just as the first axis of the Moss is developed from a parent-cell of the pro-embryo (germinal spore-filament of Nägeli), in the same manner in *Phascum*, for instance, the formation of new axes from certain capillary radicles (germinal bud-filament of Nägeli), whilst other similarly-formed roots do not appear to possess this formative power, and are therefore the only true roots, according to Nägeli's views.—Bruch and Schimper, now in conjunction with Gümbel also, have published the genera *Schistidium*, *Grimmia*, and *Racomitrium*, in four parts of their history of European Mosses (Bryologia Europæa. Fasc., pp. 25-8. Stuttg., 1845, 4to). Hampe has commenced an illustrated work, with the following title: Icones Muscorum Novorum v. minus Cognitorum (Dec. 1-3. Bonn., 1844-5, 8vo). C. Müller has given a review of *Macromitrium* (Botan. Zeit., 1845, Nos. 32, 33).—New genera: *Garckea*, C. Müll. (ibid., p. 865), from Java; from Chili, *Leptochlæna*, Mont. (Cinq. Centurie de planches cellulaires exotiques nouv., in Ann. Sc. Nat. iii, 4, p. 105), *Aschistodon* (ib. p. 109), *Diplostichon* (ib. p. 117), = *Pterigynandrum longirostrum* Brid., and *Encamptodon* (ib. pp. 120-36. pl. 14); from Lord Auckland's islands, *Sprucea* W. J. Hook. = *Holomitrium*, Brid., and *Lophiodon*, W. J. Hook. = *Cynodon*, Brid. (Antarct. Voy.)

HEPATICÆ.—Of the 'Synopsis Hepaticarum,' edited conjointly by Gottsche, Lindenberg, and Nees v. Esenbeck, the second and third part appeared in 1845, and the fourth part, which brings this important work to a conclusion, in 1846, excepting a supplement which remains to be added (Hamburg, viii, 624 pp.) The following new genera are distinguished in them: *Acrobolbus* N.

from Ireland; *Gottschea* N. = Junc. Sect. nemerosæ aligeræ; *Sphagnœcetis* N. = *J. Sphagni* Dics. and others; *Liochlaena* N. = *J. lanceolata; Micropterygium* = *J. Pterygophillum*, &c.; *Polyotus* G. = *Junc. sp.* Hook. and Tayl., from the South Sea; *Thysananthus* Ld. = *Trullania* Sect. *Bryopteris; Omphalanthus* = Jung. sp. American, &c.; *Androcryphia* N. = *Noteroclada* Tayl., *Carpolipum* N. = *Carpobolus* Schwein.

Lichens.—Montagne describes the new genus *Stegobolus* from Cumming's collection from the Philippine Isles (Lond. Journ. of Bot., 1845, p. 4). The new *Myriangium*, described by Montagne and Berkeley, belongs to the Collemaceæ; it is found in the Pyrenees, Algiers, and the Swan River (ib. p. 72); it forms a transition to the Fungi, resembling externally a *Dothidea*.

Algæ.—After the tetraspores of Florideæ had been pointed out in the Fucoideæ, Montagne was the first to find them also in one of the Confervæ, the genus *Thwaitesia* Mont. discovered in Algiers by Durieu, which only differs from *Zygnema* in this character (Compt. rendus, 1845, Oct.); however, the genus has since been rendered doubtful by the discovery of tetraspores also in several other Zygnemeæ (Revue Botan., 1846, p. 469). Decaisne and Thuret have been engaged in the investigation of the antheridia of the Fucoideæ, and have shown that the contrast between them and the sporangia is as great as in the *Charæ* or Mosses (Ann. sc. nat. iii, pp. 5-15, pl. I, II).—C. Müller has investigated the history of the development of the *Charæ* (Bot. Zeit., Nos. 24-7, pl. III). The large cell of the sporangium, filled with starch, should be regarded as a spore surrounded by a capsule, consisting of two layers of cells, which, on germination, protrudes through its envelopes (figs. 4, 6). Even before this a cytoblast appears in the place of the starch, hence the turbid fluid existing in a smaller cell situated beneath the spore, and inclosed with it in the sporangium (figs. 1, 2), probably plays an important part. From the commencement, the axis developes itself, although a mere cellular thread, in two opposite directions, as root and stem; this has been observed by Kaulfuss, and Nägeli has shown the same thing in the case of the cell of *Caulerpa*. At a subsequent period, "new" individuals are developed from the adventitious roots of the lower cells of the stem (turiones, according to fig. 10). At a much later period the whorled branches and cortical cells of the stem of *Chara* are developed, which the author has traced into the terminal bud; the former arise from the longitudinal subdivision of the cellular contents of the terminal cell (fig. 12), the latter from a growth of the branch just as in the Batrachosperms.—Fresenius has published a memoir upon the structure of the Oscillatoriæ, in which a critical history of the observations relating to these plants is contained (Mus. Senckenberg, iii, pp. 263, 292).—New genera of Algæ: Fucoideæ: *Cymaduse* Decs. Thur. (Ann. sc. nat. iii, 3, p. 12) = *Fucus tuberculatus* Huds.; *Pelvetia* D. Th. (ib.) = *F. canaliculatus; Ozothallia* D. Th. (ib.) = *F. nodosus* L. (*Physocaulon* Kütz.), so that *Fucus vesiculosus* and *serratus* are all that remain in the genus; *Pinnaria*

Endl. Dies. (Bot. Zeit., 1845, p. 288), near *Laminaria*, from Port Natal; *Contarinia* Endl. Dies. (ib. p. 289), from the same locality, near *Scytothalia; Stereocladon* Hook. Harv. (Lond. Journ. of Bot., 1845, p. 250), from Antarctic America; *Scytothamnus* Hook. Harv. (ib. p. 531)=*Chordaria Australis* Ag., from New Zealand. Florideæ: the Sphærococcoideæ, *Dicranema* Sond., from the Swan River (Bot. Zeit., 1845, p. 56), *Sarcomenia* Sond., also from this locality (ib.), *Phalerocarpus* Endl. Dies., from Port Natal (ib. p. 290), *Acanthococcus* Hook., Harv., from Antarctic America (l. c. p. 261), and *Hydropuntia* Mont., which was proposed on a former occasion, but is now fully described, and placed in this group as an aberrant form (Voy. au Pôle Sud.—Bot., i. p. 166, pl. 1); of the Rhodomeleæ: *Lenormandia* Sond. nec Mont., *Kützingia* Sond., and *Trygenea* Sond., from the Swan River (l. c. p. 54), *Epineuron* Harv., from New Zealand = *Fucus lineatus* Turn., &c. (l. c. p. 352); *Cladhymenia* Harv. (Lomentariæ), from New Zealand (ib. p. 530); *Apophlæa* Harv. (Cryptonemeæ), from New Zealand (ib. p. 549), and *Gelinaria* Sond., from the Swan River (l. c. p. 55); the Ceramiæ, *Hanowia, Ptilocladia,* and *Dasyphila* Sond., from the Swan River (l. c. pp. 52, 53). Confervaceæ: the Siphoneæ, *Struvea* Sond., from the Swan River (l. c. p. 49), *Cladothele* Hook. Harv., from the Falklands (l. c. p. 293), *Derbesia* Solier =*Bryopsidis* sp. (Rev. Bot. i, p. 452), and *Arechougia* Meneg., one of the Confervoideæ = *Phycophilæ* Kütz. sp. et *Confervæ* auctor (Atti di vi, riunione, p. 456).

Fungi.—The illustrated work of Harzer is completed by the sixteenth part (Naturgetreue Abbildungen der vorzüglichsten, essbaren, giftigen und verdächtigen Pilze. Heft xvi, Dresden, 1845, 4to).—New genera and monographic descriptions. Pyrenomycetes: De Notaris separates the following types from *Sphæria: Venturia, Massaria* = *Sp. inquinans*, Tod., *Rosellinia* = *Sp. aquila*, Fr. *Bertia* = *Sph. moriformis* Tod. (Atti di vi, riunione, pp. 484-7, t. i). Léveillé describes *Lembosia* and *Asterina* (Champign. exotiques in Ann. Sc. nat. iii, 3, pp. 58-9); Montagne, the new Pezizoideæ, *Hymenobolus*, from Algeria (ib. iii, 4, p. 359), and *Aserophallus*, from Cayenne (ib. p. 360). Gasteromycetes: Montagne describes *Xylopodium* and *Lasioderma*, from Algeria (ib. p. 361), and Czerniaïew, the following from Ukraine, *Endoptychum* (Bull. Mosc., 1845, ii, p. 146); *Trichaster* (ib. p. 149); *Endoneuron* (ib. p. 151); *Disciseda* (ib. p. 153); and *Xyloidion* (ib. p. 154). The two Tulasnes have written a monograph upon the Tuberaceæ, *Choiomyces* Vitt., and *Picoa* Vitt. Berkeley has described *Podascon pistillaris* Fr., from the Cape de Verd Islands (Lond. Journ. of Bot., 1845, pp. 291-3, t. x). Hyphomycetes: *Sphæromyces* Mont., from Algiers (l. c. p. 365); Coniomycetes, *Polydesmus* Mont. (ib. p. 365); *Phylacia* Léveill. (l. c. p. 61), from the group of Cytisporeæ to that of the Coniomycetes; *Piptostomum* Lév. (ib. p. 65). *Podisoma macropus* upon *Juniperus virginiana* is described by Wyman and Berkeley (Lond. Journ. of Bot., 1845, pp. 315-19, t. 12).

GEOGRAPHY OF PLANTS

James S[tarr] Lippincott

GEOGRAPHY OF PLANTS.

OUTLINE SKETCH OF THE GEOGRAPHY OF PLANTS AND OF THE METHODS PROPOSED FOR DETERMINING THE AMOUNT OF HEAT REQUIRED BY THE VINE AND THE WHEAT PLANT; WITH REMARKS ON THE PRODUCTION OF NEW VARIETIES OF WHEAT, AND ON THE ACCLIMATION OF PLANTS, AND OBSERVATIONS ON THE NECESSITY OF A MORE ENLIGHTENED AGRICULTURE.

BY JAMES S. LIPPINCOTT, HADDONFIELD, NEW JERSEY.

THE most superficial observer has doubtless sometimes noticed a method in the arrangement of natural objects, and believes that they have not been distributed by chance, or left without the superintendence of a directing agency. He perceives, if his view becomes more extended, that they have not been scattered at random over the surface of the globe. He who is better informed has learned that living beings (animals and plants) are regulated in their distribution by especial laws, and that their arrangement gives to each country a peculiar aspect.

The physical conditions which seem to regulate this distribution of living beings are diversified, and are mostly included under the idea of climate. Among the agencies which are understood to influence the life, health, development, and extension of the range of animals and plants, the more important are the distribution of heat throughout the year, its periodic increase and diminution, the conditions of humidity or dryness of the air and of the soil, the amount of light, the electric conditions, and in some instances the pressure of the atmosphere, and the varying degrees of actinic power or force of the chemical rays of the sun.

There are other very important considerations affecting the distribution of plants, which are not properly included under our idea of climate. Among these may be enumerated the composition and chemical character of the soil, which are influenced by the rock whence it was derived; the mechanical conditions of the soil, whether friable or coherent, readily decomposed, or resisting decomposition in a state of fine comminution, or in coarse grains or gravel; permeable readily by heat, rain, or moisture, or otherwise; and, finally, on the amount of vegetable matter therein, the color of the soil, and its inclination and exposure to the rays of the sun. The art of the agriculturist enables him to modify all these latter-named circumstances, and to adapt the soil to his various needs; but the vicissitudes of the climate he cannot control—over the weather he has not been made master.

The most important climatic influence is temperature, which, indeed, in this association may be termed the dominant power. It is beneath the ardent rays of a tropical sun that the noblest forms of vegetation are developed, while in the circumpolar zones neither trees nor shrubs can exist, and extensive districts are entirely destitute of plants, or in favored spots alone are found a few lichens and diminutive herbs, which the short summer of these regions calls forth from the partially thawed surface of the deeply frozen ground. Between these noble products of the torrid zone and the almost barren desolation of the Arctic regions there exists every variety of vegetable form, growing gradually less diversified and generally more dwarfish as we approach the limits of existence on the north and south, and exhibiting a correspondence between the modifications of temperature observed successively in our progress from the equator to the poles, and the geographical distribution of vegetable and, generally also, of animal life on the earth.

The geographical distribution of plants is a subject of vast extent and importance. The nature of the vegetation covering the earth varies, as we have remarked, according to the climate and locality; and plants are fitted for different kinds of soils, as well as for different amounts of temperature, light, and moisture. From the poles to the equator this constant variation in the nature of the flora is a shifting scene, passing from the lichens and mosses (the lowest vegetable forms in the Arctic and Antarctic regions) to the noble palms, bananas, and orchids of the tropics by a series of regulated changes through all the multiform aspects of the vegetable kingdom. The same progress and graduated fitness is observed in the vegetation of lofty mountains under the equator, when descending from the summit to the base. From the scanty vegetation of Greenland, where the only woody plants are the Arctic willows, trees scarcely a finger length in height, we may trace the expansion of vegetation as we move southward over the lichens and mosses to the saxifrages and cruciferous plants, or those which resemble the cabbage and the turnip in their mode of flowering; then to grassy pastures, and by coniferous or fir-like trees, and amentaceous or birch and alder-like trees, to the northern borders of the United States. Extending our glance further southward, we shall perceive that we enter the region of oaks, hickory, and ash, of tulip-poplar, buttonwood, walnut, red and white cedars, sugar and other maples, sassafras, sumac, laurel, and many other trees and shrubs characteristic of the temperate regions of North America. In the districts further south we find an increase both of species and of genera, and more tropical forms show themselves, such as magnolia, Osage orange, honey locust, cypress, holly bay, wax myrtle, the cotton plant, rice, the live oak, and enter the borders of the regions of the palmetto and the orange; thence to those of the sugar-cane and pineapple, the coffee plant and cocoa-nut, and the luxuriant vegetation of the equator and torrid heats. "In this progress," as Humboldt, the father of geographical botany, remarks, "we find organic life and vigor gradually augmenting with the increase of temperature. The number of species continues to increase as we approach the equator, and each zone presents its own peculiar features; the tropics their variety and grandeur of vegetable forms; the north its meadows and green pastures, its evergreen firs and pines, and the periodical awakening of nature in the spring time of the year."

Many causes intimately connected with the aspects of our globe have an influence in modifying the conditions of climate, and thus affecting the distribution of animals and plants on its surface. The geographical forms of contour, the relief or elevation and depression of the terrestrial surface, the relations of size, extent, and position, each exert a very marked effect upon the climatic peculiarities of a district. The bearing or direction of the shores of a continent, the elevation of a mountain in one place rather than in another, the subdivision of a continent into islands or peninsulas, and other minor differences, have very important bearings upon the climate of a district. The depression of a few hundred feet over some wide areas would reduce some regions to the level of the sea, or sink them beneath its waves, or so modify the climate of the higher portions left above the waters as to render them no longer tenable by the life that once enjoyed a congenial clime. This is shown by the observation that some low islands scattered in clusters are covered with a vegetation entirely different from that of extensive plains, though lying in the same latitude. A change in the bearing of the shores would modify the currents of the ocean, which would react upon vegetation. Mountain chains have oftentimes an influence upon the prevailing winds, and their height, or the plateaus from which they arise, modify the climate, and render it temperate or arctic under the fervent heats of the torrid zone. A mountain chain extending from east to west may form a barrier between the colder regions on the north and the warmer on the south, and thus protect the northern plains from the warmer winds of more temperate regions, and increase the heat on the southern slope. This is exemplified by the Alps of Switzerland, which reduce the temperature of Germany below the mean that would otherwise prevail but for their cooling influence. Under some of the high towers of this mountain barrier against the assaults of winter, the palm, the pomegranate, the orange and the olive grow in the open air, while a few miles to the eastward, in valleys, open to the north, through which the hurricane blasts of the Borra rush with terrific force and severity of cold, often sweeping vessels from anchorage, these more tender plants cannot exist.

A few thousand feet in elevation, which is insignificant compared with the mass of the earth, changes entirely the aspect and the character of a country. For evidence of this assertion we may compare the burning region of Vera Cruz—its tropical productions and its fatal fevers—with the lofty plains of Mexico, their temperate growths and perennial spring, or the immense forests of the Amazon, where vegetation puts forth all its splendors, and where animal life is abundantly prolific, with the desolate paramos or Alpine regions of the summits of the Andes, "rude, ungenial and misty." Or imagine the interior plains of the United States, east of the Rocky mountains, to be slightly inclined towards the north, and the Mississippi river to empty into the Frozen sea or Arctic ocean, or into Hudson's bay, and the new relations of warmth and moisture incident upon this simple change of direction of the current of this river would effect the most important modifications in the conditions of the vegetable and animal world, would exert a still greater influence upon the welfare of the inhabitants, and, through them, upon the destinies of society yet to be, and, perhaps, upon the entire human race.

The climate that would result from latitude alone is greatly modified by the presence or absence of extended sheets of water; and the distribution of heat through the year, for any place whatever, depends essentially on its proximity to, or distance from, the ocean or large lakes, and the relative frequency of the winds that blow over them. The equalizing influence of large bodies of water, the temperature of which is less liable to sudden changes than the atmospheric air, is quite apparent. While in Ireland and the southwestern part of England the myrtle grows in the open air, as in Portugal, fearless of the cold of winter, the summer sun of these so genial isles does not succeed in perfectly ripening the plums and pears which grow and ripen well in the same latitude on the continent. On the coast of Cornwall shrubs as delicate as the camellia and orange are green throughout the year in the gardens, though in a latitude at which, in the interior of the continent of Europe, trees the most hardy can alone brave the winter cold. The mild climate of England cannot ripen the grape almost under the same parallel where are grown the wines of the Rhine, nor will our Indian corn ever mature or attain there, even the size it will reach on our most northern border or in Canada.

In our own country the influence of the ocean is very favorable, and is more apparent in the northern sections, where it attempers the heats of summer on the land by the sea-breezes which prevail during part of the day along the coast. Nor is this genial influence of water upon the temperature of the neighboring land limited to very extensive bodies. The inland lakes of North America certainly ameliorate the otherwise severe climate of the district in which they are located, and even the minor lakes of western New York soften the extremes of cold, and, in connexion with their larger congeners, protect the incipient vegetation of spring, and prolong the growing season by preventing the recurrence of early autumn frosts.

Under the influence of the physical agencies of heat, moisture, &c., vegetation exhibits a regular progression, whether we ascend from the equator towards the poles, or from the base of a high mountain, under the equator, towards its summit to the regions of perpetual snow. In both these directions there is a striking general agreement in the order of the succession of the phenomena, so that the natural productions of any given latitude may be properly compared with those prevailing at any given height above the level of the sea. The elevation at which any variety of plant will grow spontaneously will, of course, vary with the latitude of the mountain on which it exists. There may then be said to exist a corresponding similarity of climate and vegetation between the successive degrees of latitude and the successive heights above the sea.

There is a class of plants which a slight advance beyond the freezing point calls forth from their winter sleep, which have a peculiar stamp, and constitute a peculiar flora, termed Alpine. We find this Alpine flora in regions where snow covers the earth, where the lakes are frozen most of the year, as in northern Lapland, northern Siberia, Arctic North America, on the Alps, the Pyrenees, the Carpathians, the Caucasus, in the mountains of Norway, on Scotch and Icelandic summits, in the Himalayas, the Sierra Nevada, and on our own Mount Washington. On the latter, Alpine plants identical with those of Lapland in latitude 67° north, where they grow at a height of three thousand feet and less above the level of the sea, are found at a height of not less than six thousand feet in latitude 44° north, while below this limit, in the wooded valleys of New Hampshire, there is not one species which occurs also about North Cape. The Alpine flora appears almost ubiquitous wherever a temperature sufficiently reduced, and yet adequate to sustain life, is found. Scandinavian genera and even species reappear everywhere from Lapland and Iceland to the tops of the Alps of Van Dieman's Land, in rapidly diminishing numbers, it is true, but in vigorous development throughout. They abound on the Alps and Pyrenees, pass on to the Caucasus and Himalaya, thence extend to those of the peninsula of India, to Ceylon and the Malayan Archipelago, (Java and Borneo,) and reappear on the Alps of the New South Wales, Victoria, and Tasmania, and beyond these again on those of New Zealand and the Antarctic islands, many of the species remaining unchanged throughout. It matters not what may be the vegetation of the bases and flanks of these mountains, the northern species may be associated with Alpine forms of Germanic, Siberian, Oriental, Chinese, American, Malayan or Australian, and Antarctic types, which are all more or less local assemblages; the Scandinavian, from Sweden, Lapland, and Norway, asserts its prerogative of ubiquity from Iceland, beyond its antipodes, to its Antarctic congener of high southern latitudes.*

The correspondence between the ascending vegetation on mountain sides and the distribution of trees, especially over the whole extent of the temperate zones, is so close that it may be considered a universal law. On lofty mountains climates seem arranged, as it were, in strata, one above another, with a regularity that has strongly impressed the attentive observer. He has noticed that the trees resembling the chestnut and the walnut, or most of those that flower in aments or catkins, occur in the lower latitudes under the influence of a genial clime, but disappear entirely before he reaches a latitude or an elevation where agriculture ceases. Further north he finds a variety of poplars, willows, oaks, maples, and other deciduous trees interspersed with pines, which begin to form continuous forests until they become an almost uniform pine and birch forest, covering in unbroken continuity the more northern districts as far as trees extend. So likewise, in ascending higher and higher on

* J. D. Hooker, Flora of Tasmania.—Silliman's Journal of Science, XXIX, pp. 323, 324.

the slopes of mountains, he observes the same order of succession of poplar, oaks, maples, &c., followed by forests of pines and birch to the limit of wooded growth. Beyond this, both towards the polar regions, and higher on the mountain side, the Alpine vegetation, before referred to, succeeds. A detailed comparison of northern and Alpine vegetation would show that they agree in almost every respect, and that in general corresponding species exist under similar circumstances in different parts of the Old and New World, following each other in the same succession from south to north, as well as from the plains to the mountain summits.

Illustrations drawn from observations made in our own country are always more interesting, and, perhaps, more convincing than those brought from distant regions. Many of our readers may have had the opportunity of witnessing to the truth of the following detailed description of changes in the conditions of vegetation by increase of height and consequent increase of cold, as observed on the sides of Mount Washington and in the district at its base.

The traveller who has visited the White Mountains of New Hampshire, from the south, will observe, on ascending from the lower districts towards the head-waters of the Connecticut river, that the forest vegetation begins to assume a character differing from that lower down the main valley or that nearer the ocean. At Windsor, three hundred feet above the level of the sea, the chestnut has already disappeared. Between Windsor and Littleton the hemlock spruce, the white pine, arborvitæ, larch, the buttonwood, the beech, the white, the black, the yellow, and canoe birches, the white, the red, and swamp white oaks, the elm, white ash, aspen poplar, the linden or basswood, sugar maple, and a few other species of maples and less important trees may be seen. At Littleton the buttonwood and hog walnut or pig nut disappear, the oaks are fewer and smaller, and the mountain maple, which is not found below here, makes its appearance. From Littleton, eight hundred and seventeen feet above the sea level, to Fabyan's, which is fifteen hundred and eighty-three feet above the sea, he will notice the white spruce, balsam spruce, hemlock spruce, white pine, larch, linden, white ash, sugar maple, mountain maple, elm, the birches above-named, and some others. The cup-bearing trees, or the oaks, the chestnut and the birch have disappeared, and the pitch pine is no longer observed. This vegetation continues from Fabyan's to a level of two thousand and eighty feet, where the pines are the prevailing features of the forest. From this level the slope becomes much steeper, and the variety of trees is much reduced. Above this, to the height of four thousand two hundred and fifty feet, the vegetation consists almost entirely of white and balsam spruce,* the lofty or yellow birch and the canoe birch, which become gradually more and more stunted till at the height above named, the species which form tall and splendid trees one or two thousand feet lower, appear here as mere shrubs or low bushes with crooked branches so interwoven as almost entirely to hedge up the way, excepting in places where a bridle-path has been cut through. Above this level, which is almost as elevated as the summit of Mount Clinton, and higher than Eagle Head, near Eagle pond, of the Franconia mountains, the mountain is destitute of forest trees. Many minor plants, however, appear which remind the well-informed and philosophic botanist of the flora of Greenland, and many of which have been found growing on the sub-arctic northern shores of Lake Superior.

* Most of the high mountain tops in western North Carolina and East Tennessee are covered with Abies nigra and Abies Fraseri, the former the black spruce, the latter the balsam fir. The black spruce grows at a lower elevation than the balsam, but neither of them are often met with beneath a height of 4,000 feet. Eight degrees of lower latitude causes them to ascend to a greater height.—S. B. Buckley, Silliman's Journal, XXVII, pp. 286-294.

The summit of Mount Washington,* six thousand two hundred and eighty-eight feet high, produces several plants which have no representatives south of Labrador. These, it may be interesting to some readers to be informed, are Andromeda hypnoides, Saxifraga rivularis, Rhododendron lapponicum, and Diapensia lapponica.

We are all as well aware that the mean annual temperature in this country decreases as we proceed towards the north, as we are that we shall encounter increase of cold on ascending a high mountain. All are not equally well acquainted with the rate of variation either on the plain or on the mountain slope. In central Europe the change in mean annual temperature takes place at about the rate of one degree of Fahrenheit for each degree of latitude, or for about every seventy miles; while in the interior of the United States, in the Mississippi valley, the rate of decrease is about one degree of Fahrenheit for every forty miles; or, in other words, as we travel north or south we reach successively localities the mean annual temperatures of which are one degree of Fahrenheit's thermometer lower or higher as we pass over seventy miles in central Europe, or forty miles in the valley of the Mississippi. In the Alps of Switzerland, in ascending or descending, the same change of one degree in mean annual temperature is experienced for about every three hundred feet of vertical height, so that we can pass, within the narrow limits of between six or seven thousand feet, from the vine-clad shores of the lakes of northern Italy to the icy fields or snow-capped mountains, whose summits are never adorned by vegetation, a journey that may be made in a single day; while to descend to the level in our own land and latitude, which corresponds in mean annual temperature to that experienced at the foot of the Alps, and travel northward, we shall find in the valley of the Mississippi a diminution of one degree of temperature for every forty miles; so that to find the region where the mean annual temperature is reduced to the freezing point, we should travel over about twelve degrees of latitude. This would conduct us to districts beyond the northern borders of Lake Superior before we could pass over the same range of climatic changes as we would encounter in one day upon the slopes of the Alps.

The mean annual temperature of 32° Fahrenheit does not imply a total absence of vegetation on the plains, or in the northern districts of North America, nor the presence of eternal frost. The heats of summer extend far beyond the limits above referred to, and even Indian corn may he ripened beyond the line of mean annual frost, or of a mean annual temperature of 32° Fahrenheit in districts northwest of Lake Superior, where, in the valley of the Red river, in latitude 50° north, about sixty days of clear tropical summer occur, which are sufficient, for reasons to be hereafter cited, to ripen some varieties of this most valuable of cereal grains.

Though the correspondence between the forest vegetation on mountain sides and the distribution of trees is thus marked, there exist many circumstances which show that climatic influences alone, or as at present existing, however extensive, will not fully account for the geographical distribution of plants and animals. Their various limits do not agree precisely with the outlines indicating the intensity of physical agencies upon the surface of the earth. The limit of forest vegetation does not, as before remarked, coincide with the isotherm or line of the mean annual temperature of 32° Fahrenheit; nor is the limit of vegetation in altitude on mountain sides strictly in accordance with the mean temperature; nor are the plants and animals distributed under the different zones of climate the same in their respective zones in the northern and

* The above-cited heights are taken from Guyot's "Appalachian Mountain System," a very full detail of recent accurate measurements of most of the more elevated portions of the mountains of the Atlantic border of the United States.—See Silliman's Journal, (1861,) XXXI, pp. 158 to 187.

southern hemispheres, nor even in the same zone in the Old and in the New World, certain classes of plants being exclusively circumscribed within certain regions, such as the magnolia and cactus in America, and certain animals, as the kangaroos in Australia, and the elephant, rhinoceros, and hippopotamus in Asia and Africa, which are limited to their respective districts.

Although there exists an intimate correlation between climate and vegetation, the temperature and other influences which constitute climate do not reveal all the causes which are operating or have produced these differences as they are repeated under the isothermal lines, or lines of equal heat, between the eastern and western shores of the Old World in the same order as along the eastern and western shores of North America. The similarity of the climates of eastern North America and eastern Asia, and of western North America and western Europe, originating in similarity of position as respects latitude, exposure to ocean currents, protection by mountain barriers, &c., does not excite surprise; but that there should exist in Japan many genera and species of plants unknown except to the eastward of the Rocky mountains in North America—that the district of the lower Amoor river, on the shore of the Pacific, should produce several species identical with peculiarly eastern North American kinds, and many others nearly allied, if not identical therewith, which are not found in northwestern America nor elsewhere on the globe, is more worthy of remark; and when on further inquiry we learn that the Canary islands and Azores possess American genera not found in Europe nor in Africa, and that the lofty mountains of Borneo contain Tasmanian (Van Dieman's Land) and Himalayan representatives, that the Himalayas contain Andean, Rocky mountain, and Japanese genera and species, and the Alps of Victoria (a district of Australia) and Tasmania, assemblages of New Zealand, Fuegian, Andean, and European genera and species, the interest becomes greatly enhanced, and the query arises, How shall all these seeming anomalies of a similarity of product, under similarity of climate, yet wide disseverance of species identical with each other, over districts between which there seems to exist no possible means of communication, be satisfactorily explained? Causes now in operation cannot be made to account for a large assemblage of flowering plants characteristic of the Indian peninsula, being also inhabitants of tropical Australia, while not one characteristic Australian genus has ever been found in the peninsula of India. Still less will these causes account for the presence of Antarctic and European species in the Alps of Tasmania and Victoria, or for the reappearance of Tasmanian genera on an isolated lofty mountain in Borneo.*

The above-cited and a multitude of analogous facts have led to the study of the agencies which may be reasonably supposed to have had an influence in determining the distribution of plants. Among these agencies may be named supposed changes of climate, which geological research seems to exhibit, arising either from direct cooling of the entire mass of the earth, or from new positions and elevations of the surface, which we cannot doubt have occurred at various epochs of the history of the progress of the earth. We shall not pause to dwell upon these most interesting problems which are tasking the learning and acumen of the most profound and acute philosophers, and which have not yet found a satisfactory solution, nor enlarge upon the nature of the facts and reasoning brought up from the study of extinct plants and animals, whose remains, entombed in the solid rock, bear witness of their former life, and of the conditions demanded for its support, but will pass to more practical applications of the general law of distribution under the regularly recurring changes of temperature now influencing the life of the globe.

---

* Dr. J. D. Hooker, Tasmanian Flora in Silliman's Journal of Science, vol. XXIX, pp. 19, 21.

THE DEFINITE AMOUNTS OF HEAT REQUIRED BY PLANTS.

When considering the coincidences existing between the distribution of plants, according to their respective zones of climate on plains and on zones of elevation on mountain sides, we have had especial reference to the mean annual temperatures prevailing in the respective regions observed. These, it has been remarked, are in general correct as regards the distribution of forest trees, which brave the extreme cold of winter, and require but a few months of moderate heat for foliation and for maturing their fruit. The discrepancies found to exist between the isotherms in latitude and isotherms in elevation are more especially noted when we come to consider the arrangement of annual plants under these respective relations. Originally the mean annual temperature was alone observed, and the polar limits of plants, it was presumed, could be thereby determined. But opinions on the subject of polar limits, or isothermals, bounding on the north in the northern hemisphere the extension of certain species of plants, have changed with the progress of physical geography. More recently it was taught that the mean temperature of seasons is of more importance than that of the year, and that in general two similar climates may be distributed in fractions very dissimilar, which may neutralize one another in the estimate of their mean annual temperatures. Thus it has been observed that countries in which the summers are short but very warm, and the winters very long and cold, have a vegetation different from that of those which enjoy seasons more equable, and where the march of temperature from month to month is more gradual, or the changes less sudden and violent, although the mean annual temperature of both may be the same. It is, therefore, believed that to the relative distribution of heat over the seasons rather than to the absolute amount received through the year that we are to attribute the fitness or unfitness of a region for the growth of certain kinds of vegetation. In other words, that in a knowledge of the mean temperature of seasons, or of monthly temperatures, we may find an explanation of the causes that have influenced certain species of plants in their choice of home. Also that they have advanced from their original centre over a continent to a certain line which passes through points having an equal temperature during some period of the year, which may vary with various plants, and have occupied all the region to the border thus indicated, unless their extension has been arrested by a climate too dry or too humid, or by the impassable ocean.

In the progress of knowledge, it has become apparent that these lines of equal temperature or isotherms for certain seasons do not answer all the requirements of each case; that they do not limit even the extension of annuals, though their vegetation is confined wholly or mostly to the three months of summer.

It appears that the conditions which define the limits of a plant require that we should know the degree of temperature at which its vegetation begins and ends; that at which it will flower and will mature its seeds or fruit; and also the sum of the mean daily temperatures during these periods respectively. The hypothesis that a definite amount of heat is required in order to develop each plant in its progress from one stage of growth to another was first advanced by Reaumur, better known in America from the thermometer which bears his name than through the scientific labors which added largely to the wealth of his native France. This philosopher proposed to calculate the amount of heat demanded by a plant, by multiplying the number of days required to pass through its growth by the mean temperature of the period.

To Michael Adanson, a French naturalist of comprehensive mind, great acuteness and perfect independence of thought, but too far in advance of his contemporaries to be rightly appreciated, we are indebted for the hypothesis that, by adding together the mean temperatures of each day from the commencement of the year, it will be found that when the sum shall have reached a certain figure the same phenomena of vegetation will be exhibited, such as foliation, blooming, and maturation of the fruit.

Boussingault, that "prince of agronomical chemists," in whom combine the successful farmer and the profound philosopher, whose "Rural Economy" should be studied by every farmer, and whose new "Agronomie" promises yet more brightly to illumine his name, and the path of the scientific seeker for "light—more light," revived the hypothesis of Reaumur and enlarged its application. Many years' residence in South America, engaged in scientific observation and research, where vegetation upon mountain sides appears under almost every aspect and condition, combined with experience in the pursuits of agriculture on his farm at Brechelbronn, in Alsace, had taught him that if we multiply the number of days—the length of time a summer plant endures—by the mean temperature of this period, the product will be the same in all countries and in all years.

Baron Quetelet, of Brussels, styled by his countrymen "the Belgian Arago," has experimented upon the amount of heat required by plants, and proposes that it should be measured not by the simple product of the temperature of the several days by the number of days required, but that the squares of the mean temperatures should be employed in lieu of the daily mean. The younger De Candolle has, however, sought in vain in the researches of Quetelet and others for any positive facts showing the direct advantages of the variations of heat over continuous even temperatures, or for the evidence fully determining that one day having a mean of 20° centigrade, is of equal value with four days having each a mean of 10° centigrade, as is assumed by Quetelet in the Bulletin de l'Academie Royale de Bruxelles, v. XIX, pt. I, p. 543.

Babinet, a distinguished physicist, has proposed a method for determining the heat required by plants for the performance of certain functions. He compares the action of temperature to that of a force, such as weight, which produces effects proportioned to the intensity of the cause and the square of th time. He has not, however, made experience the basis of his hypothesis, and it is rejected by Quetelet as unsound. According to the latter philosopher, by the hypothesis of Babinet, if one day at 20° centigrade produces a certain effect, two days having a mean of 10° centigrade should produce four times 10° or the effect of 40° centigrade, and four days at 5° centigrade should exert an influence equal to sixteen times 5° or 80° centigrade—results which the general experience of horticulturists will not permit us to accept as true.

Count Adrien de Gasparin, a minister of Agriculture and Commerce, and Peer of France, who, by his numerous memoirs addressed to the societies of the departments as well as to the Academy of Sciences, attained an honorable place among contemporary agronomes, or scientific agriculturists, has also given much attention to the question of the amount of heat demanded by plants.* In his excellent Cours d'Agriculture, or Course of Agriculture, he suggests that the mean heat of the day should be derived in part from the direct heat of the sun, and not alone from that of the air, as is in general measured by meteorologists, because the motive power which induces the circulation of the sap is the heat derived from the atmosphere and the soil in conjunction with the direct rays of the sun. The rate of decomposition of the carbonic acid absorbed from the air must be measured by the activity of the chemical rays of the sun, and the growth of the plant is accelerated, we are aware, by exposure to its full measure of sunshine. This method cannot, however, be readily verified in

* Count Adrien de Gasparin was father of Count Agenor de Gasparin, whose virtues and generous appreciation of the importance of the struggle that now racks our nation have been so fully exemplified in his enlightened labors in our behalf among his fellow-countrymen in Europe.

the actual state of meteorological knowledge. This distinguished observer followed the vegetation of one variety of vine growing near Orange, in France, from foliation to maturity, noted the minima of heat for each day in the shade, and the maxima shown by the thermometer in the sunshine, but protected by a slight covering of earth. The mean between these minima and maxima give, according to Gasparin, a more satisfactory number than that derived from other processes; and when multiplied by the number of days during which vegetation is influenced by this particular mean, results in a sum total of heat which varies but little from year to year at the locality where the observations were made. So nearly do these sums agree that the presumption is strengthened that the process may be the correct one, and deserving of much more attention than has been awarded to it. The result of de Gasparin's experiments in 1844 was a sum of 4195°; in 1845, 4203°; in 1846, 4057°; and in 1847, 4100° centigrade.

Among the contributions of these inquirers into the mystery of heat as adapted to the wants of plants, none have hitherto elicited more interest than those of Boussingault. His method, proposed by Reaumur, appears to be logical and precise, giving results sufficiently satisfactory for determining the heat necessary for maturing annual plants, particularly the cereals in spring. He has been very happy in determining the conditions of a good vintage in his district of Alsace, and has shown that certain mean temperatures are necessary during certain eras of vegetation, and especially that a definite amount of heat is needed to insure the elaboration of a due proportion of sugar in the grape.

The principle that we must combine the values of temperature and time in our inquiries into this subject cannot be controverted, for all must perceive that heat acts proportionally as regards its duration and its force. Boussingault, therefore, asserts that if a plant has required twenty days to ripen its seed, numbering from the period of flowering, and the mean temperature during those twenty days has been fifty degrees, it will be found that the plant will have received one thousand degrees of heat. The same number of degrees of heat might have influenced the plant during a lesser number of days had the mean temperature been proportionally higher. This is well illustrated by the rapidity with which some annual plants germinate in Arctic regions on the return of midsummer heats. In these northern regions, where for a short time plants are subjected to an intense heat, often as high as 109° Fahrenheit in the shade, and which enjoy a longer continuance of the sun above the horizon than in more southern latitudes, the growth of some vegetables is said to be so rapid under assiduous culture and in genial situations that their progress may be traced from hour to hour. In Norway, in latitude 70° north, peas grow at the rate of three and a half inches in twenty-four hours for many days in summer, and some of the cereals, probably barley and oats, grow as much as two and a half inches in the same time. Not only is the rapidity of growth affected by the constant presence of the sun's heat and light, but those vegetable secretions which owe their existence to the influence of the actinic force on the leaves are all produced in far greater abundance than in more southern climes; hence the coloring matter is found in greater quantity, the tints of the colored parts of vegetables are deeper, the flavoring and odoriferous matters are more intense, though in saccharine properties the plants of Norway are not equal to those of the south.

The successful cultivation in northern countries, by artificial means, of plants naturally demanding the high temperature and long seasons of more southern latitudes do but combine the duration and degree of heat. Nor need it excite our surprise that so well defined are the laws regulating the temperatures necessarily accorded to each variety of plant undergoing these artificial climatic conditions, and so accurately determined are they by practiced gardeners, that they have become experts to that degree that they succeed in producing the flower or the matured fruit on a given day.

While there can be no doubt that different plants require different amounts of heat from the time of sprouting to full maturity, though the time through which this may be furnished may be different in different instances, and that a great heat may produce the same effect on plants which is produced by a lower degree operating during a longer term, another principle of much importance must be observed in order to the successful cultivation of plants under natural or artificial circumstances.

This second principle is that each species requires for each one of its physiological functions a certain minimum of temperature, or, as has been well said, each species of vegetable is a kind of thermometer which has its own zero or lowest degree at which it will vegetate. A temperature above a certain minimum of heat is found necessary for germination, another for one chemical modification, and a third for flowering, a fourth for the ripening of seeds, a fifth for the elaboration of the saccharine juices, and a sixth for the development of aroma or bouquet. A certain intensity of light is also demanded to render green the tissues, and a due supply of humidity in the air and in the soil to furnish a vehicle for the materials of growth and prevent undue desiccation. A plant is thus not only under all the influences which affect the thermometer, but is likewise acted upon as is a hygrometer by humidity and dryness.

From the study of special examples among plants, under the combined influences of temperature, light, and moisture, a later inquirer believes he has reached most interesting results regarding the limiting effects of these causes towards the north, and on mountain sides. From the point of view taken by this later investigator botanical geography ceases to be a simple accumulation of facts, takes a place among the sciences, and assumes to explain by the study of the distribution of living plants the actual conditions of climate, as well as those that prevailed in former ages. It is to Alphonse de Candolle, who, in his Geographie Botanique Raisonnée, has presented to the scientific public one of the most important works that has of late appeared, and one of the most generally interesting, both on account of the subjects it treats of and the signal ability and thoroughness with which most of them are handled, that we are indebted for a clearer insight into the obscurity which has shrouded the subject of the geography of plants. "This work," which, in the language of Dr. Asa Gray, "is one that addresses and will greatly interest a much broader circle of scientific readers than any modern production of a botanical author, and which will probably long be esteemed the standard treatise upon a wide class of questions highly and almost equally interesting to the botanist, the zoologist, the geologist, the ethnologist, and the student of general terrestrial physics," is not to be found in our bookstores, and appears upon the shelves of only one library in Philadelphia, the magnificent natural history collection of the Academy of Natural Sciences. "Truly we are an enlightened people!" A. de Candolle has proved himself a son worthy to bear the name and continue the renown of his father, Pyramus de Candolle, who was the only botanist since Linnæus that embraced all branches of the study with an equal genius, and whose works mark a new era in the progress of his science. The limits which we have assigned to this paper forbid an extended notice of this work, and we will confine our remarks to a few illustrations of the results arrived at, and refer the inquiring reader to Dr. Asa Gray's notice, in Silliman's Journal of Science, vol. XXII, pp. 429, 432, and to Hooker's Journal of Botany, spring and summer of 1856, for a more extended critique.

ON THE RANGE OF CERTAIN ANNUAL PLANTS.

It has been the custom to attribute the disagreement between the limits of plants and the lines of equal temperature to errors of observation on the locality of species, to uncertainty respecting the thermometric mean, or to varying degrees of dryness and humidity. But certain facts and calculations upon the heat requisite for culture in different countries have awakened doubts of the truth of these explanations, and have led to more minute investigations.

As each species of plant can bear a definite range of temperature, indeed, requires a certain amount during a given period of time to enable it to perform all its functions properly, the only true indication of climatic adaptation is, that the plant, under the conditions assumed, perfect its seed and produce its various secretions; and more recent research has demonstrated that neither the annual mean, nor that of the summer season, accurately indicate the northern limits of the extension of the range of our *annual* plants. An extended series of comparisons of the range of many European plants, with the tables of monthly temperatures and seasons over the districts where these plants naturally appear, has shown that in no case does the limit of a species exactly coincide with a line of equal temperature for any one period of the year. Also, which will appear more worthy of remark, that the limits of annual plants in the plains of Europe cross one another with considerable frequency, and that the limits of perennial and ligneous species also cross each other in different directions, and both are far from being parallel when they do not thus cross.

The lines of mean temperature are derived from a consideration of fixed periods whether of a month or of a season, while the vegetation of annuals lasts during periods which are variable, the vegetation of the plant having its commencement and completion at different epochs of the year dependent upon the condition of greater or less excess of heat when compared with the mean of the growing season. There can obviously be but small agreement between these two classes of facts which have no necessary connexion or relation.

A. de Candolle has taught that if we would estimate the heat really useful to a plant we must consider in our calculation those values only which are above a certain degree of temperature, which degree varies with the species. This he adds is a field hitherto unexplored, and proceeds to explain the application of the above principles in combination in European climates, by which are brought about similitudes or dissimilitudes to which the means ordinarily employed furnish no key.

London and Odessa are certainly not under the same lines of temperature. The mean summer heat at London is 61° Fahrenheit, at Odessa 70° Fahrenheit; while in winter the difference is rather greater. In the monthly means these two climates have but imperfect analogy. The monthly means at London and Odessa appear to be as follows, in degrees of Fahrenheit:

	January.	February.	March.	April.	May.	June.	July.	August.	September.	October.	November.	December.
	°	°	°	°	°	°	°	°	°	°	°	°
London	37.2	40.1	42.5	46.9	53.5	58.7	62.4	62.1	57.5	50.7	44.0	40.4
Odessa	25.2	27.6	33.1	46.4	57.6	66.2	73.3	70.8	59.7	52.3	40.0	29.4

For greater convenience we will convert the above into degrees of centigrade, or the French mode of dividing thermometers, which shows but 100 degrees between the freezing point and the boiling point of water, a graduation it is greatly to be regretted not yet of universal acceptance with our countrymen.*

Thirty-two degrees of Fahrenheit becomes their zero, and as the degrees below this point or that of freezing are of no account to us in this consideration, we may very properly note those only above freezing or the zero of the centigrade. We have accordingly reduced the above Fahrenheit degrees to those of centigrade, and the series of temperatures for the respective months read as follows:

	January.	February.	March.	April.	May.	June.	July.	August.	September.	October.	November.	December.
	°	°	°	°	°	°	°	°	°	°	°	°
London	2.83	4.50	5.83	8.28	11.94	14.83	16.89	16.72	14.17	10.39	6.67	4.67
Odessa	—3.78	—2.44	0.61	8.00	14.22	19.00	22.94	21.56	15.39	11.28	4.44	—1.44

Now, if we consider the time at which the temperature of 4.5° centigrade or 40° Fahrenheit commences and terminates in each of these cities, and the sum of the mean heat between these two limits, as shown in the above table, we shall find nearly the same result. At London the mean of 4.5° centigrade (40° Fahrenheit) commences on the 17th of February and terminates on the 15th of December. Between these two periods, we find by taking the monthly means and fractions for the parts of the months named—that is, 5° centigrade for the thirteen days of February, and 5° centigrade for the fifteen days of December, which may be considered allowable since the heat of the latter part of February and earlier part of December exceed the remaining portions, we find an aggregate for London of 115.50° centigrade. At Odessa the temperature of 4.5° centigrade (40° Fahrenheit) commences later, from 2d to 3d of April, and terminates sooner, from the 17th to the 18th of November; but as it is warmer during summer, the amount of heat determined as above is about equal to that of London, being 116.39° centigrade. Hence it is apparent that a plant which would require 4.5° centigrade to commence vegetating with a certain activity, and in order to reach a certain stage or condition would require in all an amount of heat of 115° centigrade, rated upon the monthly means, might advance in a northwest direction to London, and in a northeast to Odessa. If a plant should require more or less than 4.5° centigrade as a minimum, or more or less than 115° centigrade in the whole as above calculated, the climates would no longer correspond, and the limit of species would be otherwise established. Our author has made use of the daily means in his calculation, and discovered that a plant commencing its growth at 4.5° centigrade, or 40° Fahrenheit, and requiring a season extending to the return of the same temperature in autumn, would demand at London 3,431° of centigrade heat, and at Odessa would receive between the same minima 3,423° centigrade. The above is a much more accurate method of comparison; but as we have not records of daily means for these places at command, we have used the monthly means, which, in this case, appear to approximate pretty nearly to the same result. It thus appears how two climates which differ, when considered as regards their respective mean monthly temperature, may yet be identical under certain combinations of the two causes which affect the life of the species existing therein.

* The centigrade scale offers many facilities to calculation, and may be readily understood. To convert Fahrenheit degrees, the common graduation, into centigrade, observe the following formula: x° Fahrenheit $= (x^\circ - 32^\circ)\,\frac{5}{9}$ centigrade, which means that to convert Fahrenheit into centigrade, deduct thirty-two degrees therefrom, divide the remainder by nine, and multiply the quotient by five, for the amount in centigrade degrees.

For the purpose of discovering these correspondences of climates De Candolle calculated the days on which the temperature 1°, 2°, 3°, &c., up to 8° centigrade commence and end at certain European localities; these all being above the freezing point, and correspond, respectively, to 33.8°, 35.6°, 37.4°, &c., up to 46.4° Fahrenheit. At the lowest of these temperatures some forms of vegetable life are called forth from their winter sleep. By associating with each locality noted, the product indicating the heat received in excess of each of these degrees, and applying these figures to the facts of vegetation, highly satisfactory results are obtained. One instance we may adduce in illustration of this method.

The *Alyssum calycinum*, an annual plant, grows on the eastern coast of Britain, as high as Perth. It is not found on the western coast of England, nor in Ireland, in Jersey, in Guernsey, nor in Brittany, which may be attributed to the constant humidity of those regions, for this plant loves a dry atmosphere, and a plant that grows in Scotland cannot be said to find the heat of Brittany insufficient for its existence. On the continent, the *Alyssum calycinum* spreads to the northwest as far as Holstein and the Baltic sea, but not into Scandinavia, and on the northeast as far as Moscow, but is not found in Kazan in about the same latitude, though further east, and reappears in the governments of Pensa and Simbrinsk, but little south of Kazan, and again in the Steppes between the Volga and the Ural. The limits where its extension may be thought to be determined solely by temperature stretches, therefore, from near Perth, in Scotland, under the parallel of 56½ degrees of north latitude, passes along the 54th degree in Holstein, and thence oscillates in Russia between the 55th and 56th degrees. This line varies widely from any line founded on equality of temperature for any season or month of the year. On examining a table of daily temperatures, it appears that at or near Perth, in Scotland, the temperature of 7° centigrade, about 45° Fahrenheit, or upwards, continues from the 18th of April to the 31st of October, and that, during this time, the product of the days by the mean temperature amounts to 2,281° centigrade. At Kœnigsberg, on the Baltic, the temperature of 7° centigrade and upwards is of shorter continuance, but the summer being hotter, the product amounts to 2,308°. As the limit of the species is about sixty miles to the north of Kœnigsberg, this figure must be reduced, and thus more nearly resembles that found for Perth, in Scotland. At Moscow the mean of 7° centigrade commences on the 22d of April, and terminates on the 5th of October, the product in consequence of the heat of summer rises to 2,473°. This is more than is necessary for the *Alyssum* which probably flourishes more than one hundred miles north of Moscow. At Kazan the figures fall to 2,196°, so that it is not surprising that the *Alyssum calycinum* has disappeared. Thus the hypothesis of 7° centigrade for an initial temperature or minimum at which this plant commences to vegetate, and 2,280° to 2,300° for the quantity of heat demanded, accord completely with the facts.

Another instance may be adduced of conformity of facts with figures derived from temperature. The *Euonymus Europæus*, or European spindle-tree, nearly related to the burning-bush of our gardens, has for its northern limit the north of Ireland, Edinburgh, in Scotland, the north of Denmark, the south of Sweden, the isle of Aland in the Gulf of Bothnia, latitude 60° north, Moscow, southeast to Penza, in Russia, in latitude 53°, through a latitude of seven degrees. The mean annual and mean summer temperatures vary several degrees at these places when they are respectively compared, nor do the monthly temperatures agree much more nearly. The *Euonymus* requires a temperature of 2,480° centigrade, the sum of the daily temperatures between the two epochs of the year when the curve of mean temperature ascends above 6° centigrade. At Edinburgh this is found to be 2,482°. At Stockholm, beyond the limit, the sum is but 2,268°. At St. Petersburg, likewise beyond it, it is too small. At Aland there may exist a more elevated temperature due to the surrounding water. At Moscow, where the plant appears, the sum exceeds the supposed condition, and the *Euonymus* doubtless extends here further north. Finally, at Kazan this figure is reduced to 2,250°, and here the plant does not appear. The values thus found along the limit in its neighborhood, and beyond it, accord as nearly as could be desired with the double hypothesis of a minimum of 6° centigrade, and an aggregate temperature for the growth and maturation of *Euonymus Europæus* of about 2,480° centigrade.

De Candolle considers himself justified, by the results derived from the investigation of the limits of the northern extension of nearly forty species of plants, to enunciate the law of nature that "every species having its polar limit in central or northern Europe advances as far as it finds a certain fixed amount of heat, calculated from the day when a certain mean temperature commences to the day when that mean terminates." The apparent exceptions to this rule may be explained by two circumstances which restrict its application. The eastern and western limits of a species may be determined by the humidity or dryness of portions of the region over which it would otherwise extend, or perennial and woody plants are sometimes arrested in their extension towards the north by the absolute minima of temperature. In tracing a limit of a species from east to west, if the law as stated ceases to be applicable, the species may be supposed to have encountered the excluding influence of severe cold or excessive drought or humidity, and it is often difficult to decide which of these causes operates as an obstacle.

THE TEMPERATURE AFFECTING THE VINE.

There are few crops that are so much at the mercy of the atmosphere and its varying conditions as that derived from the vine. Even in vineyards that are most favorably situated, it is rare that wines of equal quality and flavor are produced in two consecutive years; and in districts upon the verge of the productive limits of the vine under extreme climates, where it exists only in virtue of hot summers, its produce is still more variable, more inconstant.

The limits to the culture of the vine in Europe are generally fixed where the mean annual temperature is from 50° to 52°. Under a colder climate in Europe no potable wine is produced. To this meteorological datum must be added the fact that the mean heat of the cycle of vegetation of the vine must be at least 59° Fahrenheit, and that of the summer from 65° to 66° Fahrenheit. Any country which has not these climatic conditions cannot have other than indifferent vineyards, even when its mean annual temperature exceeds that above indicated. It is impossible, for instance, to cultivate the vine upon the temperate table lands of South America, where they nevertheless enjoy a mean of from 62.6° to 66.2° Fahrenheit, because these climates are characterized by constancy of temperature, never rising to the higher heats necessary to the process of sugar-forming, and the vine grows, flourishes, but the grapes never become thoroughly ripe. Even in Europe, where the climate partakes of a regularity that to us is remarkable, the varying temperature of the summer mean from year to year produces a corresponding variation in the quality of the wine. This quality may, in good measure, be tested by the amount of alcohol contained therein, as this is directly as the amount of sugar elaborated by the heat of the summer months, or by the epoch of its growth, during which the sugar-forming process takes place.

A series of observations, from 1833 to 1837, inclusive, made in Flanders, shows the following results, which may be better understood if placed in a tabular form:

	MEAN TEMPERATURES.			PRODUCTS.	
	Temperature of cycle of growth of vines.	Temperature of summer.	Temperature of beginning of autumn.	Wine, per acre, in gallons.	Percentage of pure alcohol.
	°	°	°		
1833	58.46 Fahr.	63.14 Fahr.	52.52 Fahr.	311	5.0
1834	63.14	68.54	62.60	314	11.2
1835	60.44	67.10	54.14	621	8.1
1836	60.44	70.70	53.96	544	7.1
1837	59.40	65.66	53.42	184	7.7

We here perceive that the mean temperature of the period, during which the growth and maturation of the grapes take place, exercises a very remarkable influence. The temperatures of the summer and autumn of 1834, and consequently that of the entire growing season, were decidedly greater than that of any others noted above, and in the same year the strongest wine was produced. The temperature of the beginning of autumn—September, of 1835—is next in order, as higher than that of the remaining years, and though the summer of 1836 was warmer, we find the wine of 1835 superior to that of 1836. This clearly indicates that a summer heat, prolonged into the autumn, is of more value than high summer heats, with a lower temperature, later in the season. The value of the wines produced in the remaining years is nearly in accordance with the temperature of the beginning of autumn, or the month of September, and that for 1833, when the growing season was the lowest, and that for September reduced to a minimum, the wine, it is said, was scarcely drinkable. The mean temperature, therefore, of the summer is not a safe criterion for judging of the adaptation of a district to vine culture. A mild autumn must be regarded as one of the essential conditions, and, consequently, the mean for the entire period of growth where the higher heat is near the termination of that cycle, becomes a better evidence of fitness of any region for producing wine of the highest quality. This truth is strongly indorsed by the observations made in 1811, so remarkable over Europe for the quantity and excellence of its wines. This year, known as the "comet year," was distinguished by the high temperature of the early autumn. Though the summer mean was lower than several of the years above tabulated, and about an average, the months of September and October were maintained at 59° Fahrenheit, while the usual temperature of these months had not been higher than 52.7° Fahrenheit, or nearly seven degrees lower. Boussingault, to whom we are indebted for much of the data on the meteorological conditions necessary to the production of wine of the best quality, has expressed his conclusions in the following terms:

"That in addition to a summer and an autumn sufficiently hot, it is indispensable that at a given period—that which follows the appearance of the seeds—there should be a month the mean temperature of which does not fall below 66.2° Fahrenheit."

The following table illustrates the cultivation of the vine in Europe, and also the depreciation of its produce according to climatic relations. Cherbourg in Normandy, in northwestern France, Dublin, and London, show, in a remarkable manner, how, with annual mean temperatures about the same, or higher than those of several places in the interior of the continent, and a winter temperature much milder, yet the summer heat falling several degrees below, and the temperature of the hottest month differing more widely, the grape cannot be cultivated in Ireland, Britain, or Normandy. The more general prevalence of a clouded sky at the above-named places may in a measure account for the lower temperature for the summer months as observed in the shade, and strengthen the credibility of the results of the observations of Dr. Daubeny, who appears to have proved that the ripening of fruits depends more on the illuminating rays than on the calorific and chemical rays of the sun.

A table illustrating the temperatures required for the production of wine of good and excellent quality at certain celebrated localities in Europe.

Locality.	Annual mean.	Summer mean.	Autumn mean.	Mean of hot month.	Hot months.	Remarks.
	°	°	°	°		
Dublin	50.0	59.6	53.0	61.0	July and August	Vine not cultivated.
London	50.0	62.0	51.0	63.2	July	Do.
Cherbourg	52.0	62.0	54.4	63.2		Do.
Kœnigsberg	43.2	60.6	44.4	62.6		Never ripens thoroughly.
Berlin	48.1	64.5	49.2	65.8	July	Wine, very poor.
Paris	51.3	64.5	52.2	65.6	July and August	Wine, not good.
Frankfort	50.3	65.0	50.0	66.0	July	Acid wines.
Würzburg	50.2	65.7	49.4	(?)	....do	Leistenberg wine, fine.
Vienna	51.0	69.4	51.2	70.0	....do	Better than Rhine wines.
Toulouse	55.2	69.1	56.5	71.5	August	Strong wines.
Dijon	53.0	69.6	53.3	72.5	....do	Fine Burgundy.
Bordeaux	57.0	71.0	57.9	73.1	July and August	Very good clarets.
Lisbon	61.4	70.0	62.5	72.0	....do	Very good.
Cadiz	62.0	70.4	65.3	72.8	August	Do.
Madeira	65.5	70.0	67.7	72.1	....do	Excellent.
Marseilles	58.3	73.0	59.2	76.0	July and August	Very strong wines.
Naples	60.3	74.4	61.4	76.3	....do	Super-excellent.

In the plains near the Baltic, in northern Germany, a wine is produced which is very acid and scarcely potable. The summer mean between 61° and 64.5° Fahrenheit, being too low, and even the temperature of the hottest month falling below the limit of 66.2° Fahrenheit.

The mean summer temperature of Würzburg, in Franconia, is 65.7°, but one degree above that of Berlin, while its hottest month cannot be more than two degrees higher than the hottest in the latter district; and when the different qualities of the wines produced in Franconia, and in the countries around the Baltic, are compared with the mean summer and autumn temperatures, respectively, we are almost surprised to find a difference of only about two degrees.

While the wines of Berlin are very inferior, and indeed unworthy of the name, those of Franconia, whose chief city is Würzburg, are held in high esteem. The Leistenwein, one of the kinds here produced, is esteemed with justice the second finest wine of the south of Germany, and the whole produce of the small space on which it is grown is secured for the table of the King of Bavaria, and is scarcely known, and can seldom be purchased. Doubtless there are local influences affecting the quality of the product, or which elevate the temperature above that of the surrounding region and assimilate it more closely to more southern districts, where wines stronger in alcohol are made. It is, however, acknowledged that the choicest variety of grape is grown, and the utmost care exercised in the selection of the most perfect berries, and that it is only by managing both the culture of the vine and the manufacture of its juice in the most skilful manner, and preserving the result until it shall have acquired the highest degree of perfection it can attain by age, and selling as the product of this celebrated vineyard only such vintages as are calculated to acquire or maintain its celebrity that a reputation has been obtained and preserved.

Even in the land of the vine all seasons are not alike propitious. An examination of a tabular view of the vintages of four of the most diverse and celebrated wine countries, extending from almost the western to the most eastern points, where famous wines are produced in Europe, exhibits the uncertainty attending the production of a good or fine vintage even in the favored land of the vine. This table extends from 1775 to 1842, and by it we find that for the sixty-eight years noted, that there were of the following varieties, viz:

Port.	Claret.	Rhenish.	Tokay. (1783 to 1831.)
23 years, very fine, fine, or good. 27 years, middling. 18 years, inferior, bad, or very bad.	9 years, first-rate, very good, and good. 59 years, either middling, bad, or none produced.	16 years, very good, or good. 13 years, middling. 39 years, inferior, or bad.	12 years, good, &c. 20 years, middling. 17 years, bad.

Seldom did three good or middling good years occur consecutively, while in the Rhenish provinces three, four, or five bad years have thus occurred. The expression "good" refers to quality only.

A reference to the Wine Chronicle for 432 years, from 1420 to 1852, exhibits the remarkable facts, that out of this long series of more than four centuries there were but eleven (11) years that were eminently distinguished for the superior quality of the wine produced; twenty-eight (28) very good years; one hundred and eighteen (118) pretty good ones, producing a good wine; seventy-six (76) that showed a middling quality of wine; and one hundred and ninety-nine (199) inferior. These four centuries and a third exhibit one hundred and fourteen (114) years of ample yield; eighteen (18) of middling product; ninety-nine (99) of poorer, and two hundred and one (201) failures, that did not pay the expenses of labor, &c. Thus the crop was seriously injured three years out of four. Think of this, ye complaining vignerons in the eastern sections of the United States, who are discouraged if but a few leaves drop from mildew; who talk of "rot," and "shrivel," and insects, as if they only were not the favored ones, and they only endured a climate where grape-growing cannot be made profitable. Here are wide districts of country whose staple crop and main dependence is the vine, which fails one-half of the time, and for three years out of four is seriously injured. How will the assertion that our country is not adapted to the growth of the vine be sustained, when we know that we have districts east of the Mississippi where it has not for eighteen years, probably the entire period since it was planted, failed to mature a good crop of grapes, and a wider district in California where no report of a season of failure has reached us, even by the mouth of tradition.

On a further examination of the preceding table of annual, summer, and other means, we observe that as we proceed towards the south, other circumstances being favorable, the temperature of the hottest month increases, and with it the quality of the wines. The wines produced in the region around Vienna, in Austria, are better in general than those of the Rhine; those of Burgundy have long been famous. These districts enjoy a month having a temperature of 70° and 72° Fahrenheit, respectively.

The vineyards near Bordeaux, which produce the highly prized clarets, enjoy a high summer mean of 71°, and two hot months of 73.1°, July and August, and a September of 67° Fahrenheit. The mean of June is above the minimum of 66.2°, being nearly 70°. These are circumstances very favorable to the production of superior wine.

The wines of Languedoc, Provence, and Roussillon, in the southeastern part of France, are remarkable for fulness of body, but they want the fine odor or bouquet of the wines of the Rhine. Their summer temperature is the highest enjoyed by any districts of France—that at Marseilles exhibiting a summer mean of 73°, while July reaches nearly to 76°, and August to 72°.

Lisbon, Cadiz, Madeira, and Naples enjoy a very high measure of summer heat and a hot August, the most favorable for the development of saccharine matter, and the wines of the adjoining districts are celebrated the world over.

In the most northern point of the district where the grape is grown, as at Kœnigsberg, it ripens only in warm summers, and is deficient in sugar, containing only a glutinous muco-saccharine matter; while in the southern regions the sugar is actually crystallized, and the grape is devoid of those acids which are requisite in order that the wine may possess flavor and those other qualities which distinguish wine from a mixture of syrup and alcohol. These extreme limits of the wine region thus offer extreme conditions of the product dependent almost solely upon the relative degrees of summer heat or the temperature of July or August.

THE SELECTION OF VARIETIES AND THEIR ADAPTATION TO THEIR RESPECTIVE DISTRICTS.

In the wine countries of Europe the utmost regard is paid to the proper selection of varieties adapted to the respective districts, yet with all their art and long experience the years of failure outnumber those of success. On the Rhine, the Little Riessling *(Der Kleine Riessling) Vitis vinifera pusilla*, of Babo and Metzger, is generally cultivated. This variety produces the famous Johannisberger, as well as many kinds of wine much inferior. The vineyard of Prince Metternich, to whom the Johannisberg belongs, is protected by the castle wall and a stone wall ten feet high, which occupied ten years in building. This greatly promotes the steady progress towards maturity by securing a quiescent state of the air, which is known to be extremely beneficial. The wine of Lugisland and the Liebfrauenmilch owe their superiority over that of the neighboring vineyard to the protection afforded by the town wall of Worms. The advantages of protection against agitation of the air are well understood in the Rheingau, or the "Rhine country," a valley of Nassau, celebrated for its rich vineyards, and the belts of vineyards which clothe the height of Hockheim produce very different wines, according to their position. One morgen, about one acre, close to the bed of the river Main, brings in the market two thousand florins; a higher morgen brings one thousand florins, and one at the summit only five hundred. The difference between Leistenwein and Steinwein, both the produce of the banks of the Main, is explained by the fact that different kinds of grapes are grown in the two vineyards; but why these two wines should differ from all others in the district has not been satisfactorily ascertained.

While we would not advise a close imitation of many practices common in Germany and France, such as laborious and expensive terracing the precipitous heights of rocky steeps and planting vines in baskets holding the earth upon ledges among the rocks, as we have seen in many instances along the Rhine, nor loading the backs of women with baskets of manure with which to climb the hills called the Coté d'or, in Burgundy, as has been our painful lot to witness a common practice, we may yet learn much from the centuries of experience gathered by the observant and laborious vine-dressers of German-land.

Among other points worthy of attention, let us not fail to remark that the selection of varieties of the vine is always made with special reference to the locality. Those that will bear a lower degree of heat and ripen are selected for the more northern latitudes and exposures; and as we progress towards the south, towards warmer summers and longer continued heats of early autumn, more tender varieties are grown. We have said that the Kleine Riessling is cultivated on the Rhine, and that from this the finest wines are made. There are other varieties in this section of nearly equal merit, as the Traminer *(Vitis vinifera Tyrolensis)* and the Elbling grape, *(Vitis vinifera albuelis ;)* the white Traminer, called Franken, by some Gutedel, *(Vitis vinifera amœna ;)* and, finally, the Hermitage, brought from France. Vineyards planted with the first two varieties are generally superior to those which contain the Guetdel and Elbling. Rulænder, Black Clævner, and Sylvaner are grown either in inferior localities or for their early ripening and productiveness.

In the Burgundy district, on the Coté d'or, or Golden Hills, which extend upwards of eighty miles from N. NE. to S. SW., and are about 200 to 300 feet high, the variety of grape grown is the "Pineau," small and black, with a small bunch and a light product per vine, and from 8 to 12 barrels per acre. They also grow the "Gamai" grape, which much resembles the "Pineau," but the bunches and berries are larger and the product three times as much as that of the "Pineau," but a much inferior wine.

The wines of the Medoc, the clarets of St. Julien, Margaux, Lafitte, &c., are made from the grape called "Cabernet" principally.

The sweet wines from the south of France termed Frontignac, Lunel, and Riversaltes, are from the muscat grape, which on the Rhine will ripen only just sufficiently to furnish a grape for the dessert. Further south in France, as we approach the shores of the Mediterranean, we find the vine flourishing and displaying its choicest fruits under circumstances which it would not endure in the departments of the north. Instead of a methodical arrangement of stakes and the close-cut appearance of a plantation of hops on poles shortened to half their usual height, which the vineyards of the north most nearly resemble, we find in the south the long branches of the vine ranging over the dusky green olive trees, or on high espaliers, or intertwining its shoots with the almond or the elm.

The wines of Spain and Italy possess qualities derived from the increased heat of their more southern latitude. In Spain the "Pedro Ximenes" is the most commonly esteemed. The wines of Spain are mostly rich and sweet, at least those preferred by the people, such as Malaga and Alicant; while they export those which contain more alcohol to foreign consumers. Among the the latter is Sherry, which is made in Andalusia, near Cadiz, on the west coast of Spain, on the hot and chalky hillsides of this district.

Finally, the principal wine grown near Naples is the "Lacrimæ Christi," a luscious wine which holds a place in the foremost rank of the first class produced by any country; and truly can we indorse the opinion, if a judgment formed by thirsty climbers up the jagged cinders of the cone of Vesuvius, upon whose summit we have imbibed draughts of this peerless vintage grown on the warm volcanic ashes of the mountain side, can be deemed sufficiently disinterested.

We have seen from the foregoing notices of the popular varieties of European wine grapes, that certain kinds are adapted to certain districts of the wine climate or zone in Europe, where Riessling, Traminer and Rulænder, Pineau and Gamai, Cabernet, &c., may be the representatives of our Delaware, Diana, Clinton, Concord, our Norton's Virginia, and Catawba, and many others, each of which finds its appropriate location in a district whose warmest month occurs at the period of growth best fitted to develop the sugar, and whose autumn is sufficiently long for the perfection of the fruit. We have seen that by the table of comparative temperatures for summer, and the hottest month, that those places in the wine regions on the north, having a mean in July below 66° Fahrenheit, produce no good wine; that as the temperature for July rises, and the highest heat occurs in August, as we proceed south, the wines, as those of Burgundy and Lisbon, improve in quality until we reach the highest temperature in July and August near Naples, and with it the highest excellence.

This hasty sketch of the chief varieties of grapes grown for the production of the best wines of Germany, France, Spain, and Italy, with notices of the climatic peculiarities adapted to each class, has been made for the purpose of impressing more strongly upon our incipient wine-producers, who seem bent upon diverting this fruit from its true purpose, the dessert and the cuisine, to the production of a fluid of questionable utility in its more diluted alcoholic condition, certainly injurious in its stronger states—the importance of selecting each for his own district those kinds only which are adapted by constitution to the temperature and other peculiarities of each zone or grape region of our Atlantic States. That such zones exist has already been made sufficiently clear to those who have brought an intelligent understanding of the subject to the study of an article on the "Climatology of American grape vines" in the report of the Department of Agriculture for 1862, to which the reader is referred for further information.

Different climates impress upon the grape peculiarities easily distinguished in the wines produced by the same kind of grape grown under different influences. Thus the German Hock grapes yield a wine possessed of distinct qualities when grown along the Main or the Rhine. The same sort of grapes grown near Lisbon yield the Bucellas wine, which retains some of the peculiarities of the original, while the same grapes grown at the Cape of Good Hope yield Cape Hock bearing scarcely any resemblance to the true Rhenish; and the Sercial of Madeira, produced by the same sort of grapes, though a delicious wine, has scarcely any qualities, except durability, like that of the original.

The experience of the European vigneron of the effects of climate in modifying the quality of grapes and their product is paralleled in our own country. In the above we may find an explanation of widely diverse views of growers in different sections, and the varying estimates placed by them upon the value of Concord and other American varieties.

A cultivator on the banks of the Hudson denounces the Concord grape as having a thick, acid pulp, repulsive, "fit only to sell to those who do not know it," and exhausting his enthusiasm in raptures over the "Delaware," finds his opinions indorsed by the experienced editor of a horticultural journal; while an extensive grower in Missouri asserts that the Concord grown in his grounds is almost without pulp, of fine flavor, thin-skinned, sweet, and very good, scarcely inferior to the Delaware, which he cannot succeed in growing perfectly. Were the influences of climate and location properly regarded and allowance made for the increased temperature of western localities, even in the same latitude during the ripening season, such wide diversity of opinion would be readily understood and acknowledged to be honestly entertained, and disgraceful personalities be laid aside. The genial sun of the Mississippi valley dissolves the pulp of the Concord grape, and sweetens and matures its juices beyond anything that the summers in the valley of the Hudson can exhibit, rendering it decidedly better than the Catáwba, while it is much more productive, healthy, and hardy. As the cultivation of the Concord is extended further south on the Atlantic slope, its qualities improve from the increased heat of summer in lower latitudes.

A comparison of temperatures in the valley of the Hudson may be made with that at several points in the valley of the Mississippi, in Iowa, Illinois, and Missouri, by means of the following tables. Let West Point, in the valley of the Hudson, latitude 41° 23′ north, where observations have been made for upwards of 31 years, be compared, as regards the mean temperatures for the months during which vegetation is active, with old Fort Armstrong, Rock Island, Illinois, latitude 41° 30′ north, where observations have been made for upwards of eleven years.

	March.	April.	May.	June.	July.	August.	September.	October.	Summer.	Autumn.
	°	°	°	°	°	°	°	°	°	°
West Point	37.63	48.70	59.82	68.41	73.75	71.83	64.31	53.04	71.33	53.19
Fort Armstrong	37.82	51.06	62.70	71.39	76.49	74.47	63.90	52.26	74.12	51.73

West Point and Fort Armstrong are in nearly the same latitude, and yet we observe the mean temperatures through the months of April, May, June, July, and August are about three degrees higher at the latter than at the former place. This excess of summer heat in the valley of the Mississippi becomes greater when we compare the valley of the Hudson with points in Illinois and Missouri further south. Such a comparison must be instituted in order to render more plainly apparent the cause of the discrepancy between the grape-growers of diverse opinions residing at the respective localities, whose summer temperatures we propose to compare. Taking West Point again as a representative of the valley of the Hudson, we have for—

	March.	April.	May.	June.	July.	August.	September.	October.	Summer.	Autumn.
	°	°	°	°	°	°	°	°	°	°
West Point, New York	37.63	48.70	59.82	68.41	73.75	71.83	64.31	53.04	71.30	53.20
Fort Madison, Iowa	38.20	56.80	65.30	74.20	82.50	77.50	71.50	54.10	78.10	55.30
Athens, Illinois	39.60	57.90	65.60	71.80	79.40	77.40	73.00	56.00	76.20	57.20
St. Louis, Missouri	44.40	58.30	66.40	74.00	78.50	76.50	68.70	55.40	76.30	55.00
Highland, Illinois	....	....	....	....	....	....	....	....	77.90	56.80

Here we may observe a difference of from 6 to 9 degrees of temperature in favor of western localities named over those of West Point from April to September, and that while the March mean is somewhat higher in the west, that for April is much greater, thus starting vegetation earlier and urging it more rapidly. This excess is continued through the summer until it is at its height in July; and, while West Point has a mean of but 64.31° for September, the places named in Illinois and Missouri enjoy a mean heat 4 to 9 degrees higher, or that of 68.7° to 73° Fahrenheit, a temperature adequate to the ripening of late maturing grapes for which that of West Point is not fitted.

In order more clearly to illustrate the comparative temperatures of the wine districts of Illinois and Missouri, and those of France, Spain, and Italy, we append a table of mean temperatures for sundry places of observation in the latter named countries :

	March.	April.	May.	June.	July.	August.	September.	October.	Summer.	Autumn.
	°	°	°	°	°	°	°	°	°	°
Toulouse	46.8	53.5	61.0	66.0	70.1	71.1	65.2	56.4	69.1	56.5
Dijon	48.2	51.1	60.6	66.0	70.2	72.5	62.4	53.8	69.6	53.3
Bordeaux	51.3	56.1	60.8	66.9	73.1	73.2	67.1	58.1	71.1	57.9
Lisbon	56.3	59.0	63.6	69.4	72.1	71.2	69.4	62.6	70.9	62.5
Cadiz	55.2	59.6	63.7	68.1	70.2	72.8	70.2	67.1	70.4	65.3
Madeira	62.6	63.5	64.6	67.2	70.1	72.1	71.6	67.1	69.9	67.7
Marseilles	48.4	56.1	63.2	71.0	75.9	71.9	68.7	58.7	72.9	59.2
Naples	51.2	56.7	64.8	70.8	76.1	76.3	69.3	61.9	74.4	61.4

On comparing the above table with that of the means for several places in the Mississippi valley, on page 485, it will be observed that the temperatures for March are much higher than those observed in Illinois and Missouri; those for April and May, at Bordeaux, Marseilles, and Naples, more clearly accord with those of our western wine region. June, July, and August, in the Mississippi valley, are much hotter than the hottest wine districts of Europe in nearly the same latitude; and September exhibits a parallel nearly approaching the European means, while October in the west has fallen sadly away from the high measures known throughout the wine region bordering the Mediterranean, reducing the autumn mean and shortening the season of maturation. The latter mean does not appear to be higher than that of Bordeaux, and much lower than in the more favored localities where the strongest wines are produced; but as September is still very warm and adequate to the ripening of many tender varieties of grapes requiring a long season, the wine regions of Illinois and Missouri may be esteemed comparatively well adapted to the production, in favorable situations, of wines of very good quality.

REASONS WHY THE EUROPEAN WINE GRAPE CANNOT SUCCEED IN THE EASTERN UNITED STATES.

De Candolle, assuming that the vegetation of the vine commences at 10° centigrade or 50° Fahrenheit, and employing this minimum as the limit to vegetation both in spring and autumn, finds, by comparing the aggregate temperatures enjoyed at several places during the existence of this term and upwards, both within and without the region of successful vine culture, that it may be conducted with good promise of remuneration in Europe on slopes well exposed, and at those localities which exhibit the sum of 2900° centigrade, from the day when growth commences with the mean of 10° centigrade in the shade to that on which this mean ceases, provided the number of days of rain does not exceed twelve days per month. An advantage is found to arise in taking account of the number of days of rain rather than of the quantity, for we include in this manner a very important circumstance, the mean condition of the sky as more or less cloudy. As the number of rainy days increases, the heat received by the vine will be diminished, and the chemical influence of the direct rays of the sun be much impaired. Thus this varying element, which is neglected in making up the sums of temperature in the shade, and which it is impossible to estimate at many localities, is recognized in another and definite form by stating the number of days of rain. Unfortunately our meteorological records do not furnish us with these desirable facts.

The reasoning which appears satisfactory as regards the influence of heat and number of rainy days in the European wine districts does not as fully apply to the cultivation of the grape in the United States east of the Mississippi. DeCandolle has sought for the causes which have prevented the successful culture of the grape in the middle regions of the United States, and has failed to discover the reason why it has been abandoned and its place supplied by native varieties. He concludes that there is nothing either in the means or the sums of the temperatures or the number of days of rain that can be shown to be injurious to the vegetation of the vine, but suggests that in abundant rains, especially at the period of blooming, which are known in Europe at those places where the vine has not prospered, combined with other circumstances which have eluded his research, such, perhaps, as diurnal variations of temperature or the humidity of the soil, we may find the true obstacle to the successful culture of the European wine grape in the United States.

It is doubtless to the variable quantity of atmospheric humidity which characterizes the climate of the eastern United States that we ought to attribute the failure of attempts to cultivate the wine grape. Periods of excessive heat and saturation continuing from two to eight days are common in the American summer, but are rare in Europe. The maximum of 100° in the shade is frequently observed in Canada, and south of Boston the mercury may be confidently expected to make one or two visits to the region of 95° and 98° at one or more times during "the heated term." This high heat of 95° is sometimes attended by saturation with moisture, and at such times becomes destructive to human life by causing exhaustion or depression of the vital powers through the joint action of heat and humidity. When the highest measure of 100° to 104° which have been observed occur, when the air is very dry, equally injurious consequences must follow to tender vegetation from the destructive drying of the tissues. In the eastern United States there often occur periods of several days which are burdened with excessive moisture alternated with periods as marked by excessive dryness to which we find no parallel in the wine districts of Europe. Our average of atmospheric humidity is, however, very low, as is indicated by the prevailing forms of vegetation. Our forests do not abound in mosses except in elevated districts. The grasses furnish evidence of aridity which cannot be readily overlooked, and fail to cover the earth with perennial verdure, as in the moister climate of England and western Europe. South of the parallel of 38° north latitude the introduced grasses cannot be cultivated.

Though the uplands of Georgia and the interior northward to Pennsylvania, districts in which the cultivation of the European wine grape has been attempted without success, have many points of resemblance with those of France as respects capacity for cultivation, in the extremes of dryness and humidity they present no features in common. It is to these extremes we must ascribe this wide disparity, for in our country it may be said to be alternately too dry or too wet, too warm or too cold, to conform to the corresponding periods of the life of the vine as experienced in France.*

It is exceedingly difficult to obtain observations which may be properly compared because of the different local conditions of the experiments. A decisive mode of comparing the humidity present in the atmosphere of America and England may be found by measuring the quantity of water naturally evaporated from an exposed surface. The following will serve to show the difference between the evaporating powers of the air in England and in the United States:*

---

*Blodget's Climatology of the United States, pp. 226, 229.

Quantities of water evaporated in inches of depth.

	January.	February.	March.	April.	May.	June.	July.	August.	September.	October.	November.	December.	Year.
Whitehaven, English mean of six years..	0.88	1.04	1.77	2.54	4.15	4.54	4.20	3.40	3.12	1.93	1.32	1.09	30.03
Ogdensburg, N. Y., one year	1.65	0.82	2.07	1.63	7.10	6.74	7.79	5.41	7.40	3.95	3.66	1.15	49.37
Syracuse, N. Y., one year	0.67	1.48	2.24	3.42	7.31	7.60	9.08	6.85	5.33	3.02	1.33	1.86	50.20

The first series of quantities above given for Whitehaven, in one of the most humid districts of England, was carefully observed from 1843 to 1848 by evaporating from a shallow copper vessel one and a half inch deep, filled daily and protected from the rain. The annual fall of rain at this place was 42.25 inches, and the amount evaporated was in excess of the rain fall; but generally much less is evaporated than falls each year. In the United States the evaporation is always much greater from a vessel thus supplied and protected from rain than the actual depth of rain precipitated, and we may assume the evaporation from a reservoir surface, when compared with the quantity of rain to be, during the summer months, as two to one, or twice as great. The absorbent power or capacity of the air of England is thus strongly contrasted with that of the United States, the air of the former, under similar conditions, taking up but half the quantity evaporated in the latter. And though the case cited may be an extreme one, and not fully illustrative of the conditions of atmospheric capacity for moisture as it exists in France and Germany, yet the difference will be found one of degree merely, and the average dryness of the American air will prove still largely in excess of that of western Europe.

That it is to excessive atmospheric humidity, alternating with aridity, and both combined with high temperatures, that we must ascribe our failure to cultivate the European wine grape, is attested by the success with which our American varieties are grown under circumstances which combine proper temperature with more uniform humidity, and our ill success when we attempt the extensive cultivation of even our native sorts in localities which are subject to the above extremes. In the districts where the Isabella and Catawba are largely and profitably grown, the vineyards are either surrounded by water or so influenced thereby as to enjoy a comparatively humid atmosphere combined with more moderate summer maxima and long continued autumn heat. Thus the Croton Point vineyards, of Dr. R. T. Underhill, are nearly surrounded by water. The vineyards on Kelley's island, Lake Erie, where the Isabella and Catawba of superior quality are grown very extensively, and with almost uniform success from year to year, are directly under the influence of the lake waters, whose equalizing effects upon the humidity of the air upon its borders must be as marked as is that it exerts upon the mean temperature of summer and its lengthened autumn.*

* An illustration of the influence of the lake in regulating the phases of the vegetation in the region round about may be found in the following, furnished by George C. Huntington, an intelligent meteorologist and vine culturist, of Kelley's island. He has observed that the Isabella and Catawba grapes are uniformly in full bloom on the 20th of June, and accounts for this remarkable uniformity by ascribing it to the influence of the lake, the temperature of whose waters scarcely vary at this season, having, on the 30th of June, exhibited a mean of sixty-nine degrees, Fahrenheit, through a series of five years.

THE TEMPERATURES REQUIRED BY THE AMERICAN GRAPE-VINE.

The following table will exhibit the time of leafing, blooming, and ripening of twelve of the new varieties of native American grapes at Waterloo, New York, latitude 42° 55′ north, longitude 76° 50′ west, in the summer and autumn of 1862, which appears to have been a favorable season, unattended by any unusually unpropitious circumstances. This table, also, shows the number of days from leafing to blooming, the average temperature of each period, and the sum of the mean temperatures for the same, derived from a neighboring meteorological station; also, the number of days between blooming and maturity, the actual average temperature, the sum of the mean temperatures for the same, total number of days from leafing to ripening, average temperature of the entire epoch of growth and aggregate of heat required to perfect the fruit. We have superadded a column of time of stoning, based upon the time of coloring, dating three weeks anterior to the latter appearance, which is very nearly correct. From the latter epoch we have calculated the temperature for the month following, from actual observations, and have appended remarks upon the quality of the product.

The attentive observer of this table will find much therein to interest and instruct. He will remark the order in which the varieties are arranged is not based upon their respective early leafing or blooming, but upon the order of ripening and the aggregate amount of heat required for the entire cycle of growth; that while the average temperature for the term from leafing to blooming remains about 59°, the number of days required by the different kinds varies but slightly, except in the Clinton and Isabella. The amount of heat received during this stage ranges from 1893.33° in the Clinton to 2678.66° in the Delaware; also, that the period comprised between blooming and ripening varies much more widely, from 77 days, required by the Delaware, to 111 days, which were inadequate to the maturation of the Anna. He will perceive that the varieties first in order enjoy during this period the favorable temperature of about 68°, which appears adequate to their ripening, and that the sum of heat received during this period is generally much less than that required by those lower on the list, which do not indeed properly mature, and whose mean temperature is lower because of the prolongation of their period of growth into a cooler season.

Proceeding further, we note the total number of days required from leafing to maturity to vary from 122 days in the Delaware and Hartford Prolific to 154 and beyond in Anna, which did not ripen. The average temperature for the entire period of growth varied but little, but the entire amounts of heat are widely different. Finally, that those which attained the highest perfection enjoyed the highest heat during the month following the commencement of the stoning process, and that those whose temperature at this time fell to 67° or thereabouts were of but indifferent quality, and that those which were below 66° did not ripen at all.

Dates of leafing, blooming, and ripening of sundry varieties of native grape-vines, as observed at Waterloo, New York, in 1862, *by F. C. Brehm, and reported in "The Horticulturist," volume XVIII, pages* 52 *and* 53, *with average temperatures and aggregates of heat required during each period of growth, derived from daily and monthly means reduced from observations made at a neighboring station.*

	Temperature at date of leafing.	Date of leafing.	Days from leafing to blooming.	Date of blooming.	Average temp. from leafing to blooming.	Sum of heat required to blooming.	Date of stoning, (calculated.)	Temp. of month of stoning process.	Days from bloom to ripening.	Date of ripening.	Average temp. from bloom to ripening.	Sum of heat required from bloom to ripening.	Number of days from leafing to ripening.	Average temp. of entire period from leafing to maturity.	Aggregate heat for entire period.	Quality, &c.
	°				°	°		°			°	°		°	°	
Delaware	52.41	May 14	45	June 28	59.52	2678.66	Aug. 1	69.00	77	Sept. 13	68.17	5249.00	122	64.98	7927.66	First quality.
H. Prolific	52.41	May 14	44	June 27	59.03	2607.66	Aug. 9	67.14	78	Sept. 13	68.32	5319.66	122	64.98	7927.66	Ordinary.
U. Village	51.66	May 12	45	June 26	58.95	2653.00	July 21	68.58	80	Sept. 14	67.95	5436.66	125	64.71	8089.66	Good.
Clinton	51.66	May 12	32	June 13	59.16	1893.12	July 24	69.43	95	Sept. 16	66.52	6319.66	127	64.67	8213.00	Good for wine.
Diana	51.40	May 10	44	June 23	58.66	2581.66	Aug. 1	69.00	89	Sept. 20	67.31	5990.66	133	64.45	8572.66	Next to Delaware.
Concord	52.71	May 15	42	June 26	59.13	2483.66	Aug. 9	67.14	91	Sept. 25	67.30	6124.66	133	64.72	8607.66	Poor, foxy.
Isabella*	51.40	May 10	34	June 13	58.73	1999.00	July 24	69.43	104	Sept. 25	66.19	6884.00	138	64.37	8883.00	Good.
Rebecca	53.13	May 16	42	June 27	59.19	2486.00	July 24	69.43	96	Oct. 1	66.91	6414.66	138	64.49	8900.66	Best, (poor bearer.)
Y. Madeira	52.41	May 14	44	June 27	59.26	2607.66	Aug. 8	67.14	96	Oct. 1	66.91	6414.66	140	64.44	9022.33	Unfit for table.
Catawba	52.71	May 15	42	June 26	59.13	2483.66	Aug. 15	65.62	100	Oct. 4	66.65	6665.66	142	64.43	9149.33	Good, but unripe.
To Kalon	52.31	May 13	44	June 26	58.97	2594.66	Aug. 8	67.13	100	Oct. 4	66.65	6665.66	144	64.30	9260.33	Good, liable to rot.
Anna	52.71	May 15	43+	June 27	59.32	2551.00	Sept. 5	50.61	111+	....	65.58	7280. +	154+	....	9831. +	Unripe October 16.

* The Isabella, above observed, was in a sheltered position, and leafed and bloomed earlier than it would otherwise have done. In the open vineyard it appears to require the same initial heat as does the Catawba, and leafs and blooms on the same days.

The only anomaly apparent in the table respects the date of blooming and ripening of the Isabella. The first by comparison with several warmer localities is antedated, perhaps, one week, and the date of maturity may safely be assumed as quite two weeks too early. Dr. Underhill, of Croton Point vineyards, on the Hudson, informs us that the Isabella blooms from the 14th to the 25th of June, and though the fruit is gathered for market from the 14th to the 25th of September, it continues to ripen for five or six weeks. At Kelley's island, Lake Erie, another celebrated locality where the Isabella is successfully cultivated and ripened, as we are informed by George C. Huntington, the Isabella and Catawba almost uniformly bloom on the 20th of June, and that the first does not mature even under a higher heat than that known at Waterloo, New York, until the 15th of October, and that they improve even until the 15th of November.* Both Croton Point and Kelley's island are more favorably situated for early blooming and ripening than is Waterloo, and yet at the former the Isabella blooms and ripens later. The observed temperature is not, therefore, in this instance, a correct index to the wants of the Isabella, because the vine observed was exposed to reflected heat, which raised its temperature 10°, and assimilated its surroundings to those of more southern localities. As an evidence of the value of shelter, this illustration is worthy of remark. The fruit is reported perfectly ripe under the observed temperature, which could not have matured it without the increment derived from shelter.

Observations made on the vine at Brussels show that it pushed its first leaves at a mean date of April 25 for a series of years from 1841 to 1850, and that during this term the heat of the atmosphere in the shade was a mean of 10.25° centigrade, or 50.45° Fahrenheit. This mean was preceded by a sum of temperature which may be esteemed uninfluential, that for March having been almost as low as 6° centigrade, or 42.80° Fahrenheit.

The minimum temperature at which the vine commences foliation, according to Count de Gasparin, is 9.5° centigrade, or 49.10° Fahrenheit. Boussingault admits that the 1st of April may be considered the point of departure for the commencement of the growth of the vine in the department of the Lower Rhine, where his experiments were made, and that in order to favor its growth this period should have a mean temperature of 7.5° centigrade, or 45.50° Fahrenheit, according to observations made at Strasburg. De Candolle believes that the minimum temperature at the time of commencing vegetation in the spring may be that of 8° centigrade, or 46.40° Fahrenheit. He, however, inclines to the belief that 10° centigrade, or 50° Fahrenheit may be considered the temperature at which the vegetation of the vine properly commences, and such we find to be the case in our own experience with our native varieties. At this date (April 24, 1864) the buds of the Concord vines are much swollen; some of them have opened, and several other varieties are in various stages of enlargement, but without showing the leaf. The daily mean temperature for the past seven days has been 11.5° centigrade, or 53.76° Fahrenheit, and for the past two weeks 10.25° centigrade, or 50.42° Fahrenheit, and yesterday the peach blossoms expanded, and to-day the first bloom of the pear appeared

* At Kelley's island Isabella and Catawba leaf May 5 and 6, respectively, bloom June 20, ripen October 15 and November 1, respectively. By known mean temperatures and aggregates of heat calculated therefrom we learn that the Isabella requires 10,639 degrees, and the Catawba 11,307 degrees, Fahrenheit, from leafing to ripening. These are evidently the sums of heat required in the open vineyard, and are proved to be perfectly in accordance with the results of observation made at Croton Point and Poughkeepsie, New York, and at Haddonfield, New Jersey.

Since the above was written, F. C. Brehm, of Waterloo, informs that he has destroyed nearly every Isabella in his vineyard, as they do not succeed on his rich soil. He also informs that the Diana ripens two weeks earlier than the generality of Isabellas, by which we may fix the date of ripening of the Isabella in the open vineyard at Waterloo at October 4 and thereafter, which is more in accordance with experiments elsewhere.

The mean temperature of the first week of April has been 6.3° centigrade, or 43.30° Fahrenheit; that of the second week of April, 8.3° centigrade, or 46.99° Fahrenheit; and of the third week, 9.15° centigrade, or 48.49° Fahrenheit—numbers which agree very closely with those observed in France and Belgium, as demanded by the vine before and at the commencement of vegetation.

An examination of the table of leafing, blooming, and ripening of sundry American vines, as observed at Waterloo, New York, favorably situated a few miles distant from Cayuga lake, exhibits the fact that the Diana and Clinton began the leafing process at the temperature of 51.40° to 51.66° Fahrenheit for ten to twelve days previous, or from the 1st of May, and that the other varieties opened their leaves under a mean of from 52.41° to 53.13°, beginning from the same date. The leaves did not, however, expand until a day had been experienced having a mean temperature of 60° or upwards; and the same has been our observation to-day, when the first leaves of the Clinton appeared.

De Candolle, in order to determine the possibility of vine culture at various places, employs a table of aggregate temperatures derived from the date when the daily means of 8° and 10° centigrade above freezing first appear, respectively, until the day when the same mean heats are last experienced in autumn. This is based upon the assumption that the entire term is required for the maturation of the various kinds of grapes that will grow and ripen in the latitude noted. Such has not been observed to be the fact in America. Several varieties of native vines ripen their fruit in August or early in September, when the high summer means prevail, and this even near the northern border of grape culture, as we find by the foregoing table. Several mature from the middle to the last of September, while the temperature of this month at Waterloo, New York, for 1862, was 63.04°, being 13° above the mean at which vegetation commenced, and which mean was not reached until the middle of October, when the first frost occurred.

By deducting the average temperature during the period from leafing to blooming from the mean temperature at and prior to leafing, we obtain the available heat necessary to advance the plant to the era of bloom. Taking the Delaware as an example, we find that it commences vegetating at about 52.41° Fahrenheit for 14 days previous, and blooms at a mean of 59.52° Fahrenheit, a difference of 7.11° for 45 days, (the time required,) giving us an average of 319.95° available degrees of heat, according to Fahrenheit's scale, necessary to conduct the Delaware vine to this condition. The average temperature from blooming to maturity is 68.17°, and the difference between this and the initial temperature of 52.41° is 15.76°, which continued for 77 days, the time required, gives us an aggregate from blooming to maturity of 1191.52° Fahrenheit. The latter sum, added to that required from leafing to blooming, 319.95°, exhibits a final aggregate of heat demanded by the Delaware above 52.41° of 1533°.47. This is equal to a mean of 12.57° above the initial for 122 days, the entire period required, or a mean temperature for the season of growth of 64.98°, or 65° nearly.

Waterloo, New York, appears to be very favorably situated for vine-growing, though near the northern border of successful cultivation. Its summer temperature being 66.05°, and the mean for the season from foliation to ripening of the earlier varieties 64.98°, and for those which require a longer season, and yet mature there, a mean of 64.37°. These means accord very closely with those determined as necessary in Europe for the growth and maturation of the hardier varieties of the wine grape; for, says Boussingault, "any country which has not a summer mean of at least 64.4° to 66.2° Fahrenheit, can have but indifferent vineyards." An indispensable requisite to perfect maturity is adequate heat during the later stages, and this is received at Waterloo in September, which enjoys a mean heat of 63° Fahrenheit. But the most important epoch in the life of the vine is that which follows the appearance of the seeds, and at this period it is absolutely necessary that there should exist a month, the mean temperature of which does not fall below 66.2° Fahrenheit, a fact sufficiently demonstrated by the table of European localities, where failure or success attend attempts to cultivate the wine grape at certain places, respectively, as set forth on page 480 of this volume. These results are again confirmed at Waterloo, New York, where, by examination of our table on page 490, the natural product of calculations from the data in our possession, without effort to force an agreement, we find that the two varieties which did not ripen at Waterloo were the Catawba and the Anna, whose month of stoning fell below 66.2° Fahrenheit, or to 65.62°, and 50.61°, or to at least 62°, as shown by another calculation, respectively.

By judicious application of the principles deduced from the foregoing, the extent of a growing season may be readily, and, perhaps, accurately defined, and all those varieties of grapes and other fruits whose demands for heat fall within the limits designated, other circumstances being favorable, be successfully grown. Tables of aggregate temperatures might be constructed either on the principle observed in the table of leafing, blooming, and ripening, or on that noticed above as employed by De Candolle, which would serve as valuable guides in practice.

GENERAL RULES FOR DETERMINING THE FITNESS OF A DISTRICT IN THE UNITED STATES FOR THE GROWTH OF CERTAIN VARIETIES OF VINES.

The following rules derived from a close study of the requirements of the various native grapes will be found very useful to many inquirers who are at a loss to know the fitness of their district for their cultivation, and in connexion with the following tables may serve instead of an isothermal and hyëtographic or rain map prepared to accompany a former article on the "Climatology of American Vines," but not published.

1. Those places which have a summer temperature of 66.5°, a hot month of 70°, and a September of 60°, will ripen the Delaware, Clinton, Perkins, Logan, King, and some other very hardy varieties. The temperature of their growing season corresponds to a mean of 65° and upwards, and an aggregate of heat of about 8,000° Fahrenheit. This district includes many parts of New England and New York, northern Pennsylvania, northern Michigan, Wisconsin, and Iowa.

2. Those places which have a summer of 70°, a hot month of 72°, and a September of 63°, will ripen the Concord, Hartford Prolific, Diana, Crevelling, &c. Their season of growth corresponds to a mean of 67°, and an aggregate of 8,500° and upwards. This district covers part of the southeast and south coast of New England, valleys of Hudson and Mohawk, neighborhood of the minor lakes in western New York, southern border of Lake Ontario, southern Michigan, southern Wisconsin, &c.

3. Those places which have a summer of 72°, a hot month of 73°, and a September of 65°, will ripen the Isabella. Their growing season corresponds to a mean of 70°, and an aggregate of 10,000° of heat. They are not found in the State of New York, except in the southeast extremity, lower valley of the Hudson, and near some of the minor lakes, but appear on the southern border of Lake Erie, in northern Indiana and northern Illinois.

4. Those places which enjoy a summer mean of 73°, a hot month of 75°, and a September of 65°, will ripen the Catawba. Their growing season corresponds to a mean of 72°, and an aggregate of 11,000°. They are not found north of New York city and vicinity, or the southeastern counties of Pennsylvania, middle New Jersey, or southern Ohio, Indiana, Illinois, and Missouri.

5. Those places which bask under a glowing summer of 74°, a hot month of 75°, and a September of 75°, as at Los Angeles, in California, other circumstances being favorable, may ripen the most tender European wine grapes to perfection.

The exceptions to the above rules are found at places influenced by water, whose high or moderate September mean may extend into October without intervening frost.

By consulting the following tables almost any one interested may determine for himself what varieties of grapes he may grow on a small scale. It would not be wise to enter upon the culture without more minute information and larger study respecting the fitness of a locality for grape-growing. Some places are too elevated that otherwise by position might be eligible. The extensive tables of temperature accompanying the very valuable bi-monthly reports of the Agricultural Department will assist some inquirers in determining the mean temperature of their summer, their hot month, and of September, though not as reliable as the means for many years. These reports have been widely disseminated, and can or should be readily accessible in almost every county in the land. The material for the construction of the following tables have been derived from Blodget's extensive list of temperatures in his "Climatology of the United States," and "Results of Meteorological Observations, 1854 to 1859," a report of the Commissioner of Patents, &c.:

Places which are beyond the limit of vine-growing; the normal temperature required for ripening being, for summer, 66°.5; *for hottest month,* 70°; *and for September,* 60° *Fahrenheit.*

Locality.	Summer mean.	July mean.	August mean.	September mean.	Length of observation.
	°	°	°	°	
Charlotte, Prince Edward's island	66. 1	70. 5	67. 7	59. 5	One year.
Quebec, Canada	65. 3	66. 8	65. 5	56. 2	Ten years.
Fort Kent, Maine	61. 7	62. 5	63. 5	51. 6	Four years.
Eastport, Maine	60. 5	62. 3	62. 4	57. 3	Twenty-five years.
Castine, Maine	62. 0	64. 8	64. 7	58. 4	Forty years.
Bath, Maine	64. 8	68. 5	64. 6	59. 2	Ten and a half years.
Dover, New Hampshire	66. 9	70. 1	66. 7	58. 8	Do.
Hanover, New Hampshire	62. 8	64. 4	62. 3	55. 0	Three years.
Craftsbury, Vermont	63. 6	67. 2	62. 5	55. 8	Six years.
Princeton, Massachusetts	65. 8	70. 3	64. 9	58. 6	Two years.
Williamstown, Massachusetts	65. 3	68. 3	63. 4	58. 5	Five years.
Pomfret, Connecticut	66. 1	69. 2	65. 4	59. 4	Six years.
Cherry Valley, New York	65. 4	67. 0	65. 6	57. 8	Fifteen years.
Cazenovia, New York	66. 6	68. 7	62. 5	57. 5	Two years.
Utica, New York	66. 5	68. 5	66. 7	58. 4	Twenty-three years.
Lowville, New York	63. 7	67. 8	62. 1	56. 8	Two and a half years.
Potsdam, New York	66. 3	68. 4	66. 7	57. 4	Twenty-one years.
Penetanguishene, Canada	65. 2	70. 4	68. 5	53. 2	Two years.
Mackinac, Michigan	62. 0	64. 5	64. 1	55. 1	Twenty-four years.
Fort Brady, Michigan	62. 0	64. 9	62. 9	54. 6	Thirty-one years.
Fort Howard, Wisconsin	68. 5	71. 5	67. 9	57. 2	Four years.
Kenosha, Wisconsin	65. 3	68. 6	65. 7	59. 6	Three years.
Fort Ripley, Minnesota	64. 9	67. 3	64. 7	56. 7	Six years.
Sandy Lake, Minnesota	64. 8	67. 6	65. 5	58. 3	Two years.
Forest City, Minnesota	68. 9	71. 7	67. 8	56. 4	Do.
Fort Snelling, Minnesota	70. 6	73. 4	70. 1	58. 9	Thirty-five and a half years.
Steilacoom, Washington Territ'y	62. 9	64. 2	63. 8	57. 7	Five and two-thirds years.
Astoria, Washington Territory	61. 6	61. 6	63. 0	58. 7	One and one-sixth year.

Places near the northern border of vine culture.

Locality.	Summer mean.	July mean.	August mean.	September mean.	Length of observation.
	°	°	°	°	
Montreal, Canada East	70.8	73.1	70.8	60.6	Twenty-seven years.
Saco, Maine	68.8	70.8	69.4	60.9	Eight years.
Concord, New Hampshire	68.2	71.2	67.2	60.8	Five years.
Manchester, New Hampshire	70.2	74.7	68.6	60.9	Four years.
Burlington, Vermont	67.9	69.9	68.0	59.6	Twenty-one years.
Amherst, Massachusetts	68.6	71.0	69.0	60.0	Fourteen years.
Mendon, Massachusetts	68.4	71.9	68.7	61.0	Seventeen years.
Worcester, Massachusetts	68.3	71.2	67.7	60.8	Six years.
Providence, Rhode Island	68.1	70.6	68.7	60.9	Twenty-three years.
Auburn, New York	67.2	69.8	68.2	59.4	Twenty-two years.
Oswego, New York	66.1	69.8	67.3	61.0	Four years.
Rochester, New York	67.6	69.9	67.9	60.3	Twenty-four years.
Penn Yan, New York	66.7	70.8	66.5	61.4	Five years.
Detroit, Michigan	67.6	69.7	67.5	60.0	Thirteen years.
Milwaukie, Wisconsin	67.3	69.8	67.5	61.2	Eight years.
Chicago, Illinois	67.3	70.8	68.5	60.1	Five years.
Emerald Grove, Wisconsin	68.5	71.3	68.3	60.2	Three years.
St. Anthony's Falls, Minnesota	70.1	73.3	69.4	60.7	One year.
Princeton, Minnesota	71.1	75.7	69.5	59.1	Three years.

Places within the limit of vine culture, but not necessarily, from that cause alone, successful. If atmospheric humidity, &c., exist in a favorable degree success may be more probable.

Locality.	Summer mean.	July mean.	August mean.	September mean.	Length of observation.
Salem, Massachusetts	70.1	72.5	70.5	63.0	Forty-three years.
New London, Connecticut	69.3	71.5	70.1	63.3	Eleven years.
West Point, New York	71.3	73.7	71.8	64.3	Thirty-one years.
Albany, New York	70.0	72.1	70.0	61.4	Twenty-eight years.
Geneva, New York	68.8	72.9	66.4	62.2	One year.
Skaneateles, New York	66.0	67.7	69.0	63.0	Do.
Fredonia, New York	68.4	70.9	68.8	61.3	Eighteen years.
Madison, Ohio	69.9	72.4	69.0	63.4	Three years.
Cleveland, Ohio	70.2	72.7	69.1	64.2	Five years.
Kelley's Island, Ohio	69.9	73.2	72.1	62.3	One year.
Collingwood, Ohio	70.4	72.8	69.8	62.9	Four years.
Ann Arbor, Michigan	70.5	72.9	70.0	63.8	Do.
Battle Creek, Michigan	71.6	74.7	71.2	63.8	Six years.
Grand Rapids, Michigan	69.4	73.5	68.2	62.2	Do.
Janesville, Wisconsin	71.2	74.6	70.4	62.9	Do.
Prairie du Chien, Wisconsin	72.3	75.2	72.0	61.5	Nineteen years.
Dubuque, Iowa	72.1	75.2	72.0	66.3	Two years.
New York city, New York	73.0	76.7	73.1	66.6	Four years.
Easton, Pennsylvania	70.5	72.5	69.7	63.6	Five years.
Trenton, New Jersey	70.7	72.8	71.6	63.4	Do.
Philadelphia, Pennsylvania	71.0	72.8	71.5	64.1	Four and three-fourths years.
Chambersburg, Pennsylvania	70.8	77.8	74.3	61.8	Two years.
Gettysburg, Pennsylvania	73.2	76.7	71.9	61.8	Six years.
Bedford, Pennsylvania	72.6	75.0	71.7	64.1	Five years.

Places within the limit of vine culture, &c.—Continued.

Locality.	Summer mean.	July mean.	August mean.	September mean.	Length of observation.
	°	°	°	°	
Baltimore, Maryland	74.3	76.7	74.7	67.8	Twenty-four years.
Frederick, Maryland	74.5	77.7	73.1	66.2	Six years.
Sykesville, Maryland	71.6	73.8	70.5	65.3	Five years.
Washington, District of Columbia.	76.3	78.3	76.3	67.7	Thirteen years.
Berryville, Virginia	70.0	75.4	72.0	66.0	Two years.
Richmond, Virginia	75.4	77.6	74.8	67.1	Four years.
Lewisburg, Virginia	71.8	74.4	71.3	64.5	Six years.
Poplar Grove, Virginia	73.0	75.7	72.7	66.1	Four years.
Kanawha, Virginia	74.2	77.2	73.6	66.1	Two and a half years.
Marietta, Ohio	72.9	76.2	73.1	69.9	One year.
Portsmouth, Ohio	75.1	78.3	73.7	66.1	Do.
Cincinnati, Ohio	75.8	79.8	76.3	69.2	Five years, 1855 to 1859.
Cincinnati, Ohio	74.0	76.5	74.2	66.0	Twenty years, 1835 to 1855.
Richmond, Indiana	72.4	74.7	72.5	66.1	Five years.
Peoria, Illinois	74.6	75.6	74.2	69.9	Four years.
Carthage, Illinois	74.9	79.1	75.5	67.8	One year.
St. Louis, Missouri	76.3	78.5	76.5	68.7	Twenty-three years.
Manhattan, Kansas	74.9	76.1	78.6	73.2	One year.

The following will, we hope, clearly illustrate our mode of applying the foregoing principles to practical vine-growing.

By examination of tables of daily mean temperatures before us, we find that at Providence, Rhode Island, a temperature of 52.50° had existed from 1st of May to the 13th, and that on the 8th of September an aggregate of heat had accrued equal to 7,915.7°. This, according to results at Waterloo, New York, is about adequate to the ripening of the Delaware and the Hartford Prolific, which commence leafing at a temperature of 52.41°, and require 7,926.66° of heat. The time required to accumulate the above amount of heat was 118 days, being four days less than at Waterloo. The average temperature for the entire period was 66.23°, and for the month of stoning 68.6°, both of which are amply sufficient for the requirements of these varieties. They may, therefore, be grown with good success at Providence, Rhode Island, other circumstances concurring to favor the attempt.

If the Diana leafed at Providence May 12 at the temperature of 51.40°, as required at Waterloo, and an aggregate of heat of 8,572.66° was demanded, this variety also may be grown. For the sum reached at Providence on September 17, 129 days from date of appearance of the needed initial of 51.40°, was 8,588°, with an average of 66.57°, both of which exceed the demands of the Diana.

The Clinton requires about the same temperature as the Diana, leafing at a lower degree than most other varieties, and consequently earlier. Its initial temperature is 51.66°, which was reached on May 12, 1854, at Providence. The Clinton will ripen for wine with a low measure of heat, 8,213°, which had accumulated on the 11th of September, when 8,223.5° had accrued in 123 days, having an average temperature of 66.85°. At Providence, Rhode Island, the Clinton may, therefore, be successfully grown.

The Concord expands its leaf at an average temperature of 52.77°, which was attained up to the 14th of May, 1854, at Providence. It requires for maturation 8,607.66°, which accrued by the 21st of September, or thereabouts, when 8,581.3° were found to have been reached by addition of the daily means for 129 days, at an average of 66.51° for the entire term. Even with this aggregate it was reported at Waterloo "poor and foxy." The Concord may then be readily grown at Providence, Rhode Island.

Rebecca, leafing at 53.13° on 14th May, required 8,900.66° to advance it to maturity. On the 26th of September the sum of 8,906.9° was attained at Providence in 135 days from date of foliation. If the Rebecca was fully ripened at Waterloo, it might be equally well grown at Providence on a warm soil and under a favorable aspect.

If we suppose the Isabella to bloom at Providence on the 15th of May, when the initial for the Catawba, 52.71°, has been reached, which is highly probable, from comparison of observations at Kelley's island and elsewhere, and that it may mature as early as the 15th of October, which is later than the date of the average appearance of frost, and yet not an erroneous assumption, we find that between these dates a sum of heat will have accumulated of 10,451°. From this we learn that in favorable years when the autumn is warm and the frosts very late that the Isabella may be grown, but that its ripening must in general be uncertain and irregular.

The Catawba and Anna open their leaves at the same temperature as the Delaware, or thereabouts—52.71°—but require more than 11,000° Fahrenheit to ripen them—a sum of heat which is not reached at Providence before the appearance of frost, except in very favorable years. As the temperature declines rapidly after the middle of October, and killing frosts occur near the beginning of the month, not unfrequently, in the average of twenty-eight years on the 11th of October, the latter-named varieties cannot in general be properly ripened at Providence.

From the preceding application of the principles derived from the table and comparison of observations made elsewhere from similar data, on which our limits forbid further enlargement, we conclude that the Delaware and Clinton and Hartford Prolific may certainly, and the Concord, Diana, and Rebecca may generally be grown at Providence, Rhode Island; that Isabella cannot always be perfectly ripened; and that the Catawba and Anna can never be there matured, unless in the long warm autumn of extremely favorable years.

The general conclusions are, that the Delaware, Hartford Prolific, and Clinton belong to a class of grape-vines which requires for ripening about 8,000° of heat; that Concord and Diana demand 8,500°; Rebecca needs about 9,000°; Isabella 10,000°; and Catawba and Anna must be grown where they shall receive from foliation to maturity a sum of heat calculated from the daily mean temperatures which must at least amount to 11,000° of Fahrenheit's scale. Unfortunately extensive tabulated reductions of daily mean temperatures are not of ready access. Those made by Professor Caswell at Providence, Rhode Island, we have been enabled to consult while preparing this paper, but much regret that we have not been able to obtain similar valuable data for sundry places over a wide district. The records we have been enabled to consult have, however, been ample for our general purpose—to establish the principle that we can employ the elements of initial temperatures and duration of heat—combine the values of temperature and time, for the determination of the possibilities of culture in any district, of plants as yet there unknown or untried.

On comparing the table of American grapes exhibiting their dates of leafing, &c., with that in which we have noted the temperature of various places in Europe within and without the wine districts, we observe many points of similarity between our native vines and their foreign congeners. By the table of foreign wine localities we learn that at all those places where the mean for the hottest month falls below 67°, the wine is not good or the fruit never ripens, and that those enjoying a mean for the hottest month above 67° produce superior fruit and wines. This is precisely paralleled in our table of American varieties grown at Waterloo, New York, where we have observed that those which could hasten their growth so that they could take advantage of the high mean of 69°, during the month following the stoning process, were of good or of superior quality; those which stoned so late as to have only 67°, or thereabouts, were inferior, and that those which found a temperature of 65°, or thereabouts, did not ripen.

The short warm summer of northern latitudes with its longer days advances the vine as rapidly as the high temperature in places further south; but the diminished heat of September precludes the proper elaboration of the juices and maturation of the seed and saccharine in the late-growing sorts. By an examination of the table in which is instituted a comparison between the mean temperatures for the growing seasons at West Point, in the valley of the Hudson, and several points in the valley of the Mississippi, we may observe that, while at the former place the mean temperature for September falls to 64.31°—a mean nowhere exceeded in the State of New York from the year 1826 to 1850, as shown by the "Reports, &c.," except at Buffalo, where it was only eight-hundredths of a degree higher, while the mean for all the places observed in the State was 59.87°—our table shows that at Fort Madison, Iowa, the mean for September stood at 71.5°; at Athens, Illinois, 73°; and at St. Louis, 68.7°; while the mean of the hottest month varied between 73° and 82.5°—heats adequate to mature, with the favorable high temperatures of September, vines of the most "southern proclivities," or such as require the longest season and the warmest skies.

Here a comparison between this part of the valley of the Mississippi and the south of France, the vine regions of Portugal, Spain, Italy, and Madeira, will show that the temperature of the warmest month in the Mississippi valley is identical with that of the latter-named places, so famous for their genial summers and generous wines, while the autumn months are equally propitious. The Concord, the Catawba, and Anna, which find the cold of September in middle New York unfavorable to their ripening, when removed to the warm regions of Illinois and Missouri, under their high and long-continued heats, develop qualities so surpassing those they exhibit in the northeastern States, as to be scarcely recognized as the product of the same varieties, respectively, grown in the latter localities.

Those who have attentively perused the foregoing attempt to determine the amount of heat needed by the American wine grape during its various stages of growth will, we doubt not, perceive therein strong proof of the importance of close attention to the demands of each variety when selecting for culture at various localities, and will regard with interest the meteorological peculiarities of their place of habitation. To all such we would say that we believe that if they would add to the knowledge of the mean temperatures of their vineyard, the general fact of adequate atmospheric humidity—for in the present state of meteorological knowledge we cannot determine for each place, by aid of any deductions from instrumental observations yet made and tabulated, any special rules for its determination—and will be guided by the instructions for ascertaining the isotherms bounding the regions of successful grape culture in the United States east of the Mississippi, as enunciated in the Agricultural Department report for 1862, all which our recent research, as developed in this paper, but confirm, they will not have reason to doubt the worth of our suggestions and their practical value.*

*A typographical error of much importance occurs in the article here referred to, which this occasion offers the only opportunity for correcting. On page 195, first line, "America" should read Armenia.

NOTES ON THE WHEAT PLANT, ITS RANGE, ETC.

Although the plants which the earth produces spontaneously appear to be confined in range within certain districts, and few of them would survive a wide change of circumstances, creative Providence has, nevertheless, endowed many of those most essential to man with a greater flexibility of structure, so that the limits of their production may be extended by culture beyond what have, in many instances, been assigned to them by nature. The grasses that yield grains are especially favored in this respect, their culture having been much extended through the knowledge and industry of man. The *Cerealia* also affords remarkable examples of very numerous varieties derived from one species. In Ceylon alone there are one hundred and sixty varieties of rice; of *Panicum*, or millet, very many kinds; and of *Triticum*, or wheat, we have a few well-defined species, which have been increased until, by different modes of cultivation, and the influence of soils and climate, more than two hundred varieties have been produced. From the well-defined species accepted by botanists, and which may be described in general terms as the hard wheats, the soft wheats, and the Polish wheats, all the innumerable varieties known to agriculturists have descended. The hard wheats are the product of warm climates, such as Italy, Sicily, and Barbary. The soft wheats are grown in the northern parts of Europe, as in England, Belgium, Denmark, and Sweden. The Polish wheats grow in the country from which they derive their name, and are also hard wheats. The hard wheats contain much more gluten than the other varieties. This valuable ingredient is a tough, viscid substance, very nutritious, and which, as it abounds in nitrogen, readily promotes fermentation in the dough, and is essential to good light bread. The quantity of nitrogen varies with the soil and climate from five per cent. in some soft wheats to thirty per cent. in the hardest and most transparent. It is the higher proportion of gluten that exists in Italian wheats that fits them for use in the preparation of macaroni and the rich pastes that form so large a portion of the food of the people of that land. The softer wheats contain a larger proportion of starch. The latter are usually grown in England, and require to be well dried and hardened before they can be readily ground into flour.

We have said that many of these varieties of wheat have resulted from influences derived from the soil. Some soils are remarkable, far and wide, for producing good seed, and it is equally well known that this seed degenerates in other soils, so that the original is resorted to for fresh supplies of seed. This is so well known in England that the produce of a certain parish in Cambridgeshire is sold for seed at a price considerably above the average. It is not, however, the experience of all that the finest wheat makes the best seed, but in the choice of seed the nature of the soil upon which it is to be sown must be considered as well as that upon which it grew.

It has been asserted that all the various noted seed-wheats, when analyzed by the chemist, are found to contain the different elements of which they are composed in nearly the same proportion, especially the starch and gluten. For bread, that which contains the most gluten is preferred; but to produce a perfect vegetation there should be no excess of this substance, and no deficiency, and the seed should have arrived at perfect maturity. Moreover, it has also been stated, and with great apparent probability of its truth, that if we wish to grow any peculiar sort of wheat, and find by our preparation of the soil or its original composition that we produce a wheat in which the gluten and starch are in different proportions from that of the original seed, we may conclude that this is owing to more or less nitrogenous matter in the soil—that is, more animal manure—or proportionally more vegetable humus; and by increasing the one or the other, we may bring our wheat to have all the properties of the original seed. By selecting seed from those ears which appeared superior to the others in a field of ripe wheat, sowing them in a garden or in a part of the field, the variety which may have been produced by some fortuitous impregnation, or by some peculiarity of the soil of the spot where it grew, may be perpetuated. By carefully adapting the seed to the soil, and by a careful and garden-like cultivation, and adding those manures which are found to be best adapted to favor its perfect vegetation, crops of wheat have been raised which at one time would have been thought miraculous, and in Great Britain, where only is its culture regarded as its importance demands, and the highest skill, the result of enlightened inquiry into the requirements of this invaluable grain, been persistently applied, the average product has been greatly increased on all soils.

To original defect in the soil or inadequate fertilization we may reasonably ascribe the deterioration observed to follow cultivation of varieties of wheat which at first appeared well adapted to the locality and to the climate. The demand for new seed-wheat to repair the loss arising from continued decline in the product from year to year, should induce cultivators to seek for a cause for this deterioration, either in the condition of the soil and its constituents, or in an unwise culture and indifference to the selection of the best product for continuing the crop. We have many recorded instances of the very valuable results of care in selecting the largest and heaviest grains for seed. In some instances the crop has been quadrupled in quantity and quality by the use of the choicest seed selected from that with which the rest of the field was sown.

When importing seed-wheat and any other seed of new or superior varieties of plants, attention should always be directed to the peculiarities of the soil and climate under which they originated, and those under which it is proposed to grow them. English varieties of spring wheat are sown in February or early in March, have the benefit of early spring growth, and of a milder and moister summer than a spring-sown wheat can have in the eastern United States. The failure that has attended recent attempts to introduce English varieties of wheat is no new thing, such having been the almost universal result for many years past.

The distinction between winter and spring wheat is one which arises entirely from the season in which they have usually been sown, for they can readily be converted into each other by sowing earlier or later, and gradually accelerating or retarding their growths. If a winter variety is caused to germinate slightly, and then checked by exposure to a low temperature, or freezing, until it can be sown in spring, it may be converted into a spring wheat.

The difference in color between red and white wheats is owing chiefly to the influence of the soil, white wheats gradually becoming darker and ultimately red in some stiff, wet soils, and red wheats losing their color and becoming first yellow and then white in rich, light, and mellow soils. The grain changes color sooner than the chaff and straw, hence we have red wheats with white chaff, and white wheats with red chaff. The blue-stem, long cultivated in Virginia, was formerly a red, but is now a beautiful white wheat, says General Harmon, in the Patent Office Report for 1845.

If it be true that each variety of grain is adapted to a specific climate in which it grows perfectly, and where it does not degenerate when supplied with proper and sufficient nourishment, may not the consideration of the origin of each variety we propose to sow be of more importance than has yet been accorded to it in the selection of minor varieties, the product of our own country? The varieties of wheat that have originated, apparently by accident, (for there are no accidents in nature,) or from peculiar culture, do not enjoy all the surroundings necessary for perfect continuous product. Causes yet unexplained are ever at work modifying the germ of the new growth, and the guardian care of man is needed to preserve unimpaired, or to perfect the already improved sorts. In most soils, we are aware that wheat degenerates rapidly if the seed be sown year after year where it was produced. Nor is it sufficient to prevent degeneration that the seed be taken from a different field, but that grown on a soil of different quality is to be preferred, and if from a different climate, but not widely diverse, it is found that the product is increased in quality and in quantity.

English-grown seed when sown in Ireland generally comes to maturity ten days or two weeks earlier than the native grown seed. In general, plants propagated from seed produced on a warm, sandy soil will grow rapidly in whatever soil the seed is sown, and plants from seed produced in a stiff, cold soil are late in growing, even in a warmer soil. On limestone soils, which are often heavy, wheat-seed, the product of sandstone regions, generally succeeds best. The experience of a Kentucky farmer shows that seed-wheat obtained from a northern locality has failed with him, owing to late ripening and consequent injury from rust. The experiment was tried with three varieties of northern-grown seed, and with the same result in each case. When wheat from a southern locality was sown by the same experimenter, his crop ripened early, was free from rust and disease, and improved in sample over the original; while the main crop, in the same district, was ruined by rust and other diseases. This experience was corroborated by the result of four seasons of growth; and the southern-grown seed, because of its early ripening, is rapidly superseding all the later wheats in the district referred to. The kind of wheat introduced from the more southern region of Tennessee, or, perhaps, northern Alabama, is the "Early May," which, though small, possesses superior flouring qualities, and is now the ordinary wheat of some northern counties of Kentucky, where it does not deteriorate, but improves in quality. The controversy that was originated by the introduction of the Tennessee "Early May" wheat into northern localities appears to have settled into the belief that the selection of southern-grown, early-ripening varieties is judicious where it is necessary that the grain should attain early maturity. To escape the rust so destructive in the lower valley of the Ohio and Mississippi, and the midge, that ravager of the late ripening wheats in western New York and New England, early ripening sorts have been successfully introduced, none, it appears, with better promise of usefulness than the "Early May," wherever the soil is sufficiently protected by snow during the winter in more northern districts. The experience of judicious farmers in Ohio, Pennsylvania, New York, and New Hampshire, testifies to the early ripening properties and valuable milling qualities of this variety, of which may be found ample proof in the pages of that excellent agricultural journal, "The Country Gentleman," vols. XIV to XXII.

The "Mediterranean" is an early ripening southern wheat, which it is said was introduced in 1819 from Genoa, Italy, by John Gordon, of Wilmington, Delaware. It is still an early ripening and very valuable wheat, adapted to many districts where the more tender varieties subject to the attacks of the Hessian fly, midge, or the rust, have rendered resort to this kind necessary. The introduction of the Mediterranean has proved an invaluable boon to many districts. Many other valuable kinds, noted for early maturity, &c., are of southern origin. The Rochester, or original White Flint, is said to have been of Spanish origin. The Turkish White Flint is not affected by fly, rust, or midge. The China or China Velvet wheat ripens at the same early date as does the "Mediterranean;" as also does the Malta, or White Smooth Mediterranean. The "Early Japan" wheat, from seed brought by Commodore Perry, is also from a warmer region than our own, and ripens early. So valuable has this variety been deemed by one grower, that he asserts that had Commodore Perry brought many bushels, it would ere this have paid the expenses of the expedition from the increased productiveness through early ripening and adaptation to the wants of the country.

All attempts to ripen wheat early by sending further north for seed have signally failed, says a Kentucky farmer. The experiment of sowing Canada-grown wheat in Pennsylvania resulted in a ripening of the crop two weeks later than that grown from native seed. As the cereals, which, as we have said, possess great flexibility, and are readily subject to the influences of soil and climate, we might naturally expect to find that wheat grown for a long time in southern Tennessee or northern Alabama, where the mean temperature of March equals, if it does not surpass, that of April in northern Kentucky and southern Ohio, would acquire a tendency to early vegetation, which it would retain when removed to more northern localities, and the plant be thus enabled by early maturity to escape the high heats of early summer, and the insect enemies which appear at the period of the late ripening of northern-grown wheats. Though it may be advisable to use southern-grown wheat for seed, the rule, we fear, will not apply if such seed has grown more than two or three degrees further south. All northern planters who have experimented with southern-grown seed-maize have learned that they cannot ripen the crop if the seed has been brought from a few degrees of lower latitude. This arises from the sudden decline of the temperature of September and October, and the early access of killing frosts, which shorten the period of growth to which the large and rank-growing southern kinds of corn have been accustomed, though the summer heats may have been the same as they had known in their native place. In the case of the southern wheats removed to a northern soil, the variety is not more rank or strong-growing, does not appear to require a longer season, but has had impressed upon it a proclivity to early vegetation by the influence of the early heats of March and April, which are not known in the north until April and May, respectively.

The first successful attempts to grow wheat in the West India islands, in more recent times, appears to have been made in 1835, in Jamaica and New Providence. The varieties used in these experiments were the Victoria wheat, which had been grown for two centuries by the Spanish farmers of Caraccas, and an English white wheat, and an English red wheat, also the Victoria wheat of English growth. In these experiments perfect success attended the use of the Caraccas wheat when sown either in Jamaica, at a height of 2,000 to 4,000 feet above the sea, or at New Providence, but just removed above the ocean level. This is ascribed to the long acclimation of this variety to the heats of a tropical winter which may be taken at nearly 70° Fahrenheit, even in the elevated regions of Jamaica and at New Providence in the lower districts. But while the productiveness of the Caraccas wheat was increased by its approach to a lower level, the English seed-wheat sown proved a total failure. Even the English-grown Caraccas was highly productive in New Providence, though the varieties, long inured to the cool, humid climate of Britain, scarcely produced stems, and in one instance spread into a turf refusing to form straw or head under the new aspect of affairs. Thus we perceive that the wheats transferred from a colder to a warmer region did not succeed; and that the Caraccas wheat grown one season in Britain had not lost its peculiarities by the voyage, but was nearly, if not quite, equal in product to its native but untravelled brother, the original Victoria of Caraccas growth. Surely there existed in the latter variety some inherent virtue which even the dull skies of England could not destroy.

THE PRODUCTION OF NEW VARIETIES OF WHEAT.

The production of new varieties of wheat either by careful selection of superior kinds, the spontaneous growth of nature, or by judiciously crossing varieties which possess properties it is desirable should be combined and perpetuated, offers a fitting field for the exercise of the skill of those vegetable physiologists who would advance the welfare of a nation by devoting themselves practically to the increased production of this staff of life. The extraordinary improvements effected in our esculent vegetables and fruits of almost every kind are known to all, and it may be just cause for surprise that the most important growth of temperate climates, the cereals, or wheat, rye, barley, oats, maize, and the grasses generally, should have shared to but a limited extent in these improvements.

That cross-fecundation and hybridization are possible has been fully proved by the results of experiments made by Maund and Raynbird, whose "Hybrid Cerealia" received the prize medals at the industrial exhibition in London in 1851; and that success has attended judicious efforts to improve upon the ordinary wheat by continued careful selection of seed, is evident from the product of the "Giant wheat" and "Pedigree wheat," grown by F. F. Hallett, of Brighton, England. By selecting from year to year not only the best heads of wheat, but the best kernels of the finest ears, and using them for seed, this gentleman has produced a variety possessing great fecundity of grain, extraordinary strength of stem, and uniformity in the size of the ear. Some of the heads of these new varieties measured seven inches in length, and were proportionably thick. In some instances one kernel has produced seventy-two heads, containing six thousand four hundred and eighty grains, and a maximum product was obtained of sixty and sixty-two, and in one instance seventy-two bushels per acre. The highest results on the farm of Mr. Hallett were six quarters or fifty-six bushels per acre, which appears to have been produced, not upon a chosen garden spot, but upon several acres. The large numbers named above need not excite surprise or doubt of their probability, since Schuyler county, Illinois, has produced wheat heads six and a half inches long, and Talbot county, Maryland, has exhibited a field of nearly thirty acres which in 1860 yielded very nearly fifty-five bushels of wheat of sixty pounds, each, to the acre, and nine of which produced sixty-four and a half bushels upon each acre. This last was a smooth-headed wheat brought from North Carolina a few years before. William Hotchkiss, of Niagara county, New York, exhibted, at the industrial exhibition in London in 1851, the product of six acres in 1849–'50 which averaged sixty-three and a half bushels to the acre, weighing sixty-three pounds to the bushel. This extraordinary yield was, however, exceeded in the summer of 1853 by Thomas Powell, of the same county, whose field of seven measured acres averaged within a small fraction of seventy bushels to the acre—namely, four hundred and eighty-nine bushels of wheat.—(See the report of Heman Powers, of Lewiston, New York, in the Patent Office Report, agricultural division, for 1853.)

Mr. Hallett describes the system by which he produced his "Pedigree wheat" as follows:

"The best plant is called 'the selection' of the year (say 1861) in which it is thus obtained, and consists of numerous ears containing many hundreds, and even thousands, of grains, which are planted separately, those of each *ear* being kept quite distinct, as, although the best grain of any plant is nearly always found to lie in its best ear, it may be otherwise, and the successive parent *ears* must be preserved. At the following harvest (1862) the best plant forms 'the selection of 1862,' and its produce is continued on the experimental ground, while that of the remaining plants furnishes the annual seed for the farm in the autumn of 1862, and the crops are in 1863 offered to the public. Thus the selection sold is that of 1861, or in any year that of two years before: the latest selection, that of the year immediately preceding, is not sold, being solely employed as the home seed."

The extraordinary improvement induced by Mr. Hallett it is not to be expected would continue to be maintained for many seasons under ordinary field treatment. It is also charged against his wheat that in so far as he has been successful, it has been in fixing the excessive coarseness which their seeding tends to produce, and that his system has not only failed to improve, but has produced great deterioration. This is the opinion of a practical writer and farmer who may have over-estimated this deterioration, as we observe an improved wheat highly lauded and a competitor for public favor which bears the name of the practical farmer above referred to. That there is, however, ground for the belief that the improvement in the quality of the "Pedigree wheat" has not kept pace with size and productiveness, we have testimony in a comparison of sundry new wheats reported in the "Gardener's Chronicle," and reprinted in "the Pennsylvania Farmer and Gardener" for June 1864, where the bread made from "Hallett's Pedigree wheat" is described as "of most inferior description." The error of this improver thus appears to have been that he has disregarded quality in the pursuit of extraordinary size and productiveness. That all these merits are compatible may yet be shown in the productions of other experimenters who shall properly regard them in combination. Be this as it may, it is quite probable that Mr. Hallet's system may furnish valuable hints to those engaged in growing seeds for market, and incite them to closer attention to the importance of securing clean and perfect seed.

A new variety of wheat introduced into a district has in some instances proved of very great value. It is said that the product of one quart, of a variety brought from North Carolina in 1845, had in nine years benefited the farmers of Preble county, Ohio, alone, more than $100,000 by the gain over what would have accrued from the continued use of the old varieties. From a few heads uninjured by rust or midge, found in the midst of a field seriously injured by both maladies, a careful and observant farmer grew the famous Lambert wheat. This variety ripens even earlier than the Mediterranean, and as into the composition of its chaff a larger proportion of silicious matter appears to enter, it probably for that reason escapes the attacks of the midge. A superior variety of spring wheat, known as the China or black-tea wheat, originated from a few kernels found in a chest of black tea. Hunter's wheat, one of the oldest and most esteemed varieties in Scotland, was discovered half a century ago by the roadside in Berwickshire. Through long culture and want of care this variety has greatly deteriorated. The Fenton wheat, a very valuable variety, which yields heavily on very strong soil, such as that on which it originated, was derived from a few ears found growing among the rubbish derived from a quarry of basaltic rock. Piper's thick-set, a wheat which yields largely on meadow soils, having produced sixty bushels to the acre, but is deficient in straw product, was derived from a remarkable ear found in a wheat-field and its product carefully cultivated.

The above-cited examples clearly indicate the importance of careful observation and the judicious selection of material for the production of new or improved kinds. The varieties thus obtained are, however, very apt to deteriorate, and can only be kept pure by continued care in selection. The average wheat crop of England has been estimated at thirty-six bushels per acre. In the United States the average may be placed at or below fifteen bushels per acre. Many English farmers annually raise fifty bushels per acre. This large product is the result of judicious cultivation and care in the choice of seed rather than of the influence of climate, for the same success has attended wheat-growing in America where every attendant circumstance was properly regarded and the demands of the wheat-plant fully supplied.

The principal climatological source of injury to wheat arises from the extreme humid tropical heats to which the warmer temperate zone, both in the Mississippi valley and on the Atlantic slope, is liable, whereby are originated rust, mildew, &c. During the colder months wheat may be grown within the tropics but little removed above the level of the sea, but where the humid heats intrude into the cooler zone, as along the Atlantic coast as high as Norfolk, and over much of the interior plain of the Mississippi, below the Ohio, wheat cannot be successfully grown.

In the months of May and June, 1858, at Marietta, Ohio, south, southwest, and southeast winds prevailed. During May there fell twelve and a half inches of rain, with maximum heats of 84° and 99° for the respective months, and in May twenty-four cloudy days were observed, and in June ten days of obscure sky. This excess of rain was not confined to Ohio, but was felt in all the western States. The consequences of this excess of moisture, obscured sky, and high heats, were the general destruction of the crops of wheat and oats, by a blight or rust on the wheat and a mould on the oats. The last experienced a total failure in three-fourths of all the fields in the valley of the Ohio, those only that escaped were on high ground or had been sowed early. Many fields of wheat were not reaped, but were left to decay, and were ploughed under for the benefit of the succeeding crop.*

From the study of many facts connected with the growth of wheat in temperate and tropical climates, we are led to believe that in the United States the only positive limit to its cultivation on the south is the northern bounds of humid tropical heats, while on the north it may be grown to the utmost northern extremity of Maine, latitude 47° 30′, where the mean temperature for May, June, and July is but 56.11°, and that for the warmest month, July, 62.30° Fahrenheit. The northern boundaries of the United States are, however, by no means the northern limit of its culture. Wheat is grown on the Saskatchewan river, in the territories of the Hudson Bay Company, near Lake Winnipeg, in latitude 54° north, where the months of May, June, and July have a mean temperature of 57.11°, and July enjoys a mean of 61.8°.

On the plains of Bogota, where the temperature rarely exceeds 59°, wheat is reaped in 147 days, having been sown in February and harvested the last week of July, the mean temperature for the entire period having been 58° and 59°. At Quinchuqui the vegetation of wheat begins in February and ends in July, at a mean temperature of between 57° and 58° Fahrenheit. In England the summer of 1853 was in many places at too low a temperature to ripen wheat, and the deficiency was general, and said to have been from one-third to one-half of the usual product. The mean for the months of July and August were 57° and 59°, respectively, or two degrees below the usual average. From the above comparison of data it appears that wheat cannot be successfully grown where the warmest month of its growth, or that wherein the grain is maturing, falls below 60°, or the mean of its period of growth below 56°.

The author of a paper on the "Effect of Temperature on Cultivation," read at a meeting of the London Central Farmers' Club, a resumé of which appears in "The Country Gentleman" for June 11, 1863, states that in England the wheat required, in 1860, 145 days from the first growth to maturity, March 28 to August 20, of which 133 days were actual growing days, having a mean temperature above 42°. In 1861, 130 days were required, during which period no mean below 42° occurred, this being, in his opinion, the minimum at which the vegetation of wheat will continue active. The experiments of this writer, continued through several years, result in the opinion that when the temperature of the soil, during the period included between the time of earing and maturity, falls below 58° to 60° no progress can be made, and unless 60° is exceeded the crop never fairly ripens.

The above appear to accord closely with the requirements of the wheat-plant in the United States. Those places where the mean for May generally falls below 58° and 60° are the New England States and New York, except a few places in the valley of the Hudson and in the southeastern extremity of the State, most of Pennsylvania and New Jersey, except the southern parts, all of Minnesota, Wisconsin, Iowa, Michigan, northern Illinois, northern Indiana, northern Ohio, and in all these districts the wheat crop cannot mature in May. The June mean for most of the districts named is, with few exceptions, such as Washington county, Maine, which is in the extreme eastern part and much exposed to fogs, also some sections of Vermont, Minnesota, and

* Silliman's Journal of Science, N. S., vol. XXVII, pp. 216–217.

Wisconsin, is above 61°, and the process of heading, which in lower and warmer districts takes place early in May, throughout the northern sections of the United States probably occurs late in May or early in June, and the crop is harvested from the middle of July to the first of August.

Those places in the northern States which do not enjoy a temperature of 61° in June have this mean in July, and can therefore grow spring wheat, which is sown from the 10th of April to the 10th of May, and is sufficiently advanced to receive the benefit of the mean of 61°, which is sometimes barely reached in July only, that for August often falling below it, and the wheat is harvested from the 10th to the 20th of September. In Aroostook county, in the extreme northeast of Maine, the mean for July and August are about three degrees higher than in Washington county, at the eastern extreme of the same State, at points where observations have been made, and the wheat is sown in the middle of May and harvested from the last of August to the 1st of September.

In the southern counties of New Jersey, Pennsylvania, Ohio, Indiana, middle Illinois, Missouri, and Kansas the mean of 61° is reached in May, and when not grown in elevated regions the wheat is ready for the reaper sometimes in June, and generally not later than the 10th of July. In districts further south, where not retarded by elevation above the general level, the harvest occurs in June, and in Cabarras county, North Carolina, the "Early May wheat" has ripened in the month from which it takes its name, and in Montgomery county, Alabama, near the southern limit of wheat-growing east of the Mississippi, it ripens about the end of May.

The following observations on the amount of heat required by the wheat plant in different countries and under different degrees of cloudiness may be of interest in this connexion.

Under the comparatively clear sky of Germany, at Arnstadt, experiments show that wheat requires from flowering to maturity 53 days, having a mean temperature of 63°, making an aggregate of 3,339° Fahrenheit.

In the neighborhood of Richmond, Virginia, the "Japan wheat" headed 30th of April, 1860, and was reaped 14th of June, 1860, a period of forty-six days. During the thirty-one days of May the mean temperature was 66.3°, an aggregate of 2,056°.23; for the fourteen days of June, 73.56°, an aggregate of 1,029°.84; from which we learn that from heading to maturity 3,086.07° are required at Richmond, Virginia, when the early variety, known as the "Japan wheat," is sown.

At Haddonfield, New Jersey, early-sown "Mediterranean" wheat headed on the 18th of May, 1864. For seventeen days the temperature had been at a mean of 64.30°. From the 18th to the 31st of May, inclusive, a mean daily temperature of 67.66° was observed, making an aggregate of 947°.27. The thirty days of June enjoyed the high mean temperature of 69°.229, which formed an aggregate of heat of 2,076.89°. The wheat was cut on the 30th of June, in forty-four days from heading, and required an aggregate of heat derived from daily mean temperatures, as observed by a thermometer in the shade, of 3,024.16°, having been hastened to maturity by the almost cloudless skies from the 10th to the 30th of June, combined with intense heats of the last week, and long-continued drought.

In Monroe county, New York, the wheat headed on the 10th of May, 1859, and was reaped on the 8th of July. To ascertain the amount of heat required from heading to maturity, we observe that for the twenty days of May, which enjoyed a mean temperature of 57.17°, there results 1,143.40°; for the thirty days in June, with a mean temperature of 64.72°, an aggregate of 1,941.60°; and for the seven days in July, with a mean temperature of 68°.21, results 477.47°, which combined. forms a final aggregate for fifty-six days from heading to maturity of 3,562.47°.

The above resulting aggregates of heat, derived from the very meagre data our extensive research has brought to light, do not entirely agree with the amount which appears to be required for maturing the crop in Britain. For western New York, where the climate is more moist and cooler than that of New Jersey or Virginia, the aggregate accords very well; but, in the dry districts of the south, the early varieties of "Japan wheat" and "Mediterranean" appear to demand smaller aggregates of heat, as we would reasonably expect they would require from their quality of early ripening.

In England the periods of growth varied from sixty-nine days in 1860, when, for ten days, the thermometer did not rise above 58° Fahrenheit, and fifty-nine days were thus required at an average daily mean of 61½° and an aggregate of 3,629° were received, to fifty days in 1861, when an average was noted of 64½°, and an aggregate of 3,225° were required to ripen the grain. Again, in 1862 fifty-six days, at a mean of 61°, and an aggregate 3,406°, were required to produce the same result. As the wheat in England did not head until the 12th of June, and this process commences here from the 1st to the middle of May, and later, the time as observed in Monroe county, New York, though it nearly corresponds to that of Britain, ripens the crop about one month earlier. The differences observed in the above results may be ascribed to different degrees of cloudiness, which, while it affects the direct heat of the sun, cannot be indicated by the thermometer with precision.

The foregoing observations indicate that, during the period that the wheat plant is passing from the bloom to the ripe fruit, the temperature must be above 60°, or that spring and winter wheat should be sown at such a time that they can take advantage of this mean for the heading process, which commences in the northern States generally about fifty to fifty-five days before maturity.

To the English wheat-grower the most important researches of the English experimenter are those which respect the temperature of the soil at the time of sowing in autumn. As the temperature of the soil does not differ perceptibly from that of the mean temperature of the air for a series of days, observations on the soil may be dispensed with. The results of our own experiments indicate that such is the case for short terms. The observer considers it unsafe to sow autumn wheat until the temperature of the earth has been reduced to 50°, a mean temperature which is seldom reached in the United States in latitude 40° north until the second or third week of October, or until a frost has occurred. His reasons for the opinion are founded on the experience that the best crops of wheat have been grown from seed that remained in the ground upwards of thirty days, which term was required in order to enable the seed to root properly before sprouting luxuriantly, and employing its strength in forming leaves at the expense of the roots. The foregoing observations on the proper time for sowing wheat, we apprehend, are not immediately applicable to our climate, but we have presented them in the hope that they may induce corresponding observations on our own soil and crops, which possibly may enable us to select the proper time for committing the seed to the ground with the best promise of success.

THE RANGE OF THE WHEAT PLANT.

Meyen, in his Geography of Plants, asserts that wheat requires a mean annual heat of 39° Fahrenheit, a summer heat of 56° Fahrenheit, and that much inferior mean heats suffice in the extreme climates of sub-arctic America, provided the summer heat of one hundred to one hundred and twenty days be enough.

At Sitka, in Russian North America, latitude 57°, the mean of 39.7° for the year has been observed, and a summer mean of 51.5°, yet wheat will not ripen, because the warmest month reaches but to 59° Fahrenheit, though two months were maintained at nearly this mean heat through ten years of observations.

We append a list of the names of places beyond the limit of wheat cultivation as evidence of the law of heat influencing the wheat plant; also a series of localities where it is successfully cultivated, with the period of sowing and harvesting, from the northern limit to the southern bound, and from the extreme eastern, in Prince Edward's island, to the southern border in New Mexico, and the western, in California. The interest awakened by the study of this table will, we hope, induce many observers to note with method and accuracy many points concerning wheat culture of which we are at present ignorant, and advance our knowledge of the demands of special varieties.

The following named places lie beyond the boundary of successful wheat cultivation, and in the latitudes respectively appended. They have not a mean of 61° Fahrenheit for the hottest month, though they lie further south than Fort Liard, on Mackenzie's river, latitude 60°, which probably enjoys this mean, for we find it attained at Fort Simpson, on the same meridian, though nearly two degrees further north.

Beyond the north polar limit of successful wheat culture.

Sitka, Russian America	Latitude, 57° N	August temp., 59° F	Wheat does not ripen.
Fort York, on Hudson's bay	do.	July temp., 60°, summer, 54°, year, 26°.	Do.
Edmonton, on Saskatchewan river	Latitude, 53° 40′ N.	1,800 feet high	Often destroyed by frost.
Carlton House, on Saskatchewan river	Latitude, 52° 51′ N.	1,100 feet high	Do.
Fort Liard, on McKenzie's river	Latitude, 60° N.	July temp., 61° ?	Grows occasionally.
St. John's, Newfoundland	Latitude, 47° 33′ N.	August temp., 58°; summer, 54°.	Wheat not grown.

Within the limits of wheat cultivation.

BRITISH NORTH AMERICA.

Locality.	Latitude north.	Temperature.	Remarks.
	° ′	°	
Fort Fraser	54 30		Longitude, 121° west; west of the Rocky mountains.
Cumberland House, on Saskatchewan river	53 57	July, 61.8	Sown May 8; reaped in August.
Red River Settlement	50 00	July, 67.0	Wheat grows luxuriantly.
Fort Francis, Rainy Lake district	48 36		Sown May 1; reaped late in August—120 days.
Quebec, Canada East	46 49	August, 67.0	Wheat succeeds.
Prince Edward's Island	46 12	August, 70.5	Extensively grown.
Frederick, New Brunswick	46 00	August, 70.0	Wheat succeeds.
Pictou, Nova Scotia	45 34	July, 66.0	August mean, 65°; wheat succeeds.

Within the limits of wheat cultivation—Continued.

UNITED STATES.

Locality.	Latitude north.	Temperature.	Sown—	Reaped—
	°	°		
Aroostook county, Me	46 to 47	July, 65.0	Middle of May	September 1.
Franklin county, Me	45°00′	July, 65.0	May 20	September 20.
Penobscot county, Me	45 00	Aug., 70.0	May 25 to June 1	Middle of August.
Somerset county, Me	45 00	Aug., 70.0	May 25 to June 1	Aug. 20 to Sept. 20.
Washington county, Me	45 00	Aug., 63.0	April 10 to May 10	September 10 to 20.
St. Lawrence county, N. Y	44 40	July, 68.0	April to June	August.
St. Lawrence county, N. Y	44 40		September 1 to November	July.
Windsor county, Vt	43 30		September 18	July 25.
Oshkosh county, Wis	44 00	June, 70.0	September 1	August 15.
Walworth county, Wis	43 00	June, 70.0	September 1 to 15	July 20.
Hillsdale county, Mich	42 00	June, 70.0	September 10 to 25	July 10 to 20.
Wayne county, Mich	42 15		September 5 to 25	July 5 to 15.
Washtenaw county, Mich	42 15		September 1 to 20	July 8 to 20.
Genesee county, N. Y	43 00		August 15 to September 15	Late in July.
Livingston county, N. Y	42 45		Middle of September	July 20 to August 20.
Ontario county, N. Y	42 45		August 20 to September 25	July 15 to August 1.
Monroe county, N. Y	43 00		(May wheat;) September 25	July 8.
Seneca county, N. Y	42 45		September 10 to 20	July 20.
Seneca county, N. Y	42 45		(Dayton wheat;) September 2	July 13.
Ulster county, N. Y	41 45		(Mediterranean wheat;) September 1 to 20.	July 10 to 20.
Steuben county, N. Y	42 15		August 25 to September 10	July 25.
Hampshire county, Mass	42 00		Early in September	Late in July.
Madison county, Iowa	42 00		September	July 6 to 18.
Scott county, Iowa	42 00		(Spring wheat;) April	July.
Henry county, Iowa	41 00		September	July 1.
Marion county, Iowa	41 00		(Spring wheat;) April 1 to 20	July 15.
Marion county, Iowa	41 00		(Winter wheat;) August 15 to September 20.	Early in July.
Lee county, Iowa	40 45		September	July 5 to 12.
Menard county, Ill	40 00		October 1 to 15	June 28 to July 10.
St. Clair county, Ill	38 30		Last of September to October 18.	
Howard county, Mo	39 00		September 1 to last of October	Last of June.
Delaware county, Ind	40 15		September 15	July 1 to 15.
Rush county, Ind	39 30		September	June 25 to July 5.
Wayne county, Ind	39 45		September 1 to October 15	June 25 to July 7.
Harrison county, Ohio	40 15		(Soulé wheat;) Sept. 1 to 20	July 1 to 10.
Athens county, Ohio	39 30		Early in September	July 1.
Clinton county, Ohio	39 20		(Rock wheat)	July 4.
Lawrence county, Ohio	38 40		(May wheat;) October	Last of May.
Mahoning county, Ohio	41 00		(Mediterranean wheat)	July 15.
Fairfield county, Ohio	39 45		(Mediterranean wheat)	June 15 to July 1.
Westmoreland county, Pa	40 30		September 10 to middle of Oct	Early in July.
Fayette county, Pa	40 00		(Mediterranean wheat;) September 1 to 20.	First week in July.
Mifflin county, Pa	40 30		September 10 to October 1	July 1.
Dauphin county, Pa	40 30		September 1 to October 1	July 4 to 15.
Berks county, Pa	40 30		(Blue stem wheat;) Sept. 10 to 15	July 4 to 20.
Philadelphia county, Pa	40 00		(Mediterranean wheat;) middle of September to middle of Oct.	Middle of July.

Within the limits of wheat cultivation—Continued.

UNITED STATES.

Locality.	Latitude north.	Temperature.	Sown—	Reaped—
	° ′	°		
Bergen county, N. J	41 00		October 1	July 5 to 15.
Gloucester county, N. J	39 45		October 1	July 1 to 10.
Salem county, N. J	39 30		(Mediterranean wheat;) last of September to October 7.	June 25 to July 1.
Newcastle county, Del	39 45		(Mediterranean wheat;) September 20 to October 10.	
Dover county, Del	39 00		September 20 to October 10	June 15 to 23.
Sussex county, Del	38 50		Last of Sept. to middle of Oct.	Last of June to July 1.
Harford county, Md	39 45		September 1 till frozen ground	June 25 to July 15.
Jefferson county, Va	39 15		September 25 to October 15	June 25 to July 1.
Jefferson county, Va	39 15		(Mediterranean wheat;) September 4 to 23.	Third week of July.
Richmond county, Va	37 50		(Japan wheat;) Sept. 16, 1859	June 14, 1860.
Richmond county, Va	37 50		(Early Conner wheat)	June 2.
Richmond county, Va	37 50		(May wheat)	May 26, 1842.
Franklin county, Va	37 00		October 1 to December 15	June 20 to July 10.
Buckingham county, Va	37 50		October 1 to November 15	June 15 to July 4.
Mason county, Ky	38 30		(Early May wheat;) September 1 to October 15.	June 2.
Clark county, Ky	38 00		September 15 to last of October.	June and July.
Logan county, Ky	37 00		October and November	June 10 to 30.
Cabarras county, N. C	35 30		(Early May wheat;) November.	June 1 to 10.
Bedford county, Tenn	35 30		Middle of Sept. to middle of Nov.	June 1 to 14.
Habersham county, Ga	34 45		September 15 to December 1	June 15 to July 15.
Cherokee county, Ala	34 15		October 1 to December	June 1 to 15.
Montgomery county, Ala	34 15			End of May.
Guadaloupe county, Tex	30 00	Mar., 61.0	January 1	June 1.
Santa Fé, New Mexico	36 30	June, 68.0	April	August.
Albuquerque, New Mexico	35 30	May, 63.0	February and March	Last of July.
Doña Aña county, N. Mexico.	32 30	May, 71.3	Middle of January	August.
Utah Territory	43 00	May, 65.0	September 1 to May 1	June to September.
Stanislaus county, Cal			(White Chili wheat;) November.	June 1.

In connexion with the foregoing the following table of the periods of sowing and harvesting wheat in several districts on the eastern continent may not be found devoid of interest:

Locality.	Sown—		Mean time of harvest.		Time of growth.
Delta of Egypt	Nov.	—	May	—	
Malta	Dec.	1	May	13	One hundred and sixty-three days.
Palermo, Sicily	Dec.	1	May	20	One hundred and seventy days.
Naples	Nov.	16	June	2	One hundred and ninety-five days.
Rome	Nov.	1	July	2	Two hundred and forty-two days.
Alps, 3,000 feet	Sept.	12	Aug.	7	Three hundred and twenty-nine days.
Alps, 4,000 feet	Sept.	8	Aug.	14	Three hundred and forty days.
Central Germany	Nov.	1	July	16	One hundred and thirty-seven days.
South of England			Aug.	4	
Middle of Sweden			Aug.	4	

We have seen that success attended the attempts to grow the acclimated Caraccas wheat even in New Providence under a mean temperature of 72°, which prevails during the winter and spring of the Bahamas, when the humidity is much diminished, and that varieties adapted to the region grow freely on the banks of the Saskatchewan in a high northern latitude, where the highest mean temperature of a summer month is but 61.8°. We have shown that a reduction of two degrees only from the mean temperature of each day, for one or two months in Britain, curtailed a loss of one-third to one-half of the crop of wheat.* This diminution in the product brought upon thousands the fear of famine, and upon all increased expenditure, and upon the nation an aggregate loss of millions of pounds sterling. This prospective deficiency was clearly foreshadowed at the close of July of that year, (1853,) and known or could have been known by those who gave attention to observations of the temperature received at the time from England and northern Europe, with some degree of clearness in anticipation of their realization.

If the diminution of the mean temperature of each day by about two degrees for one or two months over western Europe appear to be a trifling incident, its effects were of vast magnitude to commerce. Mark one of the results in stimulating the export of wheat flour and Indian corn and meal from the United States. In the preceding and following years the export of breadstuffs amounted to the values annexed thereto, as follows:

	Indian corn and meal.	Wheat and wheat flour.
In 1851 we exported	$2,300 000	$11,500,000
In 1852 we exported	2,000.000	14,500,000
In 1853 we exported	2.000,000	19,000.000
In 1854 we exported	7,000 000	40 000 000
In 1855 we exported	8,000,000	12,000,000

We perceive that though the amount of export had been increasing gradually, that in the year ending June, 1854, it trebled the corn and doubled the export of wheat and wheat flour; and that though the demand for corn increased the following year, 1855, that for wheat and wheat flour fell below the average of former years. We may perceive in the above a striking illustration of the vast results attendant upon a seemingly trifling modification of the mean temperature during a critical period in the life of a plant; realize our daily dependence on the wisdom and benevolence of creative Providence, of which every truly enlightened mind perceives fresh evidence wherever it may turn to consider the works of Him who can bless and can blast.

In concluding this part of our subject we cannot refrain from the remark, that, in contemplating wheat culture from the point of view which we have chosen, we have been renewedly impressed with evidences of the wisdom and beneficence of the Almighty Father who watcheth over his children for good. Not only do we remark the flexibility of the cereal plants, so admirably fitted

* Since the above was written we have met with the following in a notice of "An Essay on the Meteorological Conditions which Determine the Profitable and Unprofitable Culture of Farm Crops in Scotland," by Mr. Buchan, 1860:

"The dependence of the Scottish agriculturist on the climate is significantly proclaimed by the statement that 'the chief peculiarity of the climate of Scotland, with regard to the cultivation of the corn crops, consists in the mean summer temperature being within *two degrees* of the minimum temperature required for the perfect maturity of wheat and barley crops.' The prudent Scottish agriculturist will do well to bear in mind that a variation in summer temperature, to the extent of two degrees, may blight all his hopes.—Edinburgh Journal of Agriculture, 1861."

for the food of man, and without which he could not exist a civilized being, in that they are capable of adaptation, within certain limits, to the varying climates extending from the equator almost to the Arctic circle; but also, and with renewed cause for admiration, do we perceive how He who made the earth and poised it in space has commanded it should present that aspect to the source of heat and light and life which shall kindly temper his burning rays and divide the seasons, giving to wider zones the moderate springs that shall with increasing mean advance towards the north, bearing with them the warmth, and just the warmth, adequate to mature the wheat, the rye, the oats, and maize, spreading a table for millions of men where, but for this simple obliquity of the axis of the earth, a waste, frozen and inhospitable, would spread its barren desolation to the remorseless skies.

REMARKS ON THE ACCLIMATION OF PLANTS.

It has often been asserted, and is generally believed, that tender exotics may by degrees be accustomed to a colder climate, and thus become acclimatized. Some have believed that by sowing the seeds of such plants, in the first instance, in a warm, temperate region, then collecting the seeds produced by these plants and sowing them in colder districts, the species may eventually be rendered hardy. Many plants, it is imagined, have become hardy, which were originally so. We are apt to suppose that plants which grow in tropical countries must necessarily be tender, and adapted only to hot-house culture, not reflecting that they may have grown in very elevated and cold regions in those countries. Such is the case with many species introduced from Japan, Chili, and Nepal, which appear to be hardy in England. Of the dahlia, the heliotrope, the potato, the Lima bean, it may be said long culture has done nothing to increase their hardiness.

Sound views respecting the geography of plants would correct the prevailing errors regarding acclimation. The acclimatizing of plants, or, as it is supposed to be, inuring them to lower temperatures than those they have been accustomed to or have required in their native habitats, does not appear to be a possibility. It has been satisfactorily determined that a plant must receive the same amount of heat for the proper performance of certain processes necessary to the production of leaves, flowers, and fruit, whether in places to which it is indigenous, or far removed therefrom in more northern latitudes. The definite degree which it has demanded during certain epochs of growth is still required wherever it may grow; but the aggregate of heat may be received during a shorter term in high latitudes because of the greatly increased length of the day, and the processes be hastened and maturity attained at an earlier date. This is well illustrated in the growth of maize or Indian corn, which is said to be remarkably accommodating, though it must have a semi-tropical heat wherever grown, if only for a few weeks, and this heat it obtains even beyond the northern limits of the United States. It is well known that the varieties of maize grown near the northern limit of its cultivation ripen earlier than those which are esteemed valuable further south. Man has applied to his purposes the property possessed by many plants of adapting themselves to the new conditions, and the many varieties of maize, wheat, &c., attest the possibility of change within certain limits. Thus the season required by the maize varies from six months in the elevated plains of Santa Fé, in South America, to four months in the middle United States, and two and a half months in the Rainy Lake district, northwest of Lake Superior, a seemingly extraordinary transformation of its natural habits and associations. This plant must, however, enjoy for a few weeks, wherever grown, a semi-tropical mean heat of 67°, and the period required for its growth is proportioned to the abruptness of the temperature curve, or the rapid increase and high degree of heat reached during its growth. The several varieties of maize which have acquired the peculiarity of hastening their period of maturity when judiciously planted, enable the farmer to extend the cultivation over every district in which the summer temperature reaches a certain point, however brief the duration of heat, provided the definite aggregate adapted to each variety shall have been received. The extraordinary high temperature experienced in certain northern localities remote from the ocean, and the intense calorific and chemical action of the sun's rays, enhanced by the extreme length of the days of summer, enable this plant to mature in high northern latitudes. From the success that has attended the growth of many varieties of maize, it is supposed that many other plants may be acclimated, and that those whose epoch of growth would extend over a period of low temperatures in northern localities may, like the maize, be successfully carried towards the pole, if but new varieties adapted to the new conditions could be produced. Such hopeful experimenters forget that the maize is receiving on its further northern border the same, or nearly the same, aggregate of heat it demands in its most favored clime, is exposed to the influence of the same high mean during its most critical period, and, consequently, can elaborate its juices and mature its seed as well in the shorter hot season of the north as in the more temperate longer season of the south.

The principle above enunciated is, however, subject to some modifications as regards its application to certain physiological changes which have been observed to result from long-continued efforts to cultivate plants under unnatural conditions. Plants are, without doubt, capable of modification within certain limits. Witness the numberless varieties of grains and fruits, some of which appear to be more hardy than their progenitors. Many of these which appear to, or really do, thrive in more northern districts than those from which they were derived, have merely acquired a greater susceptibility to the influences of light and heat, and are thence aroused into earlier action, quickened in their vital functions, and mature under a lesser aggregate of heat as measured by the thermometer, though for the critical periods of their existence they demand the same mean temperature as the original from which they were derived.

It is in the power of man to fix these peculiarities when observed, and in a measure to produce them by the selection of those which promise well, and continuing the selection with adequate care through several generations. Vilmorin, by skilfully applying the principles which influence plants in their tendencies to sport into new varieties and directing them into the desired channel, has almost created a new race of beets containing twice as much sugar as their ancestors, and promising to be readily perpetuated.*

Acclimating the tender plants of the tropics, and inuring them to the cold seasons of the north or temperate latitudes, is, therefore, impossible, though some minor modifications upon those of short growth during the periods of fervid heat of the northern summer have been made. The olive and the orange have not been rendered more hardy, and the peach appears to be still endowed with the same tenderness of bud it has always shown.

The difficulties of acclimation may be illustrated by the fact that certain vegetable products can be grown in particular latitudes. while others. though they may attain considerable size, cannot be grown with any useful result. For instance, in England the vine will never yield grapes capable of making wine even of a quality equal to champagne; nor will tobacco ever acquire that peculiar principle which gives it, in the estimation of many, so great a value when grown in some other countries; and yet both the vine and the tobacco plant flourish in the soil of England. The botanist and the meteorologist can

* Silliman's Journal of Science, vol. XXII, p. 448, new series

explain why this is so, and thus prevent the commencement of speculations which must end in loss and disappointment.

It is of great importance to be able to define accurately when a plant may be said to suit a particular climate. It is not enough that it live and send out leaves: it must be able to produce flowers and seeds and to elaborate the peculiar secretions and products on which its qualities depend. Indian hemp has grown in England, even, to the height of ten feet, with thick stems, vigorous leaves, and abundance of flowers, but it did not produce the resinous matter upon which its supposed value as a medicinal agent depends.

The rhubarb of Turkey, which, as regards size and vigor of the plant, thrives well in England, does not produce a root of any medicinal value, or of the same quality as that grown in Chinese Tartary, from which, though known as "Turkey rhubarb," it is derived.

The leaves of the tea plant are harmless, or but slightly stimulating in certain latitudes, while they become narcotic and unwholesome in others. This fact can be explained by the study of the connexion which exists between climate and vegetation—a question to be solved by the botanist and meteorologist. It is science only that can explain the failure of attempts to cultivate the tea plant in Madeira and in the Indian archipelago, while a variety of the Chinese plant is now cultivated in the upper districts of India with great success.

The introduction of tea culture into India—a culture which is likely to become of vast importance—had its origin in the observation made by English residents that a kind of tea plant was indigenous in Assam. The discovery had been made ten years previous, but from the absence of competent botanists, and consequent ignorance of the true character of the plants, it was not turned to profitable account. It is to the published opinions of Dr. Royle, who, from a consideration of the similarity of latitude, climate and vegetation, as far as any information could be obtained, was of the opinion that tea could be successfully cultivated in the Himalayan mountains, we are to ascribe the practical adoption of plans that have been, after a long and arduous struggle, attended with complete success. Basing his opinion upon the known varieties in climate corresponding to varying elevations, he suggested the propriety of experimental planting in certain districts in the mountains elevated about 5,000 feet above the sea.

The success attending the experiment, when fully established on a sound basis, produced a remarkable demand for land, and applications for grants poured in, and were complied with by the government on most liberal terms. In Assam alone there are now upwards of 160 plantations owned by sixty companies and individuals, which occupy 20,000 acres of land actually under tea cultivation, bearing an estimated annual crop of 2,000,000 pounds of tea; while grants of land to the amount of 123,000 acres for tea-growing had been made up to the end of 1863. In Cachar, an adjacent province, more than 6,000 acres have been brought under cultivation, and the crop is estimated at nearly 400,000 pounds of tea. In the northwest provinces, and in the Punjab, where the Chinese tea plant is cultivated, the rate of progress appears to be most remarkable, where the statistics, which may be called marvellous, prove the movement to be a national one. The native princes, the Maharaja of Cashmere, the Raja of Nadawn, and several other Rajas, English and native planters, among the Himalayas, are entering largely into the culture, and the demand for seed and plants has been so great that in 1862 eighty-nine tons of the former and two million four hundred thousand of the latter were distributed gratis by the government, yet the demand was not supplied.* All that has been done and is doing for tea culture in India is but a feeble foreshadowing

* Tea cultivation, cotton, and other agricultural experiments in India—a review by W. Nassau Lees, LL.D

of an enormous cultivation and trade. We are told that tea plants are now thriving over 4½° of latitude and 80° of longitude, a tract containing 35,000 square miles, capable of producing, were the area all planted, upwards of 900,000,000 pounds of tea, and with high cultivation of doubling this high figure.

While the valley of the Brahmapootra and the slopes of the lowest range of the Himalayas are being clothed with tea plantations, the cultivation of coffee has advanced with rapid strides in the highlands of southern India. The introduction of the plant is by tradition ascribed to a Mahomedan pilgrim, who brought seven berries of coffee from Mocha, about two centuries ago, and planted them near his hermitage in the wild hills of Mysore. But it is only of late years that the produce has been considerable. Under improved administrations, and regulation of the modes of taking up and holding land, the plant is extending over tens of thousands of acres. The number of native planters is estimated at nearly four thousand, and is fast increasing in numbers. In Mysore, where but little available land remains, the mountain and forest wastes have been turned into rich productive gardens. From being the most wild and desolate parts of Mysore, these districts have become prosperous, and the people have been raised from poverty to comfort, and in many instances to wealth.

There is another plant which promises to take a high place in point of interest and value among the products of India. The credit of first suggesting the transplantation of cinchona trees, or those producing quinine, into India, is due to Dr. Royle, who pointed out the Neilgherry hills as suitable sites for the experiment. Here are found a climate, an amount of moisture, a vegetation, and an elevation above the sea, more analogous to those of the cinchona forests in South America than can be met with in any other part of India. There, accordingly, the large body of plants conveyed in Wardian cases from Peru and Ecuador, where they were collected by Mr. C. R. Markham, under many difficulties and discouragements, were finally planted, to become the parents of millions of future denizens of this delightful region. The many varieties of cinchona, each requiring its peculiar climate as respects warmth, humidity, and protection, as well as soil and elevation, were here each provided with the requisites for healthy and successful growth. Already upwards of 150,000 plants of the eleven species of cinchona are growing, and permanent plantations established to the number of 40,000 trees, and the prospect for the extensive cultivation by private enterprise is already very encouraging.*

Under the combined influences of liberal grants of land, wise expenditure for improving the modes of intercommunication and increasing on a magnificent scale the means of irrigation, and the recent planting among the agriculturists of India, the productive cultivation of the staples above referred to, the torpor of centuries is fast giving way before the quickening influences of science applied to the practical purposes of life. In the language of the Edinburgh Review: "The wheels of progress are fairly set in motion, and it demands but scanty powers of observation to see that society is moving onward at a pace almost Anglo Saxon in its rapidity."

The above afford very striking examples of the value of an acquaintance with the botanical peculiarities of plants combined with accurate knowledge of their climatic requirements. Those only who are acquainted with botanical geography could point out the most efficient methods of conducting the experiments to a successful issue. They also furnish evidence of the grand results that may follow the wise application of the power of government when directed by accurate knowledge. Agriculture, trade, fortune, food, population, health, may all be powerfully affected by the transfer of a little packet of seed,

* Edinburgh Review for October, 1863.

or by one of those modern contrivances known as "Wardian cases," which have so facilitated the interchange of the vegetable productions of the globe. It is only when such efforts are directed by competent knowledge that they can succeed; and the example of the decided success attending the culture of tea, coffee, and cinchona in India should not be lost upon those who direct similar attempts in our own land. All that has been done for India within the last ten years, it is with regret we say it, and much more than has been done, might have been reached many years ago had scientific inquiry and liberal and enlightened government gone hand-in-hand, co-workers, as they should always be, for the benefit of humanity. To transfer from foreign countries vegetable products, whose introduction may serve to increase the activity of trade, foster new forms of industry and enlarge the measure of human comfort and the common weal, is one of the noblest duties of a government; and the efforts of our own in this cause, though not always wisely directed, should continue to receive the encouragement of all who have it in their power to facilitate its means of doing good, and who have the prosperity of the nation at heart.

The attempts to introduce the tea plant into the United States have not been as successful as those made in India, nor even as those made in Brazil. The government attempts in the latter country were indeed a failure; but there exist extensive tea plantations in Brazil, and large quantities of tea, that could not be distinguished from the Chinese preparation, have been there made. In the United States the single private enterprise, but indifferently managed and early abandoned, is the only attempt to introduce tea culture of very recent date. The experiment of Junius Smith, in South Carolina, in 1848, furnishes no evidence of the incompatibility of our climate with its successful culture. It was, indeed, attempted in 1772, in Georgia, with no better result. The tea plants appear to have grown, and to have been perfectly hardy, but were neglected upon a barren soil, to improve which no effort was made, and the culture gradually languished. It may perhaps be proper to ascribe the indifferent success to the peculiar social condition of the southern population, semi-barbarous, with but few educated dominant minds, averse to improvement or to change of any kind, and especially to the introduction of any culture that would tend to render independent the class of poor landless whites whose elevation would weaken, if not eventually destroy, the power of the "lords of the rice tierce and cotton bale, sugar-box and human cattle." Our more enlarged and accurate acquaintance with the climate of the hilly slopes of southern Virginia, the Carolinas, and Georgia, ought to enable us to compare their peculiarities with those known to be favorable to the tea plant in China and India; and we do not entertain a doubt of the perfect feasibility of introducing the tea culture into those districts, when the sound of conflict shall have ceased, order have been restored, and the unworthy population of the region displaced by northern patriots, with heads and hands competent and worthy to fulfil the mission that awaits them. That mission we believe to be to restore the fertility of the south, redeem the land from the thraldom of ignorance and barbarism, plant therein the tree of liberty and all manner of trees that the climate will sustain, and rear upon the soil, blessed with a genial air and kindly sunshine, the homes of industry, Christian piety, and purity, and happiness, a gracious Providence designed should there arise for His obedient children.

Our space will not permit us to enlarge upon this very interesting branch of our subject, already extended far beyond the limits originally designed. Those who desire to acquaint themselves with the history of the cultivation of tea in India will read with interest a work on "Tea cultivation, cotton, and other agricultural experiments in India," by W. Nassau Lees, LL.D., already referred to; "Two visits to the tea countries of China and the British tea plantations of the Himalayas," by Robert Fortune; "Wanderings in China," by the same; "Brazil and the Brazilians," by D. P. Kidder and J. C. Fletcher, American clergymen, twenty years resident in Brazil; articles on "tea culture" and the experimental garden at Washington, D. C., &c., in the volumes of the Patent Office Report, agricultural division, for 1857, 1859, &c.

Ignorance, if not always the parent of fraud, is ever apt to become the prey of the unscrupulous and designing. Ignorance may itself mislead with the best intentions, for presumption is often but the measure of want of knowledge. An instance is at hand of very recent date. Had the true character of the tea plant and its climatology been well understood, we would not have had to lament the mistake that has been made by some whom we charitably hope were well-meaning but erring enthusiasts, who have persuaded themselves that the genuine tea plant has been found growing in wild profusion, indigenous, in the mountain districts of northern Pennsylvania. A moderate acquaintance with our botany would not have permitted them to imagine the common *Ceanothus Americana*, or New Jersey tea, to be identical with their favorite *Thea viridis*, or rather with a variety of it, the *Thea Assamica*. So far from being identical, these plants are not found in the same genus, as we perceive, nor in the same order, nor even in the same sub-class of Dicotyledons; nor does the New Jersey tea contain, as shown by the analysis made by Dr Gibbs, of the Lawrence scientific school at Cambridge, Massachusetts, any Theine whatever!

It may seem to many who advocate the use of tea as a beverage that the extended cultivation of a plant seemingly so valuable should be encouraged. To such we would suggest, in the absence of direct experiments with the genuine tea plant, the propriety of seeking among our indigenous plants for those that may contain the principle of Theine, to which the virtues of tea are ascribed. There are doubtless several growing wild in the United States which will produce this substance, and perhaps those most likely to furnish it may be found in the natural order *Ternströmiacea*, to which the Chinese tea plant belongs. Linnæus declares that species of the same genus possess similar virtues—a doctrine admitted to be incontrovertible, and which reduces the whole history of medical or economical uses of the vegetable kingdom to a comparatively small number of general laws, whereby the properties of a species we have never seen before may be determined by what we already know of some other species to which it is related. The above order *Ternströmiacea* contains the Gordonia and Stuartia, both growing in this country; the former, the holly-bay of the south, is native in the latitude of Charleston, South Carolina, and the latter is found on the mountains of Virginia and Tennessee, and is quite hardy in the neighborhood of Philadelphia, at the Bartram Botanic Garden. There are few trees or shrubs whose flowers can compare in beauty with those of the *Stuartia pentagynia*. As plants of the same natural order, as above remarked, often possess properties closely allied, sometimes identical, the above-named *Gordonia pubescens* in the south, and *Stuartia pentagynia* in more northern districts, might, on examination, prove at least much more valuable substitutes for the *Thea viridis* or *Thea Assamica* than the New Jersey tea, which, says Dr. Darlington, "must have sorely tried the patriotism of our grandmothers."

The foregoing outlines of the laws regulating the effect of heat, &c., upon plants will, we hope, lead to some valuable practical results. The climate we cannot modify to any very important extent; but the injurious influences we may shun by judicious choice of locality, seasons of growth, more fitting selection of varieties of plants, superior and better adapted seeds, and by improving, through our physiological knowledge, the various vegetable products upon which we must operate.

CONCLUDING REMARKS ON THE NECESSITY FOR A MORE ENLIGHTENED AGRICULTURE IN THE UNITED STATES.

We do not flatter ourselves that we have developed anything in the foregoing pages particularly novel or especially valuable. We shall be satisfied if, by giving to our agricultural friends an outline of the vegetation of the globe as affected by climatic peculiarities, accompanied by some illustrations of the value of a right application of a knowledge of botanical geography, we succeed in inviting their attention to those which influence our own land, and so interest them that they may be induced to observe and study for themselves the phenomena of meteorology and vegetable physiology, &c., in connexion with the progress of their crops and the various processes of vegetation. If we succeed in awakening some indifferent ones to a just conception of the importance of the subject, in order to a correct understanding of American agriculture, and in exhibiting to them and others how they may aid in placing it upon a scientific basis, the only sound basis upon which anything dependent upon the influences of natural phenomena can be built, we shall not have occasion to regret our effort.

Though man has lived upon the earth several thousand years, and has tilled the ground and reaped its products, he yet knows but little of the art, still less of the science of agriculture. What he has learned is merely the result of groping in the dark, the fruits of conjecture and experiment unguided by a knowledge of principles. Were the practitioners of every other art as ignorant of the principles upon which their processes are founded, we would not have emerged from the darkness of past ages, and almost every modern discovery and invention would be unknown. It is to modern science that we are indebted for our progress in physical comforts and the vast accessions of enjoyment which have been added to the lives of the cultivated classes, and which are partaken of, in measure, by even the least intellectual. Very few of these great discoveries have been made by chance and by ignorant persons, much fewer than is by some supposed. They have generally been made by persons of competent knowledge, who have sought for them, and who have met their reward through much pains-taking investigation. The facts that have gradually accumulated, the observations recorded by patient observers, have been applied by those who were more skilled, and who have been led to extend their inquiries beyond the depths already known, and to bring up therefrom new truths which add to the pleasure or the well-being of mankind.

They who are themselves unacquainted with the principles of their art must be incapable of employing opportunities for experiment, must be unable to observe aright or to investigate, and can never strike out anything new that may be useful in art, or curious or interesting in science. They, on the other hand, who are familiar with the reasons for the processes they employ may, perhaps, become improvers of the art they work at, or even discoverers in the science connected with it. They are daily handling the tools and materials with which new experiments are to be made, daily witnessing the operations of nature, are always in the way of perceiving what is wanting or what is amiss in the old methods, and have a better opportunity for making improvements, and, if they possess the requisite ability, can take advantage of them; for, as has been well said by an able writer on the chemistry of agriculture: "He that is master of principles is like a watchman in a belfry—the whole city is before him. He who cannot rise above detail is like a man in an alley, who does not know what is passing in the next street."

It is by no means necessary that a man should do nothing else but study known truths, or devote himself to investigation, in order to become skilled in the knowledge of the principles of his art or profession. Some of the greatest men of science have been engaged in the pursuits of active life, devoting to study only a modicum of time beyond that required by the demands of business. One of the most successful farmers of Europe is also one of the most profound of philosophers, and has written one of the most valuable works on the economy of agriculture extant.* The time has passed when it was deemed absurd to consider these callings incompatible, and he who now deems the application of science to agriculture a theme for ridicule has himself become a fair subject for the pity, if not for the contempt, of well-informed men.

Every farmer knows that the quantity of produce which may be obtained from a given space of ground varies much, other things being equal, according to the skill of the cultivator; but every one is not aware that skill is in reality merely the application of the rules of vegetable physiology to each particular case. This application is frequently made by those who are unconscious of the derivation of the rules. We are apt to overlook the causes that have influenced a course of practice, and to ascribe the improvement we witness to a mere advance in art, not considering that the advance must have had a cause, and that the cause can only be the work of some master-hand, which is afterwards blindly followed by the community.

The introduction of new or improved varieties of fruits has doubled and trebled the produce in many districts of our country. But how many who profit by the change, who are enriched by the increase, consider that they owe all this to the patient skill of the vegetable physiologist and horticulturist? The new varieties of the potato, many of which greatly surpass in productiveness the older sorts, are greedily sought by the farmer, and the names of Garnet, Chili, Cuzco, and Pink-eye Rusty-coat, *et id omne genus,* are in every mouth. But who considers the skill and care, the knowledge of the laws of vegetable growth, the research into the chemistry of vegetation, or the patient, laborious application of knowledge derived from a profound study of botanical principles, all of which are involved in the production of a new sort, or even turns to ponder the name of him who has proved a benefactor to his species? †

How many farmers know that to Thomas Andrew Knight, an English gentleman of wealth and education, and the best practical gardener of his day, we owe those experiments in vegetable physiology which have resulted in the production of a vast number of new varieties of apples, pears, plums, cherries, &c., of great value, both in England and in this country? How many are aware that the principles this philosopher advanced have led to still greater improvements among our own horticulturists, and seem destined to produce others, of the magnitude and importance of which he forms no conception! Who that cultivates the turnip, a crop yet to be largely extended in this country, and which is the most important in England, where it is the great foundation of all the best systems of cropping, remembers that it is to an English statesman, Viscount Charles Townshend, able and honest in an intriguing government, chief minister to George I, we owe the introduction of this invaluable root into England? ‡ In this seeming simple act of urging upon the farmers of Britain the cultivation of this apparently contemptible root, he has done more for his fellow-countrymen than whole galleries of famous politicians, who, instead of being benefactors, rather merit the denunciation of the poet, as—

"Destroyers rightlier called, and plagues of men."

---

* Boussingault's Rural Economy.

† Since the above was written we have with pleasure remarked the effort now making by many farmers of the State of New York towards obtaining means for a testimonial fund, for presentation to C. E. Goodrich, as an evidence of their appreciation of his long, laborious, and unremunerated labors for the improvement and culture of the potato. Such actions merit unqualified praise, and are as creditable to the donors as to the worthy recipient.

‡ This has been questioned; but if he did not originally introduce, he aided greatly in extending its cultivation and usefulness. His sobriquet of "Turnip Townshend" strengthens our position that he introduced the turnip into districts where it was unknown.

A long series of the names of benefactors of the farmer might be given—men who have expended fortunes in experiments, or given the labor of their lives in elevating the calling, by studying the improvements found in other countries, and, amid obloquy and ridicule, planting them among their own people.

Sir Richard Weston, who introduced clover into England about 1645, was a diplomatist and ambassador of James I to the Elector Palatine.

John Evelyn, whose "Sylva" gave a great impetus to rural pursuits among the landed gentry of England, was employed in public service.

Jethro Tull, who introduced the drill husbandry, was a barrister, yet was the first to recommend and practice horse-hoeing, and that but a century and a half ago.

Smith, of Deanston, whose system of draining has brought millions to the English agriculturists, was a proprietor of cotton works.

Arthur Young, who carried on experiments of every kind regardless of expense, "and in fact, by his enthusiasm, outfarmed himself," who saw through his own losses his country's gain, and by his tours through England and Wales and on the continent, and his numerous publications and observations for the general good, conferred a vast benefit upon agricultural interests, "was just the man not to farm profitably," but the very man destined to give more practical and cautious men a sufficient impetus to lift them out of the stagnation in which they had been bred, and in which, were it not for the influence of such men, they would ever remain.

Coke, afterwards earl of Leicester, must conclude our list of the great benefactors of the farmer and elevators of the agriculturist above the dull, dead level of ancestral routine. Such innovators have almost invariably arisen from outside the profession, the barrister, the retired manufacturer, the statesman, the nobleman, and the prince having each in turn aided in advancing the cause of improved culture of the soil, in which all have a common interest. The improvements made by Mr. Coke were extraordinary. By skill, capital, and enterprise he became the founder of the agriculture of immense estates, which he transformed from blowing sand and flinty gravel to a fertile domain. Tracts that were not worth to a tenant three shillings an acre became under his management, and that of those to whom his judicious leasies gave an enduring interest, capable of producing eighty bushels of barley per acre, which was double the average product of the county of Norfolk, in which his estates lay, and treble that of many counties of the kingdom to this day.

The names of many others might be added to the above names, *that, by the great example of what they have accomplished, encourage us to have faith in improvements yet to come.* These are the genuine heroes, the conquerors of ignorance and prejudice, who have crowned themselves with ennobling honors, heroes whose memory the world should not willingly let die.

No country can boast a greater number of intelligent farmers than can the northern section of the United States; nowhere else are agricultural periodicals and standard publications more largely patronized; in no other country are the discoveries and experience of the better-informed, or more skilful, or more curious sooner made available for the benefit of all; nowhere else exist a people enjoying so free intercourse, or so prone to use the opportunities locomotive and epistolary; no country exhibits a more varied surface or more varied produce; in fine, no other people possess more largely all the materials for the promotion and exhibition of an enlightened agriculture; but though, as farmers, we surpass those of every other country, England and Belgium, perhaps, excepted, we are still very far from possessing that enlightened acquaintance with its principles the avocation demands and our opportunities afford. Unfortunately the sum of mental training enjoyed by the sons of farmers is generally but the meagre teaching of the common school, which, though valuable as it is in comparison with that afforded to the people of other lands, Prussia excepted, falls far below the actual demands of the age and of the profession. There are, however, many hundreds of young farmers who have received an academic or collegiate education, and have, by self-instruction based thereon, (which is, indeed, the only education truly valuable, for the academy and the college but teach how to learn,) made themselves masters of the general principles of their art. On these devolve the duty to aid their brothers by putting them forward on the track of improvement. To all such we would hold out words of invitation to come forward and engage in this useful and honorable work. A wide field is open to all willing laborers, who, with observing faculties, trained by knowledge of the elements of natural and physical science to note the phenomena occurring around them daily, can with judgment record and report the results of their observations to those who, by condensing and tabulating, shall make their multifarious notes available for the instruction of their fellows.

"Whatever is individual and peculiar in our country must be studied by ourselves; we cannot expect the savans of Europe to make the investigation for us. Many practical questions in American agriculture demand answers, and success cannot be expected without the union of science with practical experience."

The climate of the United States is a peculiar one, marked by extremes of heat and cold, of wet and dry. It is as essential to study its characteristics as it is to determine the properties of our various soils. A want of meteorological knowledge and consequent want of adaptation of our industry to the laws of climate, both general and local, is a frequent source of loss to the farmer. A system of observations is efficiently carried on by the Smithsonian Institution, at Washington, D. C., aided by numerous correspondents in almost every State and Territory, through which it is hoped the climate of the United States will yet be as thoroughly understood as its geology and its geography. This kind of information is one of the essential elements on which to found a system of scientific agriculture adapted to the various local climates of the different portions of our extended country. The degree of heat and cold and moisture in various localities, and the usual periods of their occurrence, together with their effects upon different agricultural productions, are of incalculable importance in aiding research into the laws by which the growth of such products is regulated, and will enable the farmer to judge with some degree of certainty whether any given article can be profitably cultivated. Many wide sections of our country are, however, left without an observer, and we are in ignorance of the peculiar climate of districts almost in our midst. The amount of rain which falls along the sea-shore of New Jersey, from Cape May to Sandy Hook, and in the interior, midway between the river and the ocean, has never been observed or reported, nor do we know the mean temperature of any place throughout this entire region for any season. That it differs from the climate of southeastern Pennsylvania is apparent, but wherein, and to what degree, has not been determined. The foregoing is but one of many regions that might be named wherein observations might be profitably instituted. To aid in this good work should be the desire of every educated farmer. He may enter upon a series of observations on the weather with proper instruments at command, in which employment he will find not only pleasure, but positive benefit from the knowledge acquired and the habits of method and close attention the practice will induce.

There is a wide field open to all, and in which the skilful and persevering may calculate upon success. In the efforts at improving our fruits, grains, and vegetables many will find both pleasure and profit. This field is but partially occupied, and is one in which there is ample room for many experimenters, who, if they bring an ordinary amount of attention and knowledge of plants and their physiology, will be certain to reap a harvest. Those who propose to experiment in the improvement of wheat by selection and cultivation may consult with profit an article on the subject, taken from the "North British Agriculturist," republished in "The Country Gentleman," volume XVIII, page 219, (1861;) also Mr. Hallett's description of his mode of procedure, an extract from his published results, may be found in "The Annual of Scientific Discovery," for 1864, edited by D. A. Wells, M. D., pages 307, 308. Experimenters in hybridization or cross-fertilization may find some light thrown on the process followed by Mr. Maund in the production of his hybrid wheats, exhibited at the Industrial Exhibition in London in 1851, by consulting the Patent Office Report, agricultural division, for 1855, pages 181, 186; also "The Wheat Plant," by John H. Klippart, pages 81-85, a work abounding in valuable information, interesting to every farmer, and especially interesting to the wheat-growers of the great State of Ohio, from whom its statistics were mainly derived.

In an industrial point of view the production of a new prolific variety of any of the cereal grains—wheat, oats, rye, or maize—is of immense national importance. A new variety which will yield a few bushels more per acre over the ordinary kinds would greatly increase the aggregate yield of our soils.* In this direction an extensive and inviting field is open to experimenters. The same may be said of many other products of the soil, which, by judicious selection or careful hybridization might be greatly improved. The success which has attended the efforts of many amateur cultivators in their attempts to produce new varieties of fruits should encourage renewed efforts. Dr. J. P. Kirtland, of Cleveland, Ohio, has produced no less than twenty-eight varieties of excellent cherries, which were esteemed by his pomological friends as surpassing most of the old and best varieties. Strange as it may seem, and unwilling as we may be to believe, that one experimenter could in a short time originate so many superior varieties of cherries, it is nevertheless true that such have been produced, and, also, that time but confirms the good opinion early formed of them, and the belief that they will take precedence over all other kinds yet introduced. The experiments in hybridization instituted by Edward Rogers, of Salem, Massachusetts, which, it is believed, have resulted in the production of several excellent varieties of grapes, will incite others to enter this line of experiment with hopes of similar results.

There is much room for improvement in our fruits by the selection of seed of good varieties for the production of hardier kinds, or better adapted to their place of origin. Hon. Marshal P. Wilder remarks:

> "The immense loss sustained by American cultivators who have attempted to introduce varieties of fruits not adapted to the soil and climate, suggest the importance of raising from seed native sorts which, in most instances, possess some peculiar advantage. It is now generally conceded that the plants indigenous to a country will flourish at home better than those that have had their origin in foreign localities."—Proceedings of American Pomological Society, 1854, p. 12.

The United States possesses a very varied flora, which has not been fully illustrated. There are many sections whose herbs, shrubs, and trees are not fully known. As every educated farmer, it is presumed, has made some acquaintance with botany, and unless he has, he can make but minor claims to education, as this science is of the first importance in his profession, he can readily observe and record the names of all the forest trees at least; and if he can catalogue the common minor plants, or all that are met with in his township or wider district, his contribution would become so much the more valuable, as more full and minute. Such catalogues, returned from every section of the

---

* Dr. Vœlcker, in a recent lecture before the Royal Institution, London, stated that in the county of Norfolk the average produce of wheat was, in 1773, fifteen bushels per acre; in 1796, twenty-eight bushels per acre; in 1862, thirty two to thirty-six bushels per acre, the increase being due to drainage, tillage, and to the *growth of improved varieties.*—Annual of Scientific Discovery, 1864, p. 202.

Union, would prove of great value to those who have occupied themselves in attempts to thoroughly elucidate the productions, climate, and physical geography of our country.

Every farmer, even the uneducated, may contribute to the progress of his art, by observing the dates of the chief periods in the vegetation of plants. Let him make notes of the time of seeding and of germination, the periods of the revival in spring, the date of blooming of various grain-producing plants their maturity, &c., regarding especially in each case the peculiar *variety*. Let him observe the dates of leafing, blooming and ripening of the various kinds of peaches, pears, apples, grapes, &c., noting, particularly, the especial variety, for it is upon this that the value of such notes will depend. If the observer combine with these notes of the changes in vegetation well-ascertained facts of mean temperature and humidity derived from daily observations of the air, &c., his contributions will become of much higher value. It is a knowledge, of minute facts, some of which might be regarded by many as too insignificant for notice, that is occasionally found to afford the safest guide in many questions of difficulty in agriculture as in every other art. Every farmer may also aid the cause by assisting the agent of the Commissioner of Agriculture in each county of every State, in his inquiries into the condition of the crops, by collecting statistics, &c., &c., and by encouraging the work he has begun, and strengthening his zealous labors in our behalf.

Every cultivated mind of healthy activity has felt the need of an object upon which it can concentrate its powers. In this connexion, the language of the distinguished Secretary of the Smithsonian Institution appears especially instructive, and deserving the attention of the educated sons of farmers:

"It is scarcely possible to estimate too highly, in reference to the happiness of the individual, as well as to the promotion of knowledge, the choice in early life of some object to which the thoughts can be habitually turned during moments of leisure, and to which observations may be directed during periods of recreation, relative to which facts may be gleaned from casual reading and during journeys of business or pleasure. It is well that every one should have some favorite subject of which he has a more minute knowledge than any of his neighbors. It is well that he should know some one thing profoundly, in order that he may estimate by it his deficiencies in others."—Report of Dr. Joseph Henry, Secretary of Smithsonian Institution, 1857, p. 25.

It should not be matter for surprise that more farmers become insane in proportion to their number than the members of any other profession, when we consider the extent to which their muscles are tasked at the expense of the brain. "Their life is a dull routine; there is a sameness and tameness about it, a paucity of subjects for contemplation most dangerous to mental integrity."* The proper remedy against the sad effects of a plodding routine existence is increased mental activity, a more harmonious exercise of muscle and of mind. It is a sad truth that the majority of young farmers grow up like ill weeds, untrained, uncultivated, untaught in the natural and physical sciences, for studying which they have so ample opportunities. Declining too often on the vulgar level of debasing pleasures, they refuse the sweets of literature and the delights of knowledge until they are left without an aspiration beyond the acquisition of wealth or the mere continuance of animal existence. Never taught to respect their profession, what marvel that they should not esteem it the most important as well as the most interesting and the most delightful that can occupy the attention of a mind rightly constituted and properly educated; that the study of its principles and their judicious application to practice afford scope for every faculty to the fullest extent; that it is in truth a profession worthy the mental powers of any man be he ever so intellectual. In England the study

* See an excellent article on the "Health of Farmers' Families" in Report of the Agricultural Department, 1862, an article that should be read again and again in every farmers' family in the land.

of the principles and practice of farming engage the attention of the most advanced minds, and the educated gentry enter with zeal into agricultural pursuits. No problems which task the powers of the mathematician, no points of law which plague the lawyer, no questions of medicine which bother the doctor, are half as difficult, half as instructive, or half as important, as are the problems of agriculture. Instead of regarding it as a drudgery and unworthy the attention of the clever sons of farmers, if they could view it aright, they would see in their profession a wide and promising field for scientific research, an elevated arena in which to exercise and task their mental powers, and one in which distinction may be attained, and that more worthily than through the real drudgery of professions of very questionable utility.

If to the eye of taste a waste uncultivated field is repulsive, what to the eye of educated judgments is a waste neglected mind? What opportunities for benefiting the possessor, his neighbors, his country and his age, lie fallow and unimproved. Here are energies but stimulating the growth of noisome weeds; there, the barrenness of death. Such do we daily see around us among the young men—the young farmers of our land. They know not or heed not the opportunities for improvement that surround them, that press upon them, nay, that solicit their acceptance of the boon of culture. Alas, how many of them have yet to learn that "an enlightened people understand that in our age culture is the only true distinction among nations," and equally among individuals which compose the nation.

"Why should agricultural wisdom continue to be a groping backward and downward instead of forward and upward; a digging among old fossils, rather than penetrating originalities?" Why should not farmers become intellectual and progressive? Does not superior intelligence in farming as in every other calling make the better workman? Does not the well being and progress of the country depend upon the farmers, and their wise application of the teachings of advancing science? "A man may be a farmer, no doubt, in most unblessed ignorance of the physical phenomena occurring around him, but ignorant he cannot be without experiencing the baleful effects of ignorance. And seeing that all terrestrial arrangements, however apparently unconnected, do eventually mingle and combine in the production of the results ordained by the intelligence of the Supremely Wise, we believe it most injurious to the interests of agriculture that farmers should have but a limited acquaintance with those complex agencies which must be in harmonious co-operation before even a blade of grass can spring from the soil."

No art or calling requires for its full comprehension and perfect practice the aid of as many branches of systematized knowledge as does that of the farmer. No profession claims a wider range of wants or calls more largely on the man of science, or sooner exhausts his attainments. Witness already the applications of chemistry which have caused a growth beyond its ability to supply adequate information in explanation of new difficulties which the acute practical mind is daily meeting with. It ought not to be expected of science that it should explain every difficulty as it arises, or make plain every anomaly in cultivation, or render the art of farming as certain and as easily understood and practiced as the manufacture of cloth or iron. The perfect practice of agriculture is a subject infinitely more difficult and profound than any other art, mechanical or manufacturing. "Agriculture, in its true sense, is an encyclopedia in itself—requiring great knowledge, fine powers of observation, high mental cultivation, assiduous thought and study." The intelligent and educated young farmer, who enters upon the business imbued with the right spirit, will find that when the torch of science shall have illumined the field of his labors, he will discover attractions in the many operations of the farm to which he is now a stranger. The dew, the rain, the sun, the winds, the clouds, the clear sky, the floods and the droughts, the rocks, the soils, the native plants and the introduced weeds, the chemistry and physiology of vegetation, and all the myriad operations that are daily occurring around him, will become like so many volumes of a mighty library teeming with mysteries, but each of which he is prepared to peruse, and, in measure, to understand.

It has been laid to the charge of science that in its deductions it is often at war with itself, and such it must be, because it is progressive. It is in fact undeveloped, but struggling towards a more perfect stage. It has, perhaps, promised more than it has performed, but much more has been expected of it than should have been promised. It is much to have dissipated into "thin air" the old, erroneous notions in agriculture, and to have placed us upon the track of rational investigation. Practical knowledge of our calling is of prime importance; observation and experience must ever be the foundations on which to build the superstructure of science, and we are convinced that it is not by theoretical instruction in the halls of any college that agriculture is to advance, or that the present generation of farmers is to be rendered an improvement on the last, if they do not add to that clearer insight into the principles of their art which it is the mission of science to teach, the industry and thrift, the business capacity and judgment of their fathers, and combine therewith, through *self-instruction*, a higher and truer culture of mind and heart.

The earnest student of the very interesting branch of inquiry, a very meagre sketch of which we have attempted in the foregoing pages, will attentively peruse the following popular works, which should be found in the library of every inquiring farmer, and be studied by their sons and daughters throughout the land: Mary Somerville's Physical Geography, Guyot's "Earth and Man," Humboldt's Aspects of Nature, Schould's Earth Plants and Man, and Boussingault's Rural Economy.

A popular geographical botany, with a preface by Dr. Danberry, will be found very instructive, but for more minute and extended information he will consult Murray's Encyclopedia of Geography, volume I, pages 236–254, and volume II, pages 406–425; Balfour's Text-Book of Botany; Dr. Darlington's Agricultural Botany; De Candolle's Geographical Botany; Johnston's Physical Atlas; Agassiz's Lake Superior; Dr. Cooper on the Distribution of the Forest Trees in North America, in the Patent Office Report (Agricultural) for 1860, and Smithsonian Report for 1858; Blodget's Agricultural Climatology of the United States, in Patent Office Report (Agricultural) for 1853; Russell's Lectures on Meteorology, in Smithsonian Report, 1854; Blodget's Climatology of the United States; and various excellent articles by Dr. Joseph Henry, of the Smithsonian, in the valuable reports of that institution for 1855 and 1856, and in the Patent Office Reports (agricultural division) for 1856, 1857, 1858, and 1859.

HISTORY OF ECOLOGY

An Arno Press Collection

Abbe, Cleveland. **A First Report on the Relations Between Climates and Crops.** 1905

Adams, Charles C. **Guide to the Study of Animal Ecology.** 1913

American Plant Ecology, 1897-1917. 1977

Browne, Charles A[lbert]. **A Source Book of Agricultural Chemistry.** 1944

Buffon, [Georges-Louis Leclerc]. **Selections from Natural History, General and Particular, 1780-1785.** Two volumes. 1977

Chapman, Royal N. **Animal Ecology.** 1931

Clements, Frederic E[dward], John E. Weaver and Herbert C. Hanson. **Plant Competition.** 1929

Clements, Frederic Edward. **Research Methods in Ecology.** 1905

Conard, Henry S. **The Background of Plant Ecology.** 1951

Derham, W[illiam]. **Physico-Theology.** 1716

Drude, Oscar. **Handbuch der Pflanzengeographie.** 1890

Early Marine Ecology. 1977

Ecological Investigations of Stephen Alfred Forbes. 1977

Ecological Phytogeography in the Nineteenth Century. 1977

Ecological Studies on Insect Parasitism. 1977

Espinas, Alfred [Victor]. **Des Sociétés Animales.** 1878

Fernow, B[ernhard] E., M. W. Harrington, Cleveland Abbe and George E. Curtis. **Forest Influences.** 1893

Forbes, Edw[ard] and Robert Godwin-Austen. **The Natural History of the European Seas.** 1859

Forbush, Edward H[owe] and Charles H. Fernald. **The Gypsy Moth.** 1896

Forel, F[rançois] A[lphonse]. **La Faune Profonde Des Lacs Suisses.** 1884

Forel, F[rançois] A[lphonse]. **Handbuch der Seenkunde.** 1901

Henfrey, Arthur. **The Vegetation of Europe, Its Conditions and Causes.** 1852

Herrick, Francis Hobart. **Natural History of the American Lobster.** 1911

History of American Ecology. 1977

Howard, L[eland] O[ssian] and W[illiam] F. Fiske. **The Importation into the United States of the Parasites of the Gipsy Moth and the Brown-Tail Moth.** 1911

Humboldt, Al[exander von] and A[imé] Bonpland. **Essai sur la Géographie des Plantes.** 1807

Johnstone, James. **Conditions of Life in the Sea.** 1908

Judd, Sylvester D. **Birds of a Maryland Farm.** 1902

Kofoid, C[harles] A. **The Plankton of the Illinois River, 1894-1899.** 1903

Leeuwenhoek, Antony van. **The Select Works of Antony van Leeuwenhoek.** 1798-99/1807

Limnology in Wisconsin. 1977

Linnaeus, Carl. **Miscellaneous Tracts Relating to Natural History, Husbandry and Physick.** 1762

Linnaeus, Carl. **Select Dissertations from the Amoenitates Academicae.** 1781

Meyen, F[ranz] J[ulius] F. **Outlines of the Geography of Plants.** 1846

Mills, Harlow B. **A Century of Biological Research.** 1958

Müller, Hermann. **The Fertilisation of Flowers.** 1883

Murray, John. **Selections from *Report on the Scientific Results of the Voyage of H.M.S. Challenger During the Years 1872-76.*** 1895

Murray, John and Laurence Pullar. **Bathymetrical Survey of the Scottish Fresh-Water Lochs.** Volume one. 1910

Packard, A[lpheus] S. **The Cave Fauna of North America.** 1888

Pearl, Raymond. **The Biology of Population Growth.** 1925

Phytopathological Classics of the Eighteenth Century. 1977

Phytopathological Classics of the Nineteenth Century. 1977

Pound, Roscoe and Frederic E. Clements. **The Phytogeography of Nebraska.** 1900

Raunkiaer, Christen. **The Life Forms of Plants and Statistical Plant Geography.** 1934

Ray, John. **The Wisdom of God Manifested in the Works of the Creation.** 1717

Réaumur, René Antoine Ferchault de. **The Natural History of Ants.** 1926

Semper, Karl. **Animal Life As Affected by the Natural Conditions of Existence.** 1881

Shelford, Victor E. **Animal Communities in Temperate America.** 1937

Warming Eug[enius]. **Oecology of Plants.** 1909

Watson, Hewett Cottrell. **Selections from *Cybele Britannica.*** 1847/1859

Whetzel, Herbert Hice. **An Outline of the History of Phytopathology.** 1918

Whittaker, Robert H. **Classification of Natural Communities.** 1962